Instrumentation Systems

—Fundamentals and Applications—

Tasuku Senbon, Futoshi Hanabuchi (Eds.)

Instrumentation Systems

Fundamentals and Applications

Springer-Verlag
Berlin Heidelberg GmbH

Ohmsha

Tokyo · Osaka · Kyoto

Instrumentation Systems—Fundamentals and Applications.

First published in Japanese in 1987 by Ohmsha, Ltd. Tokyo as 「Keisou Shisutemu no Kiso to Ouyou」

Originally published by Springer-Verlag Berlin Heidelberg New York in 1987

Exclusive worldwide distribution by:

ISBN 978-3-662-12091-0 ISBN 978-3-662-12089-7 (eBook)
DOI 10.1007/978-3-662-12089-7

PREFACE

This book, though small, contains a wealth of technical information on control engineering and instrumentation engineering for industrial quantities, on control-system component elements (sensing, conversion, control, monitoring, and actuation), and on the system-design approaches (system engineering) used in process automation (PA) and factory automation (FA), discussing them based on examples of their applications, and covering everything from basics to applications.

Process automation has a long history, with automation of individual functions having begun as early as the 1920's. The feedback control techniques that constitute its basis grew into an indispensable core technology along with the rapid advance of control theory and control devices from the 1960's onward. Today we are progressing further towards system-scale optimal control technology. One of the influences that spurred major innovation along the way was the birth of microprocessor-based digital computer control in the 1970's. This enabled the realization of batch and sequential control together with feedback control in the same processor thus allowing an intimate interlinkage among them all. Technology for communication between multiple processors was also introduced, fostering rapid advances in functional sophistication and installation density. Moreover, this did not stop with process automation, but also spread to total factory automation covering entire plants. This included factory automation aimed at discrete processes.

This book begins with a discussion of control theory. It moves on to discuss the product hardware and software that implement the theory, and then proceeds to describe instrumentation examples and the system-design approaches (system engineering) suitable for a variety of production processes. Thus, we believe it to be ideally suited for use as a college-level textbook on instrumentation and automation for undergraduate or graduate students, or as a reference book for practicing instrument engineers in industry.

Since the subject matter deals with extremely specialized technology, the responsibility for the authorship has been undertaken by Yokogawa Electric experts continually involved in these areas. The Yokogawa Electric Training Center has undertaken the task of editing and compiling these writings into a text.

At the same time that we express our gratitude to the authors of the many works used for reference, we would also like to offer our deepest thanks to the staff of our publisher, Ohmsha, Ltd., for their hard work and earnest cooperation. We hope that this book will be of assistance to our readers in their study of instrumentation and control systems.

September, 1987

Hisashi Tamura, Senior Vice President
Director, SBU Administration
Yokogawa Electric Corporation

ABOUT THE ENGLISH-LANGUAGE EDITION

Since its publication in 1987, the original Japanese-language edition of *Instrumentataion Systems* has already gone through several printings. This is due to its wide readership among those responsible for instrumentation and control in Japan. There is a significant relationship between the expanding number of readers of this book and the continuing rapid growth of Japan's industry and economy, with process automation and factory automation as two of its driving forces.

Today as the barriers between East and West crumble away, we hope that an even wider international dissemination of this book will lend support to the world's movement toward global industrial and economic development. The authors and editors have felt this to be one of their missions. A necessary condition has been the creation of this English-language edition.

This opportunity to carry the English-language version to realization with the full cooperation of Ohmsha, Ltd. and Springer-Verlag has been a source of great pleasure to the authors and editors. We wish to extend our thanks for the assistance of those who undertook the translation and editorial supervision.

It is the hope of all those involved that this book will be widely read and found useful by members of the instrumentation and control community all over the world.

Akio Yamamoto, General Manager
Yokogawa Electric Training Center

LIST OF CONTRIBUTORS

EDITORS

Tasuku Senbon
Futoshi Hanabuchi

AUTHORS (alphabetical order)

Naoki Asakawa
Yoshio Fukai
Katsuhiro Hikasa
Kiyokazu Ishii
Tadamichi Kai
Isao Katsuoka
Hiroshi Kawai
Toshio Kimura
Takane Kudo
Hidesada Kurioka
Tetsuro Matsumoto
Kiyoshi Matsunaga
Teruyoshi Minaki
Yoshiaki Murakami
Yoshio Nagasaki
Shinobu Nagase
Eizo Oku
Kazuo Omori
Yoshikatsu Sakai
Makoto Sekiya
Jun Shiozawa
Fuso Takamura
Shin-ichi Takigishi
Akira Tanaka
Yukio Tanaka
Katsuaki Tokunaga
Masahito Tsukamoto
Hideo Tsurumaki
Masahiko Ushioda
Sadahito Watanabe
Shigehiko Yamamoto
Shigeru Yamamoto
Michio Yoshioka
Tsuneo Zeniya

EDITORIAL ASSISTANCE

Akio Yamamoto
Sumiaki Nishikata

CONTENTS

Chapter 3 DETECTION AND CONVERSION OF INDUSTRIAL VARIABLES

Chapter 1

INDUSTRY AND INSTRUMENTATION

1.1 The Word "Instrumentation"

Within the various topics covered in this book, a number of compound words and expressions appear utilizing the word "instrumentation," such as "Instrumentation System," "Instrumentation Engineer" and "Instrumentation Technology." Although the usage here is purely technical, it's interesting to note that dictionaries also define "instrumentation" as a musical term meaning the "composition of musical instruments in an orchestra," or in other words, the technique of selecting an appropriate musical instrument makeup for an orchestra so as to achieve optimum results for performing a particular piece of music.

This definition, as it turns out, can serve as a fitting analogy to the industrial use of "instrumentation." If we replace the words musical instrument, orchestra and piece with industrial instrument, manufacturing plant and manufacturing process, we can define industrial instrumentation as the technique of selecting appropriate measurement devices for a manufacturing plant so as to achieve optimum results for a particular manufacturing process. In this case, the results include quality of products, cost of production, ease of operation, and so on.

However, as words very often seem to have a life of their own, their meaning can change from generation to generation, and technical terms in particular seem to evolve quite rapidly. A case in point is the word "instrument." As applied to instrumentation within American industry of the 1950's, it then referred to relatively simple measuring instruments, but with time has come to include very complex and sophisticated industrial instruments as well. Moreover, with the advent of information processing tools based on computer and communication technology, it has also become necessary to include computerized systems when talking about instrumentation. In addition, the range of object processes to which instrumentation is applied has also quickly expanded over the years.

Definition of Industrial Instrument:

According to JIS (Japanese Industrial Standard) Z 8104, "industrial instrument" is defined as "measuring/controlling equipment used in production processes in industry." Here, "measuring/controlling equipment" is in turn defined as apparatus which indicate and/or record quantities or physical properties, as well as having computing, controlling or alarm functions, thus including detectors, transmitters, and the like.

1.2 The Development of Instrumentation

The automatic control of the measurement of industrial quantities, such as temperature, flow and pressure, first began in the 1920's in American oil refining processes. This period was characterized by local instrumentation in which large-size mechanical controllers were installed in the process area. The subsequent development of instrumentation technology in following periods came about as the needs of various industries and the advancement of industrial instruments became closely intertwined. The development of instrumentation technology in postwar Japan and corresponding background events in society are listed in Table 1.1.

[1] *The 1950's and 1960's*

During the 1950's, instrumentation technology experienced major development together with Japan's economic restoration centered in its petroleum, steel and textile process industries. Instrumentation in this period was mainly characterized by control systems consisting of a number of controllers each of which performed analog operation processing for one loop. In the beginning, pneumatic controllers driven by air pressure were used, but with the advancement of electronics and the shift toward largescale processes, control equipment progressed from pneumaic-operated to electronic-operated models.

On entering the 1960's, computers for use in process control first appeared in the field of instrumentation. At first, they were mainly used for the monitoring and record taking of process operations (data logging). In addition, by making use of the computer's computational ability, they were used for computing optimum process conditions or safe operation conditions with calculated setpoints passed on to control-

Table 1.1 Development of instrumentation technology in Japan

Year	Social events	Instrumentation Technology
1950	• Recovery period after World War II • Technology introduction from abroad • Prosperity of synthetic textile industry • Germanium transistor	• Standardization of transmission signal (3 to 15 psi) • Practical use of electronic tube self-balancing instruments
1955	• Construction of petroleum industrial complex • Construction of power station • Computer • Silicone transistor	• Practical use of electronic control systems • Graphic panel
1960	• Construction of blast furnace • Investment for labor saving	• Computer control systems
1965	• IC • Scalp up and integration trends of plant construction	• DDC
1970	• Microprocessor • Investment for environmental maintenance (Environmental pollution problem) • First oil crisis	• Standardization of transmission signal in electronic control systems 4 to 20 mA DC (IEC) • Hierarchy systems • CAD • Robot • Office automation • Automation of machine tool
1975	• Second oil crisis • Popularization of microcomputer	• Distributed control systems • Package type control systems
1980	• Low economical growth period • Diversification of social needs	• Single loop DDC • Factory automation (FA)

lers. This is known as Supervisory Process Control or Setpoint Control (*SPC*).

In time, however, thought was given to replacing the functions performed by analog controllers and computational units by using the increasing computational power of the computer. Accordingly, direct digital control by computer, or *DDC*, came to be realized. At this time, centralized DDC was employed in which many loops up to several hundred were controlled by one computer unit. As a consequence, however, since an unexpected computer problem could bring plant operations to a halt, the incorporation of redundant design elements such as CPU duplication, backup devices, etc., came to be necessary, resulting in increased costs. As a result, due to economic considerations at this time, centralized DDC did not fully replace analog control systems.

[2] *The 1970's to the present*

The introduction of the microprocessor in the 1970's brought about revolutionary changes to many areas, and the instrumentation field was no exception. With its lower price and higher performance, control systems, which up to then could only employ the one-computer centralized-control technique due to processor cost, could now implement a "distributed instrument control system." In these systems, microprocessors are distributed amongst each function or process area, and each microprocessor communicates with centralized CRT-based terminals for process monitoring and operational control. From 1975 onward, manufacturers around the world, including Japan, expanded the application of the microprocessor to batch and sequence control.

Another major point related to this development of instrumentation technology is the adoption of a standard interface. When connecting multiple devices to each other and configuring an instrumentation loop or constructing a large scale instrumentation system, the standardization of interface signals between the devices is extremely important. Work on this standardization first began in 1950 with SAMA*[1] in the United States, which established a standard pneumatic signal of 3 to 15 psi (0.2 to 1.0 kg/cm^2); this standard eventually came to be used worldwide.

In addition, in 1970, an electric current signal of 4 to 20 mA DC was standardized by IEC*[2]. With this standardization, analog industrial instruments produced by different manufacturers could be interconnected freely, thus contributing to the overall development of instrumentation. However, as modern industrial instruments are rapidly becoming digitalized and as the interface between devices moves from analog signals to mass-information-carrying digital signals, further standardization becomes even more important. In this regard, IEC has been working on a bus standard for distributed instrumentation control systems called "PROWAY"; however, the current situation in industry still has many independent bus systems from various manufacturers. In addition to PROWAY, MAP (Manufacturing Automation Protocol) and other field busses are currently in the process of standardization, but more time is needed before these standards are implemented throughout the instrumentation field.

1.3 Trend toward Total FA

[1] *From continuous processes to discrete processes*

The general process flow of manufacturing operations from raw

*[1] SAMA (Scientific Apparatus Manufacturers Association)

*[2] IEC (International Electrotechnical Commission)

materials to finished products is shown in Fig. 1.1. As shown here, going from upstream to downstream, the objects of control may change from fluids to solids, and likewise the corresponding production processes may change from continuous processes handling fluids to *discrete* (non-continuous) *processes* handling solids.

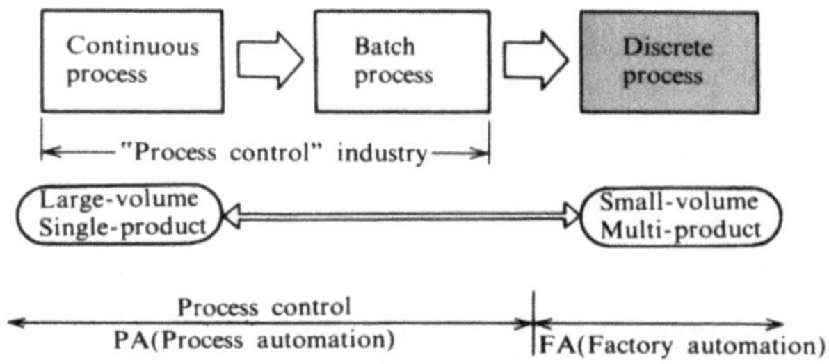

Fig. 1・1 "Upstream" and "Downstream" processes

For example, in the overall paper manufacturing process, the processes employed from pulp cooking up to paper making are continuous processes, while those employed after paper finishing and cutting, such as rolling the paper into reels and packaging for shipping, are non-continuous discrete processes.

If we apply this concept to a single factory, we can see that even for process industries whose main production is performed by a continuous process, there usually are still discrete processes required such as product inspection, packaging, transport, storage and shipping.

When speaking of process instrumentation or automation, we generally refer to upstream continuous-process control as *process automation* (PA). On the other hand, the automation of downstream discrete processes has come to be called *factory automation* (FA). Since the "tools"required here are different than those for PA, FA has been considered up to now as being in a different dimension than instrumentation. However, in order to optimize sequential production processes, discrete process instrumentation should also be standardized.

The above has described a point of view involving upstream and downstream processes when considering the factory as a flow of materials or products. If we take a point of view based on the flow of infomation, we can consider a hierarchical division of levels from the upper to the lower, as shown in Fig. 1.2. For example, at the management level, production plans are developed for the work of each manufacturing department within the factory, and specific manufacturing instructions are sent to each control system of these departments.

Accordingly, the control systems of each manufacturing department control their various production processes according to the manu-

Fig. 1・2 Instrumentation system hierarchy

facturing instructions received and produce products according to the upper level production plans. In addition, information such as production status are transmitted on-line to the management level. Based on production data, new plans or decisions can then be fed back to the control systems.

Although the prior description relates to manufacturing, it should be mentioned that in addition to manufacturing processes in a modern plant there are also automated processes within the departments of technical and business affairs, namely, LA (laboratory automation) and OA (office automation).

[2] *Achieving total FA*

Today, in order to survive the intensified market competition between various enterprises, it has become necessary to accurately determine customers' diversified needs as well as to plan for quality improvement, labor savings, energy savings and more efficient multi-product small-volume production. For this reason, a total FA system must be developed in which individually developed PA, OA, LA and FA processes are combined "organically." The enterprises manufacturing systrm can then be advanced to achieve optimal production in terms of the whole factory. In this context, instrumentation in manufacturing industries must realize a real time, flexible, consistent production system extending from the entrance of the factory to its exit.

1.4 Classification and Use of Instruments

As part of what is generally termed "instruments," this section considers the roles and features of industrial instruments.

Along with the expansion of the instrumentation field and the development of instrumentation technology, the types of instruments have been increasing and their classification has been changing as well.

Figure 1.3 shows instrument classification. This particular example describes items related to instruments as taken from the classification included in the Machinery Statistical Annual Report[1] issued by the Ministry of International Trade and Industry (MITI).

Within the electric measuring instruments field, electric instruments and electrical measuring instruments are most often used for laboratory development and product inspection on a one-unit basis.

In contrast, industrial instruments are characterized by use in factory production processes for interconnecting a system composed of multiple units. These units consist of sensors, transducers, controllers and actuators, and system operation is based on mutual relationships between the component units. Note that computer related equipment

Electric measuring instruments
- Electric instruments (indicating instruments, recorders, watt-hour meters etc.)
- Electrical measuring instruments (voltmeters, am meters and wattmeters, IC test instruments, wavefrom measuring instruments etc.)
- Industrial instruments (process use industrial instruments, process use analytical instruments, process monitoring control systems etc.)

Measuring equipment
- Measuring equipment (calipers, micrometers, water-meters, analytical instruments, instruments for pollution measurement etc.)
- Testing equipment (metallic material testing machine, Non-metallic material testing machine etc.)
- Surveying equipment (transits, levels etc.)

Electronic application equipment
- X-ray equipment (medical use etc.)
- Ultrasonic application equipment (fish finders, washers, welders etc.)
- Computers and related equipment (digital computers, control computers, peripheral devices etc.)

Fig. 1 · 3 Classification of measuring instruments

Industrial instruments
- Process use industrial instruments
  temperature instruments, flow instruments other transmitters, indicators, recorders, controllers, auxiliary instruments, manual stations
- Process use analytical instruments
- Process monitoring control system
- Other industrial instruments

Fig. 1 · 4 Classification of industrial instruments

are classified under electronic application equipment in Fig. 1.3. In addition, process analysis instruments for use in testing and inspection of products and materials, as well as process-monitoring control systems are also included in industrial instruments, as shown in Fig. 1.4.

Since industrial instruments are used in a much more continuous fashion compared with electric instruments and electrical measuring instruments, their availability factor must be high. Because of this, as well as a sometimes harsh usage environment, durability and reliability are particularly required. Moreover, when industrial instruments are installed in a dangerous environment containing combustible materials, special consideration must be taken to prevent accidental explosions that can originate in faults or breakdowns of the instrumentation system.

The number of electric measuring instruments produced for the period 1980 to 1985 according to the Machinery Statistical Annual Report[1)] issued by MITI is shown in Fig. 1.5, and that for types of industrial instruments for the year 1985 is shown in Table 1.2.

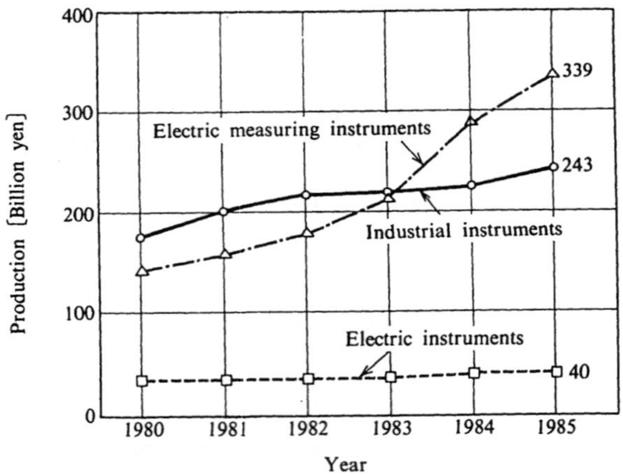

Fig. 1・5 Trends in production of measuring instruments

Table 1.2 Production of industrial instruments (1985)

Total industrial instruments	243 [Billion Yen]	100 [%]
Process use industrial instruments	117	48
Process use analitical instruments	5	2
Process use monitoring control systems	60	25
Others	61	25

REFERENCES

1) MITI, Research and Statistics Department: Machinery Statistical Annual Report (1986) (in Japanese).

Chapter 2

PROCESS CONTROL

The use of "automation," whether it be factory automation (FA), laboratory automation (LA), office automation (OA), or home automation (HA), has come to pervade almost every major field in modern society. Moreover, process automation (PA), as employed by process industries such as the petroleum, chemical, petrochemical, and steel industries, has seen the introduction of various forms of real automatic control both early and recently in its history, thus reflecting the development of process control through the years. This chapter describes the essential elements of process control beginning with feedback control, the core of process control, and leading up to sequential control, which has come to perform a fusion with feedback control via recently developed distributed control equipment.

2.1 Fundamentals of Feedback Control

2.1.1 Configuration of a control system

(a) Feedback control

"Feedback control" is defined as "control in which a comparison is performed, based on feedback, of a controlled variable and a desired value, and the subsequent corrective action taken so as to make the two values agree."

For example, in an air-conditioning system, the room temperature (controlled variable) is detected and compared with the set temperature (desired value). If a difference (deviation) between the two exists, it must be brought to 0 by turning the power (manipulated variable) on or off, or in other words, by rotating the compressor or stopping it (corrective action).

A block diagram of this air-conditioning system is shown in Fig. 2.1. As can be seen, air-conditioning equipment is configured in a closed loop.

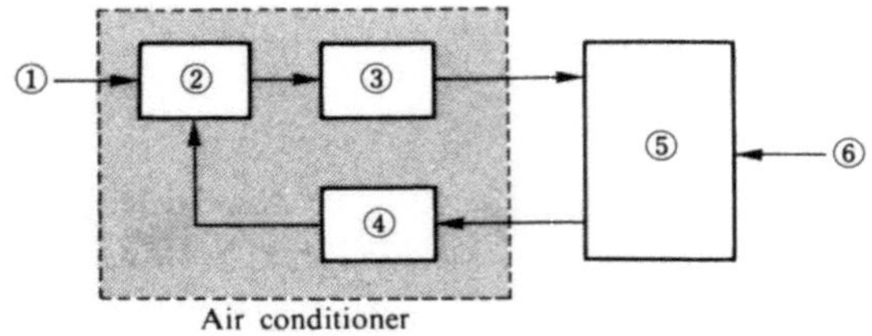

① Setting temperature
② Comparison
③ Corrective action
④ Temperature detection
⑤ Room
⑥ Outdoor temperature

Fig. 2・1 Block diagram for air conditioning

Although the above is an example of the "automatic control" of room temperature, even manual control in which people measure the room temperature themselves, compare it with the disired temperature and turn the air-conditioning power on or off, can be called a feedback control system, since a control loop is formed by a human-assisted closed loop.

(*b*) *Feedback control system*

A block diagram of a generalized feedback control system is shown in Fig. 2.2.

Figure 2.3 shows a block diagram of a generalized feedback control system as used in process control. In process control, the value of the controlled variable is obtained at a detection element and is passed

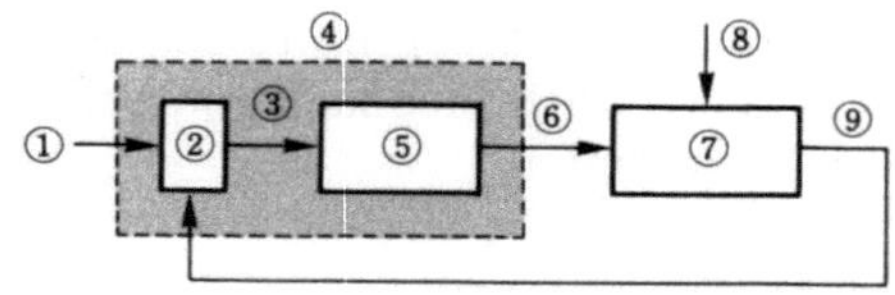

① Desired value
② Comparison
③ Controlled deviation
④ Control device
⑤ Control unit
⑥ Manipulated variable
⑦ Controlled system
⑧ Disturbance
⑨ Controlled variable

Fig. 2・2 Feedback control system

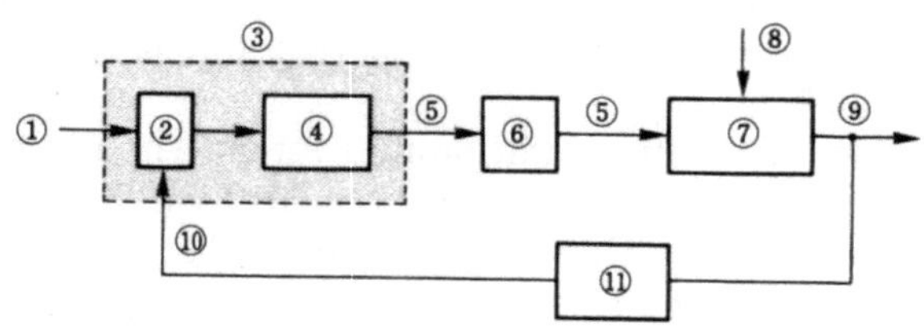

① Setpoint value (SV)
② Comparison
③ Controller
④ Computation
⑤ Manipulated variable (MV)
⑥ Control valve
⑦ Process
⑧ Disturbance
⑨ Controlled variable
⑩ Process variable (PV)
⑪ Detection・Conversion

Fig. 2・3 Block diagram for feedback process control

to the controller as a process variable (PV). The controller in turn compares the value of this variable with the setpoint value (SV) and the deviation is computed by a control element to produce a manipulation variable (MV), which is forwarded to a control valve corresponding to a final control element. The control valve then opens or closes depending on the value of the signal, thus adjusting the manipulated variable. In the above fashion, control deviation generated by outside disturbance can be corrected. Note here that the above valve is opened or closed to decrease the amount of deviation, or in other words, negative feedback is performed.

2.1.2 Characteristics of a control system

(a) Block diagram

Control systems are evaluated by considering such factors as stability and speed of response, and in order to understand these factors, it is necessary to investigate the characteristics of control systems. In this regard, block diagrams are utilized in which the constituent elements of a control system are indicated by blocks and their interconnection or signal flow by lines and arrows. Components of these block diagrams are shown in Fig. 2.4 and described below.

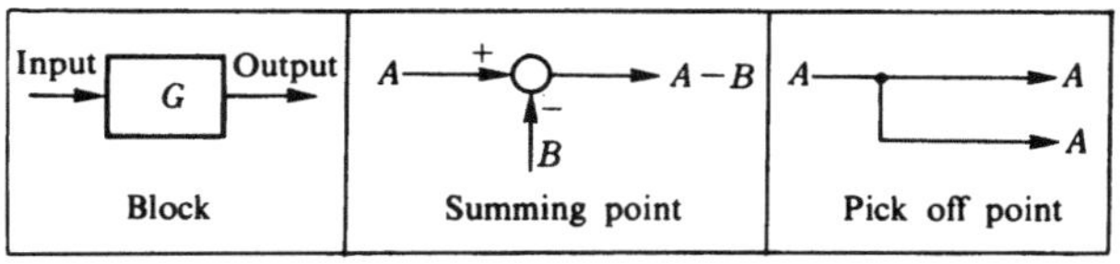

Fig. 2・4 Symbols used for block diagram

(1) *Block* Indicates a signal transfer element having particular transfer characteristics between input and output; blocks are used in conjunction with arrows to indicate input and output.

(2) *Summing point* Indicates summation of two signed signals; the dimension of the two signals must be the same.

(3) *Pick off point* Indicates signal branching.

(b) Transfer function

These are commonly used to express a signal transfer element, and are defined according to the Laplace transform. Specifically, taking the Laplace transform of input signal $x(t)$ and output signal $y(t)$ at an initial value of 0 and obtaining $X(s)$ and $Y(s)$ respectively, $G(s)$ is then expressed as the ratio of the input and output signals: $Y(s)/X(s)$, and is called a transfer function. This relationship is shown in Fig. 2.5(a).

A transfer element can also be expressed as a time function, as

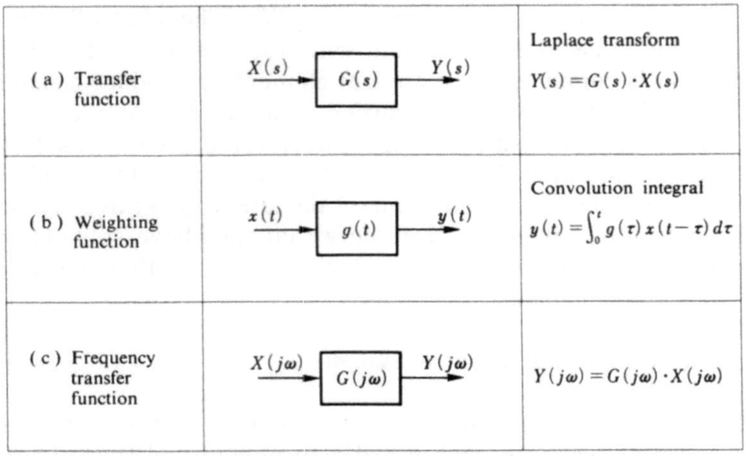

Fig. 2·5 Various transfer elements

shown in Fig. 2.5(b). In this case, output signal $y(t)$ is expressed as the convolution integral of input signal $x(t)$ and transfer element $g(t)$. Since $g(t)$ is considered to be a weighting function here, $g(t)$ is equal to $y(t)$ when $x(t)$ is a unit impulse $\delta(t)$*. To put it another way, the convolution integral of two time functions can also be expressed as the product of the Laplace transform of these two functions.

Convolution integral:

As shown in Fig. 2.6, input signal $x(t)$ is approximated by time interval partitions $\Delta\tau$. At $\tau=0$, or in other words, at the point where $t \to t-\tau$, if the input signal has a width of $\Delta\tau$ and a height of $1/\Delta\tau$, the output signal at $t=t$ nearly becomes $g(t)$, since $g(t)$ is a weighting function. Since the height at $\tau=0$ is $x(t-\tau)$, the output signal is $g(\tau)x(t-\tau)\Delta\tau$. Accordingly,

$$y(t)=\sum_{\tau=0}^{t} g(\tau)x(t-\tau)\Delta\tau$$

and as $\Delta\tau \to 0$, we obtain the integral

$$y(t)=\int_0^t g(\tau)x(t-\tau)d\tau$$

---

* Unit impulse $\delta(t)$ is defined as the following function with $\delta(t)=0$ when $t \neq 0$:

$$\int_{-\infty}^{\infty} \delta(t)dt=1$$

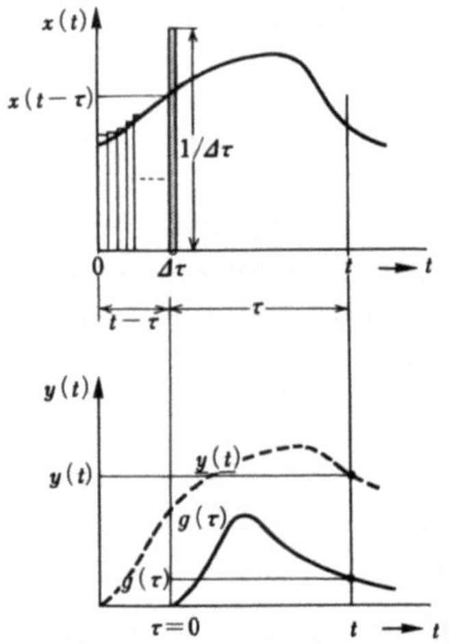

Fig. 2・6 Convolution integral

(c) *Laplace transform*

Since transfer functions express the characteristics of transfer elements in the frequency domain, elements possessing specific characteristics in the time domain also exhibit specific characteristics in the frequency domain.

To perform a conversion between a time function $f(t)$ and a frequency function $F(\omega)$, the well known Fourier transform is used, as shown below.

Fourier transform:

$$F(\omega)=\int_{-\infty}^{\infty} f(t)e^{-j\omega t}dt \tag{2.1}$$

Fourier inverse transform:

$$f(t)=\frac{1}{2\pi}\int_{-\infty}^{\infty} F(\omega)e^{j\omega t}d\omega \tag{2.2}$$

Since control engineering deals with transient responses in which $t<0$ is not considered, the Laplace transform which treats all signals as having an initial value of zero is quite appropriate. If time function $f(t)$ for $t<0$ is 0 and if $\int_0^{\infty}|f(t)|dt<\infty$, then Equation (2.1) becomes

$$F(\omega)=\int_0^{\infty} f(t)e^{-j\omega t}dt$$

If we now let $j\omega = s$, we can transform $f(t)$ to a function of s by the following formula:

$$\mathcal{L}[f(t)] = F(s) = \int_0^{\infty} f(t)e^{-st}dt \tag{2.3}$$

where $\mathcal{L}[f(t)]$ denotes the Laplace transform. In addition, the inverse transform of the above function for converting back to $f(t)$ is denoted as $\mathcal{L}^{-1}[F(s)]$ and is expressed as the following formula:

$$\mathcal{L}^{-1}[F(s)] = f(t) = \frac{1}{2\pi j}\int_{-j\infty}^{+j\infty} F(s)e^{st}ds \tag{2.4}$$

Laplace inverse transform:

In the case that $\int_0^{\infty}|f(t)|dt > \infty$, convergence factor e^{-Ct} must be multiplied in. The value C here is made larger than the value σ in which $\int f(t)e^{-\sigma t}dt < \infty$. At this time, s becomes a complex frequency expressed as $s = C + j\omega$, and the Laplace inverse transform becomes

$$\mathcal{L}^{-1}[F(\omega)] = \frac{1}{2\pi j}\int_{C-j\omega}^{C+j\omega} F(s)e^{st}ds$$

Based on the Laplace transforms and theorems for representative signals and transfer elements listed in Appendix 3, the s–transforms for basic electric circuit elements are shown in Fig. 2.7.

Electroric circuit elements		S-Transform
Resistance	R (resistor symbol)	$Z(s) = R$
Capacitance	C (capacitor symbol)	$Z(s) = \frac{1}{Cs}$
Inductance	L (inductor symbol)	$Z(s) = Ls$

Fig. 2·7 S-Transform for basic electronic circuit elements

(*d*) *Transient response*

In the event that a steady state condition as possessed by an input signal changes to another state, the characteristics present until the output signal attains a steady state are called transient characteristics, and they appear as transient response. Various forms of transient responses, such as step response, impulse response and ramp response, can be obtained depending on the type of input signal; however, the most often used signal is a unit step signal $u(t)$* for generating a step response.

Figure 2.8 shows an example in which the form of a transient response is sought when a voltage unit step signal is applied to a first-order lag circuit consisting of a capacitor C and resistor R (initial voltage of the capacitor is 0).

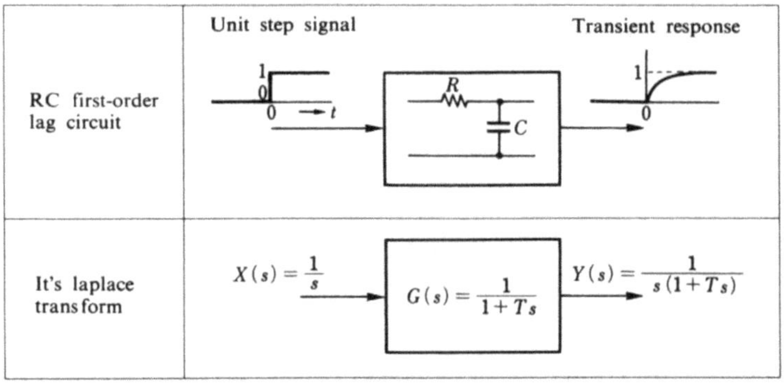

Fig. 2・8 RC first-order lag circuit

Although the Laplace transform for a unit step signal is listed in Appendix 3 as $1/s$, this can be easily obtained by directly solving the Laplace transform formula for this function. Since the unit step signal is 1 for $t>0$, we get:

$$F(s)=\int_0^\infty e^{-st}dt=-\frac{1}{s}[e^{-st}]_0^\infty=\frac{1}{s}$$

Next, we determine the Laplace transform for the CR first-order lag circuit by using the s-transforms for electric circuit components in Fig. 2.7 and solving, as follows:

$$Y(s)=G(s)X(s)=\frac{1/Cs}{R+1/Cs}X(s)$$

$$\therefore \quad G(s)=\frac{1}{1+CRs}=\frac{1}{1+Ts}$$

Here, $T=CR$ (Time constant).

* The unit step function is equal to 0 for $t<0$ and to 1 for $t>0$, *i.e.*, it is non-continuous at $t=0$.

In addition, this first-order lag circuit can be expressed as a differential equation, where $x(t)$ represents the input signal, and $y(t)$ the output signal:

$$T\frac{dy}{dt}+y(t)=x(t)$$

where again $T=CR$.

If we now apply the Laplace transform of a derivative as indicated by the theorems listed in Appendix 3, we get

$$TsY(s)+Y(s)=X(s)$$

$$\therefore \quad G(s)=\frac{Y(s)}{X(s)}=\frac{1}{1+Ts}$$

which is the same transfer function as above.

Accordingly, the output signal is then

$$Y(s)=G(s)X(s)=\frac{1}{1+Ts}\cdot\frac{1}{s}=\frac{1/T}{s(s+1/T)}$$

Although the transient response can now be determined by taking the inverse transform of the above, direct solving of Eq. (2.4), the formula for finding an inverse, can be time consuming. For the example here, however, we can again refer to Appendix 3 and use the function corresponding to the transform $1/s(s+a)$ to obtain the following expression for the transient response:

$$y(t)=1-e^{-t/T}$$

In this way, the inverse transform of function $F(s)$ having an algebraic function denominator as shown above can be found by expanding $F(s)$ into partial fractions, and then summing the inverse time functions obtained from the Laplace transform table for each factor.

(*e*) *Frequency response*

The above has described how to obtain the transient response of a transfer element through use of the Laplace transform. Also of importance is knowing the frequency response of a transfer element. An output signal in steady state resulting from the application of a sine wave input signal exhibits the characteristics of amplitude gain and phase shift with respect to the input signal. These characteristics which change depending on the frequency of the input signal are called frequency characteristics, and they appear as frequency response. In order to express frequency characteristics, frequency transfer functions are used, and specific frequency transfer functions can be obtained by substituting the s in ordinary transfer functions by $j\omega$. A transfer element expressed as a frequency transfer function is shown in Fig. 2.5(c).

As an example, to find the frequency response of a first-order lag circuit, the frequency transfer function $G(j\omega)$ can first be expressed as follows:

$$G(j\omega)=\frac{1}{1+j\omega T}=\frac{1}{1+\omega^2T^2}-j\frac{\omega T}{1+\omega^2T^2}$$

$$=\frac{1}{\sqrt{1+\omega^2T^2}}\angle-\tan^{-1}\omega T$$

And for methods of expressing the resulting frequency response, vector locus plots and Bode diagrams are commonly used.

In vector locus plots, frequency ω is used as a parameter on a complex coordinate two-dimensional surface, and the frequency response is indicated by the locus drawn out by the tip of a vector whose value changes along with ω.

In Bode diagrams, the frequency ω is assigned on a logarithmic scale to the horizontal axis, and the gain and phase shift are plotted separately along the vertical axis so as to produce a set of two plots; the units for plotting the gain are usually decibels (dB: $20\log_{10}n$), and those for phase shift are degrees or radians.

Table 2.1 summarizes the transient response and the frequency response as expressed by vector locus plots and Bode diagrams for the main types of transfer elements with respect to unit step signals.

2.1.3 Feedback control and stability

(a) Loop transfer function gain

A block diagram obtained by simplifying the feedback control system for processes and by treating transfer elements as proportional elements is shown in Fig. 2.9. If we now solve for process variable PV and control deviation DV (omitting the s symbol), we get:

$$\mathrm{PV}=\frac{K_pK_c}{1+K_cK_p}\mathrm{SV}+\frac{K_p}{1+K_cK_p}\mathrm{DS}$$

$$\mathrm{DV}=\underbrace{\frac{1}{1+K_cK_p}}_{\text{Term for setting}}\mathrm{SV}-\underbrace{\frac{K_p}{1+K_cK_p}}_{\text{Term for disturbance}}\mathrm{DS}$$

As can be seen, the influence of disturbance due to feedback control turns out to be $1/(1+K_cK_p)$. Here, the product K_cK_p is called the overall transfer function, and if this value becomes large enough, we

Table 2・1 Characteristics of transfer elements

	Transfer element	Transfer function	Step response	Vector locus	Bode diagram
a	Proportional element	K	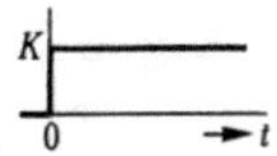	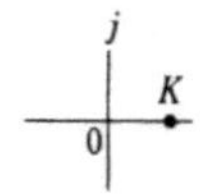	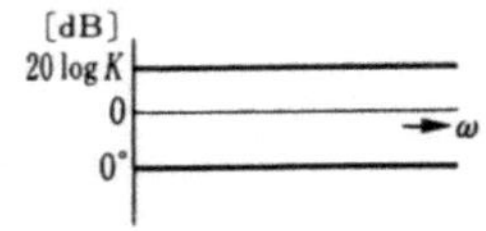
b	Fist-order lag element	$\frac{1}{1+Ts}$	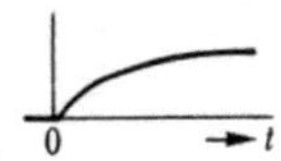	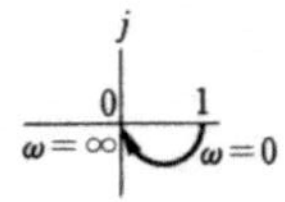	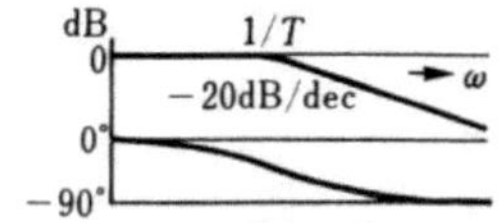
c	Second order lag element	$\frac{1}{(1+T_1s)(1+T_2s)}$	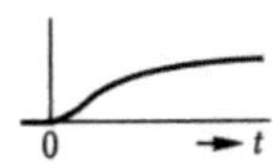	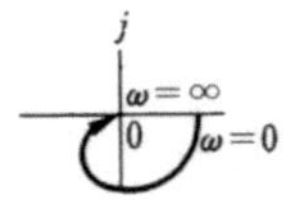	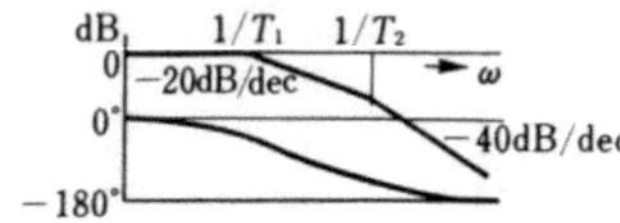
d	Dead time element	e^{-Ls}	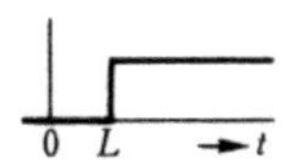		
e	Integral element	$\frac{1}{T_1s}$	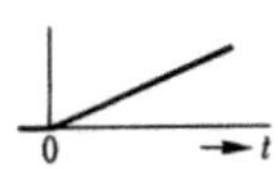		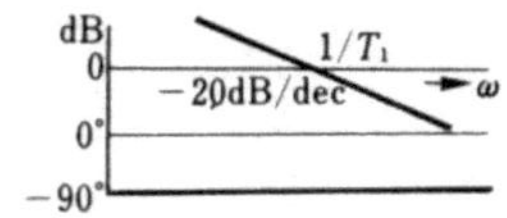
f	Derivative element	T_Ds	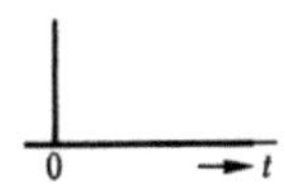		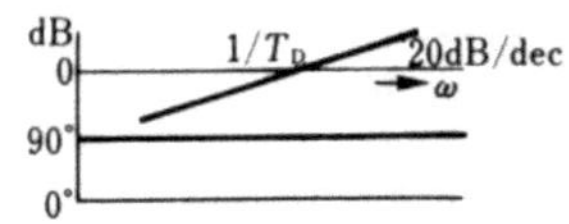

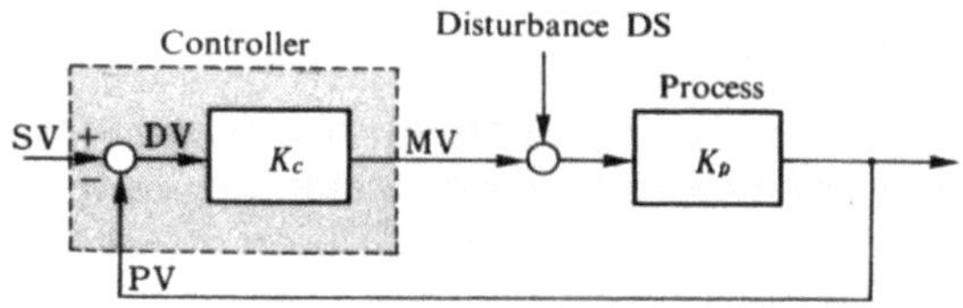

Fig. 2・9 Simplified process control loop

obtain:

$$PV = SV + \frac{1}{K_c} DS$$

$$DV = -\frac{1}{K_c} DS$$

Since in process control, controller gain K_c is made large if the overall transfer function gain is large, the above becomes

$$PV = SV \qquad DV = 0$$

In other words, if the controller gain is sufficiently large, the control variable agrees with the desired value, and the influence of disturbance becomes 0. However, due to an actual delay in processes, stability becomes a problem and insufficient rise in controller gain occurs frequently. The controlled variable is called stable in the case of damped oscillation, is considered to be at stability limit in the case of fixed amplitude continuous oscillation, and is called unstable in the case of divergent oscillation.

(b) Characteristics equation

A block diagram for a generalized feedback control system is shown in Fig. 2.10.

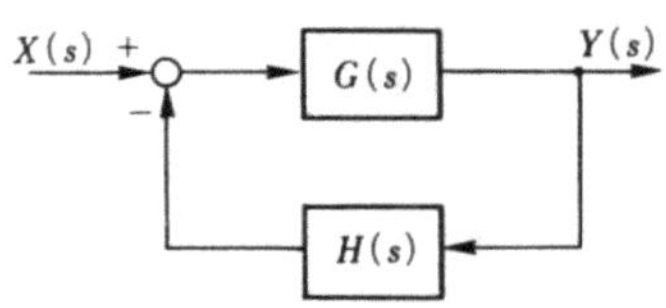

Fig. 2・10 Generalized feedback control loop

This system is expressed as

$$Y(s) = \frac{G(s)}{1 + G(s)H(s)} X(s)$$

The equation

$$1+G(s)H(s)=0 \tag{2.5}$$

is called the characteristics equation of the control system, and its roots (characteristic roots) can be used as stability criterion. If we indicate these roots as s_1, s_2, ..., s_n, the real part of each characteristics root must be negative for the system to be stable*[1].

(1) *Routh/Hurwitz stability determination method* Finding the roots of the above characteristics equation is not always simple, and Routh and Hurwitz have independently proposed methods for determining stability from the coefficients of the characteristics equation. The two methods have been combined and are now known as the Routh/Hurwitz stability determination method, as described below.

If we denote the characteristic equation as $a_0s^n+a_1s^{n-1}+...+a_{n-1}s+a_n=0$, then the following conditions must hold for stability:

(1) $a_0,a_1,...,a_n$ must all exist and be positive (in the case of first and second orders, only this condition is required)

(2) For higher orders, the following expressions must exist and be positive:
Third order: $a_1a_2-a_0a_3$
Fourth order: $a_3(a_1a_2-a_0a_3)-a_1^2a_4$
Fifth order: $a_1a_2-a_0a_3$, $(a_1a_2-a_0a_3)(a_3a_4-a_2a_5)-(a_1a_4-a_0a_5)^2$

(2) *Nyquist stability determination method* This method makes use of the vector locus plot of the overall transfer function $[G(j\omega)H(j\omega)]$. If, when incrementing from $\omega=0$ to $\omega=\infty$, the point $(-1,0)$ is to the left of the vector locus, the system is stable, while if the point is to the right of the vector locus, the system is unstable*[2]. Figure 2.11(a) shows a vector locus plot indicating stability, and Figure 2.11(b) one indicating instability.

(*c*) *Gain margin and phase margin*

In the Nyquist stability determination method, when the vector locus is drawn with the point $(-1,0)$ to its left, the closer it approaches this point, the closer it approaches instability. Furthermore, if the locus intersects this point at the frequency ω_0, it has reached the limit

---

*1 Characteristics equation and stability determination: System response is expressed as $c_0+c_1e^{s_1t}+...+c_ne^{s_nt}$, where c_0, c_1, ..., c_n are constants determined by initial conditions and the input signal. In order for the system to be stable, each of the factors e^{s_1t}, e^{s_2t}, ..., e^{s_nt} must decrease with time.

*2 Nyquist stability criterion: In vector locus plots, the vector with origin at $(-1, 0)$ is called the vector of the characteristic equation. In the s-plane, when ω changes from 0 to ∞, then in a clockwise direction from $-\infty$ to 0, if a root is located on the right side plane, its vector rotates around a characteristic root in a clockwise direction. This corresponds to rotation of the vector of the characteristic equation around the point $(-1, 0)$.

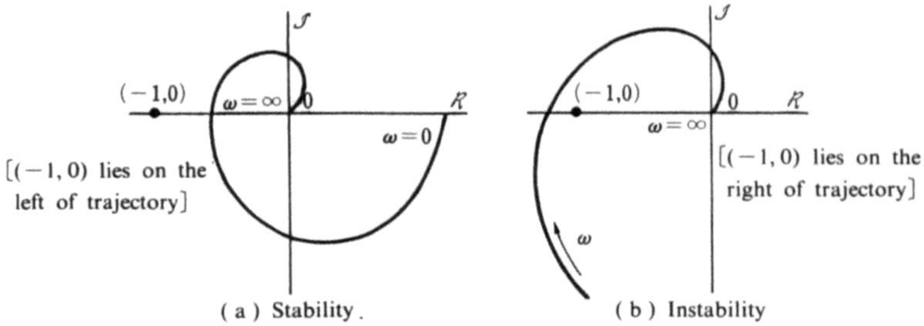

Fig. 2·11 Nyquist stability criterion

of stability and the gain becomes 1 and phase delay 180° (positive feedback) with continuous oscillation of the frequency ω_0. Accordingly, the degree of stability can be indicated by the distance from the above limit of stability.

Gain Margin (GM): The value indicating the margin between a gain of 1 and the gain at the frequency in which the phase becomes −180°.

Phase Margin (PM): The value indicating the margin between a phase of −180° and the phase at the frequency in which the gain becomes 1.

Gain margin and phase margin are shown in Fig. 2.12 using a vector locus plot and a Bode diagram.

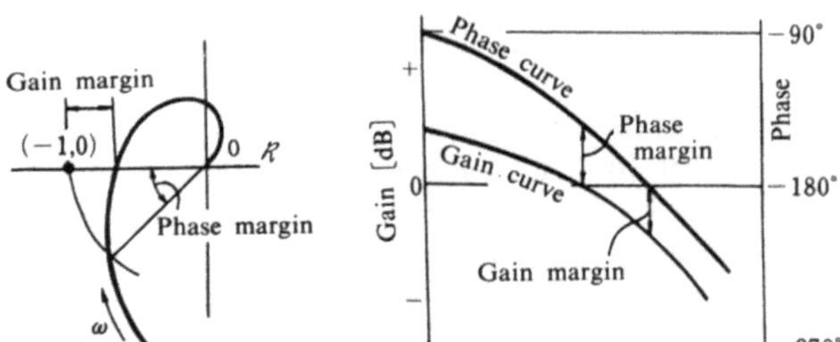

Fig. 2·12 Gain margin and phase margin

2.2 *Process Characteristics*

2.2.1 *Process degrees of freedom and controlled and manipulated variables*

For processes which are in an equilibrium state, the number of process variables which can be changed independent of each other is called the degrees of freedom of the process. For example, the process shown

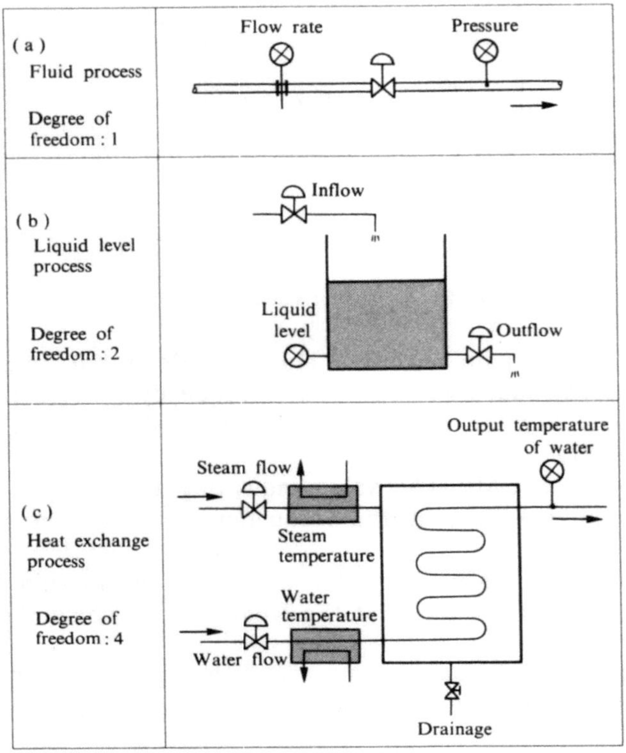

Fig. 2・13 Process and degree of freedom

in Fig. 2.13(a) characterized by the flow rate and pressure of a fluid flowing in a conduit has one degree of freedom, while that in Fig. 2.13(b) characterized by liquid level has two degrees of freedom (inflow and outflow), and while that in Fig. 2.13(c) characterized by water output temperature has four degrees of freedom (steam flow, steam temperature, water flow and water temperature). In other words, the degrees of freedom is equivalent to the number of available manipulated variables in the control system.

In order to control a controlled variable, a manipulated variable obviously is required. In the example of Fig. 2.13(a), the controlled variable may be either flow rate or pressure, while the manipulated variable must be flow rate. In this kind of a situation, it is possible to control only one controlled variable; using the same manipulated variable to configure more than one control loop leads to mutual interference. In response to such a situation, override control (see Sec. 2.3) can be used in which, depending on process conditions, any one of controlled variables is selected and controlled. In the Fig. 2.13(b) example, where the controlled variable is liquid level and the manipulated variable is either inflow or outflow, control can be performed through either of the latter. However, if we introduce a buffer tank here so that the inflow is no longer a manipulated variable and the degrees of freedom becomes 1, override control becomes necessary to control either outflow or liquid level. In Fig. 2.13(c), the controlled variable is the output water temperature, and although the manipulated variable may be selected from any of four, in practice, steam flow is made the manipulated variable. Note here that in cases where the number of controlled variables is less than that of manipulated variables, variation in the non-selected manipulated variables creates disturbance in regards to the selected controlled variable, and should thus be kept constant.

2.2.2 Process characteristics

Process characteristics can be largely divided into static characteristics and dynamic characteristics, as described below.

(a) Static characteristics

This refers to steady-state characteristics when step signals of various sizes are applied as input signals. In self-regulating processes, the property describing the size of the controlled variable with respect to that of the manipulated variable is called the process static gain. Since the controlled variable and the maniplated variable are not necessarily of the same dimension, the dimension of the transfer element would be adopted in such cases.

Self-regulation, or self stabilization, is possessed by many processes such as first-order lag systems and heat-exchangers. In contrast, an example of a process not having self-regulation is the constant outflow process shown in Fig. 2.14. Here, if the inflow = outflow, the liquid level does not change, but if the inflow > outflow, the liquid level continuous to increase, and inversely if the outflow > inflow, it continues to decrease, resulting in an integrative process. In addition

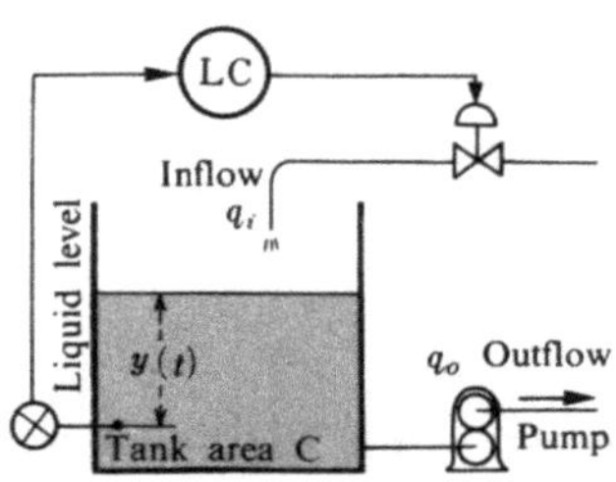

Fig. 2・14 Am example of the process without self-regulation (Liquid level process with constant outflow)

to integrative processes, another example of non-self-regulating processes are those involving exothermic chemical reactions such as polymer reactions. In order to drive such processes, control is absolutery required, and in comparison to self-regulating processes, it is more difficult to perform. Moreover, though many processes possess nonlinearity, there are many times in which they can be treated as linear at a point about the equilibrium point.

(*b*) *Dynamic characteristics*

Since dynamic characteristics ave characterized by an output signal $y(t)$ corresponding to an input signal $x(t)$, signal relationships can be expressed by a convolution integral in the time domain, and by a transfer function or frequency transfer function in the frequency domain. In addition, as for ways of portraying dynamic characteristics, transient response for time and frequency response for frequency can be effectively used as previously described.

(*c*) *Disturbance*

In processes, external influences upsetting the state of a system, *i.e.*, disturbance, typically exist. There are various forms, sizes and entry points of disturbance, and they can often be represented by a unit step signal applied to the process.

In the heat-exchanger example of Fig. 2.13(c), examples of disturbance would be changes in any of the three quantities other than the steam flow selected as the manipulated variable: steam temperature, water temperature, and specifically water flow corresponding to changes in the load. Factors such as outside air temperature must also be considered as external disturbance.

2.2.3 Process models

A process can also be viewed as one transfer element. Although representative transfer elements and their characteristics have been shown in Table 2.1, the characteristics of process models are investigated below.

(*a*) *Proportional element*

In the example of Fig. 2.13(a), the fluid flowing in the conduit can be regarded as a proportional element if we ignore the delay caused by inertia. The amount of flow rate change K occurring when the valve is opened or closed by a unit amount is referred to as proportional gain (Table 2.1(a)).

(*b*) *Dead-time element*

The conveyor shown in Fig. 2.15 possesses a dead-time element. If we specify the distance between the hopper outlet and the weighing scale as l, and the speed of the conveyor belt as v, the time from when certain particles are released up to the point when they reach the

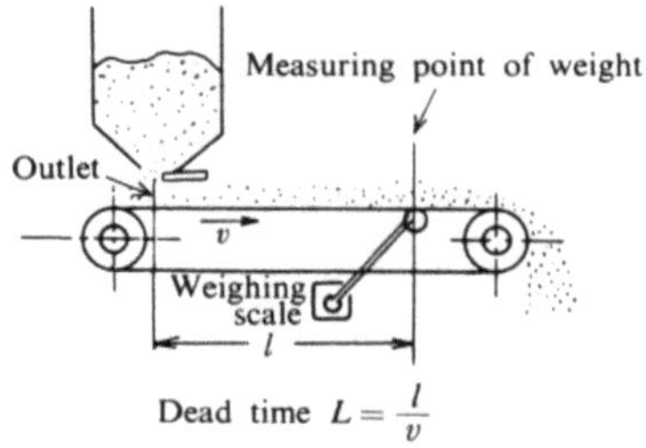

Fig. 2·15 An example of dead time process (Belt conveyer)

weighing scale is $L=l/v$, where L is the symbol for dead time. The Laplace transform for dead time L is e^{-Ls}.

As can be seen from the Bode diagram in Table 2.1(d) for dead-time elements, the gain is always constant and is unrelated to frequency. The phase, on the other hand, is delayed infinitely with increase in ω. It can thus be seen, without considering gain and phase margin, that control of a dead-time element is difficult.

(*c*) *Integral element*

The liquid level system in Fig. 2.14 consisting of the constant outflow tank is an integral element process. If we designate the inflow as q_i, the outflow as q_o, the liquid level as $y(t)$, and the tank cross-sectional area as C, we obtain the expression

$$C\frac{dy(t)}{dt}=q_i-q_o$$

If we let q_o be constant, and $q_i-q_o=x(t)$, we then get

$$y(t)=\frac{1}{T}\int x(t)dt$$

where $C\equiv T$ (Time constant).

If we now express the above as a transfer function, we get

$$\frac{Y(s)}{X(s)}=\frac{1}{Ts}$$

As shown by the Bode diagram in Table 2.1(e), the gain in this case is 1 at a frequency where $\omega T=1$, becoming a -20 dB/dec* negative slope straight line; moreover, it is infinite at $\omega=0$. The phase has a 90° delay and is unrelated to the frequency.

* -20 dB/dec: this signifies -20 dB (deci-Bel) per decade, or in other words, if the frequency is increased by 10 times, the gain turns out to be 1/10. Likewise, -6dB/oct siginifies -6 dB per octave, or if the frequency is doubled, the gain becomes 1/2.

(*d*) *First-order lag element*

The liquid level system shown in Fig. 2.16 characterized by natural down flow is assumed to be a first-order lag element process.

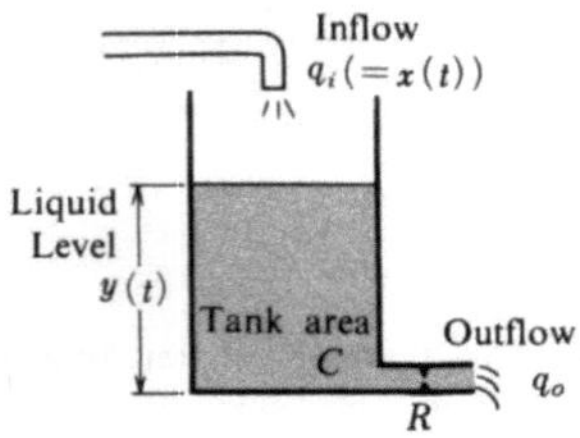

Fig. 2 · 16 An example of first-order lag process (Liquid level process)

It is expressed as

$$C\frac{dy(t)}{dt}=q_i-q_o$$

where inflow $q_i=x(t)$.

If we now designate the outflow resistance as R and let the amount of outflow be proportional to the liquid level, then

$$q_o=\frac{y(t)}{R}$$

$$C\frac{dy(t)}{dt}=x(t)-\frac{y(t)}{R}$$

If we let $CR=T$(Time constant), then we get

$$T\frac{dy(t)}{dt}+y(t)=Rx(t)$$

Expressing the above as a transfer element:

$$\frac{Y(s)}{X(s)}=\frac{R}{1+Ts}$$

The step response of a first-order lag element is shown in Fig. 2.17. At $t=0$, if we take a tangent to the response curve and extend it to the curve's final value, the corresponding time interval is called the time constant. In addition, the value of the curve at time constant T is 63.2% of the curve's final value. As shown by the Bode diagram in Table 2.1(b), the gain is characterized by a crossover point at the frequency where $\omega T=1$. At frequencies below this point, the gain is constant along the asymptotic line, while at frequencies above this point, the gain drops off to the right as -20 dB/dec. The gain at the crossover point is -3 dB; the phase at the crossover point frequency has a 45° de-

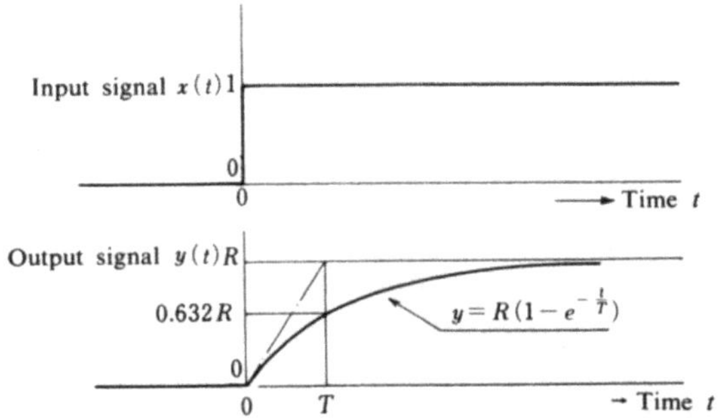

Fig. 2 · 17 Step response for first-order lag

lay, and approaches a 90° delay with increasing frequency.

(*e*) *Second-order lag element*

As shown in Fig. 2.18(a), a second-order lag element can be formed by connecting two first-order lag elements in a two-stage sequential fasion, and its transfer function is simply the product of the transfer functions for each first-order lag element. Such a sequential structure, in which following stages have no influence over previous stages, is called a cascade connection.

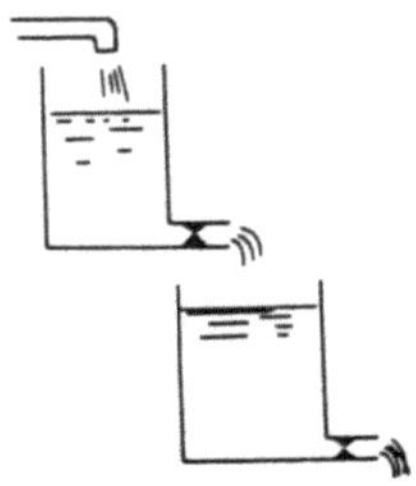

(a) An example of non-interacting second-order lag process (Dependent two tanks in series)

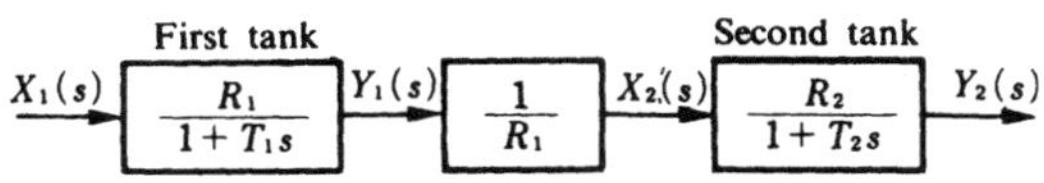

(b) Block diagram

Fig. 2 · 18 Cascade connection of tanks

The transfer function of the first-stage-tank has been previously determined as

$$\frac{Y_1(s)}{X_1(s)}=\frac{R_1}{1+T_1s}$$

Moreover, since the input $X_2(s)$ of the second-stage tank is the outflow of the first-stage tank, $X_2(s)=Y_1(s)/R_1$. The entire transfer function can then be expressed as

$$\frac{Y_2(s)}{X_1(s)}=\frac{R_2}{(1+T_1s)(1+T_2s)}$$

The block diagram for this process is shown in Fig. 2.18(b).

The step response for a second-order lag system is shown in Table 2.1(c). For lag systems of second or higher orders, the slope of the response curve at $t=0$ is zero, and since the response curve has an inflection point, the delay can no longer be expressed by a single time constant. Looking at the gain diagrams, there are crossover points at the $1/T_1$ and $1/T_2$ frequencies, and at frequencies above the higher crossover point, the gain becomes a $2\times(-20\ \mathrm{dB/dec})$ straight line. The phase is delayed up to a maximum of $2\times 90°=180°$.

If two tanks are connected as shown in Fig. 2.19, the liquid level of the second-stage tank here does influence the first-tank, which means a non-cascade connection. The transfer function here becomes

$$\frac{Y_2(s)}{X_1(s)}=\frac{R_2}{T_1T_2s^2+(T_1+T_2+R_2C_1)s+1}$$

The new time constant R_2C_1 in this second-order lag system indicates interference between the two tanks.

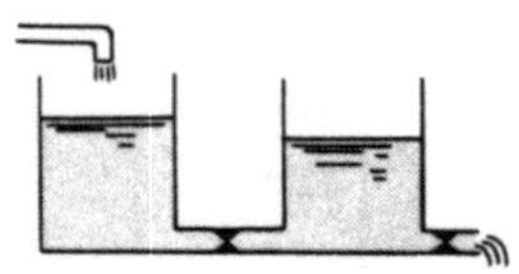

Fig. 2 · 19 An example of interacting second-order lag process (Interacting two tanks in series)

(f) Higher-order lag systems

A system formed by connecting multiple first-order lag systems sequentially is called a higher-order lag system. Step responses for cascade-connected first-order lag systems having identical time constants is shown in Fig. 2.20. As can be seen, as the order of the lag system increases, the time until the rising edge begins lengthens, indicating similar characteristics as the addition of a dead-time element. Figure 2.21 illustrates a characteristic of higher-order lag systems in which a tangent taken at the inflection point is often used to approximate systems with dead time and a time constant. Here, the dead time is called the equivalent dead time, and the time constant the equivalent time constant. Expressing this as a transfer function, we get

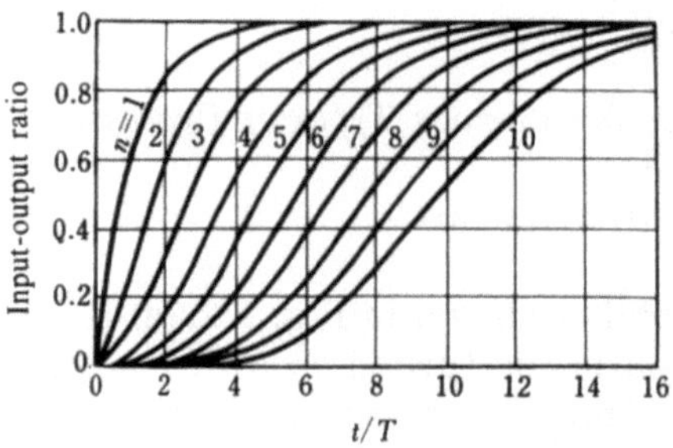

Fig. 2 · 20 Step response for $\{1/(1+Ts)\}^n$

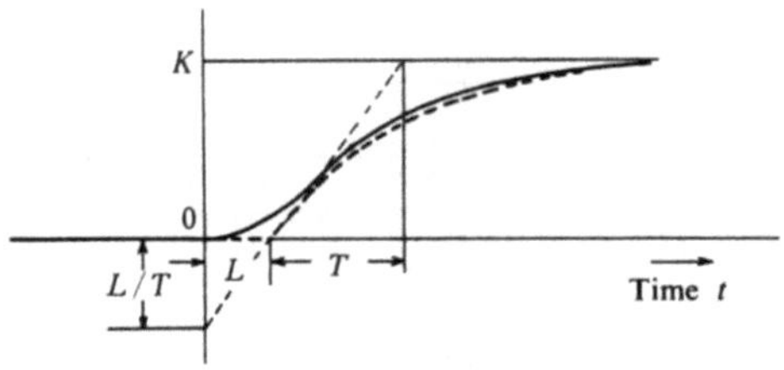

Fig. 2 · 21 Simulation of higher-order lag response

$$G(s)=\frac{Ke^{-Ls}}{1+Ts}$$

The value L/T which increases as the order of the lag system increases is used as an indicator of the relative difficulty of control. The values of equivalent dead times and equivalent time constants for the first five higher-order lag systems with identical time constants are shown in Table 2.2.

Table 2.2 Equivalent dead time and equivalent time constant

Order of lags	Equivalent dead time	Equivalent time constant	Equivalent dead time/Equivalent time constant
First-order lag	0	T	0
Second-order lag	0.28 T	2.7 T	0.103
Third-order lag	0.81 T	3.7 T	0.218
Fourth-order lag	1.43 T	4.5 T	0.321
Fifth-order lag	2.10 T	5.1 T	0.410

In regards to the frequency response, if $T_1>T_2>\ldots>T_n$, then the gain asymptote for $0<\omega<1/T_1$ is 0 dB/dec, for $1/T_1<\omega<1/T_2$ it is -20 dB/dec, and for $1/T_n<\omega$ it is $-20n$ (dB/dec). For the phase, successive delays of up to $n \times 90°$ occur.

2.3 Control Formats for Various Types of Processes

2.3.1 Single loop control systems

(a) On–off control

The on–off controller shown in Fig. 2.22(a) is called a two–position controller since its output is turned on or off in response to the polarity of the deviation. In addition, since the manipulated variable at on time or off time will be either too big or too small for maintaining the controlled variable at the desired value, cycling occurs.

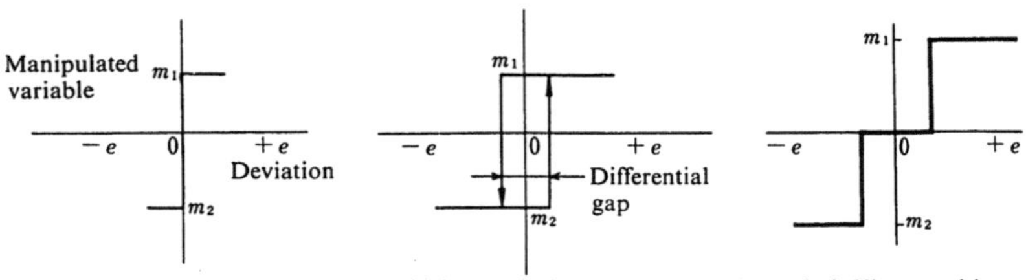

(a) Theoretical ON-OFF controller (b) Practical ON-OFF controller (c) Three-position controller

Fig. 2・22 ON-OFF controllers

In actual on–off controllers, however, a differential gap (hysteresis) is employed, as shown in Fig. 2.22(b). Without such a differential gap, the controller will rapidly turn on and off continuously in the area of the desired value, and the lifetime of the on–off mechanism will be greatly shortened. In some cases, as in a bimetal thermostat, a differential gap is inherent, while in others, such as through an on–off controller, a differential gap may be purposely implemented. In Fig. 2.22(c), a three–step controller is shown, in which an intermediate step has been designed in.

If there is a differential gap, the cycling period will become longer, and the amplitude as well will become larger. In addition, in an on–off control system, if the value of the manipulated variable just between that at on time and off time does not bring the controlled variable to the desired value, the average value will then deviate from the desired value, and offset (see *Proportional control* below) results.

(b) PID control

(1) *Proportional control* Control action in which the size of output is propotional to that of the input is called proportional action (P–action), and is expressed by the basic equation:

$$Y(s)=K_cX(s) \tag{2.6}$$

where K_c is the proportional gain. In actual controllers, however, the following property called the proportional band (PB) is used instead of K_c:

$$PB = \frac{1}{K_c} \times 100 \ [\%]$$

In other words, as shown in Fig. 2.23, the proportional band indicates in percentages the width of change in input necessary to produce an effective change in output from 0 to 100%.

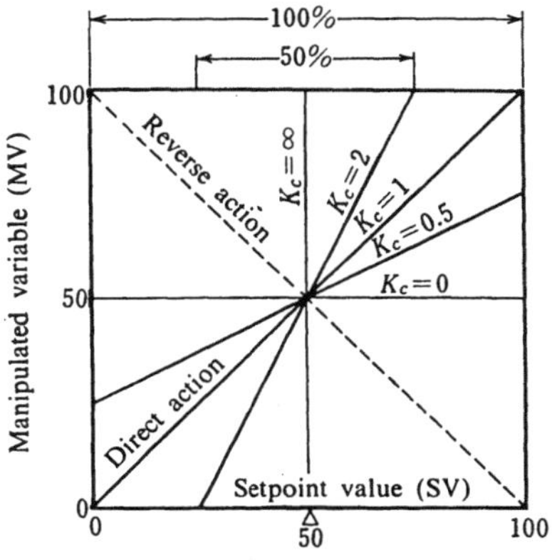

Fig. 2・23 Proportional band

Note that the transient response and frequency response of proportional action is the same as that for the proportional element shown in Table 2.1(a), in which proportional action is a variable gain proportional element.

For step response in a proportional control system, controlled deviation whose value after a sufficient amount of time becomes fixed is called an offset or steady-state deviation. In proportional control, since the controller output value takes on a specific value corresponding to the control deviation, offset will occur as long as the controlled variable resulting from the manipulated variable (output value) when the deviation is zero is not equal to the desired value.

In Eq. (2.6) above, since the output becomes 0 with an input of 0, a bias M is addid as follows:

$$Y(s) = K_c X(s) + M$$

If this bias is made variable, controller output can then be changed regardless of the control deviation and the controlled variable can be manually reset to the desired value. This manual reset function is normally provided in proportional controllers.

(*c*) *Integral action*

In integral action, also called I–action or reset action, the size of the output is proportional to the time integral value of the input, as follows:

$$Y(s)=\frac{1}{T_{I}s}X(s)$$

where T_1 is the reset time.

Since the output continues to be increased or decreased until there is no more deviation, the offset generated in proportional control can be eliminated—thus the term reset action. The strength of integral action is indicated by the reset time: the shorter the reset time, the stronger the integral action. Time units normally employed are minutes or seconds.

The transient response and frequency response of integral action is the same as that for the integral element shown in Table 2.1(e). At $\omega=0$, the theoretical gain becomes ∞. Although in proportional control, the offset decreases with rise in proportional gain, the gain increase across the entire frequency band will cause instability. Accordingly, it is considered that in integral action, offset can be eliminated by raising the gain only in the lower frequency band. On the other hand, since a phase delay of 90° across the entire frequency band is not preferable to control stability, integral action is normally combined with proportional action (PI–action) as follows:

$$Y(s)=K_c\left(1+\frac{1}{T_{I}s}\right)X(s)$$

The step response and Bode diagram for PI–action are shown in Figs. 2.24 and 2.25. In the step response, the time up to the point when the output due only to proportional action and that due only to integral

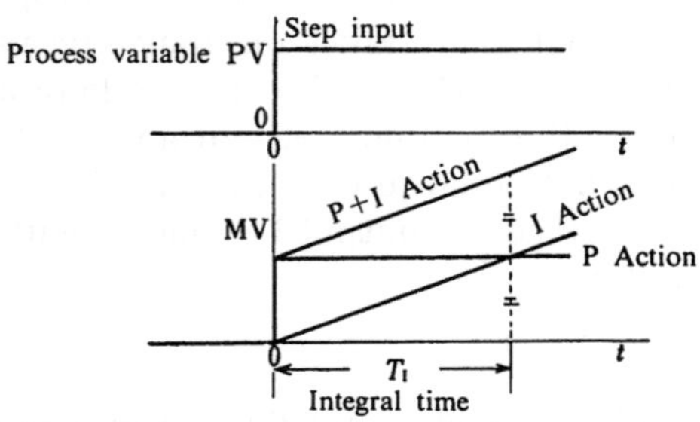

Fig. 2・24 Step response for proportional plus integral action

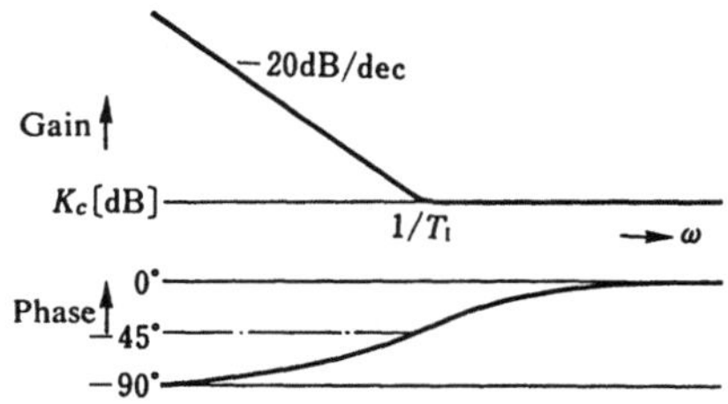

Fig. 2 · 25 Bode diagram for proportional plus integral action

action become equal is the reset time. From the phase diagram, it can be seen that at $\omega = 1/T_I$ the phase delay decreases to 45°, and it approaches zero with increasing frequency.

Reset Windup:

In integral action, since the time integral value for the deviation is output, if a state having deviation continues for a long period of time, the output due to integral action becomes saturated, similar to a control halt period in batch control. This situation is called reset windup (reset action saturation). In Fig. 2.26, control for a reactor batch process is shown. After the process is started, the deviation continues to decrease with the rise in reactor temperature, and eventually becomes zero, while the output due to integration remains saturated. Next, as the deviation polarity begins to change, the output likewise begins to decrease. However, since in general, controller output will exceed the 0 to 100% range to some extent (to ensure complete closing of the controller valve), a further delay will occur until the control valve begins

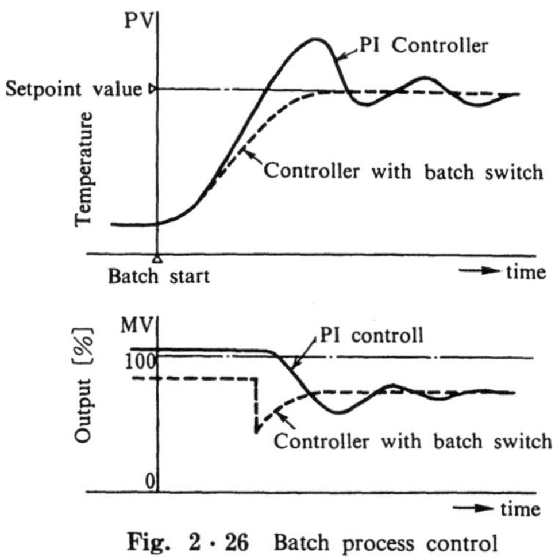

Fig. 2 · 26 Batch process control

to function. Consequently, a large reset overshoot occurs. In response to this situation, batch control has come to use controllers with a batch switch for switching output when needed as well as controllers performing proportional control in which integral action can be temporarily stopped.

(*d*) *Derivative action*

In derivative action, also called D–action or rate action, the size of the output is proportional to the time derivative of the input value (the rate of change of input), as follows:

$$Y(s) = T_D s X(s)$$

where T_D is the rate time.

As derivative action is characterized by output which is proportional to the speed of change in input, it cannot be used exclusively by itself but must be combined with proportional action or proportional and integral action. The ramp response and Bode diagram for combined proportional and derivative action are shown in Figs. 2.27 and 2.28. In the ramp response, the time up to the point where the output due to proportional action and that due to derivative action become equal is the rate time, and the longer the rate time the stronger the derivative ac-

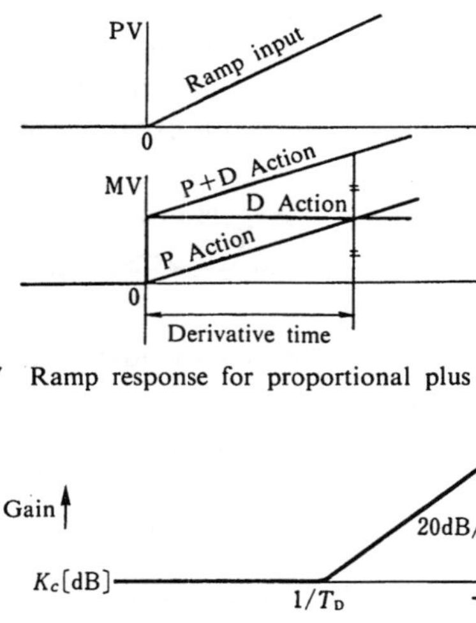

Fig. 2・27 Ramp response for proportional plus derivative action

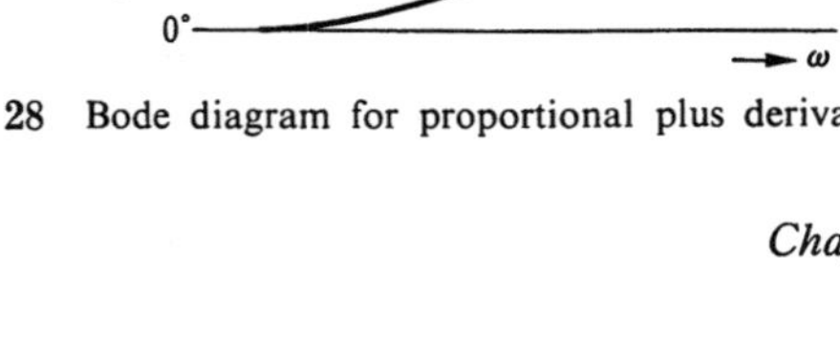

Fig. 2・28 Bode diagram for proportional plus derivative action

tion. The time units normally used are minutes or seconds.

As shown by the phase diagram, the phase advances to 45° at $\omega=1/T_D$, and advances to 90° with increasing frequency. Consequently, the process phase delay can be compensated for and control stability improved. On the other hand, above the crossover frequency, the gain increases along the 20 dB/dec asymptote. As a result, the output can undesirably change significantly in response to a small change in the setpoint value, measured value, or noise. Thus, actual controllers employ an incomplete derivative, in which a first-order lag is added to a derivative system. Extracting the differential term, we get

$$Y(s)=\frac{T_D s}{1+T_d s}X(s)$$

where the first-order lag time constant T_d is called the derivative time constant and T_D/T_d is the derivative amplitude. A value of about 10 is normally chosen for the derivative amplitude.

The step response for incomplete derivative action and the Bode diagram for combined proportional and incomplete derivative action are shown in Figs. 2.29 and 2.30. In the step response, the ratio be-

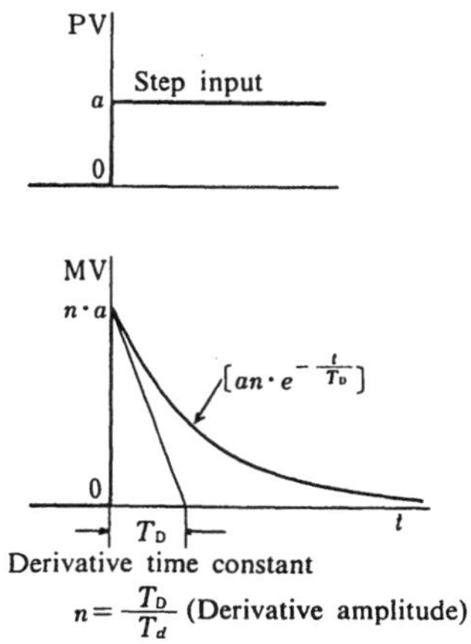

Fig. 2 · 29 Step response for incomplete derivative

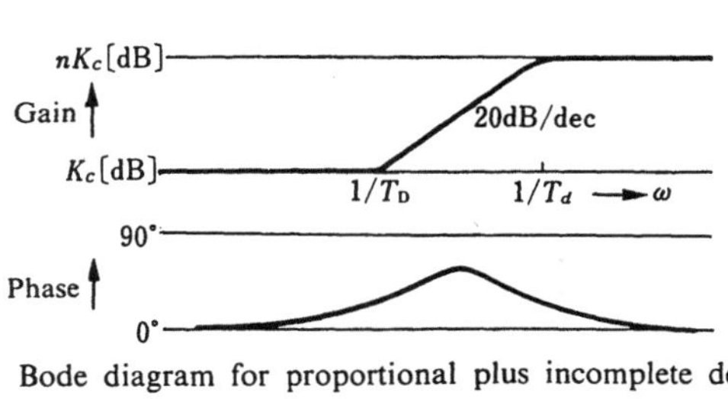

Fig. 2 · 30 Bode diagram for proportional plus incomplete derivative tween the amplitude of the input step signal and the maximum amplitude obtained from derivative action is the derivative amplitude, and the time constant indicated by the response curve is the derivative time constant. Note also that as shown by the Bode diagram, the phase never reaches 90°, though the gain levels out at a multiple of the derivative amplitude.

PV Derivative: Up to now, although we have been tacitly assuming a controlled deviation for $X(s)$, a configuration in which the measured value (PV) is applied to the input signal is called a PV Derivative or a Derivative Ahead. A block diagram for PV Derivative type PID is shown in Fig. 2.31. In regards to disturbance, this type of configuration functions in the same way as the deviation derivative type, while the output, in response to a change in the setpoint value (SV), does not change suddenly; this enables changes in the setpoint value to be performed easily.

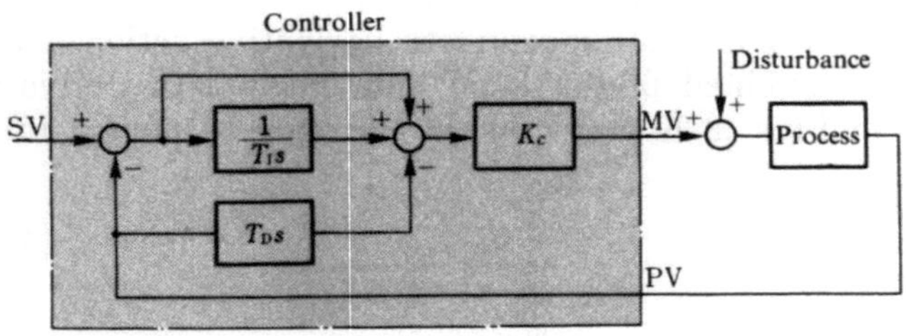

Fig. 2·31 Block diagram for derivative ahead

In addition, digital controllers employ algorithms in which the measured value can be applied not only to the derivative term but to the proportional term as well. Moreover, a PID with two degrees of freedom is adopted in which effectiveness of the above is adjustable. In regards to a change in the setpoint value, optimal algorithms in conjunction with the control mode (auto or cascade) are selected. (Refer to Subsec. 4.2.3 (b) for more details.)

(*e*) *PID action*

The combined three type of actions, proportional, integral and derivative, are referred to as PID action. The basic equation for PID action is expressed as follows:

$$Y(s)=K_c\left(1+\frac{1}{T_I s}+\frac{T_D s}{1+T_d s}\right)X(s) \tag{2.7}$$

An example of a step response for a PID controller is shown in Fig. 2.32.

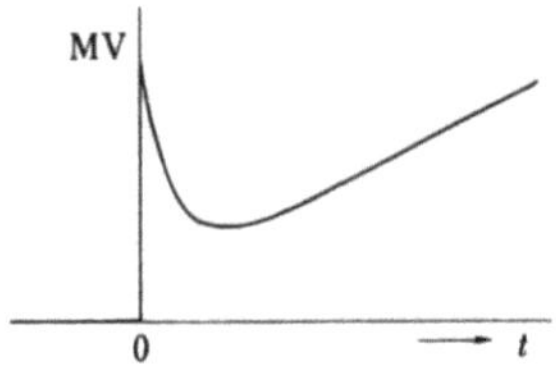

Fig. 2・32 Step response of PID controller

2.3.2 Compound loop control system

(a) Cascade control

In a feedback control system, the control which is performed by having the output signal from one control device (primary controller) determines the desired value of another control device (secondary con-

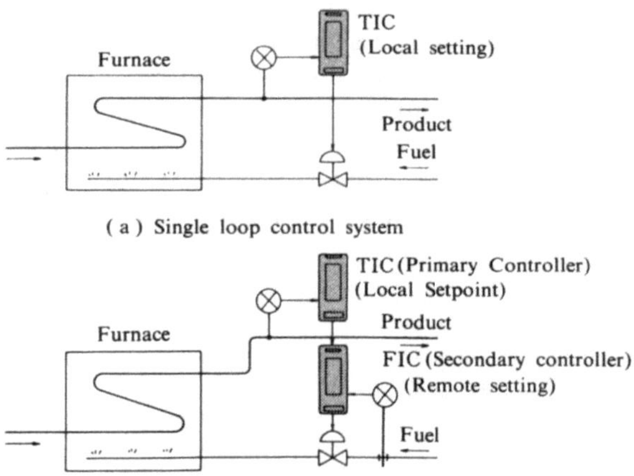

(a) Single loop control system

(b) Cascade loop control system

Fig. 2・33 Furnace control system

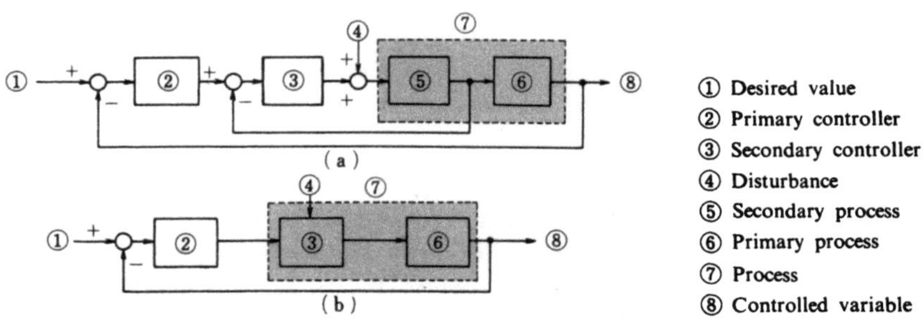

Fig. 2・34 Block diagram for cascade control system

troller) is referred to as cascade control. Figure 2.33 shows examples of a furnace in a single-loop control system and one in a cascade-loop control system. Block diagrams for a cascade control system are shown in Fig. 2.34. In particular, Figure 2.34(b) is a modification of (a), in which the secondary control system becomes one part of the process, as seen from the primary control system.

The purpose for configuring a cascade control system is to eliminate, through use of a secondary controller, the influence of disturbance on the primary process, which enters through the secondary control loop. In this furnace example, the fluctuation in the amount of fuel is suppressed by the secondary control system. In addition, since the phase delay of the secondary process is improved by the secondary control system, the response of the primary control system becomes effectively faster and the settling time shorter. Moreover, the non-linearity of the secondary process decreases.

In order to construct an effective cascade control system, the period of natural frequency of the secondary control system shoud be 1/3 or less than that of the primary control system. Figure 2.35 shows an example of temperature-temperature cascade control for a polymerization reactor. Since there is not much difference between the primary and secondary periods here compared with temperature-flow control, the secondary controller, tuned by mainly proportional control action, makes the response of the control system as fast as possible.

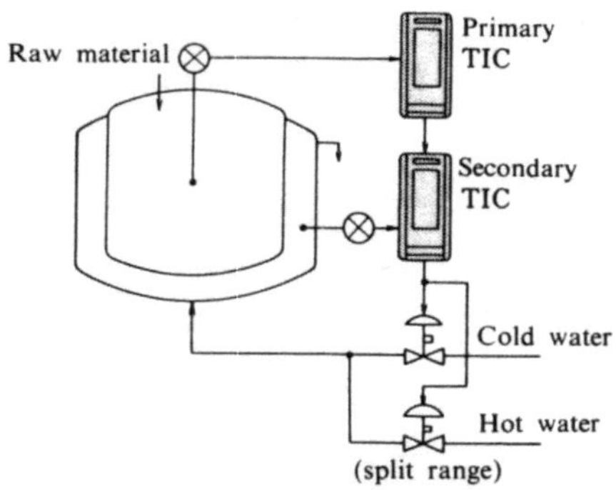

Fig. 2 · 35 Cascade control of polymerization reactor

If the secondary control system possesses non-linearity, the loop transfer function gain in the primary control system will undesirably fluctuate. For example, if the flow rate is measured with an orifice, the transmitter output signal is proportional to the second power of the flow rate. Thus, since convertors and transmitters are located on the feedback side of the control loop, the amount of feedback at small

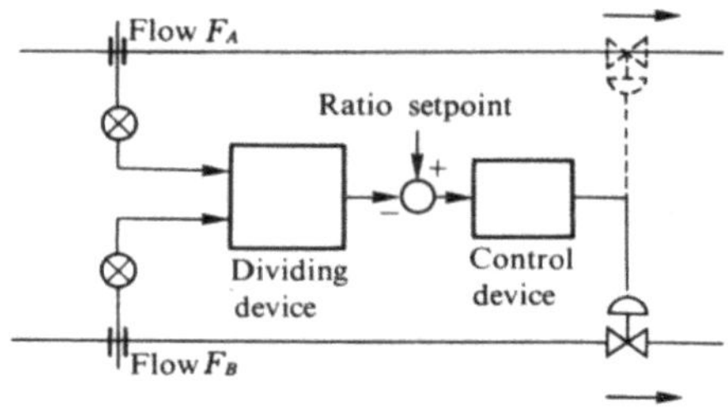

Fig. 2 · 36 Concept of ratio control

flow rate decreases, and the gain of the secondary control system increases. Accordingly, when the flow rate at times like startup is small, the control system will tend towards instable; a square-root extractor device to provide linearity is required.

(*b*) *Ratio control*

Control which attempts to preserve a ratio relationship between two or more variables is called ratio control. A typical example would be air-fuel ratio control in a combustion furnace, in which the ratio between two flow values is controlled. The fundamental concept of ratio control is shown in Fig. 2.36. As shown, the ratio between two flow values is determined by a divider and passed to a controller in which the desired ratio has been set. The controller in turn manipulates one or the other flow value.

However, in this method, since the divider is situated within the loop, if we place the control valve on the flow line corresponding to the ratio numerator, the loop gain due to change in PV corresponding to the denominator also changes, and alternatively if we place the control valve on the flow line corresponding to the denominator, the loop gain due to change in MV changes creating an undesirable non-linearity. In actuality, the configuration shown in Fig. 2.37 is employed. Here, one of the flow values is multiplied by the ratio as set by the manual ratio setting device, and the result is used as the setpoint value for the controller governing the other flow rate. In contrast to this se-

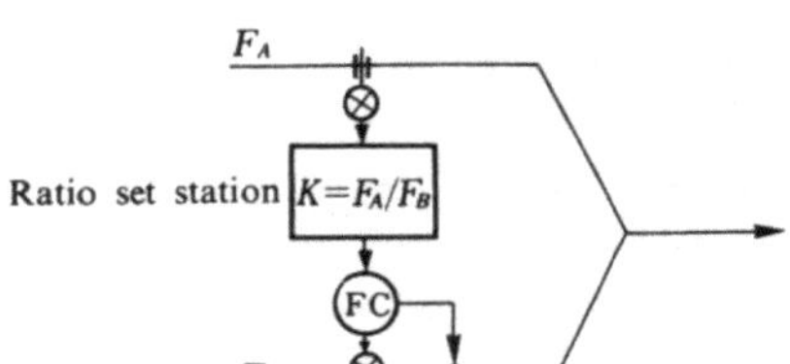

Fig. 2 · 37 Ratio control (series)

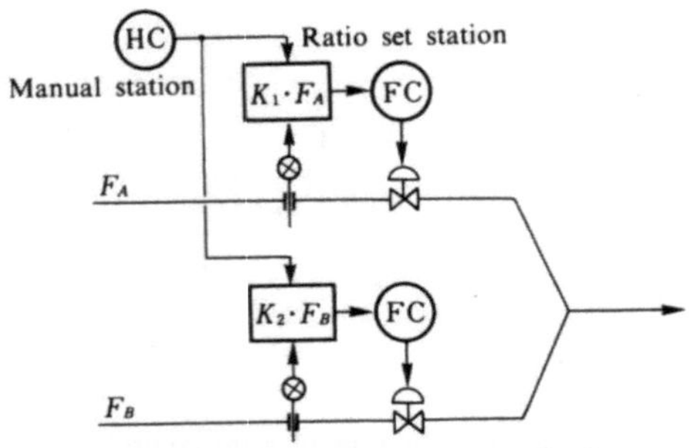

Fig. 2 · 38 Ratio control (Parallel)

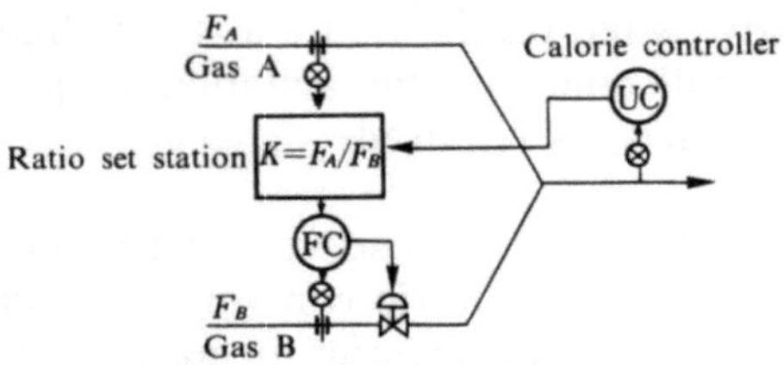

Fig. 2 · 39 Remote ratio setting

ries setting format, a parallel setting format is shown in Fig. 2.38, in which the flow delay on the controlled side in a series setting can be eliminated.

The setting range of a ratio set station for an analog unit is about 0.3 to 3.0. When using orifices or flow meters having square-law characteristics in which the square of the set ratio will become the actual ratio, a scale range from about 0.6 to 1.7 is used. In addition, ratio set stations incorporating microprocessors provide a wide setting range from about 0.0 to 8.0, as well as a square-root extraction function. Note that the ratio to be set on the ratio set station is not an actual flow rate ratio but that when each flow range is supposed to be 0 to 100%.

For ratio set stations employing microprocessors, external ratio setting is possible. In the example of Fig. 2.39, where the calorie value needs to be constant, the ratio in the set station is being remotely set.

In addition to ratio control of instantaneous flow, an integrating flow ratio control system can be used, called blending control. Here, flow meters such as PD meters and turbine meters which can accurately generate pulses are used (see Subsec. 4.2.5).

(*c*) *Selector control* This form of control includes selection of measured values as well as selection of output from either of two controllers sharing a common manipulated variable, and having different con-

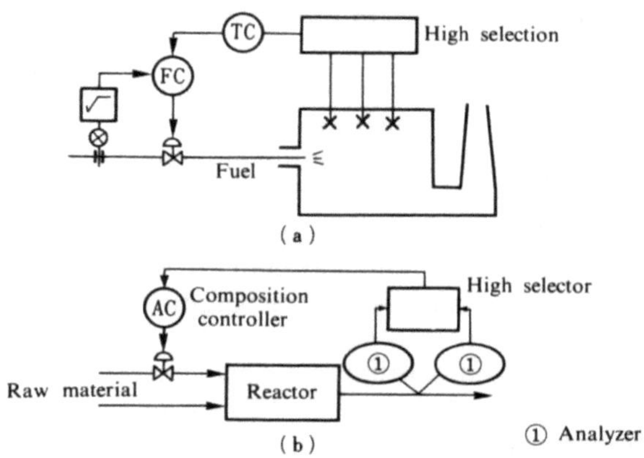

Fig. 2・40 Selective control of process variables

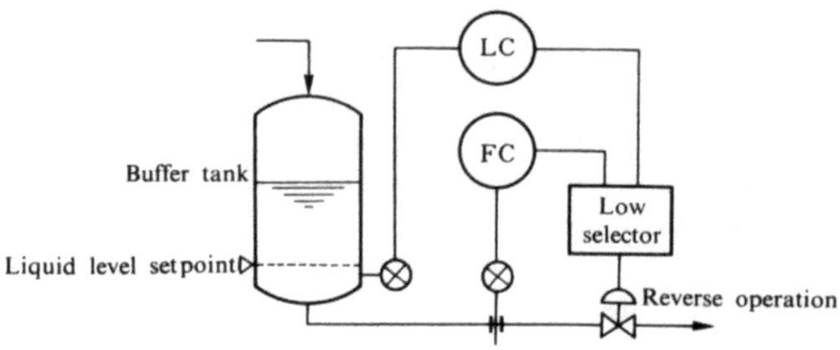

Fig. 2・41 Override control of buffer tank

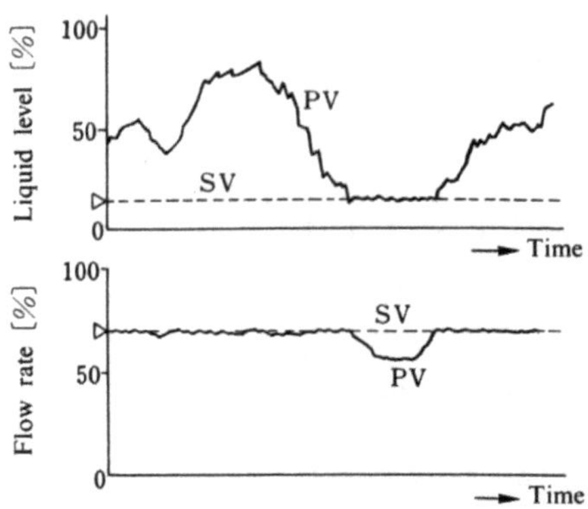

Fig. 2・42 Override control

trolled variables.

Figure 2.40 shows examples of measured value selection control. In (a), the object is to protect the furnace, while in (b), the object is to avoid the influence of analyzer failure by incorporating redundant analyzers.

Figure 2.41 shows an example of buffer tank override control. The object here is to maintain a constant feed to the next process as long as the tank is not empty. If the liquid level is above the setpoint value, the flow control system is selected and the feed amount is kept constant, while if the liquid level falls to the setpoint value, the liquid level control system is selected, and control is performed to prevent the liquid level from dropping below the setpoint value. This process is illustrated in Fig. 2.42. In this example, a "reverse operation" (air to open) control valve is used, so that the flow controller is to be "reverse action" and the liquid level controller "direct action," and the selector becomes a low selector (if the control valve is of direct operation, the selector is then a high selector).

Since there is deviation in the non-selected control loop, reset windup will occur due to integral action of the PI controller. To prevent this, an external feedback type of controller can be used, as shown in Fig. 2.43. As indicated by the following equations, the integral term of the non-selected controller can be substituted by the output from the selected controller, and the non-selected controller can thus function as a proportional controller.

Selected controller: $Y_F(s) = -K_c\left(1+\frac{1}{T_I s}\right)X_F(s)$

Non-selected controller: $Y_L(s) = K_c X_L(s) + Y_F(s)$

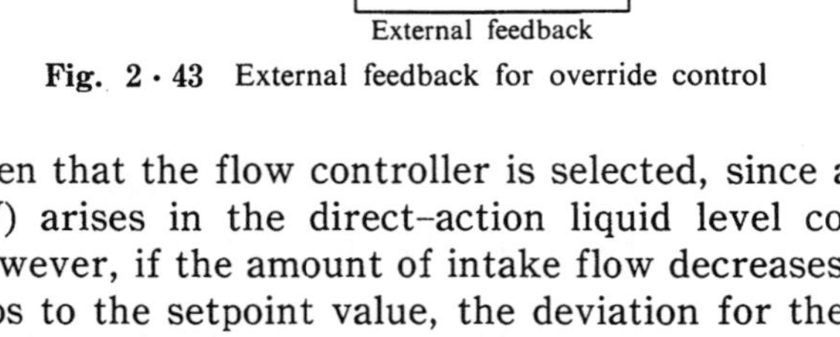

Fig. 2・43 External feedback for override control

Now, given that the flow controller is selected, since a plus deviation (SV<PV) arises in the direct-action liquid level controller, $Y_L(s) > Y_F(s)$. However, if the amount of intake flow decreases and the liquid level drops to the setpoint value, the deviation for the liquid level controller becomes 0, so that $Y_L(s) = Y_F(s)$. Moreover, if the liquid level drops below the setpoint value, the polarity of the deviation reverses,

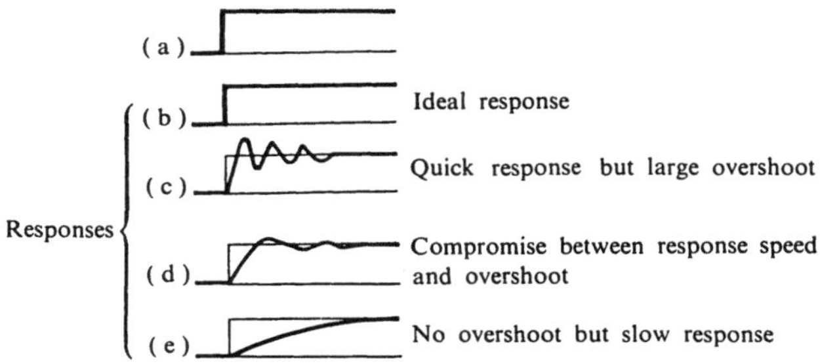

Fig. 2・44 Control responses for setpoint change

i.e., $Y_L(s) < Y_F(s)$, and a "bumpless" transfer is made to the liquid level controller.

2.4 Optimal Adjustment of Control Systems

(a) Control system evaluation

Control systems are evaluated by such characteristics as stability, speed of response and size of offset. In Fig. 2.44, four examples of responses to a step change in the setpoint value are shown. The step change is shown in (a), and the four responses in (b) through (e). Although (b) is the ideal response, it cannot actually be realized. Response (d), which lies between responses (c) and (e), is usually the most sought after; evaluation criteria, however, may differ depending on the type of process.

(b) Evaluation method

Response waveforms and their related terminology are shown in Fig. 2.45. Although there are various criteria for evaluating such waveforms, they can be roughly divided into the following three types:

(1) *Amplitude damping ratio* This is a value indicating the way in which the response waveform undergoes damping, and is determined as shown in terminology example 5.

(2) *Control area* Three ways of integrating the control area are available, as shown in terminology example 4.

(3) *Transient overshoot and response time* Since the transient overshoot (terminology example 1) and the response time (terminology example 7) possess opposing properties, response time for a given overshoot can be used as evaluation criteria.

(c) Effects of P, I and D

Figure 2.46 shows the resultant control responses when applying external disturbance in the form of a step signal, equivalent to 30% of an input signal, to a P, PI, PD or PID controlled third-order lag system having three first-order lags with a 10 second time constant and a

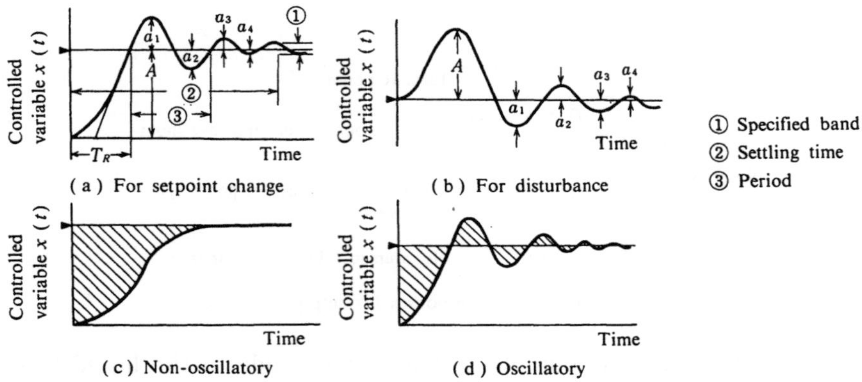

Terminology : —

1. Overshoot $\frac{a_1}{A}\times 100$ [%]
2. Specified band ±2 to 5 % of span
3. Settling time Time required for the output to enter and remain within a specified band centered
4. Control area Shaded portion of figures (c) and (d)

 Integral of absolute value of error $A_1=\int_0^\infty |x(t)-x(\infty)|dt$

 Integral of square of error $A_2=\int_0^\infty \{x(t)-x(\infty)\}^2 dt$

 Integral of time multiplied by absolute error $A_3=\int_0^\infty K|x(t)-x(\infty)|dt$

5. Amplitude damping ratio $d=\frac{a_3}{a_1}=\frac{a_4}{a_2}$
6. Period $=\frac{2\pi}{\omega}$
7. Response time T_R

Fig. 2 · 45 Terminology for transient response

gain of 2. From these results, the indivisual effects of P, I and D actions can be summarized as follows.

(1) Effects of proportional action: As the proportional band narrows,
 —Offset decreases
 —Amplitude damping ratio becomes large (oscillatory), eventually diverging
 —Oscillation period becomes short

(2) Effects of integral action: As the reset time shortens,
 —Offset disappears, and the time until reset becomes short
 —Amplitude damping ratio becomes large, eventually diverging

(3) Effects of derivative action: As the rate time lengthens,
 —Amount of offset remains unchanged
 —Amplitude damping ratio becomes small, and then large again
 —Oscillation period becomes short

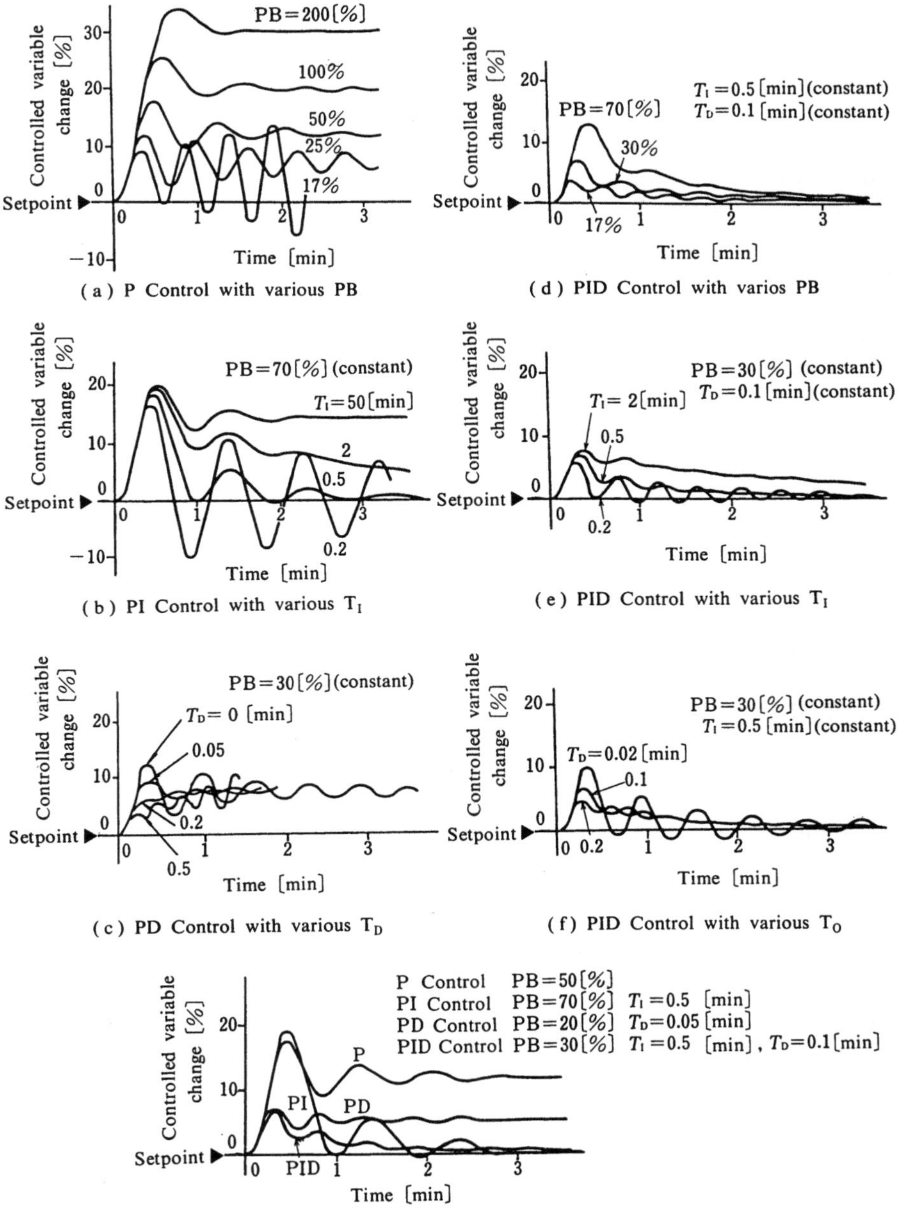

(a) P Control with various PB

(d) PID Control with varios PB

(b) PI Control with various T_I

(e) PID Control with various T_I

(c) PD Control with various T_D

(f) PID Control with various T_O

(g) Control with optimum settings

Fig. 2 · 46 Responses for PID control (for disturbance)

From the above effects and from the comparison of PID values for control under optimal conditions in Fig. 2.46(g), we see that for proportional action, offset remains; if we add in integral action, however we can eliminate the offset, although stability is degraded to some extent. If we then add in derivative action, stability increases, and the response becomes faster as well.

(*d*) *Optimal adjustment*

As control responses vary depending on the above PID values, determining those PID values which satisfy evaluation criteria for a particular process is referred to as optimal adjustment, optimal setting, or tuning. Two main methods are used in this regard:

(1) Determination from closed loop characteristics: PID values are determined from the response in a closed control loop, with amplitude damping ratio as common evaluation criteria.

(2) Determination from process characteristics: PID values are determined from investigating the process response in an open control loop, with control area or response time frequently used as evaluation criteria.

In the optimal setting methods described below, derivative and proportional action are assumed to operate on deviation for either case of disturbance or setpoint change. When PD derivative or PID with two degrees of freedom is employed, the optimal setting for the setpoint change may differ.

(1) *Determination method from closed-loop characteristics* The most well-known method is the Ziegler-Nichols ultimate-sensitivity method. Under proportional control conditions, as the proportional band is gradually narrowed from a sufficiently large value, the response determines 1) the proportional band causing fixed-amplitude continuous oscillation (ultimate proportional band PB_u) and 2) the corresponding oscillation period (ultimate oscillation period P_u). From these values, PID values can then be calculated using Table 2.3. Although these PID values in effect provide 25% damping with respect to disturbance or a change in the setpoint value, some adjustment is necessary depending on the process.

Table 2.3 Ultimate sensitivity method by Ziegler-Nichols

Control	PB	T_I	T_D
P	2 PB_u	—	—
PI	2.2 PB_u	0.83 P_u	—
PID	1.7 PB_u	0.5 P_u	0.125 P_u

Table 2.4 Transient response method (*Trans. ASME, etc.*)

Proposer	Type*1	Control mode	Optimum setting*2 PB	T_I	T_D	Optimum condition*3
Ziegler Nichols (1942)	A,B	P	$100\ K_PL/T$	—	—	25 % damping
		PI	$110\ K_PL/T$	$3.3\ L$	—	
		PID	$83\ K_PL/T$	$2\ L$	$0.5\ L$	
Takahashi	B	P	$110\ K_PL/T$	—	—	Minimum control area
		PI	$110\ K_PL/T$	$3.3\ L$	—	
		PID	$77\ K_PL/T$	$2.2\ L$	$0.45\ L$	
Chien Hrones Reswick	A	P	$333\ K_PL/T$	—	—	No overshoot and minimum response time
		PI	$286\ K_PL/T$	$1.2\ T$	—	
		PID	$167\ K_PL/T$	T	$0.5\ L$	
Chien Hrones Reswick	A	P	$143\ K_PL/T$	—	—	20 % overshoot and minimum response time
		PI	$167\ K_PL/T$	T	—	
		PID	$105\ K_PL/T$	$1.35\ T$	$0.47\ L$	
Chien Hrones Reswick	B	P	$333\ K_PL/T$	—	—	No overshoot and minimum response time
		PI	$167\ K_PL/T$	$4\ L$	—	
		PID	$105\ K_PL/T$	$2.4\ L$	$0.4\ L$	
Chien Hrones Reswick	B	P	$143\ K_PL/T$	—	—	20 % overshoot and minimum response time
		PI	$143\ K_PL/T$	$2.3\ L$	—	
		PID	$83\ K_PL/T$	$2\ L$	$0.42\ L$	
Fujii Yoshikawa	A	P	$100\ K_PL/T$	—	—	Minimum control area
		PI $L/T\leq 1$	$167\ K_PL/(T+L)$	$T+L$	—	
		PI $L/T\geq 1$	$250\ K_PL/(T+2\ L)$	$2\ L$	—	
		PID $L/T\leq 1$	$133\ K_PL/(T+(1/3)L)$	$0.5(T+L)$	$0.125(T+L)$	
		PID $L/T\geq 1$	$200\ K_PL/(T+L)$	L	$0.25\ L$	

(Notes) *1 Type A: for setpoint change; Type B: for disturbance.
*2 T, L, K_P: obtained from transient response.
*3 Refer to Fig. 2.45.

(2) *Determination method from process characteristics*

(i) *Transient response method*: If we apply a step signal to a process, then as shown in Fig. 2.47, we can measure the process gain K_p, the equivalent time constant T, and the equivalent dead time L. From these values, PID values can then be calculated. Since control valves, detectors, transmitters and the like are used as input–output devices to measure process transient response, their characteristics are inevitably included in the process characteristics. Table 2.4 shows various transient response methods.

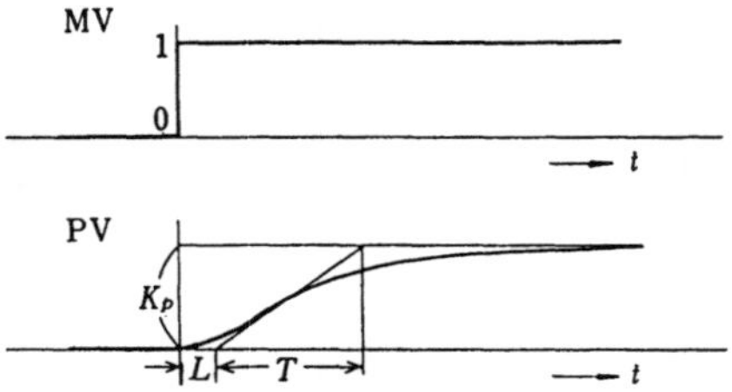

Fig. 2・47 Step response of process

(ii) *Frequency response method*: If the frequency characteristics of a process are known, the frequency characteristics of a loop transfer function including a controller can be easily obtained. Since the frequency characteristics of the controller are determined by PID setting values, it is possible to set PID values so that, for instance, gain margin and phase margin become favorable values.

(*e*) *PID for representative control systems*

Process control systems such as in temperature control and pressure control each have their own particular characteristics, which is necessarily reflected in their PID settings. Table 2.5 list PID characteristics for some representative control systems.

Table 2.5 PID settings for common process control loops (Shinsky[5])

Property	Flow and liquid pressure	Gas pressure	Liquid level	Temperature and vapor pressure	Composition
PB(%)	100–500 *50–200	0–5	5–50	10–100	100–1000
Integral	Essential	Unnecessary	Seldom	Yes	Essential
Derivative	No	Unnecessary	No	Essential	If possible

* Liquid pressure

2.5 Sequential Control

Although the material discussed up to now has been concerned with continuous control processes, there are many cases in which sequential control forms a part of process control along with continuous control. This section describes the meaning and elements of sequential control.

2.5.1 *Meaning of "sequential control"*

Sequential control is defined as "control which successively advances to each control level according to a previously determined order." Examples of sequential control in our daily life are fully automatic washing machines and elevators, while in actual process control, some examples are polymerization and crystallization batch processes, and cleaning of a water supply filter bed. In addition, even in continuous processes, sequential control must be performed at times like startup and shutdown, regardless of whether the operation is done manually or automatically.

2.5.2 *Types of sequential control*

Sequential control is devided into the following two types:

(1) *Program control* (process control)* In this case, control proceeds according to a previously determined program (from process to process).

(2) *Conditional control (monitor control)* Here, internal and external conditions are monitored, and control is performed in response to these conditions.

An example of program control would be the fully automatic washing machine mentioned above. After placing laundry and soap into the washing machine and then opening the water faucet and pushing the start button, the processes of water filling, washing, rinsing and draining are all performed sequentially according to a previously set program. This program advances in response to water level, timer and counter signals, and finally ends by sounding a wash-over chime.

For conditional control, the elevator is a fitting example. External conditions would be calling for an elevator and specifying the desired floor, while internal conditions would be those concerned with the elevator cage, *i.e.*, the current floor, cage stopped or moving, existence of passengers, etc. In addition, in the event of a multi-elevator system, the cage conditions of another elevator must also be considered, and in response to all of these conditions, an elevator can then be driven up or down.

There are also many cases in which sequential control as a form of

---

* Program control: In addition to program control narrowly defined as "control in which the desired value undergoes a previously established change," we also have this more broader form of program control as one type of sequential control.

process control will be a combination of both program control and conditional control.

2.5.3 *Sequential control description*

The following five formats are commonly used for describing the actions of sequential control.

(1) Relay circuit: Since sequential control relay circuits were traditionally realized by such elements as relays and timers, relay circuit diagrams are still used as a descriptive aid.

(2) Logic circuit: This form of description makes use of logic signals, in particular, those circuit signals as standardized by JIS and MIL.

(3) Flowchart: This description format makes use of the flowchart concept as applied to computer programming.

(4) Time chart: In this kind of chart, the mutual interaction between sequential control devices is shown according to the passage of time.

(5) Decision table: This format shows the operations corresponding to different conditions in a matrix table fashion.

Each of the above description formats has its advantages and disadvantages. In general, flowchart and time chart formats are appropriate for program control, and relay circuit and logic circuit formats are appropriate for conditional control, while decision tables can be applied to both. In fact, it is common to adopt a particular description format according to the program format of the sequential control device being used.

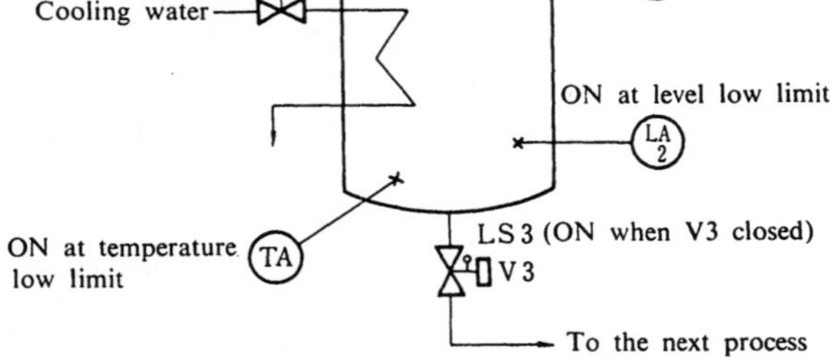

Fig. 2・48 A simple cooling process

Next, taking as an example the simple cooling process shown in Fig. 2.48, the relay circuit, flowchart and decision table description formats will be described in more detail.

The cooling process is performed as follows:

(1) Sequential control start: If no liquid is in the tank, *i.e.*, lower level limit switch LA 2 is on, sequential control can begin by pushing the PB 1 start button.

(2) Tank filling operation: After confirming that solenoid valve V 3 is closed, solenoid valve V 1 is opened and the tank is filled with liquid from the previous process until upper level limit switch LA 1 turns on, at which time solenoid valve V 1 is closed. In the event that solenoid valve V 3 is initially open, a valve closing operation is first performed before proceeding as described above.

(3) Cooling operation: After completing the tank filling operation, solenoid valve V 2 is opened and cooling is performed until lower temperature limit switch TA turns on, at which time solenoid valve V 2 is closed.

(4) Liquid transfer operation: After completing the cooling operation, solenoid valve V 3 is opened, and cooled liquid is supplied to the next process until lower level limit switch LA 2 turns on, at which time solenoid valve V 3 is closed again.

(5) Sequential control repeat or termination: If, at the point of liquid transfer completion, repeat SW 1 is on, sequential control is repeated again from the tank filling operation. If repeat SW 1 is off, however, sequential control is terminated.

Operation conditions and corresponding actions are shown for this cooling sequential control in Table 2.6. Since this sequential control is a program control type, if we implemented it with relay circuits, we

Table 2.6 Operations for cooling process

Operation	Description	Conditions for operation
Filling tank	Open V 1	PB 1 ON, LA 2 ON and LS 3 ON or SW 1 ON, LA 2 ON and LS 3 ON
Cooling	Close V 1, Open V 2	LA 1 ON
Discharging liquid	Close V 2, Open V 3	TA ON
Liquid discharge end	Close V 3	LA 2 ON

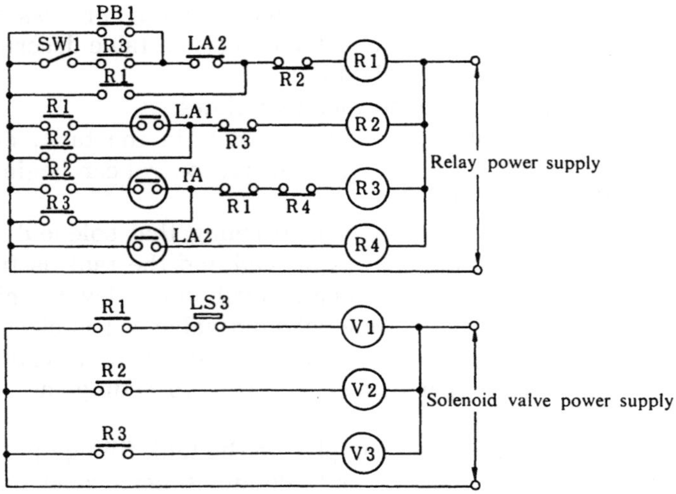

Fig. 2 · 49 Relay ladder diagram for cooling process

would need process memory relays R 1 to R 4 for recording the occurrence of each process, as shown in Fig. 2.49. In addition, as shown in Fig. 2.50, if we describe the same with the use of a flowchart, sequential control can be easily expressed in terms of each separate process.

Next, let's try expressing the above in terms of a decision table. A typical format for a decision table is shown in Fig. 2.51. Conditions are listed in the upper half of the table and actions in the lower half, with the right side of both halves divided into rule number columns. To establish a certain condition, Y is entered to indicate that sequence elements should be ON, and N for OFF; if the condition has nothing to do with the elements, the entry is left blank. Likewise for action description, Y is indicated if an ON action signal is output, and N is indicated if an OFF action signal is output, if no signal is output, the entry is left blank. Figure 2.52 shows a relay circuit described with such a decision table. In this figure, the symbols in the SYMBOL column are sequential control elements used by decision tables. Process memory relays R 1 to R 4 are substituted for internal switches which are sequential control elements. Based on the flowchart described in Fig. 2.50, the sequence is indicated in the decision table shown in Fig. 2.53. As can be seen, the internal switches for process recording are substituted by step numbers and step transfer conditions.

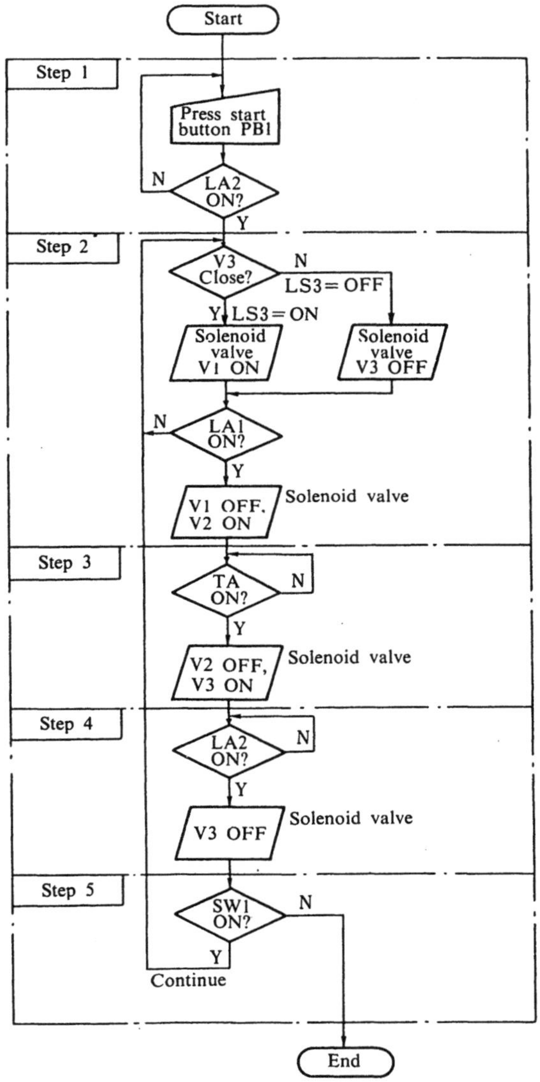

Fig. 2・50 Flow chart for cooling process

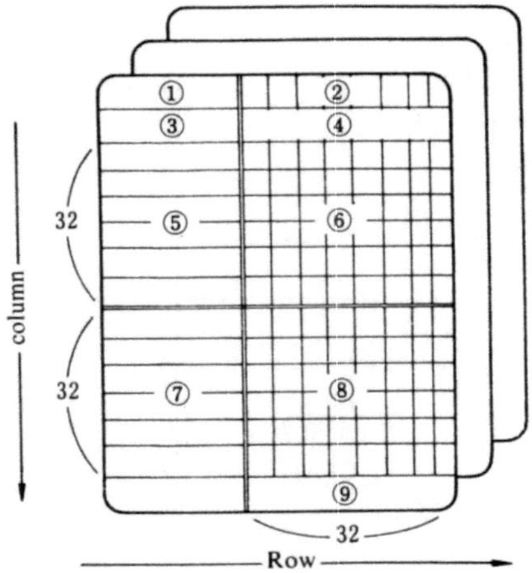

① Table number
② Rule number
③ Symbol
④ (Step number)
⑤ Condition signals
⑥ Condition rules
⑦ Action signals
⑧ Action rules
⑨ (Next step number)

Fig. 2·51 An example of decision table

TABLE NO. | 00 | ST002

DESCRIPTION: Ⓣ Cooling process 1a

			01	02	03	04	05	06	07	08	09	10	11
01		RULE NO.	01	02	03	04	05	06	07	08	09	10	11
02	SYMBOL	COMMENTS / STEP NO.											
11	NDI016	PB1	Y										
12	NSW204	R4						N	N				
13	NSW202	R2	N	N	N		Y	Y				Y	
14	NDI005	SW1		Y									
15	NSW203	R3		Y		N	N		Y				Y
16	NSW201	R1			Y	Y		N	N		Y		
17	NDI001	LA1				Y							
18	NDI000	TA						Y					
19	NDI002	LA2	Y	Y						Y			
20	NDI004	LS3									Y		
41													
42													
111	NSW201L	R1	Y	Y	Y								
112	NSW202L	R2				Y	Y						
113	NSW203L	R3						Y	Y				
114	NSW204L	R4								Y			
115	NDO010L	V1									Y		
116	NDO011L	V2										Y	
117	NDO012L	V3											Y
11[illegible]													
141													
142													
191	NEXT STEP	THEN											
192		ELSE											

CONTINUE TABLE NO.

193	

Fig. 2·52 Entry example of decision table corresponding to the relay ladder diagram

TABLE NO: | 00 | ST003 |

DESCRIPTION: Ⓣ Cooling process 1b

			01	02	03	04	05	06	07	08	09	10	11
01		RULE NO.	01	02	03	04	05	06	07	08	09	10	11
02	SYMBOL	COMMENTS / STEP NO	1	2		2	3	4	5				
11	NDI016	PB1	Y										
12	NDI002	LA2	Y					Y					
13	NDI003	LS3		Y	N								
14	NDI001	LA1				Y							
15	NDI000	TA					Y						
16	NDI005	SW1 (ON to repeat)							Y				
17													
18													
19													
41													
42													
111	NDO010H	V1		Y		N							
112	NDO011H	V2				Y	N						
113	NDO012H	V3			N		Y	N					
114													
115													
116													
117													
118													
141													
142													
191	NEXT STEP	THEN	2			3	4	5	2				
192	NEXT STEP	ELSE							1				

CONTINUE TABLE NO: | 193 | |

Shows correspondence to the relay ladder diagram 1a: R1 (rules 02–04), R2 (rule 05), R3 (rule 06)

NDOnnnH: Latched type switch

Fig. 2 · 53 Entry example of decision table using step numbers

2.5.4 Devices for sequential control

Sequential control devices originally consisted of relay control panels made up of relays, timers and the like. With the coming of the transistor age, however, such devices came to be substituted by those employing logic circuits made up of transistors and diodes. Also used at that time were sequencers in which a control program previously fixed by wiring could be set as needed through a pinboard.

The above period was in turn followed by developments in semiconductor technology, which gave rise to the stored-program PLC (Programmable Logic Controller) characterized by the use of ICs and memory elements. This stage was soon followed by the development of the microprocessor, which along with the parallel development of peripheral devices such as CRTs, enabled an even further expansion of functions, referred to as PC (Programmable Controller). Although descriptive formats such as flowcharts or logic circuits are used in programming PLCs or PCs depending on the particular kind of machine, the relay circuit format is the most generally used.

On the other hand, in DDC systems (see Chapter 1) employing minicomputers, the computational ability available could be applied to incorporate sequential control programs as well as continuous control programs as needed. With distributed control systems made possible by the development of the microprocessor, however, a high level of automatic process control, which is a fusion of continuous control and sequential control, came to be realized. In the DDC era, sequential control programs were written in either assembler or list formats based on flowcharts, but with distributed control systems, the application range has broadened, system comprehensibility has become easier, and documentation simplified with the use of descriptive formats such as decision tables.

In addition, through the use of digital input/output and the logic processing of programmable single-loop controllers incorporating microprocessors, simple sequential control has been made possible.

PRACTICE QUESTIONS

2.1 Determine the step response to a unit step signal for the circuit shown in Fig. 2.54, for the three cases:

(a) $G(s)=\frac{1}{1+Ts}$, $H(s)=1$, positive feedback

(b) $G(s)=n\gg 1$, $H(s)=\frac{1}{1+Ts}$, negative feedback

(c) $G(s)=1$, $H(s)=\frac{1}{1+Ts}$, positive feedback

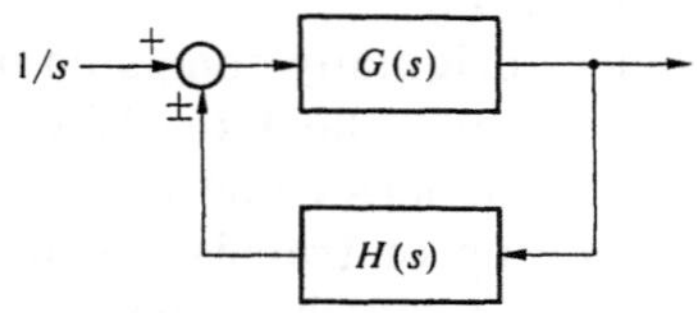

Fig. 2・54

2.2 For the case of a constant outflow liquid level process (integral element) under proportional control, determine the offset for both a change in the liquid level setpoint value and a change in the amount of outflow.
2.3 Explain any problems when using a PI controller in a batch control system.
2.4 In the buffer tank selection control example, devise a 3-value override control system in which an upper liquid level limit is incorporated to prevent overflow.

ANSWERS TO QUESTIONS

2.1 Setting up the problem with the expression, $G_0(s)=G(s)/1\pm G(s)H(s)$

(a) $G_0(s)=1/Ts$ or in other words, integral characteristics; transient response is the same as shown in Table 2.1(e).

(b) $$G_0(s)=(1+Ts)/\left(\frac{1+n}{n}+\frac{T}{n}s\right)\approx(1+Ts)/\left(1+\frac{T}{n}s\right)$$

or in other words, a proportional gain of 1, and consisting of proportional and derivative action with derivative amplitude n; response is as shown below.

(c) $G_0(s)=(1+Ts)/Ts$ or in other words, proportional and integral action with a proportional band of 100%; see Fig. 2.27 for the transient response.

2.2 Indicating tank level as $H(s)$, level setpoint value as $H_o(s)$, inflow as $Q_i(s)$, outflow as $Q_o(s)$, tank cross-sectional area as A, proportional gain as K_c and deviation as $E(s)$, the following block diagram can be drawn.

Expressing $E(s)$ in terms of $H_o(s)$ and $Q_o(s)$:

$$E(s)=\frac{1}{1+Kc/sA}H_o(s)+\frac{1}{Kc+sA}Q_o(s)$$

For a unit step change, this becomes:

$$E(s)\frac{1}{s(1+Kc/sA)}+\frac{1}{s(Kc+sA)}$$

As $t\rightarrow\infty$, the deviation, or offset can thus be expressed as:

$$S\cdot \underset{s\to 0}{E(s)}=\frac{1}{1+Kc/sA}+\frac{1}{Kc+sA}=0+\frac{1}{Kc}$$

or in other words, in regard to a change in the setpoint value, offset is zero, while for a change in outflow, offset appears.

2.3 Please refer to the discussion on integral action and reset windup.

2.4 For the override control system shown in the figure:

If the flow rate control system is selected: $M_H<M_F<M_L$

If the upper level limit control system is selected: $M_F<M_H<M_L$

If the lower level limit system is selected: $M_H<M_L<M_F$

If, however, air-to-close is used for the control valve, the above relationships will be reversed. In either case, an M (intermediate value) selector is used.

REFERENCES

1) T. Ishii: Process Control Instrumentation Systems, Denki Shoin (1973) (in Japanese).
2) The Society of Instrument and Control Engineers: Automatic Control Handbook, Devices and Applications Volume, Ohm Sha (1983) (in Japanese).

3) K. Matsunaga: "Documenting Process Control Sequences by Decision Tables," *Yokogawa Tech. Rep.*, **26**, 3 (1982) 44–50 (in Japanese).
4) J. G. Holbrook: Laplace Transforms for Electronic Engineers, Pergamon Press (1959).
5) F. G. Shinsky: Process-Control Systems, 2nd Edition, McGraw–Hill (1979).
6) N. A. Anderson: Instrumentation for Process Measurement and Control, 2nd Edition, Chilton Book (1972).
7) T. Tohyama: "Guide to Understanding of Control Engineering," *Instrumentation and Automation*, **3,** 4–9 (1975) (in Japanese).
8) Y. Kasai and M. Terao: Electrical Automatic Control, Denki Shoin (1975) (in Japanese).
9) Y. Ohshima: Encyclopedia of Terms used in Automatic Control, Ohm Sha (1971) (in Japanese).
10) JIS B 0155–1986: Glossary of Terms Used in Industrial Process Measurement and Control (in Japanese).

Chapter 3

DETECTION AND CONVERSION OF INDUSTRIAL VARIABLES

In the petroleum, chemical, iron, textile, and paper industries, raw materials are supplied to production facilities, energy is added, and the raw materials are physically and chemically processed into finished products. In these processes, a variety of variables related to each part of the process, such as temperature, pressure, flow rate, liquid level, composition, and quality are measured and controlled.

In machine industries like automobile and electrical equipment production, which stress mechanical processes such as material processing, assembly, inspection and so on, the processed objects are solids, and variables such as location, shape, dimensions, and position are measured and controlled. These types of measurements made in the production process and in related activies such as acceptance of raw materials and product shipping inspection are called *industrial measurement*. The variables of industrial measurement and control are called industial variables, and can be broadly classified according to their characteristics in the following way.

(1) Process state variables such as temperature, pressure, flow rate, liquid level, humidity, heat content, viscosity, and density
(2) Mechanical variables such as length, angle, displacement, position, and shape
(3) Dynamic variables such as mass, energy, time, revolutions, speed, and vibration
(4) Composition variables of gases, solutions and solids
(5) Electrical variables such as voltage, current, wattage, frequency, and magnetism.

There are various kinds detecting elements for these industrial variables. In this chapter we will describe some typical examples of devices currently in use.

3.1 *Measurement of Industrial Variables*[1)–5)]

3.1.1 *Methods of measurement*

There are a number of methods of measurement, that is to say, methods of comparing measured variables with some reference value. The major methods are described below.

(*a*) *Direct and indirect measurements*

Direct measurement involves comparing the measured variable with a reference value of the same type. Examples of direct measurement include measuring length with a ruler, and measuring electrical resistance by comparison, using a Wheatstone bridge circuit for reference values.

Indirect measurement, on the other hand, is done by measuring some other variable that has a fixed relationship with the variable to be measured. For example, by applying an electrical voltage E to a resistance R and measuring the current I, the resistance can be obtained by using the relationship $R=E/I$. Determining the temperature in an oven with a thermocouple, and obtaining flow rate from the pressure differential across an orifice in a pipe are other examples of indirect measurement.

There are a great many industrial variables, which range from temperature, flow rate, and pressure to solution concentration and gas composition. Measurement of these variables employs detection methods that make use of various kinds of physical law and effects. Consequently, indirect measurement methods are most common by far.

(*b*) *Deflection methods and zero methods*

When using a spring balance to measure weight, as in Fig. 3.1(a), the measured weight is read from the position of a pointer which indicates the displacement of a spring resulting from placing the measured object on the balance. A measurement method that in this way directly

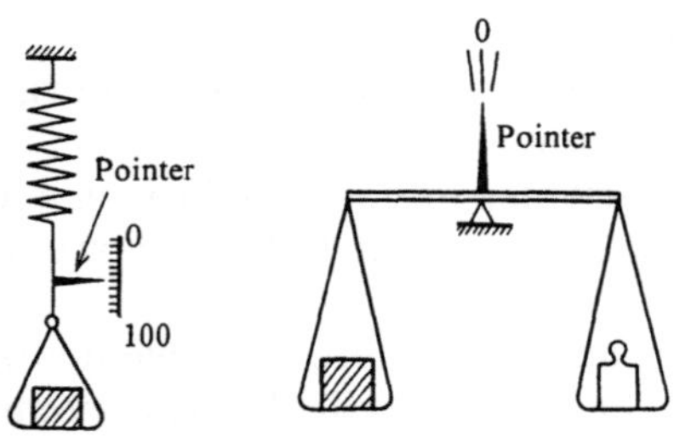

(a) Spring balance　　(b) Zero balance

Fig. 3・1 An example of deflection method and zero method translates something like a displacement or angle which has a fixed relationship to the variable to be measured into a measurement value is called a *deflection method.* The Bourdon tube pressure gauge and the movable coil voltmeter and ammeter are deflection method measuring devices. The deflection method is direct and the measuring mechanism is simple, so this method is widely used. However, since movement of the indicator takes enegy from the measured object, accurate measurement cannot ordinarily be hoped for.

As shown in Fig. 3.1(b), weight can also be determined with a zero balance. In this method, the measured weight is balanced with a known weight(in this case pieces of brass). By adjusting the known weight such that the measuring instrument indicates zero, the weight of the object being measured can be known. This type of method is called a *zero method.* Typical examples of this method are electrical potentiometers for measuring voltage, and methods for measuring electrical resistance and impedance that use a Wheatstone bridge.

When the manual balancing operation in zero method measurement is replaced with automatic balancing by servo-motors, it is called a *self-balancing method.* The zero method has the special feature that if the balancing is not complete, the difference between the measured variable and the reference value can be detected, and a feedback operation can correct the reference value.

Measurement of industrial variables mostly uses self-balancing methods for the following reasons.

(1) The balancing operation uses a separate energy supply, so the operation is strong and measurement error from friction and so on is avoided.

(2) Measurement of good accuracy can be expected bacause the measured quantity is compared with an exact reference quantity.

(3) The variable being measured is not subjected to outside disturbances, since measurement is accomplished in equilibrium.

(*c*) *Potentiometer and Wheatstone bridge*

In industrial measurement, the variables to be measured are often converted into electrical quantities such as voltage resistance and impedance. These converted quantities are often measured by potentiometer or Wheatstone bridge circuits.

(1) *Direct current potentiometer circuit* When D.C. voltage is measured with a moving coil voltmeter, as in Fig. 3.2, the current I_M flows in the measurement circuit, a voltage drop $I_M R_S$ occurus in the measured object side of the circuit, so the value E_x cannot be measured exactly. To measure D.C. voltage accurately, the zero method potentiometer mentioned above can be used. The principle is illustrated in

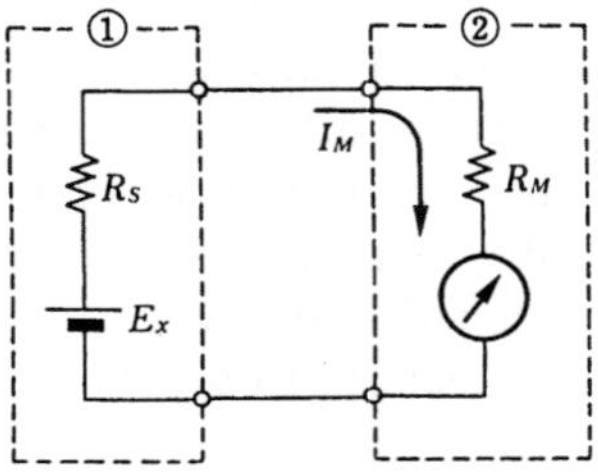

① Measured object
② Voltmeter

E_x : Unknown voltage
R_S : Internal resistance
R_M : Internal resistance of voltmeter
I_M : Circiut current
$I_M = E_x/(R_S + R_M)$

Fig. 3・2 Voltage measurement by voltmeter

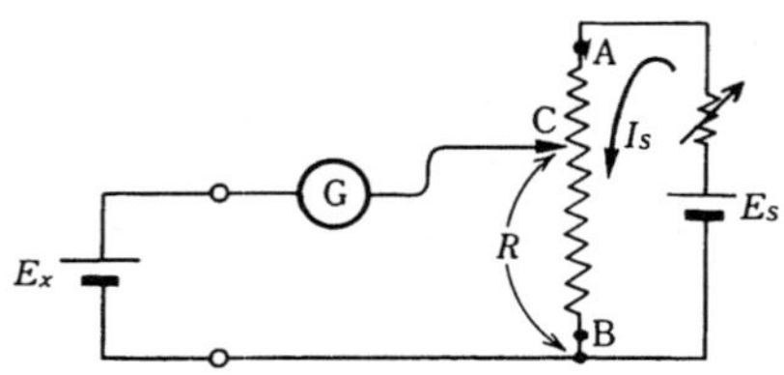

Fig. 3・3 Potentiometric circuit

Fig. 3.3.

Reference voltage E_S produces a fixed current flow I_S through a variable resistor. An unknown voltage E_x and galvanometer G are connected, and point C is sought, such that the galvanometer reads zero. If the resistance between points C and B is expressed by R, then the equation $E_x = I_S R$ holds. Thus, the value of E_x can be read directly from a voltage scale placed on the variable resistor. When the circuit is in balance, no current flows through E_x, and the internal resistance of the object of measurement is not affected. Thus, if the voltage E_S and the variable resistor are exact, precise measurement of voltage can be made.

(2) *Wheatstone bridge* The Wheatstone bridge circuit is often used for zero method measurement of resistance and impedance. In the Wheatstone bridge circuit shown in Fig. 3.4, R_x is the input side unknown resistance. If the bridge is balancd by moving the brush of the variable resistor so that the galvanometer G indicates zero, the relationship $R_x R_2 = R_1 R_3$ holds. Since R_1, R_2 is determined by the position of the variable resistor brush, if R_3 is a known value, the value of R_x corresponds to the brush position.

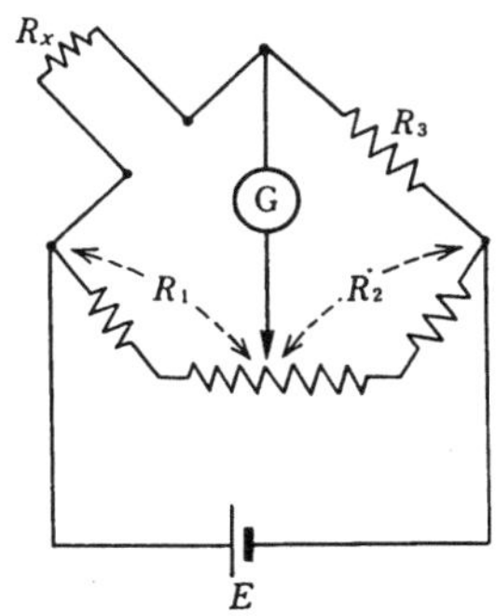

Fig. 3・4 Wheatstone bridge circuit

(d) Self-balancing methods

In the *self-balancing method*, the operation of balancing the measurement circuit so as to obtain a zero reading (such as in the methods illustrated in Figs. 3.3 and 3.4) is done automatically. The principle of a self-balancing potentiometer is shown in Fig. 3.5. Here, the input measurement voltage E_x and the voltage across the variable resistance up to the brush position E_S are compared and the voltage difference e is detected by an amplifier rather than the galvanometer. Depending on the polarity of the voltage difference e, the output of the amplifier either causes a clockwise or counterclockwise rotation in the servo-motor. Since the servo-motor and the brush are mechanically linked, the servo-motor will move the brush until the unknown voltage E_x and the voltage at the brush E_S are equal, and then automatically stop at this equilibrium position. At the time $E_x = E_S$, the brush position indicates the measured value. If the brush is mechanically linked to an indicator, the value of E_x will be displayed. Figure 3.6 shows the principle of a resistance bridge self-balancing measurement system.

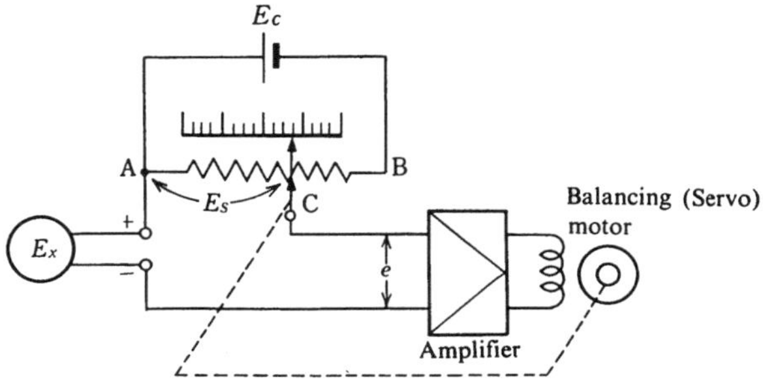

Fig. 3・5 Principle of self-balancing potentiometer

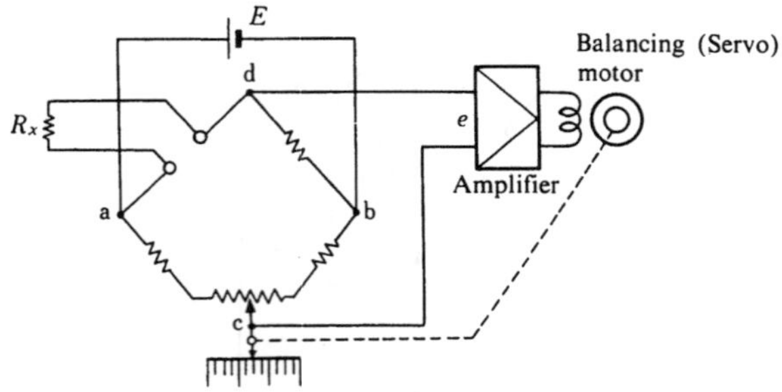

Fig. 3・6 Principle of self-balancing resistance bridge

3.1.2. *Accuracy of measurement*

The degree of correctness of the results of measurement is generally called *accuracy*. Looking at it another way, accuracy is an index representing how little *error* there is in the measurement, and is usually represented by limits of error under some appropriate conditions.

(*a*) *Errors and their causes*

Measured values always include some error. If M represents a measured value, and T the true value, then the error of measurement e represents the difference between the measured value and the true value, or, expressed symbolically, $e=M-T$.[1] The percentage of error can be expressed by $(e/T)\times 100$. The true value is an ideal value that cannot actually be known. So in order to express measurement error as a numerical value, the value measured by a standard device or instrument of the best accuracy is taken as the true value.

The sources of error can be classified in the following way.[3]

(1) *Theoretical errors* These are errors related to the principle or theory of measurement. For example, when measuring the level of a liquid in a tank with a level gauge that works on the principle of a pressure differential, there is a difference in measured pressure that is caused by a difference between the density under actual use conditions and the design density. The error of measurement resulting from this pressure difference is an example of theoretical error. It is possible to eliminate this error by correcting for the difference in density.

(2) *Instrument error* This type of error originates in the measuring instrument itself, and results from mechanical or electrical deviations in the components of the instrument such as mechanical clearance, error in resistance values, and so on. For this class of error, the measured value can be corrected by calibrating the instrument.

(3) *Individual human error* These errors result from the manner in which the measurer habitually makes the measurements. For example, a person might consistently take low readings when reading a scale.

These errors can be eliminated to a certain extent by care on the part of the measurer.

(4) *Environmental error* These are errors caused by influence of the environment or the conditions under which a measurement is made, such as ambient temperature or pressure, the position in which the measuring device is set, improper insertion of a temperature sensor, etc. If these environmental factors change, it will exert a generally irregular effect on the measurements.

(5) *Anomalous error* These errors result from defective measurements caused by abnormalities in the measuring instrument, differences in readings taken by the human measurer, and so on.

(*b*) *Accuracy*

Accuracy has always been basic term used to express the characteristics of measuring instruments. Going by the Japan Industrial Standard measurement terminology (JIS Z 8103–1984), the term accuracy is defined as "a quality that includes the correctness and precision of a measurement or a value displayed on a measuring instrument."It is used in a statistical sense, assuming that errors in measurement fall into a normal distribution.

However, the term accuracy is used differently by manufacturers of measuring instruments. What they call ordinary accuracy, or accuracy rating in instrument specifications is a definition of the limits of error in output (indicated values) permitted according to the specifications for that model, and based on the specified conditions. This is shown in Fig. 3.7. These specified conditions are ranges within which effects from fluctuations in outside factors to which the instrument is exposed, such as temperature, humidity, power supply, and vibration

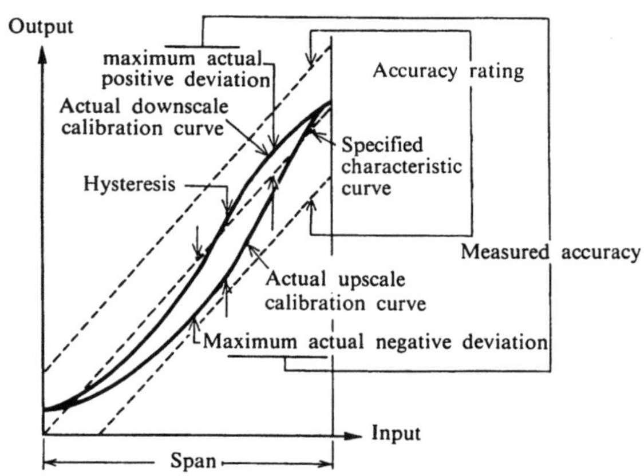

Fig. 3 · 7 Accuracy

can be neglected. For the instrument whose operating characteristics are shown in Fig. 3.7, accuracy includes the effects of hysteresis (difference in output according to the direction of change in the input), dead zone (the largest change in input that fails to produce any output), repeatability (the degree to which measurements of the same object made by the same method, under the same conditions, and repeated within a relatively short period of time, produce the same measured values; see Fig. 3.8), all of which are causes of error.

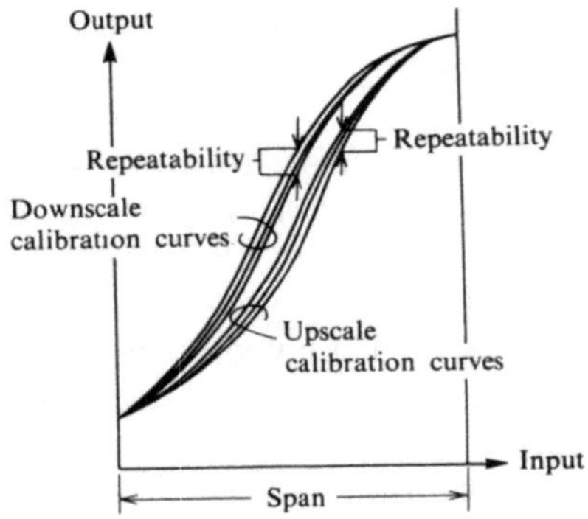

Fig. 3・8 Repeatability

The accuracy rating of industrial measuring instruments is expressed in terms of error limits, as in the following examples.

(1) A percentage of the span (*e.g.*, ±0.2% OF SPAN)

(2) A percentage of the scale length (*e.g.*, ±0.5% OF SCALE LENGTH)

(3) A percentage of the upper range value (*e.g.*, ±0.5% OF UPPER RANGE VALUE)

(4) A percentage of the reading (*e.g.*, ±1% OF READING)

The relationship between error and accuracy rating as expressed in terms of span and reading is shown in Fig. 3.9. In the case of percentage of span, the error becomes larger in proportion to the reading as the measured value decreases. Thus, when deciding on the measurement range of the instrument, use of the low part of the range should be avoided as much as possible.

Reproducibility is an expression of the stability of an instrument over time. It is the degree of consistency of readings (output) when the same instrument is used to measure the same quantity at different times, and thus represents the change in accuracy resulting from a change or deterioration of the parts of the instrument with time. Such deterioration requires that the instrument be calibrated with a frequency appropriate to the conditions of use.

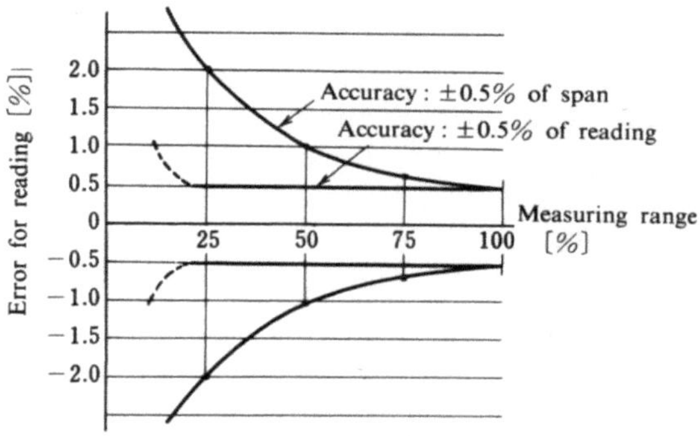

Fig. 3・9 Relation between accuracy rating and error

(*c*) *Sensitivity*

Sensitivity is the degree to which an instrument can detect a change in the quantity being measured. Its value is expressed by a sensitivity coefficient defined as

$$\text{sensitivity coefficient} = \frac{\text{change in reading}}{\text{change in measured quantity}}$$

The smallest quantity that an instrument can normally detect is called *discrimination* or *resolution*. These limits of sensitivity are simply called sensitivity. The smallest amount that an instrument can detect is also expressed as a percentage of the measuring span (*e.g.*, SENSITIVITY: 0.05% OF SPAN). Although good sensitivity is desirable, greater sensitivity increases the influence of noise, vibration, and other external conditions on accuracy. Thus it is necessary to achieve a proper balance of accuracy and sensitivity.

(*d*) *Traceability*

According to standard measurement terminology (JIS Z 8103), traceability is "a system linked to a national standard established by calibrating all instruments according to some standard or instrument of the highest quality." The accuracy of a measuring instrument is influenced by the conditions under which it is used, that is, the measurement environment, the method of use, the measurement range, and so on. Also, since the instrument will wear with use over time, accuracy will decline. Thus, to preserve the accuracy of the instrument in use and maintain the reliability of the readings, the instrument must be properly used, inspected, and calibrated as often as is appropriate to the conditions under which it is used.

Calibration involves using a standard or standard sample in place of the measured quantity as the input to the instrument, and obtaining the relationship between the output (reading) and the true value. Naturally, the standard device must be of the highest accuracy, but a standard of even higher accuracy is required in order to calibrate the standard itself. In Japan, the national standards are set according to the Weights and Measures Act, which conforms to international standards. Responsibility for maintenance and provision of the national standards rests with the Electrotechnical Laboratory (ETL), National Research Laboratory of Metrology (NRLM), Radio Research Laboratory (RRL), and other such national organs. Figure 3.10 shows the electrical standards traceability system in Japan.

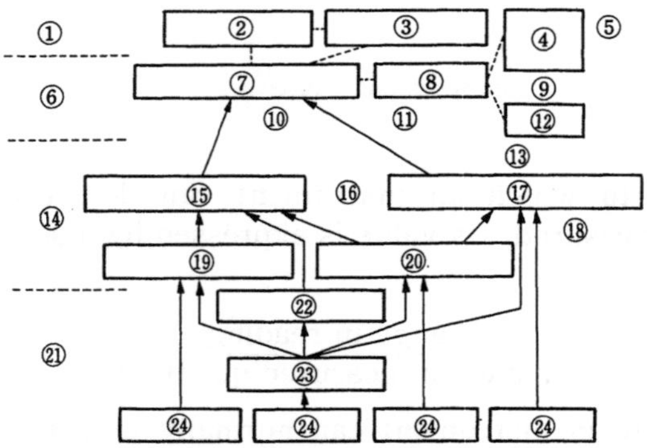

① International standards (Conveution du metre)
② International bureau of weights and measures
③ National standards laboratories of each country
④ Radio research laboratry (RRL)
⑤ Standard radio wave
⑥ National standards
⑦ Electrotechnical laboratory (ETL)
⑧ National research laboratory of metrology (NRLM)
⑨ Frequency
⑩ Electricity, light, sound, ray
⑪ Length, temperature, mass, time
⑫ Astronomical observatory
⑬ Time
⑭ Public standards agencies
⑮ Japan electric meters inspection corporation (JEMIC)
⑯ D.C. and low frequency
⑰ Japan machinary and metals inspection institute (JMI)
⑱ High frequency
⑲ Regional laboratories of JEMIC
⑳ Public industrial laboratories
㉑ Working standards user departments
㉒ Standards laboratory in the company
㉓ Standards laboratory in the factory
㉔ Each workshop

Fig. 3 · 10 Electrical standards traceability system in Japan

Figure 3.11 is an example of a measuring instrument manufacturer's basic measurement standard control system. In the standards laboratory, primary standards that have been calibrated by the national standards are maintained. These primary standards are used to calibrate the calibration standards used in every section of the company. The calibration standards are in turn used by each section for periodic independent calibration of the instruments used in ordinary work, which are called working standards.

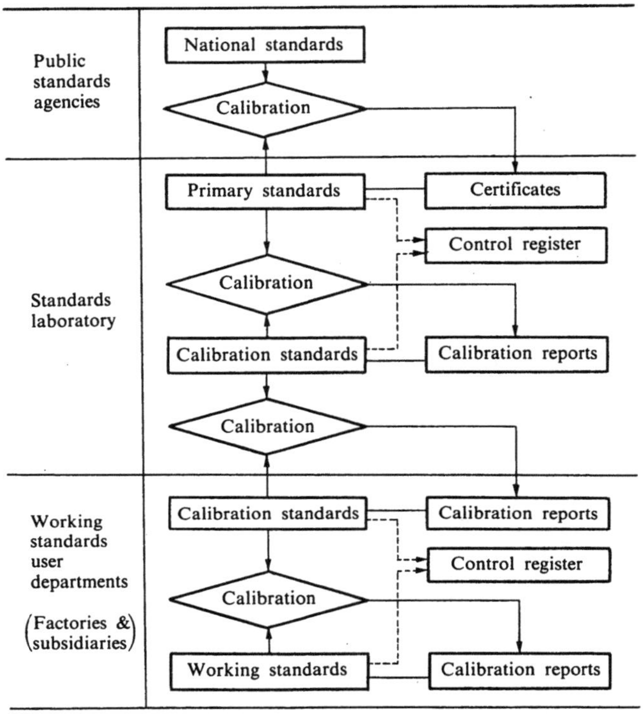

Responsibilities of each department
- Determination of calibration intervals
- Accuracy control by register
- Validity term control by label
- Storage of certificates and calibration reports
- Treatment of non-conforming standards

Fig. 3 · 11 Basic system for measurement control (Typical)

3.2 Measurement of Temperature[2),6)–16)]

Measurement of temperature, a frequent form of industrial measurement, uses various forms of sensing. The most often used are the thermoelectric thermometer and the resistance thermometer. For measurement without contact to the measuring instrument itself, there is also the radiation thermometer.

As opposed to these electrical sensors, there are also mechanical sensing devices, based on the expansion of a liquid or solid (*e.g.*, bimetal sensing element). These are not suited for transmission over a long distance, so they are most often used in local applications. Leaving mechanical sensing to other literature[2),33)], this section describes temperature measurement using electrical sensors.

3.2.1 *Thermoelectric thermometers*

(a) Measurement principle

As shown in Fig. 3.12, if two dissimilar metals, A and B, are joined at point n, and are left free at the other end, point m, the difference in temperature at the two points ($T_2 - T_1$) generates the electromotive force $E_{AB}(T_2, T_1)$. The metals A and B are called a thermocouple, point m is called a cold junction, point n is called a hot junction, and the electromotive force $E_{AB}(T_2, T_1)$ is called the thermoelectromotive force. Because there is a fixed relationship between the thermocouple materials, the cold junction temperature, the hot junction temperature, and the thermoelectromotive force, for a given thermocouple and a fixed cold junction temperature, the hot junction temperature can be determined by measuring the thermoelectromotive force.

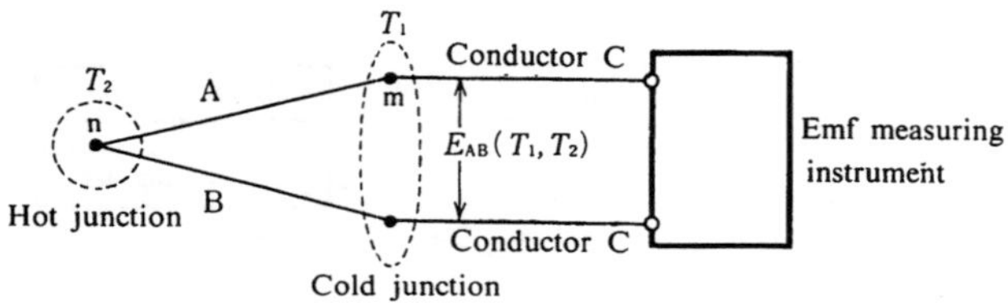

Fig. 3 · 12 Temperature measurement using thermo couple

To measure the thermoelectromotive force with almost no current flow, it is necessary to use a measuring instrument with a high input resistance.

(b) Law of successive temperatures and intermediate conductors

For a set of thermocouples, given a temperature distribution t_1, t_2, and t_3 in which $t_3 > t_2 > t_1$, if the thermoelectromotive force resulting from hot junction temperature t_3 and cold junction temperature t_1 in thermocouple elements A and B is $E_{AB}(t_3, t_1)$, then the following relation holds.

$$E_{AB}(t_3, t_1) = E_{AB}(t_3, t_2) + E_{AB}(t_2, t_1) \tag{3.1}$$

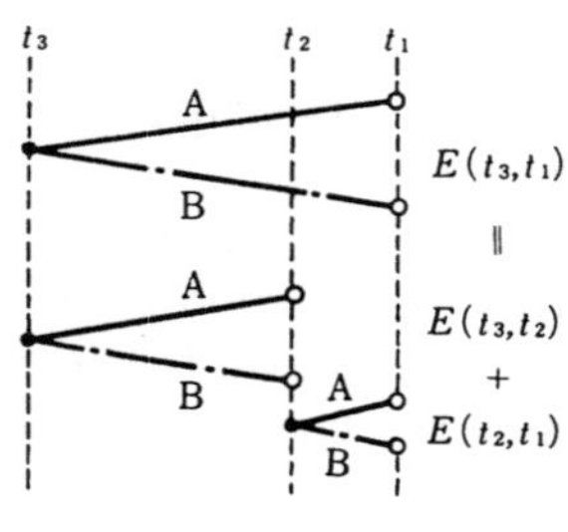

Fig. 3 · 13 Law of successive temperatures

That is, the thermoelectromotive force between t_3 and t_1 is the sum of the forces between t_3 and t_2, and t_2 and t_1. This relationship is called the *law of successive temperatures*. If, for example, $t_3=500°C$, $t_2=20°C$, $t_1=0°C$ for a K-type thermocouple, then the values

$$E_{AB}(t_3, t_1)=20\,640\ [\mu V] \qquad E_{AB}(t_2, t_1)=798\ [\mu V]$$

are obtained from the JIS C 1602 Reference Thermoelectromotive Force of K. Accordingly, from Eq. (3.1),

$$E_{AB}(t_3, t_2)=E_{AB}(t_3, t_1)-E_{AB}(t_2, t_1)=19\,842\ [\mu V]$$

The *law of intermediate conductors* applies to the effect on the thermoelectromotive force when a third intermediate conductor is inserted between thermocouples. For the hot junction temperature t_3 and the cold junction temperature t_1 as shown Fig.3.14, the following relationship holds.

$$E_{AB}(t_3, t_1)=E_{AC}(t_3, t_1)+E_{CB}(t_3, t_1) \tag{3.2}$$

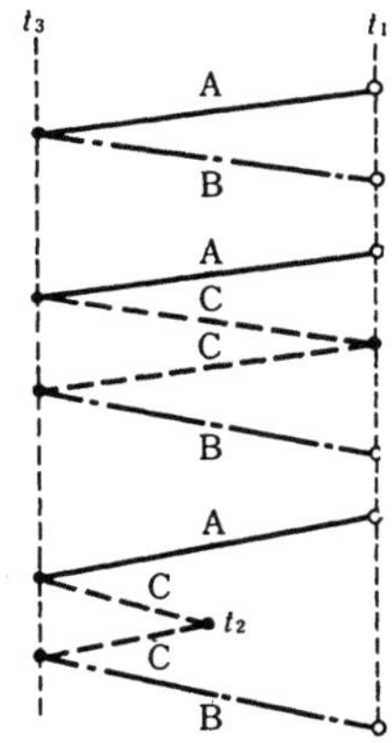

Fig. 3 · 14 Law of intermediate conductor

When both junction temperatures of the intermediate conductor are the same (in this case, t_3), the thermoelectromotive force $E_{AB}(t_3, t_1)$ is not changed.

(c) Types of thermocouples

There are many kinds of thermocoples suited for various uses, JIS C 1602 defines seven types. Table 3.1 lists their type codes, the materials from which they are constructed, measurement ranges, accuracy classes, and tolerance values. Thermocouples are broadly classified into noble metal thermocouples (type codes B, R, and S), and base metal thermocouples (type codes K, E, J, and T). The noble metal therm-

Table 3.1 Thermocouples (JIS C 1602)

Symbol of thermocouples	composition		Temperature range values	Accuracy class	Tolerance*[2]
	+Leg*[1]	−Leg*[1]			
B	Platinum rhodium alloy containing 30% rhodium	Platinum rhodium alloy containing 6% rhodium	600°C to 1 700°C excl.	Class 0.5	±4°C or ±0.5% of measured temperature
R	Platinum rhodium alloy containing 13% rhodium	Platinum	0°C to 1 600°C excl.	Class 0.25	±1.5°C or ±0.25% of measured temperature
S	Platinum rhodium alloy containing 10% rhodium	Platinum			
K	Alloy, main components of which are nickel and chromium	Alloy, main components of which is nickel	0°C to 1 000°C excl.	Class 0.4	±1.5°C or ±0.4% of measured temperature
			0°C to 1 200°C excl.	Class 0.75	±2.5°C or ±0.75% of measured temperature
			−200°C to 0°C excl.	Class 1.5	±2.5°C or ±1.5% of measured temperature
E	Alloy, main components of which are nickel and chromium	Alloy, main components of which are copper and nickel	0°C to 800°C excl.	Class 0.4	±1.5°C or ±0.4% of measured temperature
			0°C to 800°C excl.	Class 0.75	±2.5°C or ±0.75% of measured temperature
			−200°C to 0°C excl.	Class 1.5	±2.5°C or ±1.5% of measured temperature
J	Iron	Alloy, main components of which are copper and nickel	0°C to 750°C excl.	Class 0.4	±1.5°C or ±0.4% of measured temperature
			0°C to 750°C excl.	Class 0.75	±2.5°C or ±0.75% of measured temperature

T	Copper	Alloy, main components of which are copper and nickel	0°C to 350°C excl.	Class 0.4	±0.5°C or ±0.4% of measured temperature
			0°C to 350°C excl.	Class 0.75	±1°C or ±0.75% of measured temperature
			−200°C to 0°C excl.	Class 1.5	±1°C or ±1.5% of measured temperature

*1 The +leg means a leg to be connected to +terminal of an instrument to measure thermoelectromotive force, and −leg means opposite side one.

*2 Tolerance means the maximum allowable limit of such value that the temperature converted from the measured thermoelectromotive force using the reference thermoelectromotive force table minus the actual temperature of the temperature measuring junction.
The value expressed in °C or %, whichever is larger, is taken as the tolerance.

Table 3.2 Element wire of thermocouple (JIS C 1602)

Symbol of composing material	Diameter of element wire [mm]	Normal operating temperature lmit*1 [°C]	Elevated operating temperature lmit*2 [°C]	Resistance [Ω/m]
B	0.50	1 500	1 700	1.75
R	0.50	1 400	1 600	1.47
S				1.43
K	0.65	650	850	2.95
	1.00	750	950	1.25
	1.60	850	1 050	0.49
	2.30	900	1 100	0.24
	3.20	1 000	1 200	0.12
E	0.65	450	500	3.56
	1.00	500	550	1.50
	1.60	550	650	0.59
	2.30	600	750	0.28
	3.20	700	800	0.15
J	0.65	400	500	1.70
	1.00	450	550	0.72
	1.60	500	650	0.28
	2.30	550	750	0.14
	3.20	600	750	0.07
T	0.32	200	250	6.17
	0.65	200	250	1.50
	1.00	250	300	0.63
	1.60	300	350	0.25

*1 Normal operating temperature limit means the limit of temperature within which thermocouples can be continuously used in the air.

*2 Elevated operating temperature limit means the limit of temperature at which thermocouples can be used for a short time in the case of unavoidable needs.

ocouples generally have excellent resistance to corrosion, but produce small thermoelectromotive forces compared with the base metal types.

Table 3.2 lists the element wire diameter, normal operating temperature limit, elevated operating temperature limit and resistance of each type of thermocouple. As the diameter of the element wire increases, the normal operating temperature limit and elevated operating temperature increase, and the resistance decreases.

Practical use in thermocouples requires the following qualities.

(1) Good corrosion resistance, robustness against gases, etc.
(2) Large thermoelectromotive force.
(3) Good heat resistance and ability to maintain mechanical strength at high temperatures.
(4) Stable themoelectromotive force even over long periods of use, with small thermocouple loss.
(5) Interchangeability: thermocouples of the same type normally have the same characteristics.

Figure 3.15 shows the thermoelectromotive force for each type of thermocouple.

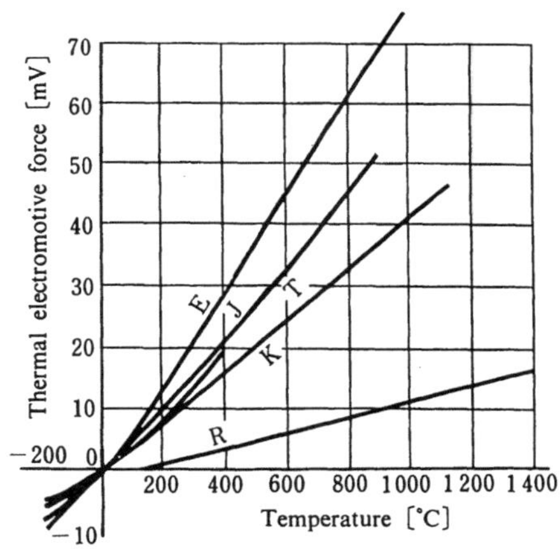

Fig. 3 · 15 Thermal electromotive force of thermocouple

(*d*) *Feature of each type of thermocouple*

(1) *Features common to the noble metal thermocouples* (*B*, *R*, *and S*)

(1) Can be used continuously in an oxidizing environment, but are not suited for use in a reducing environment
(2) Have good resistance to corrosion and chemicals
(3) Are easily degenerated by hydrogen or metal vapors
(4) Are suited for measurement of high temperatures
(5) Have small thermoelectromotive force (that of type B, is especially small)

(6) Thermoelectromotive force has good linearity.

(2) *Features of base metal thermocouples (K, E, J, and T)*

(a) *K–type* (chromel–alumel)

(1) Can be used in an oxidizing environment
(2) Deteriorates slowly in a reducing environment
(3) Strong against metal vapors
(4) Good linearity of thermoelectromotive force

(b) *E–type* (chromel–constantan)

(1) Good resistance to corrosion and oxidation
(2) Deteriorates slowly in a reducing environment
(3) Large thermoelectromotive force
(4) High electrical resistance
(5) Little measurement lag

(c) *J–type* (iron–constantan)

(1) Can be used in an reducing environment
(2) Relatively large thermoelectromotive force; good linearity
(3) Rusts easily
(4) Quality is not consistent
(5) Exhibits hysteresis thermoelectromotive force at high temperatures (800°C)

(d) *T–type* (copper–constantan)

(1) Suited for use at low and extremely low temperatures
(2) Suited for use in reducing environments
(3) Large heat transfer error

(*e*) *Standard thermocouples*

JIS C 1602 defines standard thermocouples which produce a fixed thermoelectromotive force (*reference thermoelectromotive force*) for a given hot junction temperature.

(*f*) *Thermocouples with protective tube*

The ends of two thermocouple element wires are joined by gas welding, arc welding, resistance welding and so on, to form the hot junction. In this state it is called a bare thermocouple. To insulate the two element wires electrically, they are led into an insulating tube as shown in Fig. 3.16. At this stage it is called a thermocouple with insulating tube. Because it is susceptible to corrosion and mechanical damage, it is put into a protective metal or non–metal tube, and the two element wires are connected to a terminal at the head of the tube. Fig. 3.17 shows how the thermocouple with protective tube is constructed.

(*g*) *Sheathed thermocouples*

In a sheathed thermocouple, the space between the element wires and the metal sheath is filled with a powdered inorganic insulator and sealed. It is then processed to produce a single–unit construction. JIS C 1605 defines four types of sheathed thermocoples (type symbols SK,

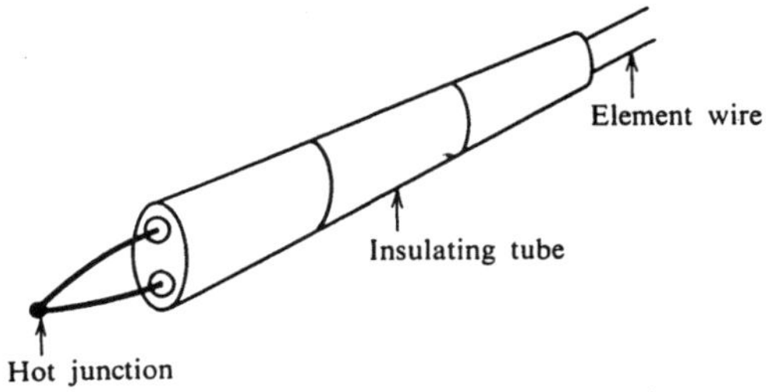

Fig. 3・16 Structure of thermocouple with insulating tube

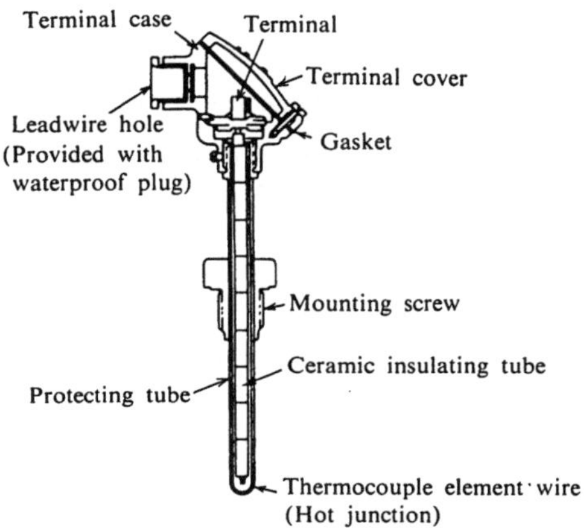

Fig. 3・17 Structure of thermocouple with protecting tube

SE, SJ, and ST). The element wire materials of these thermocouples are the same as the corresponding K, E, J, and T thermocouples.

The metal sheath is made of either austenite stainless steel (symbol A), or nichrome heat resistant alloy (symbol B). The hot junction can be either the grounded type (symbol G) or the ungrounded type (symbol U) (Fig. 3.18). The overall unit can either have a terminal plate or not (Fig. 3.19). Table 3.3 lists the symbol, measured temperature range, accuracy class, tolerance, and normal operating temperature limits for each type of sheathed thermocouple.

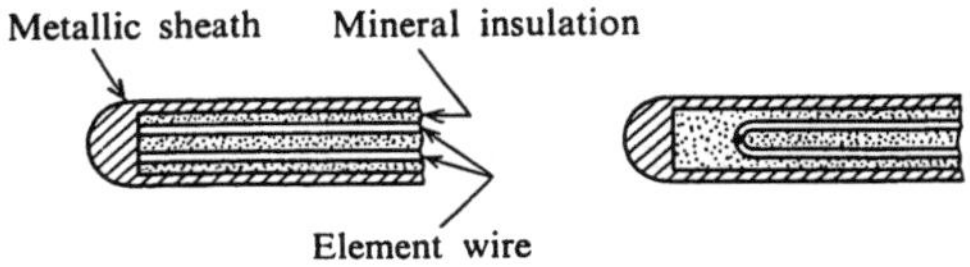

(a) Grounding type (Symbol G)　(b) Non-grounding type (Symbol U)

Fig. 3・18 Hot junction structure

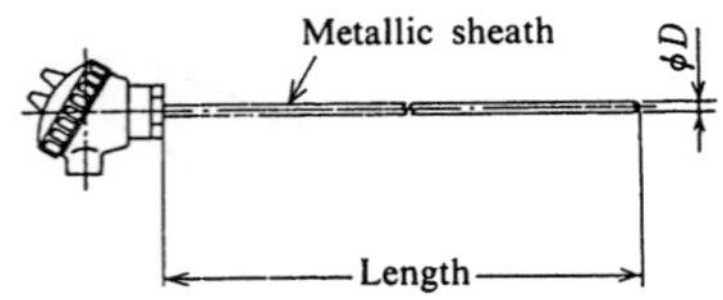

(a) Thermocouple with terminal

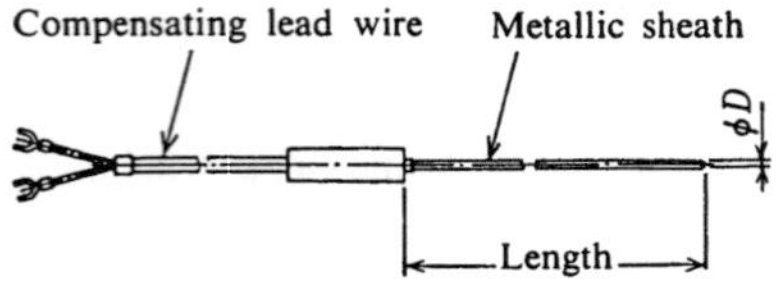

(b) Thermocouple without terminal

Fig. 3 · 19 Sheath thermocouple

Sheathed thermocouples have the following characteristics.

(1) Fast response. The metal sheath has a small diameter and the wire elements are fine, so the heat capacity is small. Also, since the sheath is bonded to the insulating material, which is in turn bonded to the element wires, heat resistance is low. Thus the response to heat is exceptionally good, and the temperature of objects having low heat capacity can be measured.

(2) Flexible: Because of the small outside diameter of the sheath,(1 mm to 8 mm) it can be curved to fit into small places.

(3) Low mechanical strength: For measurement of fast flowing liquids, or when the sheath is installed horizontally, reinforcement such as a protective tube is needed.

(4) Insulator deterioration: The insulating material (MgO) is highly hydrophilic. If it is not completely sealed, it will absorb moisture, causing deterioration of its insulating properties.

(*h*) *Compensating lead wire*

If the thermocouple is connected to the measuring instrument over a long distance using the element wires as they are, induction interference will lower accuracy, and values will become high. If a different conductor is used for the connection to the measuring instrument in place of the element wire, the junction between the conductor and the element wire will produce a contact potential difference, and changes in the temperature at that junction will generate measurement error. To avoid this, compensating lead wire is used to connect the thermocouple to the measuring instrument. Compensating lead wire is a pair of insulated conductors that, in the appropriate temperature range, including ordinary temperatures, has about the same thermoelectric characteristics as the thermocouple. Using compensating lead wire has about the same effect as using element wire to make the ex-

Table 3.3 Sheath thermocouples (JIS C 1605)

Symbol of sheath thermocouple	Measuring temperature	Accuracy class	Torelance*1	Normal operating temperature limit 〔°C〕*2							
				Sheath diameter 〔mm〕							
				1.0,1.5,1.6,2.0	3.0, 3.2	4.5, 4.8		6.0, 6.4		8.0	
				Sheath material							
				A, B	A, B	A	B	A	B	A	B
SK	0°C to 1050°C incl.	Class 0.75	±2.5°C or ±0.75% of measured temperature	650	750	800	900	800	1 000	900	1 050
	−200°C to 0°C excl	Class 1.5	±2.5°C or ±1.5% of measured temperature								
SE	0°C to 900°C incl.	Class 0.75	±2.5°C or ±0.75% of measured temperature	650	750	800	900	800	900	800	900
	−200°C to 0°C excl	Class 1.5	±2.5°C or ±1.5% of measured temperature								
SJ	0°C to 750°C incl.	Class 0.75	±2.5°C or ±0.75% of measured temperature	450	650	750	750	750	750	750	750
ST	0°C to 350°C incl.	Class 0.75	±1°C or ±0.75% of measured temperature	300	350	350	350	350	350	350	350
	−200°C to 0°C excl	Class 1.5	±1°C or ±1.5% of measured temperature								

*1 Tolerance means allowable maximum of the remainder of the temperature converted from thermoelectromotive force using the table of the reference thermoelectromotive force subtracted by the temperature of measuring junction.
Moreover, either one of °C or % whichever is larger is allowed as the tolerance.

*2 Normal operating temperature limit means the limit of temperature within which thermocouples can be continuously used in the air.

Table 3.4 Compensating lead wires (JIS C 1610)

Symbol of thermocouple to be used in combination with	Symbol	Classification by use and tolerance	Composing materials		Surface color	Tolerance	
			+Leg	−Leg		Temperature of connected point with thermocouple [°C]	Tolerance of compensating lead wire [°C]
B	BX-G	General use normal class	Copper	Copper	Grey	0 to 100	—
R	RX-G SX-G	General use normal class	Copper	Alloy; main components are copper and nickel	Black	0 to 150	+3 −7
S	RX-H SX-H	Heat-proof normal class					
K	KX-G	General use normal class	Alloy; main components are nickel and chromium	Alloy; main component is nickel	Blue	−20 to 150	±2.5
	KX-GS	General use precision class					±1.5
	KX-H	Heat-proof normal class					±2.5
	KX-HS	Heat-proof precision class					±1.5
	WX-G	General use normal class	Iron	Alloy; main components are copper and nickel			±3.0
	WX-H	Heat-proof normal class					
	VX-G	General use normal class	Copper	Alloy; main components are copper and nickel		−20 to 100	±2.5

E	EX-G	General use normal class	Alloy; main components are nickel and chromium	Alloy; main components are copper and nickel	Purple	−20 to 150	±2.5
	EX-H	Heat-proof normal class					
J	JX-G	General use normal class	Iron	Alloy; main components are copper and nickel	Yellow	−20 to 150	±2.5
	JX-H	Heat-proof normal class					
T	TX-G	General use normal class	Copper	Alloy; main components are copper and nickel	Brown	−20 to 150	±2.0
	TX-GS	General use precision class					±1.0
	TX-H	Heat-proof normal class					±2.0
	TX-HS	Heat-proof precision class					±1.0

Remarks 1. Cores of KX-G, KX-GS, KX-H, KX-HS, EX-G, EX-H, JX-G, JX-H, TX-G, TX-GS, TX-H and TX-HS are made of the same material as the thermocouple which is to be used in combination with.

2. Cores of BX-G, RX-G, RX-H, SX-G, SX-G, SX-H, WX-G, WX-H and VX-G are different materials from those of the thermocouples which are to be used in combination with, however, their thermo-electric characteristics in operating temperature range (Table 2) are the same as those of thermocouple.

3. Compensating lead wire of normal class is suitable for use in normal temperature measurement.

4. Tolerance of compensating lead wire of precision class is about a half of that of normal class and is suitable for temperature measurement which requires special precision.

5. As for the characteristics, WPR-G and WPR-H of the Former JIS may be used in combination with R and S in almost all cases.

tension. However, the thermoelectric characteristics are not completely the same, so a very small error will be generated in the compensating lead wire connection. The difference between the thermoelectromotive force of the compensating lead wire and the standard thermoelectromotive force of the corresponding type of thermocouple at the same temperature, converted into a temperature difference, is called compensating lead wire error. In Table 3.4, compensating lead wires classified according to the type of thermocouple with which they are used according to JIS standards.

Compensating lead wires are also classified according to working temperature: G for general use (−20°C to 90°C), and H for heat-resistant applications (0°C to 150°C). The ordinary accuracy class is suited for normal temperature measurements, but where precision is needed, the precision class is appropriate, since its tolerance is half that of the ordinary class. The type is identified by the color of the surface covering, and the polarity, is indicated by the color of the core insulation (red for positive, and principally white for negative).

(*i*) *Special thermocouples*

(1) *Thermocouples for high temperature use* The Pt•40%Rh–Pt•20%Rh thermocouple, which is not covered in JIS standards, can measure temperatures near 1 900°C, but has an unusually low electromotive force. The W–W•26%Re and the W•5%Re–W•26%Re are representative high melting point metal thermocouples. Tungsten is oxidized in air or in an oxidizing atmosphere, so these thermocouples are used in an inert or reducing atmosphere. They have a maximum working temperature of about 2 800°C, and a normal limit of about 2 400°C.

(2) *Dipping thermocouples* Continuous measurement of the temperature or molten metals is difficult. Thus a short–time measurement is used in which the tip of the thermocouple is opened and placed in contact with the molten metal.

3.2.2 Resistance thermometers

(*a*) *Measurement principle and characteristics*

Because there is a fixed relationship between the electrical resistance of a metal and temperature, it is possible to determine a temperature by measuring electrical resistance. Platinum (Pt), nickel (Ni), and copper (Cu) are used in this type of thermometer. The temperature and electrical resistance characteristics of these metals are shown in Fig. 3.20.

Thermometer materials must satisfy the following conditions.

(1) The relationship between temperature and electrical resistance must be continuous and have unique values within the working temperature range.

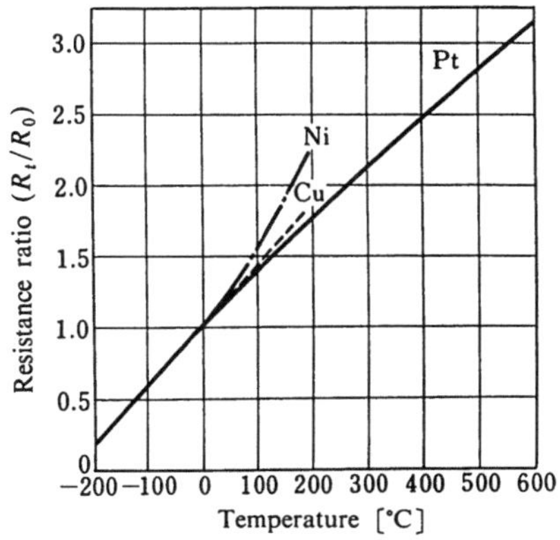

Fig. 3・20 Temperature-resistance characteristics

(2) There is no change in the resistance over time or according to other factors.
(3) There is no hysteresis effect in the relationship between resistance and temperature.
(4) The material is resistant to corrosion, thus stable.
(5) The material has a high intrinsic resistance and resistance temperature coefficient.
(6) Interchangeable.
(7) Easily worked.

Platinum has the best combination of the above properties, and the platinum resistance temperature detector is the JIS standard. Its measuring range is from −200°C to 650°C (Pt100) or -200°C to 500°C (JPt100).

The resistance temperature measurement method has the following characteristics compared with the thermocouple method.

(1) High sensitivity.
(2) Stable over a long period of time when used in a low-vibration environment.
(3) Response is slow because of its large size.
(4) Has a low maximum working temperature.
(5) Susceptible to shock and vibration because a fine resistance wire element is used.

When a current flows in a resistor, the voltage between the ends of the resistor is given by the following relationship, according to Ohm's law.

$$V = IR$$

In this equation, V represents voltage (V), I is current (A), and R is resistance (Ω). If I is fixed and V is measured, we can determine R, and thus measure the temperature of the resistance temperature detector.

(*b*) *Plantinum resistance temperature detector*

The relationship between the resistance of a platinum resistance temperature detector and temperature is standardized in JIS C 1604. The resistance value of a resistance element at 0°C is called the nominal resistance. At present nominal resistance is 100Ω. The current that flows continuously through the resistance element is called the standard current. The present standard currents are 1 mA, 2 mA, and 5 mA.

There are three working temperature grades: L for low temperature use (−200°C to 100°C), M for medium temperatures (0°C to

Table 3.5 Tolerance (JIS C 1604)

Normal resistance [Ω]	Operating temperature range	Measured temperature [°C]	Tolerance: Class 0.15 Temperature [°C]	Tolerance: Class 0.2 Temperature [°C]	Tolerance: Class 0.5 Temperature [°C]
100	L	−200	±0.45	±0.55	±1.3
	L	−100	±0.30	±0.35	±0.8
	L, M, H	0	±0.15	±0.15	±0.3
	L, M, H	100	±0.30	±0.35	±0.8
	L, M, H	200	±0.45	±0.55	±1.3
	M, H	300	±0.60	±0.75	±1.8
	M, H	350	±0.68	±0.85	±2.05
	H	400	—	±0.95	±2.3
	H	500	—	±1.15	±2.8
50	L	−200	±0.45	±0.55	±1.3
	L	−100	±0.30	±0.35	±0.8
	L, M, H	0	±0.15	±0.15	±0.3
	L, M, H	100	±0.30	±0.35	±0.8
	L, M, H	200	±0.45	±0.55	±1.3
	M, H	300	±0.60	±0.75	±1.8
	M, H	350	±0.68	±0.85	±2.05
	H	400	—	±0.95	±2.3
	H	500	—	±1.15	±2.8

350°C), and H for high temperatures (0°C to 500°C or 0°C to 650°C).

(*c*) *Internal wirs*

The conductors used to connect the resistance element to the terminals of the resistance bulb are call internal wires. As shown in Fig. 3.21, there are three types of internal wire configurations: 2-wire, 3-wire, and 4-wire. In the figure, s represents the resistance element, and A and B are the terminals.

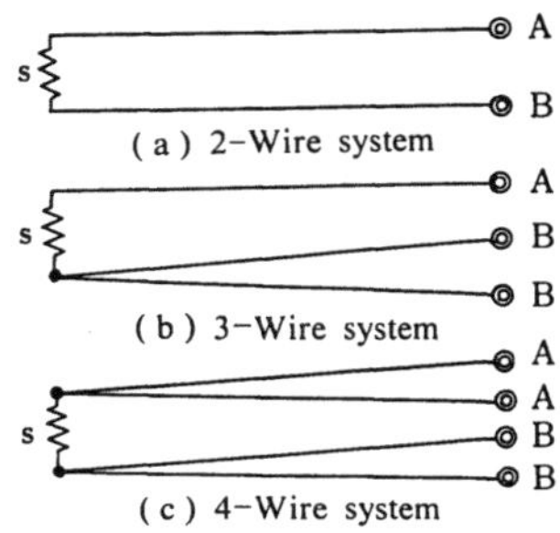

Fig. 3・21 Internal wires of resistance temperature detector

The 2-wire configuration reduces the cost of the external wiring for connecting the resistance bulb to the conversion instrument, but introduces a large error. This is because the same wire is used to supply the electric current and to detect the voltage, and as a result, the temperature of the wire directly affects the resistance. If, in order to make the influence of the external wiring resistance small, the conversion instrument is installed near the resistance temperature detector, making the instrument heat-proof and protecting it from explosion becomes a problem. For industrial measurement, the 3-wire type is most often used. If the resistances of the three wires are the same, measurement free from error is possible even if the ditstance between resistance temperature detector and conversion instrument is long and the resistance is subject to change by external temperatures. To completely eliminate this kind of error, the 4-wire configuration is used. In the 4-wire configuration, the current supply wires are independent of the voltage detecting wires, so no current flows in the detecting wires. Thus it is possible to measure the resistance of the resistance element exactly. This configuration is used in standards and for precision measurement.

(*d*) *Reference resistance element*

If R_t is the resistance of the resistance element at t (°C), and R_0 is the resistance at 0°C, then R_t/R_0 is called the resistance ratio. The resistance ratio of a metal is reduced by lower purity of the metal and by working. The purity of platinum is quite high (99.99% or more), and the standard of that purity is represented by R_{100}/R_0. In JIS C 1604, an ideal resistance element that has a constant resistance ratio R_t/R_0 for given temperatures is called the reference resistance element.

(e) Construction of the resistance temperature detector element

(1) *Mica-insulated element* An example of a mica-insulated element is shown in Fig.3.22. The element wire is wrapped around a long, thin mica plate with a width of about 3 mm to 10 mm serrated on both sides. Rather wide mica plates are mounted on both sides for insulation, and then two metal sheatks are attached. The metal sheath is bonded to the inside of a protective tube, which has the effect of guarding the element against vibration and shock, and providing good heat transfer between the protective tube and the element.

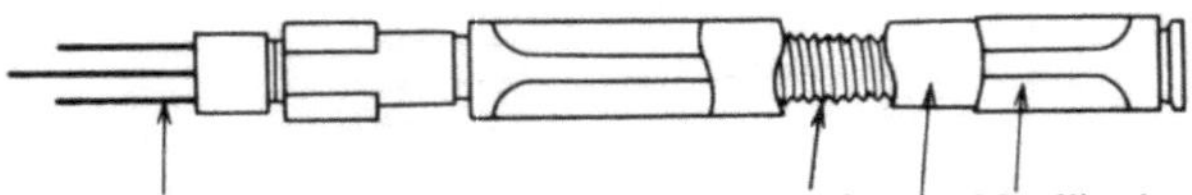

Fig. 3 · 22 Mica-insulated element

The mica-insulated element is not completely fixed in position, so distortion from heat is small and the temperature-resistance characteristic is stable. However, the unit is relatively large and thus not suitable for use in small or narrow places. Also it responds slowly to changes in temperature.

(2) *Glass sealed element* The construction of a glass sealed element is shown in Fig. 3.23. The platinum element wire is welded to platinum alloy wires and melted into a glass spool. The diameter of the resistance element is about 1 mm to 4 mm. The glass softens at about 450°C, so the maximum working temperature of this element is around 400°C.

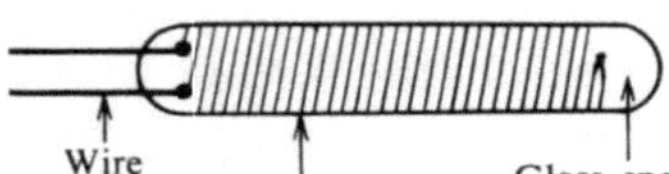

Fig. 3 · 23 Glass sealed element

(3) *Ceramic-sealed element* The construction of a ceramic sealed element is shown in Fig. 3.24. This is a rodshaped element which uses ceramic spool. Its maximum working temperature is approximately 800°C. It responds quickly to temperature changes because of its small outer diameter (1.6 mm to 3 mm), and has good insulation characteristics.

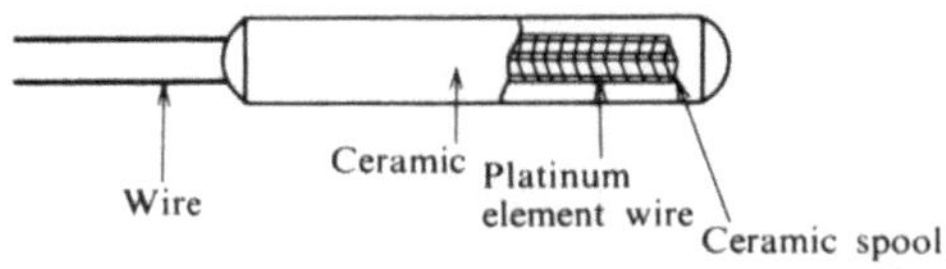

Fig. 3 · 24 Ceramic sealed element

(f) Resistance temperature detector with protective tube

A mica-insulated element inserted into a protective tube, with internal wires connected, is called a resistance temperature detector with protective tube. An internal wire material is used which does not generate a thermoelectromotive force, and which does not oxidize or degenerate at the maximum working temperature. For low to medium temperature use, silver wire covered with fluorocarbon resin or a glass tube or for high temperature use, nickel wire covered with a porcelain tube, is use. For the external wiring, the 3-wire configuration is mostly used. Moisture entering the protective tube degrades the insulation, so for low temperature use, the protective tube is packed with paraffin or filled with dry air and tightly sealed.

Figure 3.25 shows the construction of a resistance temperature detector with protective tube. JIS C 1604 specifies protective tubes with outside diameters between 3.2 mm and 15 mm, and lengths between 250 and 1 000 mm.

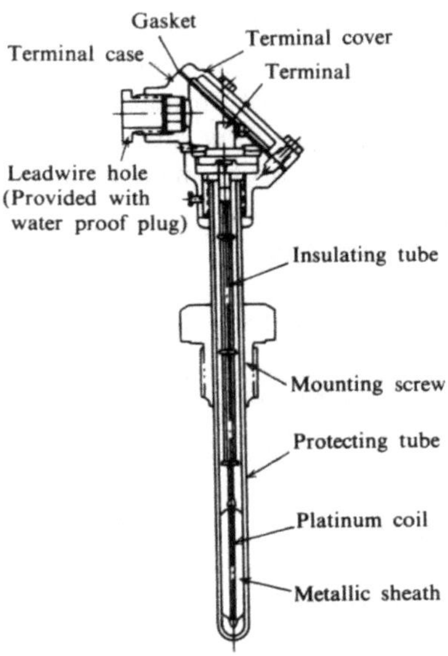

Fig. 3 · 25 Resistance temperature detector with protecting tube (*g*) *Other resistance temperature detectors*

Other than platinum, other metals such as nickle or copper are used for the resistance material. Nickel is less expensive than platinum, and its temperature coefficient is 0.6% at room temperature. However, because the thermal characteristics vary with the element meterial, the temperature coefficient is adjusted to the standard value by combining the material with constantan wire or other such material. The working temperature range is $-50°C$ to 300°C.

Copper wire of high purity can be obtained, and its temperature characteristics are uniform. Furthermore, it has better accuracy, linearity and interchangeability than nickel. However, because its intrinsic resistance is low, a long coil is needed. In addition, at around 250°C, oxidation begins to change its resistance. Thus the working temperature range is 0°C to 120°C.

(*h*) *Sheathed resistance bulbs*

Sheathed resistance bulbs are specified in JIS C 1606. The space inside the sheath and between the resistance element wires is filled with a powdered inorganic insulator and processed to produce a single-unit construction, and platinum is used in the resistance temperature detector.

Ordinary stainless steel having sufficient heat resistance and durability to prevent penetration by environmental elements is used for the metal sheath.

The inorganic insulation protects the resistance element, the internal wires and the metal sheath. This material must be stable in the appropriate temperature range, and must not affect electrical characteristics. Magnesium oxide (MgO) is often used.

The internal wire must not generate a thermoelectromotive force, and must not be susceptible to problems arising from evaporation or oxidation. Thus nickel is largely used.

The 3-wire and 4-wire configurations of internal wires are illustrated in Fig. 3.26.

The resistance element either a ceramic sealed element or a glass

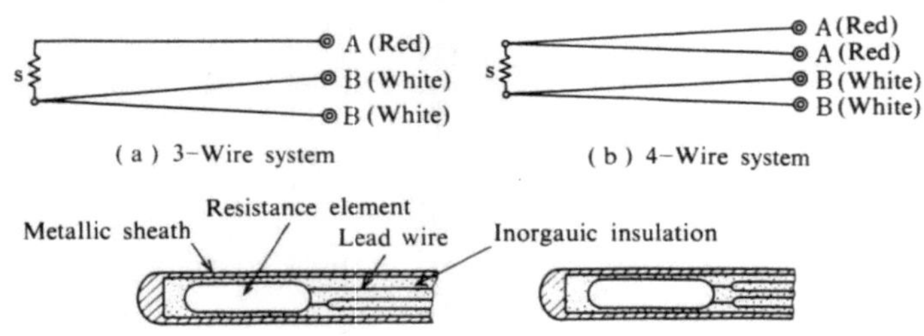

Fig. 3・26 Internal wires

sealed element. These elements are themselves hermetically sealed, and buried in the inorganic insulation inside the sheath. Thus they are not affected by the environment. They also have fast response and are strong against mechanical vibration.

3.2.3 Protective tube

Thermocouple and resistance temperature detectors are put into protective tubes during use in order to protect them from the environment. Protective tubes must statisfy the following requirements.

(1) Able to withstand the temperatures and pressures of the environment.
(2) Chemically stable.
(3) Mechanically strong.
(4) Airtight
(5) Not produce any gases harmful to thermocouples.

It is difficult to choose a material that satisfies all of these requirements, but a material that is suitable for each particular set of use conditions can be selected.

Protective tubes are classified into metal types and non-metal types.

(a) Metal protective tubes

Metal tubes are widely used because they are easily manufactured and have great mechanical strength. Typical metal tube materials are listed in Table 3.6.

Ordinary protective tubes are cut to the required length from seamless tubing. One end is sealed by welding, and the other end is threaded so that the terminal box can be attached. If a tube produced

Table 3.6 Material of typical metal protective tube

Material	Maximum operating temperature [°C]	Feature
Brass	300	Easy to work, low temperature resistance
Mild steel	600	High mechanical strength, cheap, oxidized with damp
Calorized mild steel	800	Improved mild steel on heat and corrosion resistance, low reducing gas resistance
13Cr steel	900	
20Ni 30Cr steel	1 000	
Stainless steel	850	High acid resistance
Monel	500	High sea water resistance, high dilute sulfuric acid resistance
Inconel	1 000	High strength under high temperature, applicable to oxidizing environment

by this method is not strong enough, a thick–walled tube can be made by boring out the center of a piece of rod stock. This is called a well–type or bored–type protective tube. Although well–type tubes are heavy and have bad response characteristics, they are widely used where high pressures or high–speed flows are present. Sheathed thermocouples are very often in well–type protective tubes in applications where there is severe shock or vibration.

(*b*) *Non–metal protective tubes*

Non–metal protective tubes are not as mechanically strong as metal tubes, but they have excellent resistance to chemicals, and there are many materials of use with high maximum working temperatures. Typical non–metal protective tube materials are listed in Table 3.7.

Table 3.7 Material of typical non–metal protective tube

Material	Maximum operating temperature [°C]	Features
Fluorocarbon resin	250	Good corrosion resistance
Hard glass	500	Antioxidation, preventing gas penetration, low resistance for quick heating and cooling
Quartz	1 000	High resistance for quick heating and cooling for continuous use at high temperature. Becomes brittle if exposed to high temperature.
Ceramic (PT–2)	1 300	Low resistance for quick heating and cooling
Ceramic (PT–1)	1 450	Higher heat resistance compared with PT–2
Recrystallized alumina (PT–0)	1 600	Most applicable to high temperature, low resistance for quick heating and cooling

3.2.4 *Thermistor thermometers*

Thermistors are sintered bodies of the oxides of metals such as manganese, nickel, and cobalt, whose resistance varies with temperature. The resistance–temperature coefficient of thermistors is negative; an increase in temperature lowers the electrical resistance.

Compared with platinum resistance temperature detectors, thermistors, have the following characteristics.

(1) Large electrical resistance

(2) High sensitivity

(3) Poor linearity of resistance
(4) Fast response due to small size

Measurement principle

The resistance R_T of a thermistor at temperature T is represented theoretically as

$$R_T = R_a e^{B(1/T - 1/T_a)} \quad [\Omega] \tag{3.3}$$

Where R_a is the resistance (Ω) at temperature T_a, and B is the thermistor constant (K).

If α is taken as the thermistor temperature coefficient, then because $\alpha = (1/R_T)(dR_T/dT)$, and from Eq.(3.3),

$$\alpha = \frac{1}{R_T} \cdot \frac{dR_T}{dT} = -\frac{B}{T^2} \quad [\mathrm{K}^{-1}] \tag{3.4}$$

From Eq.(3.4), we see that thermistor sensitivity varies greatly with the temperature. The value of B is usually 3 000K to 5 000K, but actually depends on temperature. The value of α is as large as $-3\%/°\mathrm{C}$ to $-6\%/°\mathrm{C}$ at room temperature.

The standard temperature characteristic of thermistors is shown in Table 3.8.

3.2.5 Radiation thermometers

(a) Black body radiation

Physical objects emit all wavelengths of electromagnetic waves, and the strength of the radiation increases with the temperature of the object. An object which absorbs completely all wavelengths of electromagnetic waves from external sources is called a black body. The spectral radiant emittance for a black body is represented, according to Planck's law, by

$$M_B(\lambda, T) = \frac{c_1}{\lambda^5} \cdot \frac{1}{\exp(c_2/\lambda T) - 1} \quad [\mathrm{W/m^3}] \tag{3.5}$$

where, $M_B(\lambda, T)$ is the spectral radiant emittance of a black body,

λ = wavelengh (m),

T = absolute temperature (K)

c_1 = first radiation constant, (3.7418×10^{-16} W·m²) and

c_2 = second radiation constant (1.4388×10^{-2} m·K).

The spectral radiant emittance for a black body is shown in Fig. 3.27.

(b) Emissivity, luminance temperature, and effective wavelength

The thermal radiation energy emitted by a physical object at a given temperature is greatest for a black body. For a wavelengh λ_e, the relationship between the spectral radiance of an ordinary object

Table 3.8 Standard temperature characterisitics for thermister (JIS C 1611)

Temperature [°C]	Connecting type	Element replaceable						Combined resistance						Ratio					
	Nominal resistance	6kΩ (0°C)	30kΩ (0°C)	3kΩ (100°C)	0.55kΩ (200°C)	4kΩ (200°C)	8kΩ (200°C)												
	Operating temperature range [°C]	−50 to 100	0 to 150	50 to 200	100 to 250	150 to 300	200 to 350	−50 to 100	0 to 150	50 to 200	100 to 250	150 to 300	200 to 350	−50 to 50	0 to 100	50 to 150	100 to 250	150 to 300	200 to 350
−50		75.36 kΩ						11.86 kΩ						3.415					
−25		19.53 kΩ						8.041 kΩ						2.677					
0		6.000 kΩ	30.00 kΩ					4.610 kΩ	3.357 kΩ					1.841	3.415				
25		2.179 kΩ	10.67 kΩ					2.853 kΩ	2.548 kΩ					1.328	2.585				
50		904.2 Ω	4.356 kΩ	13.57 kΩ				2.123 kΩ	1.748 kΩ	2.997 kΩ				1.094	1.841	3.415			
75		421.2 Ω	2.000 kΩ	6.081 kΩ					1.190 kΩ	2.328 kΩ					1.360	2.535			
100		215.6 Ω	1.017 kΩ	3.000 kΩ	6.281 kΩ				0.8659 kΩ	1.718 kΩ	4.971 kΩ				1.094	1.841	3.415		
125			560.7 Ω	1.601 kΩ	3.046 kΩ					1.267 kΩ	4.179 kΩ					1.378	2.837		
150			330.5 Ω	912.1 Ω	1.510 kΩ	13.33 kΩ				0.9702 kΩ	3.332 kΩ	865.6 Ω				1.094	2.290	3.415	
175				538.2 Ω	913.3 Ω	7.111 kΩ					2.583 kΩ	765.7 Ω					1.840	2.797	
200				330.6 Ω	550.0 Ω	4.000 kΩ	8.000 kΩ				2.007 kΩ	661.0 Ω	3.238 kΩ				1.504	2.265	3.415

225				347.8 Ω	2.356 k Ω	4.481 k Ω				1.598 k Ω	561.0 Ω	2.505 k Ω				1.264	1.841	2.809
250				230.0 Ω	1.445 k Ω	2.634 k Ω				1.316 k Ω	472.0 Ω	1.953 k Ω				1.094	1.515	2.274
275					920.1 Ω	1.619 k Ω					396.8 Ω	1.564 k Ω					1.273	1.841
300					602.4 Ω	1.030 k Ω					335.8 Ω	1.299 k Ω					1.094	1.517
325						690.7 Ω						1.120 k Ω						1.274
350						468.0 Ω						1.000 k Ω						1.094

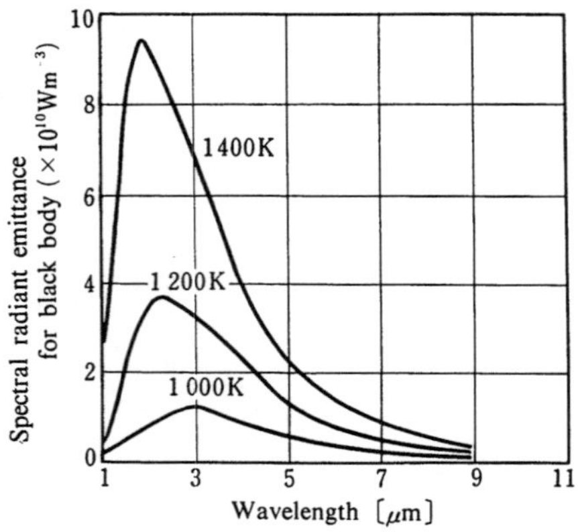

Fig. 3・27 Spectral radiant emittance for black body

$L(\lambda_e, T)$ and a black body $L_B(\lambda_e, T)$ is represented, using emissivity $\varepsilon(\lambda_e, T)$, as

$$L(\lambda_e, T)=\varepsilon(\lambda_e, T)L_B(\lambda_e, T)=L_B(\lambda_e, S) \tag{3.6}$$

Here, λ_e is called the effective wavelengh, S is the luminance temperature, T is the true temperature.

(c) Optical pyrometers

Optical pyrometers are specified in JIS Z 8706, measuring the luminance temperature for an effective wavelength of 0.65 μm. The operating principle is illustrated in Fig. 3.28. The current to the internal electric lamp is adjusted until the operator determines by sight that the filament luminance matches the emission luminance of the object being measured. The current to the lamp filament at that time is a measure of the luminance temperature. A red filter is used to restrict the measurement wave band to a narrow region which falls within the visual sensitivity of the human eye. The optical pyrometer is widely used because it has a simple construction and is relatively accurate.

For luminance temperature S (K), true temperature T (K), and ef-

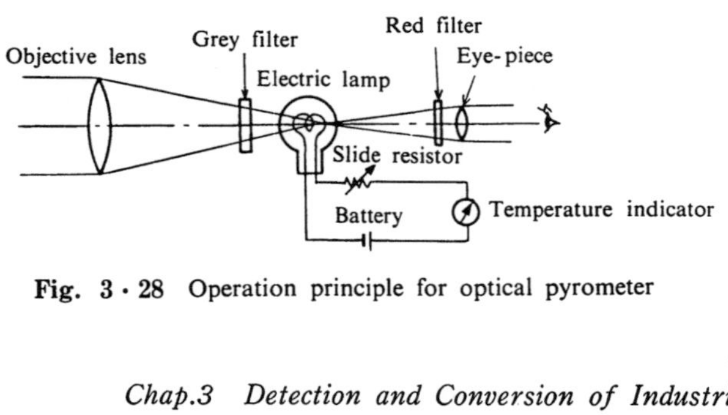

Fig. 3・28 Operation principle for optical pyrometer

fective wavelength $\lambda_e = 0.65$ (μm), the following equation holds.

$$\frac{1}{T} - \frac{1}{S} = 4.52 \times 10^{-5} \log_e \varepsilon \qquad (3.7)$$

An example of emissivity for an effective wavelength of 0.65 μm

Table 3.9 Emissivity (wavelength; 0.65μm) (JIS Z 8706)

Material and its stage			Temperature [°C]	Emissivity
Carbon			—	0.85 to 0.95
Chrome oxide			900	0.81
Alumina			900	0.18
Ceramics			1 200	0.25
			1 500	0.32
			1 800	0.38
Iron	Melted iron		1 100 to 1 900	0.4
	Melted slag		1 400 to 1 830	0.65
	Solid iron oxide		800	0.98
			1 200	0.92
	Iron non-oxide		1 200	0.37
Melted steel	Bare metal surface		—	0.35
	Oxide film		—	0.5 to 0.8
Nickel oxide			800	0.96
			1 300	0.85
Nickel			—	0.37
Copper	Melted		—	0.15
	Non-oxide	Liquid	—	0.15
		Solid	—	0.11
	Oxide		—	0.6 to 0.8
Tungsten			700	0.46
			2 000	0.43
			3 000	0.41
Platinum			1 000	0.29
			1 300	0.30
			1 700	0.32

is shown in Table 3.9.

(*d*) *Silicon radiation thermometers*

Figure 3.29 shows an example of a silicon radiation thermometer system. Light from the object of measurement is condensed by a variable focal point objective lens, and a part is sent through a half-silvered mirror to a finder for determining the position of the object and focussing. After passing through the mirror, the light is restricted by a slit to control the field of vision to a very small portion of the beam for temperature measurement. To heighten accuracy and stability, it is necessary to avoid the influence of external disturbance by visible light. For this purpose, a filter with narrow band characteristics is positioned in front of the photocell. This filter passes light of wavelengths between 0.7 μm and 1.1 μm. The silicon photocell is sensitive to light in the range of 0.4 μm to 1.2 μm. By changing the standard current of the electrical circuit, the effective emissivity is corrected. The measurement range is 600°C to 1 600°C.

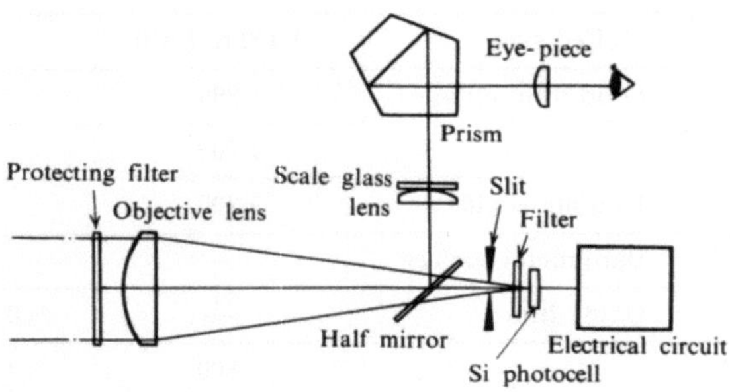

Fig. 3 · 29 System for silicone radiation thermometer

3.3 Measurement of Flow[17)-32)]

Flow measurements are often used in industrial processes. Of the methods of measuring flow, the constriction-type differential pressure flowmeter has been used from early times. There are many other forms of flow meaurement having various theoretical bases, such as the area flowmeter, the positive displacement flowmeter, the electromagnetic flowmeter, and the ultrasonic flowmeter. The type of flow measuring instrument is selected according to the required accuracy, magnitude of flow, type of fluid being measured (liquid, gas, etc.), and the properties of the fluid (temperature, pressure, viscosity, corrosiveness, etc.). The principles and characteristics of each type of flowmeter are described in the following sections.

3.3.1 Differential pressure flowmeters

Differential pressure (head) flowmeters, also called constriction flowmeters, are suitable constriction devices placed in the pipeline. The flow causes a difference in the pressure in front of and behind the constriction. This pressure differential corresponds to the amount of flow, so by measuring this pressure differential, the flow rate can be determined. These devices have a simple construction and are suitable for use with liquids, gases, and steam, so they have long been the most commonly used type of industrial flowmeter. For proper measurement of flow using constriction devices, the following JIS concerning constriction devices is widely applied.

JIS Z 8762 Measurement of Fluid Flow by Means of Orifice Plates, Nozzles and Venturi Tubes

(*a*) *General equations for constriction device flow calculations*

Figure 3.30 illustrates a constriction device installed in a horizontal pipe that has a circular cross section. The constriction device is a flat plate (called an *orifice plate*) having a circular hole that is concentric with the pipe. It is mounted perpendicular to the flow. If the fluid is incompressible and not affected by viscosity, with a steady-state flow which completely fills the pipe, then between the upstream cross section (① in Fig. 3.30) and the downstream cross section, called the vena contracta (② in Fig. 3.30), the flowing two equations hold:

$$\frac{1}{2}\rho v_1^2 + p_1 = \frac{1}{2}\rho v_2^2 + p_2 \qquad (3.8)$$

$$\rho v_1 F_1 = \rho v_2 F_2 \qquad (3.9)$$

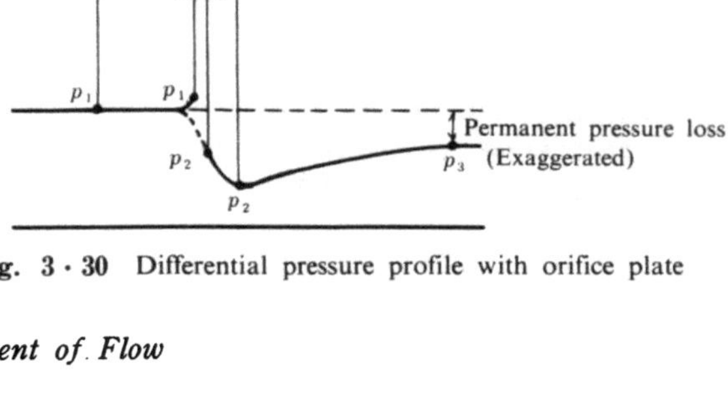

Fig. 3・30 Differential pressure profile with orifice plate

Here, v is the mean stream velocity (m/s), p is the pressure (Pa abs), ρ is the density of the fluid (kg/m^3), F is the sectional area of the flow, and the subscripts 1 and 2 refer to cross sections ① and ② in Fig. 3.30. These equations are derived from Bernoulli's equation and the flow equations of continuity.

Equations (3.8) and (3.9) lead to the following relationship between the volume rate of flow Q (m^3/s) and the pressure differential (p_1-p_2) (Pa).

$$Q=v_2F_2=F_2\frac{1}{\sqrt{1-(F_2/F_1)^2}}\sqrt{\frac{2}{\rho}(p_1-p_2)} \tag{3.10}$$

In a practical equation for real fluids which have viscosity, as well as for compressible fluids (gases), F_2 is replaced by the sectional $(\pi/4)d^2$, and the volume rate of flow Q_v (m^3/s) becomes

$$Q_v=\varepsilon c\frac{(\pi/4)d^2}{\sqrt{1-\beta^4}}\sqrt{\frac{2}{\rho_1}(p_1-p_2)}=\varepsilon\alpha\frac{\pi}{4}d^2\sqrt{\frac{2}{\rho_1}(p_1-p_2)} \tag{3.11}$$

Also, from $Q_m=Q_v\rho_1$, the mass rate of flow, Q_m (kg/s) is,

$$Q_m=\varepsilon\alpha\frac{\pi}{4}d^2\sqrt{2\rho_1(p_1-p_2)} \tag{3.12}$$

Here, ρ_1 is the fluid density (kg/m^3) for the upstream pressure measurement section, β is the ratio of the constriction orifice diameter d (m) to the pipe diameter D (m) ($\beta=d/D$), c is the coefficient of discharge, α is the flow coefficient ($\alpha=c/\sqrt{1-\beta^4}$), and ε is the expansion correction factor (for an incompressible fluid, $\varepsilon=1$).

Equations (3.11) and (3.12) are the general equations for flow rate calculation, and are suitable for use with any type of constriction device, such as a nozzle or venturi tube, in the same way as with the orifice plate. The coefficients in these equations are determined empirically. The *flow coefficient* α is given as a function of the constriction device diameter ratio β and the Reynolds number of the flow for each set of conditions including the type of constriction device and the tap position at which the differential pressures are measured (pressure tap method). The *expansion correction factor* ε corrects for the change in density of a compressible fluid (gas) after passing through the constriction device, and is determined by the type of constriction device, the specific heat capacity ratio of the gas, p_1, p_2 and β. The Reynolds number is a characteristic value of the flow conditions, and the *Reynolds number* R_{eD} with reference to the inside diameter D of the pipe is a dimensionless number given by Eq.(3.13).

$$R_{eD}=\frac{\bar{v}D}{\nu} \tag{3.13}$$

Here, $\bar{v}$ is the mean stream velocity (m/s), D is the inside diameter of the pipe (m), and ν is the kinematic viscosity of the fluid (m^2/s). Figure 3.31 shows the relationship between the flow coefficient α and the Reynolds number R_{eD} for a typical constriction device. Corresponding to each constriction area ratio, $\beta^2=(d/D)^2$, for the respective values above which they have a Reynolds number, the flow coefficient α takes on a more or less constant value. Generally, flow measurement using constriction devices is done in this region of R_{eD} in which α can be regarded as fairly constant (that is, can be treated as a mathematical constant).

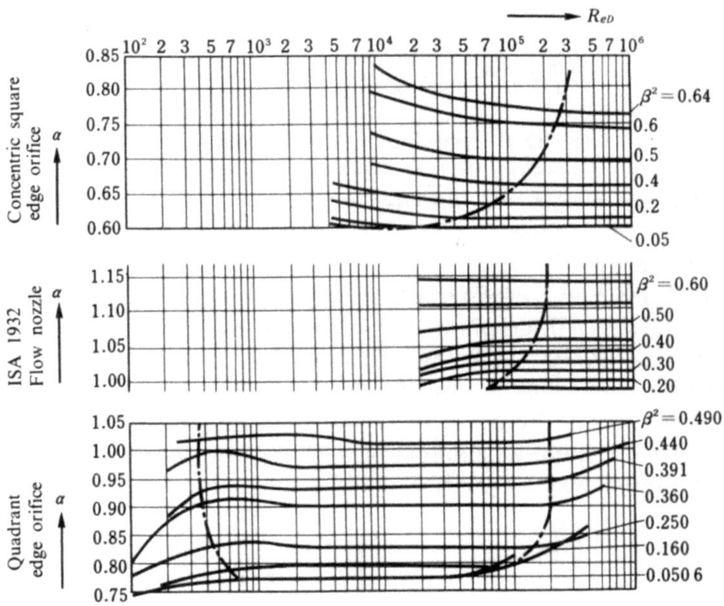

Fig. 3 · 31 Correlation between flow coeffiuent α and reynolds number R_{eD}

As can be seen from Eqs.(3.11) and (3.12), the magnitude of the pressure differential (p_1-p_2) is proportional to the square of the flow rate. When the flow rate is 1/10 of its maximum, the pressure differential is 1/100 of its maximum. Thus caution is required when measuring low flow rates, because error in measuring the pressure differential exerts a large influence on the accuracy of the indicated flow rate.

(*b*) *Temperature and pressure correction for gas flows*

As shown by Eqs.(3.11) and (3.12), flow rate is principally obtained by measuring the pressure differential created by a constriction device, taking the density of the fluid to be constant. However, the density of a gas changes with change in pressure and temperature. Accordingly, if measurement is done of a flow at temperatures or pressures different from the design values, the fluid density will also be different, causing measurement error. If we assume that the fluid is an ideal gas (taking the compressibility coefficient to be 1), then from the law of gases, the density under actual use conditions ρ_1 is

$$\rho_1 = \rho_n \frac{p_1}{p_n} \cdot \frac{t_n}{t_1} \tag{3.14}$$

where p is the gas pressure (Pa abs), t is the temperature of the gas expressed as an absolute temperature (K), and the subscript n represents the design condition.

From Eqs.(3.12) and (3.14),

$$Q_m = \varepsilon\alpha \frac{\pi}{4} d^2 \sqrt{2\rho_n \frac{p_1}{p_n} \cdot \frac{t_n}{t_1} (p_1 - p_2)} \tag{3.15}$$

This is to say, by measuring the changing pressure p_1 and temperature t_1 and the pressure differential $(p_1 - p_2)$ at the same time and then correcting the results by multiplying by the ratio with the design values for the constriction device, the correct flow rate can be obtained.

(*c*) *Constriction device shape*

There are many kinds of constriction devices, but the most typical among them are orifice plate, nozzle, and venturi tube specified by JIS.

(1) *Concentric orifice* A concentric orifice is illustrated in Fig. 3.32. The orifice plate has a simple construction and is inexpensive. For these reasons, it is the most generally used among the constriction devices for a wide range of applications. However, because the flow is abruptly dammed up, the orifice plate produces a relatively large pres-

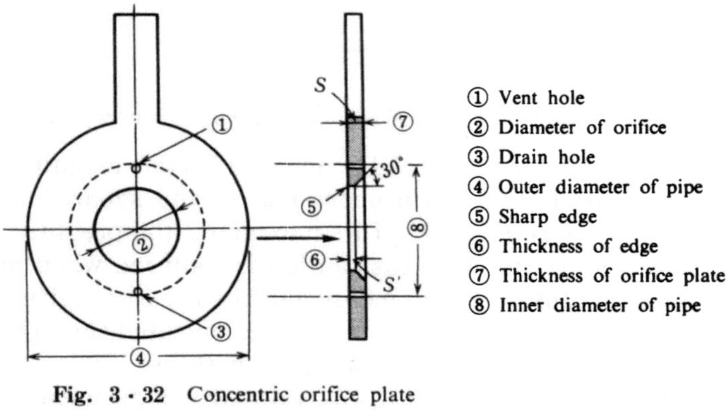

Fig. 3 · 32 Concentric orifice plate

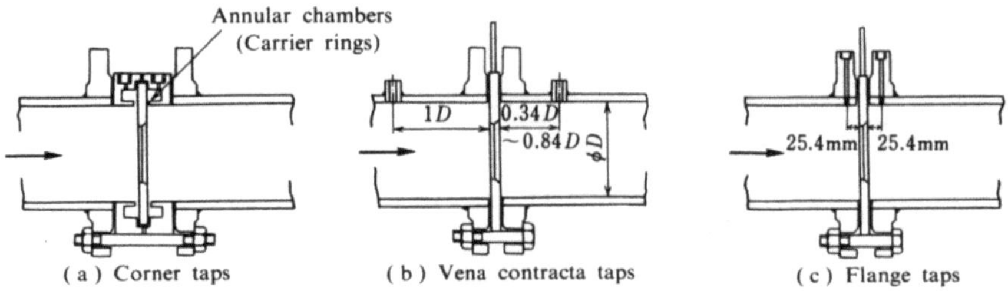

Fig. 3・33 Pressure taps

sure loss compared with the other constriction devices, and wear from abrasion at the edge of the orifice is a cause of measurement error. Thus it is necessary to avoid adulteration of the fluid by solid objects.

Figure 3.33 illustrates methods of tapping the pressure differential (pressure tapping) based on JIS.

(i) *Corner taps*: This type is mostly constructed of an orifice plate sandwiched between two annular chambers (carrier rings). The pressure differential can be tapped by reading the pressure immediately in front of and behind the orifice plate, averaged by the chambers.

(ii) *Vena contracta taps*: The pressure is tapped upstream at a distance equal to the inside diameter of the pipe (1*D*), and downstream at the vena contracta, where the cross-sectional area of the flow is the smallest. The vena contracta position has an empirically confirmed fixed relationship with β. Depending on the value of β, the vena contracta position is between 0.34 *D* and 0.84 *D* downstream from the orifice plate.

(iii) *Flange taps*: Pressure tap holes are placed 25.4 mm from the orifice plate surface on both upstream and downstream sides of the oriffice plate. With this configuration, flanges with pressure tap holes are required.

(2) *Nozzles* There are two types of nozzles: the JIS standard nozzle (ISA 1932), and the long-radius nozzle (Fig. 3.34). The flow coefficient of a nozzle α is near 1. It has a strong construction and excellent durability, and is thus suited for use with high temperature and fast moving fluids.

(3) *Venturi tubes* The JIS nozzle venturi and classical venturi are illustrated in Fig. 3.35. The constriction is gradual, so this type of constriction device has the advantages that it is difficult for solids contained in the flowing liquid to accumulate on the upstream side of the constriction, and that permanent pressure loss is small. Also, as the shape is smooth, it has good durability against abrasion. Figure 3.36 compares permanent pressure losses of venturi tubes, orifices, and nozzles.

(4) *Quadrant edge orifices* Although not specified in JIS, the quadrant edge orifice is suited for low Reynolds number flows. The flow

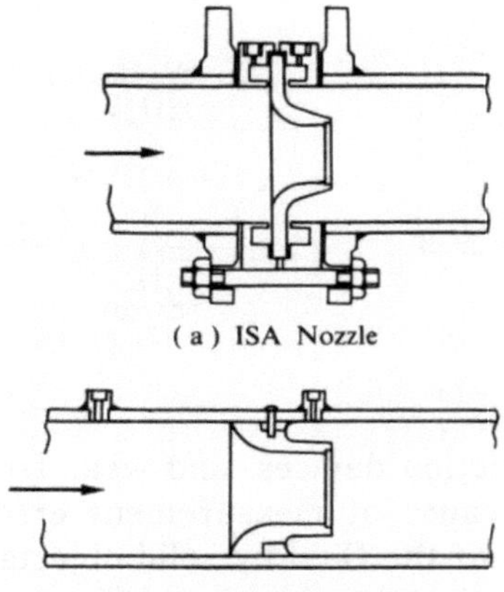

(a) ISA Nozzle

(b) Long-radius nozzle

Fig. 3・34 Flow nozzle

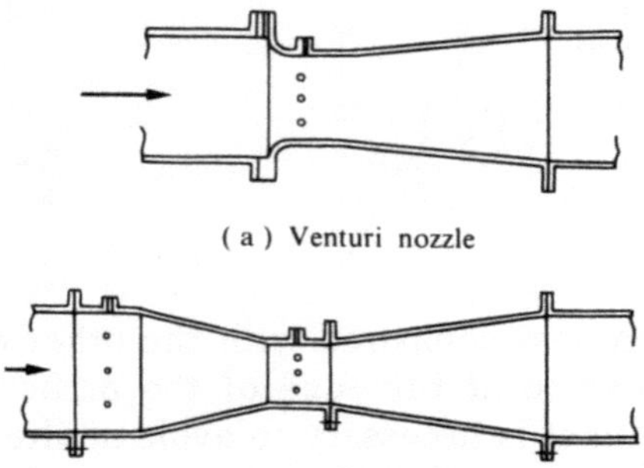

(a) Venturi nozzle

(b) Classical venturi tube

Fig. 3・35 Venturi tube

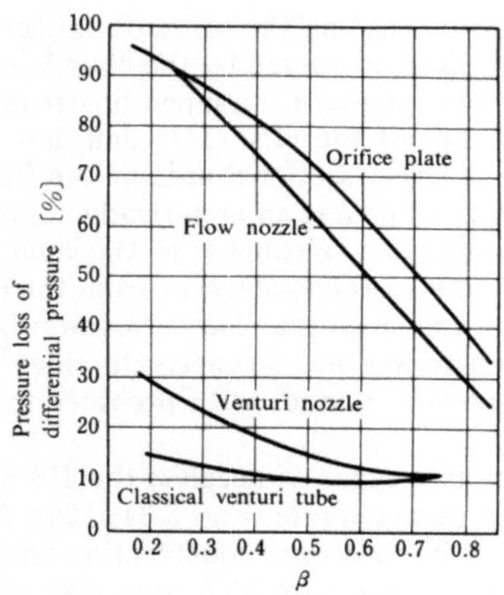

Fig. 3・36 Comparision of permanent pressure loss among primary flow elements

coefficient and characteristics of this device are given in Fig. 3.31, and the configuration is illustrated in Fig. 3.37.

(d) Standards for constriction flowmeters

As the flow coefficient of a constriction device α and the expansion correction factor ε of a gas cannot be calculated by theory, there is no other way of obtaining these values than by experiment. By making a constriction device geometrically equivalent to one on which ex-

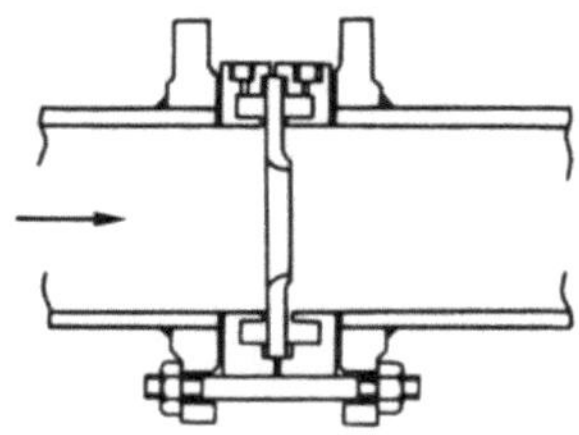

Fig. 3・37 Quadrant edge orifice

periments have been conducted, and using it under dynamically equivalent fluid conditions (the same Reynolds number), the values for α and ε previously determined by experiment can be used without having to determine them anew. Standards are experimental data reports which provide the empirical values of α and ε, and the geometric and dynamic conditions under which they were obtained. Some of the main specification organizations are DIN 1952 (Deutsche Industrie Norm, Germany), ASME (American Society of Mechanical Engineers, U.S.A.), ISO R541 (International Organization for Standardization), and AGA (American Gas Association, U.S.A.). In Japan, there is the JIS Z 8762 established on the authority of ISO.

Although at a glance there appears to be a large number of experimental data reports issued, the basic data is from DIN and ASME; other standards are nothing but converted forms of this basic data. Constriction flow meters are generelly used without calibration, using the given α and ε values as they are, to obtain the flow rate indirectly by measuring the pressure differential. Thus it is necessary to pay attention to the details of the standards.

(e) Differential pressure converters

The pressure differential generated by the constriction device is lead to a transmitter, where it is converted to an electrical or pneumatic signal and sent to a receiving measuring device. When it is not necessary to transmit the signal, the flow rate is displayed by a U-shaped tube manometer or diaphragm manometer installed at the measurement site. See Sec. 3.4 regarding transmitters.

3.3.2 Float-type area flowmeters

(a) Measurement principle

The position of a float placed in a vertical taper tube in which the fluid to be measured is flowing from bottom to top will move up and down according to changes in the rate of flow. The *float-type area flowmeter* measures the flow rate by detecting the position of such a float. Figure 3.38 illustrates this principle. The float is in equilibrium, stationary in the tube. At this time, the annular area A between the tube and the float (the flow-through area), and the volume flow rate

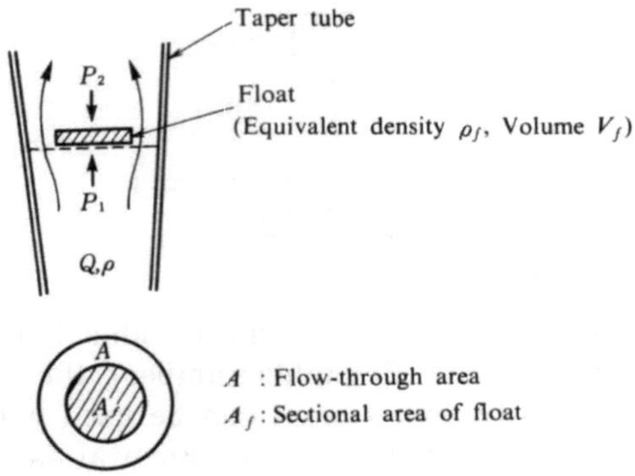

Fig. 3 · 38 Measuring principle of float type area flowmeter

through that area Q are related by the Bernoulli equation, as described earlier for the pressure differential flowmeter.

$$Q = A\alpha\sqrt{\frac{2}{\rho}(P_1 - P_2)} \tag{3.16}$$

Here, α is the discharge coefficient, ρ is the density of the fluid, and $(P_1 - P_2)$ is the upward and downward pressure differential of the float. If the float is stationary in the flow, its effective weight is balanced by the pressure differential applied to the float surface. Thus, because

$$V_f(\rho_f - \rho) = A_f(P_1 - P_2) \tag{3.17}$$

and

$$P_1 - P_2 = \frac{V_f(\rho_f - \rho)}{A_f} \tag{3.18}$$

the pressure differential $(P_1 - P_2)$ becomes a constant value. From Eqs. (3.16) and (3.18),

$$Q = A\alpha\sqrt{\frac{2}{\rho} \cdot \frac{V_f(\rho_f - \rho)}{A_f}} \tag{3.19}$$

is obtained. The discharge coefficient α is a constant obtained empirically from the shape of the float and the Reynolds number of the flow. Also because V_f, ρ_f, ρ, and A_f, are also constants, Equation (3.19) shows that the flow-through area A (that is to say, the float position) is proportional to the flow rate.

(b) Types of area flowmeters and their structures

(1) *Transparent taper tube area flowmeter* Figure 3.39 shows a transparent taper tube area flowmeter constructed simply from a tapered transparent hard glass tube and a float. The flow rate can be measured

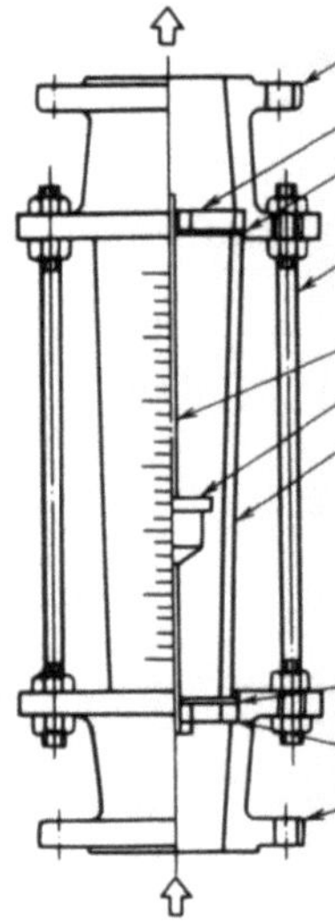

Fig. 3 · 39 Transparent taper tube type area flowmeter

by reading the float position directly on a flow rate scale printed on the taper tube. The float is made of a material such as stainless steel, Teflon®, or PVC, selected according to the fluid being measured.

(2) *Metal taper tube area flowmeter* In this type of flowmeter, the taper tube is made of metal, and the float and its spindle are made as one movable piece. The spindle then serves as an extension of the float. As seen in Fig. 3.40, the flow rate can be displayed directly by the position of the moving part viewed through a transparent straight tube mounted on the upper part of the flowmeter. Alternatively, the motion of the float extension can be detected by a magnetic coupler and transmitted to an external indicator through a lever or linkage mechanism. In either case, the metal tube has excellent mechanical strength against heat and pressure compared with the glass tube, and is also suitable for use with opaque fluids.

(3) *Metal taper tube area flowmeter with converter* As shown in Fig. 3.41, the motion of the moving part inside a metal tape tube is detected by a magnetic coupler. In addition to displaying the flow rate on an indicator, by attaching a position transducer, the flow rate signal can be transmitted to a distant location. The signal output by the transducer can be an electrical signal of from 4 mA to 20 mA DC, or a pneumatic signal of from 0.2 kg/cm^2 to 1 kg/cm^2.

The structure of a magnetic induction position sensor and position-to-electrical current converter used in an electrical position transducer is shown in Fig.3.42. The input is a degree of rotation that is pro-

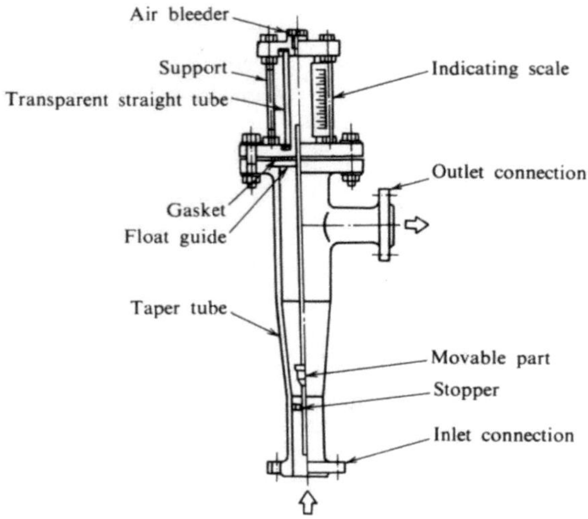

Fig. 3・40 Metal taper tube type flowmeter

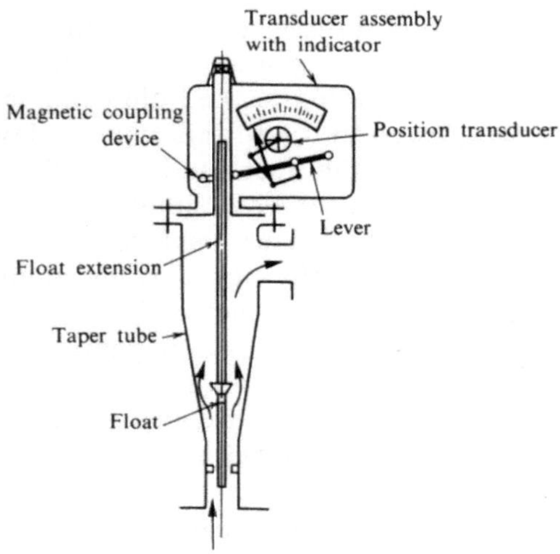

Fig. 3・41 Metal taper tube type flowmeter with converter

portional to the rate of flow. A ferrite core moves in correspondence with the rotation of the input stem, changing the connection between an alternating current exciting primary coil and a secondary coil. The exciting coil (primary coil) and receiving coil (secondary coil) are formed on a printed circuit board, and connected electromagnetically by the U-shaped core. As the core is rotated counterclockwise, the area of connection between the receiving and exciting coils increases, so the

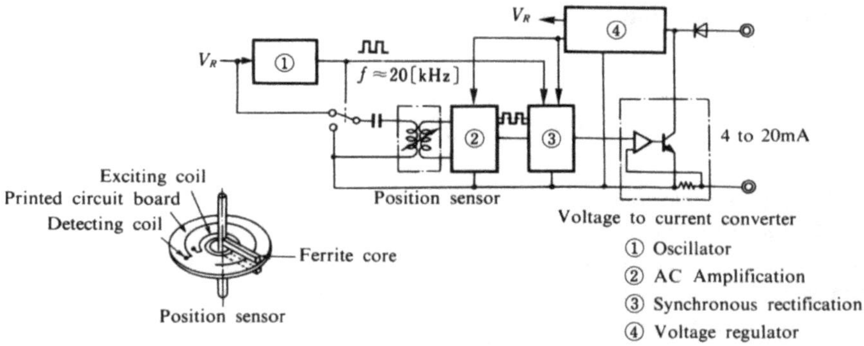

Fig. 3 · 42 Block diagram of position transducer

voltage induced in the receiving coil also increases. This voltage is amplified and rectified, and output as a direct current signal between 4 mA and 20 mA that is proportional to the degree of rotation.

(*c*) *Features of area flowmeters*

(1) Area flow meters are more able to measure small flow rates and highly viscous fluids than pressure differential flowmeters. Figure 3.43 shows the empirically determined relationship between the dischage coefficient and the Reynolds number of the flow path for typical float shapes. For shapes ② and ③, the discharge coefficient α is more or less constant until the Reynolds number becomes small.

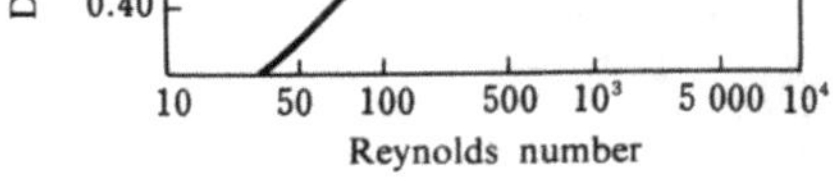

Type of float		Features
①		The dischange coefficient is large, but subject to influences from viscosity.
②	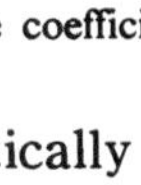	The discharge coefficient is smaller than ①, but little influenced by viscosity.
③	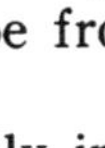	The discharge coefficient is smaller than ②, but steady until very low Reynolds number.

Fig. 3 · 43 Relation between discharge coefficient and Reynolds number

(2) The flow rate scale is a theoretically uniform scale, so the effective range of measurement can be from 10% to 100% of the maximum flow rate.

(3) Area flowmeters are theoretically influenced by changes in density, so when the density in actual use ρ' differs from the design value ρ, it is necessary to multiply the reading Q on the flowmeter

by the following correction factor to obtain the correct flow rate Q'.

$$Q'=Q\sqrt{\frac{(\rho_f-\rho')\rho}{(\rho_f-\rho)\rho'}}$$

3.3.3 Volumetric flowmeters

These devices measure volume flow rate by using the space of fixed volume formed between a rotor, piston, or other such moving part and its case as a measuring container, that is by continuously filling it up and emptying it again and counting the times this is done. This type of device is called a positive displacement flowmeter (PD meter).

(a) Operating principle

There are various types of volumetric flowmeters such as rotor types, piston types, and rotating vane types, each having different shapes measurement space formed by the moving part and its case. Here, as examples of the rotor type construction used in industrial measurement, the principle of the rotating oval gear type and the rotaing lobe type is described (Fig. 3.44). In both types the volumetric chambers are two halfmoon shaped areas formed by the space between the case and two rotors of identical dimensions. The pressure differential in the fluid between the intake and outlet sides exerts a rotary torque on the rotors, which are continuously driven in the direction shown by the arrows. The fluid trapped in the volumetric chambers is discharged through the outlet, so that with each complete revolution of the rotors, four times the volumetric chamber capacity flows out. Accordingly, by counting the revolutions, the integrated flow rate can be deter-

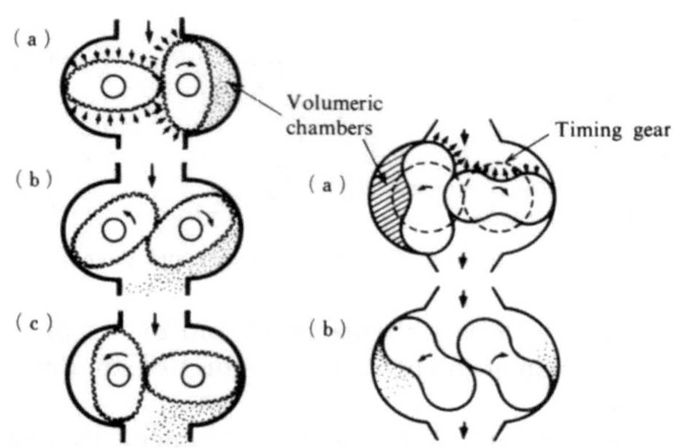

(a) Rotating oval gear type (b) Rotating lobe type

Fig. 3 · 44 Pictorial operating principle of P.D. flowmeter

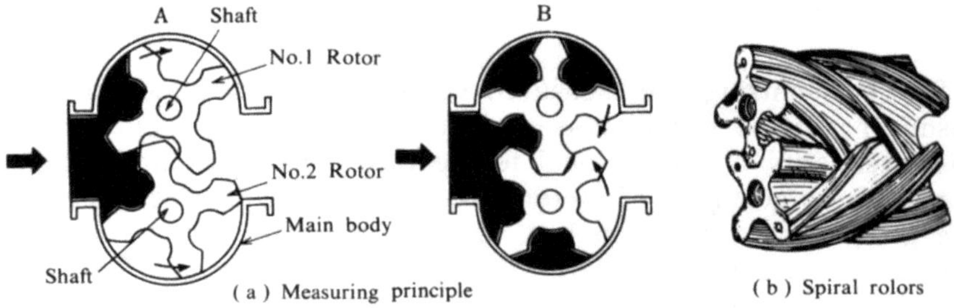

Fig. 3 · 45 P.D. flowmeter with spiral rotors

mined. Alternatively, from the speed of revolution, the instantaneous flow rate can be determined.

Figure 3.45 shows a spiral rotor flowmeter. Two spiral gears are used for rotors, so they rotate at a uniform speed. Thus the outflow does not pulsate, and little vibration or noise is generated from this flowmeter.

Oval gear flowmeters have capacities ranging from 0.2 *l*/h to 1 000 m^3/h, and are used for measuring water solutions, oils, etc. The spiral rotor type, making the most of their special properties, are used for management of relatively large volume loading and unloading of petroleum products. As their precise construction allows little seepage between the rotors and the case, both types are capable of accurate flow mesurement over the range from small flow volume to large volume.

(b) Construction and characteristics of oval gear flowmeters

As shown in Fig. 3.46, the rotation of the oval gears forming the volumetric chambers is transmitted by an electromagnetic coupling that is completely sealed off from the liquid to a transduccer located outside the case. This transducer is in turn connected to a counting mechanism (register).

A reduction gear train converts the rotation into a unit of volume (1 *l*, 10 *l*, etc.) convenient for display on the register (indicator). The register consists of a counter representing the measured amount and a

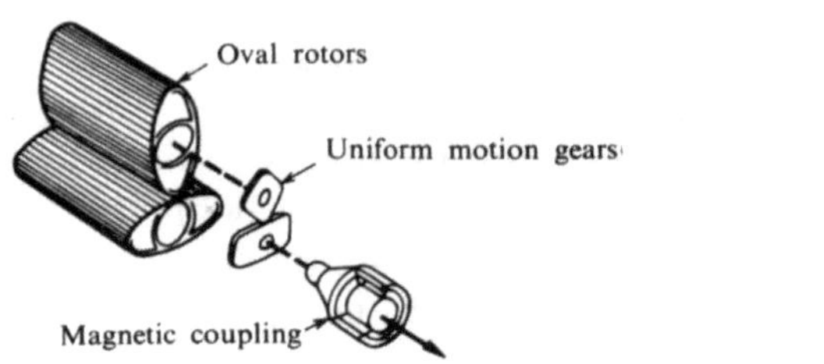

Fig. 3 · 46 Typical magnetic coupling assembly

display mechanism on which the counter value can be read directly, on site.

For transmitting the measured value to a distant location, the transducer is fitted with a pulse generator which outputs a pulse signal whose frequency is proportional to the flow rate. The output signal can be sent to any kind of meter or control device. An example of a pulse generator is shown in Fig. 3.47. A disk with slits rotates between a high-frequency osillating coil and a detecting coil. The rotation of the disk corresponds to the rotation of the oval gears in the flowmeter. The induced magnetic field between the two coils is intermittently cut, producing a square-wave pulse signal with a frequency proportional to the number of slits in the disk and the number of rotations of the flowmeter rotors.

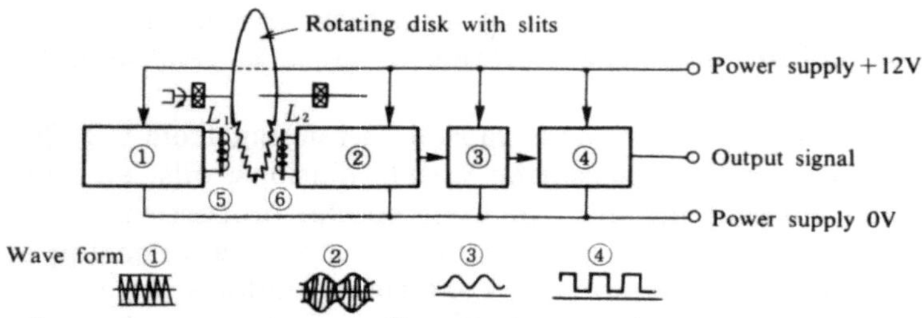

① High frequency oscillator ② Detector ③ Rectifier ④ Wave-shaping ⑤ Oscillating coil ⑥ Detecting coil

Fig. 3・47 Principle of pulse transducer, type PG30 (Oval engineering Co.)

Figure 3.48 shows the error and pressure loss characteristics for representative medium capacity oval gear flowmeters (Type 55 and Type 56). For highly viscous fluids, error values are nearly constant with respect to flow rate. However, for fluids of low viscosity, error can be seen to vary with flow rate as a result of leakage between the rotors and the flowmeter case. Pressure loss is high for highly viscous fluids because of viscous resistance, mechanical friction of the rotors, and fluid loss.

(*c*) *Features of oval gear volumetric flowmeters*

(1) These are theoretically a volume counting type of flowmeter. They are especially accurate for highly viscous liquids ($\pm 0.5\%$ of the displayed value; $\pm 0.2\%$ is possible with the precision grade), and can be used as a commercial or standard flowmeter.

(2) The flowmeter does not have to be attached to a straight part of the pipe.

(3) Because there are moving parts in the fluid stream, it is necessary to place a strainer (filter) at the flowmeter intake to prevent

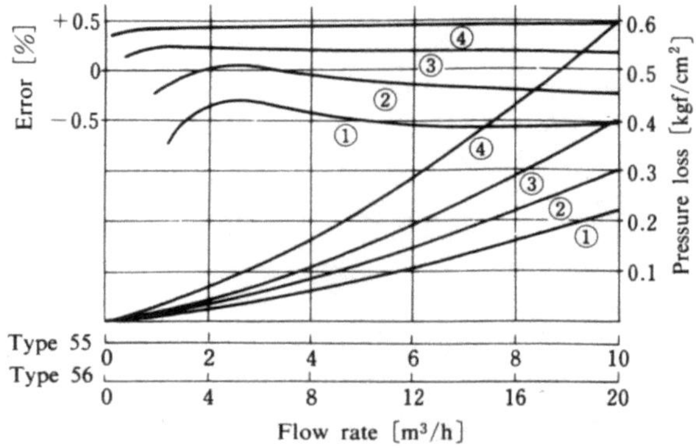

Fig. 3・48 Flow characteristics for viscosity

solid objects from being ingested by the rotors.

(4) Pressure loss is relatively large with the filter attached.

3.3.4 Turbine flowmeters

A turbine rotor placed in a round pipe will rotate at a speed that is proportional to the flow rate of a fluid flowing through it. The turbine flowmeter measures flow by detecting the rotation speed of such a turbine. Generally, in industrial measurement a highly accurate turbine flowmeter is used, which has a low friction, low flow resistance axial flow rotor (the rotor revolves at a right angle to the flow, with the center of flow as its axis).

(a) Structure of the turbine flowmeter

An example of the structure of a turbine flowmeter is shown in Fig. 3.49. Inside a pipe-shaped housing, the rotor rides on bearings held in place by supports fitted in the housing. The rotor diameter is about the same as the inside diameter of the housing. From four to ten blades (vanes) are attached to the rotating spindle at the fixed angle. The blades are magnetized to allow detection of their rotation. Low friction, durable carbide sleeve bearings are used. The upstream support has radial flow conditioning blades to direct the flow to strike the vanes of the rotor.

The number of rotor rotations is detected by a pickup coil that is electromagnetically coupled to the rotor blades, separate by the non-magnetic housing. That is, as shown in Fig. 3.50, when the rotor vane passes under the pickup coil, the magnetic flux density of the coil changes, generating an induced alternating voltage. A preamplifier amplifies the coil voltage and shapes the waveform, converting it into a

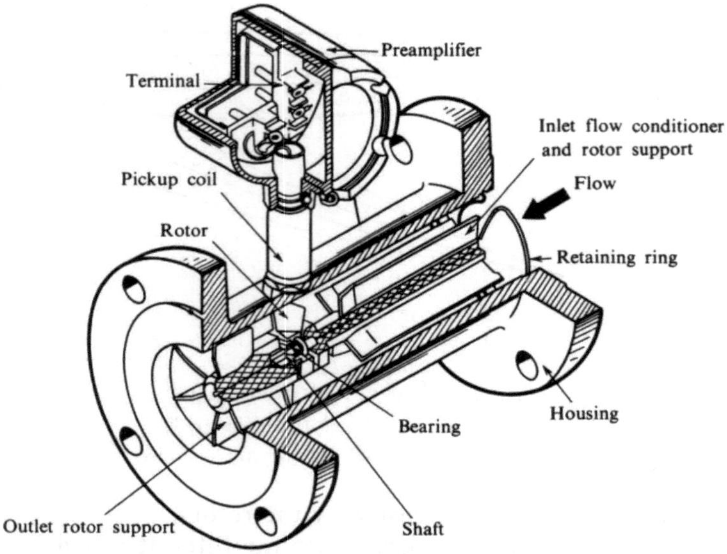

Fig. 3・49 Cutaway view of turbine flow meter (Example)

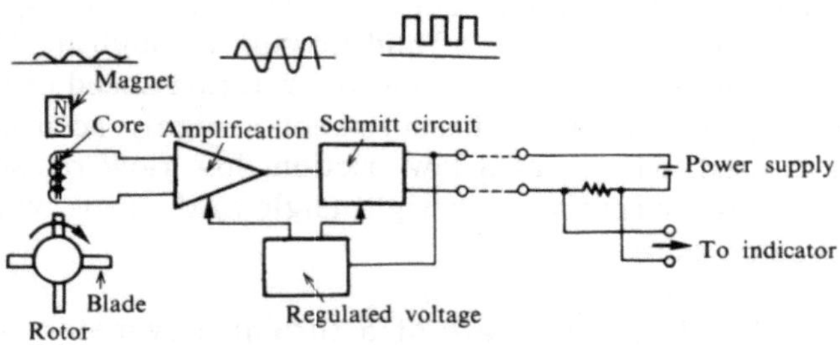

Fig. 3・50 Measuring principle of turbine slow meter

pulse signal, which is then transmitted to a display device.

Within a certain range of flow rates, the frequency of output pulse is proportional to the volume rate of flow of the fluid being measured. The constant of proportionality is called the flow coefficient. This value is accurately measured by calibration with the actual flow rate, and is recorded on the flowmeter housing in pulses per liter.

(*b*) *Flow rate characteristics*

A turbine flowmeter calibration curve is shown in Fig. 3.51. The ideal characteristic maintains a constant value as the flow rate changes, as shown by the broken line in the figure. However, the actual characteristic, represented by the solid line, varies somewhat with the flow rate, because there is a minimum flow rate required to move

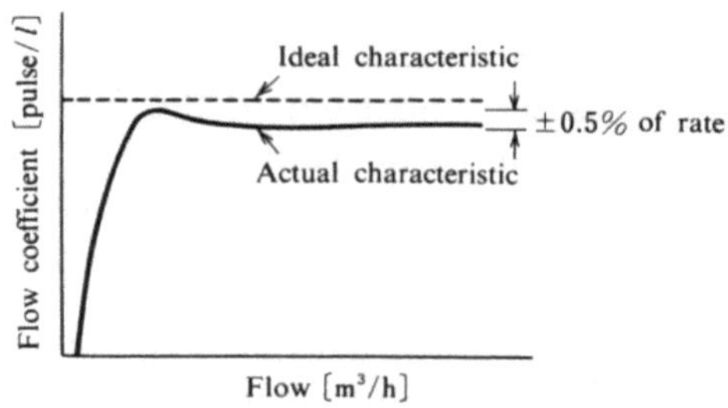

Fig. 3・51 Typical calibration curve of turbine meter

the rotors due to mechanical friction in the bearings. Normally an accuracy of ±0.5% is equivalent to ±0.5% of the width at most of flow coefficient variation. The flow coefficient is more or less constant when the kinematic viscosity of the fluid is low, but for fluids of high kinematic viscosity, the change in the flow coefficient becomes exceedingly large. Lets consider why this is so.

Let the angle between the rotor vanes and the direction of flow be θ. If a fluid of density ρ moving at a velocity of v strikes the vanes, a force proportional to $\rho\, Qv\tan\theta$ (where Q is the volumetric flow rate) is generated in the rotational direction of the rotor. The rotor rotates with the angular velocity ω, thus the rotor torque T is

$$T = c\rho Q(v\tan\theta - \omega r)r \tag{3.20}$$

where r is the mean radius of rotation, and c is the constant of proportionality.

The rotor rotates in balance with a mechanical counter-torque T_m resulting from bearing friction, and a fluid counter-torque T_f caused by fluid viscosity, turbulence and so on, thus

$$c\rho Q(v\tan\theta - \omega r)r = T_m + T_f \tag{3.21}$$

Solving Eq.(3.21) for ω/Q (this value is equivalent to the flow-coefficient in Fig.3.51 and remains constant in the appropriate use range of the flow rate, and can be referred to as the ideal characteristic) we obtain the following equation.

$$\frac{\omega}{Q} = \frac{v\tan\theta}{rQ} - \frac{T_m + T_f}{c\rho r^2 Q^2} \tag{3.22}$$

If A is the cross-sectional area of the pipe, then $v = Q/A$, and Equation (3.22) becomes

$$\frac{\omega}{Q} = \frac{\tan\theta}{rA} - \frac{T_m + T_f}{c\rho r^2 Q^2} \tag{3.23}$$

The counter-torque values T_m andT_f in the second term on the right side of Eq.(3.23) are zero, and ω/Q reduces to a constant value determined solely by the physical dimensions of the flowmeter, thus becoming the ideal flow characteristic. In reality, neither T_m nor T_f are zero, so both exert an influence on the flow characteristic.

As the flow rate increases and regains of turbulence develop, a counter-torque due to flow resistance becomes dominant. This counter-torque T_f is proportional to the density ρ and Q^2, and so can be represented by $T_f = c_1 \rho Q^2$. Accordingly, assuming $T_m = 0$, we obtain

$$\frac{\omega}{Q} = \frac{\tan\theta}{rA} - \frac{c_1}{cr^2} \tag{3.24}$$

Thus, ω/Q is a constant value not influenced by flow rate or fluid viscosity. Hewever, in laminar flow areas, the influence of a counter-torque due to viscosity appears, and at extremely low flow rates, a counter-torque due to friction becomes large, and ω/Q changes greatly.

(*c*) *Features of turbine flowmeters*

(1) Turbines flowmeters are most suitable for pure liquids of low viscosity and little fluid resistance. They have good linearity in turbulent flows, thus their accuracy is stable. Accuracy is generally $\pm 0.5\%$ of the displayed value, but $\pm 0.2\%$ are also available as commercial flowmeters. The flow rate range has been standardized as between 0.8 m^3/h (8 mm nominal bore) and 7 500 m^3/h (600 mm nominal bore).

(2) The rotor response is fast, producing a high resolution pulse signal that is proportional to the flow rate.

(3) It is necessary to have a level straight pipe at least $20D$ (D: nominal diameters of pipeline) long on the upstream side of the flowmeter for use with a flow conditioner. Also, a strainer (filter) must be placed in front of the straight pipe section to prevent foreign objects from damaging the rotors. Figure 3.52 shows an example of such an installation. Circular flow and an uneven flow speed distribution affect the torque on the turbine vanes, and lower meas-

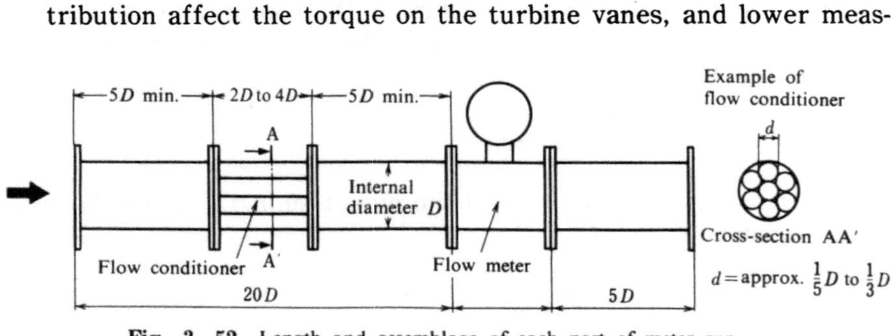

Fig. 3·52 Length and assemblage of each part of meter run urement accuracy. To guard against this, it is necessary to provide a fixed flow conditioner upstream from the flowmeter to make the flow velocity profile symmetrical around the pipe axis.

3.3.5 Magnetic flowmeters

Magnetic flowmeters apply Faraday's law of electromagnetic induction. They measure the flow rate of an electrically conductive fluid flowing through a magnetic field by detecting the voltage induced in the fluid. In industrial measurement, magnetic flow meters having nominal bores ranging from a very small 2.5 mm to a huge 2 000 mm are in practical use. Figure 3.53 shows the external appearance of some typical magnetic flowmeters.

Fig. 3 · 53 Detector and converter of magnetic flowmeter (Yokogawa: Type YEWMAG)

(*a*) *Measurement principle*

As shown in Fig. 3.54, a measuring tube having an inner diameter d (m) whose inner wall is coated with an insulting material crosses at a right angle through a magnetic field of flux density B (T). If a conductive fluid fills this tube and flows with a mean velocity of $\bar{v}$ (m/s), and a pair of electrodes are fitted opposite each other on the wall of the pipe at right angles to the magnetic field and the flow, then an electromotive force E (V) will be generated between the electrodes. This electromotive force is expressed as

$$E = kBd\bar{v} \quad [\mathrm{V}] \tag{3.25}$$

where k is a constant intrinsic to the detector and dependent on the magnetic field conditions.

For volumetric flow rate Q (m^3/s),

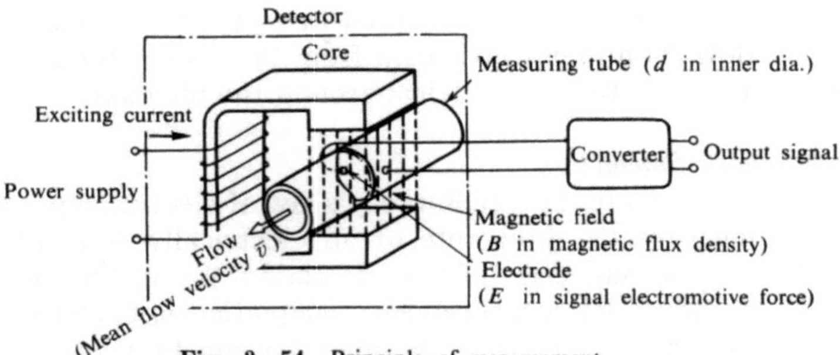

Fig. 3 · 54 Principle of measmement

$$\bar{v}=\frac{Q}{(\pi/4)d^2}=\frac{4Q}{\pi d^2}\ \text{[m/s]} \tag{3.26}$$

Thus from Eqs.(3.25) and (3.26), the electromotive force E becomes

$$E=k\frac{4}{\pi}\cdot\frac{BQ}{d}\ \text{[V]} \tag{3.27}$$

Furthermore, from Eq.(3.27), flow rate Q becomes

$$Q=\frac{1}{k}\cdot\frac{\pi d}{4}\cdot\frac{E}{B}\ \text{[m}^3\text{/s]} \tag{3.28}$$

As shown by Eq.(3.28), if the flux density B is constant, then the flow rate Q is proportional to the electromotive force E. This electromotive force is amplified and converted to a universal signal (4 to 20 mA, DC) or pulse signal by a converter and then output.

Further, it can be proven theoretically that the generated electromotive force E is proportional to the mean flow velocity $\bar{v}$ in Eq.(3.25) if the flow velocity profile in the measuring tube is symmetrical with respect to the central axis of the pipe, and the magnetic flux density is uniform.

(*b*) *Excitation method*

If a direct current magnetic field is used, the electrodes become polarized. To avoid this, an excitation method using commercial frequency alternating current had always been used in the past. With this method, the flow rate signal voltage generated between the electrodes also has the commercial frequency. Influenced by eddy currents generated by changes in the magnetic flux, and other induced noise, the zero point of the signal fluctuates. Nowadays, to eliminate the influence of various kinds of noise on the flow rate signal voltage and improve the stability of the signal zero point, a square wave of lower frequency than the commercial frequency is often used. The following are the main reasons.

(1) Using a square wave current having a low frequency that is a fraction of the commercial frequency as the exciting current, and detecting the flow rate signal at a stationary magnetic flux value, the noise based on the rate of change of the magnetic flux (magnetic flux differential noise) can be eliminated, making the flow rate signal zero point more stable. The noise is generated when eddy currents flowing in the fluid are not symmetrical about the electrode axis, and when the magnetic flux of the impressed magnetic field becomes interlinked with the signal detection loop formed by the fluid, the electrodes, and the electrode signal wires. Under these conditions, the noise is generated even when the flow rate is zero, and is thus a source of error.

(2) In the low-frequency excitation method, the signal frequency is different from the commercial frequency, so the signal is not easily affected electrostatic or electromagnetic noise induced in the lead wires carrying the signal from the detector to the converter.

(c) Detector structure

Examples of the various types of detectors are listed in Table 3.10. Detector construction varies somewhat according to the nominal bore and the application, but it basically consists of a measuring tube lined with an electrical insulator, electrodes for tapping the generated electromotive force, and a coil and core for generating a magnetic field. Figure 3.55 shows the structure of a detector of small nominal bore. Nonmagnetic stainless steel pipe, which easily passes magnetic flux to the fluid, is used for the measuring tube (spool pipe). The lining on the inner surface is either fluororesin (PFA) or polyurethane rubber, chosen according to the nature of the fluid being measured and the temper-

Table 3.10 Detectors of magnetic flowmeter (Examples)
〔*Yokogawa Tech. Rep.*, **28**, 2, p.145 (1984)〕

Symbol for size	Model	Nominal diameter 〔mm〕	General use type	Submergible type	Sanitary type	Explosion-proof type
SS	YM100	2.5～15	↑			↑
S	YM200	25～100	⇡ ↓	⇡	↕	│
M	YM300	150～400	┆	┆		↓ 200
L	YM400	500～1 000	┆	┆		
LL	YM500	1 100～2 600	⇣	⇣		

Remarks: Liner materials
⟵⟶ ; PFA (Tetra fluoroethylene perfluoro alkoxy resins)
←- - - -→ ; Polyurethan rubber

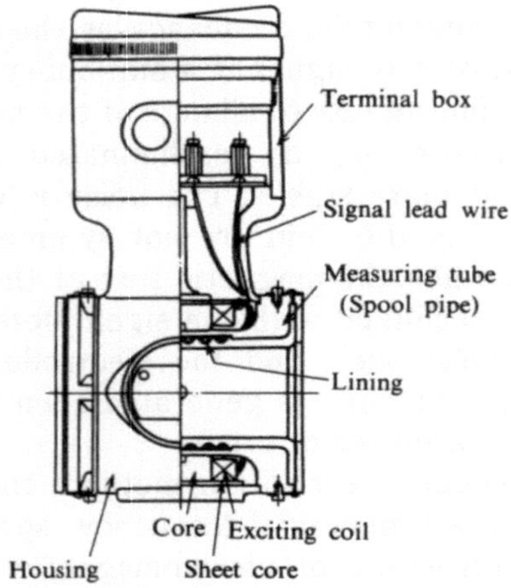

Fig. 3・55 Small size detector (Size : S)

ature. The measuring tube, housing, terminal box, and flanges are all welded together, and the inside of the detector is protected by sealing with resin or inert gas, thus it is mechanically strong and moisture-proof. For large nominal bores, after the core and coil are attached to the inside of the steel tube, polyurethane rubber is applied by centrifugal casting to produce an integrated, one-piece construction. Table 3.11 lists the characteristics of lining materials.

The electrodes must be made of non-magnetic material that is also resistant to corrosion, so SUS 316L plantinum-iridium can be used. Depending on the fluid, hastelloy C, titanium, tantalum or other such corrosion resistant metal can also be used. An example of electrode

Table 3.11 Characteristics of lining

Lining	Characteristics	Measured liquid
PFA	Mechanically strong and smooth, so free from pipe wall deposits. Resistance to heat and corrosion. Temperature limits: −10°C to +160°C (14°F to +320°F)	Permeable liquids such as hydrofluoric acid, hydrochloric acid, acetic acid. Liquids will a tendancy to solidify will adhere. Corrosive liquids (electrolytes, caustic soda, sulfuric acid, etc)
Polyurethane rubber	Good abrasion resistance to fluids containing solids. Resists weak acids and weak alkalies. Unsuitable for liquids containing organic solvents. Temperature limits: −10°C to +40°C (14°F to +104°F)	Suitable for water supply, waste water including sewage industrial water, and sea water.

construction for a PFA resin lined detector is shown in Fig. 3.56. The measuring tube lining and the electrode seals are molded as one piece, surrounding the electrode with insulating PFA resin and ensuring the insulation of the electrode from the grounding steel tube, and improving the strength of the seal against pressure from the fluid. As shown in Table 3.10, in addition to general use detectors, there are explosion-proof models that can be installed in explosive environments, sanitary models which use sanitary joints that allow easy removal for the periodic cleaning required in the food industry, and submersible models coated with tar epoxy that can be installed in places like pits where there is danger of the device falling into water.

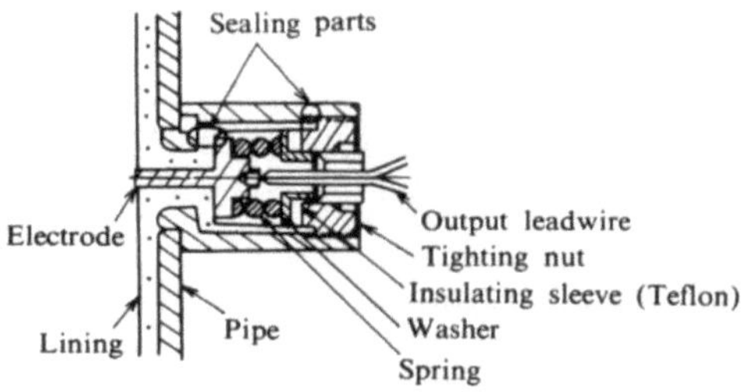

Fig. 3 · 56 Sealing and insulating construction for electrode

(d) Converters

The basic functions of the converter are to provide and exciting power source for the detector, and convert the flow rate signal voltage generated between the electrodes into a proportional output signal (4 to 20 mA DC), or a pulse signal. However as described above, while the electromotive force E generated at the electrodes is proportional to the volumetric flow rate Q, it is at the same time proportional to the flux density B. Accordingly, it is necessary to either stabilize the exciting current and make the flux density B constant, or to eliminate the influence of fluctuations in magnetic flux density by means of dividing the electromotive force E by a signal that is proportional to the exciting current (*i.e.*, proportional to the flux density). Furthermore, the electromotive force E is a low-level alternating voltage, and it is also necessary to remove the influence of induced noise as described in Subsec. (b) above.

As an example, the circuitry of a high accuracy, multi-function converter incorporating a microprocessor and used in combination with a low frequency square wave excitation method detector is shown in Fig. 3.57. The signal waveforms of each part are shown in Fig. 3.58.

The exciting coil of the dectector is driven by a three-valued (positive, zero, negative, zero) constant current from the exciting circuit

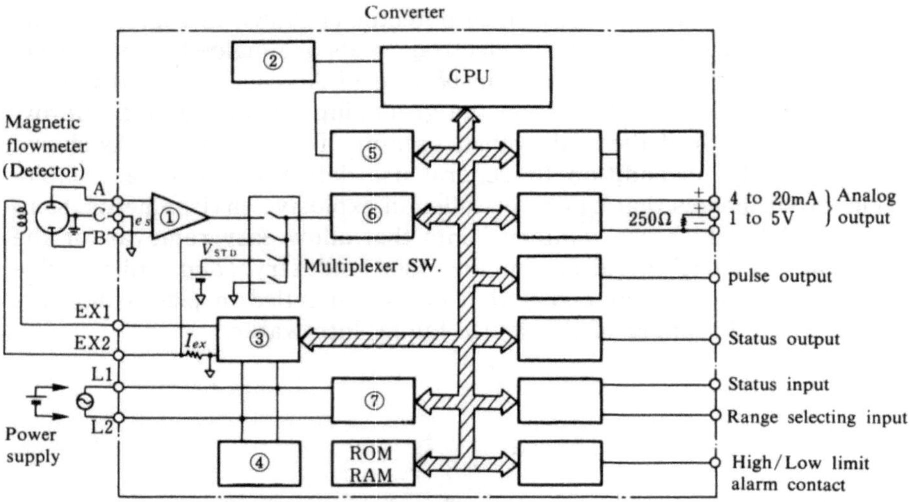

① Amplifier
② Watch-dog timer
③ Exciting circuit
④ Ragulator
⑤ Timer
⑥ A/D Converter
⑦ Power supply frequency counter
⑧ Setting display
⑨ D/A Converter
⑩ Pulse output
⑪ Status output
⑫ Status input
⑬ Alarm output

Fig. 3・57 Circuit diagram of microprocessor-based converter

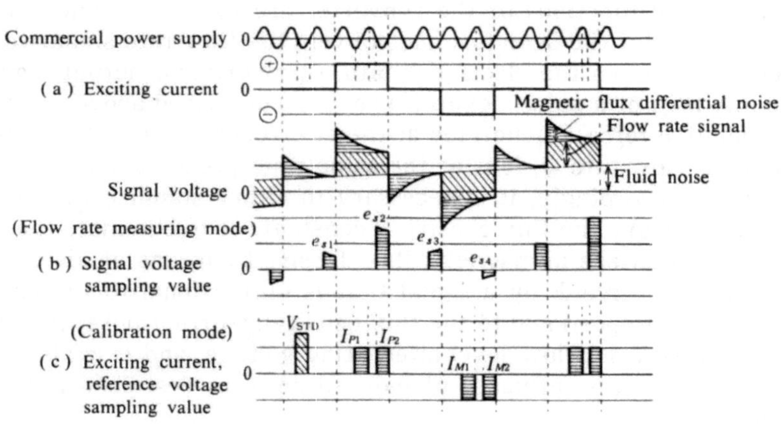

Fig. 3・58 Signal waveforms of converter main parts

(Fig. 3.58 (a)). The exciting frequency is synchronized with the frequency of the commercial power supply at either 50/8 Hz or 60/8 Hz. A signal voltage e_S that is proportional to the average flow velocity and the exciting current I_{ex} is generated between the electrodes. The signal voltage is sampled by the multiplexer switch (Fig. 3.58 (b)), converted to a digital signal, and read into the microprocessor. The exciting current is also read into the microprocessor, and the following calculation is performed.

$$e_0=\frac{K}{I_{ex}}(-e_{s1}+e_{s2}+e_{s3}-e_{s4})$$

Here, K is a constant.

Using this caluculation, the influence of fluid noise (noise cause by fluctuation in contact potential on the surface of the electrodes) and magnetic flux differential noise can be prevented. Also, the result of digital filtering of e_0 is sent to the digital-analog converter, which converts it to a pulse width signal, and generates a universal 4 to 20 mA DC signal. In addition to an electrical current output, a pulse output that is proportional to the flow rate is also generated.

The power supply uses an insulation-type swiching regulator. This regulator has the merits of using both alternating current and direct current, and having a wide control range of 20V DC to 130V DC and 80V AC to 138V AC. Figure 3.59 shows the circuit diagram of the power supply, which is the heart of the exciting circuit. A three-valued magnetic flux timing pulse controlled by the microprocessor turns transistor swiches Q_1 and Q_2 alternately on and off. While the transistor switch is off, the counter electromotive force generated in the exciting coil is stored in a condenser. When the switch is turned on, the stored energy is reused as an exciting current supply, and power consumption is reduced. Regarding the exciting current, the microprocessor controls the absolute value of the amplitude, and at the same time controls the flatness γ to a value of one, as expressed by in the following equation, thus eliminating magnetic flux differential noise.

$$\gamma=\frac{I_{P1}-I_{M1}}{I_{P2}-I_{M2}}$$

Every few cycles of the exciting current, the standard voltage V_{STD} is read and span self-checking is performed (Fig. 3.58 (c)).

For slurries and fluids of low conductivity, the electrochemical potential of the electrodes is constantly changing, which can appear as a noise in the flow rate signal. This type of noise decreases as the exciting current frequency increases. Thus, although it is accompanied by a small loss of accuracy, the noise can be reduced by raising the exciting current frequency. For this purpose, this converter has a 50/2 Hz and 60/2 Hz exciting mode in addition to the normal 50/8 Hz and 60/8 Hz setting, and the exciting mode suitable for the application can be chosen.

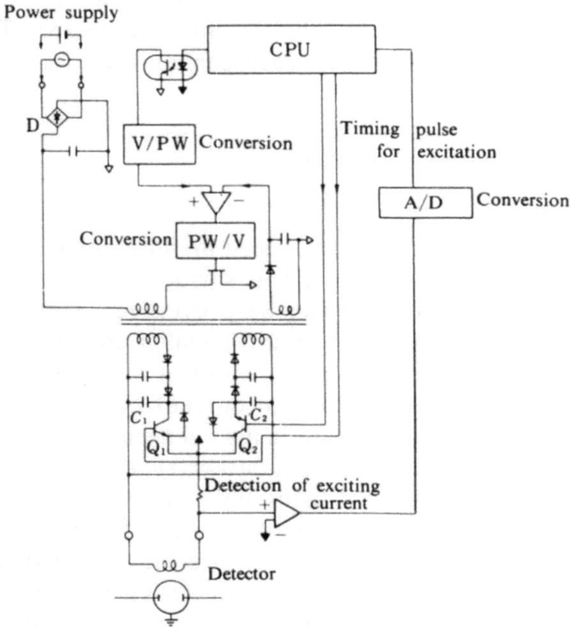

Fig. 3・59 Scheme of exciting circuit

Using the keyboard and LED display on the converter control panel, parameters such as flow rate range, nominal bore of the detector, and damping constant can be set, and various functions such as multiple range switching, forward and reverse flow rate switching, and exciting mode switching can be selected.

In addition to the flow rate indication, the display also indicates the upper and lower limit flow alarm values and the results of self-diagnosis (exciting circuit short or open circuit, abnormal input signal, abnormalities in the A/D converter or microprocessor, etc.), and at the same time outputs an external interface signal.

Figure 3.60 shows an example of an electromagnetic flowmeter having an integrated construction. This flowmeter is designed to be small and light-weight, with the converter and detector integrated into a single unit. It also has the advantage of being easy to use, because of simplified function. The converter's signal processing circuit is designed for stability of the signal zero point, using a two-valued excitation method of low frequency. It is contained in an airtight hybrid IC and mounted in a compact converter case.

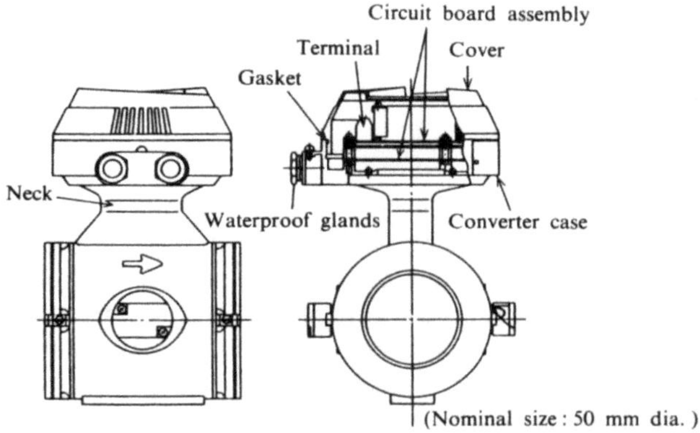

Fig. 3・60 Magnetic flowmeter with integrated converter (Typical)

(e) General characteristics of electromagnetic flowmeters

(1) Electromagnetic flowmeters are not affected by fluid temperature, pressure, density or viscosity, and can output a signal proportional to the volumetric flow rate over a wide measurement range.

(2) The detector has no moving parts and does not obstruct the flow, so if a suitable lining is used, measurement of slurries or corrosive liquids is easy.

(3) There is no pressure loss, and a shorter lengh of upstream straight pipe is need compared to other flowmeters.

(4) Flow in both directions can be measured, with no response delay.

(5) These flowmeters cannot measure gases, oils, or non-conductive fluids.

3.3.6 Vortex flowmeters

As shown in Fig. 3.61, if a columnar object (vortex shedder) is inserted perpendicular to the flow, vortices revolving in mutually opposite directions are created alternately on either side of the object. Downstream, what is called a *Kalman vortex street* is formed. The number of vortices created per unit of time (vortex frequency) is, within a certain range of Reynolds numbers, proportitional to the flow velocity. Thus flow rate can be measured by detecting the vortex frequency.

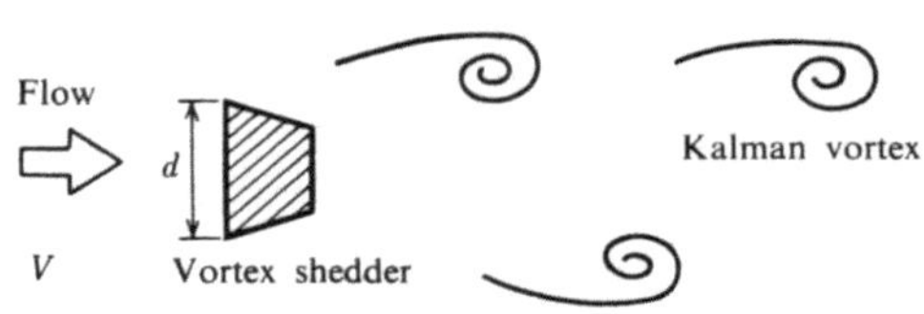

Fig. 3・61 Kalman vortex street

(*a*) *Measurement principle*

If f is the vortex frequency, d is the width of the vortex shedder, and V is the flow velocity, then the following relationship exists.

$$f = S_t \frac{V}{d} \tag{3.29}$$

Here, S_t is a non-dimensional constant called the *Strouhal number* determined by the shape and dimensions of the vortex shedder. With an appropriately chosen vortex shedder, the Strouhal number is constant over a wide range of Reynolds numbers. This means that within this range of Reynolds numbers where S_t is constant, the vortex frequency f is proportional to the flow velocity V, and is not affected by factors such as the fluid density and viscosity. Further, the Reynolds number R_e for a typical value of d can be expressed as $R_e = Vd/\nu$ where ν is the kinetic viscosity.

The relationship in Eq.(3.29) is also established for the flow in a measuring pipe of inner diameter D, so for flow rate Q and flow velocity in the constricted part $V = Q/\{(\pi/4)D^2 - dD\}$,

$$f = \frac{S_t}{\{(\pi/4)D^2 - dD\}d} Q = KQ \tag{3.30}$$

where $K = S_t/\{(\pi/4)D^2 - dD\}$. Accordingly, if S_t is known in advance, the flow rate Q can be measured by measuring f.

The vortex shedder is a dominant constituent of the flowmeter. It must produce strong, stable vortices, and at the same time, the proportional relationship between Q and f shown in Eq.(3.30) must hold over a wide range of flow rates. For this reason, much effort is concentrated on experimentally determining the most appropriate shape for the vortex shedder.

Figure 3.62 shows the cross-section shapes of several vortex shedders currently in practical use. What is common to all of these is that the surface facing the flow has sharp edges, so that the point at which the vortices are shed is well-defined, improving the linearity of the Strouhal number S_t(S_t remains constant even if the Reynolds number fluctuates). Also, the width of the vortex shedder facing the flow d, taking advantage of the vena contracta effect near the walls of the pipe, acts to average the flow velocity distribution in the longitudinal direction of the vortex shedder, so by choosing the right shedder width d in relation to the bore of the pipe D, a stable Karman vortex street can be formed even in a round pipe.

Figure 3.63 shows the relationship between S_t and the Reynolds number for a flowmeter using the vortex shedder having a trapezoidal cross section described below. For Reynolds numbers between 2×10^4

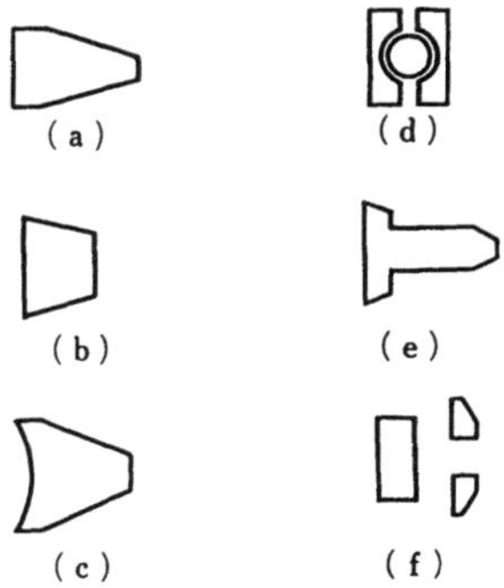

Fig. 3 · 62 Section of vortex shedders (Example)

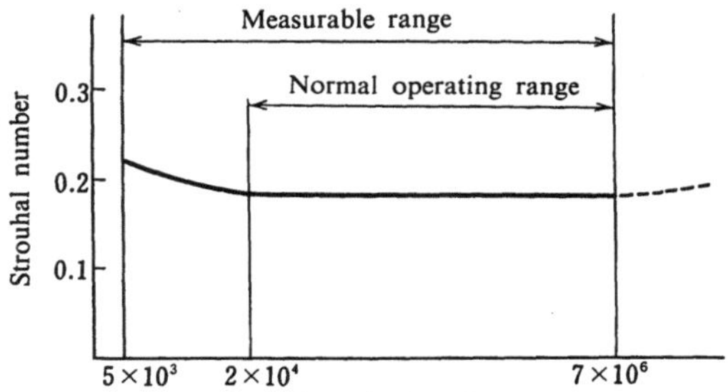

Fig. 3 · 63 Relationship between Strouhal number and Reynolds number

and 7×10^6, S_t is more, or less constant, and this represents the normal operating range of the flowmeter.

Various types of vortex frequency detectors are in use, in combination with vortex shedders. The major types of sensors are described in Table 3.12. They are broadly classified into those that detect changes in flow velocity caused by vibrations in the fluid resulting from vortex shedding, and those that detect changes in pressure.

(b) Flowmeters that use the stress detection method

As an example of an actual device, a flowmeter in which the alternating stress created in the vortex shedder by vortex formation is detected by a piezoelectric element is described. The external appearance of this flowmeter is shown in Fig. 3.64.

(1) Stress detection principle Figure 3.65 shows the vortex frequency detection principle. If the fluid flows perpendicular to the page, an alternating lift is applied to the vortex shedder that has the same frequency as the vortices shed from it. This alternating lift produces a change in stress inside the vortex shedder. The frequency of this

Table 3.12 List of sensors

Quantity detected	Detection system	Sensor
Flow velocity change	cooling the heating element	Hot wire
		Thermistor
	Detecting sound velocity change	Ultrasonic wave
Pressure change	Detecting differential pressure	Diaphragm + piezoelectric element
		Diaphragm + capacitance
		Diaphragm + inductance
	Distinguish movement	Flag + strain gauge
		Ball + inductance
	Distortion produced in vortex shedder	Strain gauge
	Stress produced in vortex shedder	piezoelectric element

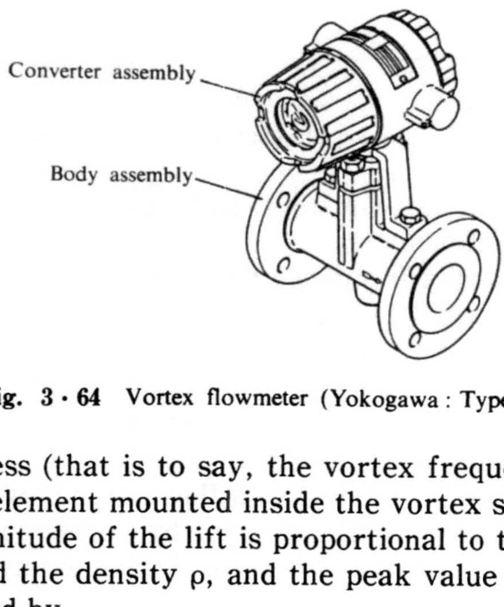

Fig. 3 · 64 Vortex flowmeter (Yokogawa : Type YEWFLO)

change in stress (that is to say, the vortex frequency) is detected by a piezoelectric element mounted inside the vortex shedder.

The magnitude of the lift is proportional to the square of the flow velocity V and the density ρ, and the peak value of the alternating lift F_L is expressed by

$$F_L = \pm \frac{1}{2} c_L \rho V^2 dD$$

where c_L is a dimensionless constant, D is the pipe bore and d is the width of the vortex shedder.

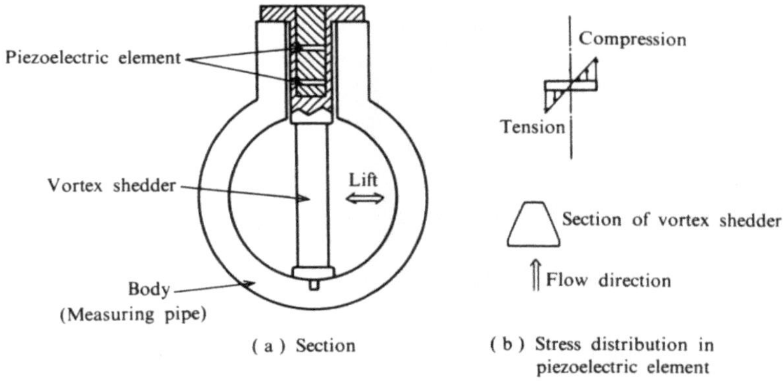

Fig. 3 · 65 Sensing principle of vortex shedding frequency

The average stress σ_M created in the piezoelectric element, and electric charge q induced in the piezoelectric element are expressed by the following equations.

$$\sigma_M = KF_L \qquad q = d_0 \sigma_M S$$

Here, K is a constant determined by the shape of the vortex shedder and support method, d_0 is the piezoelectric constant, and S is the area of the piezoelectric element. This electric charge is processed by an electronic circuit, and the vortex frequency is detected.

(2) *Structure* As shown in Fig. 3.64, this flowmeter consists of a detector body assembly made up of a section of pipe and vortex shedder with a built-in piezoelectric element, and a converter assembly with a built-in amplifier. Figure 3.66 shows the basic mechanism of a vortex flowmeter.

(3) *Characteristics* Figure 3.67 presents an example of actual airflow measurements. A volumetric flowmeter of $\pm 0.2\%$ accuracy was used as a standard. The broken line represents the Strouhal number obtained for water flow, and is in agreement with the characteristics of airflow. In addition, the linearity of the Strouhal number is good.

This flowmeter has a solid construction that integrates the vortex shedder and detector into a single unit, has no moving parts. The detector (piezoelectric element) does not come in contact with fluid, so it is highly reliable, and can be applied to the measurement of liquids, gases, and stream.

(*c*) *General characteristics of vortex flowmeters*

(1) The output signal of a vortex flowmeters is directly proportional to the flow rate, and a pulse signal can be obtained.

(2) The output signal indicates the volumetric flow rate, and is not influenced by temperature, pressure or the type of fluid.

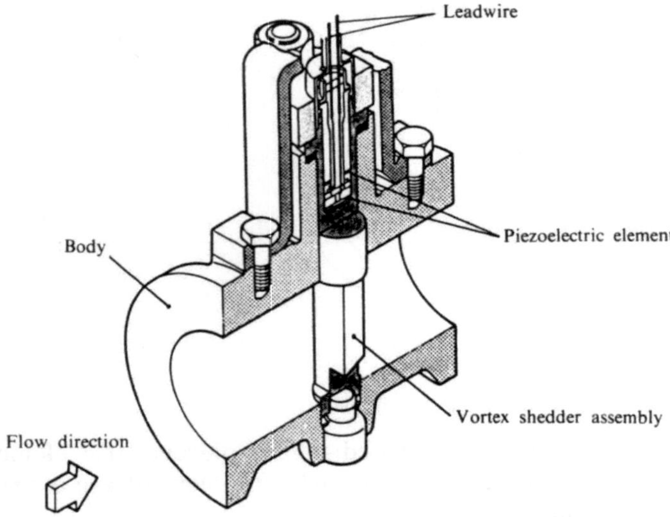

Fig. 3 · 66 Basic mechanism of vortex flow meter

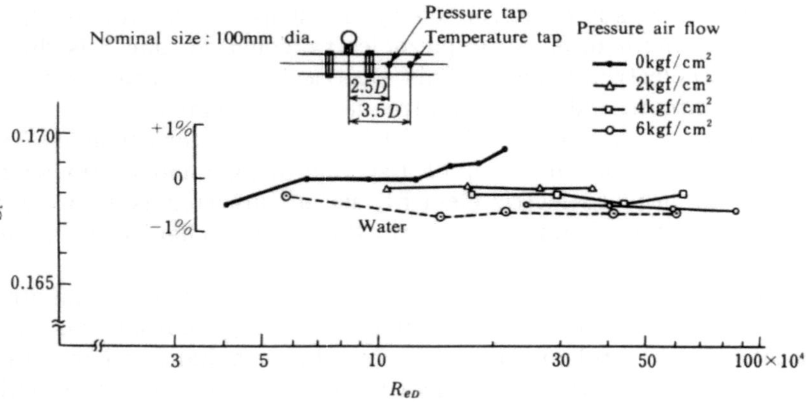

Fig. 3 · 67 Characteristics for air flow

(3) The applicable Reynolds number range is wide, and accuracy is high.

(4) The pressure loss is relatively low, but is affected by flow velocity distribution, so a straight section of pipe is required upstream, 15*D* to 40*D*, depending on pipe conditions (*D*: nominal bore of the flowmeter), and downstream, 5*D*.

3.3.7 Ultrasonic flowmeters

With ultrasonic flowmeters, ultrasonic waves are generated in the fluid from outside the pipeline, and the flow rate is determined by detecting externally the change in the transmitted waves or the reflected waves caused by the flow velocity. At this time, they are mainly used to measure the flow of water. There are two typical methods in practical use, classified broadly by the principle of measuremant they employ. One, called the transit time differential method, obtains the flow rate by measuring the difference in arrival times of ultrasonic pulses between detectors set facing each other obliquely to the pipe axis on the outside pipe walls. The other method uses the Doppler effect. By measuring the difference in frequencies of the transmitted waves and those reflected back from foreign objects in the fluid, the velocity of those objects, and thus the velocity of the fluid, can be determined.

(*a*) *Transit time differential method*

Figure 3.68 shows the basic configuration of the measuring device. A pair of detectors, P_1 and P_2 are mounted facing each other on the exterior of the pipe at an angle to the pipe axis, and in contact with the outside walls.

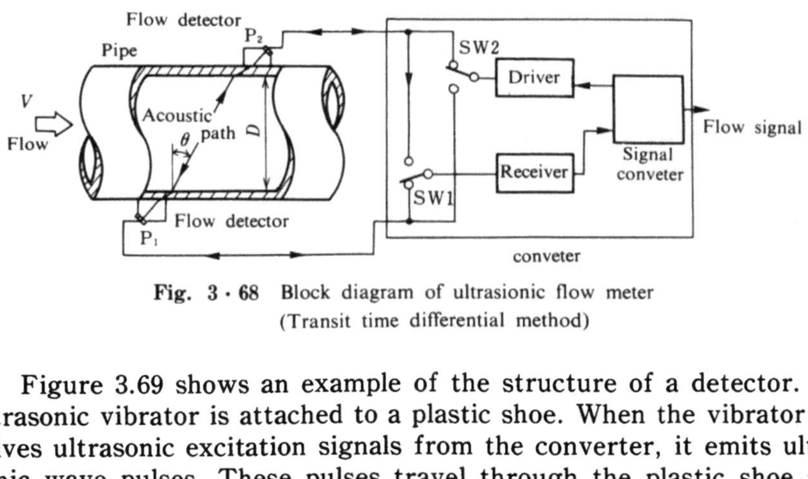

Fig. 3 · 68 Block diagram of ultrasionic flow meter (Transit time differential method)

Figure 3.69 shows an example of the structure of a detector. An ultrasonic vibrator is attached to a plastic shoe. When the vibrator receives ultrasonic excitation signals from the converter, it emits ultrasonic wave pulses. These pulses travel through the plastic shoe and through the walls of the pipe and arrive at the facing detector. As the ultrasonic vibrator, a piezoelectric element that has a resonance frequency of between several hundred kilohertz and about one megahertz is used. The material of the vibrator is usually lead zirconate.

Detectors P_1 and P_2 incorporate piezoelectric vibrators and can both transmit and receive ultrasonic waves. Ultrasonic pulses are transmitted along the paths shown in Fig. 3.68. The transmission is controlled by the opening and closing of the switches in the converter. The transit times T_2 (from P_1 to P_2) and T_2 (from P_2 to P_1) are ex-

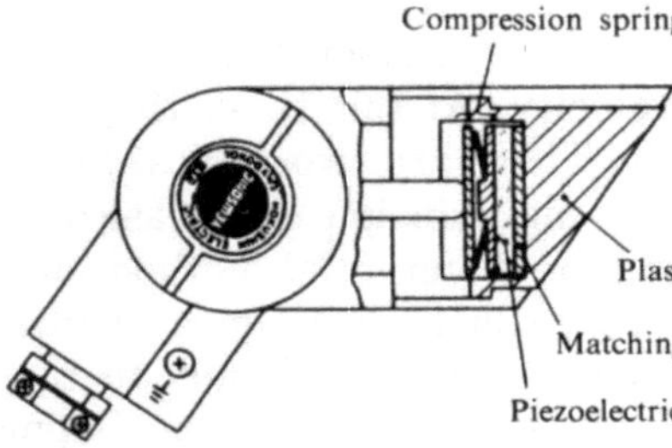

Fig. 3・69 Typical construction of detector

pressed by the following two equations.

$$T_1=\frac{D/\cos\theta}{c+V\sin\theta} \qquad T_2=\frac{D/\cos\theta}{c-V\sin\theta} \tag{3.31}$$

Here, D is the nominal bore of the measurement tube, c is the speed of sound through the stationary fluid, V is the mean flow velocity, and θ is the angle of incidence of the ultrasonic pulse. However, Equations (3.31) disregard the transit time through the pipe walls and the detector shoes.

For liquids, the speed of the sound pulse c (about 1 500 m/s) is nearly three orders of magnitude lager than the flow velocity V (several m/s), and is affected by changes in the temperature and density of the fluid. Thus correct values of the transit times T_1 and T_2 cannot be obtained using Eq. (3.31) as they are. To eliminate the term c from the equations, the inverses of the transit times (that is, the frequencies) are taken, and their difference Δf becomes.

$$\Delta f=\frac{1}{T_1}-\frac{1}{T_2}=\frac{V\sin 2\theta}{D} \tag{3.32}$$

There are various signal processing techniques available for taking the inverse of the transit times T_1 and T_2, but methods using the phased-locked loop (PLL) are often used. As shown in Fig. 3.70 the values of T_1 and T_2 are converted into the output signals of two voltage controlled oscillators (VCO). These two output signals have the frequencies of f_1 and f_2. The VCOs control their outputs such that $T-N/f$ is equal to zero, thus $f_1=N/T_1$ and $f_2=N/T_2$. Increasing the value of N improves the resolution of $\Delta f\ (=f_1-f_2)$. To obtain the frequency difference Δf, we use the following equation.

$$\Delta f=f_1-f_2=\frac{NV\sin 2\theta}{D} \tag{3.33}$$

For example, if $D=1$m, $V=2$m/s, $\theta=70°$, and $N=1\,000$, then

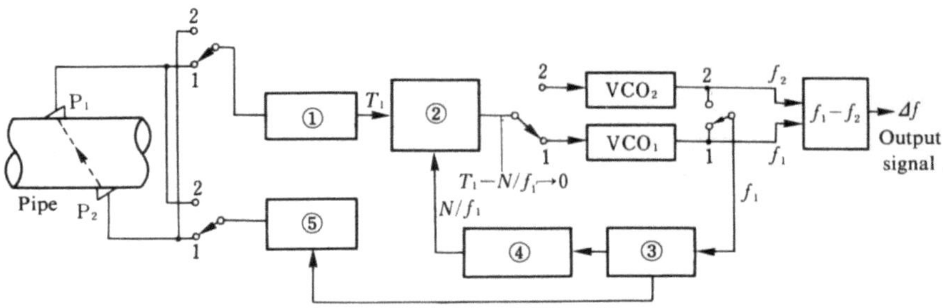

① Receiver ② Time-difference detector ③ Synchronizer ④ N Counter ⑤ Driver

Fig. 3 · 70 Block diagram of ultrasonic flowmeter (Using PPL method)

$\Delta f = 1\,285.5$ Hz. From Eq.(3.33), flow velocity V is given by

$$V = \frac{D}{N \sin 2\theta}(f_1 - f_2) \tag{3.34}$$

This equation does not include the term c, and demonstrates that given nominal bore D and the angle of incidence of the ultrasonic wave θ, the flow velocity V can be measured. The volumetric flow rate of a full pipe Q is obtained by multiplying the measured flow velocity by the cross-sectional area of the pipe. However, the velocity V representing the ultrasonic flow measurement averages the flow velocities along the path of the beam, and is different from the flow velocities averaged over the pipe cross section. If V is the flow velocity as measured by the ultrasonic method, and $\overline{V}$ is the average flow velocity over the pipe cross section, then the ratio of these two values $K = V/\overline{V}$ can be called the flow rate correction factor. The following empirical formula is widely used to express the relationship between the flow rate correction factor and the Reynolds number R_{eD} for a pipe of inner diameter D.

$$K = 1.119 - 0.011 \log R_{eD} \tag{3.35}$$

Here, $R_{eD} = \overline{V}D/\nu$, and ν is the kinetic viscosity of the fluid.

According to Eq.(3.35), even a ten-fold change in the flow velocity, that is, R_{eD}, the value of K changes by a mere 1% or so. Accordingly, from Eq.(3.34), the flow rate Q that we wish to determine is finally obtained from

$$Q = \frac{\pi}{4}D^2\frac{V}{K} = \frac{\pi D^3}{4KN \sin 2\theta}(f_1 - f_2) \tag{3.36}$$

Flowmeters using the transit time differential principle become problematical when there are particles or bubbles in the fluid that scatter ultrasonic waves, so care must be taken regarding this.

(b) Doppler method

In Fig. 3.71, assuming that minute particles in the fluid flow together at the same flow velocity V as the fluid itself, if transmitter P_1 emits an ultrasonic wave of frequency f_t at an angle of θ to the flow, then the particles approach P_1 at a velocity of $V\cos\theta$, and at the same time approach the receiver P_2 at the same velocity. Thus the frequency of the signal f_r received by P_2, given $c \gg V$, can be expressed by the following equation.

$$f_r = \frac{c + V\cos\theta}{c - V\cos\theta} f_t \approx \left(1 + \frac{2V\cos\theta}{c}\right) f_t \qquad (3.37)$$

Here, c is the speed of sound through the stationary fluid, and θ is the angle of incidence to and reflection from the reflecting particles.

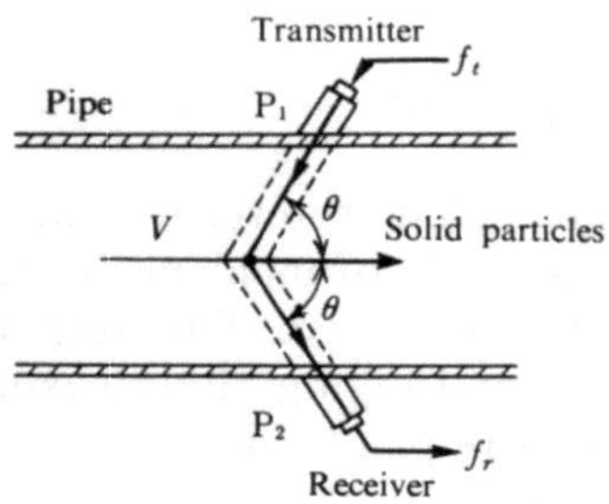

Fig. 3・71 Operating principle using doppler effect

From Eq.(3.37), the Doppler frequency f_d is

$$f_d = f_r - f_t = \frac{2f_t\cos\theta}{c} V \qquad (3.38)$$

and is thus proportional to the flow velocity V. Accordingly, flow velocity V is represented by the following equation.

$$V = \frac{c}{2f_t\cos\theta} f_d \qquad (3.39)$$

The flow rate is obtained by correcting for the difference between V and the average flow velocity.

The Doppler method requires that fluid contain suspended particles, and thus cannot measure clean fluids as can the transit time differential method.

(c) Characteristics of the ultrasonic flowmeters

(1) The detectors are attached to the outside walls of the pipe, requiring no alterations of the existing piping. Furthermore, they allow

measurement of the flow rate without stopping the flow.

(2) The detectors do not come in contact with the fluid, so there is no concern for corrosion or material adhering to them. Also, there is nothing to obstruct the flow, so no pressure loss results from their use.

(3) Disturbances in the flow velocity distribution affects measurement accuracy, so an appropriate length of straight pipe is required upstream and downstream of the place where the detectors are installed.

(4) In process industries, flowmeters for use with liquids mainly measure water flows. Based on the particular characteristics of their respective measurement principles, use of the two types of ultrasonic flowmeters is differentiated according to the characteristics of the fluid measured in the following way.

Transit time differential flowmeters: Mainly clean water, water for industrial use, raw water for water purifying plants, water for agricultural use, and other types of clean water.

Doppler flowmeters: Mainly drain water, factory effluent, and other such dirty water which contains foreign particles.

3.4 Measurement of Pressure[33)]

In process industries, pressure itself is an object of measurement, but in addition, temperature, flow rate, liquid level, and other measured variables are frequently measured indirectly by measuring pressure. Thus it can be said that pressure is the most fundamental measured variable.

The units of pressure standardized in the International System of Units (SI) are the Pascal (Pa), and the bar, which can be used together with the Pascal. Table 3.13 is a conversion table for the major units of pressure now in use, including those of the International System of Units (SI).

When pressure is detected by a pressure transmitter, it can be displayed as absolute pressure, as a pressure differential, of as a gage pressure. Absolute pressure is measured using vacuum as a reference value, and is used in equations showing the relationships among pressure and other variables. Differential pressure represents the difference from some (non-zero) reference pressure. Gage pressure is pressure measured using atmospheric pressure as the reference pressure. Thus, to show absolute pressure, the pressure indicated by gage pressure is added to the atmospheric pressure at that time. When it is necessary to distinguish absolute pressure from gage pressure,"abs"is appended to the pressure value to indicate absolute pressure, and"G" is

Table 3.13 Conversion table of pressure units

Pa	bar	kgf/cm²	atm	mmH_2O	mmHg or Torr
1	1×10^{-5}	1.01972×10^{-5}	9.86923×10^{-6}	1.01972×10^{-1}	7.50062×10^{-3}
1×10^{5}	1	1.01972	9.86923×10^{-1}	1.01972×10^{4}	7.50062×10^{2}
9.80665×10^{4}	9.80665×10^{-1}	1	9.67841×10^{-1}	1.0000×10^{4}	7.35559×10^{2}
1.01325×10^{5}	1.01325	1.03323	1	1.03323×10^{4}	7.60000×10^{2}
9.80665	9.80665×10^{-5}	1.0000×10^{-4}	9.67841×10^{-5}	1	7.35559×10^{-2}
1.33322×10^{2}	1.33322×10^{-3}	1.35951×10^{-3}	1.31579×10^{-3}	1.35951×10	1

appended to indicate gage pressure.

3.4.1 Pressure transmitters

A pressure transmitter is a device which converts the pressure of a fluid that is the object of measurement into a standardized signal, and then transmits the signal. Pressure transmitters are classified into pneumatic types and electronic types.

(a) Pneumatic pressure transmitters

Pneumatic pressure transmitters are divided into absolute pressure types, differential pressure types, and gage pressure types (generally called pressure transmitters), according to the methods of pressure display described above. The operating principle of pneumatic differential pressure transmitters is snown in Fig. 3.72.

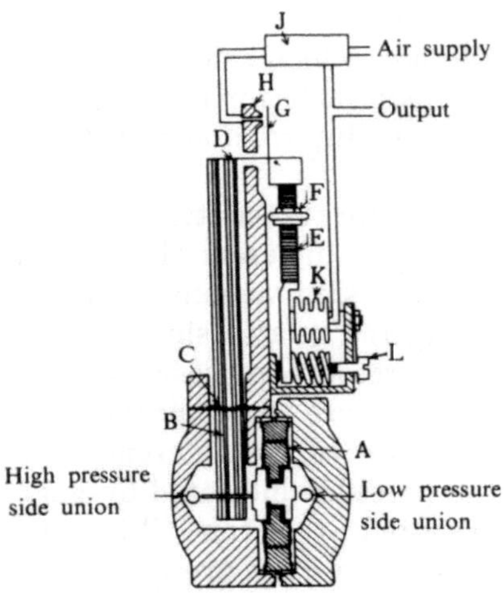

A : Twin diaphragm capsule
B : Force bar
C : Diaphragm seal
D : Flexure
E : Range bar
F : Range wheel
G : Flapper
H : Nozzle
J : Pneumatic amplifier
K : Feed back bellows
L : Zero adjusting screw

Fig. 3 · 72 Cutaway showing the principle of pneumatic differential pressure transmitter

The pressure of the fluid passing through the high pressure side union is applied to the left diaphragm of the twin diaphragm capsule, and the pressure of the fluid passing through the low pressure side union is applied to the right diaphragm of the capsule. If the pressure of the high-pressure side fluid increases or the pressure of the low-pressure side fluid decreases, then the thrust on the twin diaphragm from the left towards the right increases. This thrust moves the force bar, which rotates slightly counterclockwise, with the diaphragm seal acting as a fulcrum, and as a result, the gap between the nozzle and the flapper becomes smaller. Air normally flows out from the nozzle through this gap, but if the gap narrows, the pressure inside the nozzle increases and the output of the pneumatic amplifier increases by a larger amount. Part of the output is applied to the feedback bellows, where it is converted to a force. This force is applied to the range bar, which using the range wheel as a fulcrum, acts to pull back on the flapper, thus reducing the pressure inside the nozzle. The end result is that, compared with before the pressure differential is changed, the gap between the nozzle and the flapper has narrowed slightly, and the output has increased and stabilized. In this way the two-forces are balanced; that is, negative feedback increases the linearity of the input-output characteristics of the pressure differential transmitter, and improves the characteristics.

The zero adjusting screw is set so that the output corresponding to the minimum pressure differential is 0.2 kgf/cm^2. The position of the range wheel is adjusted so that the output corresponding to the maximum pressure differential is 1.0 kgf/cm^2. If the range wheel is set to a higher position, the amount of feedback is increased and the range widens. Figure 3.73 shows how the input-output characterstics change with the position of the range wheel.

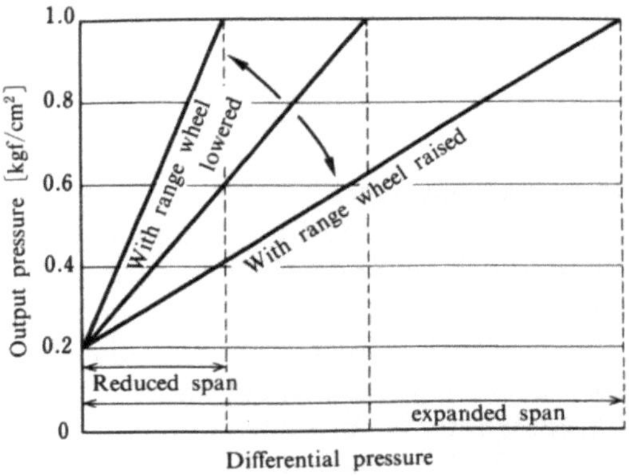

Fig. 3·73 Relationship between input-output characteristics and range wheel position (*b*) *Electric pressure transmitters*

Electronic pressure transmitters are also grouped into the same three types as are pneumatic transmitters. The operating principle of electronic pressure transmitters is shown in Fig. 3.74. Pressure is applied to the two diaphragms of the twin diaphragm capsule. This pressure is transmitted from the twin diaphragms to the sensing diaphragm through the sealing liquid. Two fixed electrodes are placed symmetrically on the left and right side of the insulator and an electrical capacitance is formed between these electrodes and the sensing diaphragm. If there is a pressure differential between the high pressure side and the low pressure side, the position of the sensing diaphragm will change in proportion to that differential, and an electrical capacitance differential will be produced. This capacitance differential is converted an electrical current and output as a signal of 4 mA to 20 mA DC, corresponding to the measured pressure differential range. Zero setting and span abjustment are done in the transmitter.

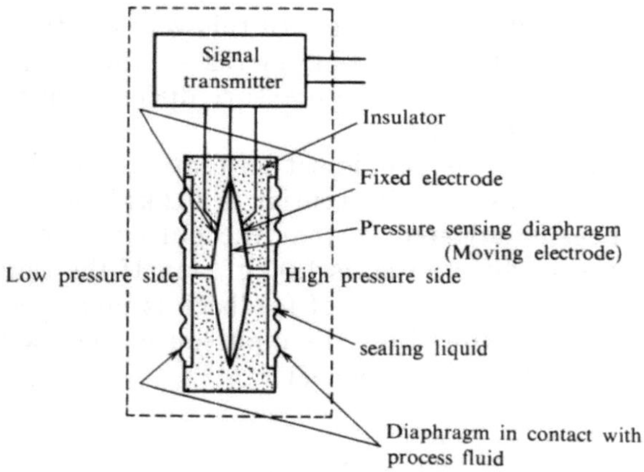

Fig. 3 · 74 Structure for electronic differential pressure sensor

3.4.2 *Types of pressure detectors*

To detect fluid pressure, the fluid is applied to a pressure sensing element (detector), which converts the pressure to a force or a displacement. In pneumatic industrial instruments, a pneumatic input signal is converted by the pressure detector to a displacement, and recorded, displayed, or used perform a control action.

In industrial instruments, pressure detectors constructed of metal whose degree of elastic deformation is proportional to the pressure applied to them are most often used.

(*a*) *Diaphragms*

A diaphragm is a thin plate fixed in place around its periphery. The plate is corrugated to increase its effective area and give it good elastic characteristics with respect to pressure. A diaphragm capsule used in a pressure transmitter is constructed as shown in Fig. 3.75. Two diaphragms are attached with a metal backup plate placed between them. The backup plate has a small cavity filled with silicone oil. This silicone oil transmits the pressure applied to the two diaphragms, and the oil in the cavity provides a damping action on the transmission of the pressure, thus increasing the stability of the entire device.

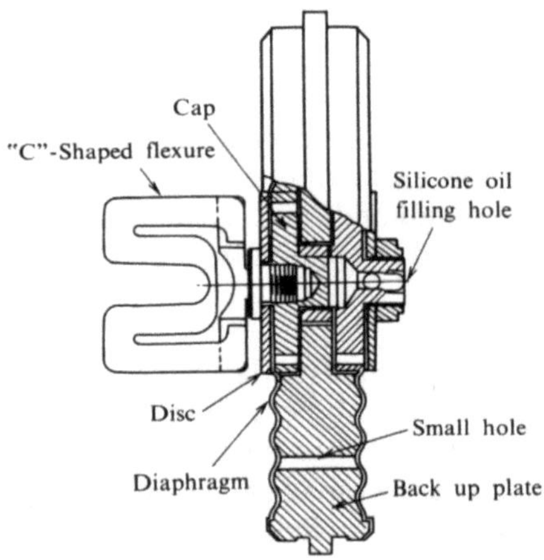

Fig. 3・75 Twin-diaphragm capsule cut away

The amout of displacement of the diaphragm is small, and normally it is used as a pressure-to-force converter. However, by combining the action of several diaphragms, thus increasing the displacement, they can also be used to drive the indicator of a meter.

(*b*) *Bellows*

Bellows are made by stamping the outer surface of a thin-walled brass, phosphorous bronze, or stainless steel pipe. The belows themselves have the characteristics of a spring, but when used as a pressure-displacement conversion device, they are used in combination with a spring as shown in Fig. 3.76 to achieve good conversion linearity. The effective cross-sectional area of the bellows to which pressure is applied given by the following equation.

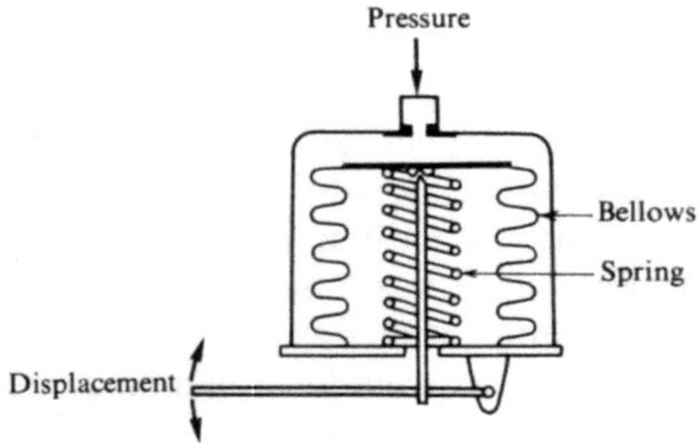

Fig. 3・76 Bellows paired with spring

$$A=\frac{\pi}{2}(R_1^2+R_2^2)$$

Here, A is the effective cross-sectional area of the bellows, R_1 is the outer radius, and R_2 is the inner radius.

If two bellows are combined, and a vacuum is applied to one, and the pressure to be measured is applied to the other, then absolute pressure can be measured.

(*c*) *Bourdon tubes*

A Bourdon tube is a seamless flat tube of bronze, phosphorous bronze, brass, stainless steel, or beryllium bronze, monel®, Ni-Span® or other such material, curved into an arc and fixed at one end. If an internal pressure is applied to this tube, its cross section will change towards a circular shape, and the free end of the tube will move outwards. This displacement of the free end is proportional to the internal pressure within the limits of elasticity. The displacement, usually from about 1 mm to 8 mm, is amplified to drive an indicator. Bourdon tubes are divided into C-shaped types, spiral types, and helical types (Fig. 3.77). Of the three types, the C-shaped type is also used as a pressure-force converter.

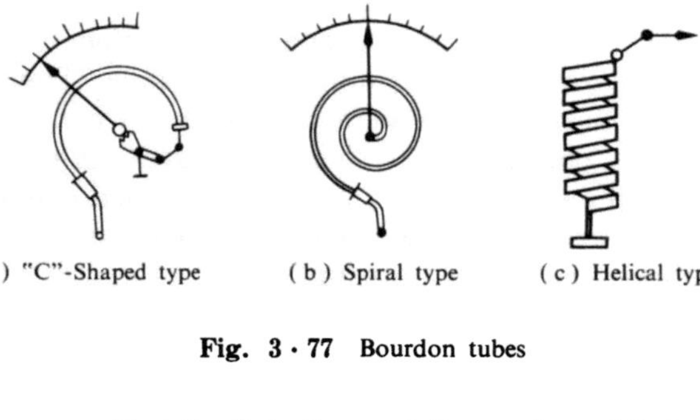

(a) "C"-Shaped type (b) Spiral type (c) Helical type

Fig. 3・77 Bourdon tubes

3.5 *Measurement of Liquid Level*[33)]

There are many methods of measuring liquid level, each having its particular characteristics. A method is selected for use according its suitability for the liquid being measured.

3.5.1 *Float liquid-level meters*

A float liquid-level meter, as shown in Fig. 3.78, measures the level of a liquid directly from the position of float resting on the liquid surface. It is sutitable for measuring the level at a dam, clean water tank, and so on. The up and down movements of the float are indicated by the degree of rotation of a pulley connected to the float by a wire cable.

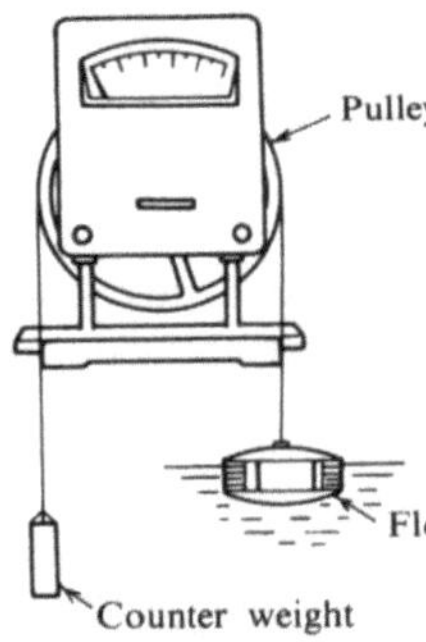

Fig. 3·78 Float liquid level meter

3.5.2 *Pressure differential liquid-level meters*

The principle of pressure differential liquid-level meters is that the static pressure at any point in a liquid is equal to the product of the distance from that point to the surface of the liquid, the density of the liquid, and the acceleration due to gravity. Thus, if the liquid density and the acceleration due to gravity are known, the distance to the surface (that is to say, the liquid level) can be determined from the pressure. The method of tapping the pressure is different for open tanks and closed tanks.

(*a*) *Open tanks*

A pressure differential transmitter or pressure transmitter is connected to the tank by a pressure tapping tube, as shown in Fig. 3.79. When a pressure differential transmitter is used, the liquid is tapped for the high pressure side, and the open air is tapped for the low pressure side.

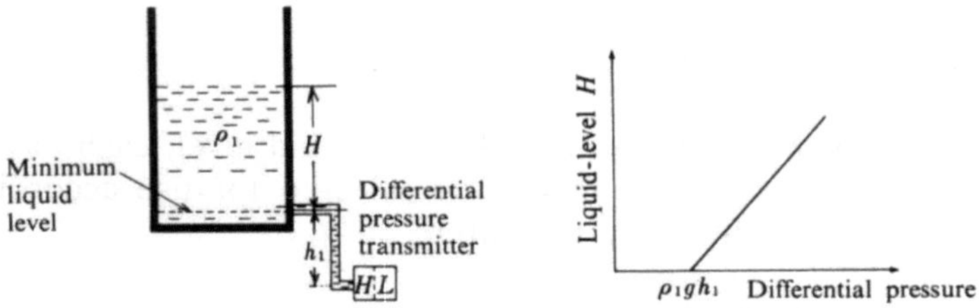

Fig. 3·79 Liquid level measurement of open tank

The following relationship exists among the pressure (gage pressure), the density of the liquid, and the acceleration due to gravity.

$$P=\rho_1 g(H+h_1)$$

Here, P is the pressure (gage pressure, Pa), ρ_1 is the density of the liquid (kg/m^3), g is the acceleration due to gravity (m/s^2), H is the plumb distance between the minimum liquid level and the surface (m), and h_1 is the plumb distance between the minimum liquid level and the pressure detector (m).

The relationship between the input and output of the transmitter is linear, so the output of the transmitter changes in direct proportion to changes in the liquid level.

(*b*) *Closed tanks*

When measuring the liquid level in a closed tank, the pressure differential transmitter is used. The low pressure tap is the pressure of the gas above the liquid in the upper part of the tank (Fig. 3.80). The pressure of the gas is also applied to the high pressure tap at the same time, so when taking the pressure differential, it cancels out and so does not affect the output of the transmitter. Thus, the level of the liquid can be known from the pressure differential.

High pressure tap pressure	$P_H=\rho_1 g(H+h_1)+P_G$
Low pressure tap pressure	$P_L=P_G$
Pressure differential	$P_H-P_L=\rho_1 g(H+h_1)$

Here, P_G is the pressure of the gas in the upper part of the tank.

If condensation from the gas in the upper part of the tank collects inside the tapping tube, the low pressure tap pressure in the tube will change and the output of the pressure transmitter will be affacted. To avoid this problem, the condensation is collected in a drain pot.

Another method involves using a relatively heavy liquid that does not easily evaporate to fill the tube, as shown in Fig. 3.81. The pressure of the gas in the tank is then applied to the pressure detector through this liquid. This method is called the wet leg method, as op-

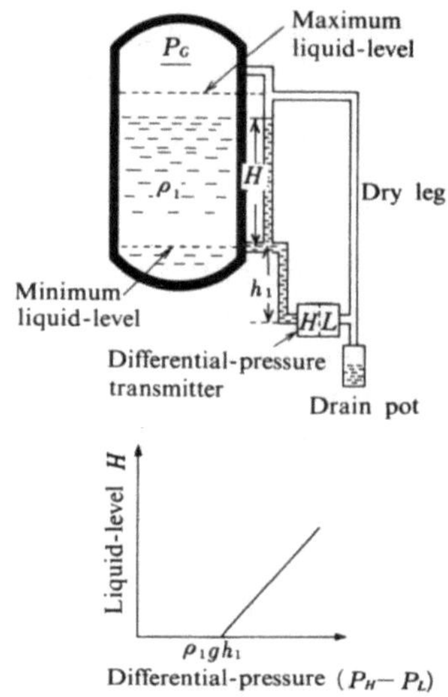

Fig. 3・80 Liquid-level measurement using a dry-leg system

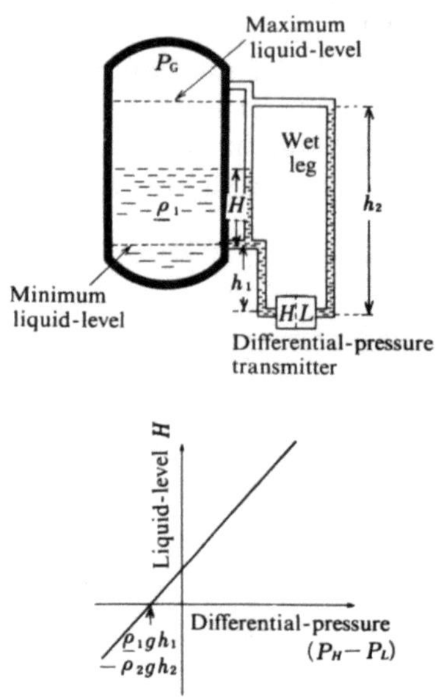

Fig. 3・81 Liquid-level measurement using a wet-leg system

posed to the dry leg method described above. In the wet leg method, the pressure is calculated in the following way.

High pressure tap pressure	$P_H = \rho_1 g(H + h_1) + P_G$
Low pressure tap pressure	$P_L = \rho_2 g h_2 + P_G$
Pressure differential	$P_H - P_L = \rho_1 g(H + h_1) - \rho_2 g h_2$

Here, ρ_2 is the density of the liquid in the wet leg (kg/m^3), and h_2 is the height of the liquid in the wet leg (m).

When using pressure differential liquid-level meters, care must be taken regarding the following points.

(1) The pressure differential transmitter must be positioned below the minimum liquid level.

(2) Corrections must be made for changes in the density of the liquid.

(3) If there is a pulsating motion in the liquid, the output of the transmitter will be unstable.

(4) The tapping tubes should be as straight as possible so as not to trap air. Also, use of points that might leak should be minimal.

3.5.3 Displacer liquid-level detectors

A displacer liquid-level detector is a round, tubular displacer combined with a force detector (Fig. 3.82). The buoyancy applied to the displacer is equal to the weight of the liquid it displaces. Thus, if the cross-sectional area of the displacer is constant along the radial axis of the displacer, the flotational force has a linear relationship with the liquid level. Accordingly, by measuring the flotational force, the liquid

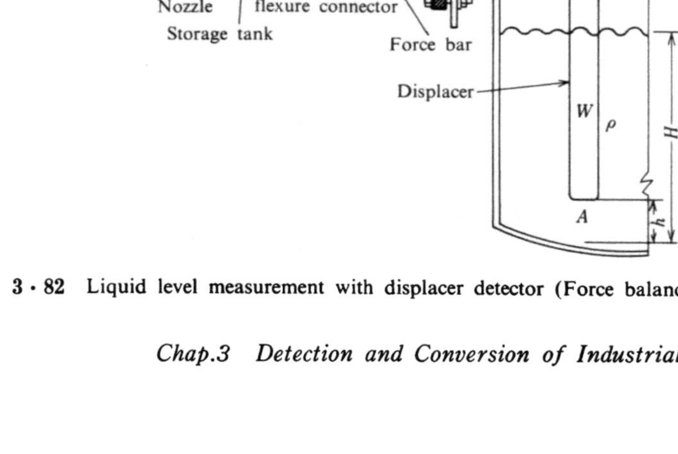

Fig. 3・82 Liquid level measurement with displacer detector (Force balance type)

level can be known.

Referring to Fig. 3.82, the force applied to the displacer can be expressed by

$$F=\rho gA(H-h)-W$$

where F is the perpendicular upward force on the displacer (N), ρ is the density of the liquid (kg/m^3), g is the acceleration due to gravity (m/s^2), A is the cross-sectional area of the displacer (m^2), H is liquid level measured from the reference position (m), h is the position of the bottom of the displacer measured from the reference position (m), and W is the weight of the displacer (N).

The flotation force applied to the displacer is transferred to the force bar, which, using the diaphragm seal as a fulcrum, transmits the force to the force detector.

The displacer liquid level dectors have the follwing special features.

(1) They can be used to measure high-temperature, high-pressure liquids.
(2) They have good sensitivity, good accuracy, fast response as well.
(3) They are suitable for short measurement spans.
(4) The measurement range can be changed by changing the length of the displacer.
(5) If impurities or other objects adhere to the displacer, its volume will change, generating measurement error.
(6) Corrections must be made for changes in the density of the liquid.

3.5.4 Purge-type liquid-level meters

Purge-type liquid-level meters are used for corrosive liquids, liquids that contain suspended objects, high viscosity liquids, and for tanks that have already been constructed, underground tanks, and so on, where installing tap tubes presents a problem.

A purge-type liquid-level measurement is illustrated in Fig. 3.83. A bubble tube is inserted into the tank, and a fixed flow of air is forced through the tube such that bubbles emerge from the end of the tube. If the air flow rate is not extremely large, the air pressure in the bubble tube can be considered to be the same as the pressure of the liquid at the end of the tube. Thus the liquid level can be measured by measuring the air pressure.

Various materials, can be used for the bubble tube, such as stainless steel, mild steel and rigid PVC. The material used is chosen accoding to the nature of the liquid. Also, when air is not suitable as a purging substance, nitrogen or some other inert gas can be used.

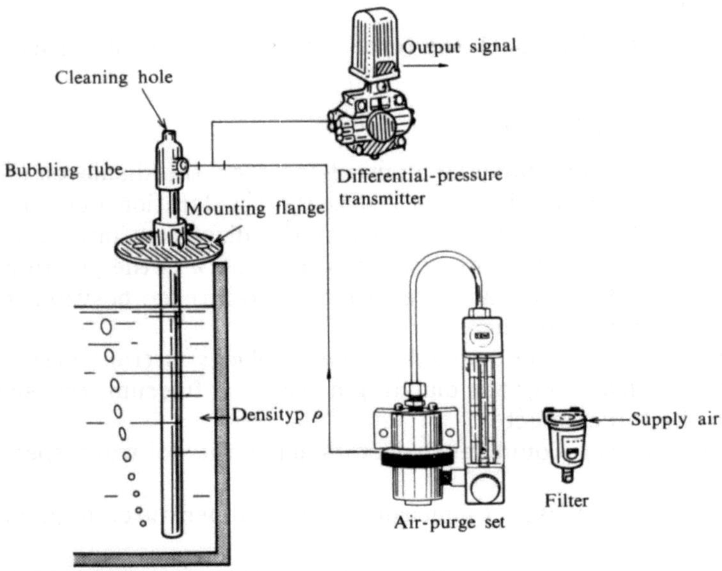

Fig. 3 · 83 Purge type liquid-level measurement system

To reduce the error associated with the air flow rate, the pressure tap is placed near the bubble tube. Corrections are necessary when there are changes in the density of the liquid.

3.5.5 Ultrasonic liquid-level meters

As shown in Fig. 3.84, an ultrasonic liquid-level meter measures the time required for an ultrasonic wave emitted by a transducer and reflected off the surface of the liquid to return to the transducer, thus determining the liquid level. The relationship between the liquid level,

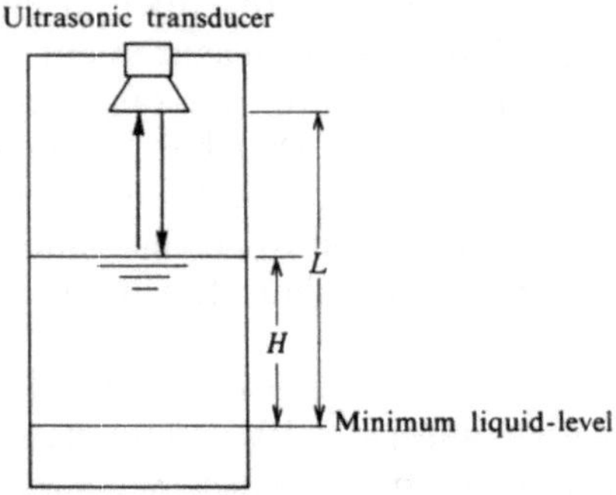

Fig. 3 · 84 Operating principle for ultrasonic liquid-level meter

speed of the ultrasonic wave, and time is given by the following equation.

$$2(L-H)=tu$$

Here, L is the distance from the minimum liquid level to the ultrasonic transducer (m), H is the liquid level (m), t is time between the emission of the signal and its subsequent reception at the transducer (s), and u is the propagation velocity of the ultrasonic wave (m/s).

The velocity of the ultrasonic wave u varies with the type of gas it travels through and the temperature, so the temperature of the gas must be measured and corrections applied according to the results.

Ultrasonic liquid-level meters have the following special features.

(1) Measurement can be accomplished without touching the liquid.

(2) There are no moving parts, so inspection and maintenance are easy.

(3) They are small and light, so installation and operation are easy.

3.5.6 Capacitance liquid-level meters

The electrostatic capacitance of two facing electrodes varies with the dielectric constant of the material separating them. As shown in Fig. 3.85, if the space between two concentric tubular electrodes contains a liquid, a change in the level of that liquid will change the capacitance between the electrodes as described by the following equation.

$$C=\frac{2\pi\{(\varepsilon_2-\varepsilon_1)H+\varepsilon_1 L\}}{\ln(D/d)}$$

Here, C is the electrostatic capacitance between the electrodes (F), ε_1 is the dieletric constant of gas (F/m), ε_2 is the dielectric constant of the liquid (F/m), H is the liquid level (m), L is the height of the electrodes (m), D is the diameter of the outer tubular electrode (m), and d is the

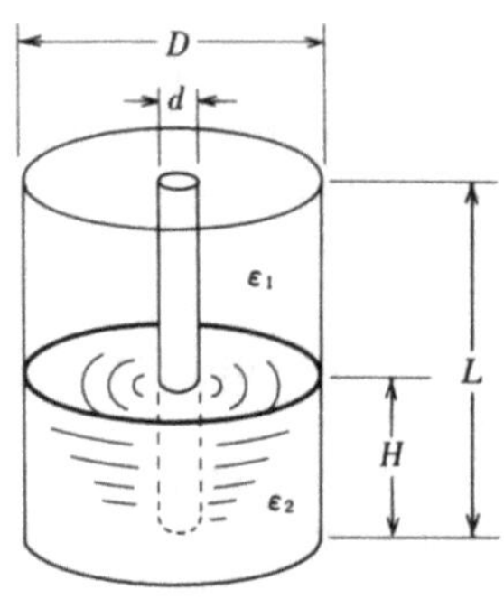

Fig. 3·85 Operating principle for capacitance liquid-level meter

diameter of the inner tubular electrode (m).

The terms L, D, d, ε_1, and ε_2 are constants, so by measuring the capacitance C, the liquid level H can be determined. The surfaces of the electrodes are coated with an insulating material that must be selected considering the nature of the liquid being measured.

Capacitance liquid-level meters have the following special features.

(1) The detector has a simple construction.

(2) They have no moving parts, and thus a long service life.

(3) Changes in the dielectric constant that are caused by changes in the temperature and density of the liquid being measured produce errors in measurement.

3.6 Measurement of Displacement and Angle[34)-39)]

Displacement and angular movement are objects of measurement in themselves, but their measurement also is used in the displacement-conversion parts of converters employed in the measurement of pressure ,flow rate and other industrial variables, and in the position-conversion part of recording device servo-mechanisms. Many methods have been developed to measure displacement and angles; some representative methods are described here.

3.6.1 Resistance potentiometer methods

In resistance potentiometer methods, displacement and angular movement are converted to changes in the brush position of a linear or circular slide resistor, thereby converting them to changes in electrical resistance. These devices have a simple construction and have been used from early times.

In wire coil-type meters, metal wire that has a low thermal coefficient of resistance, such as manganin wire, or constantan wire. However, although their linearity is good, a weak point is that noise is easily generated by poor brush contact. CP types use conductive plastic resistance elements formed by heat-shaping a mixture of conductive particles, such as carbon, and plastic pellets. These devices have the merits of high resolution, and little wear due to smooth contact. Also, it is possible to raise the resolution and stability of the wire coil-type devices by coating with this sort of conductive plastic.

3.6.2 Electromagnetic induction methods

The induction methods employ the change in inductance resulting from displacement of a ferrite core or short-circuit ring. Although there are many methods based on device structure, we will introduce only a few of them here.

(a) Differential transformer method

As shown in Fig. 3.86, a movable core is placed between a primary coil and two secondary coils positioned symmetrically on both sides of the primary coil. If an exciting alternating current flows in the primary coil, a voltage is induced in the secondary coils. However, as this voltage is differentially extracted, if the movable core is centrally positioned, the voltages in the two secondary coils cancel each other out and the output voltage is zero. When the core moves, the induced voltage increases in one of the secondary coils, and decreases in the other. Accordingly, the differential output increases. The relationship between the displacement of the core and the transformer output can be made linear by choice of coil winding size and distribution. The measurement range is from a few millimeters to several tens of millimeters, and the linearity is approximately $\pm 0.1\%$ to $\pm 1\%$.

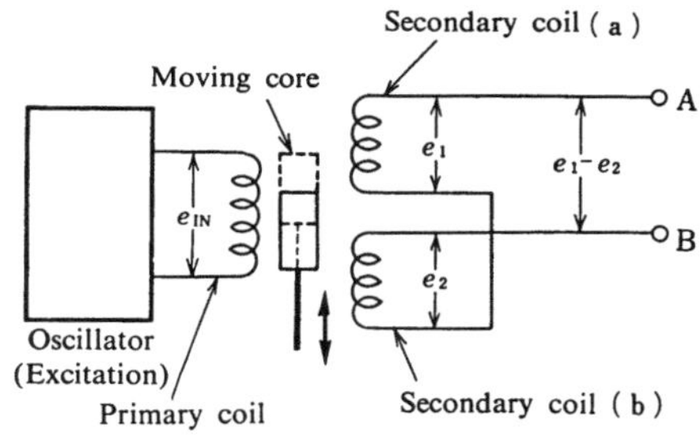

Fig. 3 · 86 Operating principle for differential transformer

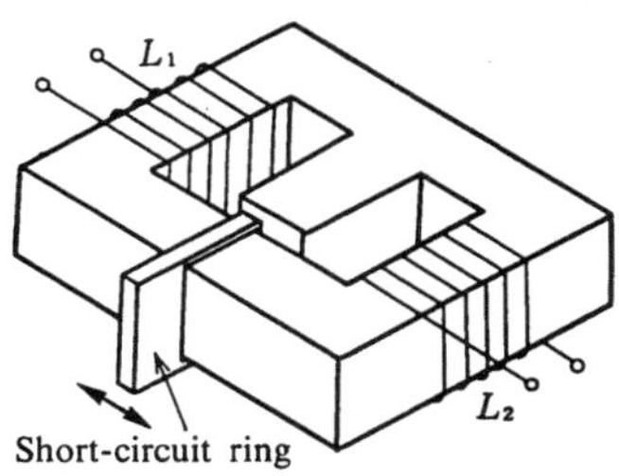

Fig. 3 · 87 Structure for short-circuit ring type displacement gage

(b) Short-circuit ring method

As shown in Fig. 3.87, a copper short-circuit ring lies in the middle of the magnetic path of two coils (L_1, L_2). Displacement of this short-circuit ring differentially changes the inductance of the two coils, and by measuring this change, the degree of displacement can be determined. If the displacement is limited to a certain range by proper choice of the two coils, L_1 and L_2, the output will be proportional to the displacement x, and the following approximation equation can be applied.

$$\frac{x}{a}=\frac{L_1-L_2}{L_1+L_2}=\frac{i_2-i_1}{i_1+i_2}$$

Here, a is the effective length of movement of the short-circuit ring, i_1 is the current flowing in coil L_1, and i_2 is the current flowing in coil L_2.

Thus, if (i_1+i_2) is held constant and (i_2-i_1) is measured, the displacement x will be obtained. Linearity is approximately $\pm 0.1\%$ for a measurement span of 1 mm.

(*c*) *Variable coupling method*

This method uses a coil formed on a printed circuit board, and detects displacement as a change in mutual inductance. The operating principle is illustrated in Fig. 3.88. If an alternating current flows in the exciting coil, and induced voltage is generated in the receiving coil through the U-shaped ferrite core. If the core moves in the direction of the X axis, the coupling area of the receiving coil increases, increasing the induced voltage. This voltage, when rectified and amplified, yields an output that is proportional to the position of the core. The relationship between the displacement of the core x and the voltage generated in the receiving coil e_2 is expressed by the following equation.

$$e_2=amx\frac{dB}{dt}$$

Here, m is a constant representing the relationship between the receiving area and the displacement ($f(x)=mx$), and a is the length of movement of the core along the X axis.

Figure 3.89 shows the construction and circuit diagram for the device when used in an angle converter. The coil is a pattern formed on a printed circuit board. For a 140° angle, the linearity is $\pm 0.1\%$ or

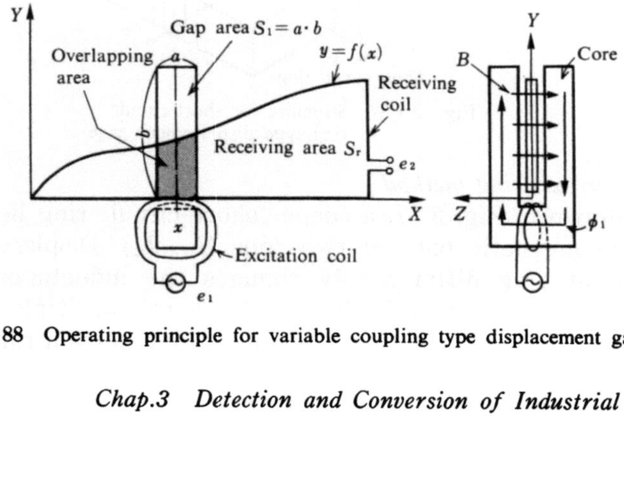

Fig. 3・88 Operating principle for variable coupling type displacement gage

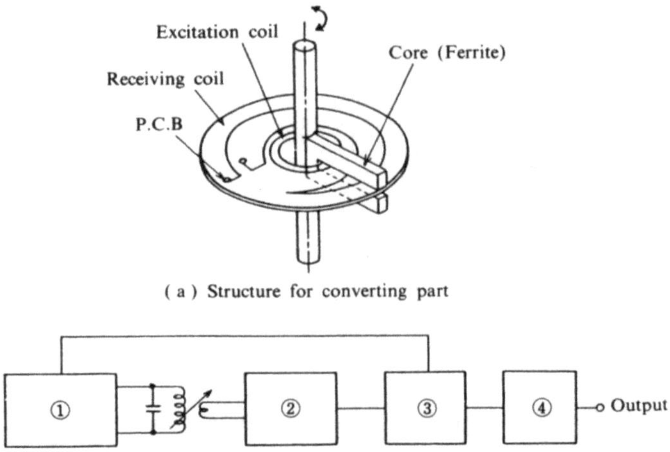

(a) Structure for converting part

① Constant amplitude oscillation
② AC Amplification
③ Synchronous rectification
④ Smoothing amplification

(b) Block diagram

Fig. 3 · 89 Configuration of angle converting part

less. Angle converters and transmitter units for showing the degree of opening of motor-drive valves are examples of the applications for this device.

(d) Eddy current displacement gages

In this method, differences in eddy currents due to displacement are detected as changes in the impedance of a coil. This principle is illustrated in Fig. 3.90. If an alternating current flows through the active coil used to measure displacement, an eddy current is generated in the surface of a conductor that is drawn near to the coil. The magnetic field from this eddy current produces a counter-electromotive force in the coil, which changes the effective impedance of the coil. If the conductivity, magnetic permeability and thickness of the conductor are fixed, then impedance becomes a function of the distance x between the coil and the conductor. If the output voltage is detected by a coil and condenser Wheatstone bridge circuit, that voltage V_0 is expressed by the following equation.

$$V_o = a \exp(-bx) + c$$

Here, a, b and c are constants.

If this output is made linear by a logarithmic amplifier, then

$$V = -bx + \ln a,$$

which represents a proportional relationship between the displacement

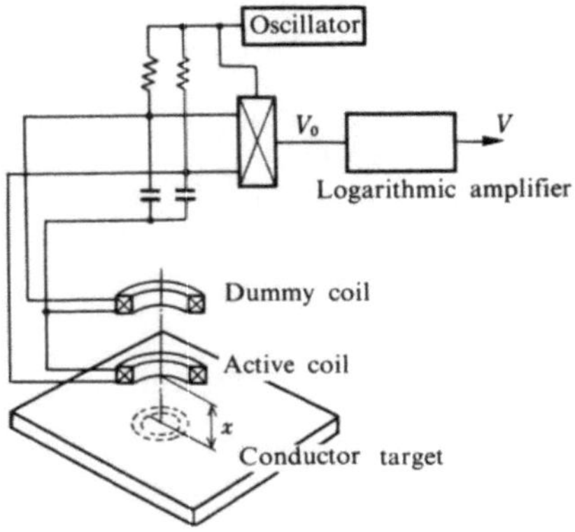

Fig. 3・90 Operating principle for eddy current-type displacement gage

and the output voltage. In actual practice, a dummy coil placed near the active coil compensates for temperature. By taking the excitation frequency above a few hundred hertz, only the reactance portion is measured. As there is no contact with the object of measurement, applications include measuring the axle displacement and vibration of rotating machinery. Linearity is $\pm 1\%$ or less for a measurement span of from 1 mm to 5 mm.

3.6.3 Magnetic balance method

In the magnetic balance method, the magnetic flux generated in a magnetic path by a change in the angular position of a permanent magnet is detected electrically. This measured magnetic flux is balanced with a magnetic flux from a feedback coil, from which the angle of the permanent magnet can be known. Figure 3.91 illustrates this principle. The stator is a saturable ferrite core of high magnetic permeability, and when the permanent magnet rotates, the magnetic flux ϕ_i changes in proportion to the angular displacement of the magnet θ.

$$\phi_i = k_1 \theta$$

If output current I_o flows through the feedback coil, a magnetic flux ϕ_f is generated, as expressed by the following equation.

$$\phi_f = k_2 I_o$$

Here, k_1 and k_2 are constants.

Thus, a magnetic flux of $\Delta\phi = \phi_i - \phi_f$ remains in the leg of the core, and if a current I_o flows such that $\Delta\phi$ becomes zero, the amount of this current corresponds to the angular displacement. In detecting the magnetic flux differential $\Delta\phi$, the direct current components of

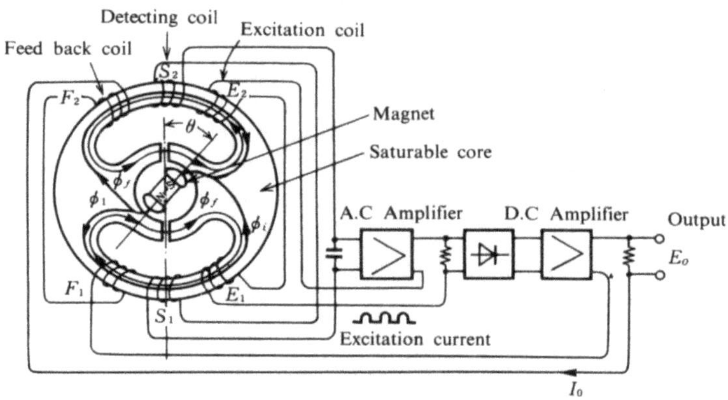

Fig. 3 · 91 Magnetic balance-type angle converter

half-wave alternating currents flowing in exciting coils E_1 and E_2 change the magnetic flux in relation to ϕ_i. Specifically, the component from E_1 adds to the flux, and the component from E_2 substracts from it. The detecting coils S_1 and S_2 are differentially connected so that only the voltage that is proportional to the angular displacement ϕ_i is obtained. This voltage is amplified and the feedback current I_o is obtained. This angle converter has a linearity of $\pm 0.2\%$ for an angle of 40°.

3.6.4 Magnetic strain method

This method was developed for use as a servo-motor recording device position converter. For details, refer to item 4.1.3 in Chap. 4, Sec. 4.1.

3.7 Measurement of Rotation[3),40)]

Tachometers are used to measure the rotational speed of moving bodies, or in other words, the number of revolutions per unit of time. The unit of measurement most often used is the number of revolutions per minute (rpm). In industrial measurement, analog tachometers using generators as rotation detectors have been widely used in the past, but for highly accurate measurement, digital counting type tachometers which employ frequency output detectors are used.

3.7.1 Measurement using tachometer generators

Tachometer generators measure rotational speed by using alternating current generators as rotation detectors, applying the principle that the voltage induced in the generator is proportional to the speed of rotation. An example of a medium-speed (500 rpm to 5 000 rpm) alternating current generator tachometer is shown in Fig. 3.92. It is a single-phase, four-pole alternating current generator, having a permanent magnet attached to the rotor as shown in Fig. 3.93. When the permanent magnet rotates, the magnetic flux interlinking with the pickup coil changes, and an alternating voltage is induced in the pickup coil. This voltage is rectified and converted to a DC voltage output.

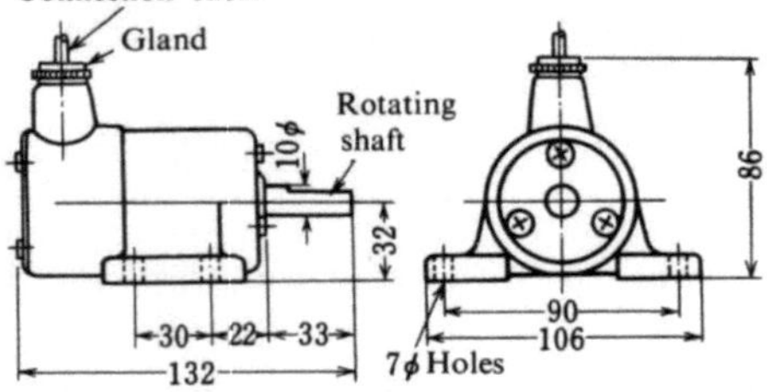

Fig. 3・92 Ac tachometer generater (Example)

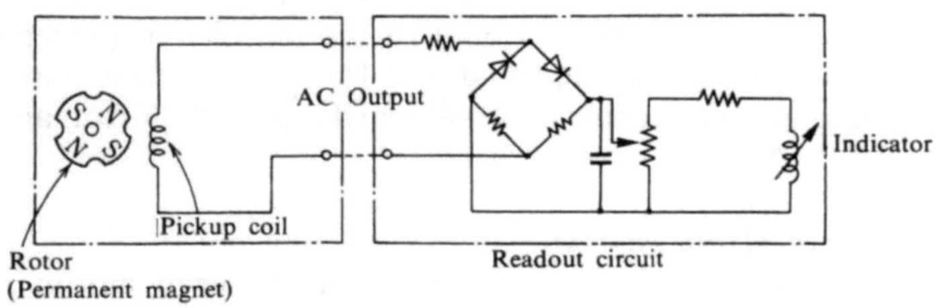

Fig. 3・93 Ac tachometer schematic

By this method, low rotational speeds produce a low-frequency output from the generator, and after rectification ripples in the DC output increase, causing the tachometer indicator to vibrate and making measurement difficult. For measurement of low-speed rotation, a low speed alternating current generator that uses a multiple-pole permanent magnet is employed. Such a tachometer can measure rotational speeds of from 0 rpm to 20 rpm. The voltage induced in the generator undergoes full-wave rectification by the built-in rectifying circuit, and direct current output voltage is obtained. This output voltage is used for remote measurement of the rotation speed.

Alternating current generators do not have brushes and commutators as do direct current generators, thus they are highly reliable and easily made explosion-proof.

3.7.2 Pulse output sensors

These are sensors which count the number of pulses (frequency) of an electrical signal that is proportional to the rotation speed of a rotor. Typical sensing methods include the electromagnetic method and the photoelectric method.

(*a*) *Electromagnetic method*

Figure 3.94 shows an example of this method. As the toothed wheel of iron rotates, the magnetic resistance between the wheel and the ferrite core varies periodically. Thus, a pulsed voltage signal that is proportional to the number of teeth in the wheel is generated in the pickup coil. At low speeds approaching zero motion, the output voltage theoretically becomes small, so these devices cannot detect low rotational speeds. However, they have the advantages of excellent durability and not requiring a power source.

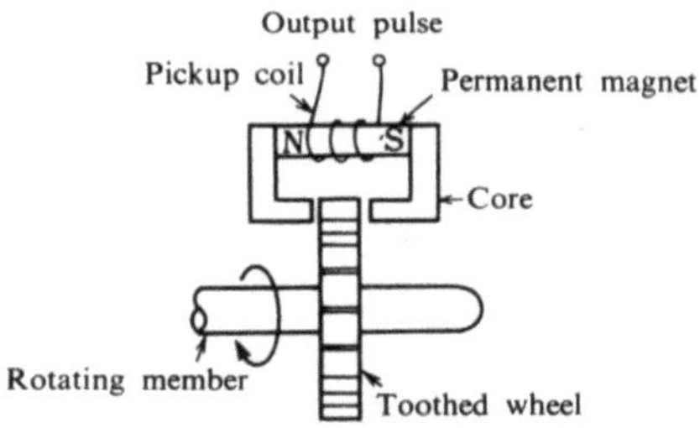

Fig. 3 · 94 Magnetic pickup-type speed sensor

(*b*) *Photoelectric method*

This method employs a light source and phototransistor or other photoelectric converter to convert rotational speed into countable pulses. The light detecting methods include the incident light method and the reflected light method.

The incident light method is illustrated in Fig. 3.95. A slotted disk attached to the rotating part rotates between the light source and the receiver photocell. This produces a pulsed signal that is proportional to the number of slots in the disk. This system has the merit that if the

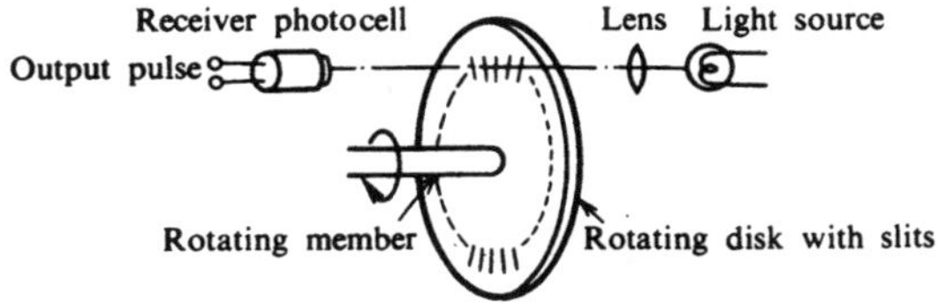

Fig. 3 · 95 Photo-electric speed senor (Incident light type)

number of slots is increased, higher resolution measurement can be made.

The reflected light method is shown in Fig. 3.96. The surface of the rotor is patterned with marks of different degrees of light reflectivity, such as a black and white pattern or highly directional reflective tape. The receiver photocell outputs a pulsed signal generated by the light reflected back to it from the marks in an on-and-off fashion.

The photoelectric sensors described above use either a light-emitting diode or an ordinary electric bulb as a light source, and a phototransistor or other such device can be used for the receiver photocell. The photoelectric sensing method allows detection of rotational speed without touching the rotor, so it is used for detecting high-speed rotation. The reflected light method places no load on the rotor, so it is suitable for measuring low-torque rotations.

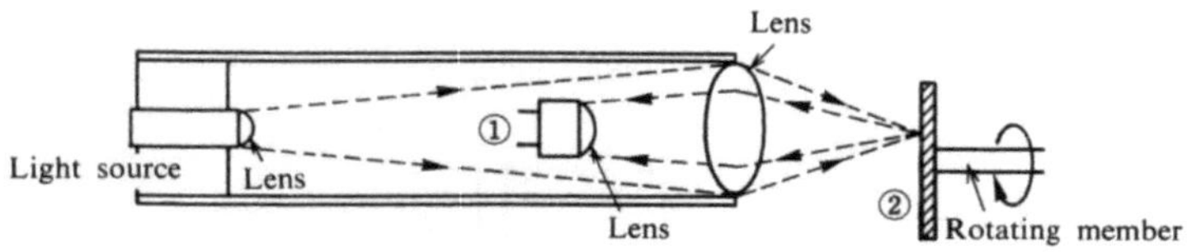

① Receiver photocell
② Rotating disc with dark and light areas

Fig. 3 · 96 Photo-electric speed sensor (Reflexed light type)

3.7.3 Digital counting tachometers

These are two representative methods of digital counting as used in measuring the number of rotations. In both methods, the pulse train signal output from the detectors is processed digitally, allowing highly accurate measurement.

(a) Pulse (frequency) counting method

This method directly counts the number of pulse signals in a fixed amount of time, yielding the average number of rotations for that period of time. A block diagram of this circuit is shown in Fig. 3.97.

The pulse signal picked up from the rotor is proportional to the number of rotations. It undergoes waveform shaping and is transformed into a pulse-train signal. It is then fed to the gate. The gate is opened for a fixed period of time by a control signal whose duration is determined by a stable quartz oscillator. The number of input pulses that pass through the gate while it is open are counted and displayed as a numerical value.

(b) Rotation period counting method

The rotation period (time) of the rotor is measured, and the number of rotations is obtained from the inverse of the measured value. As

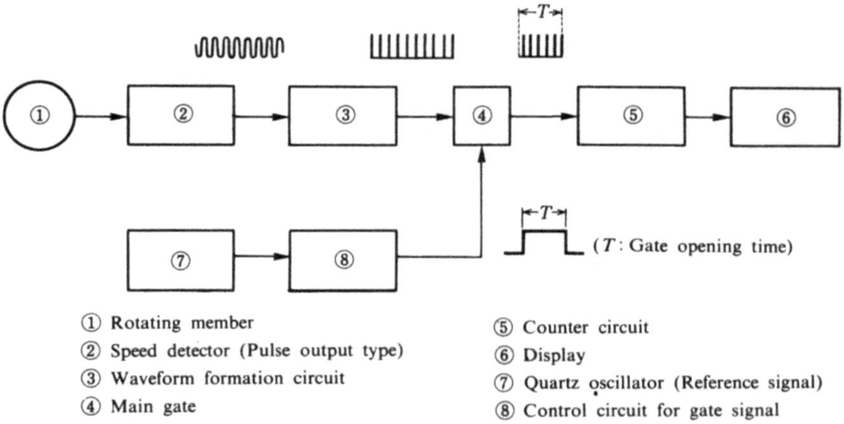

Fig. 3 · 97 Sohematic of pulce counting circuit

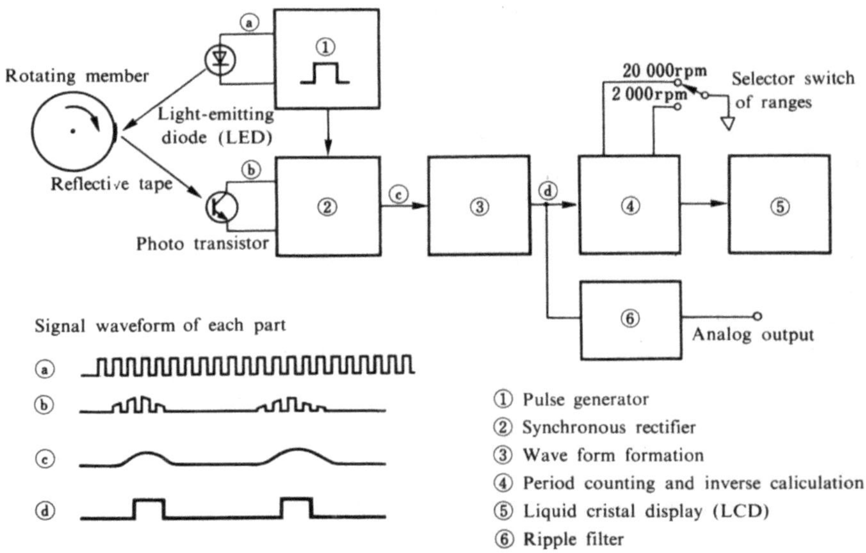

Fig. 3 · 98 Functional schematic of photo electric type tachometer

an example, the functional schematic and signal waveforms of a photoelectric tachometer are shown in Fig. 3.98.

Special reflective tape (which is covered with many small spherical lens such that incident light is reflected back to the source even if the angle of incidence is not perpendicular to the tape surface) is attached to the rotor. An easily focussed visible red light emitting diode (LED) is used as the light source. The reflected light is sensed by a phototransistor, and converted to a pulse signal.

As shown in the signal waveforms in Fig. 3.98, the LED is fired in pulses, and a pulse signal having a good signal-to-noise ratio is obtained. By synchronous rectification at the receiver side the rotation pulse is obtained. Measuring the period of this rotation pulse and taking its inverse yield the number of revolutions. This method has a short response time even for low-speed rotation, and makes high-resolution measurement (0.1 rpm) possible.

3.8 Measurement of Composition

Gases, liquids and solids are all objects of composition measurement in industrial processes. The purpose of composition measurement is to determine the actual constituents of a material. However, at times it is also used as an indirect means to measure the properties of a material when those properties cannot be measured directly. Thus, there are many types of composition to measure, and many methods of measuring them.

For length, mass, temperature, and so on, there are established international standards, and thus there is little problem with traceability. However, in the measurement of composition there are many different objects of measurement and measuring methods, so few standards exist. The current situation is that relative calibration, or sample calibration, is done by comparison with what is considered to be the most reliable reference instrument or standard. Thus composition measurement is actually more like relative measurement than absolute measurement.

This section describes the main instruments and methods used measure the composition of gases and liquids used in industrial processes.

3.8.1 Gas chromatography[47)-49)]

In gas chromatography, the components of a sample are separated and highly sensitive quantitative analysis is carried out for each component. This process can measure virtually any type of gas or any liquid that can be vaporized. This ability to analyze multiple components at one time is a special feature shared by no other analyzer, so the gas chromatograph is used in many industries.

The basic structure of a process gas chromatograph is shown in Fig. 3.99. It consists of an analyzing section and a computing control section, and is used in conjunction with a sampling valve device. The externally-attached sampling equipment produces the most suitable measuring conditions for analysis of the gas or liquid by adjusting the

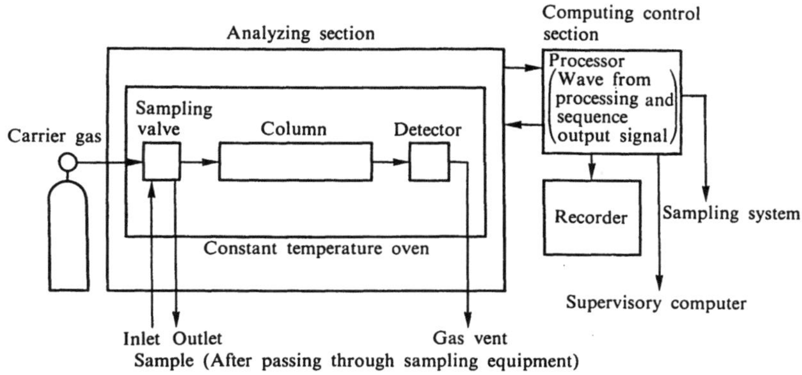

Fig. 3・99 Basic structure of a process gas chromatograph

sample pressure and temperature, removing water vapor, and filtering out solid particles. It then conveys the sample to the analyzer section. The sampling equipment also includes calibration equipment to facilitate calibration with a reference gas or liquid. These functions are essential for stable measurement.

The analysis section separates the components of the sample, and coverts their concentrations to an electrical signal which is output to the computing control section. The computing control section is made up of several components. The waveform processor amplifies and converts the signal output by the analyzing section. The digital input/output circuit and processor controls the electromagnetic valve and the constant temperature oven of the analyzing section, and other such operations. The transmitter outputs the results of the analysis to the supervisory compute or a recording device.

The basic structure of the analyzing section, as shown in Fig. 3.99, comprises the sampling valve, a column, and a detector. These components are contained in a constant temperature oven controlled to within $\pm 0.1°C$. A sample of fixed volume is taken in by the sampling valve, and infected into the column under the pressure of a carrier gas. The component separation process within the column is shown in Fig. 3.100. Although the sample consists of a gaseous mixture of various components, the alternating adsorption and desorption that occurs between the filling in the column (fixed phase) and the carrier gas (moving bed) causes the various constituents to travel through the column at different speeds due to differences in their distribution coefficiences (the ratio of the amount in the fixed phase to the amount in the moving bed). Thus the various components appear at the column outlet at different times, and the concentration signals for each component coming out of the detector over time comprise the chromatogram.

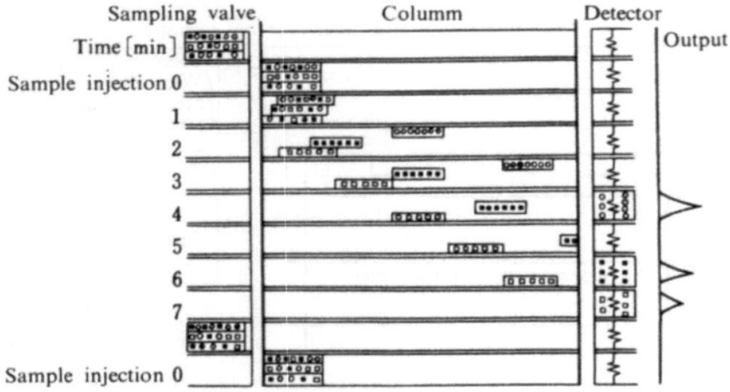

Fig. 3 · 100 Gas separating principle in column

The main structural parts of the analyzing section are described below.

(*a*) *Column and column system*

Complete separation of the measured constituents in the shortest possible time, and long-term stability are desirable features. Thus, selection of the column filler and ensuring the appropriateness of the working conditions, and selection of the column system are important matters. There are three types of column systems: pre-cut (backflush and foreflush types), regrouping, and column by-pass. The backflush type is used for measuring minute quantities among the main constituents, shortening analysis time, and preventing contamination of the column. It is the most basic type of system, and is often used in combination with other types. Figure 3.101 illustrates a backflush column system.

The column by-pass system is used when measuring a mixture of organic and inorganic gases. A separate column is used for separating inorganic gases so that it does not to suffer damage from the organic gases (Fig. 3.102).

(*b*) *Detectors*

Generally, thermal conductivity detectors (TCD) and flame ionization detectors (FID) are used, but electron capture detectors (ECD) are

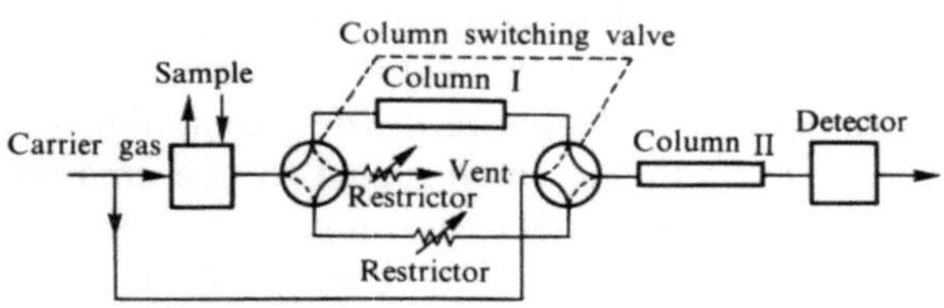

Fig. 3 · 101 Back flush column system

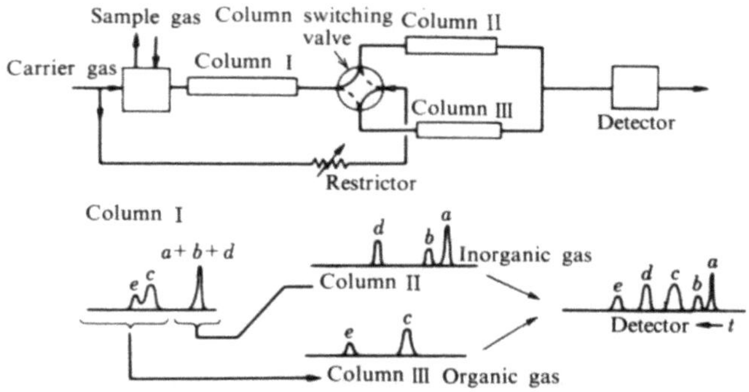

Fig. 3 · 102 An example of gas separation by column bypass system

also used. Figure 3.103 illustrates the principle of a TCD. If a carrier gas which contains the constituents is introduced to the measuring filaments, the difference in thermal conductivity of the two gases will cause a change in the temperature of the measuring filaments. This in turn changes the electrical resistance of the measuring filaments, resulting in a change in output from the Wheatstone bridge that corresponds to the amounts of the constituents of the gas. For the carrier gas, H_2, He, Ar, N_2, and so on are used, and most gases in the measurement range from 0% to 0.1% or more can be analyzed.

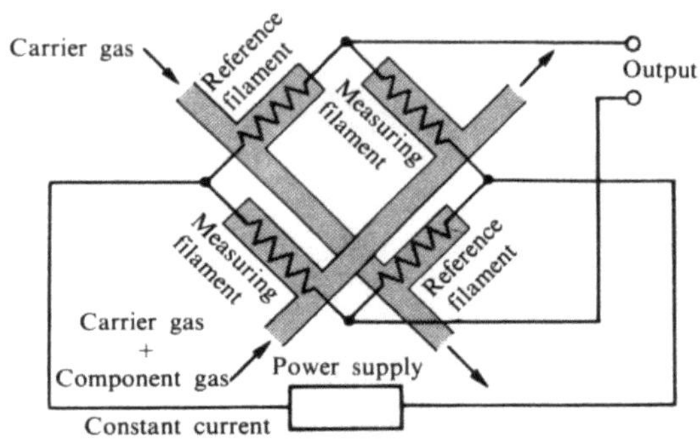

Fig. 3 · 103 Principle of thermal conductivity detector

Figure 3.104 shows the principle of FID. Using FID, organic constituents can be measured over a wide range from extremely small amounts (on the order of parts-per-million) up to 100%. Hydrogen gas, with air added, is burned in a flame. If an organic material is introduced into the flame together with the carrier gas, the carbon in the organic compound will undergo electrolytic dissociation. Further, if an electric potential of several hundred volts is applied to the nozzle and

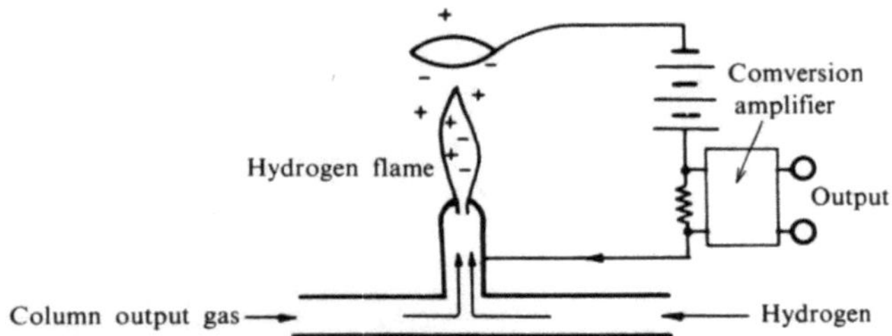

Fig. 3 · 104 Principle of flame ionization detector

to the collector electrode above the flame, a direct current of about 10^{-7} to 10^{-12} amperes corresponding to the amount of organic carbon will flow. By amplifying this current, a signal corresponding to the amount of the constituent organic gases as will be obtained. It follows that this method is not suitable for analysis of inorganic gases. Either N_2 or He are generally used as FID carrier gases.

Recently, gas chromatographs having two detectors built into the analyzing section have become practical. An example of such an application is shown in Fig. 3.105. For the column by-pass system, having two detectors has special merits such as shortening the analysis time and allowing the degree of sensitivity to be selected through choice of the proper detectors.

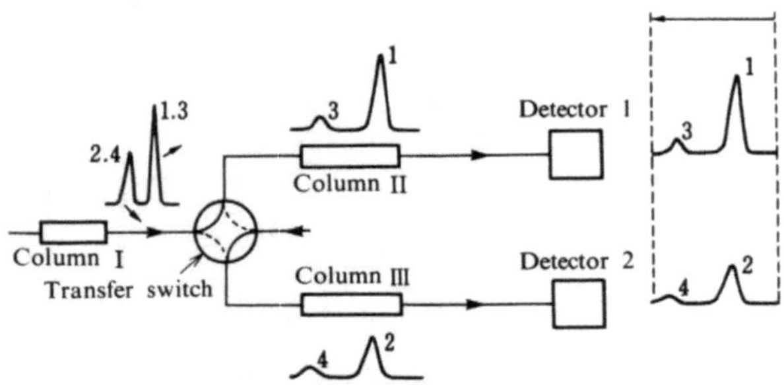

Fig. 3 · 105 Application example of double detectors gas chromatograph

As an example of analysis, Figure 3.106 show a chromatogram for LPG. The fields of application cover a variety of industries including petroleum refining, petrochemistry, organic and inorganic chemistry, the steel industry, metal industry, ceramics and glass-making, the electric power industry, the food industry, and the pharmaceutical industry. It is widely used in industrial processes involving ethylene, polypropylene, blast furnaces, poly-vinyl chloride, BTX, LPG, polyethylene, and so on.

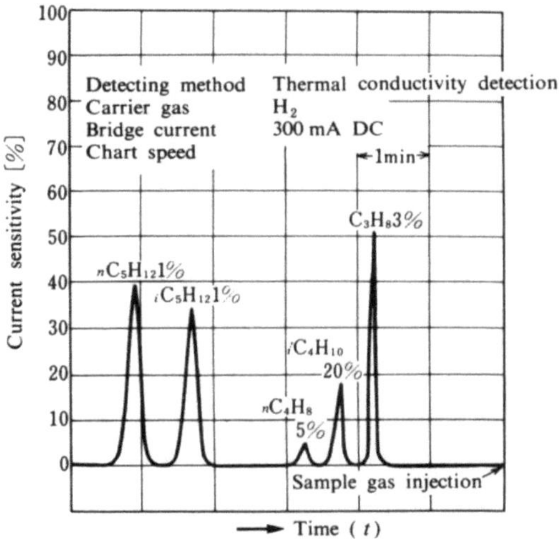

Fig. 3・106 An example of LPG (Simulation gas) chromatogram

3.8.2 *Infrared analyzers*[50)]

There are dispersion and non-dispersion types of infrared analyzers. Dispersion infrared analyzers are used in laboratories as spectrophotometers; non-dispersion analyzers are used for continuous measurement in industrial applications. In non-dispersion analyzers, the wavelength used is fixed in advance, and the concentration of the target constituent is measured continuously from the transmittance of that wavelength.

The absorption spectrum of infrared radiation absorbed by a gas, is unique to the type of gas, and the Lambert-Beer law applies between the degree of infrared absorption and the concentration of the gas. If I_0 is the strength of the infrared radiation before injection into the measuring cell, I is the strength after exiting the cell, l is the cell length, and C is the concentration of the gas, then the following equation holds.

$$I = I_0 \exp(-klC)$$

Here, k is the coefficient of light absorption. The values of k, l, and I_0 are constants determined by the structure of the cell and the type of gas. The concentration C can thus be determined by measuring the value of I.

Figure 3.107 illustrates the principle of semiconductor and microphone-type infrared analyzers. To allow the tiny (about 2 mm²) detectors to make the most use of the light emitted from the light source,

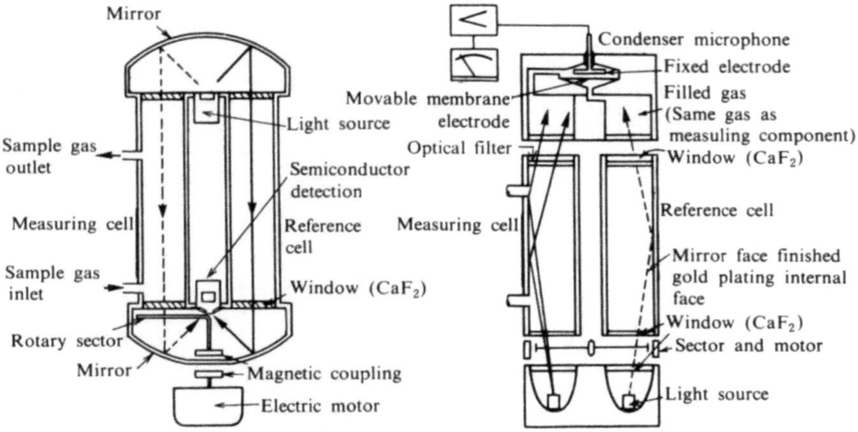

(a) Semiconductor infrared analyzer (b) Condenser microphone infrared analyzer

Fig. 3 · 107 Semiconductor- and microphone-type infrared analyzer

that light (infrared) is split into two parallel beams by the upper mirror. One beam is used for the measurement and the other is used as the reference. The reference light beam passes through the reference cell, which is filled with air or N_2, and is then reflected off the bottom mirror onto a semiconductor detector. The measurement light beam passes through the measurement cell, and is reflected onto a semiconductor detector in the same fashion as the reference light. The component to be measured in the gas sample flowing through the measurement cell absorbs some of the measurement light, thus reducing the strength of the light relative to the strength of the reference light. The two light beams are alternately cut off from the detector by a semi-circular rotary sector. This allows the detector to convert the difference in the strengths of the measurement and reference light beams into an alternating electrical signal that represents the concentration of the gas being measured.

The semiconductor detector output does not depend on the light wavelength, thus it is used with a band-pass filter so as to eliminate interference gas effects. This filter is formed by depositing alternating layers of thin-film dielectric materials of high and low refractivity. The interference effect of these layers permits only a certain range of light wavelengths to pass through the filter, thus allowing selective measurement of the target. Figure 3.108 shows the spectroscopic characteristics of a band-pass filter for use with CO, CO_2, and CH_4.

In the condenser microphone type of analyzer, the interior surfaces of the cells are first given a mirror finish and then gold-plated so as to maximize the amount of light reaching the detector. If the meas-

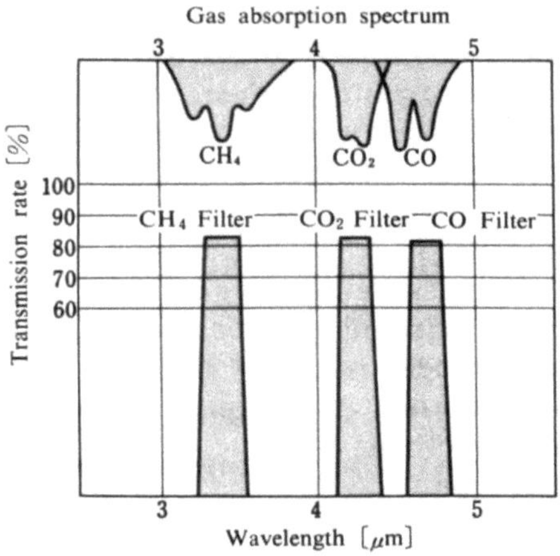

Fig. 3 · 108 Spectroscopic charateristics of a band-pass filter

urement light and the reference light are alternately injected into the condenser microphone detector, infrared radiation is absorbed by the filling gas (the same gas as the component being measured), which then expands due to heating. The movable membrane electrode is displaced by the pressure from this heat expansion. The movable electrode and the fixed electrode form a condenser whose capacitance varies according to the changes in the distance between the electrodes resulting from the displacement of the membrane. This change in capacitance modulates the frequency of a square-wave output signal.

Although any gas that has absorption characteristics in the infrared region can be measured, this method is generally used to measure the gases listed in Table 3.14. The concentration measurement range can be increased by lengthening the cells, and the sensitivity can be increased by increasing the pressure of the filling gas. Infrared gas analyzers are widely used in all industries for such purposes as the analy-

Table 3.14 Application examples of infrared gas analyzer

Measuring gas	Molecular formula	Minimum range [ppm]	Maximum range [%]
Carbon monoxide	CO	0 to 30	0 to 100
Carbon dioxide	CO_2	0 to 20	0 to 100
Sulfur dioxide	SO_2	0 to 100	0 to 100
Nitric oxide	NO, NO_2	0 to 100	0 to 100
Ammonia	NH_3	0 to 1 000	0 to 100
Methane	CH_4	0 to 200	0 to 100
Ethane	C_2H_6	0 to 200	0 to 100

sis of furnace gases, process component analysis, pollution measurement, and atmospheric gas analysis. They are especially valuable for measuring flue gases, which have temperatures in the range of 100°C to 300°C, contain large amounts of dust, and are saturated with water vapor. The analyzer is used in conjunction with sampling equipment that includes an electronic cooler/dehumidifier and triple filters.

3.8.3 Oxygen analyzers[51)]

The methods of measureing oxygen concentration are classified into paramagnetic methods and electrochemical methods.

(1) Paramagnetic methods
- Magnetic wind method
- Magnetic force methods
  - Dumbbell type
  - Pressure detection type

(2) Electrochemical methods
- Zirconia method
- Electrode methods
  - Controlled potential electrolysis
  - Galvanic type
  - Polarograph type

The magnetic methods employ the force of attraction generated when paramagnetic oxygen molecules are magnetized within a magnetic field. The electrochemical methods use oxidation-reduction reactions of oxygen, or oxygen concentration cells. At present, the zirconia method and magnetic wind method are mainly used in process oxygen analyzers. These methods are compared in Table 3.15.

(a) Zirconia oxygen analyzers

As shown in Fig. 3.109, a zirconia cell is constructed of a solid zirconium oxide ceramic electrolyte covered on two sides by gas-permeable porous platinum electrodes. The zirconia is made conductive by high temperature (600°C or higher) oxygen ions (O^{2-}), If gases with different partial pressures of oxygen are introduced to the two sides of the cell, an oxygen concentration cell is formed. The following two reactions involving oxygen and oxygen ions occur. The oxygen reaction takes place at the negative electrode, and the oxygen ion reaction takes place at the positive pole.

Negative pole: $O_2 + 4e \longrightarrow 2O^{2-}$
Positive pole: $2O^{2-} \longrightarrow O_2 + 4e$

An electromotive force expressed by the Nernst equation is generated between the two electrodes. If the temperature of the zirconia element is raised to 750°C, the electromotive force E is given by the fol-

Table 3.15 Comparison between zirconia method and magnetic method

Method / Item	Zirconia method	Magnetic method (Magnetic wind method)
Measurement range	Min. 0 to 1% Max. 0 to 100% (Principally 1 to 100ppm available)	Min. 0 to 1% Max. 0 to 100%
Cell output	Logarithm	Nearly linear
Response time (90%) (Case of gas introduced from caribration gas inlet)	1 to 5 seconds	5 to 30 seconds
Gas temperature	1 to 1 400°C	0 to 50°C
Warm up time	20 minutes	1 to 2 hours
Interference of coexistent gas	Combustible gas of CO, H_2, CH_4	High thermal conductivity gases such as CO_2 etc.
External sampling devices	Not required	required

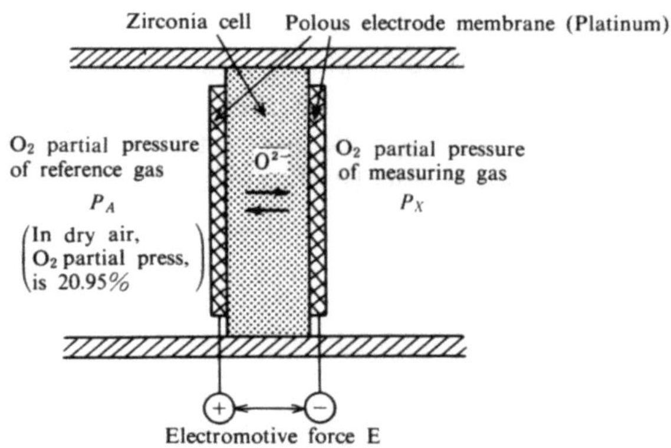

Fig. 3 · 109 Principle of zirconia oxygen analyzer

lowing equation.

$$E=-50.74 \log \frac{P_X}{P_A}=-50.74 \log \frac{P_X}{20.95} \text{ [mV]}$$

Here, P_X is the partial pressure of oxygen in the gas being measured (vol%), and P_A is the partial pressure of oxygen in the reference gas (vol%).

If dry air at 1 atm is used as the reference gas, then $P_A=20.95$ vol%. By measuring the electromotive force E, the partial pressure of

O_2 in the measured gas P_X can be obtained by calculation.

The zirconia oxygen analyzer uses a direct-feed method, so sampling equipment is not required. As a result, the measurement response is fast, and highly accurate, even for low concentrations. Thus these analyzers are often used for control of the air-fuel ratio in combustion furnaces and low combustion management.

(*b*) *Magnetic wind oxygen analyzers*

Figure 3.110 illustrates the principle of magnetic wind oxygen analyzers. The sample gas flows into the glass ring tube through the gas inlet, then flows through the tube, and out the gas outlet. If oxygen is present in the sample gas, the gas will be pulled from the left to the right within the tube by the magnetic field of the magnet poles. If this gas is heated by a heating coil, the magnetic characteristic of the oxygen will be weakened, and the gas will be pushed out to the right. In this way, a gas flow (magnetic wind) that corresponds to the concentration of oxygen in the sample will be generated in the middle part of the tube. The magnetic wind cools the upstream coil a more than it cools coil b. Electrical resistance in the coils decreases with temperature, so an unbalanced voltage is generated in the Wheatstone bridge. This voltage is amplified and converted to serve as an indication of the oxygen concentration.

Although nitrous oxide (NO) and nitrogen dioxide (NO_2) can interfere with the measurement, normally they are present in very small amounts, and because the magnetic susceptibility of oxygen is overwhelmingly large, oxygen is selectively measured.

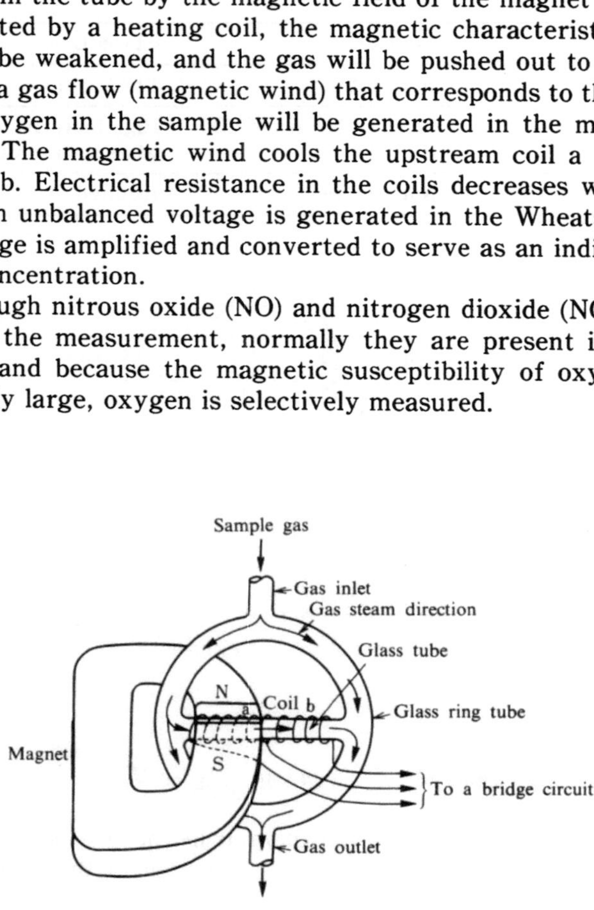

Fig. 3 · 110 Principle of magnetic oxygen analyzer

3.8.4 *pH meters and ORP meters*[44),46)]

(*a*) *pH meters*

The pH is a unit expressing the degree of acidity or alkalinity, and is defined by the following equation.

$$\mathrm{pH} = -\log_{10}\,(\text{hydrogen ion activity})$$

Here, the hydrogen ion activity is the product of the hydrogen ion concentration and the activity coefficient and represents the degree to which ions in a solution participate in chemical reactions. When the ions can move about freely, as in weak solutions, the activity is 1.0, but as the ion concentration increases, the ions restrict each other and the activity decreases from 1.0.

There are various methods of measuring pH, but the main methods are those using chemical indicators, hydrogen electrodes, quinhydrone, antimony electrodes, or glass electrodes. Here, the glass electrode type is described, as it is currently used widely in industry.

The glass electrode method is illustrated in Fig. 3.111. This method employs the electromotive force that is generated on both sides of a glass membrane when the membrane separates two different solutions. This electromotive force is proportional to the difference in the pH of the two solutions. The actual configuration of this type of device is shown in Fig. 3.112. A solution of known pH is put into electrode B, formed from a thin glass membrane. If this electrode is then immersed in solution A, the solution which is being measured, then an electromotive force is generated on the two sides of the membrane. The reference electrode is filled with a KCl solution, and an appropriate internal electrode E_1 is immersed in this. The reference electrode is connected

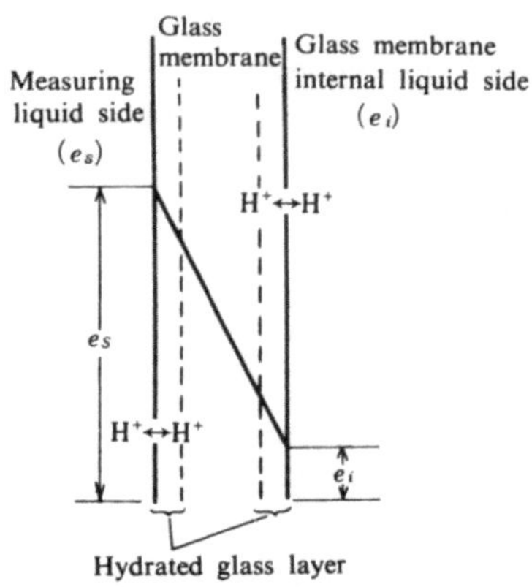

Fig. 3 · 111 Principle of glass membrane

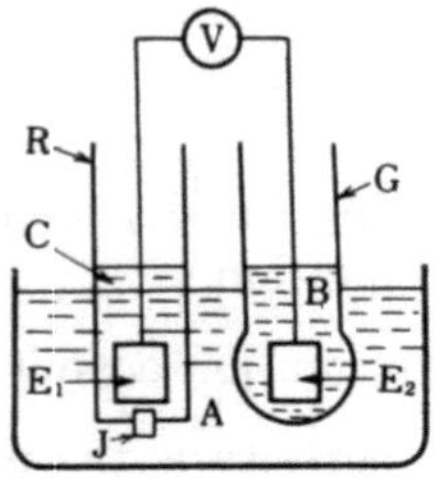

Fig. 3・112 Principle of pH measurement

with solution A through the liquid junction J. The KCl solution inside the reference electrode flows out through this porous ceramic liquid junction and comes into contact with the liquid being measured, and a fixed reference potential is obtained. This method is superior to others in that it offers a wide measurement range, short measurement time, excellent reproducibility, and easy operation. On the other hand, while the strength and high resistance (several tens of MΩ to several hundred MΩ) of the glass membrane are points that require attention, these factors are no problem at all for commercial products in actual use.

From the Nernst equation, the surface potential generated on the glass membrane can be expressed by

$$E_g = \frac{2.303RT}{F}(\mathrm{pH}_i - \mathrm{pH}_s) + E_{AS} = (54.19 + 0.198\,3t) \times (\mathrm{pH}_i - \mathrm{pH}_s) + E_{AS} \tag{3.40}$$

where R is the gas constant, T is the absolute temperature, F is Faraday's constant, t is the temperature(°C), pH_i is the normal pH 7 of the buffer solution filling the glass electrode, pH_s is the pH value outside of the glass electrode, and E_{AS} is the deviation from the theoretical value due to the asymmetry potential.

In actual practice, standard solutions are used to adjust for E_{AS} in the above equation, and for span calibration. The following JIS standard solutions are normally used.

phthalic acid salt (pH 4.01)
neutral phosphoric acid salt (pH 6.86)
boric acid salt (pH 9.18 at 25°C)

High input impedance and low bias current are desirable in the pH meter circuitry, and recently MOS FETs and junction FETs have been used.

Figure 3.113 shows an example of a recent pH meter designed for industrial use. The wide application range (corrosion resistance, tem-

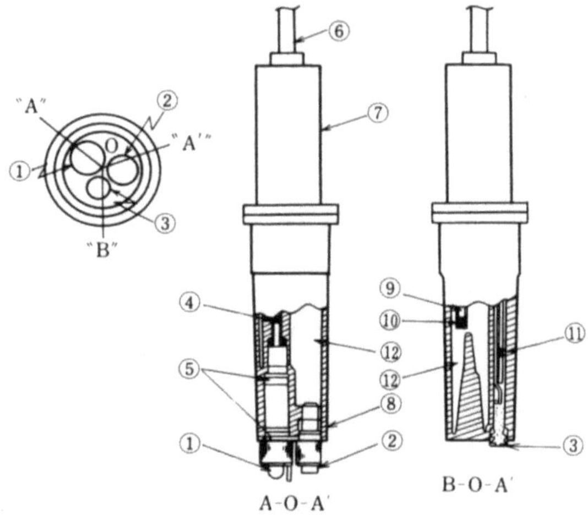

①: Glass electrode
②: Junction
③: Liquid earth
④: Conductive rubber
⑤: O-ring
⑥: Cable
⑦: Detector body
⑧: O-ring
⑨: Reference electrode
⑩: Porous plastic
⑪: Resistance temperature detector
⑫: KCl solution

Fig. 3・113 Internal construction of industrial use pH detector

perature, pressure, etc.), ease of maintenance (cleaning, calibration, etc.), and high reliability are important considerations in the design of instruments for industrial use.

Here we use equation 3.40 to calculate an actual numerical value for the electromotive force. Taking $E_{AS}=0$,

$$E_g=0.198\,4(7-X)T \text{ [mV]}$$

where X is the pH value, and T is the absolute temperature.

(b) Oxidation–reduction potential meters (ORP)

If a reference electrode and an electrode made of platinum, gold or some other chemically inert metal are inserted into a solution containing an oxidation–reduction material, an electric potential will be generated between the two electrodes. This potential is called the oxidation–reduction potential, and is determined by the strength of the acidity or alkalinity of the solution. Devices using this principle to measure the acidity or alkalinity are generally called ORP meters, deriving their name from the oxidation–reduction potential. Their main applications are in the control of cyanides in the treatment of waste water from metal plating processes, control of the amount of sodium metabisulfite added in the treatment of chromic acid waste water, and so on.

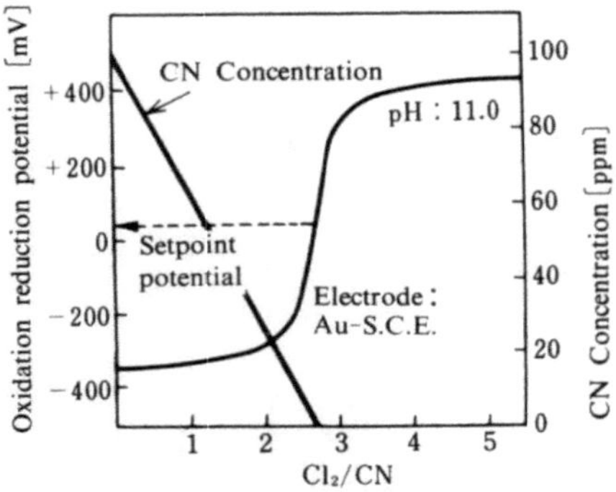

Fig. 3 · 114 Oxidation reduction potential curve and CN concentration

Figure 3.114 shows an example of this. In addition to these applications, ORP is also used in bleach liquor production processes, and the bleaching of textiles, paper, and pulp.

3.8.5 Moisture/humidity meters[51)]

Moisture/humidity meters include those employing the non-dispersive infrared method in the near-infrared region (explained in Sec. 3.9), relative humidity meters which employ microwave absorption characteristics, and for measuring moisture in gases there are the devices which detect the changes in frequency of a quartz oscillator.

Here, the zirconia humidity meter is explained. The zirconia humidity meter is for use in industrial processes to measure the moisture content of high-temperature gases (max. 600°C). The measurement

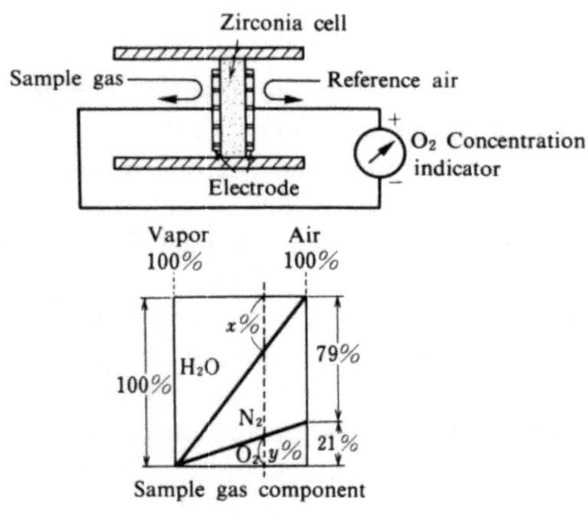

Fig. 3 · 115 Principle of zirconia humidity meter

principle of this devices is shown in Fig. 3.115.

If a solid zirconia electrolytic cell (zirconia ceramic) is operated at high temperature as described in Subsec. 3.8.3, the electromotive force E is generated between the two electrodes. The strength of E is dependent on the partial pressures of oxygen in the sample gas and the reference gas, and is defined by the Nernst equation.

$$E=-\frac{RT}{nF}\log_e\frac{y}{a}\ [\mathrm{V}]$$

Here, R is the gas constant (8.3143V·C/K·mol), T is the absolute temperature (K), $n=4$, F is Faraday's constant (96 487 C/mol), y is the partial pressure of oxygen on the sample gas side of the zirconia cell (vol%), and a is the partial pressure of oxygen on the reference gas side of the zirconia cell (approx. 21 vol%).

If this humidity meter is used only for sample gases that are mixtures of air and steam, and the humidity of this gas mixture is represented as x (vol%), then

$$y=(100-x)\times 0.21.$$

From the above two equations (but with $k=RT/2.303\ nF$), we obtain

$$E=-k\log\frac{y}{a}=-k\log(1-0.01x)\quad [\mathrm{mV}]$$

and the humidity x(vol%) can be obtained directly from E, the electromotive force generated between the electrodes of the zirconia cell.

The structure of the detector (Fig. 3.116) is the same as for the zirconia oxygen detector. This device can measure sample gases in the temperature range from 0°C to 600°C, so it can be inserted directly into the environment of the gas being measured without sampling. Response time is fast (90% of the responses are within 5 seconds), and the H_2O measurement range is from 0% to 100%, so this device has a wide

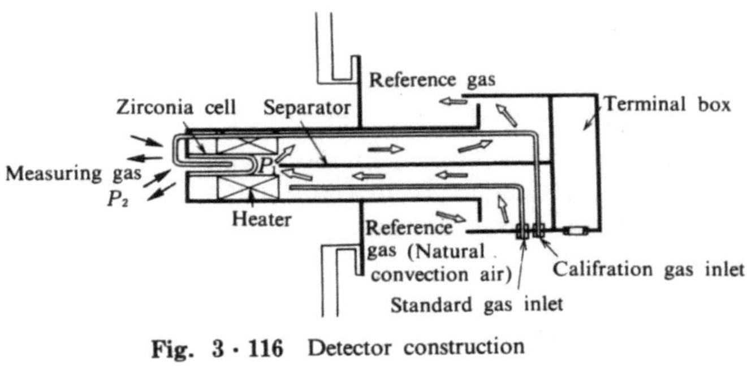

Fig. 3 · 116 Detector construction

range of applications as a humidity meter for industrial processes.

Also, zirconia detectors that measure the humidity of ordinary combustion gases rather than only air–steam mixtures have become practical. In these detectors, the gas being measured is dehumidified by an electronic cooling dehumidifier or some other such equipment, and used as the reference gas. As a result, the application field of zirconia humidity meters is expected to widen.

This device is used in all industries for the control of humidity over a wide range of temperatures in drying and heating processes. Through its use, quality and productivity are increased, and energy is saved.

3.8.6 Turbidity meters[41)]

In the treatment of clean water and waste water, the measurement of turbidity is becoming an important control indicator for manufacturing processes in every kind of manufacturing industry.

For continuous measurement, the methods available include the transmission method, transmission–scattering method, scattering method, and the integration sphere method. The basic principles of these methods are listed in Table 3.16. In the JIS K 0101 standard for turbidity, one liter of water containing 1 mg of refined kaolin is defined as having a turbidity of 1 ppm.

An example of the transmission–scattering method is as follows. The measurement principle of the transmission–scattering method of measuring turbidity is illustrated in Fig. 3.117. A long-life tungsten bulb is used as the light source. The light from this source is converted to parallel light rays by a condensing lens, and then is passed through a glass window into the measuring chamber. A measuring chamber between 5 mm and 300 mm in diameter can be chosen according to the turbidity of the fluid. In the chamber, the light is reflected, absorbed and dispersed by the material suspended in the fluid. The transmitted light and part of the dispersed light pass through the chamber and arrive at the detector. The detector consists of two photocells. The inner photocell (PC_2) detects the transmitted light, and the donut-shaped outer photocell (PC_1) captures the dispersed light. The ratio of the output from these two photocells, which depends on the relative amounts of transmitted and dispersed light coming from the measurement chamber, is determined by the calculation circuit. As a result of these calculations, influences from fluctuation in the light source and the color of the liquid being measured are eliminated. This is explained in equation form below.

If the photoelectric current for PC_1 is I_1 and that for PC_2 is I_2, then the relationship of these values with turbidily within the measure-

Table 3·16 Principle and features of turbidimeter

	Principle	Features
A	Transmission method 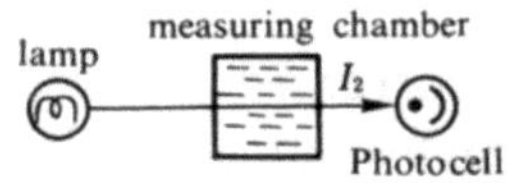	$T \propto (K - \log I_2)$ 0 to 2, 0 to 10 000 ppm Available for high turbidity Window dirts, Liquid color influential
B	Surface scattering method 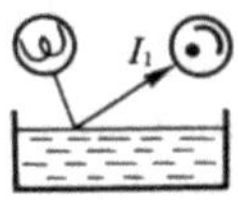	$T \propto I_1$ 0 to 1, 0 to 5 000 ppm Window dirts not influential Liquid color influential
C	Scattering method 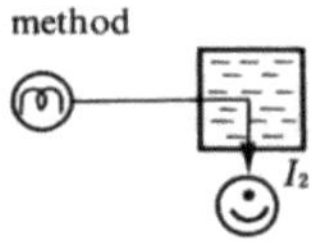	$T \propto I_1$ 0 to 0.2, 0 to 30 ppm Window dirts, Liquid color, influential
D	Scattering and transmission method 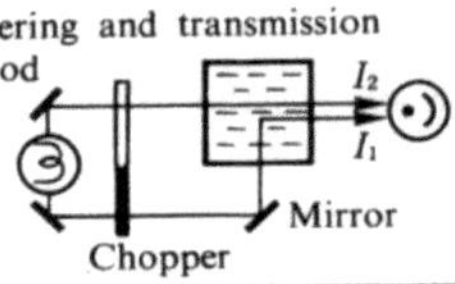	$T \propto I_1/I_2$ 0 to 0.5, 0 to 20 ppm Liquid color, not influential Window dirts, influential
E	Scattering and transmission method 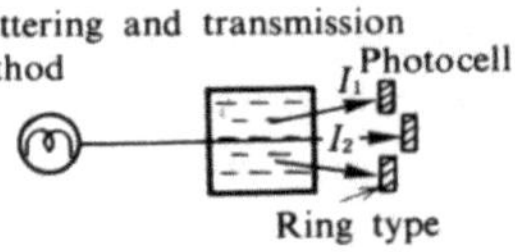	$T \propto I_1/I_2$ 0 to 0.5, 0 to 20 ppm Liquid color, not influential, Window dirts, influential
F	Integration sphere method 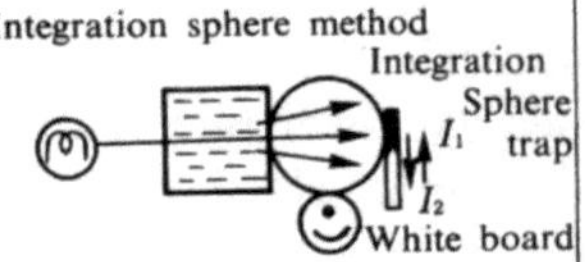	$T \propto I_1/I_2$ 0 to 5, 0 to 1 000 ppm Liquid color, not influential, Window dirts, influential

T: Turbidity, I_1: Scattering beam, I_2: Transmission beam, K: Constant

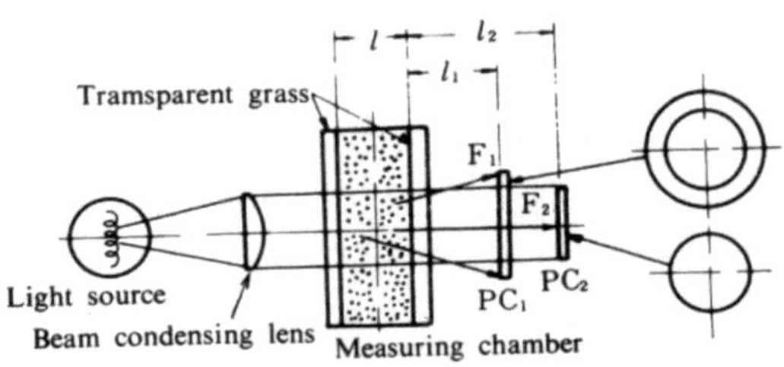

Fig. 3·117 Principle of measurement

ment range is given approximately by the following equations.

$$I_1 = K \exp(-\beta lN)(I_{1(0)} + \alpha lNI_{2(0)})$$
$$I_2 = K \exp(-\beta lN) I_{2(0)}$$

Here, N is the turbidity, $I_{1(0)}$ is the current due to the scattered light when the turbidity and color components are zero, $I_{2(0)}$ is the current due to the transmitted light when the turbidity and color components are zero, l is the effective length of the measuring chamber, α and β are constants determined by the shape of the measuring chamber, the characteristics of the photocells, the properties of the material suspended in the fluid, and other such factors, and K is a constant determined by the characteristics of the photocells, the length of the measuring chamber, and the color component.

Taking the ratio of I_1 and I_2, we obtain

$$\frac{I_1}{I_2} - \frac{I_{1(0)}}{I_{2(0)}} = \alpha lN$$

The left side of this equation is proportional to the turbidity, and is not influenced by color.

3.8.7 Conductivity meters[42]

At a given temperature, electrolytic solutions exhibit a conductivity that is a characteristic of the substance and its concentration. Thus, by measuring the conductivity, the concentration can be determined. Demineralized water, such as boiler water, has an extremely small conductivity (theoretically 0.055μS/cm at 18°C), so by measuring conductivity, the presence of impurities can be determined. In this way conductivity meters can be used as water purity meters. A particular use for these meters is the detection of NaCl, so they are also some-

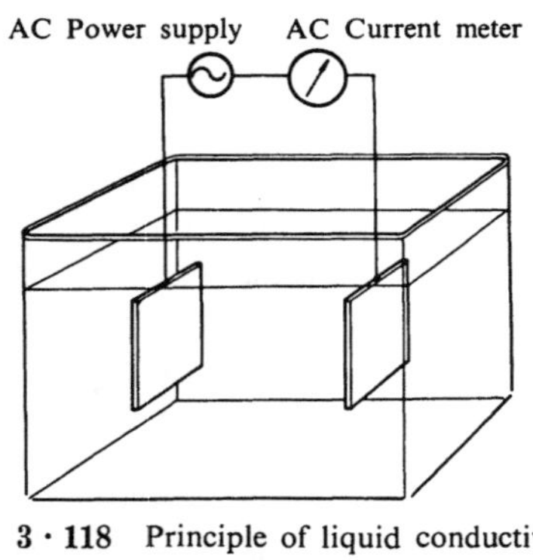

Fig. 3 · 118 Principle of liquid conductivity measurement times called salinometers.

To explain the principle of conductivity measurement, the method generally used to measure conductivity in liquids is illustrated in Fig. 3.118. Two metal plates are inserted into the liquid. These plates serve as electrodes. An alternating voltage of fixed amplitude is applied to the two electrodes, and the current flowing through the liquid is measured.

The resistance between the electrodes R_C, is inversely proportional to the conductivity K, and can be expressed as $R_C = J(1/K)$. In this equation, J is a proportionality constant determined by the shape of the electrodes, generally called the cell constant. It is expressed in units of cm^{-1}. If the impressed electromotive force between the electrodes is E, and because $i = E/R_C$, the two equations yield $i = (E/J)K$. The values of E and J are fixed, so by obtaining a value for i, conductivity can be measured.

In actual practice, however, an electrical double layer forms on the surface of the electrodes, and ions in the solution receive the electrons and electrical charge from the electrodes (electrode reaction). Thus the impedance between the electrodes becomes extremely complicated. Figure 3.119 shows a typical equivalent circuit. Figure 3.120 shows the equivalent circuit in which the impedence between the electrodes is represented as resistances and capacitances in series. To

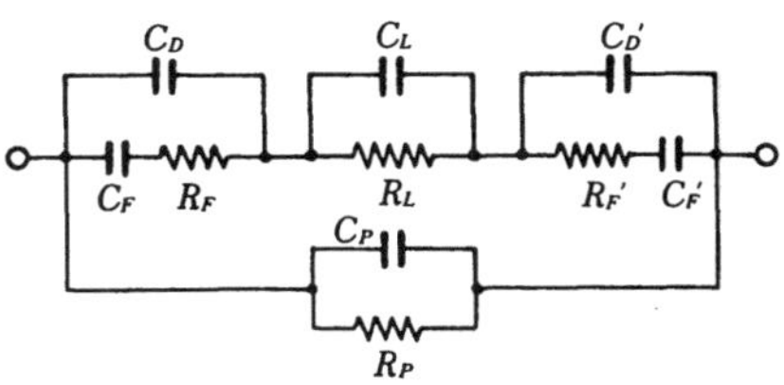

R_L : Liquid impedance between electrodes
C_L : Electro-static capacitance of liquid caused by dielectric constant
C_D : Electro-static capacitance on the surface of electrode caused by electric double layer
C_F, R_F : Electro-static capacitance and resistance on the surface of electrode caused by electrode reaction
C_P, R_P : Parallel electro-static capacitance and resistance composed by lead wires to electrode

Fig. 3 · 119 Equivalent distributed parameters between electrodes

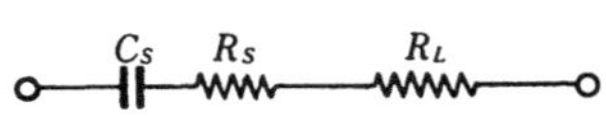

Fig. 3 · 120 Equivalent circuit of impedance between electrodes

measure R_L, it is necessary to reduce the impedance due to electrode reaction and other such factors, represented by C_S, and R_S, to a value sufficiently small compared to R_L. The values of C_S, and R_S change greatly with the material, surface conditions and the structure of the electrodes. Platinum electrodes that are covered with platinum black and set facing each other in a parallel fashion is the optimum configuration. However, because of the strength of platinum black, cleaning, and other problems, in actual practice platinum, graphite, stainless steel and other such materials are used, particularly in process applications. A fixed alternating current voltage suited to the conductivity measurement range is applied between the electrodes. The frequency of this voltage is chosen so as to reduce impedance for the purpose of avoiding polarization from electrode reaction and so on. As for the materials used in the parts other than the electrodes, ethylene tetrafluoride and hard vinyl chloride are used as insulating materials, and stainless steel is often used for the outer tube. As the conductivity of the liquid being measured has a temperature coefficient, some devices also have a temperature sensor built into the electrode, and perform a temperature compensation function. Figure 3.121 shows and example of an actual electrode for process use.

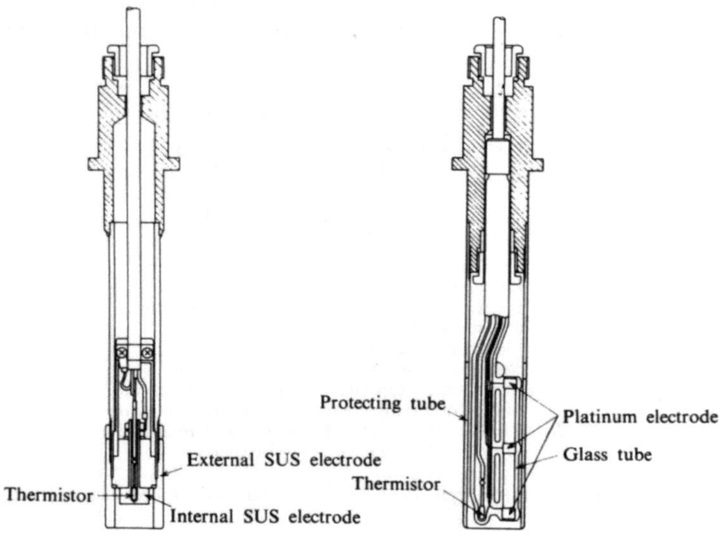

(a) Cell constant 0.05 cm^{-1} (b) Cell constant 5 cm^{-1}

Fig. 3 · 121 Electrode of process conductivity meter

(1) *Conductivity temperature coefficient for dilute solutions* The conductivity corresponding to the temperature of the liquid can generally be

expressed by the following equation, using α and β. Although the values of α and β vary with the type and concentration of the solution, for weak solutions (0.001 mol or less), the concentration factor becomes more or less constant.

$$K_t = K_{ts}\{1+\alpha(t-t_s)+\beta(t-t_s)^2\}$$

Here, t_s is the reference temperature, K_{ts} is the conductivity at the reference temperature, and K_t is the conductivity at temperature t(°C). For example, for NaCl at $t_s=18$°C, $\alpha=226\times10^{-4}$, and $\beta=84\times10^{-6}$.

(2) *Conductivity in concentrated solutions* Many highly concentrated solutions have a maximum conductivity value with respect to concentration as shown in Fig.3.122. Sulfuric acid characteristics in particular are rather complicated, having a maximum in the vicinity of 30%, and a secondary local maximum near 92%. These characteristics are for two-component systems comprising a single solute and pure water as the solvent. Adding even a minute amount of another component may cause great changes in the conductivity.

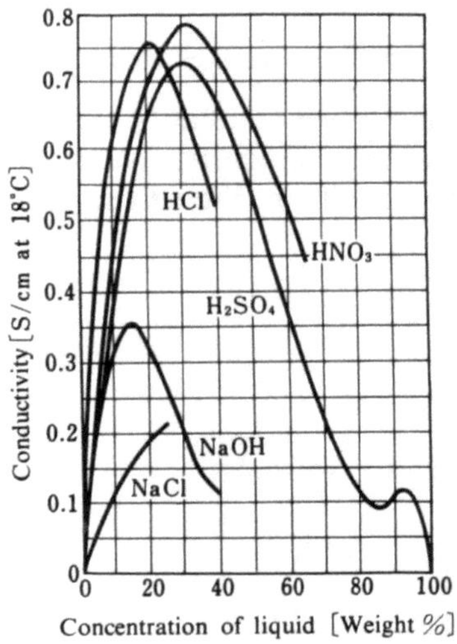

Fig. 3 · 122 Relation between concentration and conductinity of liquid

3.8.8 Other composition measuring devices

(a) Sulfur-in-oil analyzers[52)]

The strength of x-rays passing through a chamber filled with petroleum decreases according to the carbon, hydrogen and sulfur content of the petroleum. The mass absorption coefficient of sulfur is ten times as large as that of carbon and hydrogen, so the strength of the transmitted x-rays varies markedly with the sulfur concentration. Sulfur-in-oil analizers use this phenomenon to measure the sulfur concentration of petroleum.

The transmitted x-ray are detected by an ionization chamber. The current generated in the ionization chamber, I_S, is expressed by the following equation.

$$I_S = I_{SO} \exp\{-\rho t(\mu_H C_H + \mu_C C_C + \mu_S C_S)\}$$

Here, $C_H + C_C + C_S = 1$; I_S is the ionization current (A) when the measuring chamber is empty; ρ is the density of the petroleum (g/cm^3); t is the effective length of the measurement chamber (cm); μ_H, μ_C, and μ_S are the mass absorption coefficients for hydrogen, carbon, and sulfur (cm^2/g); and C_H, C_C, and C_S are the relative weights of hydrogen, carbon, and sulfur.

The values of the mass absorption coefficient of each element varies with the energy of the x-rays, but as can be seen from Fig. 3.123, at 20 keV the coefficients of hydrogen and carbon are equal, and the above equation becomes

$$I_S = I_{SO} \exp[-\rho t\{(\mu_S - \mu_H) C_S + \mu_H\}]$$

In this equation, I_{SO} and t are constants associated with the detector, and μ_S and μ_H are physical constants. Thus, by measuring the ionization current I_S and the density ρ, the sulfur concentration C_S can be obtained.

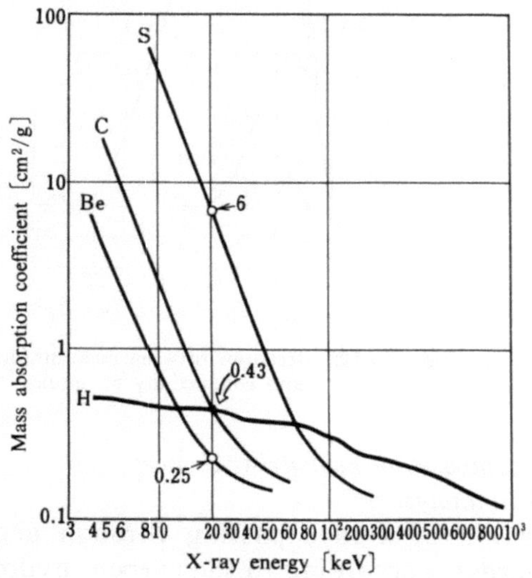

Fig. 3 · 123 Energy-dependence of mass absorption coefficient

An example of the structure of a sulfur-in-oil analyzer for process use is shown in Fig. 3.124. This device includes a transmitted radiation sulfur detector, a density detector, and sulfur converter. It performes the calculations described above, and outputs the value of the sulfur concentration C_S. Refer to Subsec. 3.8.8 (d) for information regarding the density detector.

An example of a sulfur-in-oil detector is shown in Fig. 3.125. Gamma radiation from a radioactive source (^{241}Am) strikes a silver target, producing 20 keV x-rays. These x-rays pass through the liquid being measured and enter the ionization chamber. Teflon® is used for

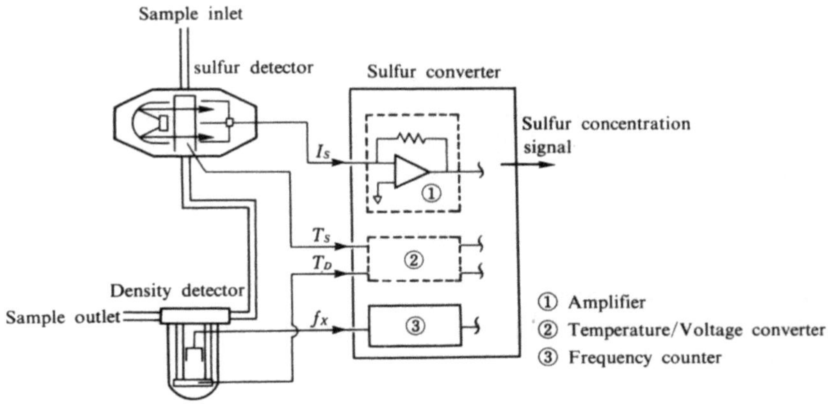

Fig. 3・124 Sulfur-in-oil analyzer system configuration

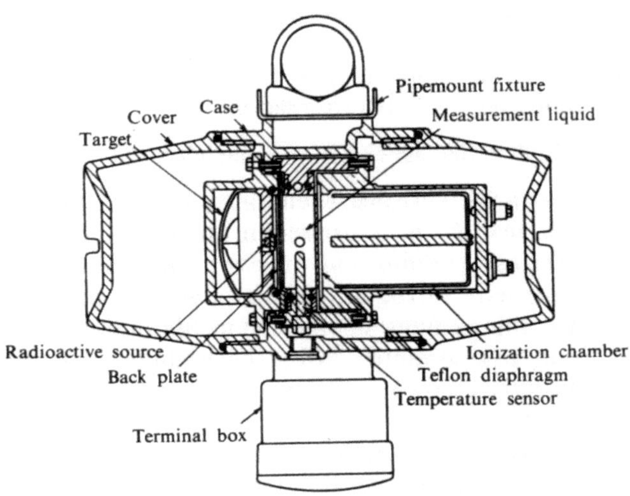

Fig. 3・125 Construction of sulfur detector

the measurement chamber because it is not easily stained. A material with a low mass absorption coefficient, such as beryllium, is used for the ionization chamber window.

The transmitted radiation type sulfur-in-oil analyzer has the following special features.

(1) Relatively unaffected by the CH ratio (± 0.005 wt% sulfur/± 2C/H)
(2) High pressure tolerance (1 to 20 kg/cm^2)
(3) Fast-response
(4) Long-term stability

In addition, this method allows a stronger measurement chamber construction than the analyzer which use low-energy fluorescent x-ray excitation.

The applications of the transmitted radiation sulfur-in-oil analyzer include crude oil and fuel oil inspection, monitoring and control of fuel oil blenders, control of the operation of sulfur removal equipment, and monitoring of sources of pollution. It is mainly used in the measurement range of 0 to 5 wt%. This analyzer uses a radioactive source, so there are legal restrictions concerning its use, and safely precautions are necessary.

(*b*) *Residual chlorine detectors*[43)]

In process water testing, residual chlorine is defined by the following equation.

$$\text{Residual chlorine (ppm)} \\ = \text{free chlorine (ppm)} + \text{combined chlorine (ppm)}$$

Generally, residual chlorine is called total chlorine to distinguish it from free chlorine.

(1) *Free chlorine* If chlorine gas is absorbed into water, Cl_2 will maintain the following equilibrium reaction.

$$Cl_2 + H_2O \rightleftharpoons HClO + HCl \qquad HClO \rightleftharpoons H^+ + ClO^-$$

This equilibrium varies with pH. Generally, free chlorine exists in water as Cl_2, HCLO, or ClO^- as expressed in the above reaction.

(2) *Combined chlorine* Pollution of rivers and so on results in the presence of ammoniate nitrogen, such as ammonia, in the water. This produces various chloramines such as monochloramine (NH_2Cl), dichloramine ($NHCl_2$), or trichloramine (NCl_3), depending on the pH of the liquid. The antibacterial power of these compounds is several tens of times lower than that of free chlorine.

(3) *Methods of measuring residual chlorine* The various residual chlorine measurement methods can be broadly classified into colorimetry methods, and polarograph methods. Here, we describe the colorimetry

and polarograph methods.

(i) *Colorimetry* (the standard used in Japanese water purification facilities): This method, which is called the o-tolidine method, is based on a reaction which exhibits a thin yellow to yellow-brown color corresponding to the free chlorine content when an o-tolidine solution is added to the water being tested. It is used together with sodium arsenite. Accuracies of 0.5 ± 0.05 ppm at 0.5 ppm and ± 0.1 ppm at 0.5 to 1.0 ppm have been recorded. However, repeatability is poor and there can be large differences among the results produced by different operators. Furthermore, the results are sensitive to iron ions, high-valence manganese, and nitrous acid ions.

(ii) *Polarography*: This method is used for almost all continuous industrial measurement. As shown in Fig. 3.126, two electrodes are placed in an electrolytic solution. One of the electrodes has very small surface area, and the other has a large surface area (about ten times as large as the small one). If a variable direct electrical potential is applied between these two electrodes, as the voltage changes a graph like that shown in Fig. 3.127 is obtained. (This graph is an example of an I_2 plateau curve.) This kind of graph is called a polarogram, and the equipment is called a polarograph. If electrolysis begins, due to the direct current voltage applied to the electrodes, concentration polarization occurs because the supply of ions from diffusion is disrupted near the smaller electrode. As a result, a saturation current that is determined by the ion concentration and diffusion rate of the liquid flows between the electrodes. This current is represented by the flat part of the graph, and is called a plateau.

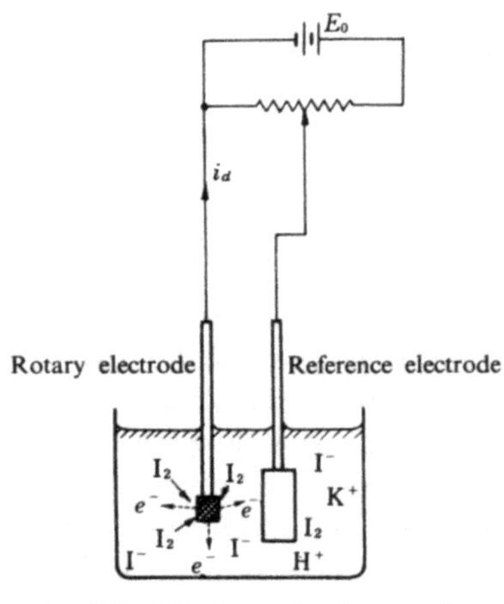

Fig. 3・126 Principle of polarograph (Example of polarogram of I_2)

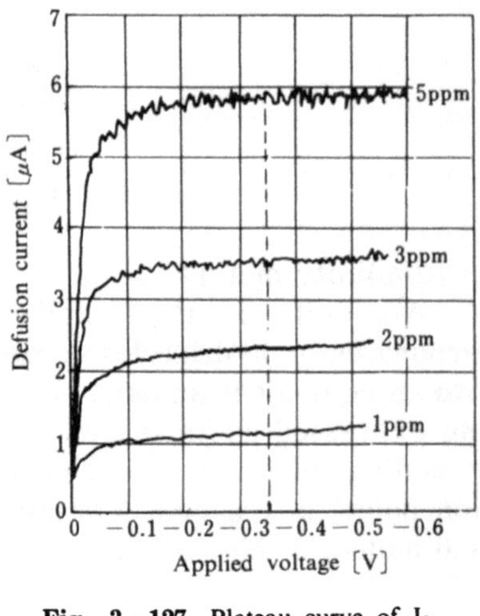

Fig. 3・127 Plateau curve of I_2

The platinum electrode method is one variation of the polarograph method. Below, a residual chlorine analyzer that uses a rotating platinum electrode is described simply.

If potassium iodide is added to the liquid to be measured, iodine atoms are liberated by reaction with the chlorine in the liquid.

$$Cl_2 + 2I^- \longrightarrow I_2 + 2Cl^-$$

By measuring the current that flows when these iodine atoms become ions again as a result of electrolytic reduction, the total chlorine can be measured. In this case, platinum electrodes are used for the working electrode and the counter electrode. In addition to the potassium iodide, the reagent includes a buffering solution composed of a glacial acetic acid and glacial sodium acetate to stabilize the pH value.

Figure 3.128 shows the relationship between chlorine concentration and diffusion current. Although the process is theoretically the same as for measuring free chlorine only, using potassium bromide allows selection of an appropriate pH value for the reaction liquid.

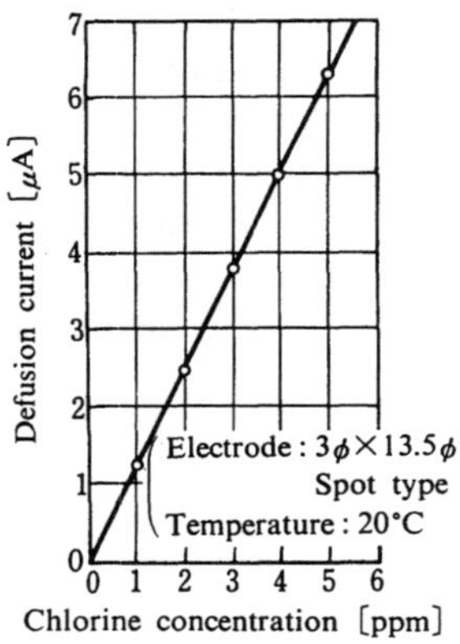

Fig. 3 · 128 Calibration curve

(*c*) Alkalinity meters

Alkalinity is the alkaline portion consisting of bicarbonates, carbonates or hydroxides contained in water converted by calculation into an equivalent amount of calcium carbonate in milligrams per liter, and expressed as parts per million (ppm). This measurement is made by adding sulfuric acid of a certain concentration until a fixed pH (titration point) is reached. The alkalinity is obtained from the amount of acid added. Table 3.17 lists typical alkalinity conversion factors. From this table we can see that the alkalinity of a liquid containing 2 ppm of OH^- and 3 ppm of $CO_3{}^{2-}$ is $2.94 \times 2 + 1.67 \times 3 = 10.89$ ppm.

Table 3.17 Conversion factor to calium carbonate of components

Components / Item	Calium carbonate $CaCO_3$	Hydroxide ion OH^-	Carbonate ion CO_3^{2-}	Bi-carbonate ion HCO_3^-
Equivalent	50	17	30	61
Conversion factor	1	2.94	1.67	0.82

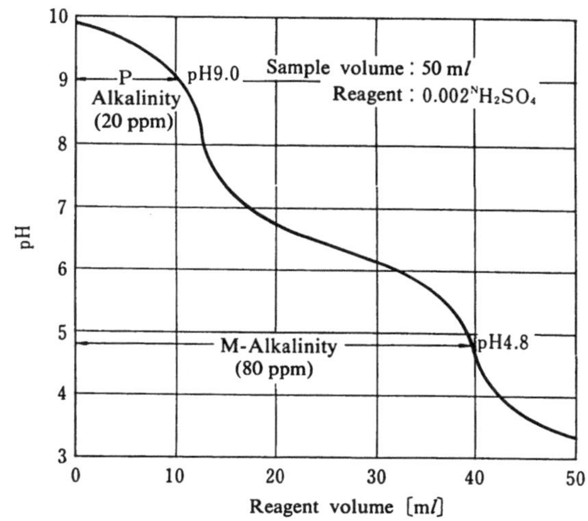

Fig. 3 · 129 Titration curve

(1) *P alkarinity and M alkalinity* While P alkalinity is represented by the amount of acid required to neutralize all of the hydroxides and one half of the carbonates, M alkalinity is the amount of acid required to neutralize all of the alkaline in the water. The titration curve for this is shown in Fig. 3.129. The titration point pH values are different, as shown in the figure. However, as the pH values for ordinary measurements of river water alkalinity are between pH 6 and pH 8, the alkalinity meters used in clean water treatment facilities are M alkalinity meters only, with no P alkalinity meters exsiting.

(2) *Alkalinity and pH* When hydroxides and carbonates are dissolved in water, hydroxide ions expressing alkalinity are generated, and the pH value becomes higher. However, when bicarbonates are dissolved in water, almost no hydroxide ions are formed, and the pH value does not rise. Thus, depending on the composition of the alkali, there are times when there is almost no relationship between pH and alkalinity.

(3) *Measurement principle of alkalinity meters* Below, an example of an actual intermittent titrator using the neutral titration method is

given.

A certain amount of liquid to be measured is drawn, and pH electrodes for detecting the end point of the reaction are inserted. The liquid is titrated until the end of reaction by adding an accurate concentration of titrating reagent (sulfuric acid) to the liquid. At this point, the chemical reaction is expressed as

$$CaCO_3 + H_2SO_4 \longrightarrow CaSO_4 + H_2CO_3$$

The following relation holds between the amount of reagent consumed and the alkalinity of the liquid being measured.

$$CaCO_3 = C_R f \frac{W_R}{W_S} \times 1\,000 \times 50 \text{ [ppm]}$$

Here, C_R is the concentoration (N) of the titration reagent, W_R is the amount of the titration reagent consumed, f is a coefficient based on the concentration of the titration reagent, and W_S is the amount of the liquid being measured (g).

This relationship is not disturbed by other components that may be in the water because the titration reagent reacts only with the alkaline components.

(*d*) *Liquid density meters*[45)]

Density is a basic variable in the determination of the nature or composition of a substance. The standard units for measuring density are kg/m^3 or g/cm^3. The methods of measuring the density of liquids include the JIS hydrometer method, pycnometer method, and displacement method. For industrial use, there is the float-type density meters, which use the change in flotational force on a float; the weight-type density meters, which use the change in weight of a liquid flowing through a U shape-tube of fixed volume; vibration-type density meters, which use change in the vibrational frequency of a pipe; and radiation density meters, which use change in the transmittance of radiation. Of theses, we will explain the widely used vibration density meter and radiation density meter.

(1) *Pipe vibration density meters* If f_x is the frequency of free vibration in the horizontal direction of a pipe of uniform cross section filled with liquid, the following equation is given.

$$f_x = \frac{K_1}{\sqrt{1 + K_2 \rho_x}}$$

Here, ρ_x is the density of the liquid being measured, K_1 and K_2 are constants determined by the inside and outside diameters of the pipe, the pipe material, the modulus of longitudinal elasticity, and the length of the vibrating part of the pipe.

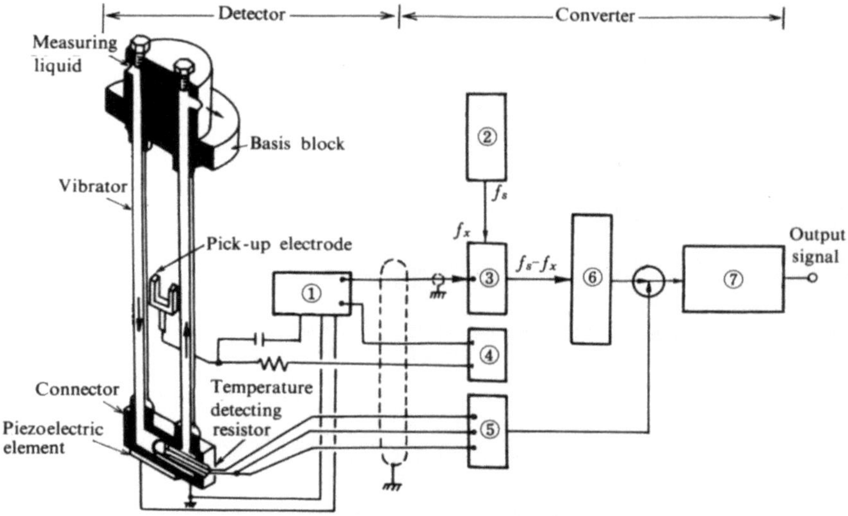

① Oscillation amplifier
② Standard oscillator
③ Mixing detection
④ Power supply
⑤ Temperature compensation
⑥ Frequency/Voltage converter
⑦ Voltage/Current conversion and linearizing circuit

Fig. 3·130 Principle of pipe vibration type density meter

From this equation, the density of the liquid ρ_x can be determined. An example of this principle applied to a practical density meter is shown in Fig. 3.130. Two pipes are used. These are symmetrical in the direction of vibration, so forces applied to the upper and lower connectors negate each other so no energy escapes to the outside (the main pipe system), and Q becomes large. The pipe walls are made 0.15 mm thick, increasing sensitivity. The vibrator becomes a kind of tuning fork. Combined with a positive feedback self-exciting oscillator circuit as oscillating element, this vibrator vibrates continuously

(2) *Ring mode density meters* Density meters which employ the circular vibration mode of pipes allow the detector to be small, and are applied in research. For these detectors, the natural frequency of vibration is given by the following equation.

$$f_x = f_0 \Big/ \sqrt{1 + \frac{\rho_x}{\rho_0} \cdot \frac{R}{t} \cdot \frac{1}{2}}$$

Here, f_0 is the natural vibration frequency in a vacuum of a round tube of radius R whose walls have thickness t, and which is made from a material of density ρ_0. The oscillation circuit for this device is shown in Fig. 3.131.

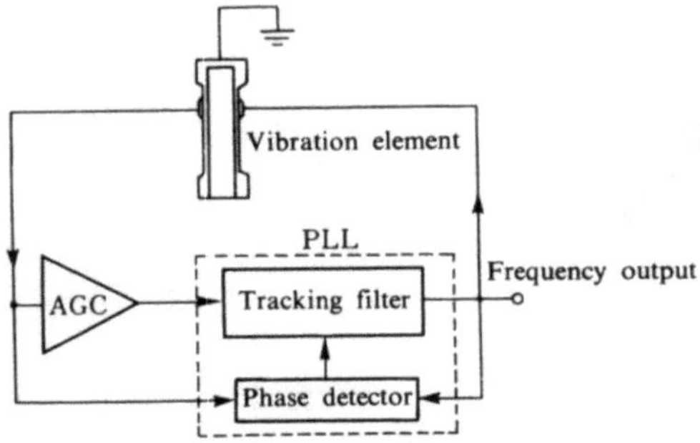

Fig. 3 · 131 Principle of self-oscillation circuit of ring vibration type density meter

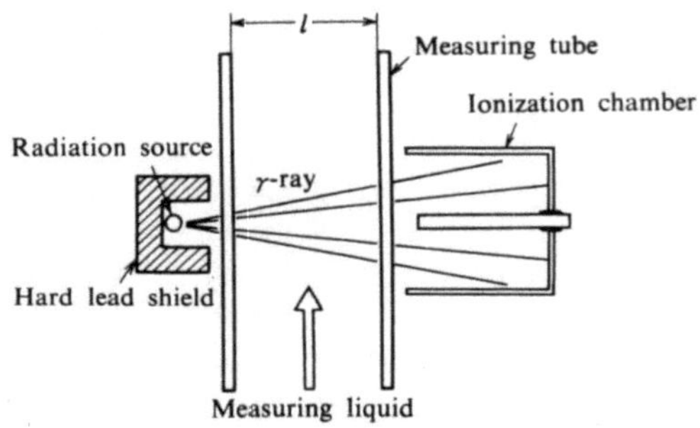

Fig. 3 · 132 Principle of γ -ray density meter

(3) *Radiation density meters* For the continuous measurement of slurries and corrosive liquids, mainly gamma-ray density meters are used. These meters are based on the principle that when gamma-rays from a radioactive isotope (cesium 137 or cobalt 60) passing through a material of fixed thickness, the radiation will be reduced as a result of absorption, diffusion, and so on. This principle is illustrated in Fig. 3.132.

3.9 B/M Systems[49),51),53)-55)]

In the paper industry, a ,measurement control system called the B/M system is used in papermaking and paper coating processes. This system is comprised of a process computer system and sensors for measuring the basic characteristics of paper, such as basis weight (weight per unit area), water content, ash content and thickness. An example of the configuration of a B/M system is shown in Fig. 3.133, but see Sec. 10.5 for more details. The intelligent frame (Fig. 3.134), which carries five kinds of sensors, including a beta-radiation basis weight meter, an infrared moisture content meter, an ash content me-

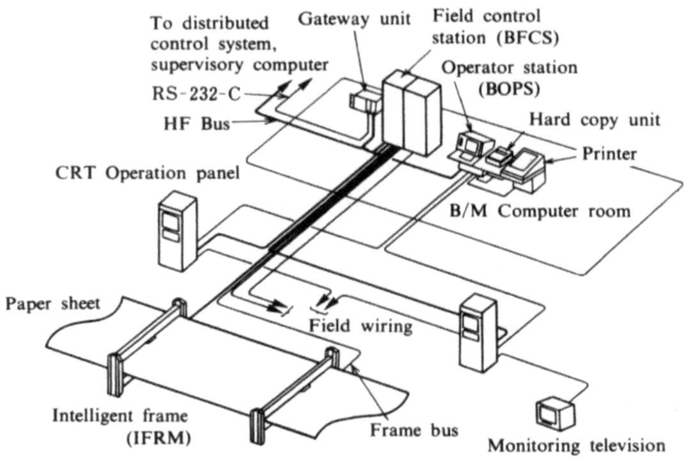

Fig. 3 · 133 System configuration

Fig. 3 · 134 External view of intelligent frame

ter, and a thickness meter, collects data on paper quality by shuttling the sensors back and forth across the width of the paper.

The B/M system must have multiple functions in order to measure the various paper parameters, and it must have high-performance sensors which do not affect the raw materials or papermaking conditions. We will describe the operating principles of the detectors now in practical use. Table 3.18 lists the measurement principles and measurement ranges of each type of sensor.

3.9.1 Basis weight sensor (B sensor)

The basis weight sensor employs a detector which detects the

Table 3.18 Measuring range and detecting principle of B/M meter

Name of detector	Measuring principle	Measuring range
Basis weight meter	β-ray transmission absorption method	Source of ^{85}Kr: Basis weight 10 to 1200 g/m^2 Source of ^{145}Pm: Basis weight 5 to 100 g/m^2
Caliper	Eddy current displacement meter (Floating target position detection)	Paper thickness 0 to 400 μm
Moisture meter	Near infrared method Microwave method	Moisture 0 to 12% (In case of basis weight 10 to 360 g/m^2) Moisture 4 to 12% (In case of basis weight 150 to 1 000 g/m^2)
Ash meter	X-ray transmission method (Ti target)	Ash 1 to 35% (In case of basis weight 30 to 200 g/m^2)
Color sensor	Spectrometer	Wavelength of 385 to 740nm measurable with 5nm resolution

transmittance or absorption of beta radiation. The mass absorption coefficient of beta radiation, μ, does not change with the atomic number of the material being measured. Thus the absorption coefficient remains more or less constant for any kind of material, so this sensor is universally suitable for measuring the area weight of whole sheets. Also, the type of radiation source can be selected according to the

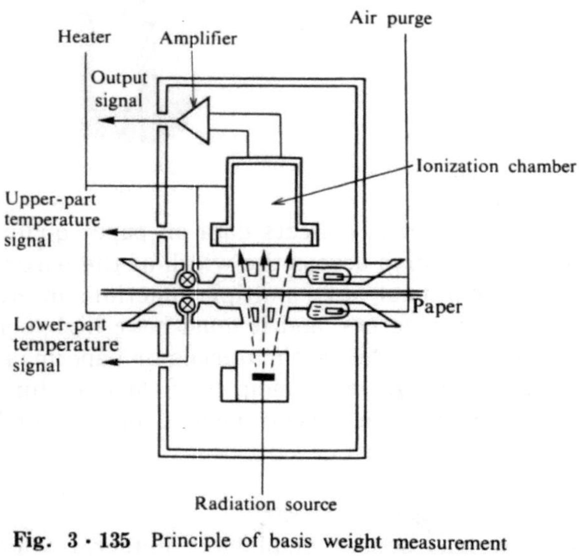

Fig. 3 · 135 Principle of basis weight measurement

basis weight of the substance being measured.

Figure 3.135 illustrates the measurement principle. When the object of measurement is placed between the radiation source and the sensor, the amount of radiation transmitted I is expressed by the following equation.

$$I = I_0 \exp(-\mu\rho t)$$

Here, t is the thickness of the material being measured (m), ρ is the density (g/m^3), μ is the mass absorption coefficient (m^2/g), and I_0 is the amount of radiation transmitted when there is no material to be measured present in the system. Accordingly, if μ is known, then by measuring the amount radiation passing through the material being measured, the basis weight $t\rho$ (g/m^2) can be obtained.

In actual practice, the signal from the basis weight meter is converted to the basis weight from a calibration curve which approximates the absorption characteristics of the individual sensor.

3.9.2 Moisture sensors (M sensors)

The combined transmission/dispersion moisture sensor (N/T beam moisture sensor) is a high-precision device whose output is almost unaffected by the basis weight or the color of the pulp. This is because it applies a calculation formula derived from theoretical analysis to the basis weight signal and four signals obtained from two light beams of different optical path length which pass through the paper. The resulting precision is higher than either the transmittance moisture sensor or the dispersion moisture sensor.

Figure 3.136 shows an example of the optical element construction. For the N optical path of the multiple dispersion optical element, light from the source enters the light detector head after repeated reflections and being dispersed between the paper and the upper and lower disperison surfaces. On the other hand, for the T optical path of the directly transmitted optical element, part of the light from the source is reflected by a mirror, passes through the same filters, is projected on the paper, and then enters the detector head. The filter wheel carries two monochromatic filters. One is the 1.94 μm M filter, the wavelength absorbed by water molecules; the other is the 1.80 μm R filter, a wavelength not absorbed by water molecules. As the filter wheel rotates, a reference light (R_N) and measuring light (M_N) which follow the N optical path, and a reference light (R_T) and measuring light (M_T) which follow the T optical path are emitted successively. The optical axis of the light detecting element is coaxial with the T optical path. Each of the lights injected into the optical path are collected at the PbS, converted into their four respective signals, and output. These

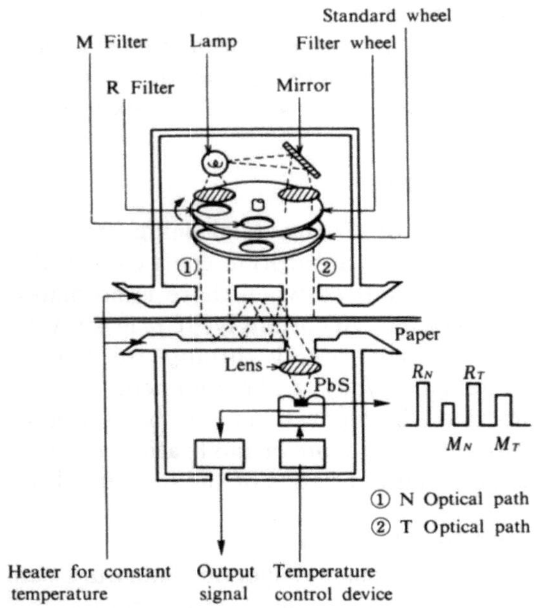

Fig. 3 · 136 Principle of moisture meter

four signals are processed numerically and displayed as a percentage moisture content.

3.9.3 Calipers (paper thickness gauges)

Calipers are a direct measurement sensor. They can measure the thickness of paper in micrometer units by a one-side, non-contact method which applies the air-micrometer principle. The principle is illustrated in Fig. 3.137.

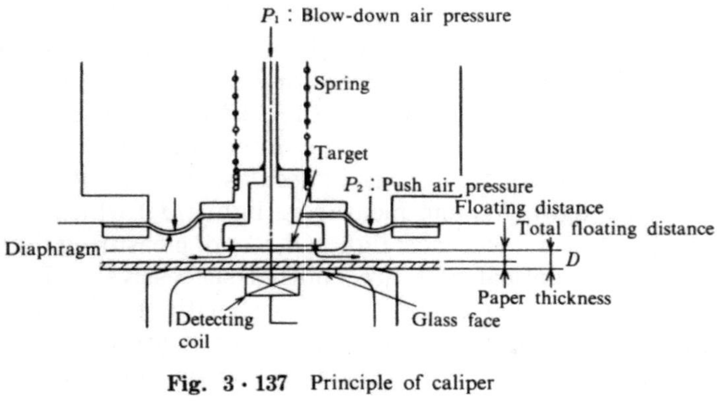

Fig. 3 · 137 Principle of caliper

Above the upper paper surface, a round metal target floats at a certain distance as a result of air pressure. On the lower paper surface, there is an eddy–current displacement meter, which measures the distance D to the target. The paper thickness is obtained by subtracting the known fixed floating distance d from the measured value.

There are several holes on the bottom of the target, through which air is forced out. When there is no paper present, these jets of air hit the glass surface and create an air curtain. The target is supported by a thin, weak diaphragm. The target is pulled up by a spring, and at the same time pushed down by a separate air pressure P_2, and maintains a fixed floating distance above the glass surface determined by the balance of these two forces. If paper is inserted between the glass surface and the target, the surface of the paper takes the role of the glass surface and the target adjusts its position to the fixed floating distance above the paper surface.

3.9.4 Moisture sensor for thick paper

These devices use the fact that water absorbs microwaves, and so are suitable for measurement of moisture in black paper and paper of large basis weight, for which near–infrared moisture sensors cannot be used.

Microwave absorption depends on moisture content MW, and when absorption by the paper material is negligible, the electrical transmittivity T_P is given approximately by the following equation.

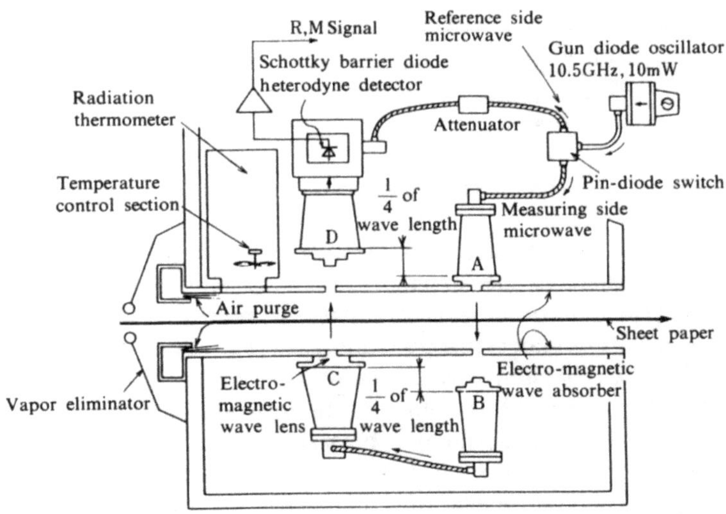

Fig. 3 · 138 Configuration of moisture detector

$$T_P = T_{PO} \exp(-\mu \times \mathrm{MW})$$

Here, T_{PO} is the electrical transmittivity when the moisture content MW is zero, and μ is the absorption coefficient.

The structure of the detector is shown in Fig. 3.138. Microwaves (10.5 GHz) generated by the gun diode are time-divided into a measurement signal (M) and a reference signal (S). Signal M is emitted from horn A. After passing through the paper two times, it enters horn D and is detected. Signal R is fed directly to the detector. The moisture content is obtained after compensating the two signals for temperature according to a radiation thermometer, because the microwave absorption coefficient varies with temperature.

3.9.5 Color sensors

Color, opacity (which is the degree of non-transparency), and brightness (which indicates color), are measured concurrently and without contacting the paper. The principle and structure of these devices is shown in Fig. 3.139.

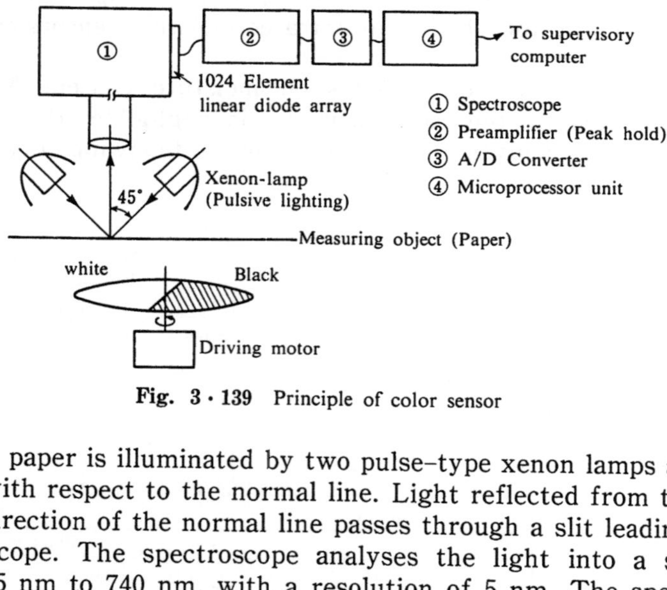

Fig. 3 · 139 Principle of color sensor

The paper is illuminated by two pulse-type xenon lamps at angles of 45° with respect to the normal line. Light reflected from the paper in the direction of the normal line passes through a slit leading to the spectroscope. The spectroscope analyses the light into a spectrum from 385 nm to 740 nm, with a resolution of 5 nm. The spectrum is then projected onto a 1 024 element linear diode array. Below the object being measured (the paper), white and black plates are alternately set. While the white plate is set, color is measured; while the black plate is set, opacity is measured. The results of the measurements are displayed as a wavelength spectrum. The units of color, brightness, hue, saturation, hunter white (representing whiteness), and opacity are also displayed.

3.9.6 Ash sensors

Using the same principle as the basis weight meter, the construction of the ash meter is about the same except for the use of the titanium target x-ray tube. The ash content of the paper (mainly oxides of Mg, Al, Ti, etc.) have x-ray absorption characteristics four to five times stronger than the pulp. Ash content is measured by measuring the absorption and transmission of 4.5 keV x-rays. Also, using fluorescent x-rays, the calcium in the $CaSO_4$ contained in intermediate paper can be selectively and quantitatively analyzed.

To enable the detectors employed in the above B/M system to be used under any environmental conditions of high temperature and humidity and paper dust flying about, internal pressure air purging is used to protect the instruments against humidity and dust.

3.10 Signal Converters[56),57)]

3.10.1 The purpose of signal converters

In process control systems, a variety of detectors are used to measure physical quantities, and these devices generate various signals. These signals must be converted into a form that allows easy connection with other elements, such as recorders, controllers, and computers. Various types of converters for this purpose are described below.

(*a*) *Signal level converters*

The analog signals generated by detectors have various voltage levels, ranging from low level to high level voltages. These signals are converted to a fixed signal level by passing through an amplifier. In process control systems, signal voltages of 1 to 5V DC and 0 to 10 V DC are often used.

(*b*) *Signal type conversion*

There are also cases where the signal is converted to another type of signal to facilitate processing. For example, converting a signal representing changes in resistance value into a voltage signal, makes the signal easy to amplify or transmit. When the distance between the detector and the receiving instrument is great, it is possible to eliminate signal attenuation during transmission by converting the signal into an electrical current signal. An electric current signal having the span of 4 mA to 20 mA is generally used.

(*c*) *Linearizing*

Detector input and output characteristics are often nonlinear. Nonlinear signals from thermocouples, resistance temperature detec-

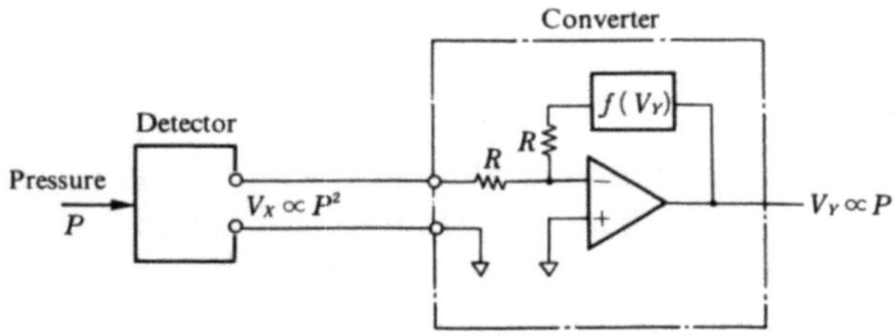

Fig. 3 · 140 Principle of linearization

tors, differential pressure flowmeters, and so on are made linear by a signal converter before being transmitted to an indicator or recorder.

Figure 3.140 shows the principle of linearization. The symbols in the figure have the following meanings.

P: The process variable being detected (pressure, for example)
V_X: The voltage signal generated by the detector
V_Y: The voltage signal output from the converter

If the input/output characteristics of the detector, for example when $V_X \propto P^2$ and the amplifier feedback characteristic is defined by

$$f(V_Y) = V_Y^2 \tag{3.41}$$

then, from the amplifier input current balance condition

$$\frac{V_X}{R} + \frac{V_Y^2}{R} = 0 \tag{3.42}$$

and accordingly,

$$V_Y = -\sqrt{V_X} \propto P \tag{3.43}$$

Which is to say that by passing through this converter, signal V_Y, which is proportional to the voltage, is obtained.

(d) Filtering

In process control systems, machinery that consumes large amounts of electric power, like electric motors and electromagnetic valves are often used together with measuring instruments that handle very small signals. Accordingly, noise that is synchronous with 50 Hz or 60 Hz electric power supplies and pulse noise is often present. Signal converters can prevent the abnormal operation of signal-receiving instruments due to these types of noise. Low-pass filters, made up of sets of capacitors and resistors, eliminate noise components above 50 to 60 Hz.

(e) Signal isolators

When processing multiple input signals or coupling signals into different systems, the signals must be isolated from each other to prevent interference. Transformers or photocouplers are used to insulate the signals.

Examples of signal converters are shown below.

3.10.2 Thermocouple signal converters

The following uses temperature sensing by a thermocouple as an example of amplification of a low-level signal. Figure 3.141 shows the electromotive force characteristic of a thermocouple. Although there are differences according to the type of thermocouple, in all cases, the electric potential signal is extremely small, ranging from an electromotive force of a few mV to several tens of mV. The impedence of the th-

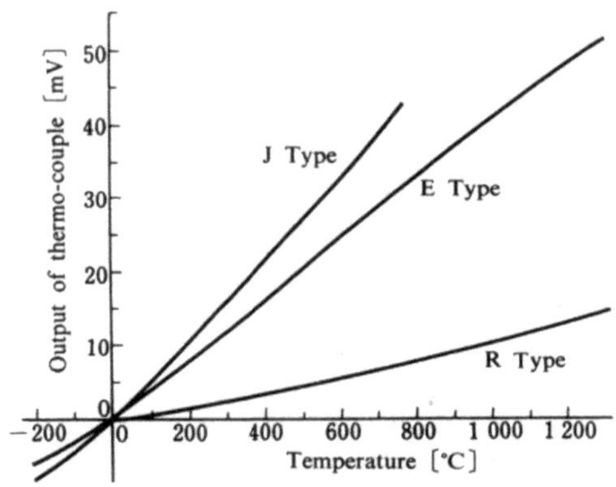

Fig. 3 · 141 Electromotive force charateristics of thermocouple

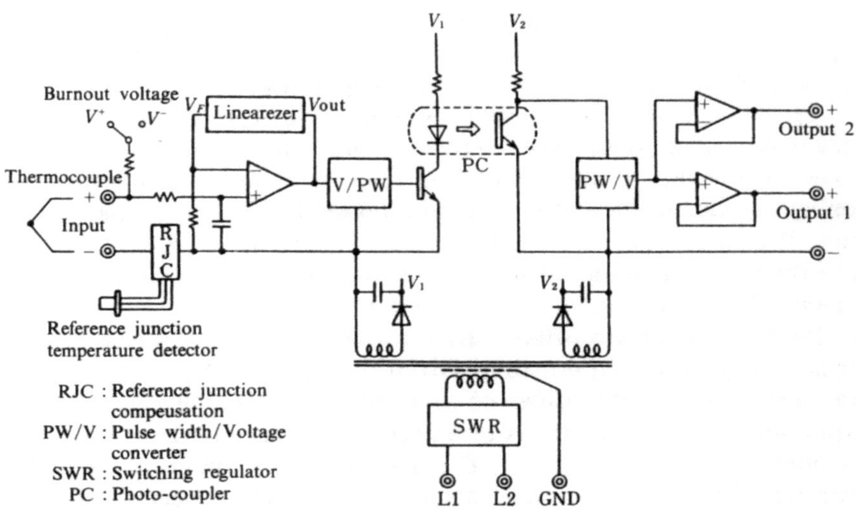

Fig. 3 · 142 Principle of thermal-electric temperature connerter

ermocouple is small, but it is necessary to perform the measurements without current flowing in the thermocouple. Thus, high-input impedance amplifiers must be used.

Figure 3.142 shows an example of a converter circuit. Conventionally, the signal from the thermocouple was converted to an alternating signal by a chopper before amplification, to avoid direct current drift in the amplifier. After amplification, the signal is converted back to a direct current. However, now low-drift monolithic operational amplifiers having high output impedence are available.

The thermoelectromotive force characteristic of a thermocouple is determined by the temperature difference between the hot junction and the cold junction (usually 0°C), thus the accuracy of the cold junction temperature is important. Maintaining the cold junction at 0°C by using the thermal balance between ice and water is one method of insuring temperature accuracy. However, in process control systems, the temperature of the cold junction is measured, and the thermoelectromotive force is compensated for the temperature. Resistors having a large thermal coefficient or transistors are used to measure the cold junction temperature. The device shown in Fig. 3.142 uses the latter method.

The relationship between the V_{BE} of the transistor and temperature is expressed as

$$V_{BE}=V_{g0}-\frac{kT}{q}\ln\frac{AT^{\gamma}}{I_C} \tag{3.44}$$

where T is the absolute temperature (K), V_{g0} is the energy gap of silicon at $T=0$ (K), k is the Boltzman constant, q is the electronic charge, A is the geometric constant of the transistor, γ is a parameter related to the base region diffusion constant, and I_C is the collector current (V_{g0}, k, q, A, and γ are independent of temperature).

Thermocouples are exposed to high temperatures, so they are quickly degraded by reactions between metal and surrounding gases. It is very important to always keep in mind this thermocouple degradation and breaks in the wiring when designing measurement control systems. Thermocouple converters include a function to make their output either the maximum or minimum value when a break in the wiring is detected. This is known as the burnout function.

Thermocouples are often used to measure temperature in electric furnaces and so on. When the insultation is degraded, the heater voltage becomes common mode noise in the thermocouple signal. If the output side of the converter is grounded, this noise current flows from the input side to the output side, and there is danger of abnormal operation of the converter and damage to the circuit. Accordingly, the input and output of the converters are generally isolated from each other. In the example shown in Fig. 3.142, the amplified signal is converted into an optical pulse signal. On the secondary side, it is converted back into an electrical signal, thus accomplishing the isolation.

3.10.3 Resistance signal converters

This is an example of signal type conversion. Figure 3.143 shows an example of conversion from resistance change of a resistance temperature detector (RTD) into an electric voltage signal. The resistance temperature detector is connected to one side of the Wheatstone bridge, and the voltage imbalance created by resistance changes in the detector is amplified.

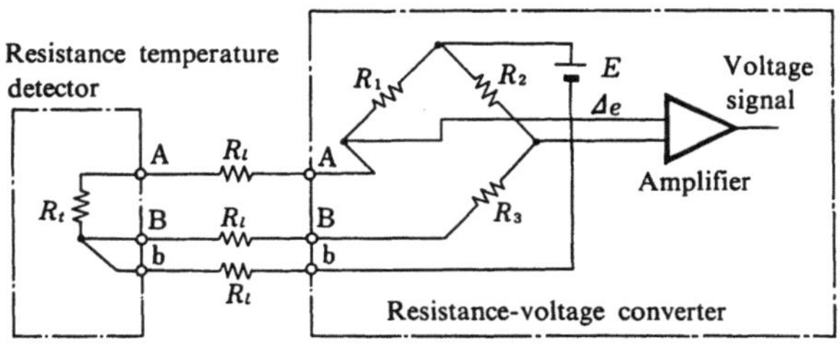

Fig. 3・143 Principle of resistance-voltage converter

In resistance-voltage conversion, sufficient attention should be paid to the effect of conducting wire resistance R_l. For the circuit in the figure, the bridge imbalance voltage is expressed as

$$\Delta e=\frac{R_2R_t-R_1R_3+(R_2-R_1)R_l}{(R_1+R_t)(R_2+R_3)+2(R_1+R_2+R_3+R_t)R_l+3R_l^2}E \quad (3.45)$$

which, if $R_1=R_2=R_a$, becomes

$$\Delta e=\frac{R_a(R_t-R_3)}{(R_a+R_t)(R_a+R_3)+2(2R_a+R_3+R_t)R_l+3R_l^2}E \quad (3.46)$$

Normally, when copper wire is used, R_l is 10Ω or less. Thus if R_a =1(kΩ), and a platinum resistance temperature detector (Pt 100Ω) is to make a measurement in the range from 0°C to 100°C, then the effect on the conducting wire resistance when there is change in the environmental temperature is a very small 0.2%/10°C. By using a three-wire configuration in which the three wires have the same characteristics, the measurement error can be made small. Figure 3.144 is an example of a resistance-temperature converter. Transistors Q_4 and Q_5 comprise a burnout circuit to drive the output to the maximum value when any of the three input wires (A, B, or b) are broken.

Platinum resistance temperature detectors (Pt 100Ω) are used in industrial measurement. The resistance change characteristic of this device is non-linear, as shown by the following equation.

$$R_t = R_0(1 + \alpha t - \beta t^2) \tag{3.47}$$

Here, R_0 is the resistance at 0°C (100 Ω), R_t is the resistance at temperature t (°C), and α and β are constants.

For example, the maximum non-linearity error in the 0 to 500°C measurement range is approximately 10°C in the vicinity of 250°C. In the circuit shown in Fig. 3.144, the non-linearity error of the Wheatstone bridge circuit is added. This error can be compensated for by proper selection of the resistance R_f, which determines the positive feedback from the output of amplifier A_1.

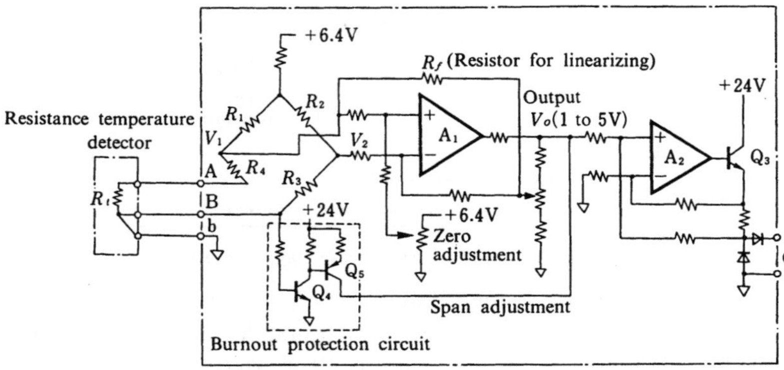

Fig. 3 · 144 Example of resistance-temperature converter

3.10.4 Two-wire signal transmission

When the distance between the detector and the converter is large, the two-wire electric current transmission method is used. Figure 3.145 shows this configuration. The two wires connect the detector with the converter, supplying power from the converter and at the same time carrying the signal component as variations in the current.

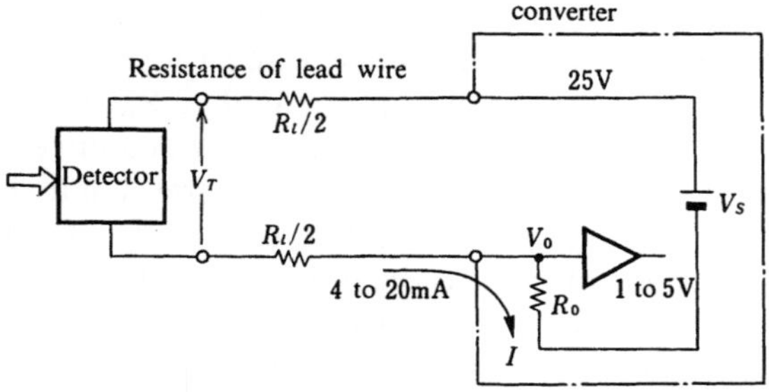

Fig. 3 · 145 Two-wire method current transmission system

Corresponding to changes in the physical quantity from 0% to 100%, the signal is converted into a DC electric current signal of from 4 to 20 mA and transmitted. When a voltage signal is required, a resistor is inserted into the path of the current, and the voltage signal is picked up at the two sides of the resistor. The allowable lead wire resistance R_l of the transmission line is given by the following equation.

$$R_l \leqq \frac{V_S - V_0 - V_T}{I_{\max}} \quad [\Omega] \tag{3.48}$$

Here, V_S is the minimum voltage supplied from the converter (V), V_0 is the internal voltage consumption of the converter (V), V_T is the minimum operating voltage of the converter (V), and $I_{\max}$ is the upper limit of the signal current (A). The electric current signal used in process control systems is being standardized at 4 to 20 mA DC, so $I_{\max}$ is taken to be 20 mA. The signal is converted to from 1 to 5 V according to the resister R_0 inside the converter, so V_0 is taken to be 5 V. The minimum operating voltage of the converter is 12 V. If V_S is taken to be 25 V, then $R_l \leq 400\Omega$.

3.10.5 Pulse signal converters

In flowmeters, devices which output pulse signals are often used. By counting the pulses in these signals, the flow rate can be determined. However, because the pulse signals from flowmeters are high-level signals, and chattering occurs because relay contact points are used, direct input to computers and so on is not possible.

Figure 3.146 shows an example of a pulse signal converter. After waveform shaping, the pulse input undergoes input-output isolation by a photocoupler. The isolated signal is then applied to the output circuit. The output is a transistor contact signal, thus compatible with

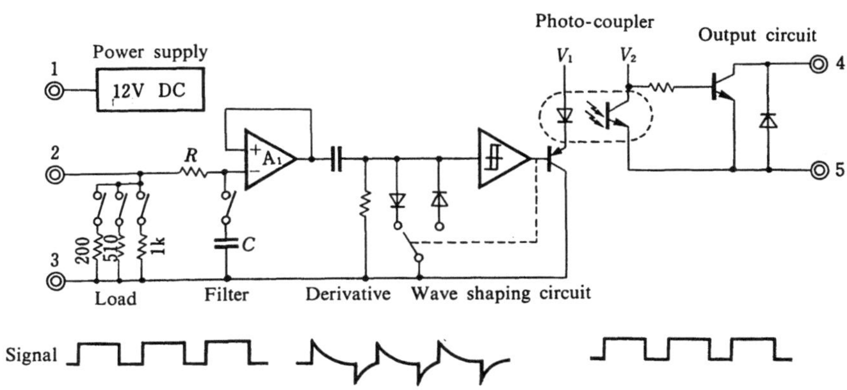

Fig. 3 · 146 Pulse signal converter

Table 3.19 An example of input pulse convertor specifications

(1) Input signal:
  Contact or voltage level pulse from two-wire transmitter system or three-wire trsnsmitter system.
 Input frequency:
  0 to 6 kHz
 Minimum input pulse width:
  ON time; 60 μs, OFF time; 60 μs
 Input level:
  Contact signal:
   Contact type: relay contact or trausistor contact (open collector)
   Contact resistance: closed, 200Ω or less; open; 100 kΩ or greater
   Contact rating: 15V DC, 15mA or greater
   Filter: insertable by switch; time constant, 10ms
  Voltage level pulse signal:
   Low level (E_L): −1 to +8V DC
   High level (E_H): +3 to +12V DC
   Swing: $E_H - E_L \geqq 3$V
   Signal source impedance: 1kΩ or less
   Load resistance on two-wire transmitter system writh power supply:
    200Ω, 510kΩ, 1kΩ (selectable by switch)

(2) Output signal:
  Transister contact (open collector)
  Contact rating: 30V DC, 30mA
  Output frequency: Same as input frequency

(3) Power supply for transmitter: 12V DC ±10%, 30mA or less

computers. This device can accept as input relay contact, transistor contact, or voltage pulse signals. For relay contact input, the filter switch is turned on and chattering noise is eliminated. The purpose of the load resistor is to convert the current pulse signal to a voltage level when the two-wire current transmission method is used. An example of input pulse converter specifications is given in Table 3.19.

3.10.6 Computer input equipment

Process control systems which use computers accept multiple analog input signals. Figure 3.147 shows a computer input processing system, taking thermocouple input as an example. The multiple input signals are switched by a multiplexer, and then undergo A/D conversion. For switching speeds of 1 to ten points per second, relay switching is suitable, but for higher speeds, semiconductor switching is used.

Figure 3.148 is an example of a flying capacitor multiplexer. The unselected channel switches are thrown to the detector side, and the input electric potentials are stored in capacitors C_1. The switch for the selected channel is thrown to the amplifier side, and the electric potential on the two sides of capacitor C_1 undergoes A/D conversion. With

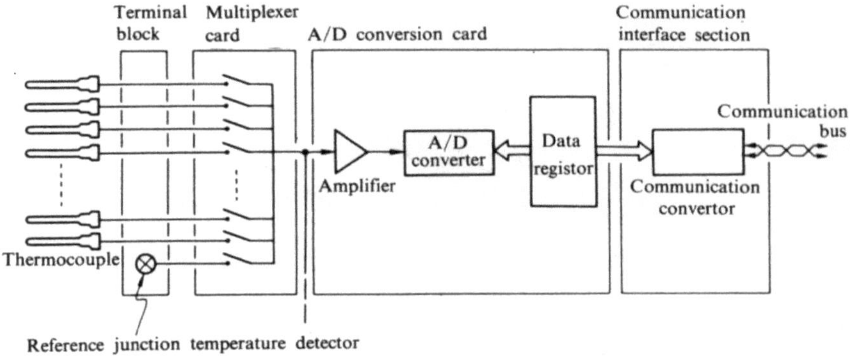

Fig. 3・147 Input treatment system of computer

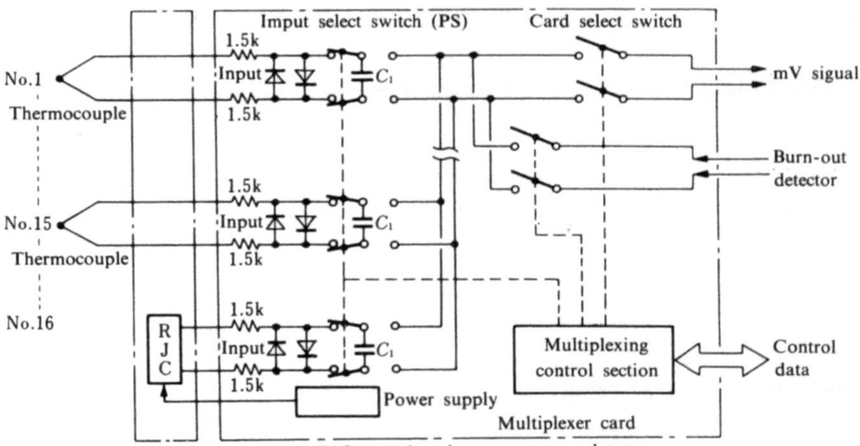

Fig. 3・148 Flying capacitor type multiplexer

this method, the input signals are isolated from each other, and at the same time analog filtering is accomplished during the conversion, because the capacitors are charged by the input voltages over a fixed time interval.

When multiple thermocouple signals are batch processed, equality of the cold junction temperatures is important. Figure 3.149 shows an example of a 15 input connection terminal board. The 15 input terminal board consists of a copper printed circuit board designed for constant temperature across the part where signal wires are fastened by screws. The temperature of the printed circuit board is measured with a resistance temperature detector, which sends a temperature com-

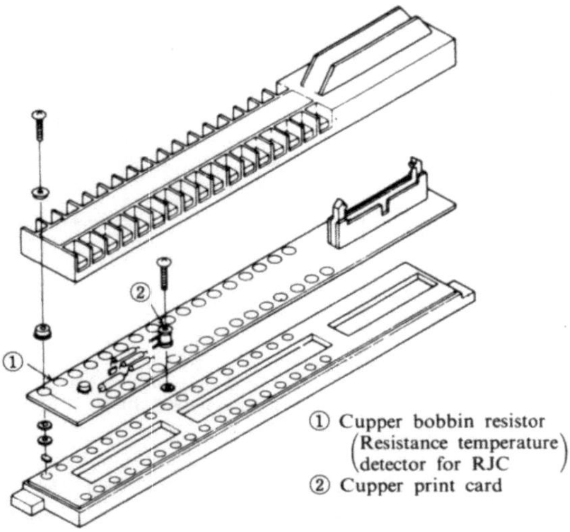

Fig. 3・149 Terminal board with integrated reference junction compensation circuit

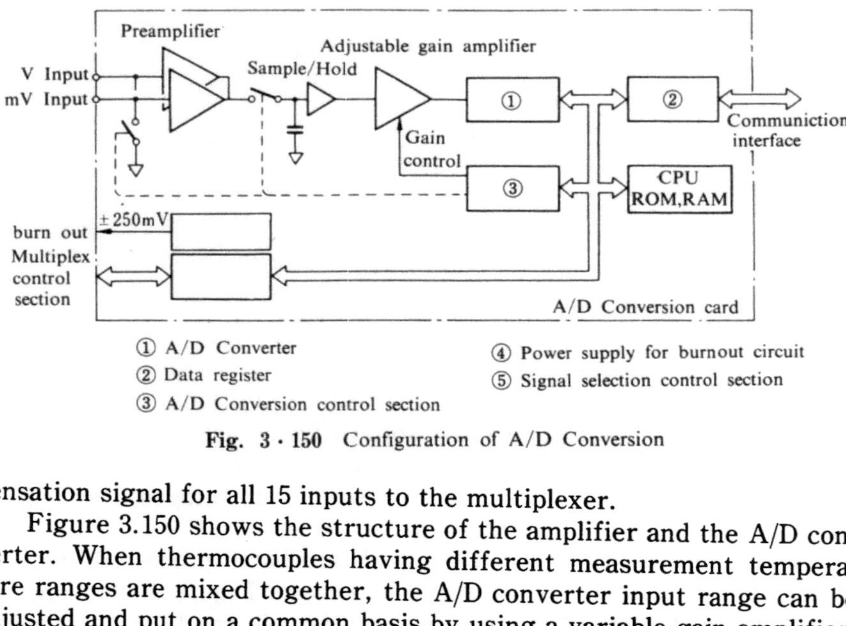

Fig. 3・150 Configuration of A/D Conversion

pensation signal for all 15 inputs to the multiplexer.

Figure 3.150 shows the structure of the amplifier and the A/D converter. When thermocouples having different measurement temperature ranges are mixed together, the A/D converter input range can be adjusted and put on a common basis by using a variable gain amplifier. When slow conversion is acceptable, an integration A/D converter is used; but when high speed conversion is required, a successive approximation A/D converter is used. The resolution required in process control is 10 to 13 bits.

PRACTICE QUESTIONS

3.1 Describe the error factors influencing the measured values of industrial variables.

3.2 Why are differential pressure type flowmeters so often used? What are their drawbacks?

3.3 Describe the features of vortex flowmeters.

ANSWERS TO QUESTIONS

3.1 See Subsec. 3.1.2. in the text.

3.2 a. Wide application for liquids, gases and steam.

b. Many types of constriction devices are available; the structure is simple so failures are rare.

c. Can generally used without calibration of actual flow, because constriction device standards are well-established.

d. High-accuracy differential pressure transmitters are available.

The disadvantages are,

a. Pressure loss is high compared to other flow meters.

b. The differential pressure is proportional to the square of the flow rate. Thus, measurement acccuracy falls in the low flow rate region (low range ability).

c. They cannot be used for some highly viscous fluids or slurries.

3.3 See Subsec. 3. 3. 6 in the text.

REFERENCES

1) JIS Z 8103-1984: Glossary of Terms Used in Instrumentation (in Japanese).

2) T. Senbon and M. Toyama: Industrial Measurements, pp.8-9, pp.240-242, Ohmsha (1965) (in Japanese).

3) T. Naito: Industrial Measurement Handbook, pp.23-26, Asakura Shoten (1976) (in Japanese).

4) JEMIS 022-1983: General Rule for Expression of Industrial Instruments Performance (in Japanese).

5) M. Kanno: Electrical and Magnetic Measurements and Instrumentation, p.30, Corona Publishing (1982) (in Japanese).

6) T. Ishii: Process Control Instrumentation Systems, pp.60-94, Denki Shoin (1973) (in Japanese).

7) The society of Instrument and Control Engineers, Temperature Measurement Devision: Temperature measurement (1981) (in Japanese).

8) The Instrumentation Control Association Japan: Industrial Instruments Management Standard III, Management of Temperature Instruments, pp.68-69 (1978) (in Japanese).

9) Daido Steel Co. LTD.: Technical Information, Portable Digital Radiation Thermometer "STAR THERMO" (in Japanese).

10) JIS C 1602-1981: Thermocouples (in Japanese).

11) JIS C 1605-1982: Metal Sheathed Thermocouples (in Japanese).

12) JIS C 1604-1989: Resistance Thermometer Sensors (in Japanese).
13) JIS C 1606-1989: Metal Sheathed Resistance Thermometer Sensors (in Japanese).
14) JIS C 1610-1981: Compensating Lead Wires. (in Japanese).
15) JIS C 1611-1975: Thermistor for Temperature Measurement (in Japanese).
16) JIS Z 8706-1980: Methods of Temperature Measurement by Optical Pyrometers (in Japanese).
17) H. Kawada, *et al.*: Flow Measurement by Constriction Devices, Flow Measurement Handbook, pp.55-65, Nikkan Kogyo Shinbun Sha (1979) (in Japanese).
18) The Instrumentation Control Association Japan: Industrial Instruments Management Standard IV, Management of Flow Instruments, pp.18-28 (1978) (in Japanese).
19) JIS Z 8762-1988: Measurement of Fluid Flow by Means of Orifice Plates, Nozzles and Venturi Tubes (in Japanese).
20) T. Kobayashi: "Field instruments," *Yokogawa Tech. Rep.*, **28**, 2 (1984) 77 (in Japanese).
21) JIS Z 8761-1977: Method of Flow Measurement by Float Type Area Flowmeters (in Japanese).
22) M. Kanayama, *et al.*: "Fundament of Flow Measurement," *Automation*, **21**, 7/12 (1976) (in Japanese).
23) Oval Engineering Co., LTD.: Training Text Book, No.A-001-1, p.35 (in Japanese).
24) F. Satori: Turbine Meter, Flow Measurement Handbook, pp.203-210, Nikkan Kogyo Shibun Sha (1979) (in Japanese).
25) JIS Z 8765-1980: Method of Flow Measurement by Turbine Meters (in Japanese).
26) N. Kayama, *et al.*: "Recent progress in electromagnetic flowmeter," *J. Soc. Instrument & Control Eng.*, **23**, 2 (1984) 20-27 (in Japanese).
27) JIS Z 8764-1980: Method of Flow Measurement by Electromagnetic Flowmeters (in Japanese).
28) JEMIS 028-1985: Methods of Flow Measurement by Vortex Flowmeters, p.7, pp.19-21 (in Japanese).
29) I. Ito: "Vortex flowmeters using stress detection method," *Instrumentation*, **34**, 8 (1985) (in Japanese).
30) H. Ohwada: "Flow measurement using ultrasonic wave," *Sensor Technology*, **4**, 2 (1984) (in Japanese).
31) Y. Yamamoto: Flow Measurement by Ultrasonic Flowmeter, Flow Measurement Handbook, pp.261-262, Nikkan Kogyo Shibun Sha (1979) (in Japanese).
32) N. Kayama: "YEWSONIC ultrasonic flowmeter," *Yokogawa Tech. Rep.*, **30**, 2 (1986) 61-63 (in Japanese).
33) Yokogawa Electric Corporation: Industrial Instrument Handbook Pneumatic Instrument Volume, 2nd Edition, Tokyo Denki University Press (1974) (in Japanese).
34) B. Saito, *et al.*: "Air-to-current type 5551," *Yokogawa Tech. Rep.*, **15**, 2, (1971) 22-26 (in Japanese).
35) Hirano, *et al.*: "Contactless transducer using a coil formed on a printed circuit board," Text Books for No.23, Training course of Automatic Control (1979) 249-262 (in Japanese).
36) S. Shindo: "Miscellaneous measuring instruments," *Yokogawa Tech. Rep.*, **25**, 3 (1981) 27-29 (in Japanese).
37) I. Ohno, *et al.*: "Magnetic displacement transducer using a self-oscillating technique," *J. Soc. Instrument & Control Eng.*, **12**, 1 (1973) 2-15 (in Japanese).
38) T. Ueda, *et al.*: "Displacement transducer using magnetostrictive delay line," *Trans. Soc. Instrument & Control Eng.*, **17**, 8 (1981) 858-864 (in Japanese).
39) H. Ohtu, *et al.*: "Programmable intelligent recorder μR100/μR180 series,"

Yokogawa Tech. Rep., **30**, 3 (1986) 15–20 (in Japanese).
40) S. Shindo: "Electronic measuring instruments," *Yokogawa Tech. Rep.*, **25**, 3 (1981) 18–27 (in Japanese).
41) H. Hirasawa: "Turbidity transmitter 'T/B cell'," *Yokogawa Tech. Rep.*, **17**, 1 (1973) 38–43 (in Japanese).
42) H.Ikegawa, *et al.*: "New electrolytic conductivity measuring equipment for process use," *Yokogawa Tech. Rep.*, **18**, 2 (1974) 63–69 (in Japanese).
43) H. Kamata, *et al.*: "Non-reagent type free available chlorine analyer 'F/C Cell'," *Yokogawa Tech. Rep.*, **20**, 4 (1976) 156–162 (in Japanese).
44) M. Tomo, *et al.*: "New series of liquid analyzers for process use," *Yokogawa Tech. Rep.*, **28**, 1 (1984) 55–62 (in Japanese).
45) T. Minaki, *et al.*: "A digital density meter," *Yokogawa Tech. Rep.*, **29**, 2 (1985) 3–8 (in Japanese).
46) Yoshimura, *et al.*: Theory of pH and Its Measuring Method, Maruzen (1968)(in Japanese).
47) Y. Fukai, *et al.*: "New gas chromatograph readout GC6P," *Yokogawa Tech. Rep.*, **24**, 2 (1980) 2–9 (in Japanese).
48) M. Maeda, *et al.*: "New thermal conductivity detector for process gas chromatograph," *Yokogawa Tech. Rep.*, **27**, 1 (1983) 27–32 (in Japanese).
49) Y. Hanaoka: "Process analyzers," *Yokogawa Tech. Rep.*, **30**, 2 (1986) 64–67 (in Japanese).
50) H. Takahara: "Type IR11, IR21 infrared gas analyzer," *Yokogawa Tech. Rep.*, **24**, 3 (1980) 16–20 (in Japanese).
51) H. Nakashima: " Process analyzers," *Yokogawa Tech. Rep.*, **28**, 2 (1984) 86–89 (in Japanese).
52) S. Muramoto:"PC6 oil calorimeter," *Yokogawa Tech. Rep.*, **24**, 2(1980) 10–16 (in Japanese).
53) S. Kiyobe: Japanese Journal of Paper Technology (Mar. 1984) (in Japanese).
54) H. Ishikawa, *et al.*: "New sensors of B/M-CENTUM System in paper plant industries—ash, caliper and new moisture sensors," *Yokogawa Tech. Rep.*, **24**, 3 (1980) 3–15 (in Japanese).
55) K. Isozaki, *et al.*: "New 'WEB GAGE' —uses radiation to measure sheet thickness," *Yokogawa Tech. Rep.*, **26**, 4 (1982) 27–31 (in Japanese).
56) S. Miyazaki: Control for Chemical Pressure, Sangyo Tosho (1971) (in Japanese).
57) C. D. Johnson: Process Control Instrumentation Technology, 2nd Edition, John Wiley & Sons (1982).

Chapter 4

RECORDERS AND CONTROLLERS

In large-scale instrumentation systems, CRT display and printout of data is often used, and accordingly the priority of recorder units is decreasing. However, recorder technology is also rapidly developing through application of electronic technology. This chapter describes the non-contact type self-balancing mechanism, which has been problematic in conventional industrial recorders. New methods of printing and so on are also described.

Controllers have developed from the early pneumatic controllers through analog electronic controllers, digital controllers to the CRT display control systems which have data communication functions. The section on controllers in this chapter is limited to single-loop controllers, and while comparing analog electronic controllers with digital controllers, discusses controller functions. Control algorithms are described in detail in Chap. 2 and Chap. 8, so discussion in this chapter is limited to the range of the basic PID form. The fine points of pneumatic controllers can be found in other works.[1)]

4.1 Recorders

Industrial recorders monitor the changes over time in process measurement signals, and provide a permanent record of those changes.

4.1.1 Types of recorders

Industrial recorders are classified in the following way according to the methods they employ.

Operation method {Self-balancing recorders; Raster scan recorders}

Recording method {Pen recorders; Dot printing recorders}

Circuit method $\begin{cases}\text{Analog recorders}\\\text{Digital recorders}\end{cases}$

As most self-balancing recorders use the zero method, the principle of that method is shown in Fig. 4.1. The input signal is converted by the preamplifier (signal conditioner) to a level suitable for transfer to the servo-amplifier. In the servo amplifier, the differential voltage signal between the output voltage from a potentiometer connected to the pen and the preamplifier output is impressed on the input voltage signal, and this differential voltage is amplified and used to drive the motor. This is currently the most often used type of recorder. Raster scan recorders will be discussed later. This section is devoted to description of recorders classified as pen recorders and dot printing recorders. These recorders are mainly manufactured in recording widths of 100 mm, 180 mm, and 250 mm, and accuracies ranging between 0.1% to 0.5% of full scale.

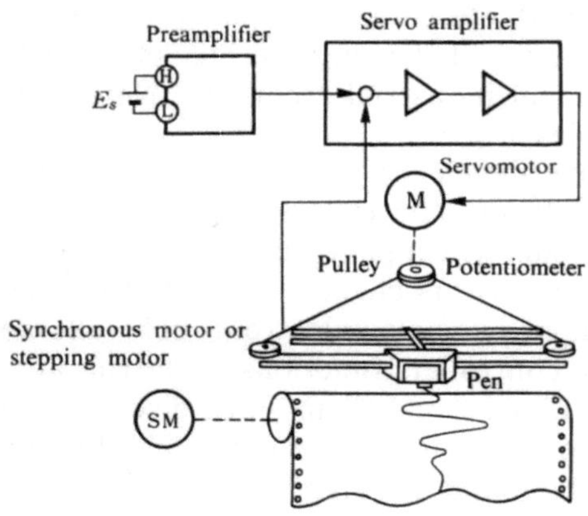

Fig. 4 • 1 Principle of self-balancing type recorder

4.1.2 Recorder functions

As shown in Fig. 4.2, recorder functions can be classified into measuring, processing and recording. In self-balancing recorders, measuring and recording are mainly accomplished by servomechanisms. Chart feed and the printing mechanism are driven by a synchronous motor. Although these mechanisms are relatively simple, if other functions such as character printing, display, computation, remote control, and so on is required, then the circuitry and mechanical components become more complex.

Analog-to-digital (A/D) conversion of the input signal allows efficient processing under control of a microprocessor. The following func-

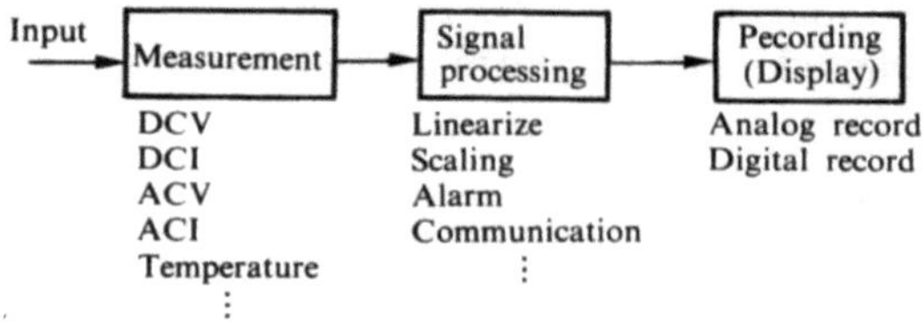

Fig. 4 · 2 Function of Recorder

tions can be readily performed.

(*a*) *Measuring*

As shown in Table 4.1, there are many types of inputs that can be measured, and the measurement range is also wide.

Table 4.1 Examples of input signal and range

Input type	Range
DC voltage	20mV to 50V
Thermocouple	TYPE R, S, B, K, E, J, T, N, W
Resistance temperature detector	Pt 100/50 Ω, Ni 100/120 Ω

(*b*) *Recording*

As shown in Fig. 4.3, the measured values, time, scale, and so on are digitally printed, and have been conveniently used to grasp long-term trends of input change in addition to the analog recording. This is referred to as hybrid recording. Recording zones for each channel can be selected and important portions can be enlarged. Also, scheduled printing, in which information on conditions such as the range of each channel, tag numbers, units, sensor alarms, the date, and chart speed can be listed at fixed intervals on the chart.

(*c*) *Display*

In addition to the digital display of the measured values for each

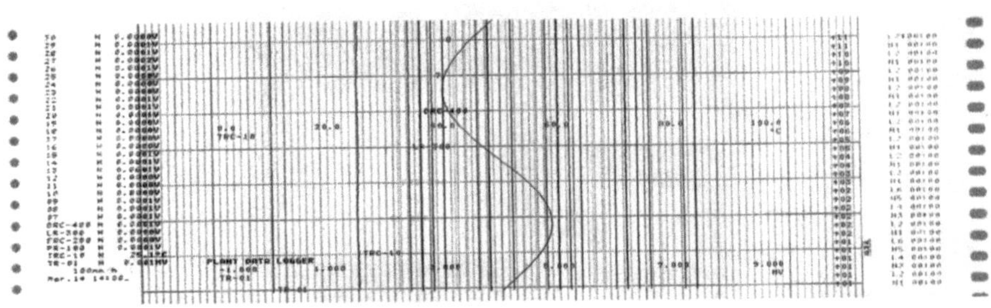

Fig. 4 · 3 Chart example (Analog + Digital record)

channel, the year-month-date, and the time, measured values, alarm settings are displayed, and at alarm times, flashing point bar graphs are displayed.

(*d*) *Computation*

Inter-channel differential computations, industrial variable conversion of input voltage signals (scaling), linearizing, and square root extractions are possible. Also, taking advantage of memory storage, the offset between pens can be compensated for. Multiple-point high speed recording and printing using the raster scan method can be performed. In addition to this, event recording, in which the data before and after any event that occurs can be preserved in a special memory area and the data can be printed on command.

(*e*) *Alarms*

From two to six alarms can be set for each channel. The types of alarms include upper limit, lower limit, upper velocity limit, lower velocity limit, difference upper limit, and difference lower limit. These alarms can be annunciated through relays as well as recorded.

(*f*) *Data communication*

These devices have GP-IB and RS-232 C communication capabilities, so communication with computers is possible. Thus they can be used as computer output devices.

The operating principle and structural components of recorders designed for multiple functions through use of microprocessors are described in the following sections.

4.1.3 Pen recorders

A typical pen recorder is shown in Fig. 4.4. Figure 4.5 is a block diagram of this device. Commercial products are mainly 1-pen to 3-pen models.

(*a*) *Operating principle*

The measurement signal, after amplification corresponding to a

Fig. 4 · 4 Pen recorder

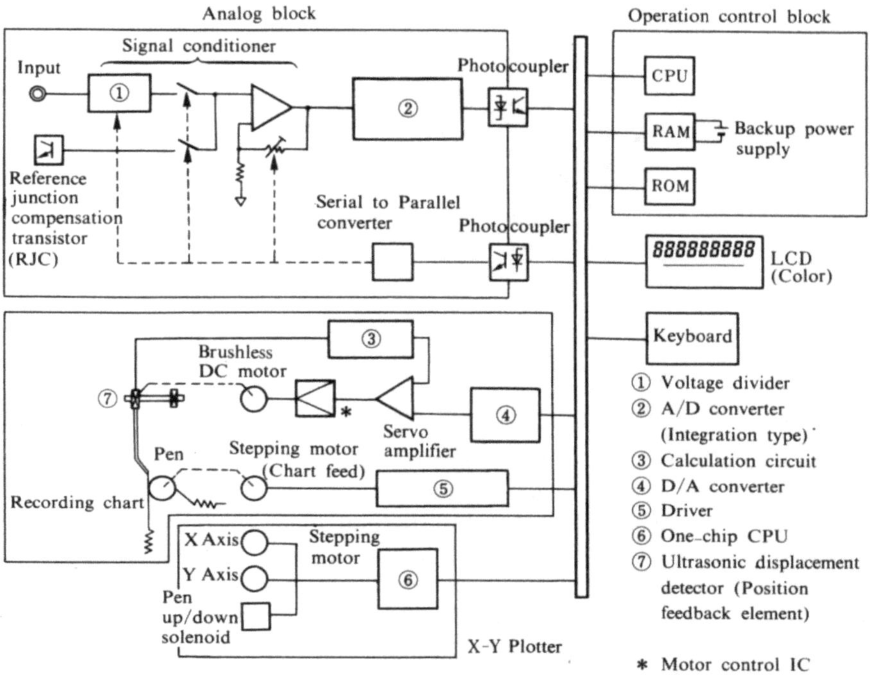

Fig. 4・5 Functional block diagram of pen recorder (1pen)

pre-set measurement range, is converted to a digital signal by an integration type A/D converter. The converted signal undergoes linearizing and alarm processing in the operation control block, and is then temporarily stored as display data in memory (RAM). It is subsequently converted to recording data compatible with the recording equipment. The recording data is converted back to analog data by a D/A converter and it is sent to the servo amplifier, where it is compared with the pen position signal obtained from the pen-position feedback element. The deviation (error) output is power-amplified by the motor control IC and drives the servomotor. The position feedback element is an ultrasonic displacement detector, and has no mechanical joints. The servomotor is a thin type brushless DC motor. These components make up a highly reliable contact-free servo system.

List printing and scheduled printing of logs is done by a small X-Y plotter that operates independently of the servomechanism. The plotter is controlled by a dedicated one-chip microprocessor.

(*b*) *Structural components*

Figure 4.6 shows the structure of the A/D converter and the servo unit.

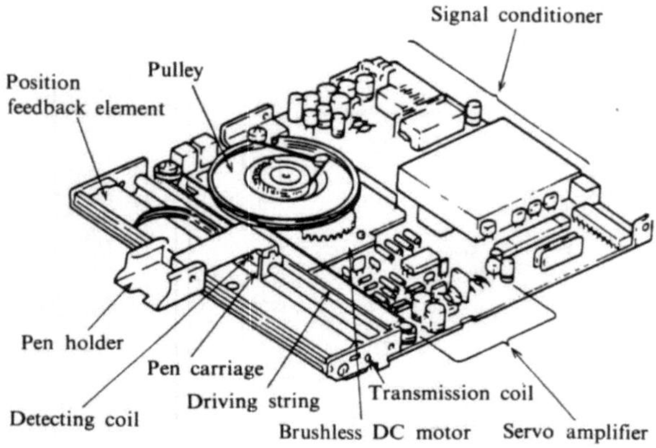

Fig. 4・6 Configuration of servo-unit and A/D converter

(1) *Signal conditioner and A/D converter* Signal conditioners are available in two models: one for DC voltage thermocouple input and one for resistance temperature detector input. The model for DC voltage thermocouple input consists of six ranges from 20 mV to 20 V, which depend on a potential divider and a variable gain amplifier. The output of the amplifier goes to an integration type A/D converter which uses the feedback-pulse-width modulation method. The obtained pulse width is isolated by a photocoupler and is applied to the operation control block. The operation control block has a high-precision reference voltage source and performs zero compensation for each range, zero compensation and full-scale compensation for the A/D converter, and other types of automatic calibration.

In the resistance temperature detector model, the input signal is converted to an electric potential by a resistance-voltage conversion circuit consisting of a fixed current of about 1 mA and a reference resistance of 300 Ω (accuracy of $\pm 0.01\%$). The resultant signal is fed to the A/D converter. The control signal from the operation control block is sent in serial form through the photocoupler to the analog block, and undergoes serial-parallel conversion to control signals for the amplifier and all switches. Typically, the resolution of the A/D converter is ± 14 bits, and its data output rate is 1/125 ms.

(2) *Servo system* The deviation of the signal from the position feedback element and the recording position signal corresponding to the measured value is amplified, and is sent to the DC motor circuit. In the D/A conversion block, the recording position signal becomes a smooth direct current signal by smoothing after pulse-width modulation by 300 Hz repetition. In the ultrasonic position detector, the pulse signal flows through the drive coil, and the direct sound waves and reflected

sound waves received in the detection coil (which moves as a unit with the pen) are converted into a position signal in the analog calculation circuit.

(3) *Brushless DC motor* For industrial recorders, tough, long-life two-phase alternating current servomotors having no frictional parts have often been used. Recently, however, small, light-weight brushless DC motors that generate little heat are coming into use. In these motors, electronic circuitry takes the place of the brushes. Figure 4.7 shows the structure of such a thin brushless motor mounted on a recorder. This is a rotating-field type DC motor, in which the armature coils are fixed on the stator (printed circuit board), and the field magnet rotates. Hall elements are used for contactless detection of the relative positions of the armature coils and the rotating field. Motor control is performed by dedicated ICs which have built-in commutation control circuits. By using this motor, a servo unit step response time (90% response) of 1 second or less can be realized.

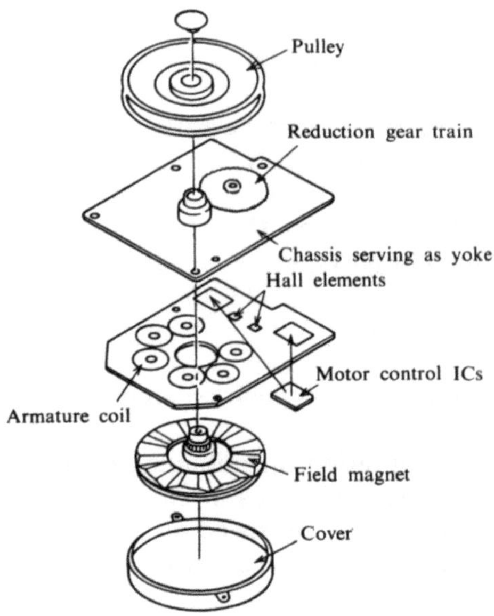

Fig. 4 · 7 Configuration of brushless DC motor

(4) *Ultrasonic position detector* The principle of the ultrasonic position detector is shown in Fig. 4.8. If a pulse magnetic field is generated at the position of N_0 by the current passing through the ultrasonic drive coil N_0, the Joule effect will result in a mechanical distortion within a ultrasonic waveguide made of magnetostrictive material, and the ultrasonic pulse will propagate. When these ultrasonic pulses (the

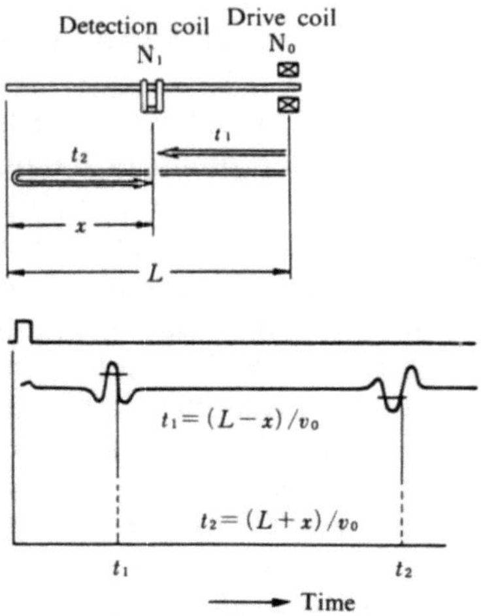

Fig. 4 · 8 Operating principle of ultrasonic pen position sensor

direct pulse from N_0 to N_1, and the reflected pulse from the reflecting end) arrive at the detection coil N_1, an electric potential pulse is generated in N_1 by the Villari effect. As shown by the propagation waveform in Fig. 4.8, by measuring the propagation time of this pulse, the position of the moving part can be obtained. Using the time width signals t_1 and t_2 for each sample cycle, the computation circuit carries out the following computation.

$$\frac{x}{L} \text{ (a dimensionless position signal) } = \frac{t_2 - t_1}{t_2 + t_1}$$

Because the computation for obtaining changes in the relative positions is done in this way, changes in environmental conditions such as temperature, which change the speed of sound v_0 do not affect the output. In this way infinitely high resolution and linearity of 0.1% or less are achieved.

Other than the contactless position feedback element for use in industrial recorders described above, there are also magnetic detectors and capacity type detectors. Also often used are contact type devices such as coiled wire potentiometers, conductive plastic potentiometers (CPP), and hybrid potentiometers in which the coiled wire slide surface is coated with conductive plastic. However, contact type position feedback elements often require cleaning of the resistor and exchanging of the contact mechanisms, so the movement will probably be toward con-

Table 4.2 Comparison between analog method and digital method for multipoint-recorder

Function	Measurement			Processing			Recording		
	Scanner	Signal conditioner	Convertor	Linealize	Alarm	Internal sequence control	Method	Print	Chart feed
Analog	Rotary switch	DC amplifier	—	Function potentiometer	Analog comparator	Synchronous motor and gear train	Servo	Ink hammer (motor and cam-driving)	Synchronous motor
Digital	Relay	Programmable DC amplifier	AD convertor	ROM table	Digital computation	Microprocessor	Raster scan	Wiredot	Stepping motor

Yokogawa Tech. Rep., **28,** 2, p.34 (1984)

tactless devices in the future.

4.1.4 Multipoint recorders

The configurations of analog and digital type multipoint recorders are compared in Table 4.2. Digital recorders are also called hybrid recorders. They have a structure near to that of data loggers, and they are difficult to distinguish by function. Figure 4.9 shows the external appearance of a hybrid recorder.

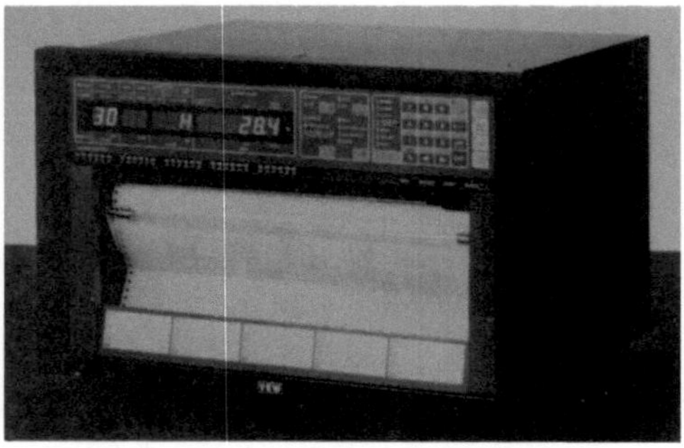

Fig. 4 · 9 Hybrid recorder

(a) Construction and operating principle

As shown in Fig. 4.10, the device is made up of an analog block, an computational and control block, a recording block and a keyboard block. The analog block consists of a multipoint measurement circuit made up of a relay scanner and programmable amplifier, and a pulsewidth-modulation, integration type A/D converter circuit that has a resolution of 4½ bits. The computational and control block consists of two microprocessors, ROM, RAM, and peripheral circuits. One microprocessor controls the analog block, the keyboard, and the communication interface; the other controls the recording block.

The recording block is made up of a wire dot hybrid printing head, chart feeding mechanism, motor driving equipment, and so on. The keyboard consists of various setting keys and an LED display.

Thirty direct current voltages and temperature input signals are selecting in turn by the scanner, and undergo A/D conversion according to pre-set measurement ranges in the analog block. This data is sent to the computational and control block, where, according to the type of input, it undergoes linearization, scaling, alarm processing and so on. It is then converted to display data and recording data, and stored in RAM. The data for recording is converted into analog and character printing data corresponding to their respective recording positions. Control up to this point is done by the first microproces-

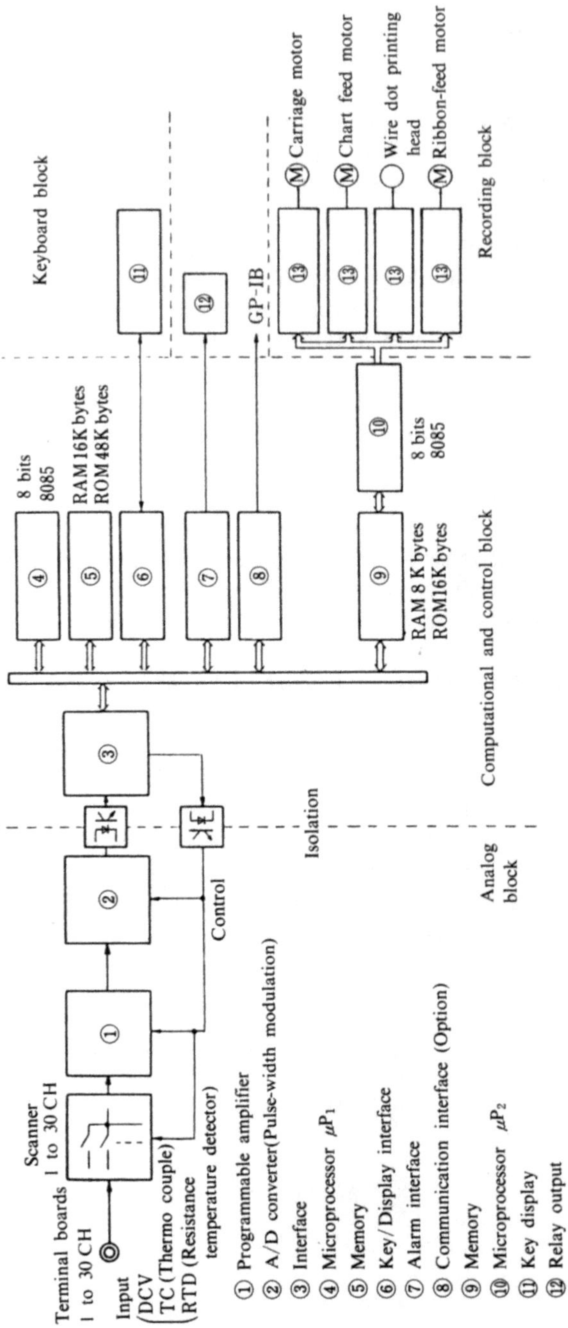

Fig. 4·10 Block diagram of hybrid recorder

sor, μP_1.

The second microprocessor, μP_2, drives the printing head. As the head moves across the chart from one side to the other in 0.1 mm steps, μP_2 refers to the recording data. When relevant data is present, μP_2 excites the solenoid of the printing head, causing a dot to be printed. This operation is repeated, and one scan is completed when the head arrives at the end of the chart. Another scan is performed as the head returns in the opposite direction to the first side. This recording method is called raster scanning. As shown in Fig. 4.3, in the hybrid recording mode, 6-color analog recording is done on 250 mm wide chart. Also, measured values and the date and time are printed on the left side of the chart. While the alarm data are printed out on the right side. The scan time is selectable, with a minimum of six seconds. In logger mode, the measured values are printed across the entire width of the chart. The scan interval can be set to one minute or any larger value.

(*b*) *Structural components*

(1) *Input scanner* In the same way as for pen recorders, there are different types of scanners for DC voltage thermocouple input and resistance temperature detector input. The former uses low thermoelectromotive force relays for switching elements; the latter uses CMOS semiconductor switches.

(2) *Wire-dot printing heads* Figure 4.11 shows the printing head structure. This is what is called the wire-dot printing head. It features a two-layer construction, with solenoid drivers for analog recording wires in the bottom layer and solenoid drivers for character printing wires in the top layer. Figure 4.12 illustrates the recording method, with the black dots in the head representing the wire array. As shown in the figure, each wire corresponds to the six-color ink ribbon. The device has wires specially for printing character, and can print one row of 5×7 dot characters in one scan.

The wires are the free-flight type. As shown in Fig. 4.13, the wir-

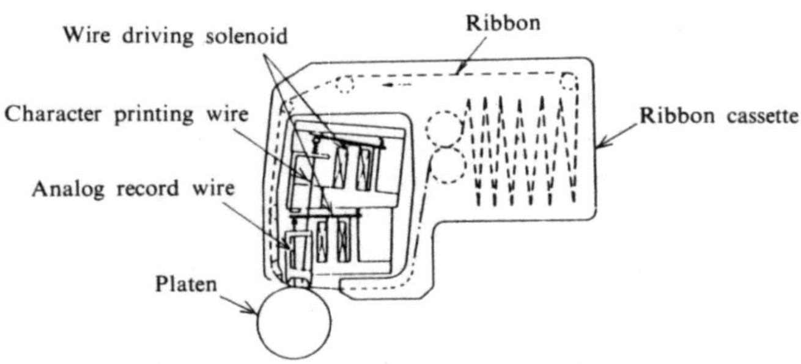

Fig. 4・11 Printing head mechanism

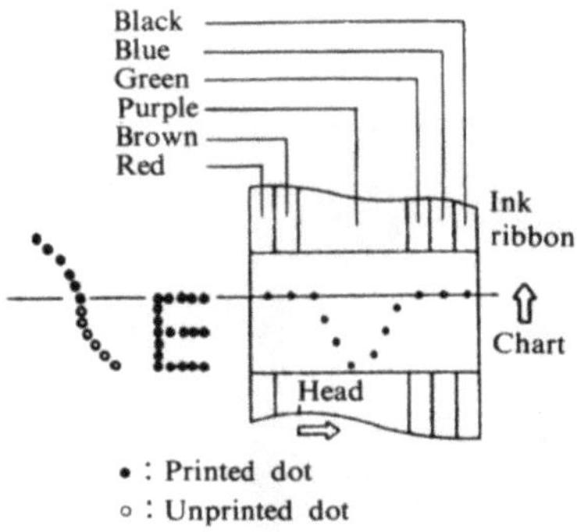

Fig. 4・12 Recording method

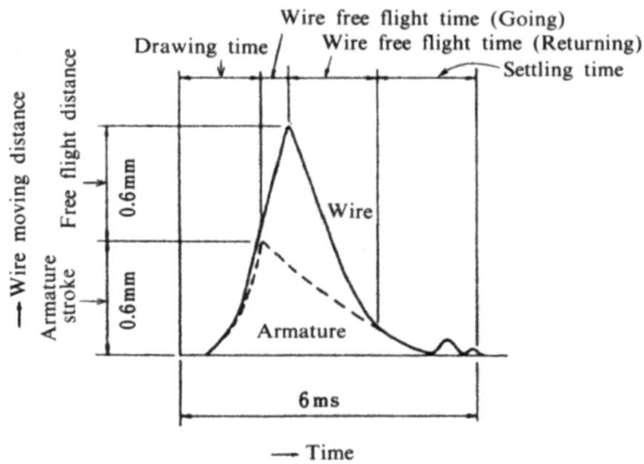

Fig. 4・13 Wire movement

es continue free flight after the armature has hit the yoke, strikes the ink ribbon and paper, and bounces back. This method can get a long wire stroke, and adjustment and maintenance are easy.

The ink ribbon is a nylon cloth cartridge type. It has six separate colors corresponding to the analog recording wires, and is moved forward by a motor with each stroke.

In addition to the color wire-dot recording method described above, multipoint recorder methods also include the multicolor dot printing pen head method, the heat-sensitive method, the color ink-jet method, and the electric discharge method.

The external appearance of a recorder that employs the multicolor dot printing pen head method is shown in Fig. 4.14. As shown in Fig. 4.15, this recorder has the special feature of recording the maximum and minimum values input during the recording period in line segments.

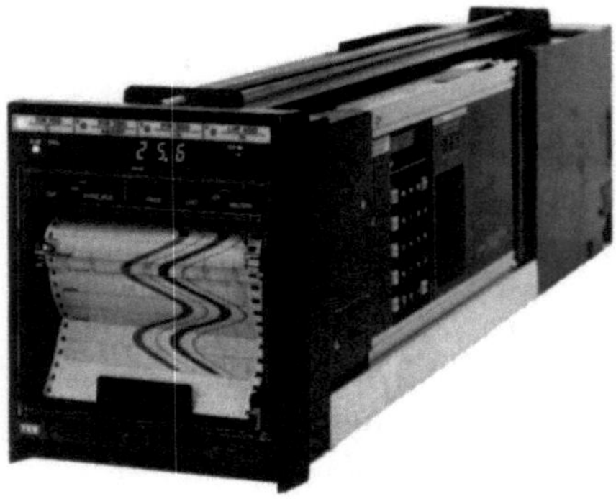

Fig. 4 · 14 External view of process-use recorder

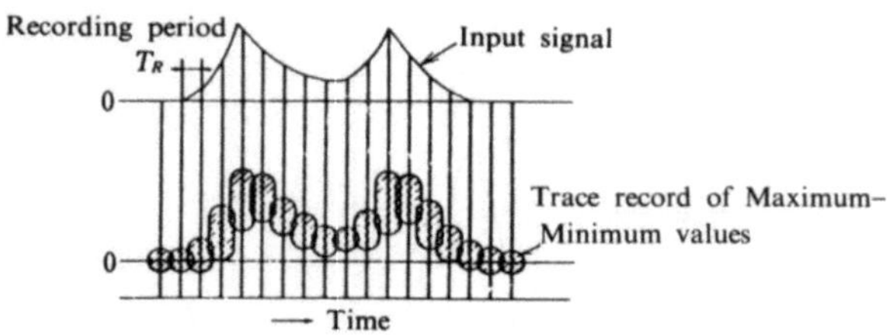

Fig. 4 · 15 Relationship between input signal and trace record of max./min. values

The dot printing recorders generally have from 6 to 30 dots, and a recording speed of from 1 dot per 5 seconds to 30 dots per 5 or 6 seconds. However, in the future, with faster A/D converters and faster processing through distribution of microprocessors to each block, even higher speed multipoint recorders will become commercially available.

4.2 Controllers

A computerized multi–loop digital control system can conveniently handle 8–loop, 16–loop or 32–loop multiple processes. However, here we will describe the structure and functions of a single–loop digital controller, which handles a single process loop. Multi–loop systems are explained in Chap. 5.

4.2.1 Pneumatic and electronic controllers

As explained in detail in the history of industrial instrument development (Sec. 1.1), development of the field of electronics was accompanied by a move away from pneumatic controllers towards electronic controllers, and as computers came to be used in process instrumenta-

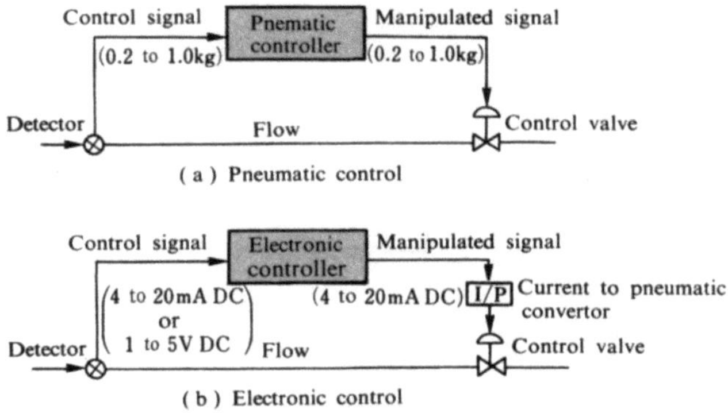

Fig. 4 · 16 Loop configuration of pneumatic and elctronic control

Table 4.3 Comparison between pneumatic and electronic control

Item	Pneumatic	Electronic
Transmission characteristics of signal	Slow	Fast
Usage in explosive gas atmosphere	Safe	Should be careful
Computer compatibility	Poor	Good
Actuator compatability	Good	Poor

tion, the trend became stronger. Figure 4.16 shows the pneumatic and electronic type loop configurations for a flow process, and Table 4.3 lists the special features of each type. Suitability for use in an explosive gas environment is the greatest advantage of pneumatic controllers. However, compatibility with computers has become an important point in recent years as central monitoring of processes using computers and digital control have become popular. Thus the trend is towards electronic controllers. This chapter is mainly concerned with electronic controllers.

4.2.2. Analog electronic controllers

Figure 4.17 is a block diagram showing the basic configuration of a controller. Controller operation can be switched among three modes: manual (M), automatic (A), and cascade (C). In the manual mode, the output signal (manipulated variable) can be directly manipulated by means of the manual control block. In the automatic mode, the controller itself compares the input signal (controlled variable) with the set-point value, and the control computation block acts to make the deviation zero. The output signal (manipulated variable) is then sent to the operation block. The only difference between the automatic mode and

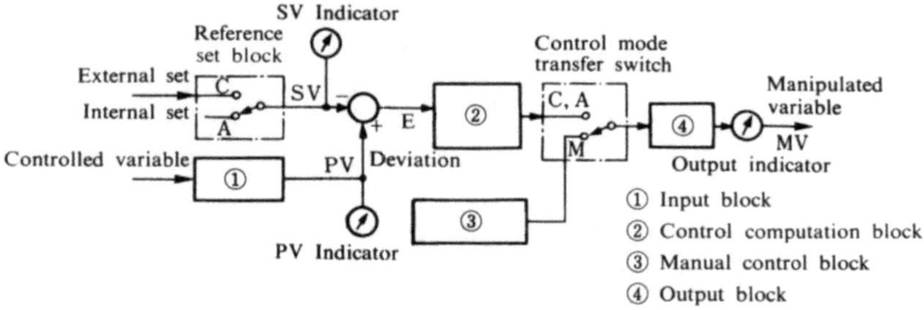

Fig. 4・17 Basic configuration of controller

the cascade mode is that in the cascade mode the setpoint is obtained from outside the controller.

The indication and operation part of this kind of controller is shown in Fig. 4.18. The indicator arrangement is generally such that the relationship between the setpoint value and the controlled variable can be easily seen at a glance. Setting of the setpoint value and manual operation is generally done from the front of the instrument, as shown in the figure, and the setting functions for other control constants are generally placed on the side.

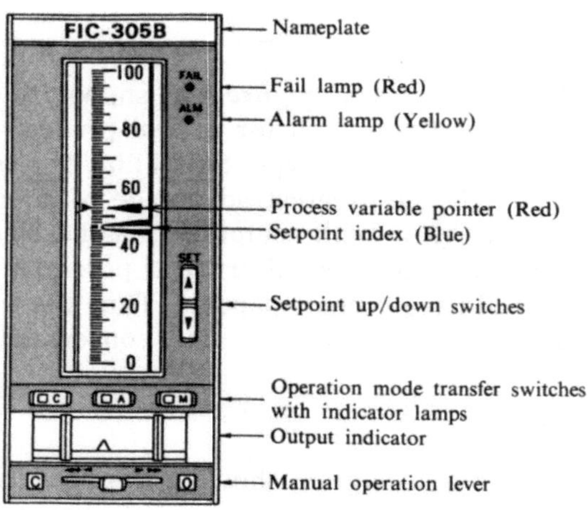

Fig. 4・18 Indication and operation section of controller

The methods for manipulating the setpoint value and output value include position form setting methods, which typically use variable resistors, and velocity form methods which change the setting relative to the current value using push-button switches. In the example shown in Fig. 4.18, as an example of any type of velocity form setting switch, there is a push-button switch for increasing the value, and another push-button switch for decreasing the value. While either of these buttons continues to be pressed, the value changes. However, there are differences in the manipulating characteristics when used for setting the setpoint value and when used for setting the output value. For example, for the setpoint value, the ability to set the value accurately is the important consideration, and there is the single-speed, 40 seconds per full-scale change rate. For setting the output, on the other hand, two setting change speeds are available. By pressing lightly on the operating lever, the change rate is the same 40 seconds per full scale; but by pressing hard, the change rate becomes a fast 4 seconds per full scale.

(*a*) *Control block configuration example*

Figure 4.19 is an example of the circuitry of a PID controller which uses an analog operation circuit. The deviation signal *E* undergoes derivative and proportional operations in operational amplifier A_1 and integral operation in amplifier A_2, and is then output. The proportional band, derivative time, and integral time are determined by the constants of the resistors and capacitors, and the variable resistors can be set to a wide range of values. Generally, however, resistance of at least several megohms and capacitors having a capacitance of several tens of microfarads are required. Thus in circuit design, care must be taken to maintain high insulating resistance.

A switchover from automatic control (A) to manual operation (M) or from manual to automatic is performed in front of amplifier A_2. Then, A_2 acts as a hold amplifier in manual operation.

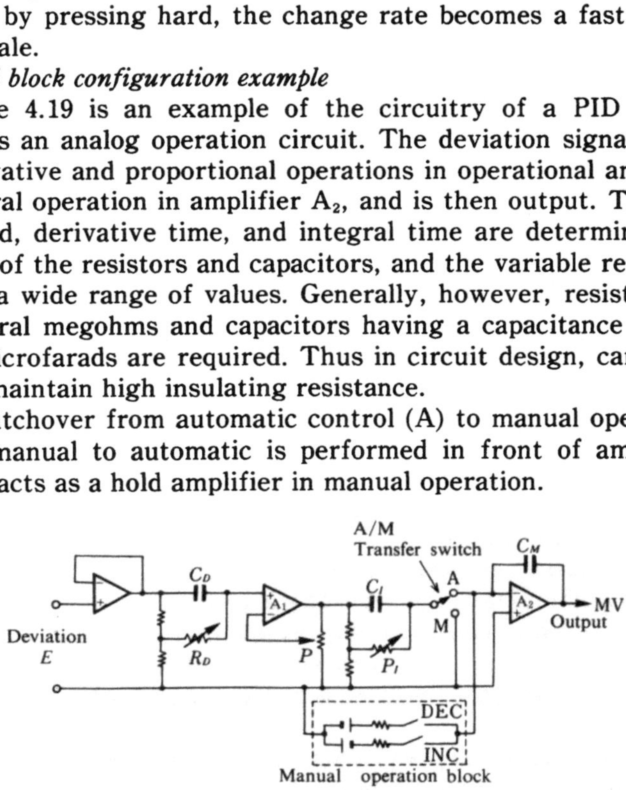

Fig. 4 · 19 Example of analog controller circuit

(*b*) *Manual operation block*

The manual operation block has the important function of allowing the operator to directly operate the valve at plant start-up or when abnormalities occur. The ability to switch from manual operation to automatic control or from automatic to manual without any sudden changes in output, known as "bumpless switching," is required function. In the circuit shown in Fig. 4.19 for example, when the A/M switch is in the "M" position, the output signal is determined by the electric charge on capacitor C_M. The output signal can be manipulated by the charging or discharging current flowing through the INC (increasing) or DEC (decreasing) switches in the manual operation block.

On the other hand, the entire output of amplifier A_1, which operates on the deviation signal E, is stored in capacitor C_I so even if the A/M switch is thrown to "A," the charge on C_M does not change suddenly. Conversely, the charge on C_M is not changed by switching from "A" mode to "M" mode. In this way, bumpless switching between the automatic and manual modes becomes possible.

(*c*) *Control action*

The simplest control action is the on–off operation. For example when using a heater to control temperature, the heater is turned off when the setpoint temperature is exceeded, and turned on when the temperature drops below the setpoint. However, both overshooting and undershooting from the desired temperature occurs, producing a zig–zag control effect. To obtain a good control without constant cycling, PID control action is often used.

The basic formula of PID controllers was shown in Chap. 2 (Eq.(2.7)), but an actual analog controller such as the one shown in Fig. 4.19 uses the following equation.

$$\mathrm{MV}=\frac{100}{\mathrm{PB}}\left(1+\frac{1}{T_I s}\right)\left\{\frac{1+T_D s}{1+(T_D/m)s}\right\}E \tag{4.1}$$

Here, MV is the manipulated variable, E is the deviation, PB is the proportional band, T_I is the reset (integral) time, T_D is the rate (derivative) time, m is the derivative gain, and s is an operator.

At plant start-up time, the values for PB, T_I and T_D can be set to the most suitable values for the characteristics of the process using the variable adjustors provided on the side of the controller. However, as seen from Eq.(4.1), the proportional and integral actions and the derivative actions take the form of a product, and create an interaction in the PID action.

In the third term, $1/\{1+(T_D/m)s\}$ is for incomplete differentiation, so normally the value of m is chosen to be between 10 and 20.

4.2.3 Digital controllers

In direct digital control (DDC) using computers for process control, a large number (up to several hundreds) of loops are controlled at essentially the same time. However, with the development of the microcomputer, economical digital controllers which control several loops or even a single loop have become practical, rivaling analog controllers.

In digital controllers applying microprocessors, ratio control, cascade control, feed-forward control, nonlinear control and other such control algorithms, which are difficult to achieve with conventional analog controllers, can be easily realized. Beyond that, digital controllers can be equipped with preprocessing of the input signal, data transfer to a supervisory system, self-diagnosis, and many other such functions.

(a) Configuration example

Figure 4.20 shows an example of a digital controller which uses a microprocessor. The entire operational function of the controller is programmed in the system read-only memory (ROM). First, an analog signal is selected by the multiplexer. It then undergoes A/D conversion and is read into the microprocessor as a digital variable. After digital processing, the signal is converted back to an analog form by the D/A converter and then output. This operation is repeated continuously in intervals 0.1 or 0.2 seconds. During the intervals, the output signal is maintained at a constant value until it is updated at the next cycle.

This method is a sampling control method, but as the sampling time is ten times or more faster than the process time constant, its control ability is about the same as that of analog controllers. Also, every type of computational and control function required of the controller can be executed by the microprocessor, so these controllers have the advantage of being suitable for diversified control methods.

It is desirable for the A/D converter used in digital controllers to

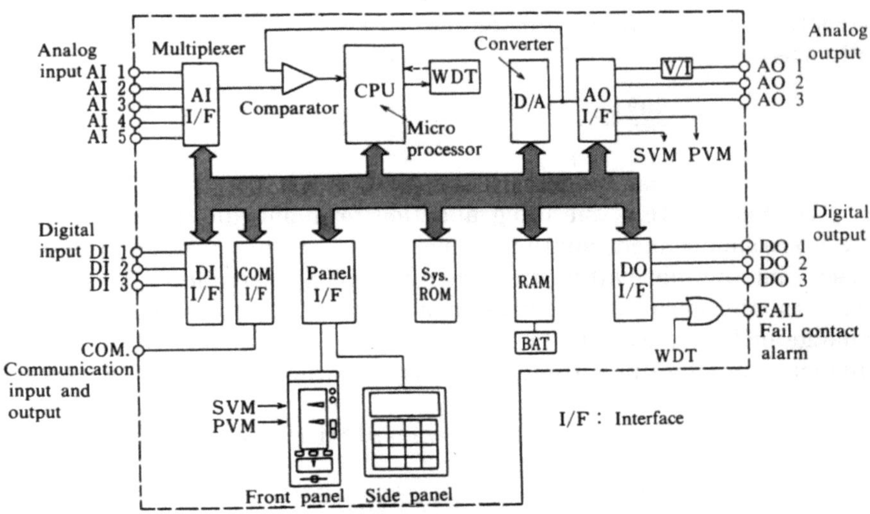

Fig. 4 · 20 Construction of digital controller

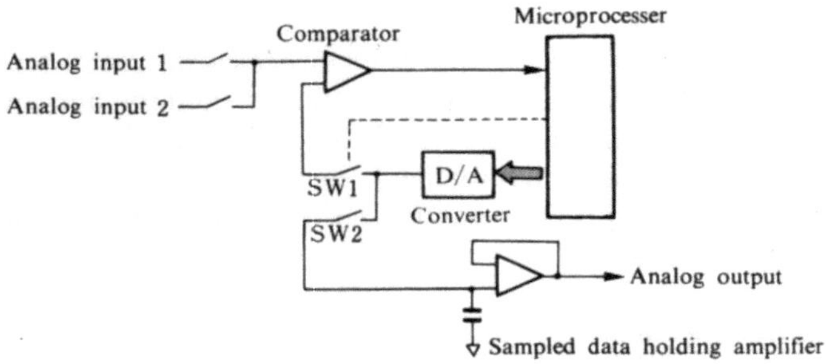

Fig. 4 · 21 Micro processor based A/D-D/A converter

have a resolution of at least 12 bits, and conversion speed of a few microseconds. Generally, the successive approximation method is used. The conversion procedure is controlled by the microprocessor, which allows simplification of the circuitry. Figure 4.21 shows an example of a single D/A converter being used for both D/A conversion and A/D conversion. First switch SW_1 is turned on, and the comparator, D/A converter, and microprocessor form an A/D converter by the successive approximation method, and the analog input signal is converted into a digital signal. When converting the computation results of the microprocessor into an analog signal, SW_2 is closed, the signal passes through the D/A converter and is output to the sample and hold amplifier. The A/D conversion method is explained in detail in Subsec. 5.5.2.

(*b*) *Controller action*

For digital controllers, any type of control algorithm can be constructed. In the automatic mode, the following PID algorithm is generally used.

$$\mathrm{MV}=\frac{100}{\mathrm{PB}}\left\{\mathrm{PV}+\frac{1}{T_{\mathrm{I}}s}E+\frac{T_{\mathrm{D}}s}{1+(T_{\mathrm{D}}/m)s}\mathrm{PV}\right\} \tag{4.2}$$

where PV is the controlled variable.

In this equation, the integral action term and the derivative action term are independent additional operations, so the interaction between reset time and derivative time seen in analog controllers does not occur. Furthermore, the proportional action is only with respect to change in PV, so even if the setpoint value (SV) is changed, there is no sudden change in the output, which is a desirable operating characteristic.

In the cascade mode, the following algorithm is largely used.

$$\mathrm{MV}=\frac{100}{\mathrm{PB}}\left\{E+\frac{1}{T_{\mathrm{I}}s}E+\frac{T_{\mathrm{D}}s}{1+(T_{\mathrm{D}}/m)s}\mathrm{PV}\right\} \tag{4.3}$$

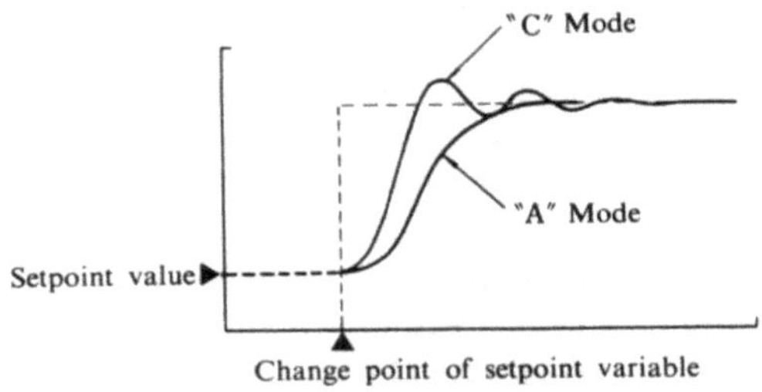

Fig. 4 • 22 PV Response to setpoint value changes

In this equation, the proportional action responds to the deviation (setpoint minus measurement) signal of the controller. Thus fast response to changes in the setpoint value is achieved. The difference in controlled variable response to stepwise change in the setpoint variable for the two kinds of control represented by Eqs. (4.2) and (4.3) is shown in Fig. 4.22. Digital controllers can in this way switch between two or more control algorithms and execute the most suitable control for the purpose at hand. The processing within the digital controller is

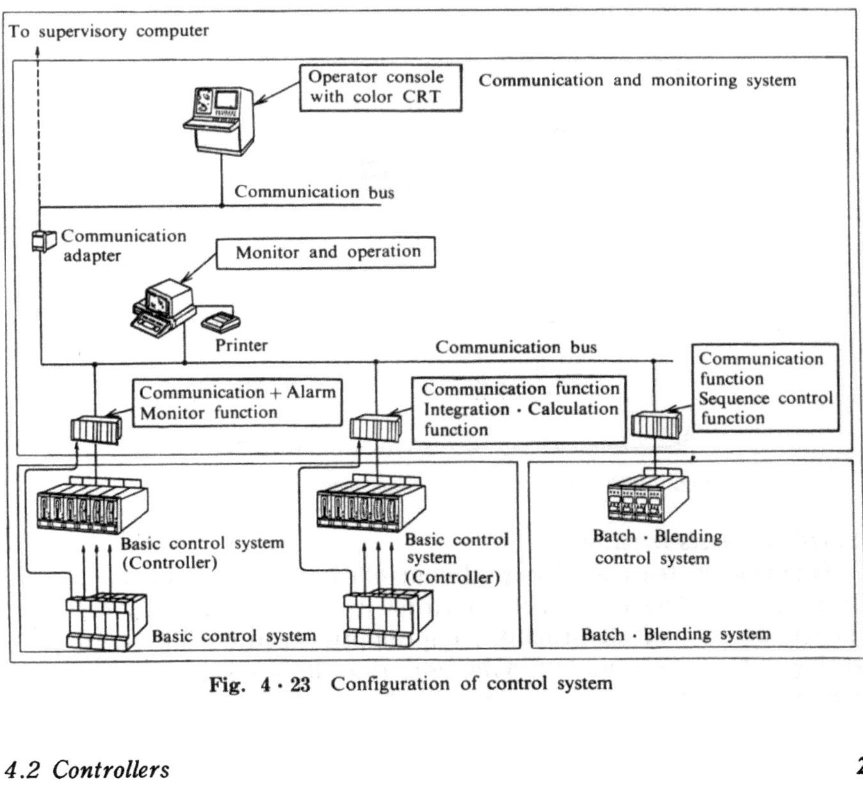

Fig. 4 · 23 Configuration of control system

a discrete in sampling interval *T*, and Equations. (4.2) and (4.3) should be expressed in discrete form. However, for ease of comparison with analog controllers, they have been expressed in a continuous form.

(*c*) *Communication with a supervisory system*

When considering the configuration of a process control system, generally the principle of "distributed control and centralized information" is followed. Figure 4.23 shows an example of a system configuration. Digital controllers are arranged in a dispersed fashion, with a CRT-equipped operator console as the heart of the system. Input data coming through the communication bus from the controllers can be centrally monitored on the CRT, and at the same time even a change in the setpoint value and output value for each controller can be manipulated from the supervisory system side. This type of configuration is called a setpoint control system (SPC) or direct digital control (DDC) system.

An example of the operating conditions of a controller displayed on the CRT of the supervisory system based on the data sent from the controller is shown in Fig. 4.24. This screen shows the input and output signals, and at the same time the high and low alarm settings of the input signals (PH and PL). Thus abnormal process conditions can be detected from the screen.

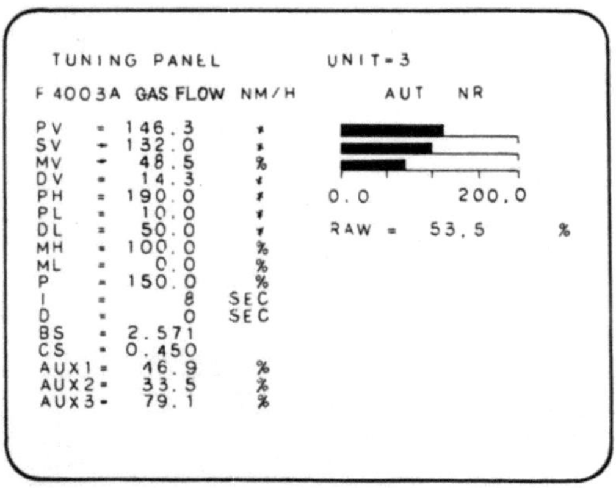

Fig. 4・24 Monitor and operation panel of controller

(*d*) *Self-diagnosis function*

One special feature of digital controllers using microprocessors is self-diagnosis. The controller confirms its own operating conditions. When it detects an abnormality, it promptly notifies the operator by lighting a lamp or sounding a buzzer, thus improving the safety and maintainability of the plant operation. The following are self-diagnosis items.

(1) Microprocessor abnormalities
(2) A/D converter and D/A converter accuracy
(3) Range over of input signal
(4) Computational overflow
(5) Break in the electric current output

4.2.4 Programmable controllers

In analog controllers, there is a one-to-one correspondence between hardware and computation functions, thus the execution of complex control calculations is difficult. On the other hand, with controllers employing microprocessors, various additional control calculations can be performed through software. Thus in addition to the basic PID algorithm, complicated control algorithms can be performed easily by assembling various software modules chosen according to the requirements.

Figure 4.25 shows a programmable controller set up for flow ratio control. In controlling flow so as to maintain a fixed ratio between line A flow rate and line B flow rate, the flow rate (pressure differential) of line A is measured, the square root is taken, the ratio is found by division, the bias added, and the result serves as the setpoint value for line B control. Accordingly, it is necessary to combine the square root

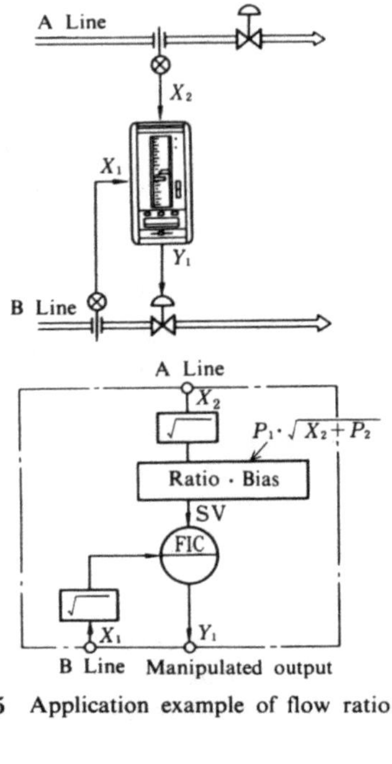

Fig. 4 · 25 Application example of flow ratio control

operation, division operation, and addition operation.

Programmable controllers can be programmed by the either the fill-in-the-form method (FIF), or the step writing method (procedure-oriented language, problem-oriented language in a narrow sense). In the former, module specification items are written into a table to assemble the program; in the latter, the functions to be executed are written as instructions. We will describe the step writing method here.

(*a*) *Computation principle*

The operation of programmable controllers is built up of three functions (input reading, computation, and output of the result), and can be written as instructions. In Fig. 4.26, the program for a two-input adder and an arithmetic register operation are shown. The computation is done completely in a common arithmetic register S. Connection of the adder signal, that is to say the S register input, is performed by a LOAD instruction (written as LD). The S register consists of stacks S_1 to S_5. By the LD instruction, the data is pushed down from S_1 to S_2 with each input. The FUNCTION instruction is used to perform operations on the input data. The operation modules required by control computation are prepared. They are written as symbols such as $+$, $-$, $\div$, $\times$, $\sqrt{\ }$, and so on. The data in the S register is used for the number required by an operation, and the results are stored in regsister S_1. To read out the computation results, the STORE instruction (written as ST) is used.

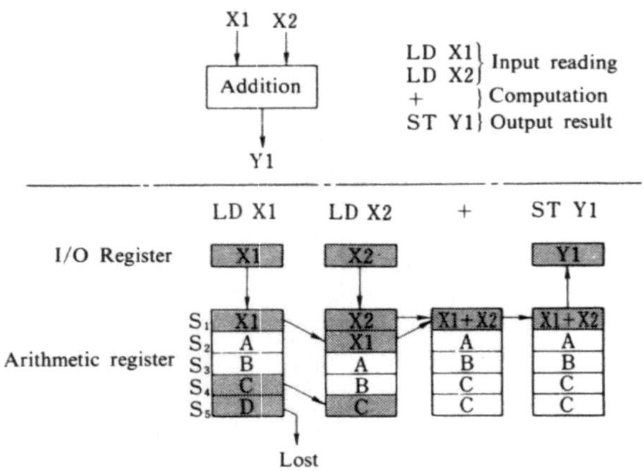

Fig. 4 · 26 Arithmetic stack register operation

(*b*) *Structure of the input/output register*

Figure 4.27 shows the structure of the input/output register. Before execution of the user program, the analog input, digital input and setting parameters are put into the X_N, DI_N, P_N registers, respec-

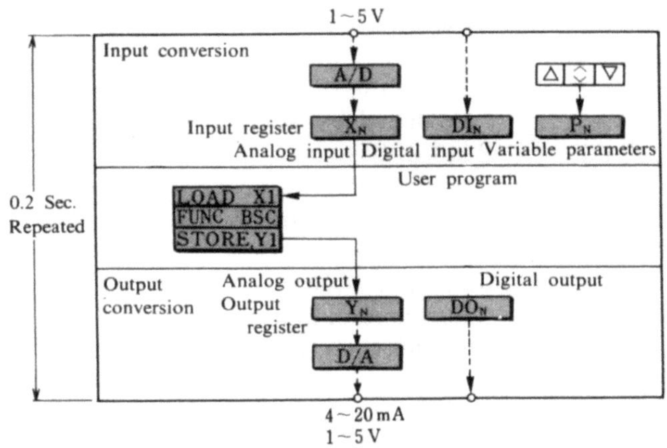

Fig. 4·27 I/O Register configuration

tively, as normalized values. In the user program, as described earlier, the required input signal and parameters are loaded from their respective registers into the arithmetic register by the LD instruction. Then the results of the calculation are connected to the output register by the ST instruction. Next, the contents of the Y_N or DO_N registers are converted to analog or digital output. The user program handles all signals as normalized digital data, so concern over process input/output conversion is unnecessary.

(c) Computational and control functions

Examples of the computational and control functions of a programmable controller are listed in Table 4.4. The basic PID control function BSC, and the cascade function CSC, in which two controller units are connected in series, are each defined as a different computational function. Thus, if the programmable controller is to be operated as a simple PID controller, then the program is simply written as follows.

1 LD X1
2 BSC
3 ST Y1
4 END

The BSC functions are shown in Fig. 4.28. As a dedicated extended function of BSC, the A and FL registers can be used by the program. For example, when it is desired to set the setpoint value for cascade, the cascade setting input is connected to the A1 register by the ST instruction.

The program for the flow rate ratio control in Fig. 4.25 is written as follows.

1	LD X2	:	Read in X2 input
2	$\sqrt{\ }$	:	Calculate $\sqrt{X2}$
3	LD P1	:	Read in the ratio
4	$*$	:	Calculate $P1 * \sqrt{X2}$
5	LD P2	:	Read in the bias
6	$+$	:	Calculate $P1 * \sqrt{X2} + P2$
7	ST A1	:	Connect to the cascade input terminal
8	LD X1	:	Read in X1
9	$\sqrt{\ }$	:	Calculate $\sqrt{X1}$
10	BSC	:	Basic control module
11	ST Y1	:	Connect the results to the operation output
12	END	:	

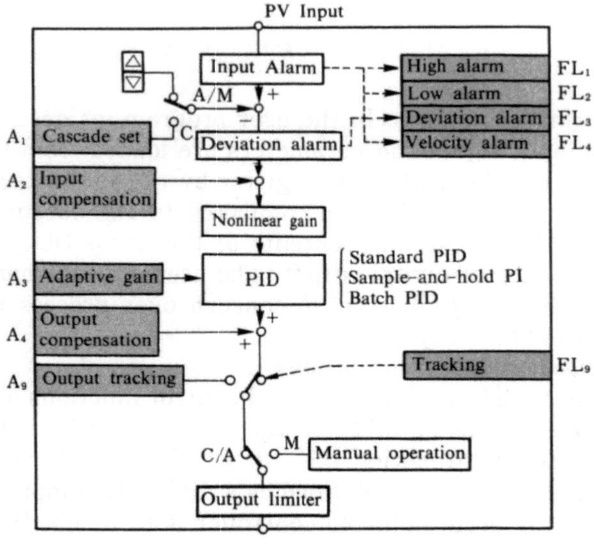

Fig. 4 · 28 BSC Function

(d) Programmer and programming procedure

The program can be written either by using the programming panel on the side of the controller to write directly to the RAM, or by using a dedicated tool called a programmer to write to the ROM. The functions of this programmer include:

(1) Program key-in,
(2) Display and set the computational constants,
(3) Test the finished program,
(4) Display the process input and output,
(5) Write to the ROM, and
(6) Print out the program listing.

Table 4.4 Computational functions of programmable controller

Function name	Instruction code	
Basic four function	+ − * /	
Signal conversion		
Absolute	ABS	(Absolue)
Square root extraction	SQT	(Square root)
Line segment	FX	($f(x)$)
Selector		
High	HSL	(High selector)
Low	LSL	(Low selector)
Limiter		
High	HLM	(High limiter)
Low	LLM	(Low limiter)
Velocity	VLM	(Velocity limiter)
Dynamic functions		
1st order lag	LAG	(Lag)
1st order lead	LED	(Lead)
Dead time	DED	(Dead)
Velocity	VEL	(Velocity)
Timer	TIM	(Timer)
Analog conditional decisions		
Comparison	CMP	(Compare)
High alarm	HAL	(high alarm)
Low alarm	LAL	(Low alarm)
Logical functions		
AND	AND	(AND)
OR	OR	(OR)
NOT	NOT	(NOT)
Branching	GIF	(Go if)
Control functions		
Basic control	BSC	(Basic control)
Cascade control	CSC	(Cascade control)
Selector control	SSC	(Selector control)
Other		
Analog I/O	LD	(Load)
Digital I/O	ST	(Store)
Branching	GO	(Go to)
End	END	(End)

Figure 4.29 shows an example of such a programmer connected to a controller.

4.2.5 Batch controllers and blending controllers

Fluid blending and batch shipment systems use dedicated controllers. Figure 4.30 shows an example of the configuration of a batch ship-

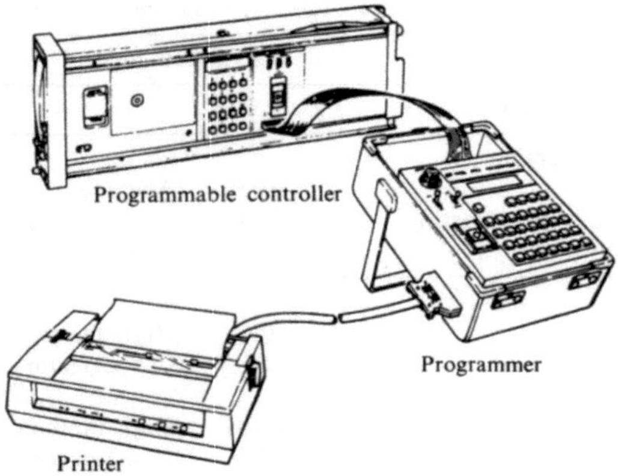

Fig. 4 · 29 Connection of programmer

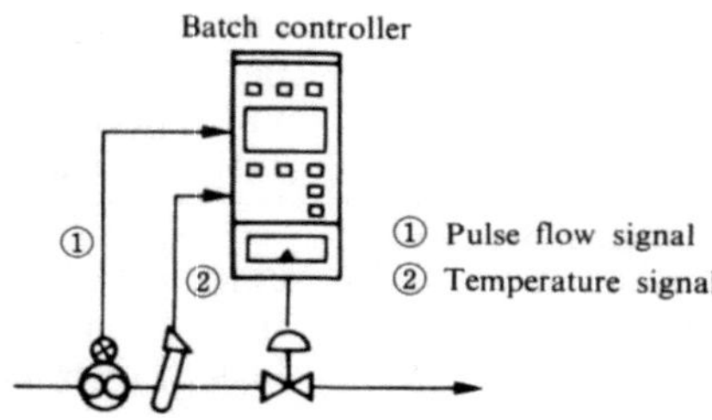

Fig. 4 · 30 Batch shipment system

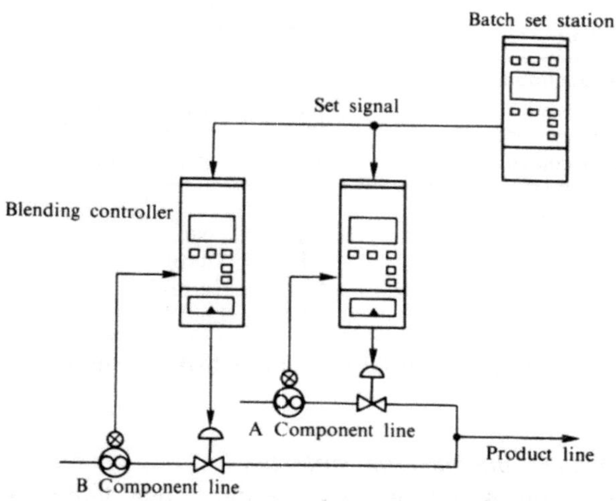

Fig. 4 · 31 Blending shipment system

ment system. Batch controlleres are PI controlleres that perform programmed control of the measured flow rate such that accmulated flow produces a pre-set batch volume. The program is written to suit the controlled object, and can be freely specified at the time.

Figure 4.31 shows an example configuration of a blending shipment system. The blending controller receives a set signal from the bath set station, and performs PI control such that the integrated values of the set signal and the measured flow rate maintain a fixed ratio.

The control algorithm for blending controllers is expressed by the following equation.

$$\begin{aligned} MV &= \frac{100}{PB}\left(\frac{1}{s}+\frac{1}{T_I s^2}+T_D\right)E \\ E &= SV-PV \end{aligned} \qquad (4.4)$$

(*a*) *Display, setting, and operating functions*

An example of the front display and operation panel of a batch controller is shown in Fig. 4.32. The batch sequence command switches are located at the top, the batch set value and the totalized value can be monitored at the same time on a two-stage digital display, and in

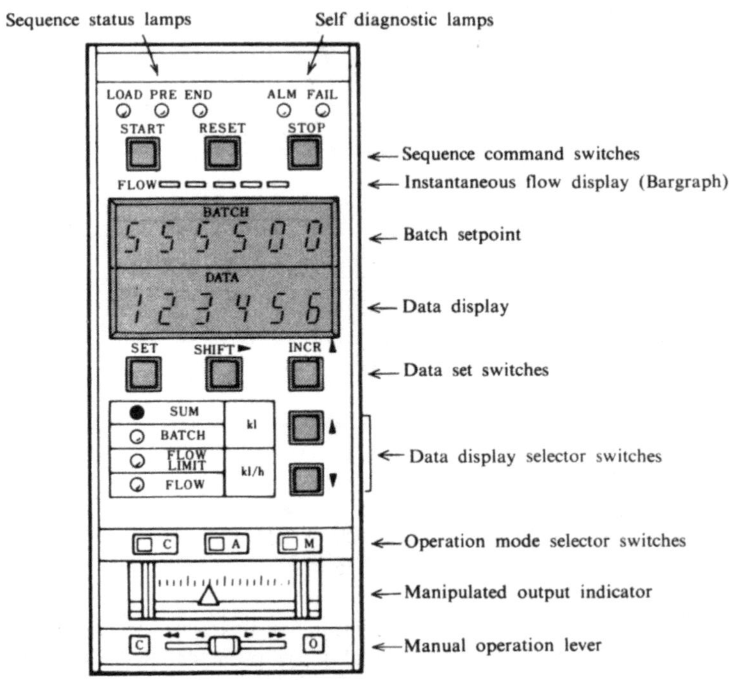

Fig. 4 · 32 Front view of batch controller

addition, all parameters can be set using this display.

(*b*) *Process input/output interface*

The inputs to a batch controller include the flow rate input, compensation input, and contact input. Also, the form of the signal from the pulse flow signal transmitter has not been standardized, so there are differences in specifications such as the power supply voltage for the transmitter, the signal amplitude, the frequency, and the load resistance according to the manufacturer. Thus it is necessary to provide interfaces matched with these various specifications, and there is some inconvenience concerning maintenance. Figure 4.33 shows an example of a pulse input circuit constructed to support a wide range of transmitter specifications. The transmitter power supply and load resistance can be switched to match the specifications of the transmitter. The specifications of this pulse input circuit are listed in Table 4.5.

(*c*) *Flow rate compensation*

In batch blending systems, flow rate compensation is an important

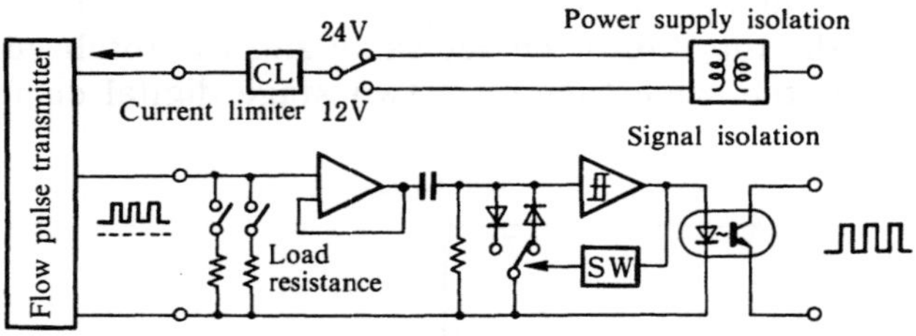

Fig. 4・33 Pulse input circuit

Table 4.5 Pulse input specifications

Item	Description
Type of pulse signal	Voltage level pulse or contact
Frequency	0 to 6.0kHz, zero elevation not available
Minimum pulse width	50μs
Voltage level pulse signal	
ON (E_H)	3 to 24V DC
OFF (E_L)	-1 to $+8$V DC
Pulse peak value ($E_H - E_L$)	3V or more
Input resistance	10kΩ or more
Power supply voltage	12V DC or 24V DC
Current	8mA or less
Contact signal	
ON	Signal source resistance 200Ω or less
OFF	Signal source resistance 100Ω or more
Signal source rating	30V DC, 30mA or more

function for accurate trade. High accuracy is required, and the compensation method must be adaptable for various types of compensation according to the type of fluid. The following three types of compensation computational formulas are used.

$$c=(1+\alpha)f(\rho, t) \tag{4.5}$$

$$c=(1+\alpha)\{1+10^{-2}\beta(t-t_0)+10^{-6}\gamma(t-t_0)^2\} \tag{4.6}$$

$$c=(1+\alpha)\{(TI_2)+0.5\}, \quad TI_2=0 \text{ to } 1.0 \tag{4.7}$$

$$F=cF_0 \tag{4.8}$$

In these equations, c is the flow rate compensation coefficient, α, β, and γ are compensation coefficients that can be set freely, ρ is the density, t is the measurement temperature, t_0 is the compensation reference temperature, F_0 is the flow rate before compensation, and F is the flow rate after compensation.

Equation (4.5) relates mainly to petroleum, and using the generally used ASTM* (specific gravity–volume conversion table), automatically calculates the compensation coefficient for the reference temperature. The only specified data is the specific gravity of the fluid. The coefficient α is for specifying compensation for instrumental error of flow transmitter, required depending on the properties of the fluid such as viscosity. It can be used for various other purposes according to the fluid being measured.

Equation (4.6) is a multiple-item compensation computation for processes other than petroleum processes, and has a wide degree of freedom. The temperature vs. density curve is approximated by a quadratic equation, and compensation coefficients β, and γ are set. Equation (4.7) is for special conditions for which the above Eqs. (4.5) and (4.6) cannot calculate compensation, and is used by connecting with an externally computed analog compensation input (TI_2). It is linearly converted over the range of 0.5 to 1.5 corresponding to a 0% to 100% change.

(d) Flow rate addition and subtraction

In the shipment of fluid of volatile fluids such as LNG, several percent of the flow changes to a vapor and returns, causing error in the shipment volume. The measured vaper flow is introduced to the flow rate addition and subtraction function through analog or pulse input, and compensates for the lost amount, thus making highly accurate shipping control possible. A block diagram of the functions of a batch

*ASTM: American Society of Testing Materials.

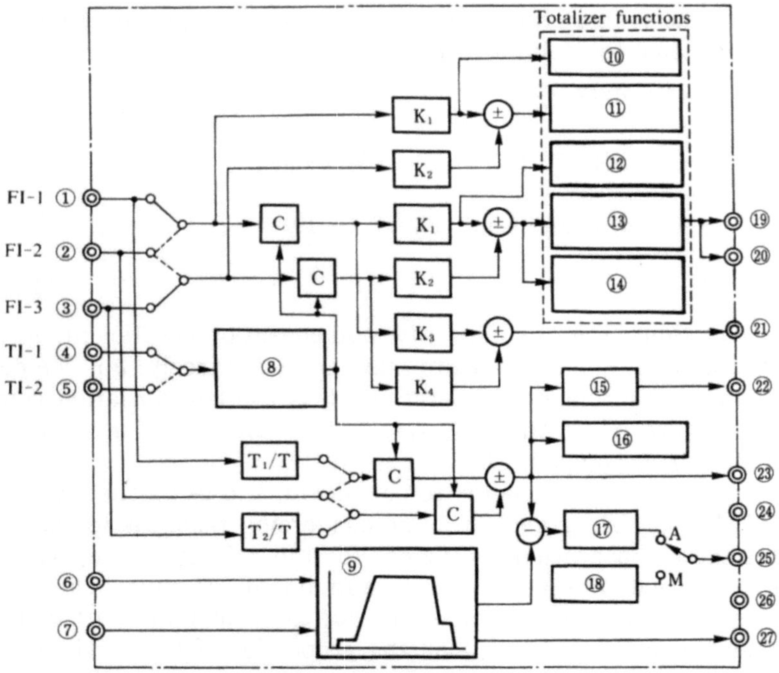

① Process variable input (Pulse signal)
② Process variable input or added/subtracted flow signal (1 to 5V)
③ Added/Subtracted flow signal (Pulse signal)
④ Compensation signal (Pt·100 Ω)
⑤ Compensation signal (1 to 5V)
⑥ Start, Stop, Reset signal
⑦ Master pacing
⑧ 1) ASTEM equation
 2) Quadratic equation
 3) General compensation equation
⑨ Flow setpoint program
⑩ Measured batch flow
⑪ Cumulative flow totalizer (With added/subtracted flow signal)
⑫ Measured batch flow (With compensation computation)
⑬ Batch flow totalizer (With added/Subtracted flow signal and compensation computation)
⑭ Cumulative flow totalizer (With added/Subtracted flow signal and compensation computation)
⑮ Alarm
⑯ Instantaneous flow display
⑰ PI Control
⑱ Manual operation
⑲ Batch end pre-alarm
⑳ Batch end alarm
㉑ Flow signal repeater (Pulse signal)
㉒ Alarm output
㉓ Flow signal repeater (1 to 5V)
㉔ Communication
㉕ Manipulation signal (4 to 20mA)
㉖ Fail signal
㉗ Reset signal

Fig. 4·34 Block diagram of batch controller

controller are shown in Fig. 4.34.

(*e*) *Batch sequence*

The batch sequence is controlled by the start, stop, and reset buttons on the front panel, or by external contacts. At the same time, the status of the controller can be known from lights or the output of external contacts. Figure 4.35 shows the relationship between batch sequence and flow rate settings.

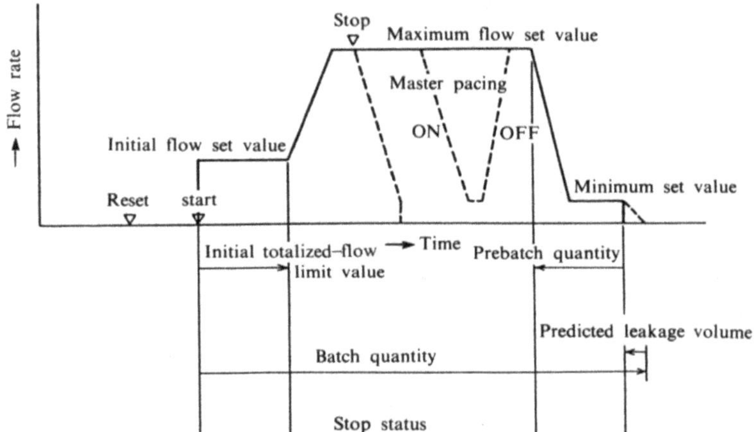

Lamps	LOAD	☼	●	●	☼		●	
	PRE				☼		●	●
	END				☼			●
Status (Contacts)	Prebatch	Open	Open	Close	Open	Close	Open	Open
	Batch	Open	Close	Close	Open	Close	Close	Open
	Reset	Close	Open	Open	Open	Open	Open	Open

Fig. 4 • 35 Batch sequence and flow rate setting

4.3 Computing Stations and Set Stations

4.3.1 Alarm set stations

An application example of an alarm set station is shown in Fig. 4.36. For safe system operation, control status is monitored by an alarm set station, which is a system component that is independent of the controller. This device monitors the absolute value of the measurement signal, the deviation between two signals, and so on. When an abnormality arises, its contact output sounds an alarm buzzer, may initiate a system shutdown sequence, or initiates some other such action. An example of the circuit of an absolute high and low alarm set station is shown in Fig. 4.37. Generally, the action mode is selectable accord-

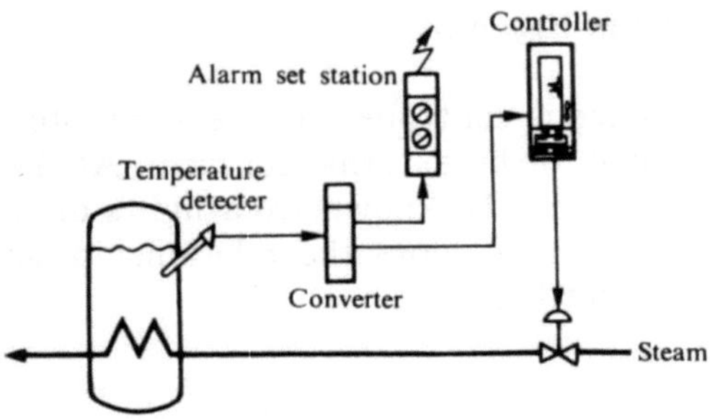

Fig. 4 · 36 Application example of alarm set station

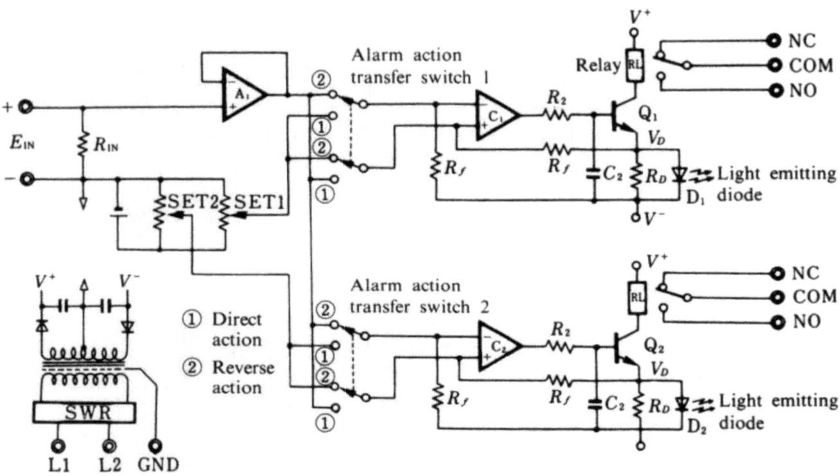

Fig. 4 · 37 Principle of absolute high & low alarm set station

ing to whether the relay is activated or deactivated when the alarm is generated, with deactivation being the fail-safe direction.

After the input signal has reached the alarm setpoint, the relay output is latched by a lockup function such that it does not switch back and forth from the on and off positions with small changes in the input signal. The relationship between the relay operation mode and the lockup action is shown in Fig. 4.38.

4.3.2 Programmable computing units

In addition to square root extraction, temperature compensation, and other such operations on the signal from the sensor, this device also performs filtering for input to a recorder or controller. Conventional analog computers were also used for this purpose, but because the computation was implemented in hardware, the accruracy of com-

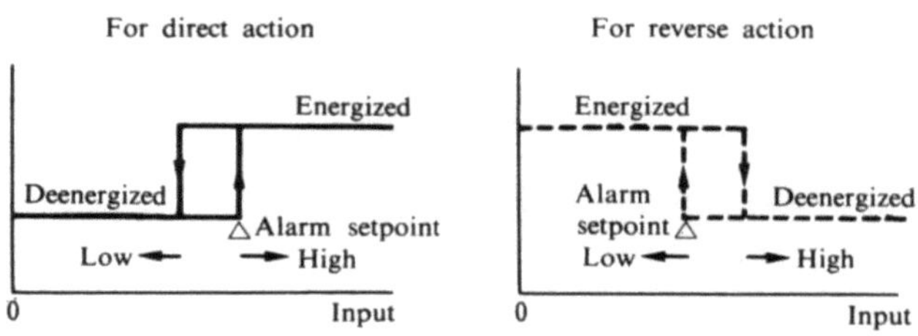

Fig. 4 · 38 Lockup action

plicated calculations was poor and it was sometimes the calculations were not possible.

In the same way as with the programmable controllers in Subsec. 4.2.4, the application of microprocessors allows complex computations to be freely performed through the use of groupings of computational software modules. Figure 4.39 shows an example of a programmable computing unit. It dose not have a display or operating panel, because it is not normally operated by an operator.

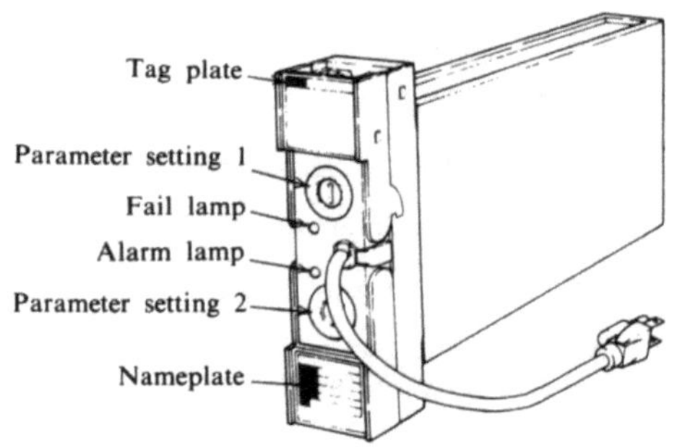

Fig. 4 · 39 External view of programmable computing station

Figure 4.40 shows the interior of the device. The program is written by the step writing method using a dedicated programmer, and written into ROM. The specifications of this type of programmable computing unit are listed in Table 4.6. This device can perform the computations of ten conventional analog units, so it is possible to simplify the rack instruments, which have become overcrowded as a result of current control requirements.

4.3.3 Manual set stations and manual operating stations

A function block diagram for the manual set station is shown in Fig. 4.41. Figure 4.42 shows the function block diagram for the manual operating station. The internal function configuration excludes the

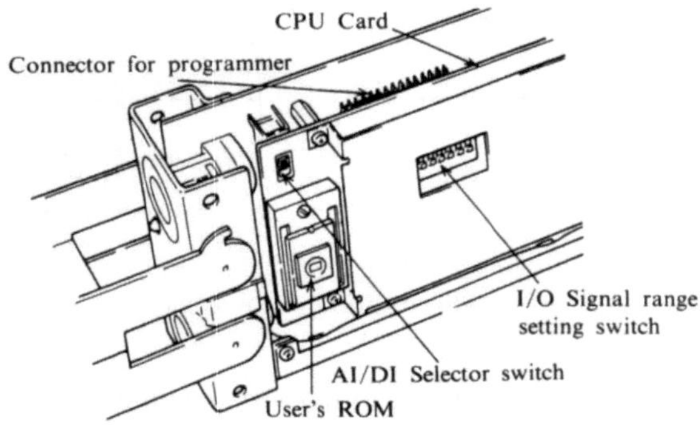

Fig. 4 · 40 Internal view of programmable computing station

Table 4.6 Specifications of programmable computing station

Item	Description
Analog input	1 to 5V DC, 3 points (4 points available with no digital input)
Analog output	1 to 5V DC, 2 points
Digital input	ON–OFF signal dry contact or voltage, one point
Digital output	Transistor open collector, one point
Adjustable constant	One turn potentiometer, 2 points (0 to 100% scale)
Fail output	Transistor open collector, one point
Calculation function	Addition, subtraction, multiplication, division, and logical sum, etc., 26 functions
Program step number	Max. 99 steps

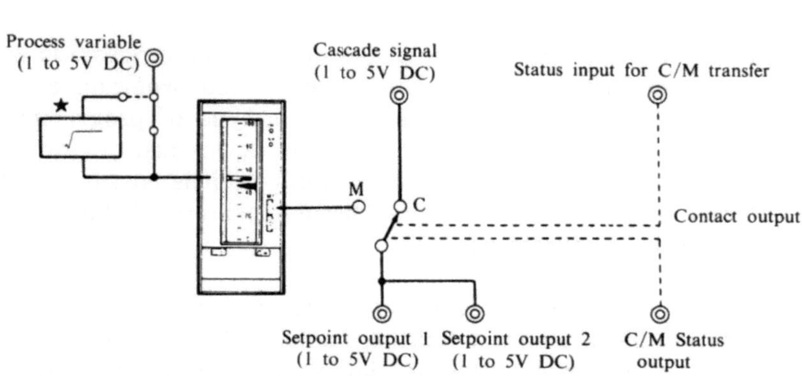

Fig. 4 · 41 Manual set station (Setpoint output)

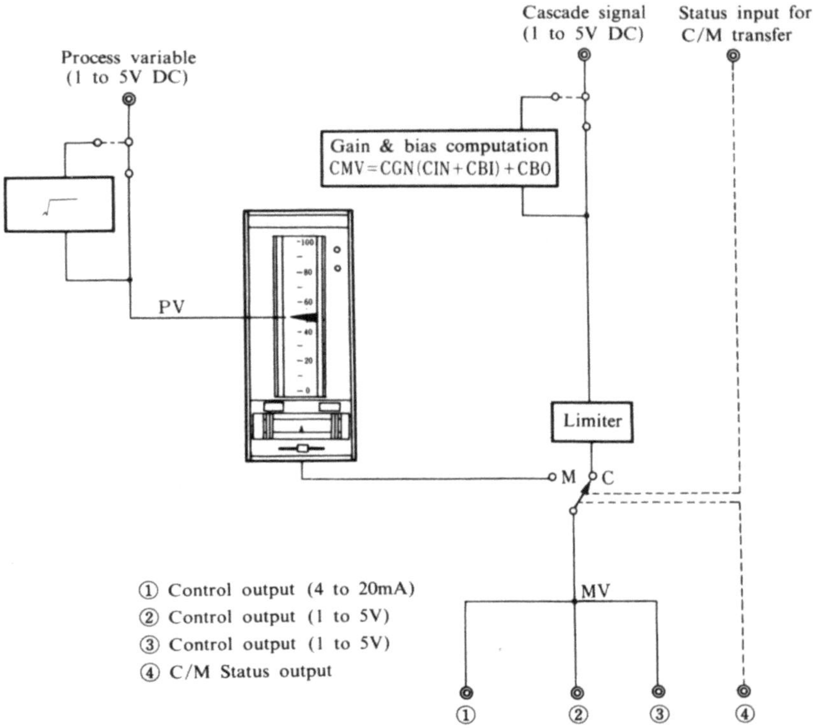

Fig. 4 • 42 Manual operating station (Manipulated variable output)

PID computation function from the controller.

The manual set station is sometimes used to provide a common set-point value to a number of controllers. On the other hand, from the system safety design point of view, the manual operating station is often used as a controller backup device. It is also used as the final control device in the application example in which the output of one controller unit drives a number of final control elements.

REFERENCES

1) Yokogawa Electric Corporation : Industrial Instrument Handbook Pneumatic Instrument Volume, 2nd Edition, Tokyo Denki University Press (1979) (in Japanese).
2) T. Utsumi : "Measuring instruments—Recorders," *Yokogawa Tech. Rep.*, **28**, 2 (1984) (in Japanese).
3) H. Ohtsu, *et al.*: "Programmable intelligent recorders μR 100 / μR 180 series," *Yokogawa Tech. Rep.*, **30**, 3 (1986) 15-19 (in Japanese).
4) I. Morishita : Process Control Digital Instrumentation Systems, The Society of Instrument and Control Engineers (1983) (in Japanese).

5) S. Narita : Digital System Control, Shoko Do (1980) (in Japanese).
6) H. Tamura, *et al.*: "Electronic control system YEWSERIES 80," *Yokogawa Tech. Rep.*, **25**, 2 (1981) (in Japanese).
7) Yokogawa Electric Corporation : Model SLPC, SLMC, SCMS, SPLR, SPRG Programmable Instruments Functions awd Applications, *Yokogawa Technical Information*, TI-1B4C2-02E.
8) Yokogawa Electric Corporation : Model SLCD (Style E) Indicating Controller Functions and Applications, *Yokogawa Technical Information*, TI–1B4C1-03E.
9) Yokogawa Electric Corporation : Batch-Blending Control System Overview, *Yokogawa Technical Information*, TI-1B4A3-01 (in Japanese).

Chapter 5

SYSTEM CONTROL EQUIPMENT

The computer was introduced to process control in the 1960's, and broad advancement in digital control techniques was achieved. After that, electronic devices such as microprocessors developed rapidly. Hand-in-hand with the development of digital communications, distributed control systems first appeared in 1975. Use of this type of system quickly spread, and today has achieved prominence among process control systems. This digitization has greatly transformed control systems, bringing broad improvements in productivity and product quality, and reducing manpower requirements. From a broad perspective with respect to industrial processes, these automatic systems are called process automation (PA).

In the area of discrete processes, where assembly work plays the leading part, factory automation (FA) has recently been developing rapidly towards automation and labor reduction through combinations of such automation machinery as computers, numerical control machines, and robots.

In recent years, the control systems of entire factories have been networked into unified systems which include production management computers. Through the integrated system design, production-related information is accurately transmitted among the system components. Direct connection of this information and the control system shortens production lead time, makes production more flexible, greatly reduces production costs, and leads in the direction of a system matched to an era of diversified manufacturing. This kind of unified production system centered around a computer is called computer-integrated manufacturing (CIM).

This chapter narrows the field of digital control, and explains the most important basic points of distributed control systems for continuous processes, production line control systems for discrete processes, and production management computer systems. The description uses actual examples as much as possible to facilitate understanding and show how everything fits into the total system. The main features of high-reliability technology, and the basic software such as real-time operating systems and control algorithms, and other software which form the basis of digital control are also introduced. These points help clarify the trends in technology.

5.1 Overview of System Control Equipment[1)]

5.1.1 Development

Instrumentation system equipment has progressed hand-in-hand with the expanding requirements of the process industries. This progress has been supported by the development of electronics and control technology. The system control equipment dealt with here is mainly for digital control systems. Figure 5.1 shows an example of the historical development of such systems.

In Japan, the automation of process control (PA) for continuous processes (fluid or continuous material flow) began around 1955, and after that achieved rapid progress. Now, the field of control systems has

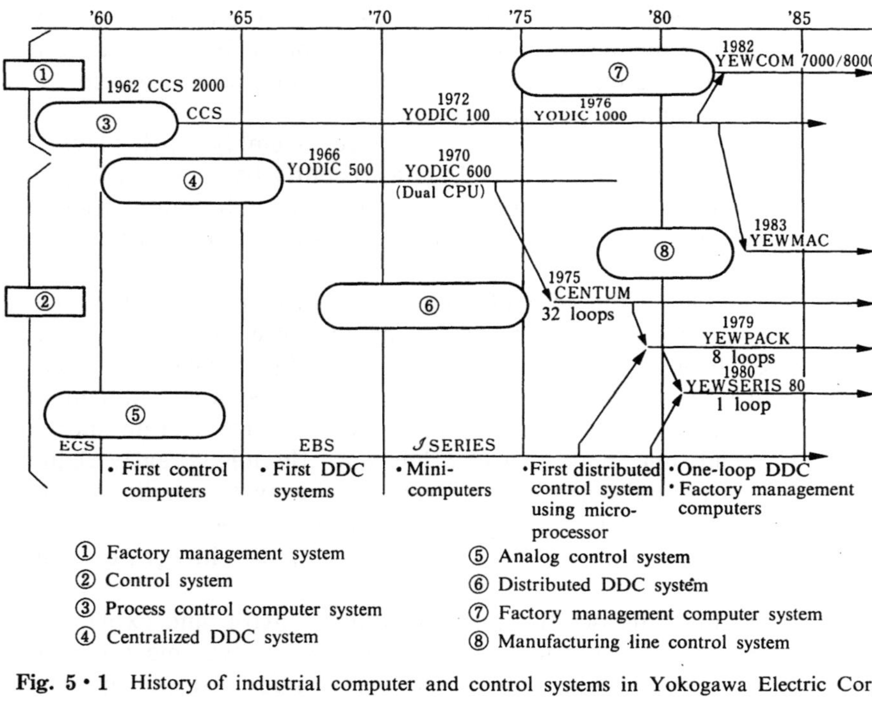

Fig. 5 • 1 History of industrial computer and control systems in Yokogawa Electric Corp.

firmly taken root, both technologically and conceptually. These systems are the backbone of plant operation, and are closely related to productivity, product quality, safety and so on. Looking at the breadth and depth of the increasing practicality of automatic control, it can be seen that the work done by humans is being given over to machines step-by-step as it becomes possible. Process control equipment is making the transition from pneumatic devices to electric or electronic types, through the introduction of computers, and on to direct digital control (DDC), distributed direct digital control, and single loop controllers. The digitization of control equipment is bringing about rapid development of all aspects of control systems.

One such aspect is controllability. There has been spectacular development in control theory, and this has had its effect on the field of process control. However, the basis of control algorithms, even now, is still the classical control theory, and the PID control method is the mainstream of process control. Still, complicated calculations are possible through the use of digital computing equipment, and with appropriate switching of PID control parameters and nonlinear control, advanced control yielding high productivity while keeping in mind safety limits is being achieved. Also, by combining sequence control and PID control, the automation of plant start-up and the full automation of batch processes has become possible.

Moreover, process control was performed by humans actively handling the control elements, thus managing the entire plant. However, as plant construction became complex and highly developed, and the pressure for reduction of labor costs became stronger, the dependence of process control operation on computers became higher. Also, the operational feature of process control changed from large instrument panels to the completely new method of the CRT. The copious flexible functions of the CRT allowed the revolutionary transition to graphic display of information, clear, simple and accurate alarm indication, and simple operation based on one-touch operation that effectively prevents operation error. Furthermore, the meaning of operation itself has been changing from the adjustment of control loop settings to management aspects including specification of product names, production amount, product quality, and so on. Figure 5.2 shows the conceptual trend in operation. The trend towards integrated control systems is expected to continue, and operation of these systems will undergo great development making use of high performance features such as display of data from other systems on CRT monitors, ultra-high density, and artificial intelligence.

The cost-performance ratio of computers is decreasing by a factor of 10 every ten years, with capacity and speed of computation increas-

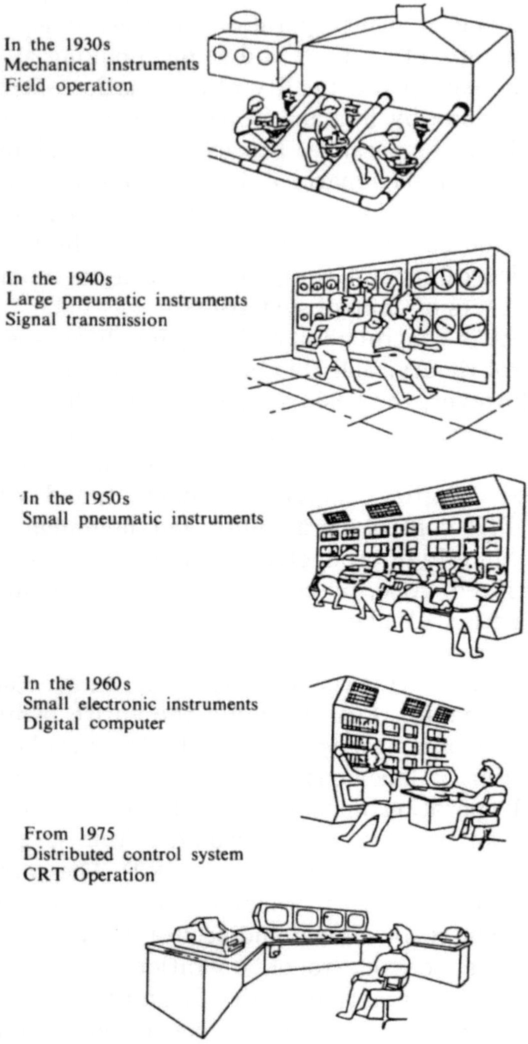

Fig. 5 · 2 History of man-machine interface (Reprinted from "Prevention review" 140 p.18 issued by Japan non-life insurance association)

ing. Powerful database management systems form the core of production management, such as production and process planning, and process line management such as optimizing calculations and process simulations. The apportionment of these various areas of management is becoming a goal of management by control, and the transformation

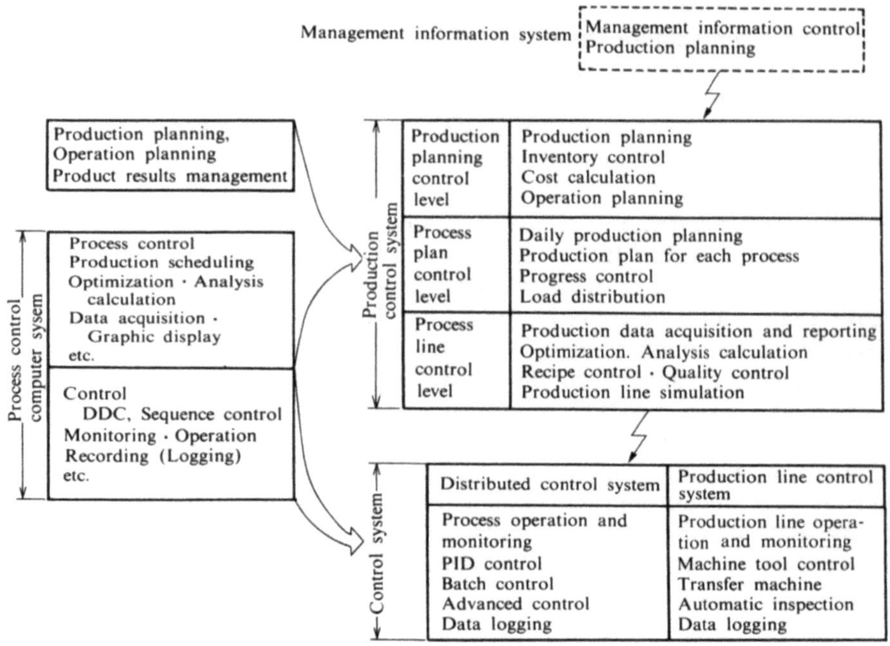

Fig. 5 · 3 Trends of role in process control computers

of the control system structure into a hierarchical form is becoming clear. (Fig. 5.3)

Taking the automotive industry as a representative case, discrete processes like machining and assembly developed, quickly improving in productivity compared to the original mass production or lot production. Regarding the control of these processes, numerical control (NC) first appeared in 1948, followed by the machining center and DNC. Then came the programmable controller (PC) in 1969, which came to play an important role. Entering the era of diversity, low-volume production of a varied line of products followed the trend of the times. To handle the business related to this, in those years the personal computer, which had become economical and spread very quickly, had begun to be used in sales management. Although the concept of factory automation (FA) had begun to be known in the 1980's, in 1983 the special features of the programmable controller and the personal computer were integrated, and the FA computer was achieved. Thereafter, FA made great and rapid progress.

Actually, however, FA has still not been fully developed. The automation of factories is not sufficient. Also, the combination of

CAD/CAM with production management is still in the planning stage. However, even though there are no large integrated systems, partially automated lines consisting of combinations of numerical controlled machines and robots, and automated local material flows, as well as the linking of assembly line and production management being steadily developed. These have reached a fairly advanced level.

5.1.2 Configuration of a total FA system

Recently, terms like "complete factory automation," and "total FA" have begun to be widely used. However, these expressions mean more than simply automating manufacturing processes. As shown in Fig. 5.4, these concepts apply the latest in information processing, control and mechatronics technologies to such production activities in the factory as design, production line, inspection, production management, and orders and shipping. The goal is an integrated factory system designed for flexible, minimum-cost production. The important point here is that the automated systems of all sections are connected together by unified communication network, managed by a central database in such a way that the required data is sent to the section that needs it at the time it is needed.

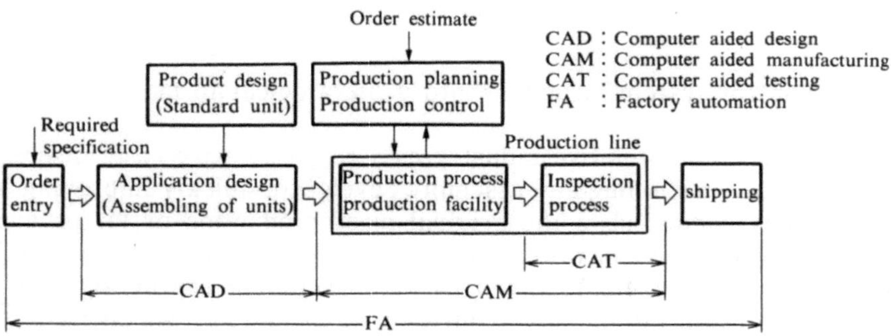

Fig. 5 · 4 Conceptual diagram of FA

Figure 5.5 illustrates the concept of the total FA network. The upstream PA (continuous batch process) and the downstream FA (assembly, inspection, and packaging) are subsystems with individual busses in a distributed control configuration, but in total FA these subsystems are connected in the trunk line of a unified factory local area network (LAN). However, although we speak of them as connected, there

* The terms "total FA," "PA" and "FA" are used. However, in this chapter, "total FA" has the broader meaning of a combined control system which includes PA. "FA" refers to a control system in the narrow sense of a discrete process.

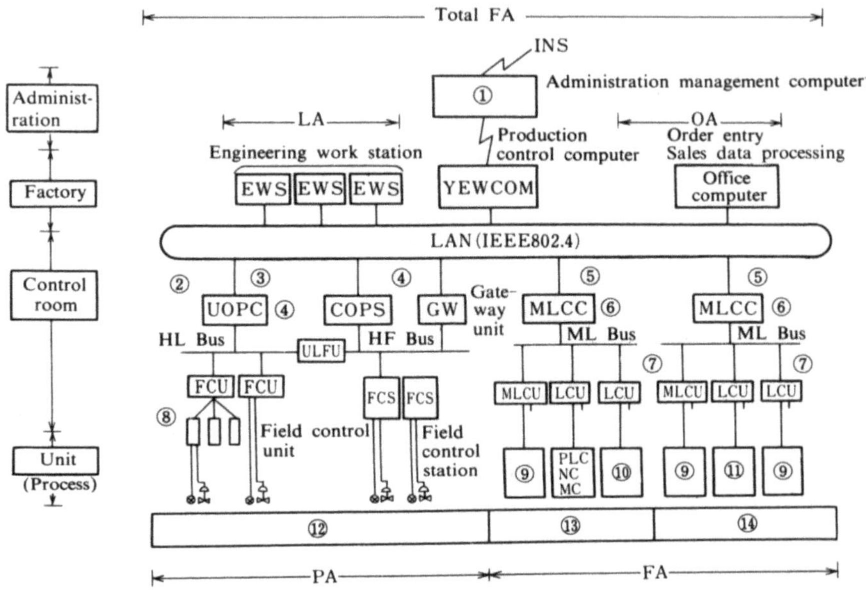

① Supervisory computer
② YEWPACK Package control system
③ CENTUM Distributed control system
④ Operator station
⑤ YEWMAC Production line control system
⑥ Line computer
⑦ Line controller
⑧ YS-80 Electronic control system
⑨ Robot
⑩ Visual inspection
⑪ Visual inspection tactile sensor
⑫ Contenuos · Batch process
⑬ Assembling
⑭ Inspection/Packing

Fig. 5・5 Schematic diagram of total FA network (Example of Yokogawa Electric Corp.)

are few examples where the PA and FA subsystems exist together in the same factory. Thus it is better to think of total FA as the connection of their respective subsystems.

When making a total system with a hierarchical structure such as shown in Figs. 5.3 and 5.5, the roles of the subsystem functions for each level must be determined. Table 5.1 lists rough standards related to the characteristics and processing time of information for each level. The upper levels require the capability to process large volumes of data and long-term data storage;in the lower levels, real-time response is the important factor. Timing and data volume information of a suitable level is exchanged among subsystems both horizontally and vertically in the hierarchy by means of the communication network.

Table 5.1 Information peculiarity in each level

System	Relevant Dept.	Managing period	Management Unit	Data acquisition cycle	Response time (Transaction processing)	Data retention time	Data bank capacity
Management information gystem	Entire company	Months to a few years	Months	Months	Several tens minutes	1 year or more	Several hundreds M bytes or more
Production planning · management level Production planning, Inventory control, Cost calculation, Order control, etc.	Factory	1 month to 6 months	Weeks or months	Days	Several tens seconds	6 months to 1 year	Several hundreds M bytes or more
Schedule control · management level Production plan for each process, Progress control, Load distribution, etc.	Production dept.	Weeks to months	Per shift or day	Hours	A few seconds	1 month to 3 months	Several tens M bytes to several hundreds M bytes

System	Relevant Dept.	Managing period	Management Unit	Data acquisition cycle	Response time (Transaction processing)	Data retention time	Data bank capacity
Process line management Monitoring, Data acquisition, Analysis calculation, etc.	Production section	For each shift and day	Minutes or hours	Minutes	2 sec. or less	Days to weeks	Several M bytes to several tens M bytes
Operation Monitoring, Data acquisition	Production line process	(Real time)		(Real time)	1 sec. or less	A few ms to 1 hour	Several tens K bytes to several hundreds K bytes
[Continuous process] Control Alarm				Scan cycle 1 to 10 sec.	100 ms or less		
[Discrete process] Control Alarm				Scan cycle 1 ms to 1 sec.	100μs to 10 ms		

5.2 *Distributed Control System*[2)-6)]

5.2.1 *Concept of the distributed control system*

The basic architecture of the distributed control system, as shown in Fig. 5.6, consists of an operator station, a control station, and a communication system. The operator station collects data relating to the process operation and can display and manipulate those data. The control station contains control functions such as for the DDC. According to the scale and type of object process, multiple stations are used. However, the degree of dispersion differs according to the number of control loops to be processed. There are 1–loop, 8–loop, and 40–loop types. The communication system exchanges data at high speed between the operator station, the control station and other stations.

The distributed control system consists of the four interfaces depicted in Fig. 5.6. The process interface is the interface between the distributed control system and the plant (*i.e.*, measurement sensors

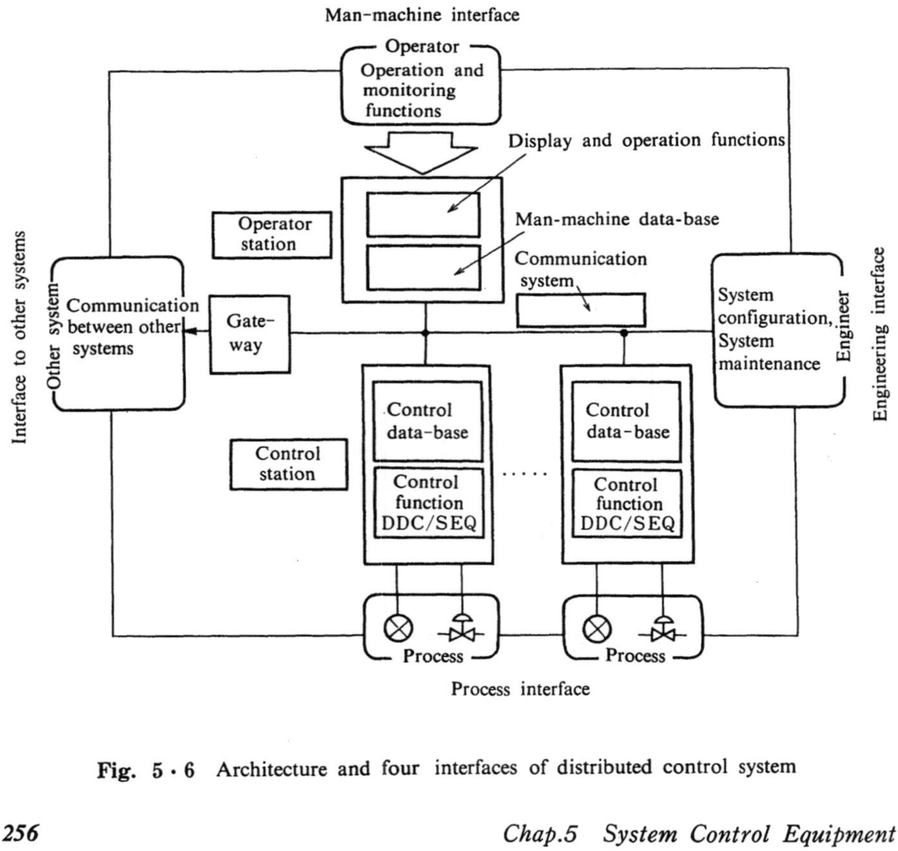

Fig. 5・6 Architecture and four interfaces of distributed control system

and final control elements), The control stations receive measurement signals from sensors such as for temperature, pressure, and flow rate, and perform control calculations in accordance with the deviations from the setpoint values. Output signals are then sent to the final control elements to perform compensatory actions.

The man-machine interface is the interface between the distributed control system and the operator. It does central monitoring of the plant and permits the operator to perform operations. The operator console is composed of a powerful microprocessor, CRT, and keyboard. Many different operations can be carried out by switching displays on the CRT. Operating each plant using the CRT as above is called "CRT operation." System operation using electronic instruments arranged together on the control panel for control loop unit, is called "board operation."

Other interfaces consist of a supervisory computer interface and control sub-system interfaces. The supervisory computer interface connects the distributed control system to a supervisory computer. It transmits control data and receives supervisory operation commands and optimal settings.

The control sub-system interface connects the distributed control system to other types of instruments such as the PLC (programmable logic controller) or a composition analyzer to integrate plant operation.

The engineering interface permits communication between the distributed control system and the engineer. It permits system build-up and software maintenance in the distributed control system.

The main system names for some representative products available at present for this type of distributed control system are listed in Table 5.2. Figure 5.7 shows a typical configuration with Yokogawa Electrics Corporation's total process control system.

CENTUM is a total distributed control system comprised of a

Table 5.2 Main distributed control systems

Manufacturers	System name
Yokogawa Electric Corp.	CENTUM (40 loops type or 8 loops type) YEWPACK (8 loops type) YEWSERIES 80 (1 loop type)
The Foxboro Company	SPECTRUM
Fuji Electric Co., Ltd.	MICREX-P
Hitachi, Ltd.	Unitrol EX Series
Mitsubishi Electric Corp.	MACTUS
Bailey Japan Co., Ltd.	NETWORK-90
Toshiba Corp.	TOSDIC
Yamatake Honeywell Co., Ltd.	TDCS-3000

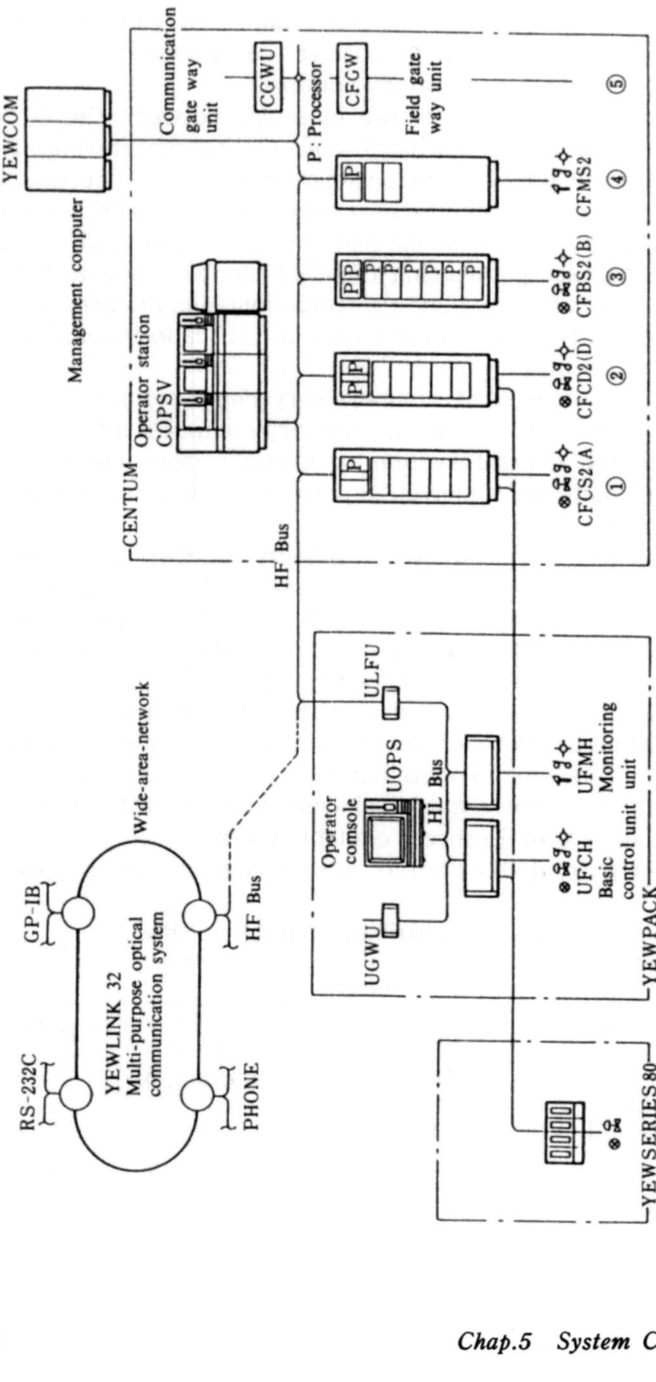

① Basic control station ③ Muliple-processor control station ⑤ Sub-control systems *e.g.* PLC
② Duplex control station ④ Field monitoring station

Fig. 5・7 Total process control system of Yokogawa Electric Corp.

powerful CRT operator station, a control station which performs with various process instrumentation, and a gateway unit. There are four types of basic control stations: a 40-loop basic control station, a duplex control station which duplicates the basic control station, an 8-loop N:1 backup-type highly distributed control system, and a field monitoring station which monitors processes.

There are two types of gateway units: a communication gateway unit for the supervisory computer and a field gateway unit for a controller sub-system. CENTUM has a wide range of uses in large scale continuous processes as well as batch processes.

YEWPACK is a distributed control system on a relatively small scale. It has 8 loops per basic control unit, and is ideal for the control of batch processes. The YEWSERIES 80 is a 1-loop board instrumentation system. It is suited to small to large scale instrumentation of continuous processes. These systems can perform monitoring and operation by communicating with the CENTUM operator station. Figure 5.8 shows a general view of several CENTUM operator stations and a control station.

Fig. 5・8 External view of CENTUM

5.2.2 Configuration of the distributed control system and its functions

(*a*) *Configuration of the control station*

As shown in Fig. 5.9, the control station consists mainly of a station control nest (SCN) and I/O nests. The SCN has a microprocessor and performs process control, communication control, and duplex control. The I/O nests contain interface cards for input and output. Signal conditioner nests and terminal boards can be installed in the rear of the cabinet, as necessary. Signal conditioner cards provide signal isolation and standardize signals from the process to 1 to 5V DC. Output conditioners provide isolation for 4 to 20 mA DC. As shown in Fig. 5.10, analog signals such as 4 to 20 mA DC outputs.

As shown in Fig. 5.10, analog signals such as 4 to 20 mA DC from

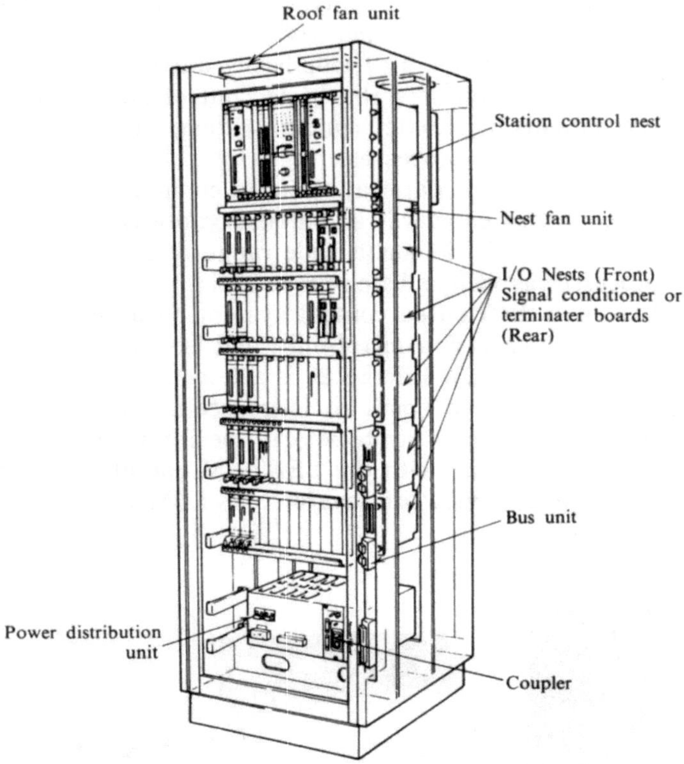

Fig. 5 · 9 Configuration of control station

the field transmitters and thermocouple inputs are linearized and/or normalized to 1 to 5 V DC by signal conditioner cards. The inputs are cabled into a group of 8 or 16 points to the I/O card. These signals are converted to digital values by the I/O cards, transmitted to the processor via the SI bus to become input signals for control algorithms, alarm monitoring, etc. The output of the control computations is returned to the I/O cards via the SI bus.

The digital control output is converted to 4 to 20 mA DC analog output signals and sent to valves and other such devices via signal conditioners.

The MAC 2 card effectively combines 8 point analog inputs and 8 analog 4 to 20 mA DC control outputs. However, failure of the MAC 2 card means failure of the 8 control outputs, creating a problem in process operation. A duplex configuration (two cards) is usually implemented to avoid this. The SI bus is an internal station bus which connects the SCN and the I/O nests. Because control of all input and out-

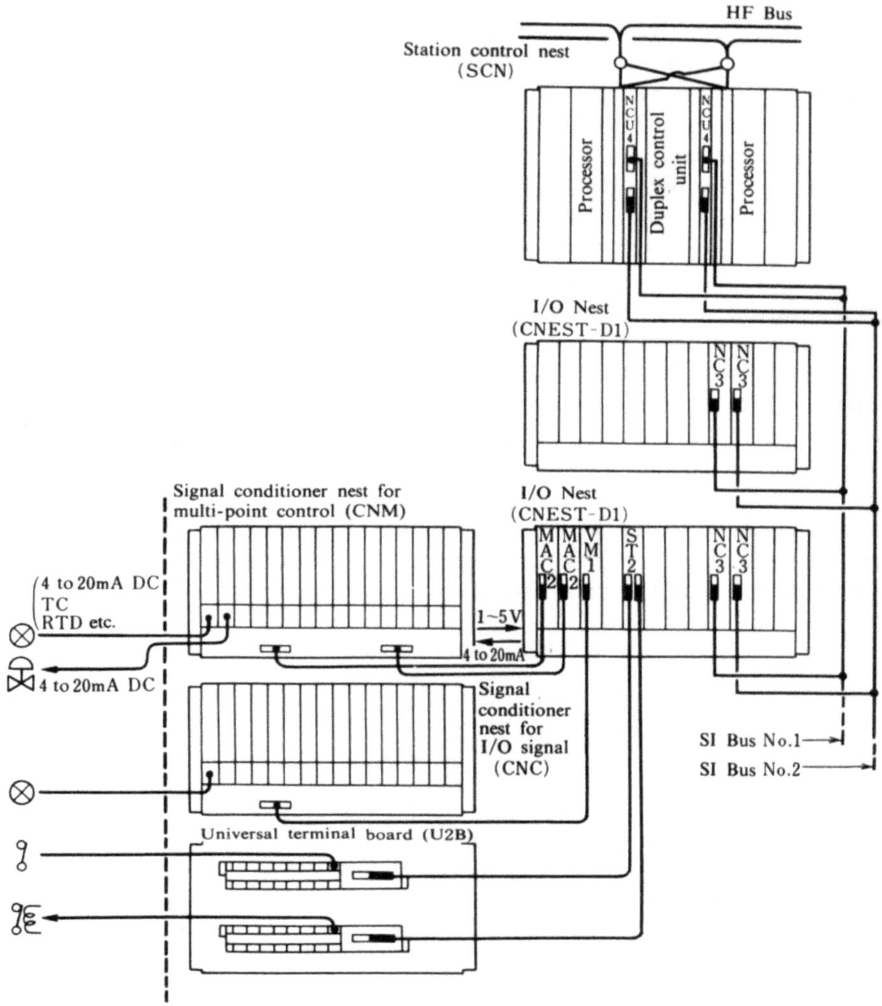

Fig. 5 · 10 Signal flow diagram of control station

put signals becomes impossible if the SI bus fails, reliability is increased by making it duplex. Card types NC 3 and NCU 4 are interface cards for sending and receiving signals between I/O cards and the SI bus, and between the SI bus and the processor. The duplex control section selects a standby processor if the processor or HF bus interface card fails.

Contact input signals connected to terminal boards are cabled to the I/O card in 16- to 32-point groups. These signals are combined into digital value at the I/O card and then sent to the processor via the SI bus. They typically become inputs to sequence control tables. The action output of the sequence control tables is again transferred to the I/O card via the SI bus. The output signals are cabled to terminal boards for connection to field circuits, in order to perform motor start-stop and lamp on-off.

Table 5.3 shows the different types of signal conditioner cards and Table 5.4 shows the different types of I/O cards. The signal conditioner cards isolate the system from noise from the field. On-line maintenance is possible because failed signal conditioner cards and I/O cards can be replaced without affecting other cards in the same nest. The processor uses a 16-bit microprocessor and memory consists of both ROM and RAM. The ROM contains programs such as for the DDC, and the RAM holds system configuration data, set and measured values, etc. The contents of RAM not lost as a result of power failure because of a battery backup system. The functions of the control station start automatically during normal power-up. Also, this processor is connected to a stand-by processor via a duplex control section. The database in the memory of the stand-by processor is continually up-dated to equal the database in the control processor.

Table 5.3 I/O card list for signal conditioner nest

	Card type	Card name
For input signal	CM1	mV input card
	CT5	Thermocouple input card
	CR5	Resistance temperature detector input card
	CS1	Potentiometer input card
	CH1	Input isolation card (1 to 5 V input)
	CH2	Input isolation card (0.2 to 1 V input)
	CH5	Input isolation card (1 to 5 V input with square root extraction)
	CA1	2-wire transmitter input card
	CA5	2-wire transmitter input card (with square root extraction)
	CP1	Pulse train input card
For output signal	CC0	Control output isolator card
	CA0	Output isolator card (4 to 20 mA output)
	CH0	Output isolator card (1 to 5 V output)
	CX1	I/O through card

(b) Process data highway

The data highway is a local network in which data is transmitted bit-serially along a single coaxial cable. The following features are included in the system, according to typical industrial requirements for

Table 5.4 I/O card list

Model	Name	Functions and specifications
MAC 2	Multipoint analog control I/O card	8-loop control I/O(1 to 5 V input; 8 points, 4 to 20 mA output ; 8 points)
VM 1	Multipoint analog input card	1 to 5 V DC input; 16 points
VM 2	Multipoint analog I/O card	1 to 5 V DC input; 8 points, 1 to 5 V DC output; 8 points
PM 1	Multipoint pulse train input card	pulse input; 16 points
ST 2	Multipoint status I/O card	Input(contact or voltage) ; 16 points, output(Tr contact) ; 16 points
ST 3	Multipoint status input card	Input(contact or voltage) ; 32 points
ST 4	Multipoint status output card	Output(Tr contact) ; 32 points
PB 5	Multipoint push-button input card	Contact or voltage status change signal; 16 points
MX 2	mV input multiplexer card	mV, thermocouple non-isolated inputs; 32 points
MX 3	mV input multiplexer card	mV, thermocouple isolated inputs; 16 points
MX 4	High level input multiplexer card	Voltage non-isolated inputs; 32 points
MX 5	High level input multiplexer card	Voltage isolated inputs; 16 points
LCS	Loop communications card	Interface for YEWSERIES 80; 8 units
LCU	Loop communications card	Interface for loop display unit; 4 units
LDI	Loop display interface card	Interface for MAC 2; 8 units

control use.

(1) *Transmission distance*: The transmission distance is several kilometers, and can be adjusted to suit the scope of the plant.

(2) *High reliability and safety*: This is the only centralized part of the distributed control system. In addition to using duplex redundancy for the transmission path and communication control interface, a system is required to prevent the propagation of the effects of errors among connected stations.

(3) *Real time*: In order to display data dynamically on the operator console, event-driven data such as alarm and fixed-period scan data are transmitted along the data highway. For example, at each station, 4 to 20 bytes of data are transmitted in less than 20 ms, as frequently as several hundred times per second.

(4) *Resistance to noise*: Transmission lines are often installed in areas of plants having high levels of electrical and RF noise. Error control for countering this noise and with error detection and recovery techniques is required.

(5) *Expandability*: The system can be easily reconfigured when expanding the plant.

(6) *High availability and maintainability*: Breakdowns are easily spotted and repairs are quick. The failed equipment can be removed and repaired while other lines carry on normally.

(7) *n:n transmission*: Each station is able to enjoy mutually effective communication with other stations. There are no master-slave relationships.

(*c*) *High system reliability*

The technology for high reliability is based on three ideas: 1) To be difficult to damage (fault avoidance), 2) Not to be unduly influenced by failure (fault tolerance), and 3) To be able to be repaired without stopping the system (on-line maintenance). Important parts which are shared by the system (parts which greatly influence the entire system when they fail) are all of duplex configuration to increase fault tolerance. System redundancy and maintainability are directly related. The more reliable a system is, the more the simpler the maintenance requirements are.

Errors detected by self-diagnosis of system components appear as maintenance information on the operator station CRT or on the printer. This makes quick diagnosis and accurate repair possible. Table 5.5 gives examples of CENTUM system redundancies.

With this highly reliable design, for example on a control station (CFCD 2), an availability factor of over 99.99% can be expected. This has been proven from field data of the CFCD 2. Figure 5.11 shows the calculation of the availability factor of the SCN section. The processor

Table 5.5 CENTUM system redundancies

System components			Applied technology
COPS			Duplexed processors
HF bus, coupler			Dual-redundant HF buses
CFCD 2			
	SCN (processor and memory unit)		Duplexed processors
	Internal bus, NC 3 card		Duplexed internal buses and control cards
	CNEST	Power supply card	Duplexed cards
		MAC 2	Duplexed analog I/O cards
CFBS 2			
	CIU		Back up by BCU
	Internal bus, NC 1 card		Duplexed internal buses and control cards
	CFCU 2	Power supply card	Duplexed cards
		Processor and memory card	$n:1$ back up by BCU
		MAC 2	Duplexed I/O cards

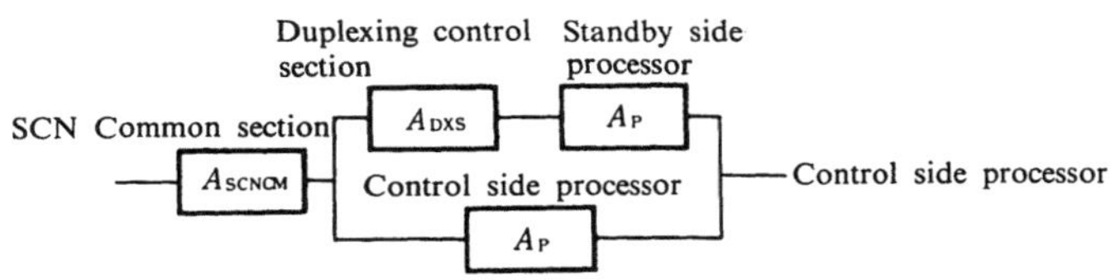

(a) Reliability model of duplex SCN configuration

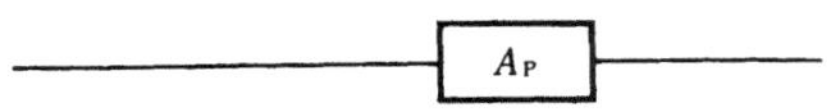

(b) Reliability model of single SCN configuration

Availability of duplex type

$A = A_{SCNCM}\{1-(1-A_P)(1-A_PA_{DXS})\}$
$= 0.999\ 999\ 459$

The availability of duplex type is 3 digits better than that of single type (0.9998).

	Failure rate [fit]	Availability	
SCN common section	60	A_{SCNCM}	0.999 999 5
Processor	24 300	A_P	0.999 81
Duplex control section	3 200	A_{DXS}	0.999 974

Fig. 5・11 Availability of duplex type SCN

section includes the processor, memory, HF bus interface card, and SI bus interface card. The SCN common section has several logic circuits which make up the control enable logic of the duplexing control section. The duplexing control section can detect failure through self-diagnosis. To enable on-line maintenance, control right can be fixed with a

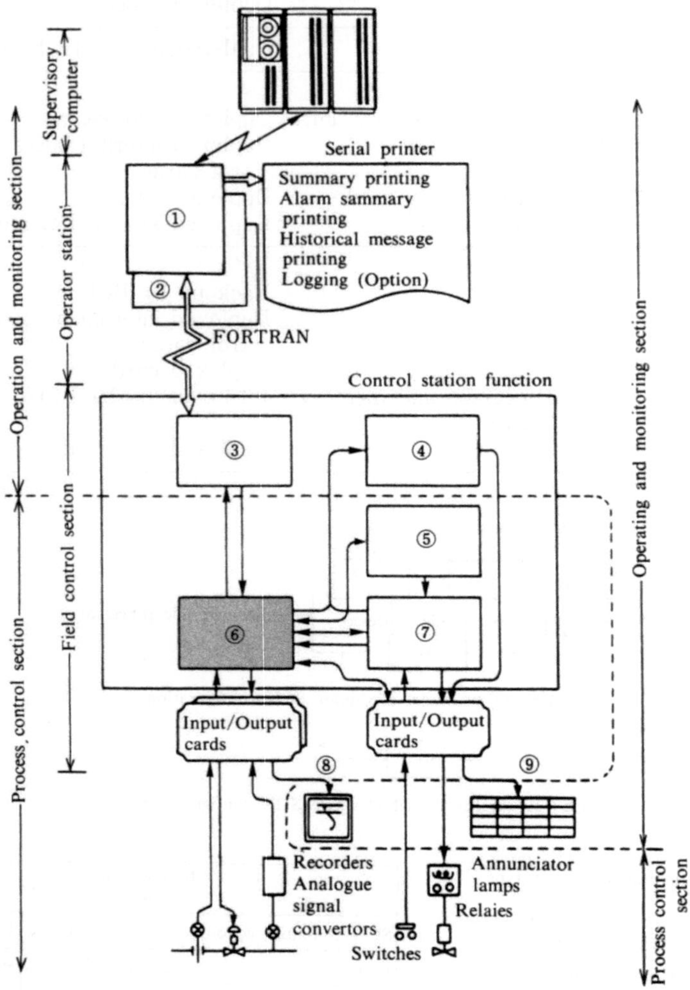

① Operator station ,Display/Setting alarm, monitoring, functions
② Builder maintenance functions
③ Communication functions
④ Annunciator functions
⑤ Computational functions
⑥ Feedback control functions
⑦ Sequence control functions
⑧ Analog input/output
⑨ Digital input/output

Fig. 5 · 12 Distributed control system function configuration selector switch. It is considered that the pass of a standby section includes the duplex control section. This makes the duplex system even more effective.

5.2.3 Feedback control

Figure 5.12 shows the functional configuration of the distributed control system. Feedback control function and sequence control function processing is done at the intelligence section of the control station.

(a) Feedback control function

The feedback control function consists of a combination of loops, as shown in Fig. 5.13. Examples are the single loop and the cascade loop. The single loop consists of a software instrument (a function unit derived from a control algorithm and function parameter, set beforehand) which processes one input and one output. The cascade loop consists of two instruments combined in a cascade. The instruments are indicated by circles in the figure. Each instrument is assigned a tag number such as FIC 100 (meaning flow indicating controller No.100) for each loop. Tags are used for loop identification. Table 5.6 is a list of instrument types.

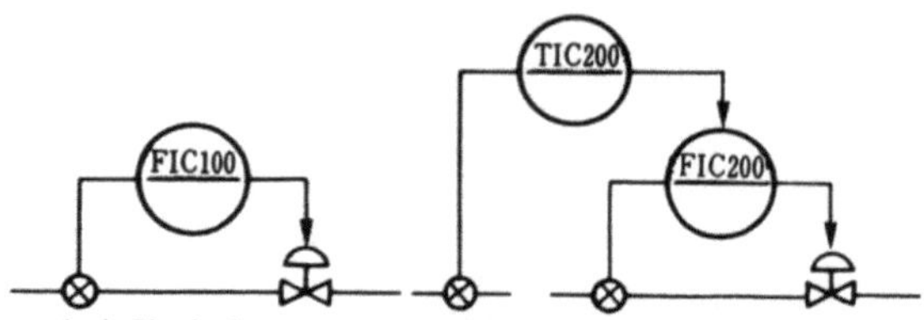

(a) Single loop (b) Cascade control loop

Fig. 5 · 13 Loop configuration

Figure 5.14 is a functional block diagram of a digital control or algorithm. The input signal conversion process linearizes a non-linear input signal of a flowmeter and other such device, and then puts it through a digital filter. A first order lag filter is applied as a digital filter. The compensation/totalization process performs flow temperature and pressure compensation, etc. In addition, flow rate can be totalized.

The alarm check function detects input signal errors and alerts the operator by sending a warning which is displayed on the monitor of the operator station, and may be output to the printer. Alarm status can also used as a condition for testing shifts in process for sequence control action. For example, it is can be used to test if a temperature or pressure has reached a certain value.

There are several types of alarm checks. The input open check

Table 5.6 List of instrument types

Instrument type	Instrument function
Indicators	Input indicator
Controllers	PID controller PID controller with dead band En-type PID control En-type PID controller with dead band PID controller with batch switch Two-position ON-OFF controller Three-position ON-OFF controller Pulse duration ON-OFF controller PD controller with manual reset Blending PI controller
Manual loaders	Manual loader Manual loader with input indicator Auto/manual station with output transfer switch
Ratio set units	Ratio set unit
Signal selectors	Auto selector Singal selector
Selector switches	Three-pole three-position selector switch Data set switch
Computing units	First order lag unit First order lead unit Union filter unit (ramp unit) Lead lag unit Dead time unit Dead time compensation unit Moving average unit Cumulative average unit Line-segment function unit Special line-segment function unit Calculating unit
Data set units	Data set unit
Program set units	Program set unit
Batch set units	Batch set unit
Batch data set units	Batch data set unit
Data acquisition units	Data acquisition unit
Station data link	Station data link

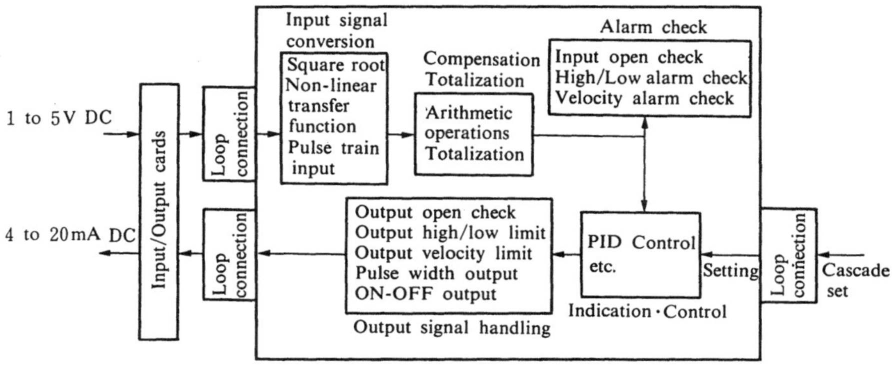

Fig. 5·14 Digital (Control unit configuration)?

tests for an open input signal. There is the high/low alarm check for input of measured values, and the deviation alarm check which checks for deviations from the set values. The velocity alarm check tests whether the change rates of an input exceed a limit value or not.

For indication and control functions, there are the PID control algorithm and the PID control algorithm with a dead band. The output signal handling function checks for errors in the output wiring (output open check), can apply a variable output limit to the controller output, and finally outputs the manipulated variable to the valve. Output limit is a function to perform the process operation safely. It consists of an output velocity limiter and output/high/low limiter. The output velocity limiter limits the incremental change in output during an output processing cycle so that changes in output are smooth. In the output/high/low limiter, if the output reaches the high or low limit value, the output value is limited to the high or low limit setting. In addition to the 4 to 20 mA analog output, there is a pulse width output operate motor-operated valves. Also, an on-off output which can be used for controlling electrical heaters, etc.

The loop connection function corresponds to the wiring of analog instruments. It specifies the instrument signal input and output destinations, and the setpoint connections of the cascade loop. Also, the loop connection function transmits the loop status at the lower loop of the cascade loop and clarifies the cascade loop open/close status for the upper loop. In this way, operability is enhanced.

(b) PID control algorithm

The PID control algorithm of the analog controller is expressed in the following formula.

$$M=\frac{100}{P}\left(e+\frac{1}{T_I}\int edt+T_D\frac{de}{dt}\right) \tag{5.1}$$

Here, M is the output value, P is the proportional band (%), e is the deviation (desired value − measured value), T_I is the integral time, and T_D is the derivative time.

The PID control algorithm for digital control is expressed by the following difference equation.

$$M_n=K\left(e_n+\frac{\tau}{T_I}\Sigma e_n+\frac{T_D}{\tau}\Delta e_n\right) \tag{5.2}$$

Here, M_n is the output value, e_n is the control deviation ($=m_n-s_n$), s_n is the setpoint value, m_n is the measured value, $\Delta e_n=e_n-e_{n-1}$, K is the proportional gain (= $100/P$), T_I is the integral time, T_D is the derivative time, and τ is the sampling period. The variable n expresses the number of sampling times.

Equation (5.2) changes to the following.

$$\Delta M_n=K\left(\Delta e_n+\frac{\tau}{T_I}e_n+\frac{T_D}{\tau}\Delta^2 e_n\right) \tag{5.3}$$

Here, $\Delta M_n=M_n-M_{n-1}$ and $\Delta^2 e_n=\Delta e_n-\Delta e_{n-1}$. Usually, Equation (5.2) refers to the positional and Equation (5.3) refers to the velocity or incremental change. Incremental output ΔM_n has to be accumulated and converted to positional output M_n. However, when compared with Eqs. (5.2) and (5.3) allows bumpless switching between manual and automatic operation by putting the manual operation value to the initial value at that point.

Also, with the positional type, even if the output value is saturated at 0% or 100%, Σe_n is calculated. Even if the sign of control deviation e_n is changed, the output recovery is delayed by Σe_n which accumulated during the preceding interval, and the tendency to overshoot arises. This is called reset windup. With the incremental type, even if the output gets caught in the output limit, the output begins to change rapidly along with the change in e_n, because the accumulated value is not in the output value itself. Reset windup and over shooting can be prevented. This integration term is only in the output value itself, and the inherent property, in which various changes in the control status appear only in the changed portion of the output value. It is favorable for advance control, which uses a variety of changes the control parameters and control modes.

When constant-value control is performed, $\Delta s_n=0$, and since $\Delta e_n=\Delta m_n-\Delta s_n$ in the incremental type Equation (5.3), this Equation (5.3) changes as follows.

$$\Delta M_n = K\left(\Delta m_n + \frac{\tau}{T_I} e_n + \frac{T_D}{\tau}\Delta^2 m_n\right) \tag{5.4}$$

Even if the setpoint value is changed in a step-wise fashion, that change does not drive the value excessively from the proportional and derivative terms. Because only the integral action responds to a setpoint step change, Equation (5.4) has the advantage of not over-driving the process. However, because there is a delay tendency of response to a setpoint value change at the secondary in a cascade loop, Equation (5.3) is often better where S_n changes moment to moment. The use of Eqs. (5.3) and (5.4) should be selected according to the situation. The correct algorithm is automatically selected in the Yokogawa CENTUM system, unless otherwise specified.

5.2.4 Sequential control

The sequential control function is used for sequential processes such as process startup and shutdown, and batch process switchovers. It is complementary with DDC and is particularly effective for automatic plant operation.

(a) Sequence description technique

There are many techniques for describing sequence operation including Boolean logic and relay ladder diagrams, however, in the design stage, the following techniques are often used (Fig. 5.15).

(1) *Logic circuit* Logic circuits are based on the symbols AND and OR. They are commonly used for interlocking processes. The symbols

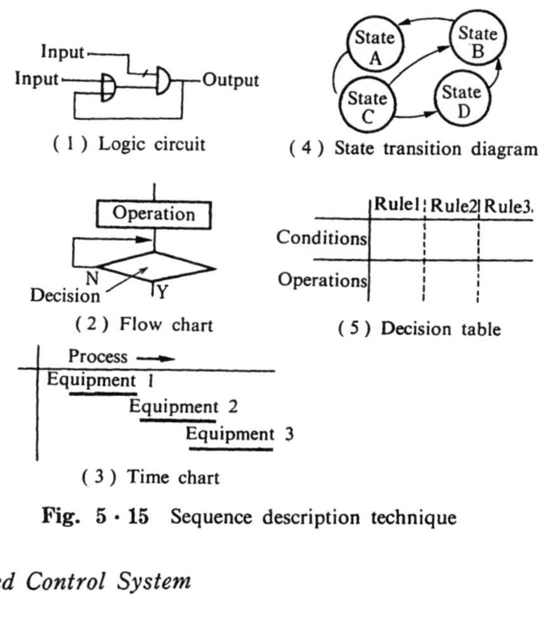

Fig. 5 · 15 Sequence description technique

used in Fig. 5.15 are prescribed by JIS and U.S. MIL 806B (Military Specifications and Standards). For use in processes, there is the ISA (Instrument Society of America) standard.

(2) *Flowchart* Flowcharts are used in computer programming and are excellent for describing operations which follow a fixed sequential procedure. For example, flowcharts are often used to describe the start-up operations of processes.

(3) *Time chart* Time charts are easy to read because they define the device operations according to time flow. Usually, time charts are used to describe the time-relevant operations of batch processes in which products are finished while moving, in order, between many small machines.

(4) *State transition diagram* Not related to time, state transition diagrams are effective for expressing what conditions several processes require for transition, or in what way a command, such as manual, automatic, stop and re-open, operates in the flow of a sequence operation.

(5) *Decision table* In the decision table, conditions and operations are lined up one by one to describe, in chart form, which operation should be performed under a combination of conditions. The decision table is suited for describing not only parallel operations and interlocking, but also sequences in general.

(*b*) *Example of sequence control function*

The following is an explanation of the CENTUM decision table, shown in Fig. 5.16, as an example of a sequence control function.

The condition signal is written into each vertical row of column B of the sequence table, and the condition signal combination, as the operation condition, is written into each vertical row of column C. To operate, enter Y (Yes) for the condition signals which as a condition must be activated, N (No) for signals which must not be activated, and leave blank for a signal which has no relation to operation. Write the action signal into section D. Use Y ("on") or N ("off") in section E, for the actions to take when the above conditions are satisfied, and leave a blank for those which will not be operated (have no relation). The columns of sections C and E correspond one-to-one. Rows F and G are used when doing stepped sequences at row A. Write either the number of the next step to proceed to in row F when the conditions are satisfied, or enter a step in row G when the conditions are not satisfied. Leave blank when the conditions have no relation.

Condition signals can use only contact input/outputs, internal switches, operation and stop statuses of timers and counters. The switch position of the switch unit, which switches the path of the signals, and alarm status of the feedback control function can be de-

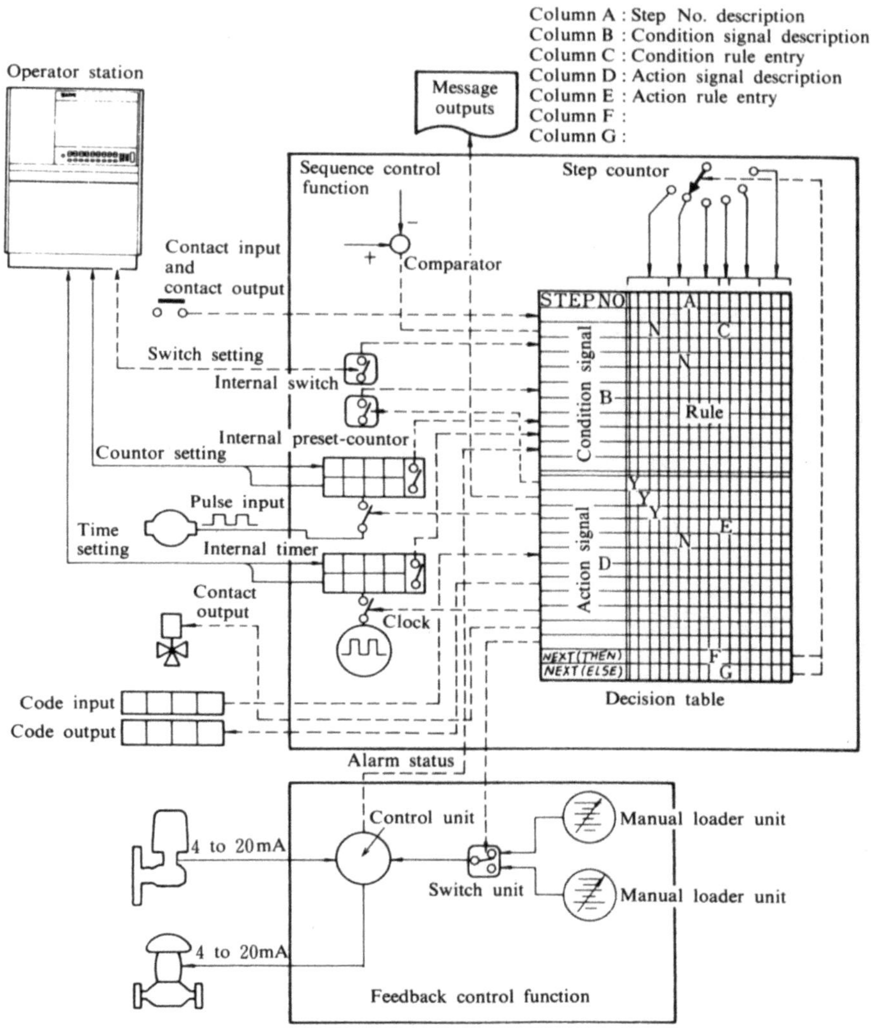

Fig. 5 • 16 Example of CENTUM decision table

scribed. For action signals, in addition to contact output and internal switch setting and resetting, and timer and counter starting and stopping, change of loop status of the feedback control function (for example MAN to AUTO) and change of the switch position no. on the switch unit are possible. In this way, by combining the sequential control and feedback control effectively, process startup and shutdown can be automated, and shifting of process in batch process control is

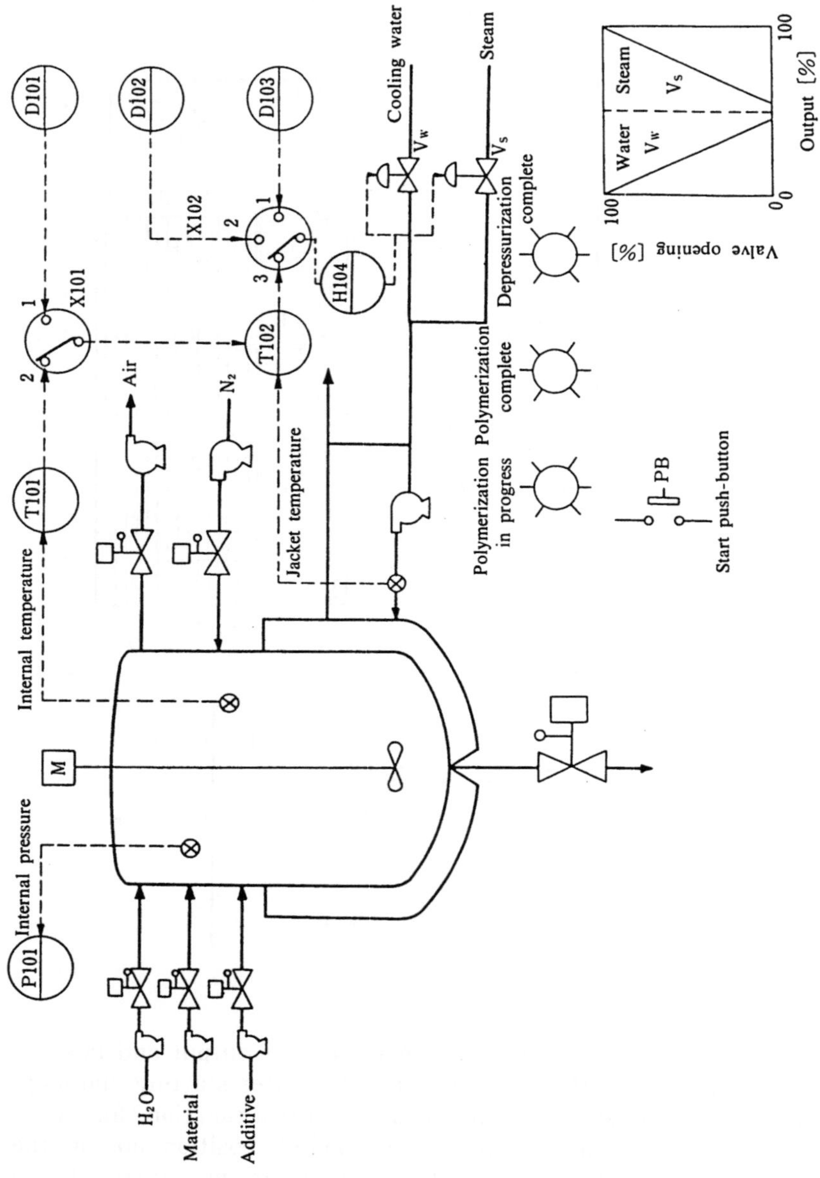

Fig. 5 · 17 Polymerization process heating sequence

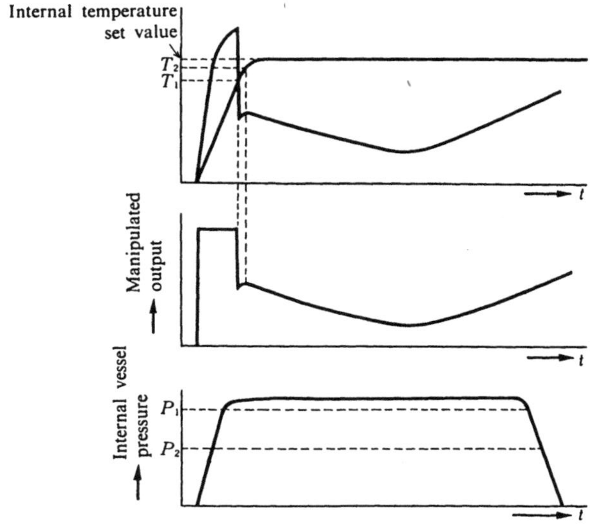

Fig. 5・18 Time chart of polymerization process heating sequence

simplified.

A descriptive example of a sequence table application is shown for the polymerization process heating sequence in Fig. 5.17. Figure 5.18 is a time chart for the process.

(1) After completion of the charging process, push the start push button to begin the polymerization process. The polymerization in progress lamp lights and the depressurization complete lamp goes out. Turn the switch (X102) to the manual station device (D103) side. The output of the D103, via the manual station with switch H104, causes the cooling water or steam valve to move. At plant side, the steam valve (V_S) opens completely and the jacket heats up quickly.

(2) When the measured value (PV) of an internal temperature controller (T101) reaches non-alarm status (NR) ($PV \geq T_1$), turn switch (X102) to the manual station (D102) side as a previous step for fixed value control of the jacket temperature. Also, start timer (U700). At plant side, the steam valve (V_S) closes completely and a cooling water valve (V_W) opens a fixed amount.

(3) After time is up on the timer (U700), turn switch (X102) to the jacket temperature controller (T102) side and turn a switch (X101) to the manual station (D101) side. Put the jacket temperature controller (T102) to AUTO and cascade connect the jacket temperature controller (T102) and the manual station (D101).

(4) When the measured value of the internal temperature controller

(T101) reaches the high alarm setting (PH) (PV$\geq T_2$), turn the switch (X101) to the internal temperature controller (T101) side and cascade connect the internal temperature controller (T101) and the jacket temperature controller (T102). Put the internal temperature controller (T102) to AUTO and start a timer (U701).

(5) After time is up on the timer (U701), when the measured value of the pressure indicator (P101) reaches the non-alarm status (NR) (PV$\geq P_1$), the polymerization in progress lamp goes out, "polymerization complete" is printed and the polymerization complete lamp lights up.

(6) When the measured value of the pressure indicator (P101) drops below the low alarm setting (PL) (PV$\geq P_2$), the polymerization complete lamp goes out and the depressurization complete lamp lights up.

Figure 5.19 shows this example in the form of a sequence table. Corresponding inputs and outputs are allotted as follows.

Status input:	Start push button PB:	N300
Status output:	Polymerization in progress:	N700
	Polymerization complete lamp:	N701
	Depressurization complete lamp:	N702
Internal switch:	Charging process end flag:	N100
	Polymerization in progress flag:	N101
Timer, printer:	Internal timer:	U700
	Internal timer:	U701
	"Polymerization complete" print-out:	U400

The "polymerization complete" print-out message is output to the operator console.

5.2.5 Man-machine interface

Plant operation by a distributed control system is as shown in Fig. 5.20. In the even of a buzzer sound which indicates an alarm, the operator gets information regarding the cause of the alarm at the operator console. He judges whether to leave the situation as is or to deal with the problem. When left as is, the system maintains the original monitoring status. To deal with a problem, he may drive various valves from the operator console or take other actions, and return to the original monitoring status. Event-driven information cautions the operator at the operator console, and it is used for notification of process alarms and operator guide messages. Buzzer sounds are used. Plant status information is provided by the operator console in response to operator requests. The information is provided visually on the CRT or a logging printer. Operator console operation refers to operation whereby the operator obtains desired status information from the operator

TABLE NO.

00– U 503 ,

DESCRIPTION

Ⓣ , ① ,

	SYMBOL	COMMENTS (RULE NO / STEP NO)	01	02	03	04	05	06	07	08	09	10	11	12	13	14	15	16	17	18	19	20	21	22
01–																								
02–																								
11–	N100	, ②	Y																					
12–	N300	, ③	Y																					
13–	T101, NR	, ④		Y																				
14–	T101. HI	, ⑤				Y																		
15–	P101, NR	, ⑥					Y																	
16–	P101. LO	, ⑦						Y																
17–	U700. HI	, ⑧			Y																			
18–	U700.IOP	, ⑨		Y																				
19–	U701. HI	, ⑩					Y	Y																
20–	U701. IOP	, ⑪				Y																		
21–	N101	, ⑫	N	Y	Y	Y	Y	Y																
22–		,																						
23–		,																						
24–		,																						
25–		,																						
26–		,																						
41–		,																						
42–		,																						
111–	N700. H	, ⑬	Y				N																	
112–	X101. 1	, ⑭			Y																			
113–	X101. 2	, ⑮				Y																		
114–	X102. 1	, ⑯	Y																					
115–	X102. 2	, ⑰		Y																				
116–	X102. 3	, ⑱			Y																			
117–	T102.AUT	, ⑲			Y																			
118–	T102.CAS	, ⑳			Y	Y																		
119–	T101. AUT	, ㉑				Y																		
120–	U700	, ㉒	N	Y																				
121–	U701	, ㉓	N			Y																		
122–	U701. H	, ㉔					Y	N																
123–	U400. 1	, ㉕					Y																	
124–	N702. H	, ㉖	N					Y																
125–	N101. H	, ㉗	Y					N																
126–		,																						
142–		,																						
191–	NEXT STEP	THEN																						
192–		ELSE																						

CONTINUE TABLE NO.

193– ,

① Polymerization temperature rising sequence
② Charging process end flag
③ Start push-button (Panel)
④ Internal temperature ≧T1
⑤ Internal temperature ≧T2
⑥ Internal pressure ≦P1
⑦ Internal pressure ≦P2
⑧ Timer, time-up
⑨ Timer, reset state
⑩ Timer, time-up
⑪ Timer reset state
⑫ Polymerization in progress flag
⑬ Polymerization in progress lamp
⑭ T102 Setting transfer (D101 side)
⑮ T102 Setting transfer (T101 side)
⑯ Output transfer (D103 side)
⑰ Output transfer (D102 side)
⑱ Output transfer (T101 side)
⑲ Jacket temp. controller auto
⑳ Internal temp. controller auto
㉑ Timer
㉒ Timer
㉓ Polymerization complete lamp
㉔ Polymerization print-out
㉕ Polymerization print-out
㉖ Depressurizing complete lamp
㉗ On polymerization flag

Fig. 5 · 19 Entry example of polymerization temperature rising sequence

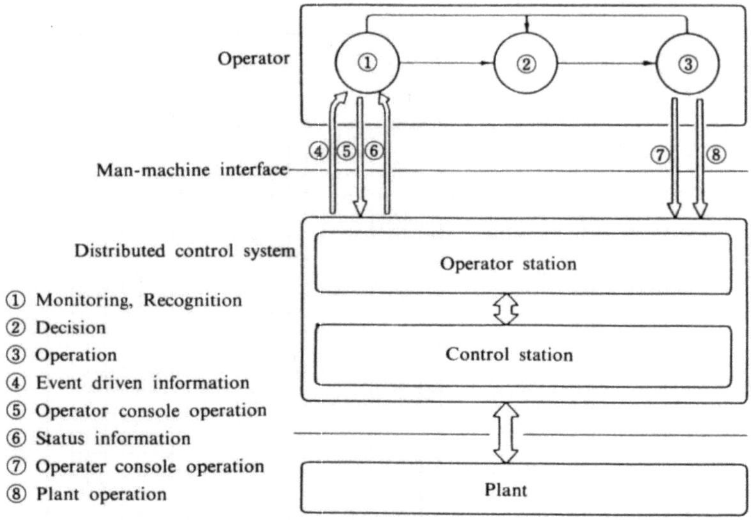

Fig. 5 · 20 Plant operation by distributed control system

console. This does not affect plant operation. The less the operator console operation done the better. Plant operation refers to actions which influence the plant such as changing setpoint values and operating valves. Although a configuration which prevents incorrect operation is necessary, it is also necessary to be able to perform an operation quickly during an emergency. The operator console is designed considering the factors described above. The CENTUM operator console is explained in the following paragraphs.

The operator console, as shown in Fig. 5.21, consists of the display section and the keyboard section. The display section consists of the system messages section, the main display panel, the sketch pad section, and the soft key labels section. The system messages section displays messages which notify the operator of information from the operator console. It is used mainly for notifying about process alarms and operator guide messages, and for messages operation error messages. On the main display panel, a function panel (explained below), which is selected through key operation, is displayed.

The sketch pad section displays data entries done by the operator, measured values, and the alarm status and the loop status of one designated loop. The soft key labels section displays the variable labels which correspond to each function panel. They are used to call up data, and allow the display of settings and other function panels, with one-touch operation. The keyboard section consists of the tuning keyboard and the operation keyboard. The tuning keyboard is mainly used

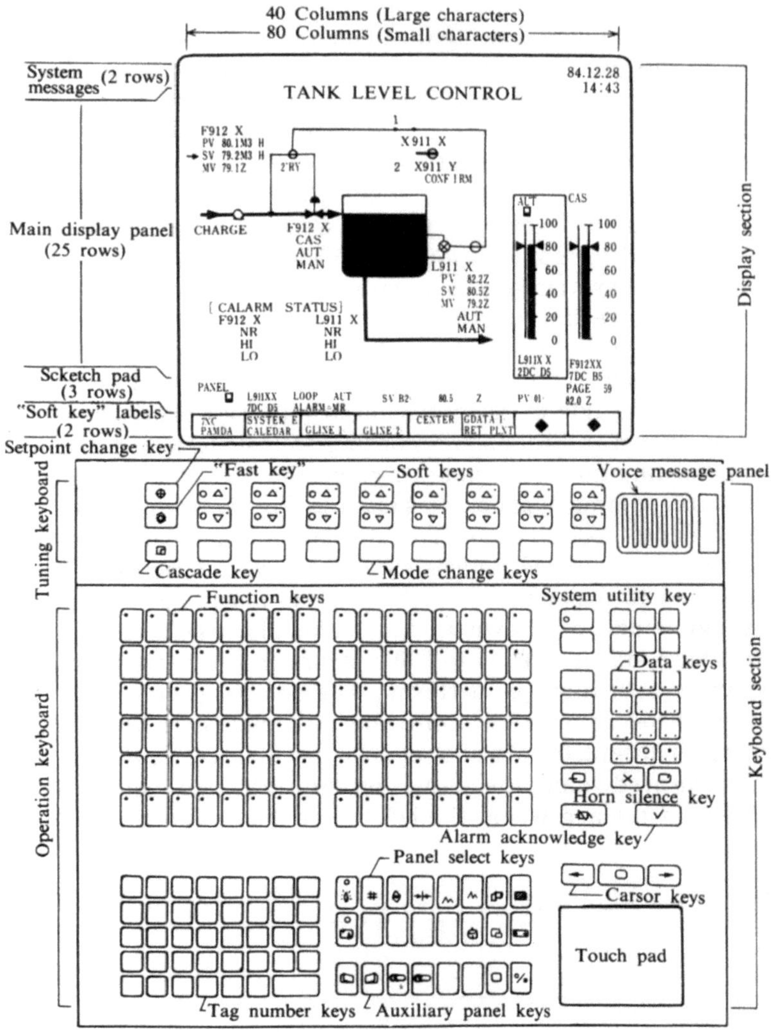

Fig. 5 • 21 Configration example of operater console

for switching loop statuses and changing controller setpoints and output values. The soft keys are labeled by the screen. The speaker is used to alert the operator by various tones of an alarm. Optionally, it can provide voice communication. The operation keyboard consists of function keys, panel select keys, auxiliary panel keys, a touch pad and cursor keys, an alarm acknowledge key, tag number keys, data keys, and system utility keys. The function keys, allow one-touch call up of user-defined panels. They also call up the printing of logs. Each key has an LED which flashes or lights upon the occurrence of specified process alarms to urge operator action. The panel select keys call up the operation monitoring panels. The auxiliary panel keys are used to scroll pages in the function panels. The touch pad and cursor keys are used to move the cursor for selecting items in the function panels. The alarm acknowledge key is used to acknowledge the occurrence of an alarm. The tag number keys are used to designate tag numbers. The data keys are used to input various data. The system utility keys are used to call up the system test, system maintenance, the system builder, and other engineering functions.

Function panels follow the format shown in Fig. 5.22. The operation monitoring panel is used for process operations and is hierarchical. The operator guide message panel outputs user pre-defined guide messages, as shown in Fig. 5.23, according to instructions from the sequence control function. It is able to guide the operator during operation. The alarm summary panel, as shown in Fig. 5.24, displays process alarms in order of occurrence. The display for an alarm disappears when normality is restored, unless an alarm lock is specified. The display is designed to make the occurrence of an alarm readily apparent. The overview panel, as shown in Fig. 5.25, displays in color on one screen the alarm statuses of up to 800 instruments. It features the ability to recognize operations as patterns. The control panel, as shown in Fig. 5.26, is made up of eight controller faceplates. In this example, eight controller faceplates are shown. The bar displays in the middle indicate the measured value, the arrows on the right side indicate the setpoint value, and the arrows on the left side indicate the controller out-

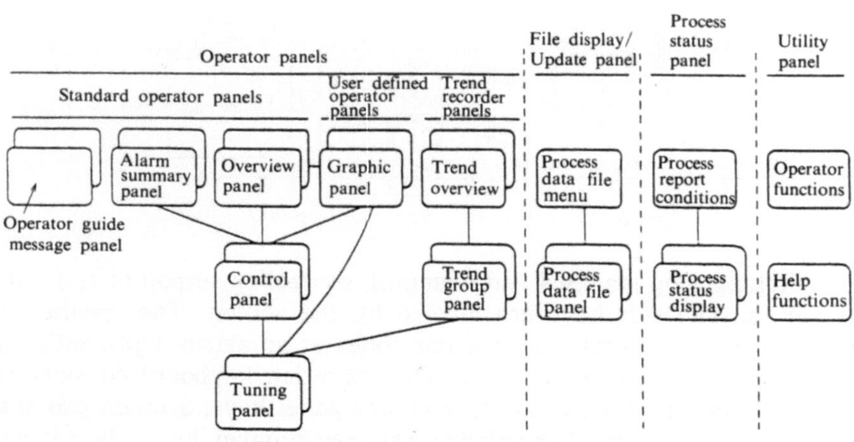

Fig. 5・22 Function panel configuration

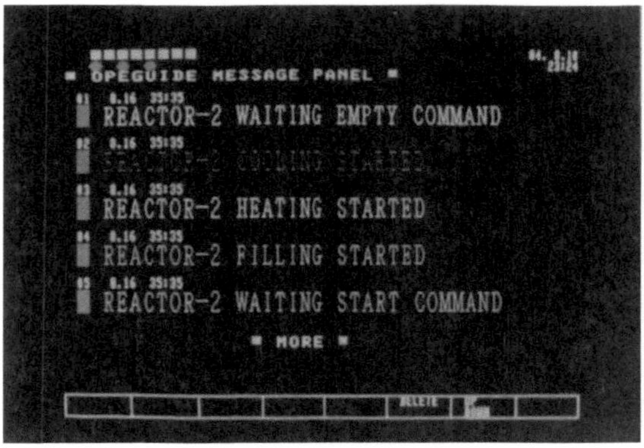

Fig. 5 · 23 Operator guide message panel

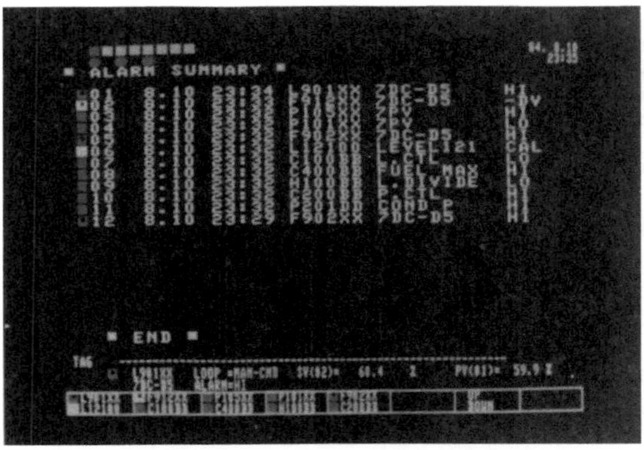

Fig. 5 · 24 Alarm summary panel

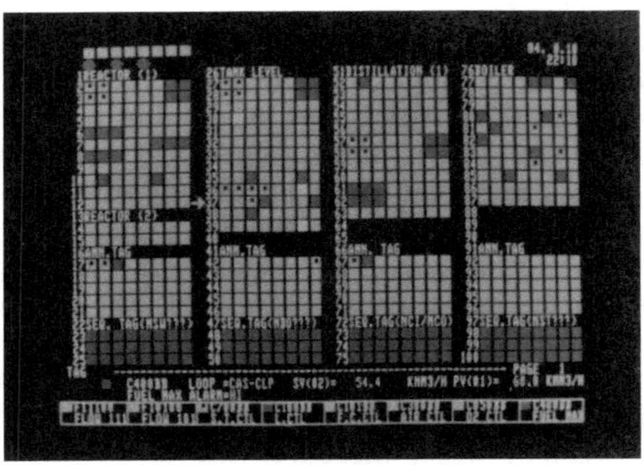

Fig. 5 · 25 Overview panel

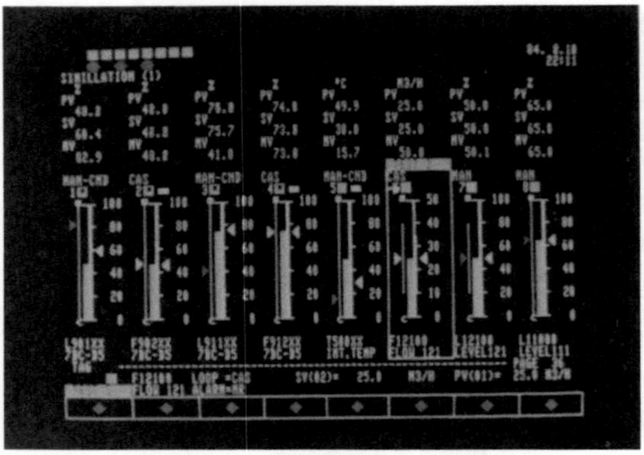

Fig. 5 · 26 Control panel

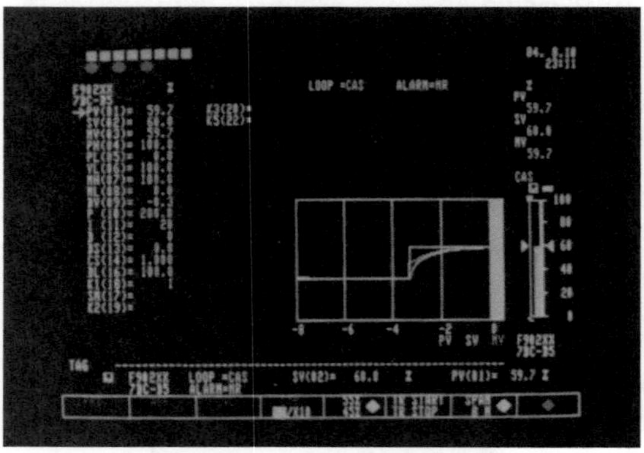

Fig. 5 · 27 Tuning panel

put value. PV, SV and MV are the corresponding digital displays.

Setpoint and output values can be increased and decreased in a gradual manner with the soft keys of the tuning keyboard, in accordance with AUTO and MAN status. Operation error from incorrect entry of data values is prevented. L901XX and 7DC–D5, etc., are displays of tag numbers and tag comments. The frame surrounding F12100 is called a tag marker. Tags signifying a test operation, etc., can be attached to the faceplate and operation can be prohibited when specified. The tuning panel, shown in Fig. 5.27, displays all loop parameters. It is used for tuning loops, changing alarm settings, and other control parameters. The parameters PH, PL, VL and DL are the alarm settings (measurement high, measurement low, velocity alarm, and deviation alarm, respectively). P, I and D are, proportional band, inte-

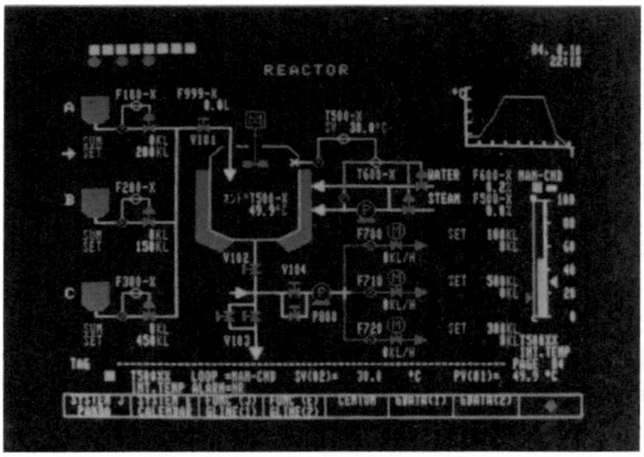

Fig. 5 · 28 Graphic panel

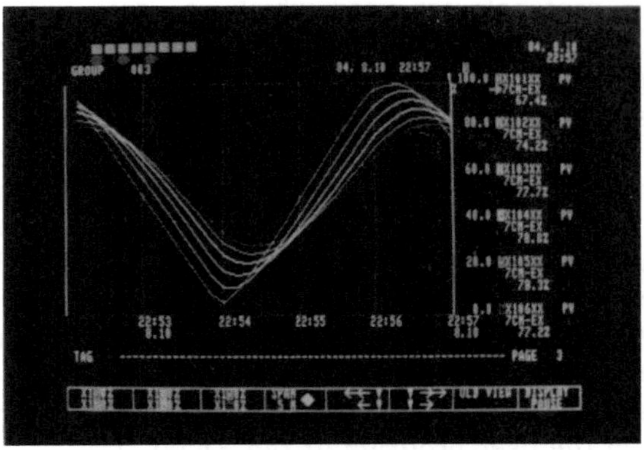

Fig. 5 · 29 Trend group panel

gral time, and derivative time. A trend display is provided for tuning.

The graphic panel is a display in which the user can follow and direct operation. It is the center of operation. An example is shown in Fig. 5.28. Colors and graphics change in accordance with alarm changes, and operation status can be made intuitive. The trend overview panel and trend group panel replace pen-and-ink recorders. Trending on a CRT brings a high degree of freedom and ease of use in the display of large amounts of record points, permits enlargements and reductions, and operation is possible while watching process changes over time. Also, data can now be easily stored on auxiliary media such as floppy diskettes. Use can also be made for trend data for quality control. Figure 5.29 shows a trend group panel. The file display change panel can be used to service simple data bases for logging printing data and printing format files, storing brand name data for brand control, and storing recipe data files. The process status report panel allows the operator to retrieve process statuses under some conditions. The utility panel displays help messages for settings of each type of system constant such as time settings, and for when incorrect operation occurs.

5.2.6 Communication with other systems

With the advance of computer communications technology, total FA systems have become a reality. Distributed control systems, in meeting the demand, have also acquired standard functions for communicating with other systems. As an example of this, the CENTUM system will be explained below.

For interfaces between computers used for control applications, as shown in Fig. 5.30, the GP–IB or RS–232C are widely used. Beyond these, standard protocols are used which use data with tag numbers and industrial units which have ASCII character strings as their base. These are unified as presentation interfaces so that communication gateway units (CGWUs) can interface with supervisory computers in exactly the same way that the operator station sends and receives character string information to the operator. Table 5.7 shows the message formats used in a controller. Figure 5.31 shows the information flow for process data acquisition by a supervisory computer, YECOM.

(1) Configuration information of the control system, tag lists and data conversion tables are downloaded from the operator console, and the communication gateway unit begins to function.

(2) The message "message omitted here" is sent to the CGWU via GP–IB from the user program of the YEWCOM.

(3) The message received at the CGWU is transferred to the command process routine via the GP–IB driver.

(4) The command process routine can call up the DDC process routine and sequence data process routine separately by message format or tag number. For this example, the DDC data process routine is called.

(5) The DDC data process routine changes the tag numbers to station numbers and loop numbers (the tag order number), the data types are changed into data numbers, and a request for data extraction is put to the HF bus driver.

(6) The HF bus driver extracts controller data via the HF bus and passes it to the request origin (DDC data process routine).

(7) The DDC data process routine converts the data format inside the controller to character–format industrial–unit data from the type of instruments and the data type.

(8) The command process routine arranges the message format and

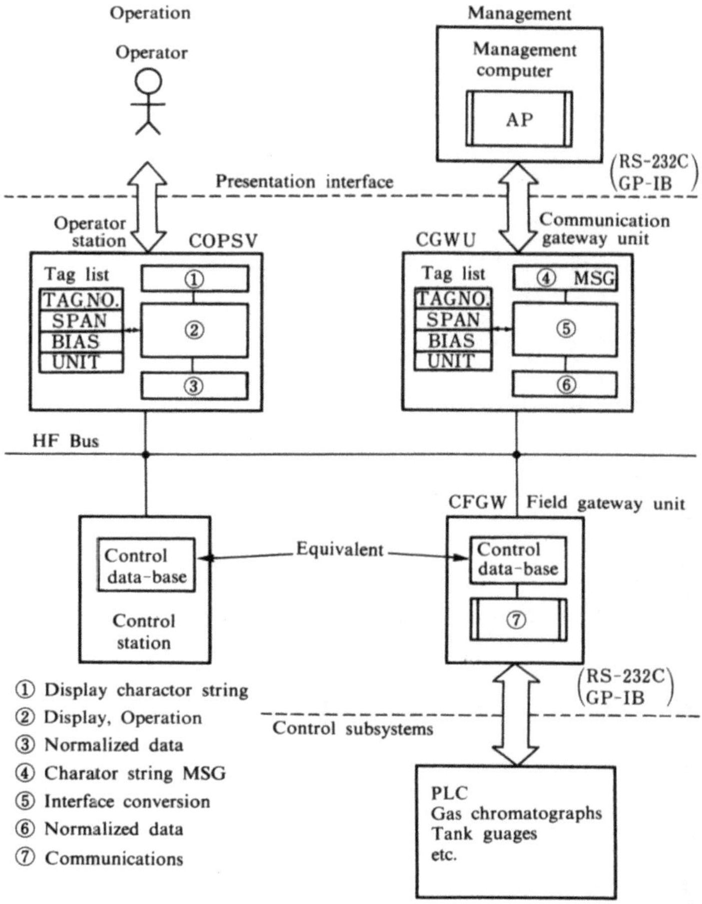

Fig. 5 · 30 CENTUM Communication gateway

passes it to the GP–IB driver.

(9) The GP–IB driver sends the message "message omitted here" to the YEWCOM via the GP–IB.

(10) The GP–IB driver of the YEWCOM passes the received message to the user program.

5.2.7 *Engineering*

The distributed control system uses computer technology to the fullest, however, it greatly differs from computer systems in the way of the program is written. Computer systems are programmed and tested using assembly language or a general–use programming lan-

Table 5.7 Message type used in a controller

Function			Message type
DDC data acquisition by Tag No.	Send	TG␣n_1	␣Tname, Dtype␣…␣ Tname, Dtype CRLF
	Return	TG␣n_1	␣Data␣……␣Data CRLF
DDC data set by Tag No.	Send	TP␣n_2	␣Tname, Dtype␣Data␣…… ␣Tname, Dtype␣Data CRLF
	Return	TP␣n_2	␣Data␣……␣Data CRLF
DDC/SEQ data acquisition by loop No.	Send	DG␣ST␣n_1	␣Dcd␣……␣Dcd CRLF
	Return	DG␣ST␣n_1	␣Data␣……␣Data CRLF
DDC/SEQ data set by loop No.	Send	DP␣ST␣n_2	␣Dcd␣Data␣…… ␣Dcd␣Data CRLF
	Return	DP␣ST␣n_2	␣Data␣……␣Data CRLF
Message from HF bus side	Send	MS	CRLF
	Return	MS	␣FM␣mno␣PAmes CRLF

n_1 ; Data number (1 to 32)
n_2 ; Data number (1 to 8)
Tname; Tag No.
Dtype; Data type
Data; Data
Dcd; Data identifier
ST; Station number
PAmes; Alarm message

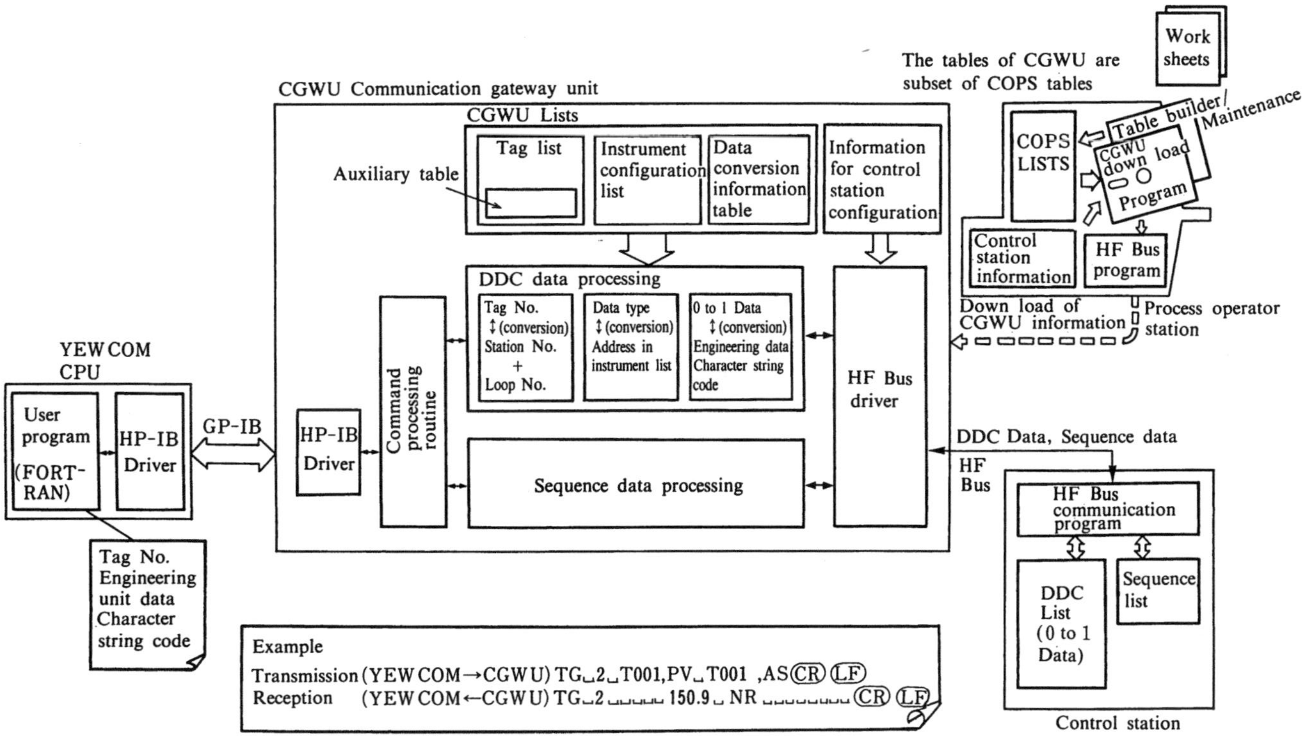

Fig. 5・31 Information flow in process data acquisition

P. /

CENTUM INSTRUMENT SPECIFICATIONS

STATION NO.

Loop Number

Tag Number

Instrument Model

Input Processing
- Scale High Limit
- Scale Low Limit
- Engineering Unit
- Ratio Set Unit Ratio Scale
- Pulse Rate
- Totalizer/No. of Computational List

Alarm Processing
- Discriminative High/Low Limit Alarm
  - Alarm Output Point
- Velocity Alarm +/−
  - Period (Sec.) for Velocity
  - Alarm Output Point
- Deviation Alarm +/−
  - Alarm Output Point

Control Processing
- Control Period (Sec.)
- Gain in Dead Band
- MV Tracking
- I/O Compensation

Output Processing
- Direct/Reverse Action
- Output Tracking
- Output Velocity Limit
- Pulse Width Output
  - Full Stroke
  - Backlash
  - Min. Output
- Time Proportioning On/Off Period
- Compensation Output Data
- Computing Module Positional/Incremental

Program Set Unit
- Time Span
- Type
- CMP-DDC

Special Processing

NL	7SW	7PG	7BD	CALC	INL	7SL

	Input Point %	Output Point %
NL, INL		
7SW	——	Data %
7PG	Time S/M	Data %
7BD	K1 – KE•	D1 – DE•

Rows 1–14.

No. of Segments on + Side 7SL (rel. posn. in list): rows 1–11

No. of Segments on − Side

Linked stn No.

Linked tag No.

(Note) •: Engineering Unit Data.
"Output Point %" column can not be specified for NL.

REV	n	REMARKS	DATE	REV. BY

ENGINEER: DR CH

PLANT NAME (JOB NAME)

YOKOGAWA ◆

WS 34B2X2-23E
3rd Edition: June, 1987
Printed in Japan (TA)

Fig. 5 · 32 Example of work sheet (Instrument specifications) guage such as FORTRAN after a system engineer consults with the user and writes the functional specifications.

In contrast to this, the distributed control system is completed with the user or sales engineer filling in worksheets which includes everything from system and hardware configuration to configurations of feedback control functions and sequential control functions. This is how basic specifications are decided. These worksheets are called a "fill-in-the-form" (FIF) style worksheets. Their sole purpose is to permit communication of system configuration in the language of an instrumentation engineer. The contents of the worksheets are input into the system using the builder/maintenance function (system creation and alteration) which follows an interactive format, and the system is created. The created system is then tested using the test functions

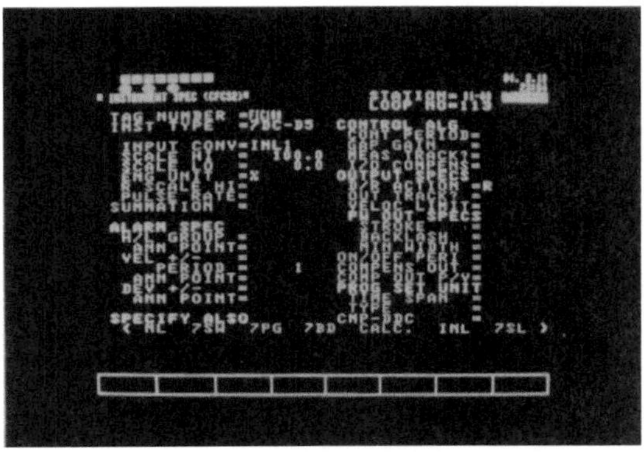

Fig. 5 · 33 Instrument specification maintenance panel

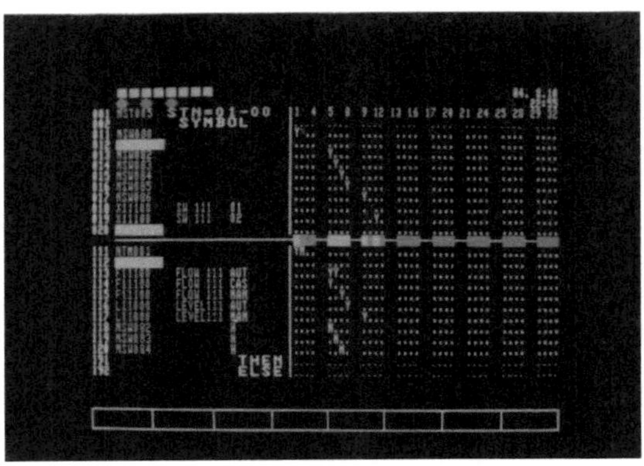

Fig. 5 · 34 Sequence table maintenance panel

such as the status display function and the integrated retrieval function. Corrections are done using the builder/maintenance function. After the tests are completed, the distributed control system is finished. Using the self-documentation function, complete documentation is printed out for the system, to simplify maintenance. Figure 5.32 shows the worksheet, Figure 5.33 shows the builder/maintenance function; Figure 5.34 shows the status display; and Figure 5.35 shows an example of self-documentation.

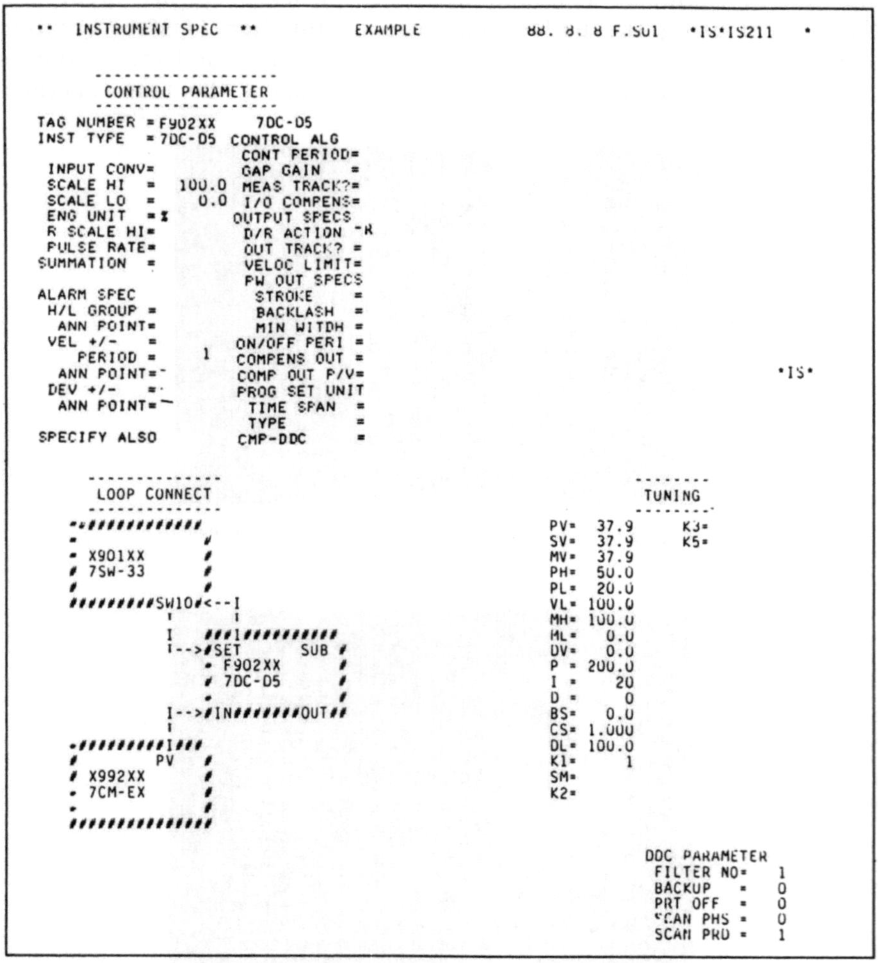

Fig. 5 · 35 Example of self-documentation

5.3 Production Line Control System[7)-10)]

The implementation of mechanization and automation of production lines, where products are worked upon and produced, are very important factors concerning plant automation. Many difficulties lie, however, in the fact that both materials and information must be handled in such systems. In this section, an explanation of the different types and the overall position of the systems for controlling production lines, and a summary of FA computers for production line control which have come into wide use lately are explained.

5.3.1 Summary of production line control systems

FA production systems, as shown generally in Fig. 5.36 form a hierarchical structure reaching from the production line level and spanning the entire company. Computer systems, which form the core, are utilized at each level, and the type of computer system used depends on what goes on at that level. At the company level, general purpose computers are used, while at the factory and production department levels, computers for production control are used. Many types of computers are used at the production line level depending on the type and scope of the production line, and the type of control system. These computers are generally referred to as "production line control sys-

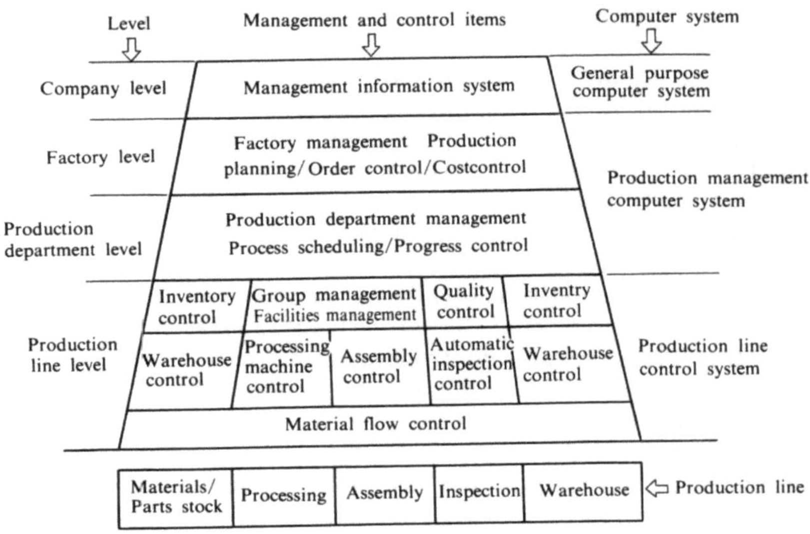

Fig. 5 · 36 The hierarchical structure of FA manufacturing system tems." A factory automation system at the production line level, which uses a production line control system, will reflect the diversity of the production line. For this reason, such systems are difficult to classify into a single group. Also, proceeding with factory automation in sequence from the actual site, a system configuration becomes desirable which starts from the factory floor, bottom up. This is the layer where physical and mechanical information from the production line are converted to logical information which can be processed by a computer. This layer, which is the basis of total factory automation, is one of the most important elements of the system and is the key to its success.

5.3.2 Types of production line control systems

Production line control systems come in many types. The span extends from computers which would be better considered controllers than computers, such as those which are embedded in every machine in the plant and those which use a PC (Programmable Controller) for main sequential control, to computers which classify more closely as production control computers which center around data processing.

A summary of a typical production line control systems in actual use as follows.

(a) Specialized controllers for FA equipment

Controllers for robots, NC machinery, automatic warehouse stacking cranes, and other such FA equipment are embedded in the body of the equipment. Most of these controllers use 8- or 16-bit microprocessors, and both the hardware and software are specially designed to optimize the performance and function of each piece of equipment. The program is generally stored in ROM. Although in some cases, the operating instruction set is fixed, normally instructions are received from a programming device, paper tape, floppy disk, or from a supervisory computer by communication. These operating instructions are stored in RAM. Usually a battery back-up power supply is used to insure that the memory contents are not lost in case of power failure.

(b) Programmable controllers

Programmable controllers (PC) are control computers that have in recent years been replacing the formerly-used relay boards in sequential control. Programmable controllers are more easily changed and expanded than relay boards. They offer great flexibility, and their high reliability facilitates maintenance.

An example of PC hardware configuration is shown in Fig. 5.37. The sequential control procedure is input in advance from any kind of terminal, and stored in the user subprogram memory. The control procedure is as follows. First, the input signal is read in from the I/O device and stored in the data memory (input processing). A microcom-

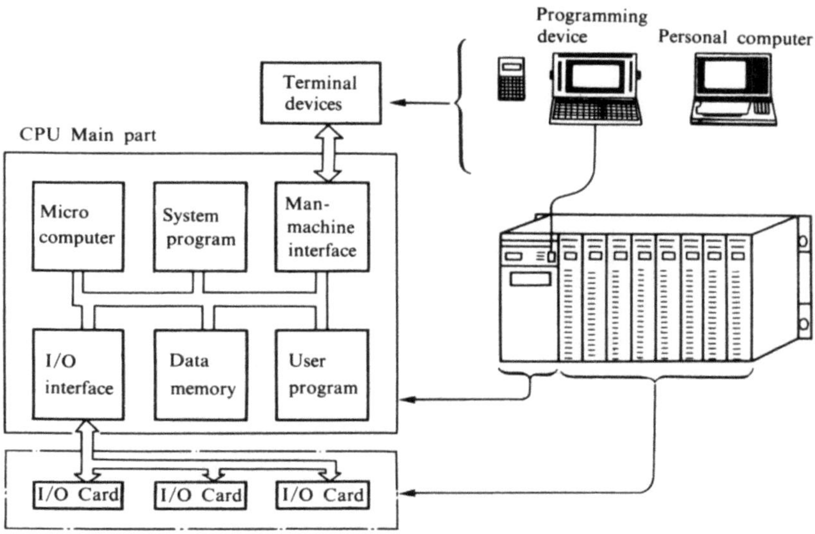

Fig. 5 · 37 Configuration of programmable controller

puter reads the user program, in order, and performs logical operations based on the status of input signals stored in the data memory. The results are then stored in the data memory (operation processing). When processing of the user program is finished, the operation result stored in data memory is output by the input/output device (output processing). This one period: input, operation, output, is called "one scan." This is executed cyclically as the sequence control program is advanced. The scan period differs according to the type of PC and number of steps in the user program. Typically, a speed of 10 to 50 ms is required, but for high-speed machine control, a speed of 1 to 10 ms is required. Ladder diagrams, which resemble operational sequence diagrams for relay boards, are often used to express user programs (Fig. 5.38). Ladder diagrams are easy for technicians experienced with relay boards to understand, but are not always the appropriate method for expressing the overall progress of sequence control. Flowcharts, Boolean algebra, and decision table methods are also used.

In addition, with FA becoming more and more widespread recently, programmable controllers often incorporate high level features such as analog control, data processing, and external communications functions, but FA computers (described later) are usually used in applications requiring such functions, and in general programmable controllers are used as small, relatively inexpensive on-site controllers.

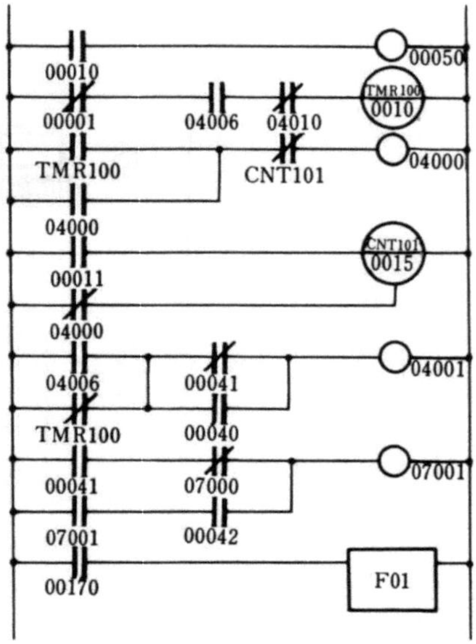

Fig. 5・38 Example of ladder diagram

(*c*) *Computer system*

Programmable controllers are widely used as inexpensive FA controllers, but as FA advances there is a demand for more general purpose data processing functions, various man-machine interfaces, communications with host systems, and management functions for all kinds of production lines. Production line control systems that use computer systems allowing the construction of more general purpose systems than programmable controllers are increasingly being applied. Computers based on existing minicomputer and personal computer systems are common. In addition to these, special purpose production line control computers, *i.e.* FA computers are also available on the market and are becoming widespread.

The features of these computers as production line control systems are outlined below.

(1) *Minicomputer based systems* Minicomputers such as the YODIC and YEWCOM from Yokogawa Electric Corporation have long been used in the field of process control (PA). They have also been used in the field of FA, mainly for production management. They are commonly used in production line control systems with a centralized type system construction for controlling several processes. They are also used as front end processors of large scale production management systems.

(2) *Personal computer based systems* When 16-bit personal computers

came onto the market they became popular in the field of office automation, and their prices dropped quickly. This contributed to their gradual appearance in the field of FA. Factors such as reliability, process input/output functions and expandability are of concern regarding the use of personal computers for control on production lines, but they are widely used for independent management in relatively small scale systems.

(3) *FA computers* These are specialized computers for application in production line control systems and have the following features.

(1) Powerful real-time functions and high reliability.
(2) A wide range of I/O interfaces.
(3) Sequential control functions.
(4) Powerful communications functions for communicating with host systems and FA equipment.
(5) Flexible system structure and easy system construction.

Following the release of the YEWMAC FA computer, by Yokogawa Electric Corporation in 1983, FA computers were released by many other companies and this new computer market quickly expanded.

5.3.3 FA computer systems

The system structure an actual FA computer developed for production line control is explained below.

(a) System structure

For total control of a production line, it is important that in addition to the transfer of production command data and accumulation of results data from each process in the vertical direction, tracking data and quality data flow between each process in the horizontal direction.

The system structure of a total production line control system using personal computers is shown in Fig. 5.39. The personal computers that control each process are connected to a host process management

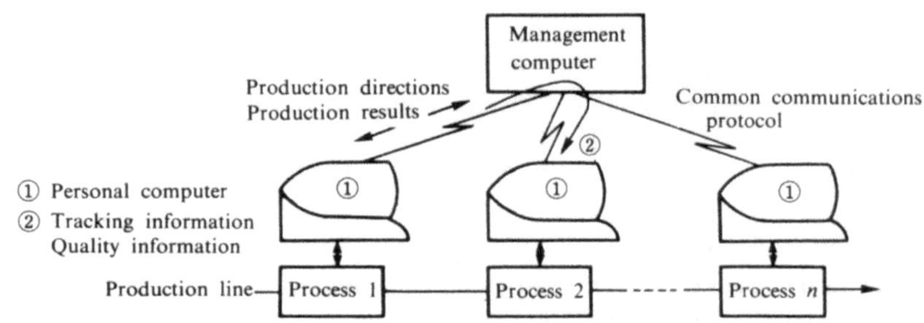

Fig. 5・39 Production line control system by personal computer

computer, and normally, transmission of data between processes is performed via this management computer using some common communications protocol. In this type of system the performance of the communications link between the computers and ease of data transfer procedures are important points.

FA computers are designed with this kind of total production line control system in mind, and in general have the distributed type structure shown in Fig. 5.40. The control units controlling the process, and the management unit that controls the total system all have their own processors, and are connected together via a high speed data bus and function as one system. Each unit can directly and simply communicate at high speed, and users can easily access production data JIT (Just In Time). This is necessary to construct high efficiency production control systems.

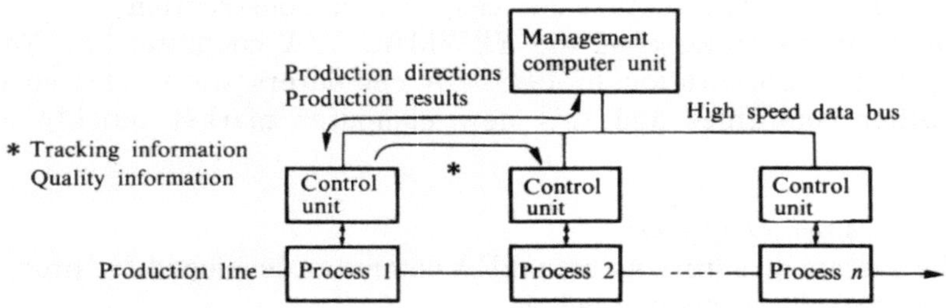

Fig. 5 · 40 Production line control system by FA computer

(*b*) *System example*

To illustrate the general concept of FA computers, Figure 5.41 shows the system structure of an inspection process employing Yokogawa Electric Corporation YEWMAC FA computer. The YEWMAC system consists of line computers which handle management functions (MLCC; Manufacturing Line Control Computer), networked via the ML bus to line controllers which handle the control functions (MLCU; Manufacturing Line Control Unit), thus forming a hierarchical distributed system. The MLCU (equipped with a 16-bit microprocessor) and the MLCC (with 16- or 32-bit microprocessor) run user programs written in distributed real time BASIC which has inhanced functions for FA use. The MLCC and MLCU are shown in Figs. 5.42 and 5.43. In the system shown in Fig. 5.41, an MLCU reads the product type of each product via a bar code reader, and controls the various measuring devices under control of the inspection program corresponding to that product type. Based on the results of the measurements, the DI/DO signals are directly controlled to control the product flow. The inspection results are also displayed on terminals in the factory to inform operators. The MLCC simultaneously stores and manages the inspection pro-

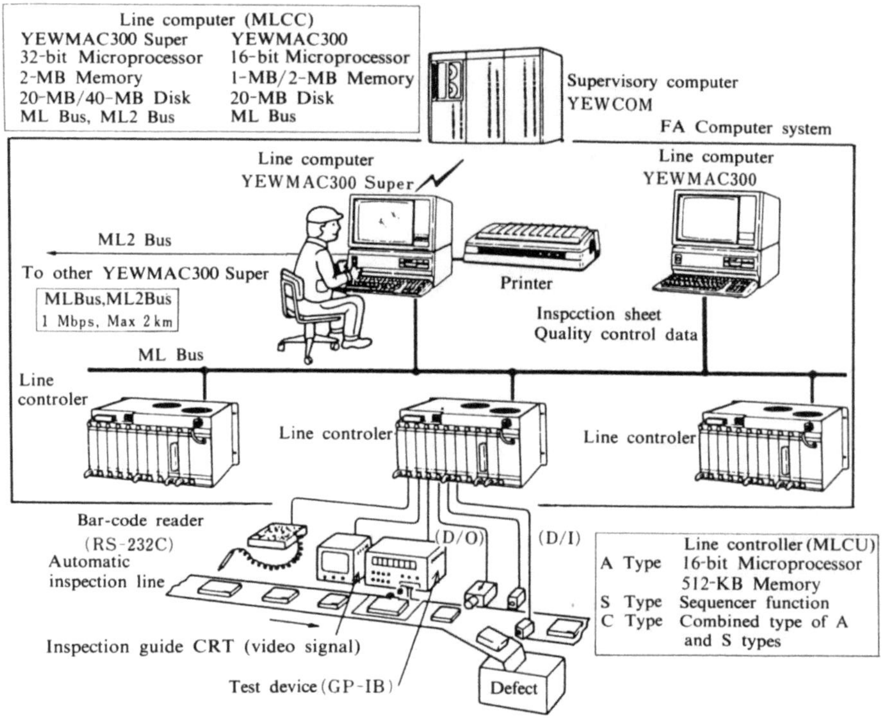

Fig. 5 · 41 Example of FA computer system (YEWMAC)

Fig. 5 · 42 External view of YEWMAC 300 MLCC

Fig. 5 · 43 External view of YEWMAC 300 MLCU(C-M Type)

gram master file and accumulates the quality management data, inspection results table data, and operation results data. It displays the data on the management terminal, outputs to the printer, and transmits data to the host computer as necessary. From this example it can be seen that the important points for an FA computer are, in addition to the status of the system structure, the production line interface (process interface), man/machine interface, communication interface, and the software to control them.

5.3.4 FA computer system hardware

(a) Process interface

A production line control system must connect to the many types of sensors and actuators used in production lines, and control the FA equipment. Table 5.8 shows the types of I/O cards and the sensors and actuators they are used with.

Features of the I/O cards for FA compared to those for PA.

(1) *A diverse variety of I/O signals* Unlike PA, in the case of FA, the I/O signals to the system are not assumed to be uniform. It is necessary to install an assortment of I/O cards to handle the variety of signals from the sensors and to the actuators.

(2) *Multi-function card* For flexibility it is desirable to have a multi-function card that includes a variety on I/O specifications on a single card.

For example, a multi range analog I/O card, a communications card compatible with a variety of communications standards such as RS-232 C, RS-422, and current loop, and a card for use with the factory terminal to handle the CRT output and keyboard input.

Table 5.8 Example of I/O card, sensor and actuator

Type of I/O cards	Example of combined sensors and actuators
Analog input	Potentiometer, Differential transformer, Strain-gauge, Thermister, Thermocouple
Analog output	Electric motor, Recorder
Digital input Interrupt signal Pulse input	Microswitch, Limitswitch, Electrophoto switch, Push-button switch, Rotary encoder
Digital output Pulse output	Lamp, Solenoid valves, Electromagnetic contactor, Servo-motor, Stepping-motor

(3) *High speed* The demand for high response speed is greater for discreet processes than for continuous processes, particularly for mechanical system control. Scan periods are normally 10 to 100 ms, and the required control periods may be 1 to 10 ms. Accordingly, to comply with the high speed, I/O cards for the production line control system should be capable of high speed. Many I/O cards have their own processors for high-speed I/O control. Process interface cards are listed in Table 5.9. General purpose I/O cards, communications cards, and man/machine interface cards are also listed in the same table.

(*b*) *Communication interface*

The FA computer system communications interface is classified into two types; host and field communication.

(1) *Host communication* FA computers receive production planning and manufacturing data from the host computer, and send back the actual production results and the production line operating status. This communication with the host computer is performed at every batch, such as a day or a shift. It is important that the production line control system be able to operate as autonomously as possible. Figure 5.44 shows an example of a control system for an NC machine. Here, the FA computer receives the required NC data for the day or shift and performs the processing locally in its own system while the machine is operating. This improves host system performance, and enables more responsive downloading of the NC data for the variety of products to be manufactured.

The communication procedure with the host is not fixed but the asynchronous procedure, BSC (Binary Synchronous Communication), and SDLC (Synchronous Data Link Control) are often used. For connection to general purpose host computers, terminal emulators, and RJE

Table 5.9 Input/output card used with MLCU

Use	Type	Function	Microprocessor
General use I/O cards	VM 1	16 points, 1 to 5 V input	
	VM 2	8 points, 1 to 5 V input/output each	
	AN 3	16 points, ±10 V input	
	AN 4	8 points, ±10 V input/output each	
	HAD 1	16 points, ±1.25/2.5/5.0 V input Resolution 14 bits 50 K samples/second	○
	PM 1	16 points, pulse train	
	S T 2	16 points, digital input/output each	
	S T 3	32 points, digital input	
	S T 4	32 points, digital output	
	P B 6	16 points, push-button input	
	G B 1	GP-IB interface	○
	T V 3	Color TV/keyboard interface	○
	L C S	YEWSERIES 80 intereface	○
	R S 2	Multiboard (4-ch) RS-233 C	○
	R S 3	RS-232 C, RS-422, 2 ports/current loop	○
Inspection use I/O cards	3568-01	10 points ±20/50/200 mV input Thermocouple, Type R/K/J/T/E	○
	3568-02	10 points, ±1/10/50/1 to 5 V input Universal counter	○
	3568-04	DC-1 MHz, pulse totalization Frequency/period/measurement of frequency ratio	○
	3568-05	2 points, 1 to 10 V programmable voltage output, with sweep function	○
Sequence control use I/O cards	AC 1	16 points, AC 100 V ON-OFF input	
	DC 1	16 points, DC 24 V ON-OFF input	
	S C 1	16 points, contact input	
	PU 1	2 points, up-and-down pulse counter	
	S R 0	16 points, solid state relay switch	
	TR 0	16 points, transistor output	
	RL 0	16 points, relay output	

(Note) ○: Incorporate microprocessor

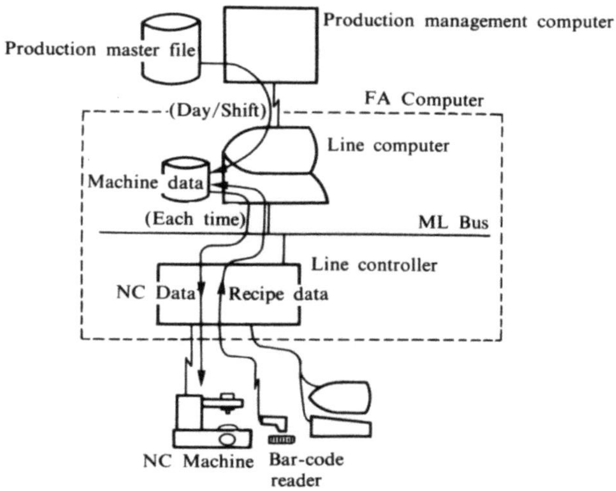

Fig. 5・44 Data flow in NC machine control system (YEWMAC's example)

(Remote Job Entry) are required. With these, communication is possible without any particular need for awareness of the communications procedure.

(2) *Field communication* Machines with intelligence such as robots, Programmable controllers, inserters, and NC machines are widely used on production lines. From now, connection to FA equipment will in general be via RS-232 C, RS-422, current loop, and GP-IB. There will be a variety of communication procedures such as the asynchronous procedure, or basic procedure, depending on the FA equipment, and the production line control system will have to be able to flexibly respond to these.

(*c*) *Man/machine interface*

At present because of technical and economic considerations, there are few examples of totally automated factories with no human operators, and in fact systems that harmonize the high level decision making ability of human operators with machines are more practical. Accordingly, the man/machine interface that the operators and technicians in the factory use to input and receive necessary data is a very important factor in the system.

(1) *Factory terminal* It is not a requirement that the display on the operator's terminal have great resolution for displaying large amounts of information. Rather, it is desirable that the information be displayed in large size characters that can be seen and confirmed from a distance. CRT and LED displays are often used, and recently demand for Chinese Character displays and color displays is increasing because

they are easy to read and understand for Japanese operaters. Keyboards with numerical keys and just a few alphabetic keys are commonly used. They are installed throughout factories, so low cost is a strong requirement. Figure 5.45 shows an example of this type of terminal.

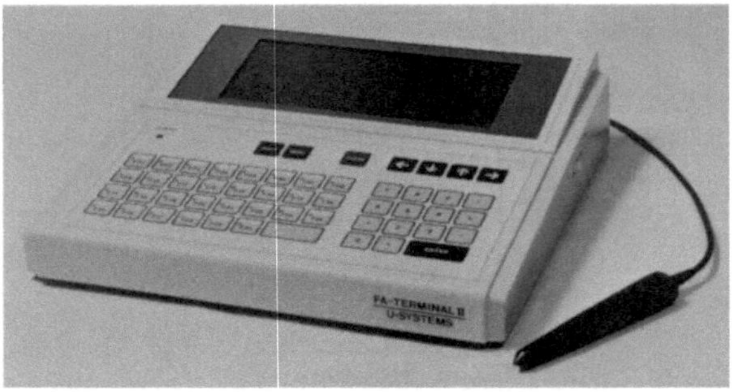

Fig. 5 · 45 Example of FA terminal

There are a wide variety of man-machine interface devices used to meet the needs of factories. These include bar code readers for reading the manufacturing numbers off parts and manufactured goods, sign board readers for single product flow type production systems, and voice recognition systems for use in sorting systems.

(2) *Management terminal* The management terminal informs production line supervisors and operators of a variety of management data relating to the production line and detailed information about the status of the production line. Due to the increasing complexity of the data to be displayed, and for human engineering reasons, displays are generally CRTs that can display in Japanese, with color and full graphics capability. Display resolution of about 640×400 dots is required and 14-inch screens are usual, although for process control use, 20-inch screens are also used. Normally, these terminals are also used for program development, and general purpose keyboards (*e.g.*, JIS specification) are generally used. Flat keyboards covered with silicon rubber are also used because of their closed nature.

(*d*) *Sequential control functions*

Sequential control is often used in production line control, and fast cycles of 1 to 10 ms are often required. To avoid placing to much load on the CPU, it is desirable to have a dedicated sequence control processor to handle these high speed sequence control tasks. Unity and coordination between the processing done on the CPU and the processing done on the sequence control processor is important. It is also necessary for the program running on the CPU to be able to quickly access the sequence control processor data, and upload and download the sequence programs. The YEWMAC can be equipped with MLCU dedicated sequence control processor cards, and access from BASIC on the CPU is easy.

5.3.5 FA computer software

In recent years the percentage of software development cost in computer system development has risen sharply. FA computer systems are not just hardware, and the software must also be carefully considered.

The software consists of the system programs supplied as standard with the system, and the programs created for the applications of individual users. The program development environment and operating environment are heavily influenced by the content of the system programs. The important operating systems, programming languages, and software packages are discussed below.

(a) Operating systems

The operating system (OS) is the nucleus program that controls the operations of the software system.

The features required in operating systems for FA computers are as follows.

(1) *Real time functions*: Fast response to external events, priority interrupt processing, efficient allocation of resources, and high speed task scheduling.

(2) *Multitasking functions*: Parallel processing of many tasks such as; data acquisition and screen display, and external communications.

(3) *Efficient I/O management system*: Efficient and consistent management of process I/O, the man/machine interface, and communication interface are required.

The operating systems generally used in FA computers can be classified as shown in Fig. 5.46.

Operating systems from computer manufacturers, which are written exclusively for their machines, have the advantage that they are designed to optimally utilize the special features of these machines, and

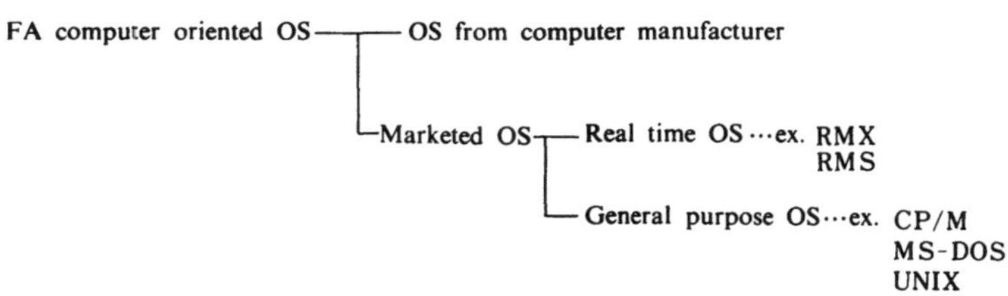

Fig. 5 · 46 Classification of FA computer oriented OS

are efficient in terms of functions and performance. On the other hand, commercially available operating systems, which have started to come into use because of the popularity of personal computers, have the advantage that the various types of software marketed for the same OS can be used relatively easily. Marketed operating systems can be classified into real-time operating systems and general purpose operating systems. However, as stated previously, FA computers require real-time operating systems. There are more software tools and computer languages for software development for the general purpose operating systems, so some systems are switched between the two types of operating system depending on the application.

(*b*) *Languages*

There are a wide variety of production lines that FA computers control, and compared to the case of process control, they are difficult to typify. For this reason, procedure-oriented languages such as assembler, BASIC, FORTRAN and PASCAL are normally used rather than problem oriented languages. It is necessary to consider the following points when adopting these languages.

(1) Ease of learning for factory technicians who are not computer experts.
(2) Efficient program development and simple debugging.
(3) Execution speed of the developed programs.
(4) Memory space required by the developed programs. (High coding efficiency)
(5) Ease of transferring the programs to other systems.

The features of procedural orientated languages for FA.

(1) *Assembler* Programs written in assembler have the advantage that they can be fast and compact. Disadvantages are that assembler is difficult to learn, program development is inefficient, and that programs are difficult to transfer to other systems.

(2) *FORTRAN, PASCAL and C* FORTRAN is widely used for scientific calculations in process control. PASCAL features rich data structures and is suitable for writing structured programs. The C language used to write UNIX* closely resembles assembler language and features a large number of operators, and the ability to perform the bit control necessary in control and measurement programming.

Program development using these compiled languages is fast compared to assembler, but more memory space is required and execution speed is slower. They are more oriented to specialists than BASIC which is described below.

(3) *BASIC* This language came into wide use as personal computers became popular and it is easy for beginners to learn. With the nor-

* Unix was developed by AT&T and is a registered trade name.

mally used interpreter type there is no need for compiling or linking, so debugging is relatively easy. Slow program execution and poor real time functions can be raised as disadvantages, but in recent years, powerful BASICs that solve these problems have appeared.

One example of such a BASIC is the distributed type real time BASIC used in the Yokogawa Electric Corporation FA computer YEWMAC. (Refer to Subsec. 5.2.3 (b) regarding YEWMAC).

The BASIC that supports YEWMAC has, in addition to the functions of normal personal computer BASIC, the following functions for production line control.

(1) Using time slicing, multiple programs can be run on the MLCC.
(2) Distributed processing functions for sending and receiving data between the MLCC and MLCU.
(3) Real time processing for timed and unpredictable events. (Refer to Table 5.10)

Table 5.10 List of real time statements

Real time statements	Function
WAIT	Waiting for execution of program
ON TIME	Branching by time
OFF TIME	Cancellation of branching by time
ON TIMER	Start timer
OFF TIMER	Stop timer
SIGNAL	Notice of event occurrence
ON EVENT	Waiting for event occurrence
OFF EVENT	Cancellation of waiting for event occurrence
DISABLE	Inhibits interrupt
ENABLE	Cancellation of interrupt inhibit
CHAIN	Chainning of programs
START	Start program
SETTIME	Set time
SETDAY	Set data
RESERVE	Declaration for use of common resources
RELEASE	Cancellation of declaration for use of common resources

(4) Dedicated statements for I/O cards and sequencer cards (32 statements).
(5) High level graphics such as closed-region filling color, and Window/View (35 statements).
(6) MLCU debugging can be performed from the MLCC.

(c) Software packages

Using software packages is one way to improve the efficiency of software development. In the world of office automation (OA), a large number of packages have already been developed, and this so called distributed software is widely used. The FA field is difficult to typify and

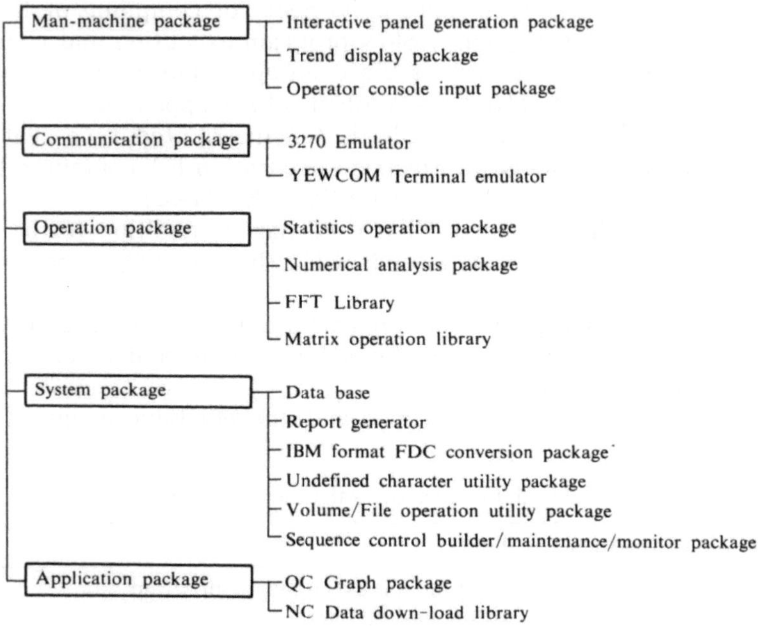

Fig. 5 · 47 Software package (In case of YEWMAC)

so there are not many software packages available compared to the number of OA packages available, but from now their use will gradually increase. Examples of package software that support FA computers are shown in Fig. 5.47.

5.4 Computer System Equipment for Production Management[11)–13)]

5.4.1 Computer components and configuration

Production management computers must generally meet the following requirements.

(1) Have a real-time operating system capable of fast response to the control system
(2) Be capable of processing and storing large volumes of data
(3) Have a redundant architecture for high reliability under round-the-clock operating conditions
(4) Be capable of highly reliable long-term storage and frequent updating of production information
(5) Possess a human interface that allows easy, highly responsive operation
(6) Have software tools such as data base management packages, pro-

cess data acquisition packages, and human interface packages, that allow easy system construction

(7) Be capable of forming a computer network or factory network for system integration

(*a*) *System configuration*

A production management computer system generally consists of a high-performance central processing unit together with large-capacity auxiliary memory devices, operator console and other peripherals, process input/output equipment, a distributed control system interface, a factory LAN interface, an interface for communication with other computers, and other such components. The YEWCOM system produced by Yokogawa Electric Corporation shown in Fig. 5.48 is an example of such a production management computer system. The detailed hardware configuration of this system is shown in Fig. 5.49.

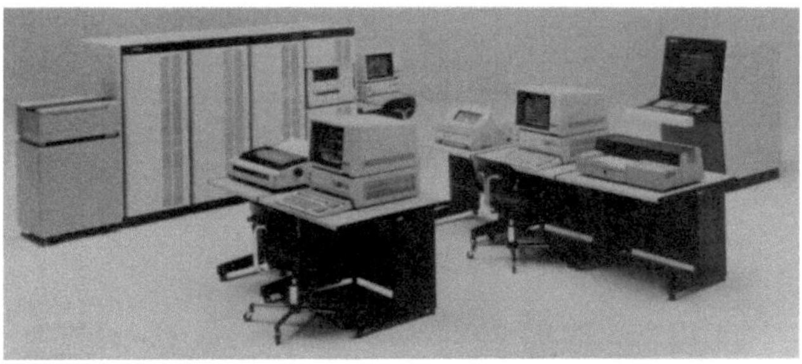

Fig. 5・48 Production management computer YEWCOM

(*b*) *Central processing unit* (*CPU*)

A processor that can be classified as a minicomputer or super minicomputer is used for the central processing unit. Formerly, 16-bit word length processors were mostly used. Recently however, 32-bit processors have come to be used as well, due to increases in the volume of information processing. Also, recent production management computers adopt the latest high-performance technologies from high-end general purpose computer systems, in addition to the real-time capabilities that have been developed on the process control computer. Previously processing performance of about 0.5 MIPS* was common, but recently performance has been raised to the 3 MIPS to 4 MIPS level. In addition, main memory capacity has expanded from a few hundred kilobytes (K) to several megabytes (M).

The following are techniques that are contributing to the increase in CPU performance.

* MIPS (million instructions per second) is one measure of processing power. It represents the number of instructions the processor can execute in one second.

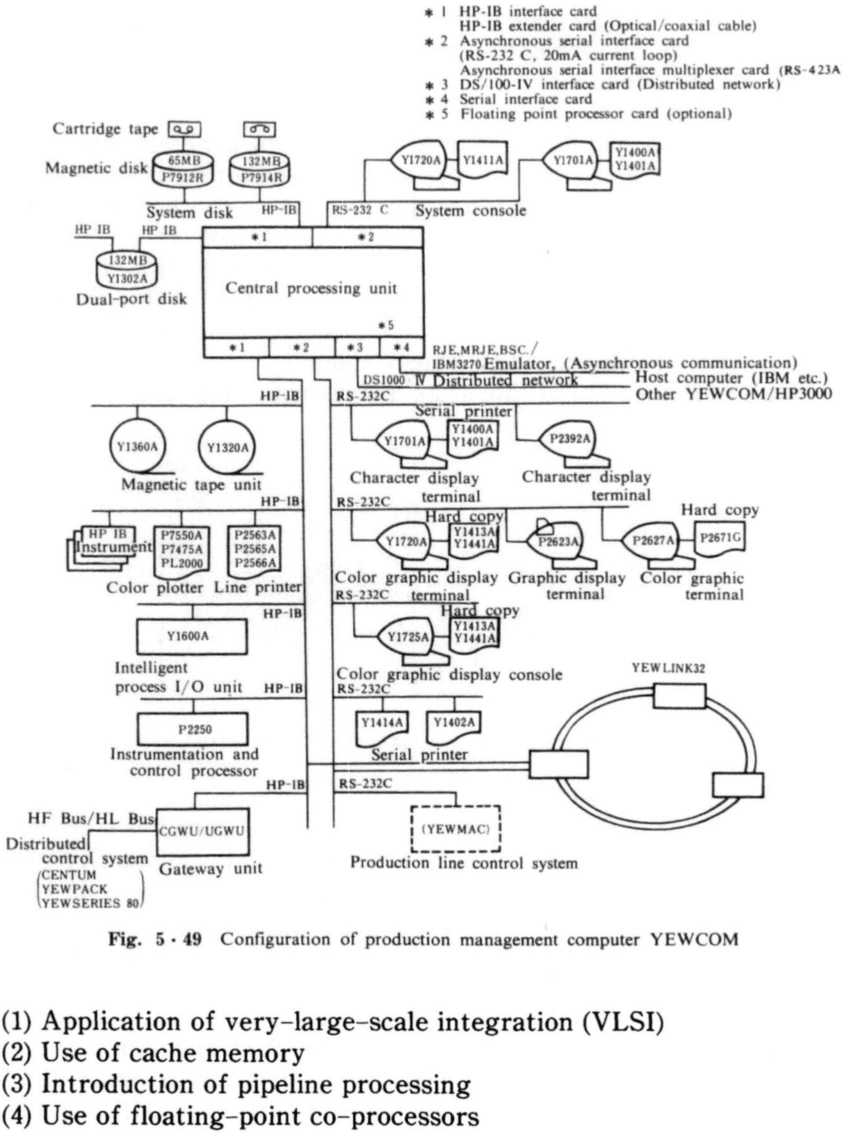

Fig. 5・49 Configuration of production management computer YEWCOM

(1) Application of very-large-scale integration (VLSI)
(2) Use of cache memory
(3) Introduction of pipeline processing
(4) Use of floating-point co-processors
(5) Use of firmware for the operating system (OS)
(6) Instruction sets corresponding to high-level programming language

(*c*) *Memory devices*

There is an inverse relationship between the operating speed of memory elements on one hand, and memory capacity and cost per unit

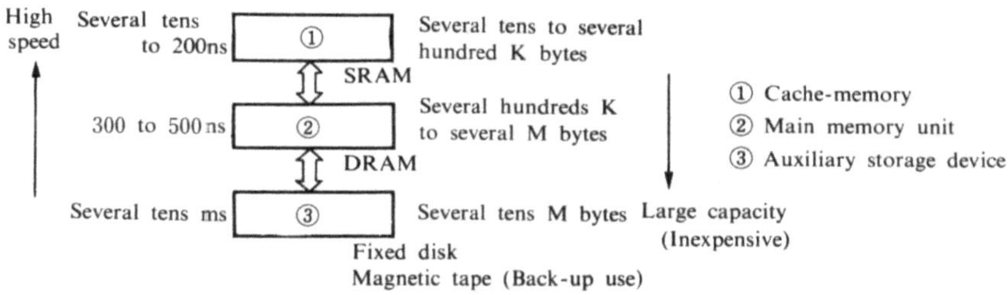

Fig. 5 · 50 Hierarchical configuration of storage devices

of memory capacity on the other. Thus a hierarchical configuration of different types of memory devices such as shown in Fig. 5.50 is generally used.

Semiconductor memory devices have become reliable and inexpensive, so dynamic random access memory (DRAM) units are used for the main memory device. Semiconductor memory requires little electric power, so battery back-up is a possible measure against the volatility of the DRAM. Also, using error correcting code (ECC), one-bit errors can be corrected and two-bit errors can be detected. Thus high reliability is achieved. Cache memory is an ultra-high speed buffer memory placed between the main memory device and the CPU in order to increase the effective speed of the main memory. It uses static random access memory (SRAM) or memory elements using the same circuit technology as the CPU, and has an access time that is equal to the cycle time of the CPU. The copies from the most recently accessed memory remain in the cache memory, and speed is increased because memory access is localized within small memory address range during a program executing period.

Fixed-head magnetic disk (hard disk) equipment is mainly used for auxiliary memory devices. The hard disks in general use have capacities ranging from several tens of megabytes to several hundreds of megabytes, although progress is being made towards smaller disks of larger capacity. Magnetic tape devices and tape streamer devices are used for data back-up and long-term storage.

(*d*) *Input/output control and intelligent input/output equipment*

The most common method of input/output (I/O) control uses a common I/O bus interface. The date transfer rate of the I/O bus ranges from 2 Mbyte/s to 10 Mbyte/s, and the trend is towards higher speeds. One method for control of data transfer is program control, in which the transmission of each word of data is controlled by a program. Another method is direct memory access (DMA), in which data is exchanged by main memory and I/O devices directly and at high speed, without depending on the CPU. In large computer systems, there is also a dedicated channel interface for high-speed I/O. As a means of increasing the I/O throughput, intelligence is being added by equipping each I/O interface with an I/O processor to perform I/O control independently of the CPU.

Figure 5.51 outlines the intelligent functions of the I/O interface and I/O devices of the YEWCOM system Since each I/O device handles the complicated processing required for its own operation, the application programs are considerably simplified. The I/O processor controls DMA transfer, interrupt processing, I/O wait, and so on. It also controls protocol among I/O devices, and processes command strings. Intelligent I/O interfaces include the byte multiplexer for asynchronous communication, HP-IB interface, the binary synchronous communication interface (BSC), and the high-level data link control (HDLC).

Along with bringing intelligence to the I/O interface, the various types of I/O devices themselves are being given intelligence by the addition of microprocessors. This allows I/O processing to be done in parallel with CPU operation. Also, because CPU overhead for I/O response no longer depends on the number of I/O devices, high overall system throughput and speed of response are achieved.

(1) *Intelligent process input/output device* (*SPACE I/O*) The SPACE I/O intelligent process I/O device shares the burden of I/O processing by downloading a program from YEWCOM. The program is written in MML (measuring and monitoring language), a specialized measurement programming language that is similar to BASIC. Figure 5.52 shows the apportioning of processing for the case of analog input. SPACE I/O gathers process inputs at fixed intervals. YEWCOM receives data that has been filtered, linearized, and converted to industrial units in SPACE I/O, thus it can devote itself to the processing of the data.

(2) *Communication gateway unit* (*CGWU*) The CGWU is an intelligent

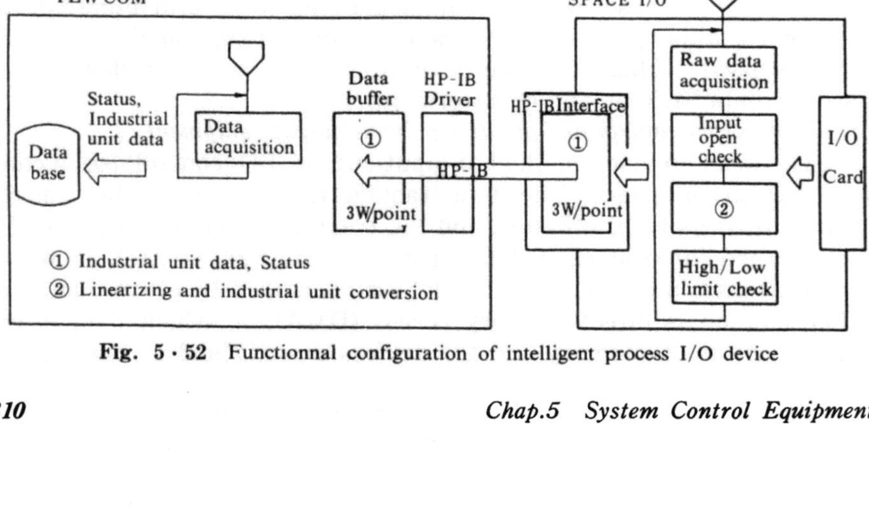

Fig. 5・52 Functionnal configuration of intelligent process I/O device

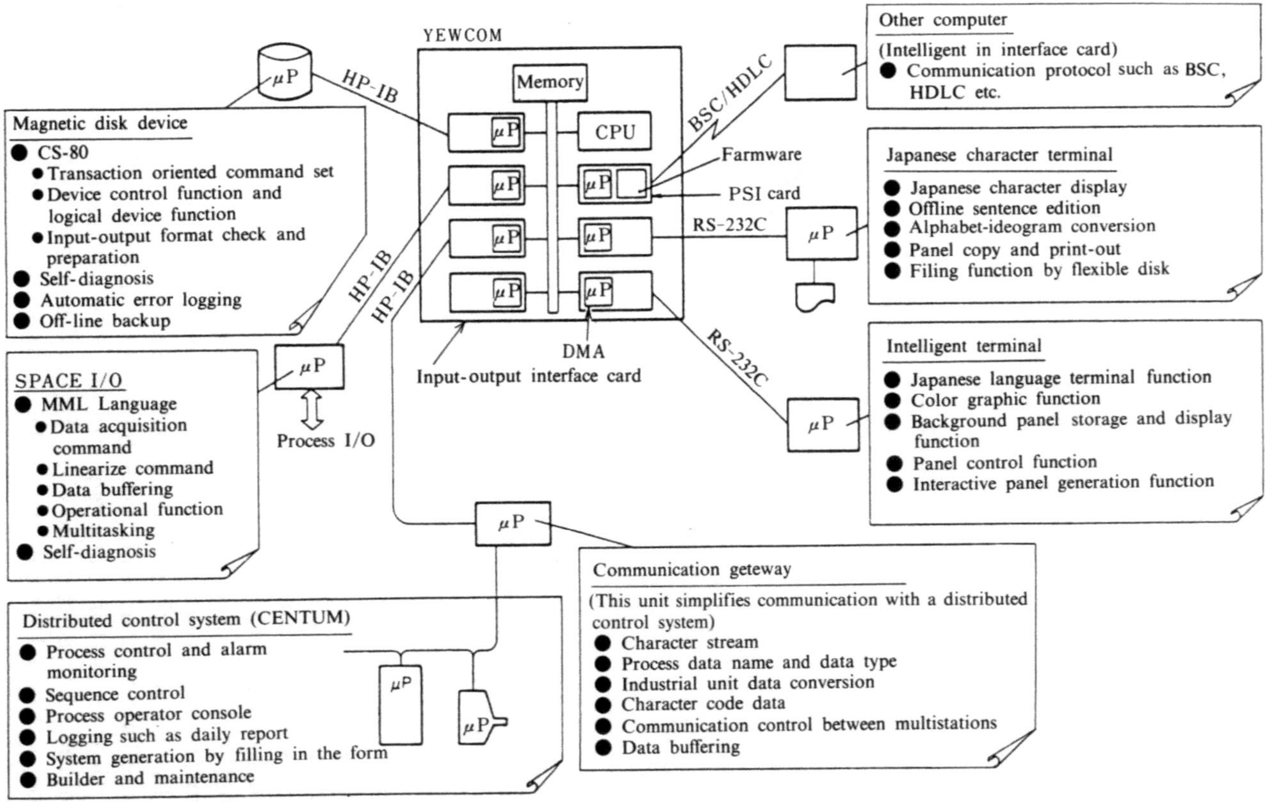

Fig. 5・51 Intelligent functions of I/O interfaces and I/O devices communication control device that connects the various kinds of stations on the HF bus of the CENTUM distributed control system to YEWCOM through the HP-IB. CENTUM data can be accessed by tag number and data type in character form, and the industrial unit data can be collected and set. As shown in Fig. 5.53, conversion between tag number and station loop number, and industrial unit conversion is handled by CGWU intelligence. Because the communication procedure is performed with completely logical data, communication can be done with no awareness at all of the arrangement or format of the data on the OENTUM side.

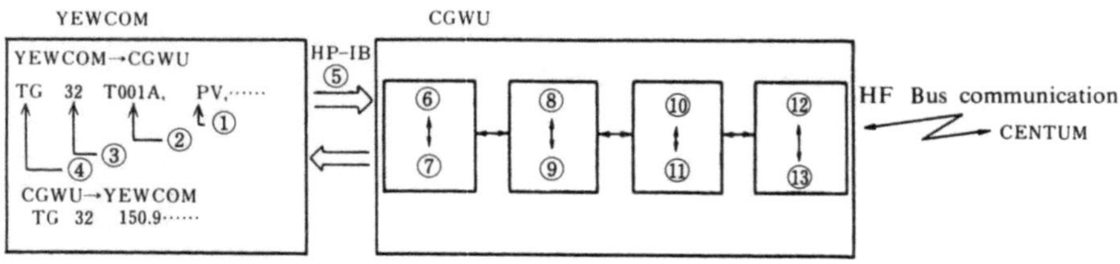

① Data type
② Tag No.
③ Data quantity
④ Command
⑤ Communication
⑥ Tag number
⑦ Station No. Loop No.
⑧ Data type
⑨ Address in instrument list
⑩ Industrial unit character string data
⑪ Industrial unit binary data
⑫ Industrial unit binary data
⑬ 0 to 1 data

Fig. 5・53 Data conversion in communication geteway unit

(3) *Intelligent terminal and operator console* In Fig. 5.54, the Y1720A is shown as an example of an intelligent terminal. An example of its display screen is also presented. In this terminal, as shown in Fig. 5.55, the required number of fixed background screens, such as process flow diagrams, are prepared in advance using the terminal's interactive screen construction program and stored in the terminal file. When the desired screen is called up from the keyboard, first the background screen is displayed. In parallel with this, the process data is received from YEWCOM, and the foreground screen is superimposed on the background screen, greatly improving response. The definition of the information required for the foreground display can also be handled by the interactive screen construction program. Moreover, terminal intelligence can realize the features of a dedicated operator console. Most of the functions required for good operability, such as input and output data editing and checking, and display of system messages can be performed by calling up the appropriate screen by means of a menu. Figure 5.56 compares the response time for an ordinary terminal and an intelligent terminal. For intelligent terminals, the response

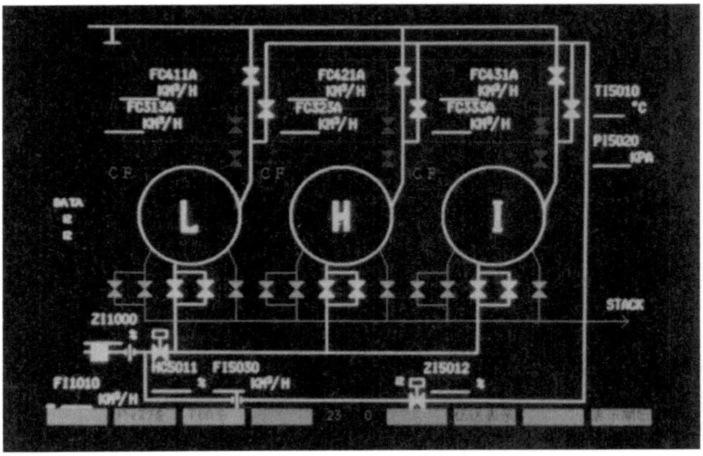

Fig. 5 · 54 Intelligent terminal (Y 1720 A)

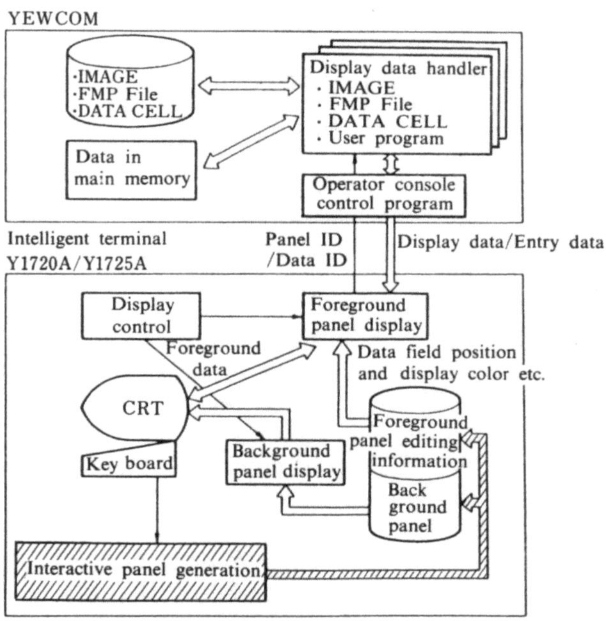

Fig. 5 · 55 Functional configuration of intelligent terminal (Y 1720 A) and operator console (Y 1725 A)
(Shaded portion shows background processing)

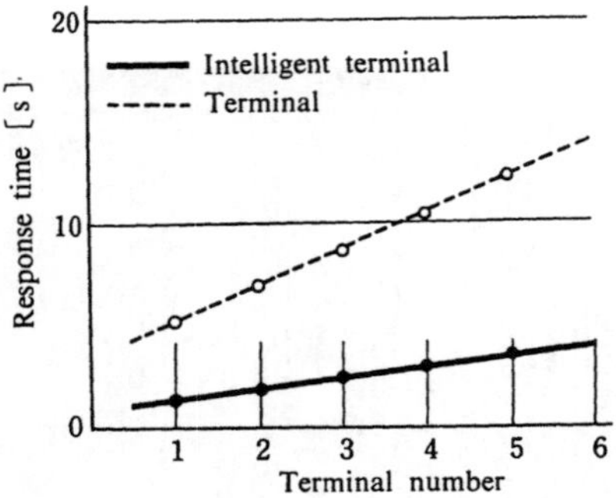

Fig. 5・56 Response time of intelligent termrnal

time increases little as more terminals are added. Display within two to three seconds in any system configuration is a great advantage of intelligent terminals.

(*e*) *Duplex system configuration*

The three types of redundant system configurations shown in Fig. 5.57 are representative of high-reliability systems designed to improve reliability.

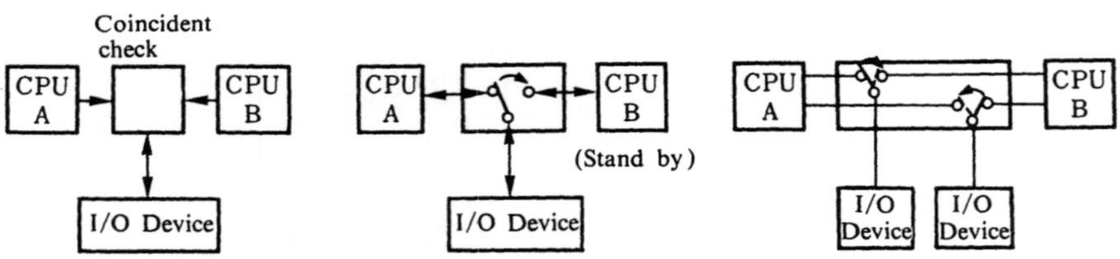

(a) Dual configuration (b) Duplex configuration (c) Loadsharing configuration

Fig. 5・57 Duplex system configuration

(1) *Dual configuration* In this configuration, two CPUs with synchronized clocks perform the same processing. When an inconsistency occurs, each of the CPUs performs self-diagnosis. The normal processor that passes the diagnosis continues processing and the other is cut off. When one of the CPUs goes down, complete continuity of processing is attained. However, this method requires special hardware design, and thus has less flexibility than other methods.

(2) *Duplex configuration* In the duplex system, one CPU does the processing, and the other CPU stands by. If the active processor goes down, the processing is automatically switched to the stand-by unit. Although with this method processing stops for just an instant, the availability factor can be greatly improved. This method is implemented by adding a simple duplex control device to two CPUs, and can thus

be widely used.

(3) *Load-sharing configuration* In this method, the processing load is apportioned between two CPUs. If one of the processors goes down, its share of the processing is taken up by the normal processor. During normal operation, the system has about the same performance as two CPUs. However, during failure, engineering for degeneration of function becomes important.

Figure 5.58 shows the duplex system configuration used in YEW-COM. This configuration can apply either the duplex method or the load-sharing method. The duplex control section monitors both CPUs. When it detects failure in the executing CPU, it switches the I/O device group over to the stand-by processor. The dual-port disk holds common data. The dual-port disk controller also provides a means of communication between CPUs, and has a buffer and a resource management table for that purpose. Thus when there is no other means of

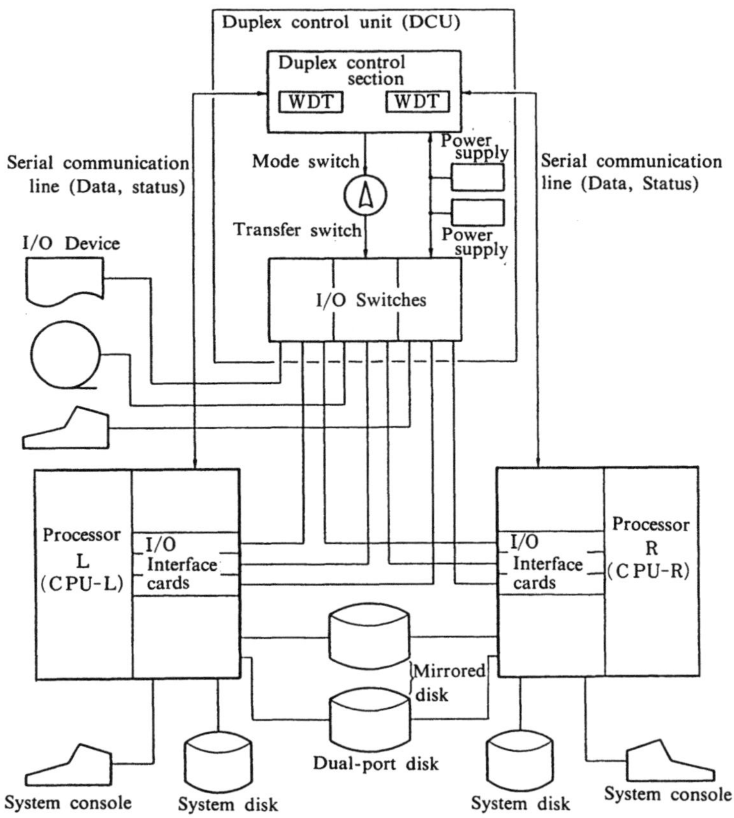

Fig. 5・58 Duplex system configuration in YEMCOM

communication between processors, the controller can perform effective common resource management.

In production management computer systems, high data reliability is required to cope with long-term storage and frequent updating. Thus data back-up capability is very important. This system uses a mirrored configuration for the dual-port disk. In this mirrored disk, data is written to two disk devices at the same time. If either disk fails, then processing continues using the other disk, and there is no loss of data. The system also supports recovery by making the data in the two disks equivalent after repairing the failed disk without disturbing system operation.

5.4.2 Software for production management computer systems

(*a*) *Software configuration*

The software configuration is shown in Fig. 5.59. The core of the configuration is an operating system that supports both real-time processing and large-volume batch processing. A development support package consisting of a language processor, interactive screen editor and various utility programs for program development provides an effi-

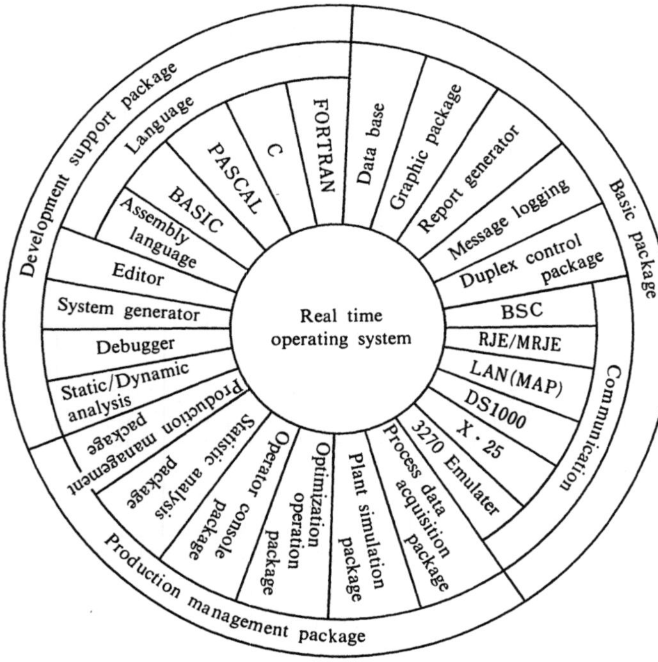

Fig. 5·59 Software configuration of production management computer cient environment for programming and system generation. The basic package contains the standard tools used for any type of application. For process management computers, the real-time data base and network communication packages are particularly important. The production management package is a set of packages for facilitating development of application software for production management systems. It consists of a human interface package, a process data acquisition package, and so on.

(*b*) *Operating system*

The operating system (OS) is a program that manages and controls all of the various system resources, including the hardware components, software, and data for the purpose of efficient operation. The following are the basic requirements for a real-time operating system.

(1) High throughput
(2) Fast response
(3) Ease of use
(4) Reliability, availability, and serviceability (RAS)
(5) Integrity
(6) Security

(1) *Task management and multiprogramming* One of the most important functions of real-time operating system (OS) is the parallel execution of multiple programs. This is called multiprogramming. In real-time OS, the object of each executing program is called a task, or a process, and system resources such as CPUs, I/O devices, programs, and data, are allocated using the task as the basic unit of management.

(i) *Interrupt control* Interrupt control refers to a system design in which the CPU is notified when certain events occur, causing the currently executing task to be interrupted to allow execution of a higher priority task. The following are some of the different types of interrupt.

(1) *I/O interrupt*: notifies of the completion of an input/output operation.
(2) *Timer interrupt*: causes an interrupt at a specific time or a specific time interval.
(3) *Power stop/restore and error interrupt*: notifies when power stops or is restored, and when a CPU error occurs.
(4) *Program interrupt:* Notifies of an error in the machine language code.
(5) *SVC interrupt*: generated when a user program executes a SVC (supervisor call) instruction call to the OS.

(ii) *Resource management* In multiprogramming, multiple tasks are performed at the same time, and each task requires system resources. Simultaneous requests from various tasks for resources are received, and the tasks are arranged into queues to await the availability of each required resource. When a task finishes using a resource, the resource is made available to another task from the queue. The task dispatcher allocates use of the CPU, the most important resource in the system. This task switching is called dispatching or scheduling. Tasks are generally managed by transitions among the following four states.

(1) *Run*: being executed (use of the CPU secured).

(2) *Ready*: can be executed (resources other than the CPU are secured), but is waiting while a task of equal or higher priority is using the CPU.

(3) *Wait*: unable to execute while waiting for completion of an I/O operation, a signal from another task, etc.

(4) *Idle*: no activity.

Figure 5.60 illustrates the concepts of priority processing for foreground tasks which require real-time operation, and parallel processing for I/O wait. Also, when several users are operating different terminals on the same CPU, such as for program development, and editors, compilers and so on are executed as background tasks, it is necessary to allocate CPU time fairly among them. Figure 5.61 illustrates the concept of time slice processing for fair execution of multiple background tasks.

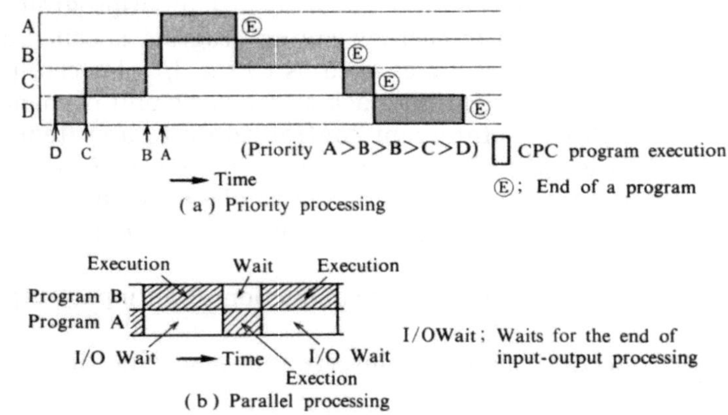

Fig. 5 · 60 Priority processing and parallel processing

(2) *Memory management* When multiple tasks are executed in parallel, it is desirable to have many programs in main memory. However, there are physical limits to main memory size. The OS, subroutines used in common by tasks, and common data are maintained in main memory. Highly urgent tasks and tasks that are very frequently used are also placed in residence in main memory. Other programs are stored on disk, and loaded into main memory as needed for execution

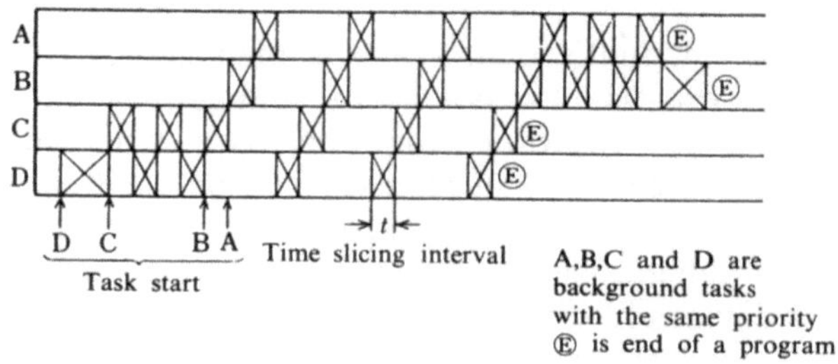

Fig. 5 · 61 Time slice processing

as non-resident tasks. In such cases, it is normally necessary to divide the program up into a number of segments according to its size, and load the segments into main memory as the processing progresses. This is called overlay processing. Virtual storage allows the programmer to regard the main memory and auxiliary memory as one large main memory unit, without concern at all for the size of the program.

(3) *File management* Files are collections of logically related records. File management is a function that provides unified access to logical files for data input/output and storage without concern for the physical configuration of memory or I/O devices or differences in physical interfaces. File access is accomplished by means of the file name. Using the file name, the file managemant finds the physical memory address and file attributes in a look-up table (directory) in memory,

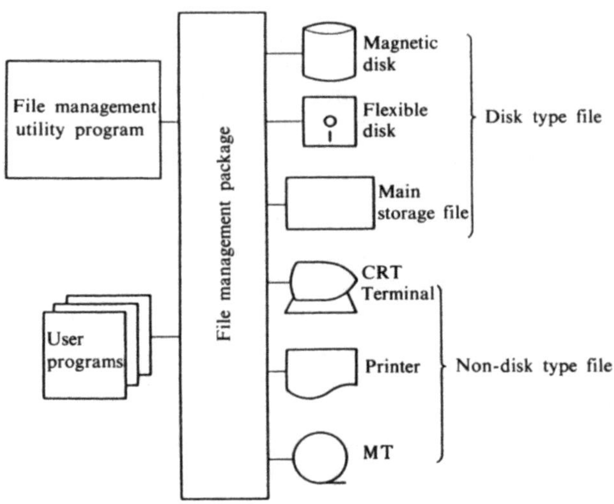

Fig. 5 · 62 File management configuration

and performs the I/O operation on the appropriate memory device.

Figure 5.62 shows the configuration of the YEWCOM file management system. All magnetic disk space is managed as files. The user program specifies the disk space it needs, and can use this space exclusively or in common with other programs. Also, when high-speed file access is needed, blocks of main memory can be used as a virtual disk device, and accessed in the same manner as a disk. File management involves sequential devices such as magnetic tape and printers as well as random access devices like hard magnetic disks and floppy disks. Each device is given a logical device number (LU). By specifying a LU, terminal input and output can also be accessed as a file. A file management package is a collection of subroutines that can be called by a user program. It consists of a group of programs for creating, opening, closing, and deleting files, and for disk file accessing. File management utility programs allow file creation, deletion, storage, copy, packing, and so on from the operator command level.

(*c*) *Development support software*

Although software development is also done on large, general-purpose computers, in recent systems it has become possible to do program development on most target systems in parallel with real-time processing. FORTRAN is the most widely used programming language. FORTRAN 77, an extended function structured language that can be used for real-time system, has become the ISO standard. In 1982 it appeared in revised form as JIS FORTRAN (X 3001), although it remained substantially the same, and is widely used. To address the serious problem of how to increase productivity in software production and insure high quality, software development support has been studied in great earnest. Various kinds of interactive utilities such as interactive screen editors, on-line relocatable loaders, on-line system generators, and symbolic debuggers, are being perfected, and utility programs such as static analysis tools, coverage measurement tools, and operability analysis tools are coming into use.

Computer workstations have become cheaper and more powerful. With advances in development and standardization of data communication systems centered around local area networks (LAN), the trend is toward a software development environment in which workstations are integrated with host development systems. The UNIX operating system and the C programming language are widely used, and are becoming the industry standard.

(*d*) *Basic software for production management systems*

Figure 5.63 shows an example of a YEWCOM production management system configuration and data flow model. The IMAGE/1000 J data base management system is the core of the system. The basic

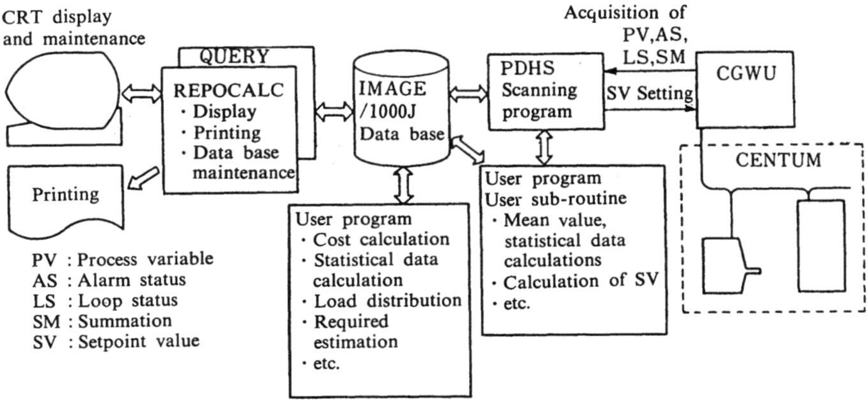

Fig. 5 · 63 Configuration of production management system

package comprises the human interface (QUERY and REPOCALC), the process data acquisition packages (PDHS), and a package for communication with various kinds of computer move over many types of software packages and user's programs are added to met the target system requirements.

(1) *Data-base management system* If, for example, a part number appears in multiple files, such as the product file, the quality control file, the price file, the production performance file, and so on, then part number data will accumulate job-by-job. This results in redundancy and is a source of possible data conflict. A data base is a collection of data structured by logical relations. The purpose of the data base is to allow the data to be shared.

The IMAGE/1000 J data-base management system (DBMS) for the YEWCOM system is a network DBMS that can handle Japanese text. This system offers data independence, data consistency, data security, and high operability. Data independence means the accessing program is not affected even if the data base is altered. Data consistency is brought about by the deletion of redundant data. Data security results from many kinds of read/write protect functions. Also, a high-level data manipulation language allows a high degree of operability. These capabilities provide the operator with large amounts of production data in a useful form. On the other hand, if data addition and deletion functions are not required after the system has been constructed, such as for a process data acquisition system, then the data structure can be simplified and a simple DBMS designed for high performance can be used. The YEWCOM data cell is an example of this. As shown in Fig. 5.64, the data cells have a two-dimensional structure, and it's records consist of several items which differ in word length and data format.

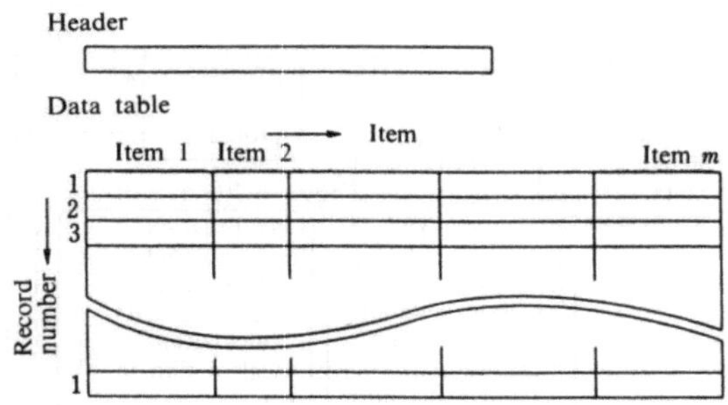

Fig. 5 · 64 Data construction of data cell

A group of these records comprises a data cell. For example, if item 1 is allocated to tag name, item 2 to PV, item 3 to SV and so on, this collection of process data would make up one data cell. The data cell access statement can be written using the item name in correspondence with the FORTRAN READ/WRITE statement. Ease of programming and high performance at execution time are achieved by using a cell preprocessor, which determines the physical offset of the item name from the beginning of the record. There are many other utilities, such as the date cell generator, which constructs data cells from a data cell definition table, and the cell scope, which allows interactive inspection and changing of cell contents.

(2) *Human interface package* In production management systems, the screen construction support package is particularly useful because programs for CRT display screens, and forms and tables have to be written very frequently. The report calculation package (REPOCALC) is a typical example of software for display of listings on the CRT or printing of forms on the printer. It has the following functions.

(1) Creation of screen specifications for any kind of listing and form printout using a high-level dedicated language

(2) Automatic continual updating of process data display by means of a recursive screen display function

(3) Hourly, shift time and daily print outs at fixed times are also possible

As shown in Fig. 5.65, an output screen can be constructed easily by creating a FORMS file using the editor. The input/output data definition section can be used to easily define search conditions, I/O editing, data checks, simple calculations, and so on.

(3) *Process data acquisition package* The process data acquisition package (PDHS) is a program which collects data from the distributed control systems CENTUM and YEWPACK, the production line control system YEWMAC, and the intelligent process input and output device

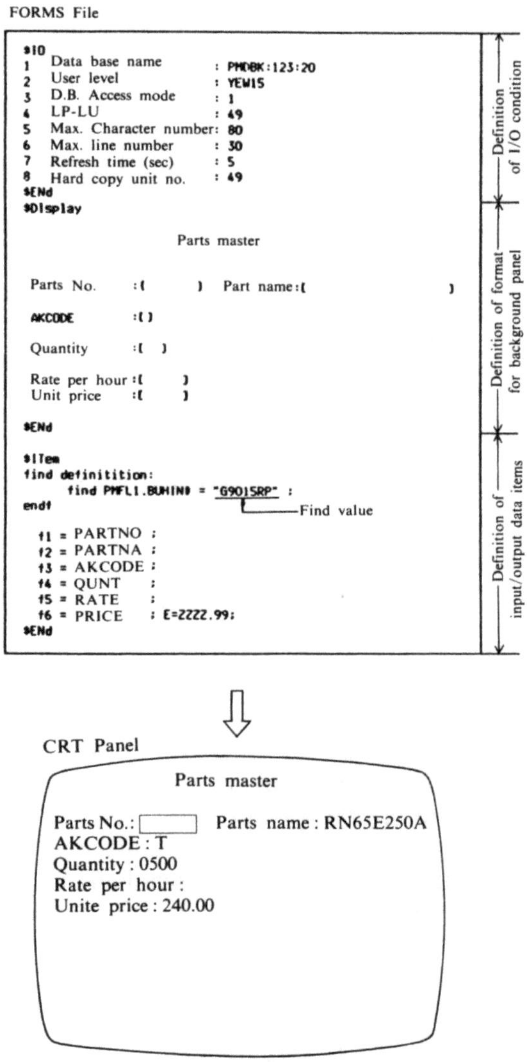

Fig. 5・65 Output panel generation by REPOCALC

SPACE I/O. PDHS stores the data in a centralized data base. The data acquisition specifications are defined by filling in blank items in a work sheet. An example of such a work sheet for defining the process acquisition method and so on is shown in Fig. 5.66. The work sheet is used to specify the acquisition interval, what process data to obtain, and where in the PDHS data base to store the data. In this example,

SCAN & USER'S PROGRAM DEFINITION PDHS FORM SU1

TAG NAME

F101 , ... , ... ; 1st blank space: Tag name

F102 , ... , ... ; 2nd blank space: Tag No of CENTUM, YEWPACK

END , ... , ... ; 3rd blank space: Data acquisition destination

SCAN & USER'S PROGRAM DEFINITION PDHS FORM SU2

PERIOD & PHASE 60 , 0 ; 1st blank space: Period, 2nd blank space: phase

SCAN DATA

PV , • , PV , ... ; 1st blank space: Item name

ASTS , ... , AS , ... ; 2nd blank space: Tag No of CENTUM, YEWPACK

SV , ... , SV , ... ; 3rd blank space: Data type name

LSTS , ... , LS , ... ; 4th blank space: Data acquisition destination

END , ... , ... , ... ;

Fig. 5 · 66 Work sheet of process data acquisition package (ADHS)

PV, AS, SV, and LS for tag numbers F101 and F102 in the CENTUM system are collected at 60-second intervals. The acquired data is stored in the PDHS of the YEWCOM system as items named PV, ASTS, SV, LSTS and having the tag names F101 and F102. In addition, by using the work sheet to define the PDHS data base and the conditions for executing a user program for average value and accumulative value calculations, a flexible process data acquisition system can be constructed easily. The data stored in the PDHS data base can be displayed on the screen and printed out by using the above-mentioned REPOCALC human interface package.

(4) *Inter-computer communication package* Communication between computers is becoming increasingly important considering progress in distributed configurations and total system integration. Figure 5.67 shows overall configuration of communication in the YEWCOM system. The communication techniques for YEWCOM are classified according to which type of computer system is involved in the communication.

(1) *Communication between YEWCOM systems (DS* 1000-*IV*): DS

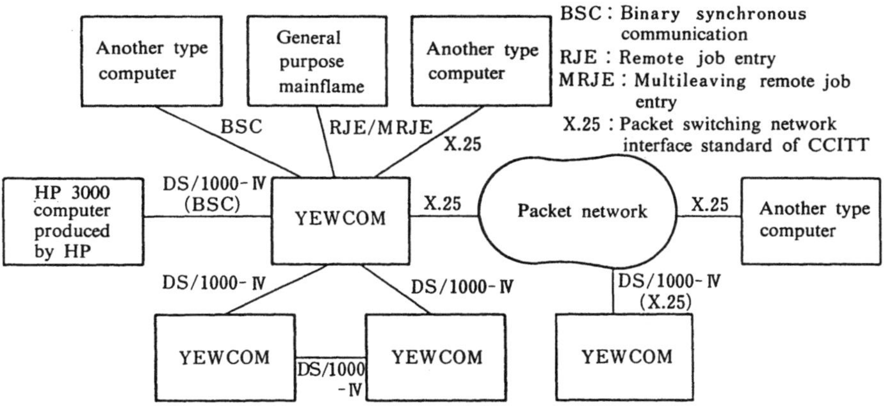

Fig. 5・67 Computer network in YEWKOM system

1000–IV is network software which realizes a distributed processing system by interconnecting a number of YEWCOM systems. This network has a topology–independent architecture, and has no preferences among line, star, or ring configurations. It features communication between programs, remote file access, remote command processing, remote data base access, and other such functions. It also possesses many functions for reliability, operability, and maintainability, such as dynamic message re–routing. In dynamic re–routing, the communication route is automatically changed when a failure occurs.

(2) *Communication between YEWCOM and other computer systems*: The communication method can be selected from among following, according to the amount of transaction, amount of data involved, and the communication function possessed by the other computer.
1) Binary Synchronous Communication (BSC)
2) Remote Job Entry/Multileaving Remote Job Entry (RJE/MRJE)
3) X.25 (CCITT packet exchange network)
4) IBM 3270 Emulator communication

The recent trend is towards total factory management, and for this reason standardization of a horizontal type of communication line among multivendor systems (LAN) is rapidly progressing. The importance of this type of network will continue to increase.

5.5 *Data Communication and Equipment*

5.5.1 *Data communication and standards*

In the application of computers to the fields of measurement and data processing and process control, the system must quickly convert from the user data input to the output processing results. In other words, the need for real-time processing is exceptionally great. Recently, progress in microprocessors and the technologies applied to them have advanced the intelligence and decentralization of measurement instruments and control devices. The result of these advances is that data communication technologies that allow accurate and real-time communication of the required data among these types of equipment, and also between these devices and computers, is becoming extraordinarily important.

Date communication can be described as the linking of intelligent equipment such as computers and control devices by means of communication lines through which digital data is mutually transferred with good efficiency according to certain procedures that are determined in advance. To send correct data to the destination system, several methods and types of specifications must be the same for both communicating systems. These include the types of communication cables and connectors, the electrical characteristics of the signal, the signal delivery procedure, and the method of expressing the data. It is advisable to follow as much as possible the standardized methods and specifications, which are known as communication protocols.

One of the most frequently observed items in standardization is the set of standards known as the open systems interconnection (OSI)

Table 5.11 OSI reference model

Layers	Function
Layer 7 Application	Provides all services directly comprehensible to application programs
Layer 6 Presentation	Restructures data to/from standardized format used within the network
Layer 5 Session	Synchronize & manage data
Layer 4 Transport	Provides transparent, reliable data transfer from end node to end node
Layer 3 Network	Performs packet routing for data transfer between nodes
Layer 2 Data link	Improves error rate for frames moved between adjacent nodes
Layer 1 Physical	Encodes and physically transfers bits between adjacent nodes
Physical medium	Transmission line

basic reference model created by the International Organization for Standardization (ISO). As shown in Table 5.11, the OSI model classifies all the functions which are indispensable for data communications into seven layers, ranging from the physical layer implemented in hardware (including the function of converting data to an electrical signal for the transmission line) to the application layer implemented in software (application-oriented functions such as methods of expressing messages and file transfer procedures). In addition, the functions of each layer are defined and serve as guidelines for standardization.

Recently, the demand for total network integration of production information systems on the factory and enterprise level has become high. For this reason, the factory communication network MAP (Manufacturing Automation Protocol), whose standardization and practical dissemination has been proceeding gradually with the cooperation of standardization organizations and many enterprises, centering in the United States and including Europe and Japan. Under the primary leadership of the General Motors Corporation, MAP is also conforming to OSI.

On the other hand, there are also many widely-used communication networks and protocols which do not entirely conform with international standards. One such example is the Systems Network Architecture (SNA), which has been the IBM Corporation standard network since 1974. It is also used worldwide.

Also, in Japan there are many examples of practical optical communications systems. This is another area in which standardization is proceeding slowly. In any case, when the use of non-standard protocols is unavoidable, it is necessary to make detailed preparations and carefully design for stability and reliability.

5.5.2 *Methods of data communications*

There are many data communication methods, but if we represent them in terms of communication distance versus communication speed, we obtain something like Fig. 5.68. The classification is not entirely clear, but the following rough categories can be distinguished.

(1) Communication between devices in close proximity
(2) Communication at intermediate distances
(3) Wide area networks

(a) Communication at close proximity

Within testing laboratories and so on, automatic measurement and data processing is often carried out with various measuring instruments connected to a computer. The IEEE-488 instrument bus, specified by the Institute of Electrical and Electronics Engineers (USA) is widely used for such purposes because it allows easy reconfiguration

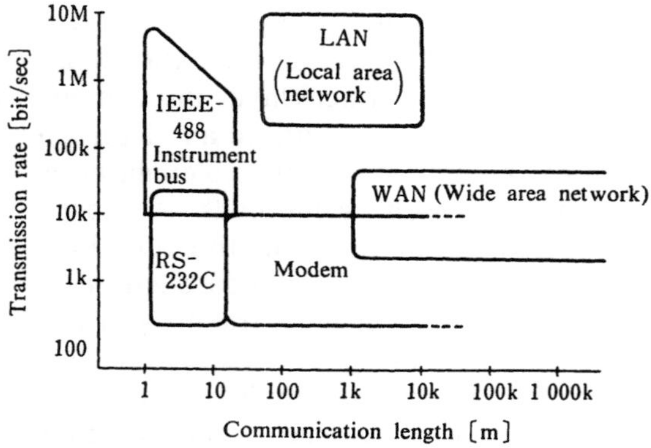

Fig. 5・68 Communication methods

of the system for various experiments and instruments can be easily mounted and dismounted. In the atomic power field, the Computer Automated Measurement and Control Equipment (CAMAC) standard has been established and is used for about the same purposes. Also, for connection of computers to terminal equipment such as CRT terminals and printers, connection of various types of measurement instruments and automation machinery (robots, NC machines, etc.), the RS-232 C interface standard set by the American organization, the Electronic Industries Association (EIA), is being widely used. Recently the RS-422 and RS-423 standards have been set as an extension of the RS-232 C to longer distance communication. These are replacing the RS-232 C, albeit slowly.

(*b*) *Intermediate distance communication*

In the field of measurement and control, PROWAY (Process Dataway) specified by the International Electrotechnical Commission (IEC) is the most important method for exchange of data among dispersed computers, control devices, terminals, and so on at distances of up to ten kilometers. The PROWAY standard is a means of highly reliable, high-speed data communication between the operator console and multiple filed control units in a distributed control system. It served as a model for the F bus, the predecessor to the HF bus of the CENTUM system. The major feature is the baton-passing method used in the CENTUM system. MAP and PROWAY are examples of local area networks (LAN), which have recently been receiving much attention as a means of integrating dispersed equipment in various environments such as offices, research centers, and factories.

The modem is often used as a simple and convenient way to make a fixed connection between two devices. By introducing a modem between devices equipped with the RS–232 C interface described in Subsec. (a) above, the transmission distance can be extended. There are different types of modems. Specialized modems for on-site communication lines are used to extend the distance between connected devices within a facility. Public communication line modems and private communication line modems can connect devices that are far apart on a one-to-one basis through lines provided by commercial communications enterprises.

(c) *Wide-area networks*

Wide-area networks are a method of connecting devices across large distances through commercial communication systems. Data is transmitted over the communication lines in the form of data blocks, called packets. Information on the destination and packet sequence number are included in the packet, thus the communication circuit utilization efficiency is very high because many users can share a circuit. This method can be used for on-line transmission of orders, production plans, and so on from the head office to distant factories.

Below, the communication methods which are particularly important in measurement and control are explained in concrete terms.

5.5.3 *The IEEE–488 instrument bus*

The IEEE–488 instrument bus is most often used to connect computer and personal computers to measurement instruments, plotters,

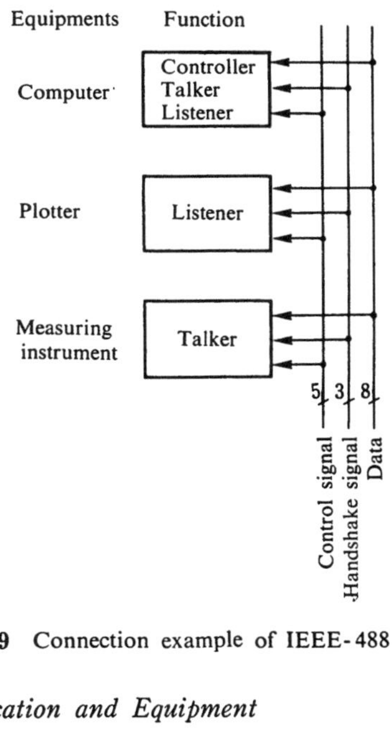

Fig. 5 · 69 Connection example of IEEE-488 instrument bus

and other such peripheral devices in testing laboratories. In this way, automatic measurement, analysis, data processing and recording can be accomplished. The functions of the connected devices can be classified into three types: controller, talker, and listener. Each device must be equipped with at least one of these three functions (Fig. 5.69).

The controller performs such management functions as specifying the talker and listener, and releasing a connection. The talker loads data onto the bus, and the listener downloads data from the bus. Generally, computers are equipped with all three functions, while measurement instruments are talkers, and plotters and other such devices are listeners. As an example of a simple system, the computer performs the controller function as required and specifies the instrument as the talker, and itself as the listner, and then receives the measurment data. After completion of data transmission, the disgnations for both talkers and listener are released. When data analysis is complete, the

Table 5.12 IEEE-488 instrument bus signal

Classification	Signal line		Function
Handshake signal	DAV	Data valid	Data is valid
	NRFD	Not ready for data	Not ready for data
	NDAC	Not data accepted	Data is not accepted
Control signal	ATN	Attention	Command designation
	REN	Remote enable	Remote control designation
	IFC	Interface clear	Reset of interface
	SRQ	Service request	Interrupt notice to controller
	EOI	End or identify	End-of-data
Data	DI 01 to 08	Data input output	Data

Fig. 5 · 70 Data transfer cycle timing

computer specifies itself as the talker and the plotter as the listener. It then sends the analysis data to the plotter for recording.

The bus has 24 signal lines, including 5 internal control signal lines, 3 handshake signal lines for data transfer timing, 8 data lines for 8-bit parallel data, and 8 common bus lines. The bus line names and functions are shown in Table 5.12. Figure 5.70 is the data transfer timing chart. The data transfer is asynchronous, and is matched to the slowest among the talker or multiple listener devices.

5.5.4 The RS-232 C interface and modems

The word modem is an abbreviation for modulator-demodulator. These devices are used to extend the communication route. They perform different functions at the sending and receiving ends. At the sending end, a serial digital signal is used to modulate an analog signal (using, for example, frequency modulation), which is sent out on the communication circuit; on the receiving end, the analog signal is detected and demodulated to reconstruct the original serial digital signal. Modems for connection to private or public communication lines must be authorized and conform to standard frequency band and transmission power specifications. They must also be designed to transmit signals while avoiding external interference from adjacent communication lines and be approved.

The interface between the modem and the devices it is connected to (computer, terminal, etc.) is standardized in terms of connectors, sig-

Table 5.13 RS-232C signal

Pin arrengement of connector

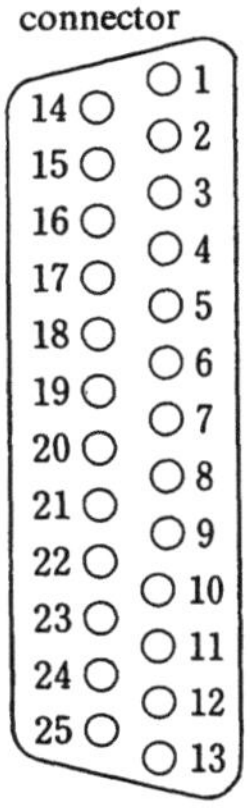

Code	Pin No.	Circuits mame	Signal direction From MODEM	To MODEM
FG	1	Frame ground	–	–
SG	7	Signal ground	–	–
SD	2	Send data	○	
RD	3	Received data		○
RS	4	Request to send	○	
CS	5	Clear to send		○
DR	6	Data set ready		○
ER	20	Equipment ready	○	
CI	22	Call indicator		○
CD	8	Carrier detector		○
ST 2	15	Send signal element timing		○
RT	17	Received signal element timing		○
ST 1	24	Send signal element timing	○	

Table 5.14 Typical electric specifications of RS-232 C, RS-422, RS-423, RS-485

	RS-232 C	RS-422	RS-423	RS-485
Connection mode	1 : 1	1 : 1	1 : 1	$n : n$
Transmitted-received circuits (Trans.) (Receiv.)	 Single end type Single end type	 Differential type Differential type	 Single end type Differential type	 Differential type Differential type
Transmitted signal voltage	5 to 15 V (plus, minus)	2 to 6 V (plus, minus)	4 to 6 V (plus, minus)	1, 5 to 6 V (plus, minus)
Received signal threshold voltage	±3 V or more	±200 mV or more	±200 mV or more	±400 mV or more
Transmission distance	15 m	1.2 km (100 kbit/sec. or less) (0.2 mm^2 twisted pair cable)	1.2 km (6 kbit/sec. or less) (0.2 mm^2 twisted pair cable)	–

nal lines, electrical specifications, data exchange method, and timing. One such standard is the RS–232 C interface. Table 5.13 shows how the signal lines are used in this standard. The typical electrical specifications are listed in Table 5.14.

Serial data is passed between devices by either the asynchronous method or the synchronous method. In the asynchronous method, data is handled in units of one character (6, 7, or 8 bits). As shown in Fig. 5.71, data is delimited by appending a start bit before each character, and a stop bit after each character. The receiving device recognizes the beginning of the character by detecting the start bit, and from that time reads each bit making up the character at fixed time intervals (determined by the transmission speed).

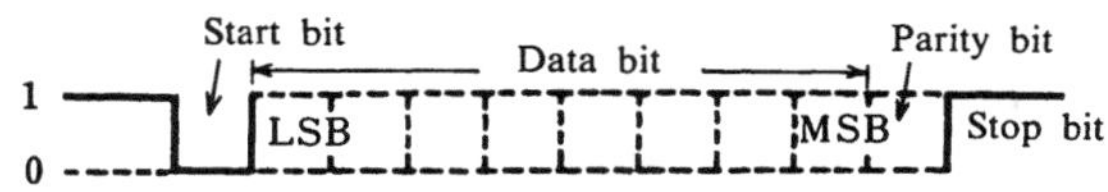

Fig. 5 · 71 Asynchronous data

Synchronous transfer, on the other hand, sends data in units of character strings, which are made up of a number of characters. As shown in Fig. 5.72, no start bits or stop bits are added as is done when the transfer unit is one character. Bit position is detected by synchronization with a special clock signal, and characters are distinguished by detecting a special synchronization character at the head or in the middle of the character string and then counting bits as the data is received.

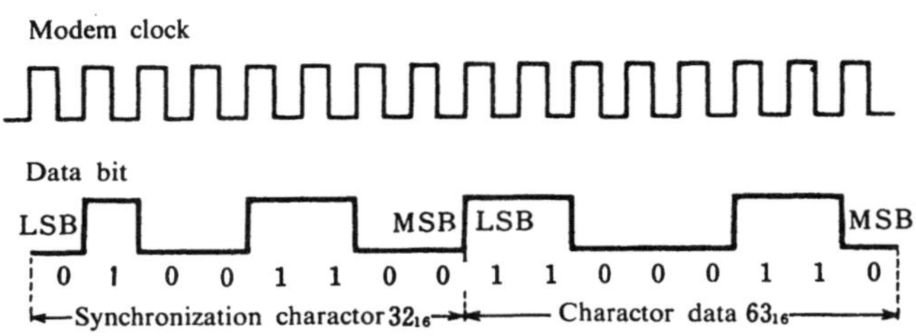

Fig. 5 · 72 Synchronous data

Recently the synchronous transfer method called the High Level Data Link Control Protocol (HDLC) has been standardized. This method uses frames of variable bit length as the data transmission unit rather than single characters or character strings. It is being used in wide area networks for connecting computers. In this method, as shown in Fig.5.73, a special bit string (01111110), called a flag sequence, is places at the beginning and end of the frame. These flags are used to

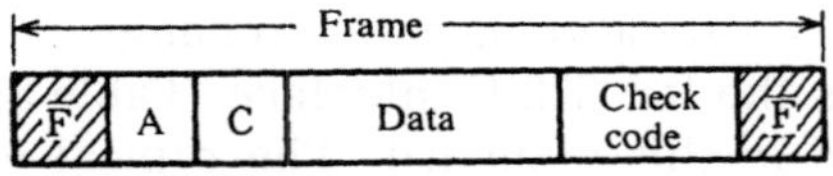

F; Flag sequence C; Control information
A; Secondary station address

Fig. 5 · 73 Data frame of HDLC

detect the frame and the bit positions. In the figure, the address represents the address of the secondary station corresponding to the main station, and the control information specifies the data frame type.

When the transmission line is long or in factory environments where there is much electrical noise, noise errors may be generated in the transmitted data. Thus, a means of detecting and correcting these errors is needed to allow transmission of correct data. In asynchronous transfer, a parity bit is added to each character unit of data; in the synchronous method, a check code that has a high error-detection power is appended to the end of each data unit (either character string or frame). Using these features, the receiving side can detect whether or not there is an error in the data. If an error is detected, the receiving side issues a request to the transmitting side for re-transmission.

The RS-232 C interface can be used for direct connection of nearby devices (within 15 meters) as well as for the widely used standard modem connection. Thus it is standard equipment on many computers and control devices. As a substitute for the RS-323 C interface for transmission without using a modem at relatively longer distances (several hundred meters or so, depending on the cable diameter), the RS-422 (differential type), RS-423 (single-end type), and RS-485 (differential multi-drop type) interfaces have been standardized, and their use is spreading. The electrical specifications of these interfaces are listed in Table 5.14.

5.5.5 Local area networks

Local area networks (LAN) were developed as a means of compact, high-speed data exchange through a single medium among a number of computers, workstations, controllers and so on located within a limited area such as in the same building, or on the same floor. Many types of LAN have been developed for various purposes, but they can be simply classified according to their physical configurations (topology). Figure 5.74 illustrates three topologies: the bus, the ring, and the star configurations. The methods for transferring data to the transmission medium (access method) include the token passing (or baton passing) scheme and the CSMA/CD method.

The bus topology has high reliability because the transmission

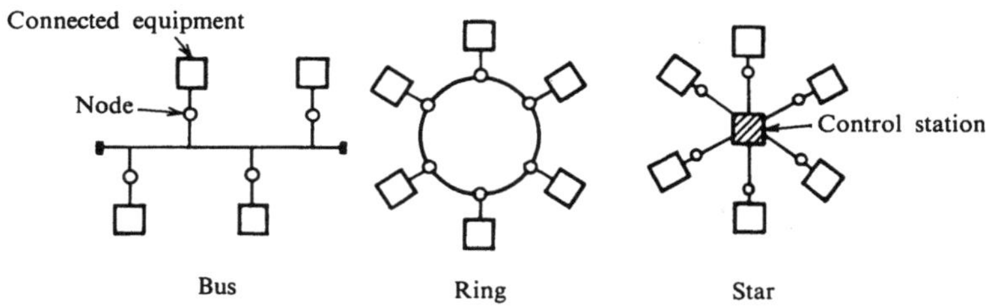

Fig. 5 · 74 Topology

route contains no active elements (transistors, LSI circuits, etc.) in series, unlike the ring and star configurations. Thus the bus topology is suitable for factory on-line networks. The PROWAY and CENTUM HF bus use the bus topology. In the ring topology, the signal is intercepted and retransmitted at the connecting point for each device (node). Optical fibers are often used as the transmission medium for LAN. The star topology is used in telephone exchanges, and seldom used in factories.

In the token pass method, a frame (called a token) makes rounds to all the devices on the LAN. Only the device that is to receive the token can send a signal to the medium. In the CSMA/CD method on the other hand, any device can send a signal at any time if there is no signal on the medium, thus signal collisions may occur. At such times, each device is made to wait for a set time, which is different for each device. The CSMA/CD method is efficient when the traffic load is light. However, when the load increases, multiple conflicts develop regardless of the waiting. As a result, efficiency is lost and response time is lengthened. Degradation of response time beyond the limit is unacceptable in a network for factory control and management, thus the token passing method is most often used for factory networks. MAP and PROWAY are examples. ETHERNET (bus-type) is a typical example of CSMA/CD. It is used extensively for light-load situations such as in offices, and in research and design facilities.

Standardization is currently proceeding in IEEE. Here, three standards have been discussed. Of these, brief specifications for the IEEE-802.4 method, which is intended for factory use and is also employed in MAP are listed roughly in Table 5.15, together with the specifications for PROWAY.

5.5.6 Optical communications

As a transmission medium, optical fibers have such advantages as insensitivity to electrical noise, the capability for high-speed transmission over great distances, and safety because they do not transmit elec-

Table 5.15 Specification summary of IEEE-802.4 and PROWAY

Item	IEEE-802.4	PROWAY
TOPOLOGY	Bus	Bus
Transmission medium	75 Ω co-axial cable	75 Ω co-axial cable
Transmission distance	—	Max. 2 km
Transmission rate	1, 5, 10 Mbit/sec	1 Mbit/sec
Node number	—	Max. 100
Modulation method	Phase continuous FSK Phase coherent FSK Broadband	Phase continuous FSK
Control method	TOKEN PASSING	TOKEN PASSING (Baton passing)
Error control	Not data error monitoring	Retransmission and recovery by immediate acknowledge monitoring
Worst case access time	In case of no transmission error, it is decided by station numbers and working parameter value.	50 ms or less

tric power. For these reasons they hold great promise for use as a transmission medium in factories, and their use is spreading. However, optical connections are difficult, the cost is high, and standardization is slow. Thus, there is poor compatibility among the products of different manufacturers. In the past optical media has been used for special environments and high-performance communications. Nonetheless, these points of difficulty will be overcome steadily, and in the near future we can expect that use of optical communication media will spread.

The optical fibers used in communications are difficult to branch, so they are mostly applied to one-to-one communications or to ring-type communication configurations. However, recently optical fibers have been introduced logically to bus and star LANs by using a multiple-branch element called the star coupler. Furthermore, RS-232 C connectors with built-in optical modems for a duplex communication are being developed, and low-cost, convenient optical communication is becoming possible. As a representative example of a high-performance ring network, the structure and main specifications of YEWLINK 32 are introduced in Fig. 5.75 and Table 5.16.

5.6 Basic Components of Digital Control [14)-18)]

5.6.1 Microprocessors

Microprocessors, which have developed from the first 4-bit device

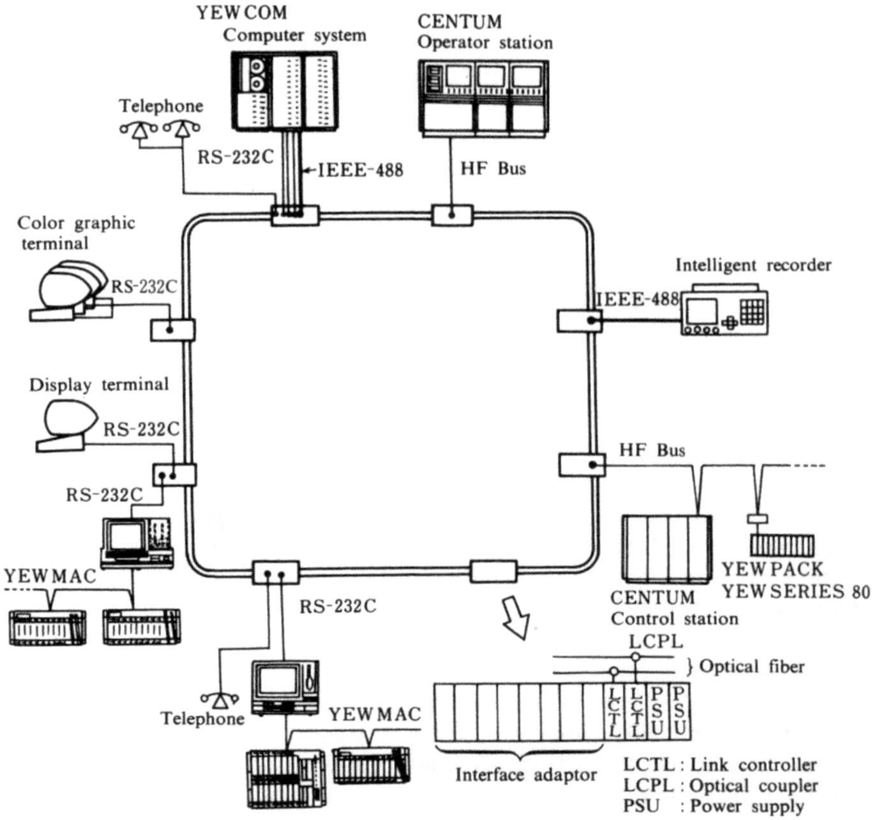

Fig. 5・75 Configuration of YEWLINK 32

Table 5.16 Specification of YEWLINK 32

Item		Specification
TOPOLOGY		Ring
Transmission path		Optical fiber cable (GI 50/125)
Transmission distance		Max. 120 km
Transmission rate		32.064 Mbit/sec
Node number		Max. 32
Fault measures		· Loop, duplexed nodes · Loop-back · Node bypass
Interface	Loading number	Max. 8
	Type	HF bus IEEE-488 RS-232 C Telephone

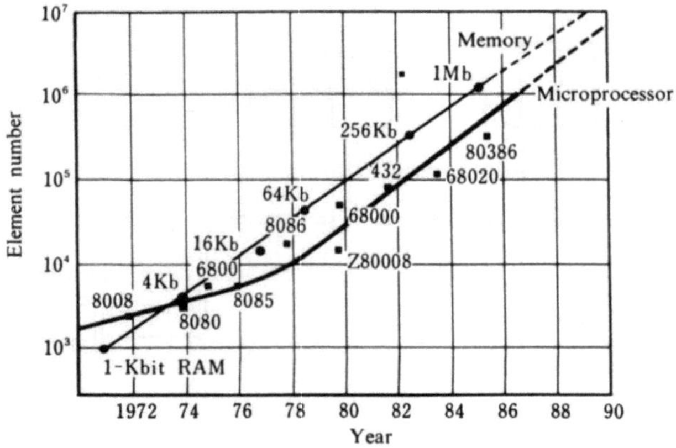

Fig. 5・76 Trends of VLSI

of late 1960 to today's 32-bit devices, are widely used in the fields of measurement and control. The course of microprocessor element development is shown in Fig. 5.76. The principle parts of the 32-bit minicomputers of the 1970's are now integrated onto a piece of silicon that is about 8 mm square.

(*a*) *Microcomputer configuration*

The architecture of microcomputer varies somewhat with the type, but it generally consists of a microprocessor chip,* which integrates the arithmetic unit, registers, and control unit, memory chips for storing programs and data, and a peripheral interface chip for control of input/output devices. The chips are interconnected by buses which obey unified electrical and logical rules. The buses include an address bus, a data bus, and a control bus. For example, an 8-bit microprocessor has an eight-wire data bus. Bus rules differ among manufacturers, but connection among chips from a single manufacturer is relatively easy. A device which has the microprocessor, memory, and peripheral interface integrated on one chip is given the special name, "single-chip microprocessor." As opposed to this, a processor comprising several chips such as shown in Fig. 5.77 is called a "general purpose microprocessor" in this book.

Figure 5.78 shows an example of an actual one-board microcomputer. This single printed circuit board also carries peripheral circuitry for flexible disk drives, hard disk drives, printers, a clock and communication control.

---

* Here, the term "chip" is used to mean integrated circuit chip (IC).

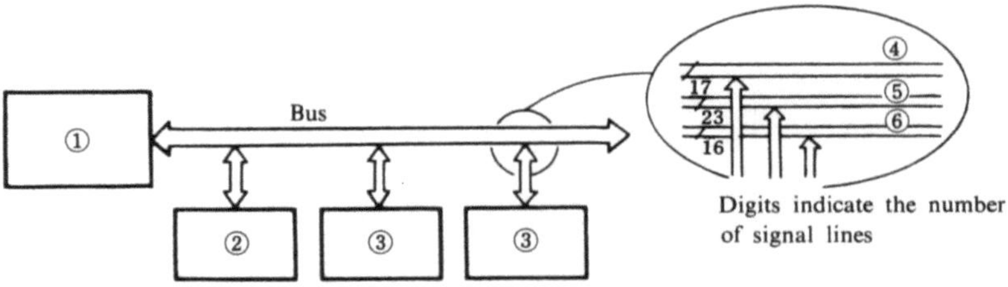

① Microprocessor ③ Peripheral interface ⑤ Address bus
② Memory ④ Control bus ⑥ Data bus

Fig. 5 · 77 Configuration of microcomputer (An example of 16-bit)

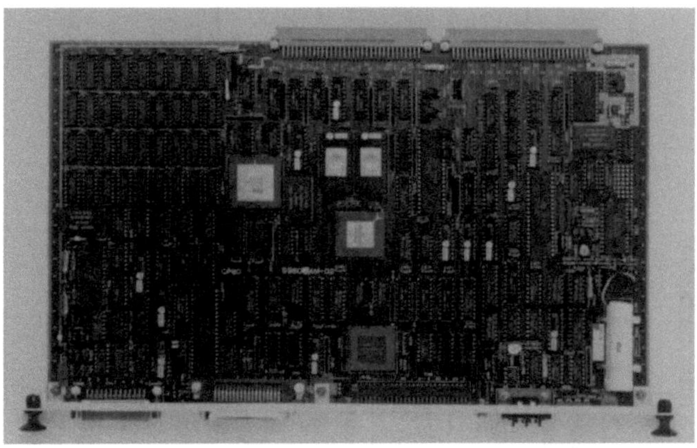

Fig. 5 · 78 Application of microcomputer
(CPU board of YEWMAC, Yokogawa Electric Corp.)

(b) Types of microprocessors

Technology for the integration of processor elements has improved and various kinds of microprocessors have been developed for diverse applications. Table 5.17 lists the broad categories of microprocessors and their main uses. The devices listed as examples are those that are well known in the market, but there are many others as well. In measurement and control, the general purpose microprocessor is most often used because it can be combined with peripheral chips in many ways. However, as integration advances and memory capacity increases, single-chip microprocessors are also coming into frequent use. There are many single-chip products, which differ according to the memory and peripheral circuitry combined in the chip. Thus selecting among the available single-chip microprocessors requires careful analysis.

Table 5.17 Classification and application of microprocessor

Classification		Main use	Examples
General purpose microprocessor	8-bit	Measuring instruments, Sequencer, Single loop controller	8085 (Intel) Z 80 (Zilog) 6809 (MOTOROLA)
	16-bit	Distributed control system, Measuring system, EWS, Personal computer	8086 (Intel) 68000 (MOTOROLA)
	32-bit	Production management computer	68020 (MOTOROLA) 80386 (Intel) V 60/V 70 (NEC)
Single-chip microprocessor	4-bit	Electrical appliance (Microwave oven, VTR, etc.)	μPD 546 (NEC) HD 44700 (HITACHI)
	8-bit	Computer I/O device, automobile engine control, Industrial instrument	8048 (Intel) 6801 (MOTOROLA) 6301 (HITACHI)
	16-bit	LAN controller, NC, Robot	8096 (Intel)
Bit-slice microcessor		Image processing, High speed sequencer	2901, 2909 (AMD) 29116 (AMD) AS 888/889 (TI)
Operational processor		Co-processor of general purpose microprocessor (floating point arithmetic)	8087 (Intel) 68881 (MOTOROLA)
Digital signal processor (DSP)		Voice recognition, Image processing, Fourier analysis	μPD 7720 (NEC) TMS 320 (TI) MB 8764 (Fujitsu)

For bit-slice microprocessors, 4-bit devices are predominant. However, their architecture allows multiple units to be connected to form 16-bit and 32-bit processors. Unlike other microprocessors, they can be custom-designed for specific purposes, such as image processors which require high processing speed, high-speed sequencers, matrix calculations, and other such applications. The operational processor performs floating-point arithmetic, trigonometric function calculations, and so on, and is often used as a co-processor for general purpose microprocessors.

In process control, digital signal processors perform real-time calculations on signals from measurement instruments. The central operation in digital processing is integrating. Digital signal processors are specifically designed for this purpose, and have several tens of times the power of 16-bit general-purpose microprocessors for special calculations. They are expected to have applications in a wide range of areas, such as voice synthesis, and voice recognition.

(c) *General-purpose microprocessors*

There are 4-bit, 8-bit, 16-bit, and 32-bit general-purpose microprocessors. However, the 4-bit general-purpose microprocessors are now seldom being used, but the trend is toward single-chip devices for use in home appliances. The 8-bit microprocessors are most suitable for handling text characters, and are applied to peripheral I/O devices such as printers. In the field of control they are used in single-loop controllers and so on. The 16-bit microprocessors have powerful calculation processing capabilities, and have broad application in distributed control systems and processors for measurement equipment. In around 1984, the 32-bit microprocessors appeared, and their power rivals the performance of the 32-bit minicomputers developed in the 1970's. Now, they will have wide application in the fields of production management, robot control and other fields requiring high performance.

Figure 5.79 shows some examples of actual 8-, 16- and 32-bit general purpose microprocessors. As the number of pins increases, the package design changes to increase the pin density. Also, although the number of chip elements increases greatly as the bit capacity increases, with improved IC technology and the application of CMOS technology, the dimensions of the silicon chip itself have not increased significantly. General-purpose microprocessors can be divided into two main series: the Intel series (8085 and 8086) and the MOTOROLA series (6800 and 68000). There are small differences between the processors of these two series, but the basic processor functions are equivalent.

Figure 5.80 shows the I/O signals of the MOTOROLA 68000 series 16-bit general purpose microprocessor. The 23 address buses allow address specification for up to 16 megabytes of memory space. A 16 data

8-bit
8085 (Intel)

16-bit
68000 (MOTOROLA)

32-bit
68020 (MOTOROLA)

Fig. 5・79 Examples of general purpose microprocessors

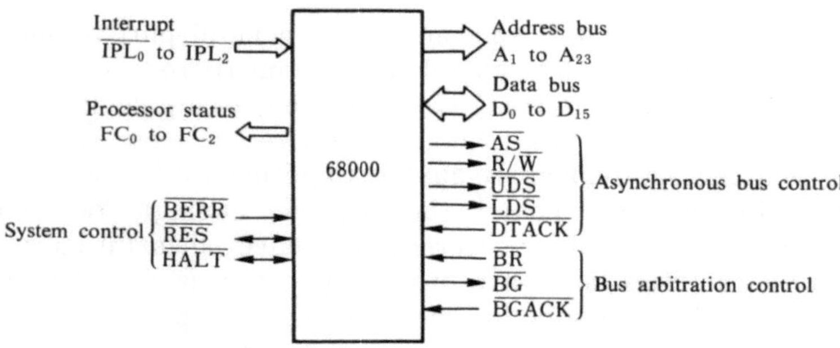

Fig. 5・80 Signal wires of 68000

buses permit 16 bits data to be read or written. Bus control employs an asynchronous method, which can match the differences in response times of memory devices and peripheral chips. In the field of measurement and control, interrupts are highly useful. The 68000 has three interrupt lines, which, when decoded, provide interrupts with seven levels of priority.

(d) Peripheral integrated circuits

Peripheral integrated circuits (ICs) are most useful in the development of various kinds of equipment. When a function is developed, whether or not there is an LSI chip suitable for the function makes a great difference in the amount and cost of the hardware involved. ICs are being developed for high-performance display, communications, and supplementary memory devices. Table 5.18 classifies peripheral ICs for 16-bit microprocessors, and lists examples of each type. The examples are broadly divided into the Intel series and the MOTOROLA

Table 5.18 Classification of peripheral LSI

Classification	Function	68000 series	8086 series
System control	DMA controller	68450	8237
	DRAM controller		8208
Display control	CRT character display	6845	8275
	CRT graphic display	63484	82720
	LCD controller	63645	
Communication control	Asynchronous method	6850	8251
	Synchronous method	6852	8251
	LAN	68824	82586
Timer	Interval/fixed cycle	6840	8253
	Watch/calendar	146818	
Digital/analog input-output	Digital input-output	6821	8255
	Analog input	46508	
	Analog output	6890	
Auxiliary storage control	FDD controller	63265	μPD 765
	HDD controller	63463	μPD 7261
General purpose bus control	GP-IB controller		8291/8292

series, but by adding some external circuitry, other series of ICs can also be connected.

5.6.2 Memory elements and storage equipment

(a) Semiconductor memory

After the appearance of the 1 kilobit MOSRAM in 1970, the bit density of memory elements increased by ten fold about every three years (Fig. 5.76), and by 1986 the 1 megabit DRAM was available for use. In parallel with this, the cost per bit dropped very rapidly, greatly improving the equipment cost-performance ratio.

Semiconductor memory is broadly classified into random access memory (RAM) and read-only memory (ROM). This classification scheme and examples of the devices are shown in Table 5.19. RAM memory is volatile, which means that the stored contents will be lost when the power supply is cut off. ROM, on the other hand, does not have this problem, thus in the field of measurement and control it is often used for storage of programs and constant values. RAM is further divided into static RAM (SRAM) and dynamic RAM (DRAM). DRAM circuit design is complicated because a refresh circuit is required to sequentially access the memory cells at regular time intervals. However, it is suitable for large-capacity memory devices. SRAM, as opposed

Table 5.19 Classification of semiconductor memory

Classification			Example of element configuration	
RAN (random access memory)	DRAM (dynamic RAM)		NMOS	256K×1-bit
			CMOS	1M×1-bit
	SRAM (static RAM)		CMOS	256K×1-bit
			CMOS	32K×1-bit
			ECL	64K×1-bit
ROM (read only memory)	Masked ROM		CMOS	128K×1-bit
	PROM (programable ROM)	Un-erasable PROM	Bipolar	8K×1-bit
		Erasable PROM		
		EPROM*1	CMOS	128K×1-bit
		EEEPROM*2	CMOS	8K×1-bit

* 1 Electrically programable ROM
* 2 Electrically erasable PROM

to this, can be connected to the microprocessor by means of simple circuitry and has the additional merits of fast read and write times. The drawback is that the cost per bit is high compared to DRAM.

Conversion to CMOS technology for the purpose of reducing the power consumption in integrated circuits is a strong trend, and this is particularly true for semiconductor memory devices. As the number of transistors in the memory device increases, this trend becomes necessary in order to control the heat generated in the chip. Reduction of power consumption in the semiconductor memory devices used in control equipment is also very important. Specifically, it would allow smaller batteries to be used in the battery backup required to preserve stored data in event of power outages. Cooling systems can also be made smaller.

Erasable-programmable ROM (EPROM) is a type of memory device whose memory contents can be erased by means of irradiation with ultraviolet light then new data can be written in its place. In contrast with this, the electrically-erasable-programmable ROM (EEPROM) can be erased electrically. Eight-kilobit EEPROM devices have reached the practical stage. As the reliability of these devices improves, they are being used more and more frequently.

(*b*) *Magnetic storage devices*

Magnetic storage devices employ the stable nature of magnetism. There are various kinds of magnetic storage devices, including flexible disk drives (FDD), magnetic disk drives, and magnetic tape devices. As shown in Fig. 5.81, magnetic storage devices are slower than semiconductor memory devices, but they are non-volatile and have large capac-

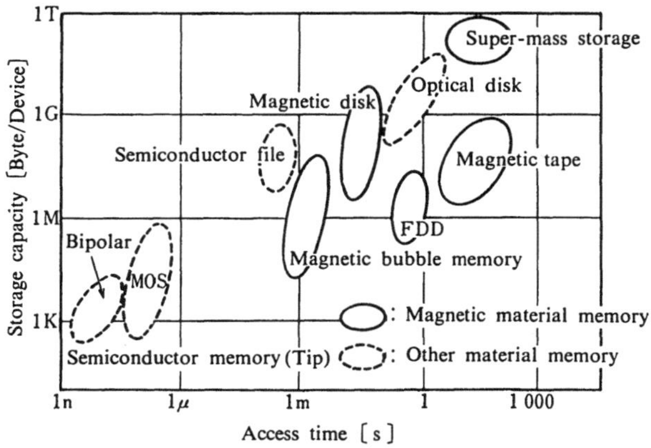

Fig. 5・81 Storage capacity and access time of storage devices

ities. Thus, they are used for auxiliary storage for computer systems. Among these devices, the flexible and hard disk drives have undergone remarkable reduction in size and cost per bit, and play an important role in the increase of system reliability and lowering of cost.

(1) *Flexible disk drive* The flexible disk drive uses a circular sheet of plastic material as a removable medium. This allow easy handling, and since its development in 1972, it has penetrated a wide range of markets to become a generally-used removable storage medium. Beginning with the 8-inch disk that could hold 400 kilobytes of data, the disk size shrank to 5.25 inches and then again to 3.5 inches. At the same time the capacity increased to the high density of 1.6 megabytes. Table 5.20 shows the various combinations of size and capacity of existing disks. Because of this variety, it is necessary to check the compatibility of the disk and disk drive. The trend is toward increased use of the 1.6 megabyte disk. The 8-inch disk is mostly used for removable media in large computer systems.

Table 5.20 Kinds of flexible disk devices

Disk size	Capacity (unformattep, Mbyte)			
8-inch	0.4 (single side single density)	0.8 (double side single density)	(high double density)	1.6 (double side double density)
5.25-inch	0.5		1.0	1.6
3.5-inch			1.0	1.6

(2) *Hard disk drives* Formerly, magnetic disk drives had disk packs or disk cartridges that were used as removable media. Recently, small, large-capacity hard disk drives (HDD) have been developed. The hard disks can be grouped together into a single unit inside the disk drive, and they have gradually become the mainstream device—at first for personal computers and now also for measurement and control equipment.

In the HDD, the disk unit is directly connected to the spindle motor. The principle of the HDD is illustrated in Fig. 5.82. The disk drive consists of magnetic disks, spindle and motor constructed as a single unit to rotate the recording medium, a head positioning mechanism, and peripheral control circuits. Head positioning is accomplished by means of either a stepping motor or a voice coil. The stepping motor method uses a simple mechanism, and is thus used for small and inexpensive drives. The voice coil method uses a servo mechanism for positioning the head, thus allowing high-speed access and high-density recording.

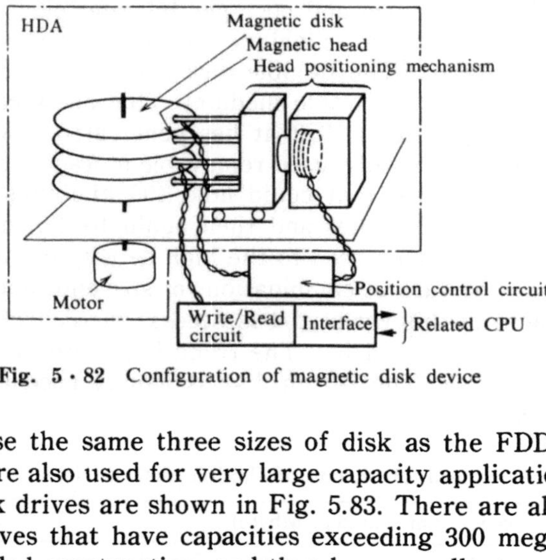

Fig. 5 · 82 Configuration of magnetic disk device

HDDs use the same three sizes of disk as the FDD. In addition, large disks are also used for very large capacity applications. Examples of actual disk drives are shown in Fig. 5.83. There are also example of 5.25-inch drives that have capacities exceeding 300 megabytes. HDDs feature a sealed construction, and thus have excellent resistance to environmental contamination compared to other types of supplementary memory devices. Recently, small environment-proof HDDs have been under development.

5.6.3 Display elements and devices

The man-machine interface allows interaction between people and

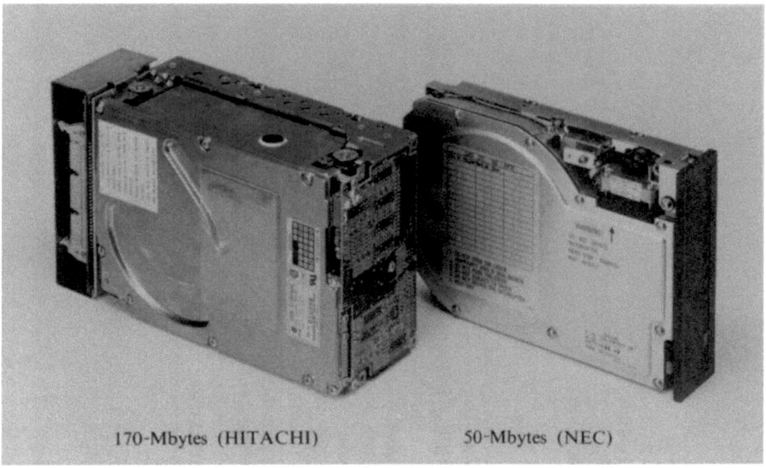

Fig. 5 · 83 Example of fixed magnetic disk (5.25 inch)

equipment. It uses various kinds of display elements and devices which are appropriate to the purpose at hand. Table 5.21 lists the types of display elements and devices, important aspects of their display characteristics, and their applications. The cathode-ray tube (CRT) has a wide range of applications, beginning with serving as a computer terminal. The plasma display is more expensive than the CRT, but it is light in weight and has good display characteristics. Thus it is beginning to be used to replace the CRT in some application areas. Liquid crystal displays are thin and feature low power consumption. They are often used for portable, small display capacity devices. The electroluminescent display (ELD) has a highly luminous light-emitting surface, and has been a subject of recent research and development. It is also used for back-lighting in liquid crystal displays. Light-emitting diodes (LED) and fluorescent display tubes are mostly used for the single purpose of alphanumeric character display devices. There are also such special devices as magnetic inversion display plates and magnetic migration displays, which are suitable for wide-area advertising.

(a) Cathode-ray tubes

Cathode-ray tubes (CRT) have been used as display devices for a long time. However, they have improved in step with progress in fine process technology, and resolution has improved to the point of being a mature technology. Presently, CRTs used as display devices are classified according to their scanning method into raster-scan types and random-scan types (Fig. 5.84). Random-scan devices are further divided into refresh-type displays, in which data must be rewritten repeatedly within the luminous persistence time of the phosphors, and storage-type displays, in which the phosphors have memory characteris-

Table 5.21 Classification of display elements and devices

Classification	Compared characteristics				Application
	Resolution	Display capacity	Display area	Thickness	
CRT	◎	◎	◎	×	Control monitor, CAD/CAM, EWS, Simulator
Crystalline liquid display	◎	△	△	◎	Measuring instrument, Portable device
Plasma display	○	○	○	○	Control monitor, Production management
LED	○	△	△	○	Instrument display, Panel display
Fluorescent display tube	△	△	△	△	Instrument display, Panel display
EL	◎	○	△	◎	Measuring instrument, Portable device
Magnetic inversion display plate	×	△	◎	△	Panel display, Traffic control display, Traffic information display, Trouble status display, Public relation, Advertisement
Magnetic migration type display	×	△	◎	△	Panel display, Public relation, Advertisement

(Note) ◎; excellent, ○; good, △; not good, ×; inferior.

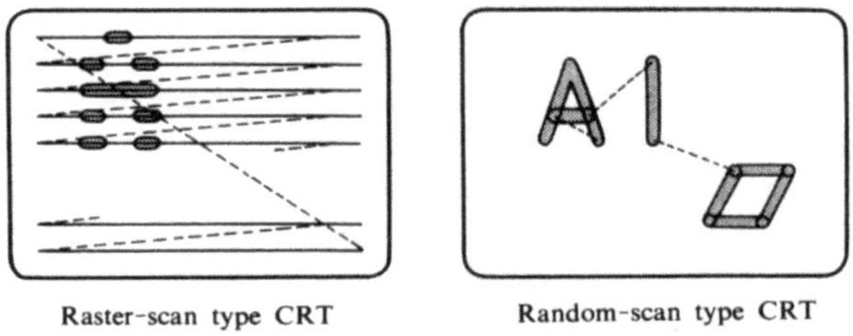

Fig. 5・84 Scanning method of CRT

tics. Random-scan type displays achieve high-quality graphics because oblique lines are free from the step-like distortion characteristic of raster-scan types of display. However, they have limited display speed and they are expensive. Now, the random-scan displays are seldom used.

On the other hand, raster-scan displays are in wide use because, as an extension of television technology, they now make use of integrated circuits, which has resulted in increased reliability and lower cost. The principle of the raster-scan color CRT display is illustrated in Fig. 5.85.

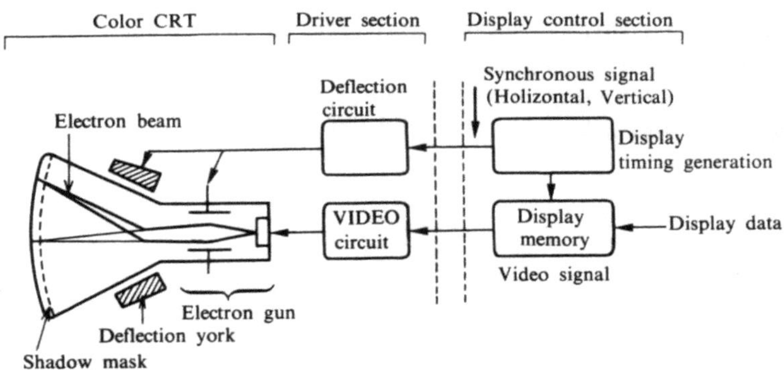

Fig. 5・85 Principle of raster scan type CRT display

(1) *Color CRTs* On a color CRT screen, red, green and blue phosphor dots are arranged in a regular, sequential pattern. There are three electron guns corresponding to these three colors (RGB). A shadow mask guides the electron beams from these electron guns such that they strike the phosphor dot array.

(2) *Color CRT drivers* To form a display on the CRT screen, external synchronization signals and a video signal are necessary. The synchronization signals (horizontal and vertical) produce deflection scanning of the electron beams. The video signal (luminance signal) causes

the phosphor dots to produce light.

(3) *Display control* The synchronization signals specify the beginning of horizontal and vertical scanning. The scanning position is accurately determined in the display control section by means of a counter. The display data corresponding to the scan position is read from the display memory, and the appropriate video signal is generated. In this way, information can be displayed on CRT screens.

(*b*) *Liquid crystal displays*

Formerly, liquid crystals were mainly used as small-capacity display elements in small devices typified by electronic calculators and digital wristwatches. However, recent progress in thin-film semiconductor technology has improved the capacity and responsiveness of liquid crystal display elements, and it has become possible for these devices to handle the large-volume display data for computer terminals and measurement instruments. Achievement of color capability is progressing faster for liquid crystal displays than for other thin display devices. Liquid crystal television screens 3 inches to 5 inches in size have already become practical, and 10-inch to 12-inch, 640-dot by 400-dot panels will soon follow.

Figure 5.86 shows the operating principle of the twisted nematic (TN) liquid crystal. The liquid crystal lies between two polarizing filters whose polarization planes are at right angles to each other. When there is no electric field applied perpendicular to the filters, the crystal molecules are twisted as shown in the figure. In this state, light passes through the filters, because the angle of light deflection is 90°. On the

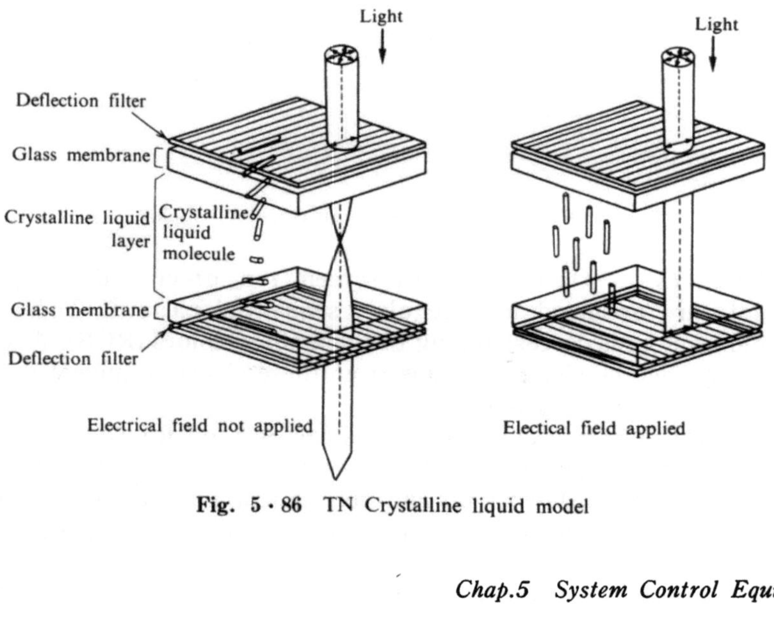

Fig. 5・86 TN Crystalline liquid model

other hand, if an electrical field is applied, the crystal molecules orient themselves parallel to the field, and the light is blocked by the polarizing filters. Employing this principle, it is possible to display information by controlling the electric field applied to individual crystals that have been formed into a pattern.

5.6.4 Analog/digital conversion

Analog/digital (A/D) conversion, which means changing an analog signal into a digital signal, is an important technology that is essential to the introduction of computers to the field of measurement and control. The output signal from sensing devices such as temperature detectors and flow rate detectors are mostly analog signals which must be converted to the digital signal form suitable for input to computers. There are many conversion methods, and the choice of method is based on conversion speed and conversion accuracy. Table 5.22 lists various conversion methods and their main applications. Figure 5.87 shows the relationship between conversion accuracy and conversion speed.

The *V–F* conversion type generates a pulse signal whose frequency (F) is proportional to the voltage of the input analog signal (V), and the voltage value of the input signal is determined by counting the number of pulses in a given period of time. The *V–T* method, on the other hand, is based on counting the number pulses generated by a standard clock within a gate time (T) that is proportional to the voltage of the analog signal (V). Both of these methods are integration methods, and have the disadvantage of being relatively slow. However,

Table 5.22 A/D conversion method

Classification		Use
Integration method	V-F conversion type ex.: Charge balance type V-T conversion type ex.: Dual slope type	High accuracy-low speed use Digital voltage meter Panel meter
Comparison method	Feedback comparison type ex.: Successive approximation type, Servo-balancing type	Medium-high speed use Data acquisition measurement Process control
	Non feedback comparison type ex.: Parallel comparison type, Serial comparison type	High speed use Image processing, Waveform memorizing device PCM communication

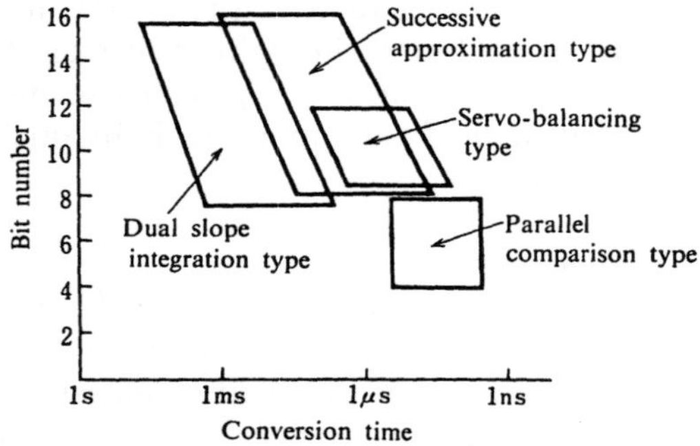

Fig. 5・87 Characteristics of A/D converters

they have the merits of being inexpensive and resistant to noise.

The comparison methods have the advantage of being easily adapted to fast IC circuitry. Of the comparison methods, the successive approximation type is often used in process control and data acquisition. There are also many ICs which can be applied in this method. The basic circuit and operation waveform of the successive comparison method are shown in Fig. 5.88. The conversion start signal clears the

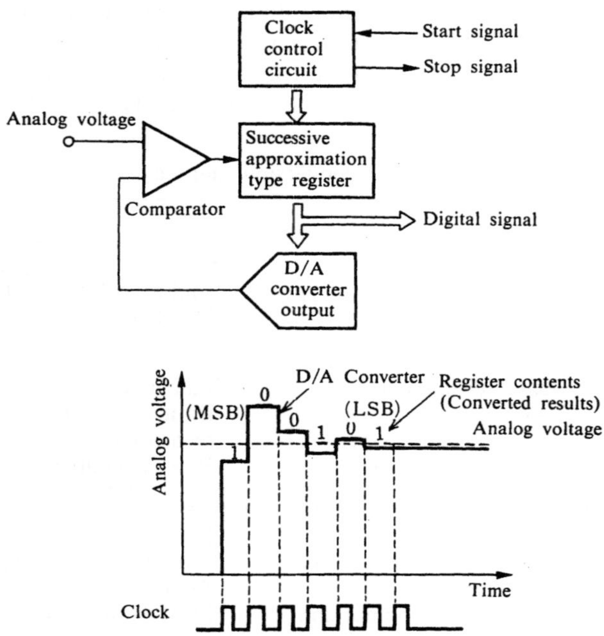

Fig. 5・88 Principle and circuit configuration of successive approximation type

register, and then the output of the D/A converter is compared with the analog input voltage bit-by-bit, starting with the highest-order bit. According to the result of each comparison, a 0 or 1 is set in the corresponding bit of the resister.

5.6.5 Optical communication elements

Optical communication, which developed out of broadband digital communication systems, has recently begun to occupy an important position as an elemental technology for the construction of total control systems in factories. The main components of an optical communications system are a light-emitting element, a light-receiving element, and an optical fiber, which serves as the transmission path. In addition, fiber splitting and coupling devices are also important. These devices are used to branch one fiber into several fibers, and conversely, to combine several individual fibers into a single fiber. Fibers are broadly classified as step-index fibers (SI), graded-index fibers (GI), and single-mode fibers (SM) , according to distributions of their refractive indices. However, in the field of control, the GI fiber is often used for reasons of economy, speed, and so on.

Light-emitting diodes (LED), and semiconductor laser diodes (LD) are used as light-emitting elements. The characteristics of these two types of device are listed in Table 5.23. There are two types of light-receiving elements. One is the PIN photodiode, which converts injected photons directly into electrons. The other is the avalanche photodiode (APD), which using the avalanche amplification effect, produces a larger electrical output for a given input light.

Table 5.23 Comparison between light emitting diode and laser diode

	Light emitting diode (LED)	Laser diode (LD)
Spectral bandwidth	350 Å	30 Å
Output power	1 mW	10 mW
Coupling efficiency to optical fiber	Several %	50%
Lifetime	One million hours	One hundred thousand hours
Use	Mainly used for shortwave band. Low, medium speed. Short distance transmission.	Mainly used for longwave band. High speed. Long distance large volume transmission

REFERENCES

1) E. Ohshima: "Transition of automation and safety in chemical plant," *for safety of High Pressure Gas Eng.*, **22**, 4 (1985) 158–165 (in Japanese).
2) Y. Wakasa, *et al.*: "Enhancing reliability of CENTUM distributed process control system," *Yokogawa Tech. Rep.*, **27**, 4 (1983) 3–12 (in Japanese).
3) S. Takigishi, *et al.*: "User oriented flexible operator station in a distributed process control system," *Yokogawa Tech. Rep.*, **28**, 4 (1984) 24–31 (in Japanese).
4) T. Ogawa: "Process data highway-configuration, functions and protocols," *Yokogawa Tech. Rep.*, **26**, 4 (1982) 46–52 (in Japanese).
5) K. Matsunaga: "Digital algorithms for process control–PID control, deadtime compensation and advanced control," *Yokogawa Tech. Rep.*, **27**, 1 (1983) 38–43 (in Japanese).
6) K. Matsunaga: "Documenting process control sequence by decision tables," *Yokogawa Tech. Rep.*, **26**, 3 (1982) 44–50 (in Japanese).
7) H. Tamura: "YEWMAC, YEW's manufacturing line control system development philosophy and overview," *Yokogawa Tech. Rep.*, **27**, 3 (1983) 24–27 (in Japanese).
8) H. Kawai, *et al.*: "Development policy and technology of YEWMAC 300/50," *Yokogawa Tech. Rep.*, **30**, 1 (1986) 13–22 (in Japanese).
9) Kojo Jidoka Jiten Henshuu Iinkai: Encyclopedia of Automization for Factories, Sangyo Chosakai (1983) (in Japanese)
10) Japan Electronic Industry Development Association: A Report of Investigation for Industrial Computer Systems, 60–A–232 (1985) (in Japanese)
11) H. Tamura, *et al.*: "How the YEWCOM computer fits into the YEW total system architecture," *Yokogawa Tech. Rep.*, **27**, 2(1983) 2–64 (in Japanese).
12) S. Yamamoto, *et al.*: "YEWCOM performance enhancement," *Yokogawa Tech. Rep.*, **28**, 4 (1984) 17–23 (in Japanese).
13) K. Ohmori, *et al.*: "YEWCOM redundant system," *Yokogawa Tech. Rep.*, **30**, 1 (1986) 7–11 (in Japanese).
14) K. Sakamura: Computer Architecture, Kyoritsu Shuppan (1984) (in Japanese)
15) I. Morishita: Hardware of Microcomputer, Iwanami Shoten (1984) (in Japanese).
16) *Infor. Process.*, **27**, 6(1986) (in Japanese).
17) Y. Nagahashi: Design of A–D/D–A Conversion Circuits, CQ Shuppan (1980) (in Japanese).
18) M. Maeda: Optical Fiber Communication, Denki Tsushin Gijutsu News Sha (1981) (in Japanese).

Chapter 6

FINAL CONTROL ELEMENTS

This chapter describes the final control elements of instrumentation systems. Featured are the control valves that operate either by signals from a controller or by manual action to control the transmission and cut-off of fluids such as liquids, gases, and steam. The controller can be considered to be the head, with the control valves acing as the hands and feet. The control valve, which receives the control signal, consists of an actuator that converts the control signal to an operating variable and a main valve body that applies the work directly to the object of control.

6.1 Types of Control Valves

There are many different types of control valves used in industry. The type used depends on the requirements of the process involved. The follwing descriptions are classified according to the types of operation signals.

6.1.1 Pneumatic control valves

This control valve functions according to an air pressure signal (0.2 to 1.0 kg/cm^2G). The air signal can be applied directly to the valve actuator, but frequently a positioner is used to improve response and minimize hysteresis. Pneumatic control valves have been used for many years, and so have become relatively standardized and very reliable. They have many advantages, such as being relatively inexpensive, easy to maintain, and equipped with essential features to prevent bursting. This type of valve is the most popular type currently in use.

6.1.2 Electrical control valves

This control valve operates by an electric current signal (mainly 4 mA to 20 mA DC). In the manipulation of motor-drive valves and electromagnetic valves, an electric pulse signal may also be used. In the case where an electric current signal is used, the actuator is manipulated by a current-to-pneumatic, current-to-current, or current-to-hydraulic positioner.

6.1.3 Hydraulic control valves

The strong points of the hydraulic control valve are its good response to the actuator and its strong operating power. It is manipulated by means of a current-to-hydraulic positioner.

6.1.4 Self-powered control valves

These are self-regulating control valves installed in the field as simple regulators. The process variable, such as pressure and differential pressure, is the valve's direct power source or the valve is manipulated by using the change in vapor pressure of the liquid enclosed in the temperature sensing element. Now, because the control valve works directly with the controlled system, it has the most important role. Thus choosing the type of device that fits the control system and the control purpose is the key toward achieving the success of the instrumentation system.

This point is important and attention is concentrated on the control valves that fit diverse control requirements and have the highest level of use as final control elements for instrumentation systems. The focus is on:

(1) A method for choosing control valves that fit the instrumentation system
(2) Types of valve bodies and their characteristics
(3) Types of actuators and their characteristics

Figure 6.1 illustrates the construction of a pneumatic control valve with a globe valve, which is widely used as a basic valve body subassembly. Particulars about globe valves will be taken up in Sec. 6.3, but this shows a representaive example of a control valve.

6.2 Choice of Control Valves

6.2.1 Various conditions affecting choice

The control valve is impotrant in the process operation. Not only are the specifications of the valve itself important, but it is also necessary to sufficiently consider other matters related to the control valve in order for it to function as required. The following are the main points to keep in mind when specifying a control valve.

(a) Target process

It is necessary to completely understand the control system, in-

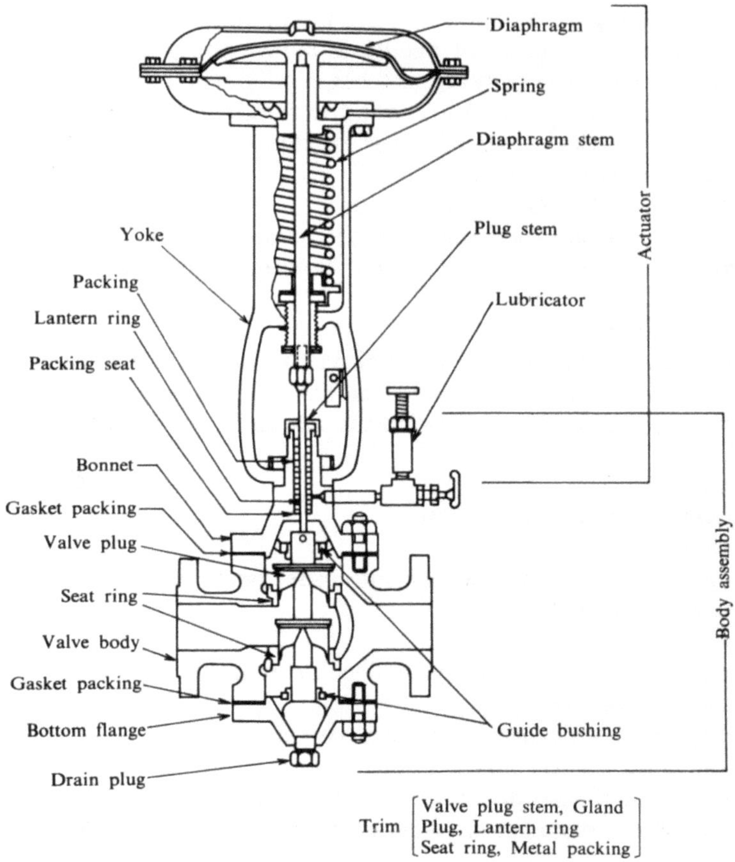

Fig. 6 · 1 Construction of control valve

cluding the control valve. Additionally, one should sufficiently understand the start-up and shut-down of the process itself, including proper conduct in an emergency situation.

(*b*) *Purpose of use*

Excluding manipulation of the process variable of the fluid itself, there are control valves that control the cut-off and release of fluids, mix two fluids, separate the flow into two directions, or exchange fluids. There are also valves that control a pressure drop from a high-pressure system to a low-pressure system. Also, there are valves that incorporate two or more of the above functions. Therefore, the most appropriate control valve is chosen after determining the purposes of a particular valve.

(*c*) *Response time*

In responding to a change in the manipulation signal, the control valve experiences a period of deadtime before the plug stem can overcome friction from the packing and begin to move. There is also a period of operating time needed to move the required distance. It is necessary to consider the effect of these factors on the controllability and safety of the entire system.

(*d*) *Specific characteristics of the process*

Determine in advance the presence or absence of self–equilibrium, the range of variation in the required flow rate, the speed of response, etc.

(*e*) *Fluid conditions*

The various conditions of the fluid can be obtained from the process data sheet, and these become the basic conditions for the selection of the control valve. The following are the main conditions that will be used.

(1) name of fluid
(2) components, composition
(3) flow rate
(4) pressure (at both the inlet and outlet ports of the valve)
(5) temperature
(6) viscosity
(7) density (specific gravity, molecular weight)
(8) vapor pressure
(9) degree of superheating (water vapor)

In addition to the above data relating to normal operating conditions, one should also determine their respective values at the beginning and end of operation as well as their maximum and minimum values.

(*f*) *Fluidity, special characteristics*

One should determine the presence of possible dangers regarding the nature of the fluid, corrosiveness, or slurry.

(*g*) *Rangeability*

In the case where one control valve cannot provide the necessary rangeability (Subsec. 6.2.4), consideration of the use of two or more valves becomes necessary.

(*h*) *Valve differential pressure*

As the rate of the valve's differential pressure decreases relative to the overall pressure loss of the entire system, the installed flow characteristics shift away from the inherent flow characteristics. (Refer to Subsec. 6.2.3.) The rate of control valve pressure loss in a piping system is a complicated problem. Although it is impossible to generalize, a value for P_R between 0.3 and 0.5 is usually chosen.

(*i*) *Shut-off pressure*

The highest value of the differential pressure at the control valve shut-off time is important data to be used in the choice of the actuator and in ensuring a sufficiently strong design for each part of the control valve. Designs in which the intake pressure is set equal to the maximum shut-off pressure are numerous, but this method might result in over-specification of the valves. Thus it is necessary to consider actual use conditions when determining the shut-off pressure.

(*j*) *Valve-seat leakage*

Determine what quantity of valve-seat leakage can be tolerated at the time of valve shut-off. It is also necessary to know the frequency with which the valve shut-off condition occurs. Generally, the degree of leakage is indicated by a percentage of the value of the rating C_v of the control valve. ANSI B 16 104-1976 is widely used for classification of leakage and for testing conditions.

(*k*) *Valve operation*

There are two objective in the operation of the valve; the first of these is fail-safe operation and the second is operation according to the valve input signal. Regarding the first objective, the movement of valve operation is in a safe direction of the process in case the input signal and the power supply is lost. The operation is classified as "air failure close," "open," or "lock." Regarding the second objective, the opening and closing direction of the valve is adjusted in according to whether the signal input to the valve increases of decreases, but the operation is not necessarily the same as the fail-safe operation. When the valve closes as a result of input increase, this is called *direct action*. When the valve opens as a result of input signal increase, this is called *reverse action*. Figure 6.2 shows combined valve action in cases where the actuator and plug show direct and reverse action in a globe valve

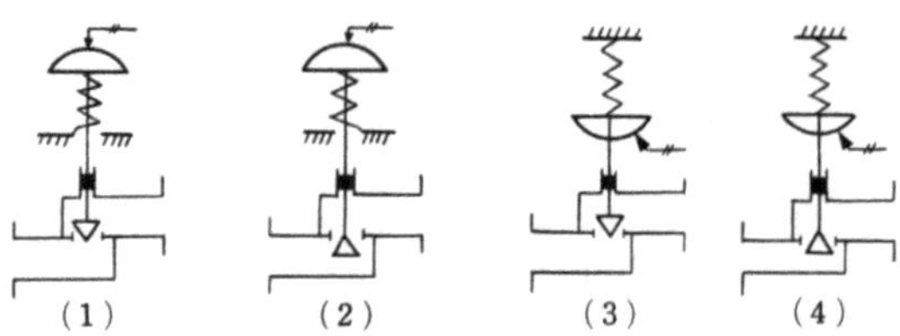

Figs.	Actuator	Plug	Valve operation
(1)	Direct	Direct	Direct action
(2)	Direct	Reverse	Reverse action
(3)	Reverse	Direct	Reverse action
(4)	Reverse	Reverse	Direct action

Fig. 6 · 2 Value action

with a diaphragm actuator.

(*l*) *Explosion-proofing*

Electrical equipment used in conjunction with the control valve must have the explosion-proof rating required by the location in which the valve is installed.

(*m*) *Power supply*

For pneumatic power supplies, it is important to provide clean air with water, oil and dust removed in order for parts such as the actuator and positioner to function without failure. At the same time one must determine the actuative pressure (air motor diaphragm area × air pressure) and capacity so as to secure enough actuative power.

(*n*) *Piping specifications*

Determine the specifications of the piping in which the control valve is installed. The important specifications include the diameter of the pipe, the piping standards, the quality of the material, the type of connecton to the piping, and so on.

6.2.2 Sizing

Sizing refers to choosing the appropriate valve by calculating the control valve's flow coefficient rating value C_v in accordance with the conditions of the fluid passing through the valve. There are several methods and modified methods of calculating C_v, but the one most for practical applications will be dealt with here. We will use the easily-managed FCI method (Recommended Voluntary Standard Formulas for Sizing Control Valves). However, there are some process conditions that cannot be covered by the FCI method. In those cases, one should use the calculation method that ANSI/ISA advances as a complement to the FCI method.*

The value of C_v: This is a number representing the capacity of the control valve. It is calculated by measuring the volume of flow in US gallons in one minute through a valve whose degree of opening is fixed, with the differential pressure held at 1 psi at a temperature of 60°F. For example, if 10 US gallons per minute can flow at 1 psi under the differential pressure, then the value of C_v 10.

---

* The FCI formulas were obtained empirically based on experiments using water. Errors will result when these formulas are used under conditions different from the experimental conditions. Other calculation formulas can be applied (ISA Handbook) for conditions such as fluids exhibiting cavitation or flushing, high viscosity or very small flow rate fluids, compressible fluids when there is a large relative pressure drop, and when valves such as ball valves or butterfly valves which have a high pressure recovery ratio are used with compressible fluids.

(*a*) *Calculating C_v by the FCI method*

(1) For liquids:

$$C_v=1.17Q_l\sqrt{\frac{G_l}{\Delta P}} \tag{6.1}$$

(2) For gases:

$$\Delta P<0.5P_1 \qquad C_v=\frac{Q_g}{273}\sqrt{\frac{G_g T_I}{\Delta P(P_1+P_2)}} \tag{6.2}$$

$$\Delta P\geqq 0.5P_1 \qquad C_v=\frac{Q_g\sqrt{G_g T_I}}{236P_1} \tag{6.3}$$

(3) For steam:

$$\Delta P<0.5P_1 \qquad C_v=\frac{W(1+0.001\,3T_{SH})}{13.5\sqrt{\Delta P(P_1+P_2)}} \tag{6.4}$$

$$\Delta P\geqq 0.5P_1 \qquad C_v=\frac{W(1+0.001\,3T_{SH})}{11.7P_1} \tag{6.5}$$

In these equations, C_v is the flow coefficient of the valve, G_g is the specific gravity of a gas in the standard condition relative to air in the standard condition, G_l is the specific gravity of liquid at the temperature of use relative to water in the standard condition, P_1 is the inlet pressure of the valve (kg/cm^2 abs), P_2 is the outlet pressure of the valve (kg/cm^{2}abs), ΔP is the valve differential pressure P_1–P_2 (kg/cm^2), Q_g is the volumetric flow rate of the gas (Nm^3/h), Q_l is the volumetric flow rate of the liquid (m^3/h), T_I is the intake temperature (K), T_{SH} is the superheated temperature of the steam (°C), and W is the weight flow rate (kg/h).

6.2.3 *Flow characteristics*

(*a*) *There are five basic flow characteristics as given below:*

(1) Quick–opening (saucer shape)
(2) Square root (quadratic characteristic, V–notch)
(3) Linear
(4) Equal–percentage (geomertic rate configuration)
(5) Hyperbolic

Three of these are commonly used: quick–opening, linear, and equal–percentage.

(*b*) *Inherent flow characteristics*

Refer to Fig. 6.3 for the characteristics exhibited where the pressure differential of the valve is maintained at a specific value.

With the quick–open characteristic, a small change in the opening results in a large change in flow. Therefore, it is limited to on/off use.

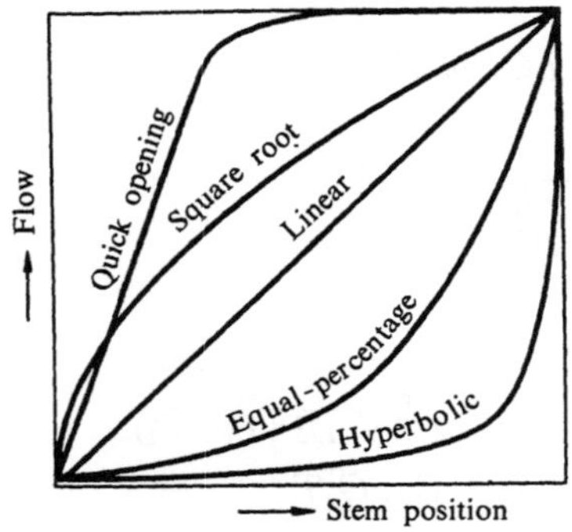

Fig. 6 · 3 Inherent valve characteristies
(ISA Hand book of control valve)

For control use, the linear, equal-percentage, or other characteristic intermediate between these is used (many of the rotating valves apply).

(c) Choice of flow characteristics

When the control valve is installed in piping, the valve pressure differential changes in accordance with change in flow, causing the installed flow characteristic to differ from the inherent flow characteristic. The installed flow characteristic changes according to the ratio of pressure drop across the valve to the pressure drop across the entire system. Examples illustrating the linear and equalpercentage characteristics are shown in Fig. 6.4.

If P_R is the ratio of the valve pressure drop to the pressure drop of the entire system (including the valve), the distortion of the installed characteristic increases as P_R decreases. For a control valve, $P_R=0.05$ is the approximate limit of tolerance. When the ratio of valve pressure drop is small, the linear installed characteristic approaches the

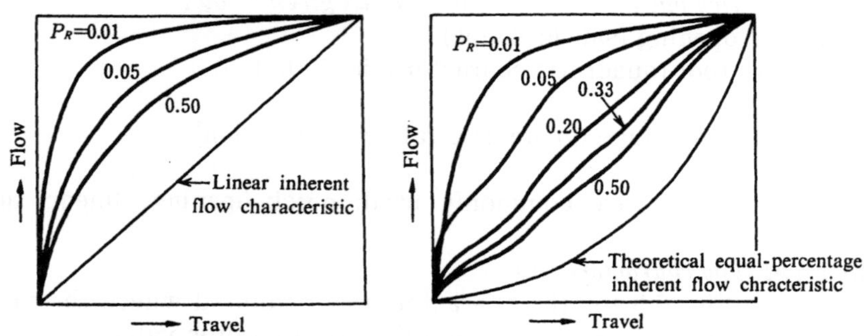

(a) Linear inherent flow characteristic (b) Equal-percentage actual inherent flow characteristic

Fig. 6 · 4 Installed flow characteristies (ISA Hand book of control valve)

quick–open characteristic, and the equal–percent installed characteristic approaches the linear characteristic.

The choice of the valve characteristic is important in compensating for the installed flow characteristic. However, determining the characteristics and distribution of the entire system is generally difficult, so as a yardstick for practical use in the following situations, the equal–percentage characteristic is chosen.

(1) When the pressure distribution is unclear.

(2) When ΔP is large for low flow rates and small for high flow rates.

6.2.4 *Rangeability*

With respect to the control valve, rangeability is the ratio of the maximum and minimum controllable flow of the valve in practical use.

(a) Inherent rangeability

With inherent rangeability, the pressure drop across the valve is maintained at a fixed level, thus

$$R = \frac{Q_{max}}{Q_{min}} = \frac{C_{v\,max}}{C_{v\,min}}$$

Generally, for globe valves $R = 30$ to 50, and for butterfly valves, $R = 20$.

(b) Installed rangeability

The rangeability of a valve actually installed in a piping system is called installed rangeability. Installed rangeability changes according to the valve's position in the system. Figure 6.5 gives an example of a system including resistance in series with the valve (series resistance), and Figure 6.6 shows an example of a system including resistance in

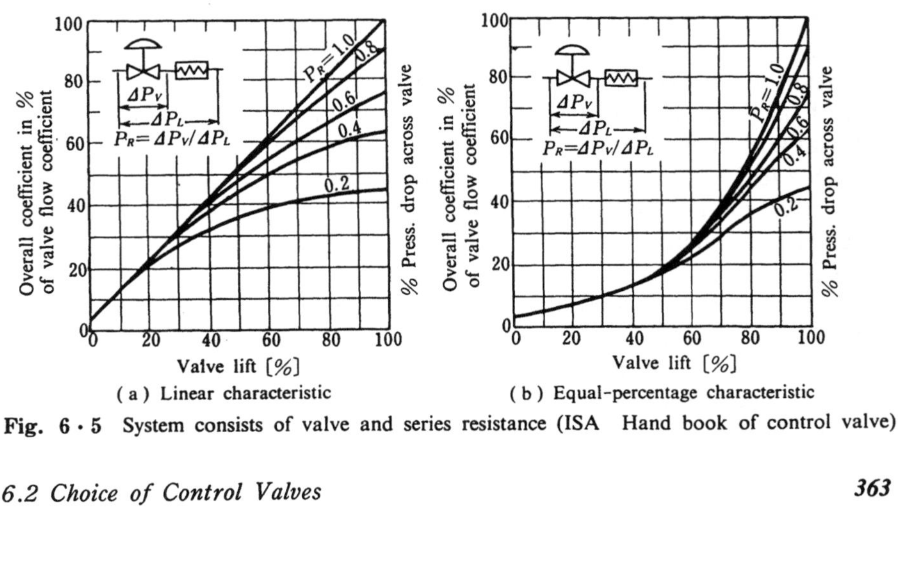

(a) Linear characteristic (b) Equal-percentage characteristic

Fig. 6・5 System consists of valve and series resistance (ISA Hand book of control valve)

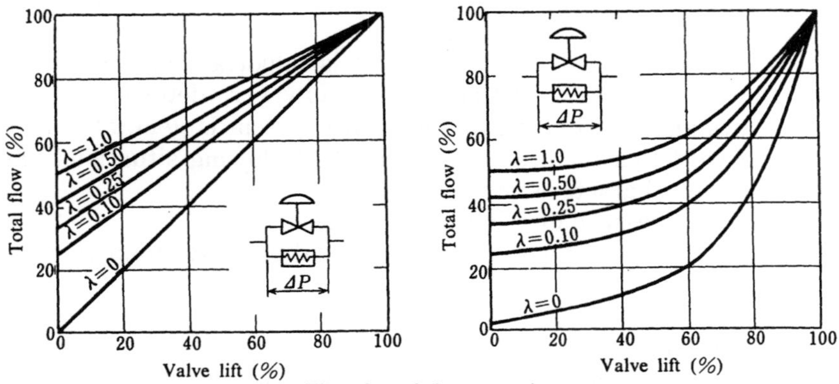

$$\lambda = \left(\frac{\text{Flow through by-pass resistance}}{\text{Flow through valve at full lift}}\right)^2$$

(a) Linear characteristic (b) Equal-percentage characteristic

Fig. 6 · 6 System consists of valve and by-pass resistance

parallel with the valve (by-pass resistance). In both cases, the rangeability of the control valve drops and influences controlability, so attention should be paid to the pressure distribution of the piping system.

6.2.5 Materials

(*a*) *Valve body materials*

The most commonly-used valve body material are listed in Table 6.1.

(*b*) *Trim materials*

Trim comes in direct contact with the fluid and is a changeable part. Materials on level higher than those used in the valve body are used to maintain trim function over a long period of time. Materials in standard use are SUS 304 and SUS 316, but in cases such as when high differential pressure or high temperatures, abrasion due to slurry, or cavitation are expected, one must choose materials that have a high degree of hardness. Two methods are mainly used, namely using a hard material and facing with a hard material. These are described below.

(1) *SUS 440C* This is made of 17 Cr martensitic steel, and has a relatively high degree of hardness due to heat treatment. It is a superior material for resisting cavitation for use with hot water.

(2) *SUS 630* This is made of precipitation-hardened stainless steel (also known as 17-4 PH). It resists corrosion to the same extent as SUS 304, and is often used because of its resistance against to wear.

(3) *Stellite®* As an alloy composed mainly of cobalt, this material is excellent because of its strength at high temperatures and its resistance to corrosion. It is widely used as a hard facing material.

Table 6.1 Value body materials (JIS B 8243-1981)

Material		Code		Operating temperature	Main use
		JIS	Compatible material of ASTM	(℃)	
Cast iron		FC 20, FC 25	—	0 to +250	General
Cast steel	Carbon steel	SCPH 2	A 216-WCB	−5 to +540	General
		SCPL 1	A 352-LCB	−45 to +350	Low temperature
	Cr-Mo steel	SCPH 21	A 217-WC 6	−5 to +593	High temperature and high pressure
		SCPH 61	A 217-C 5	−5 to +650	
Stainless steel casting		SCS 13 A	A 351-CF 8	−196 to +800	Corrosion resistance, cryogenic and high temperature
		SCS 14 A	A 351-CF 8 M		

(*c*) *Seal material*

Control valve seal materials are used in the gasket, where sealing is employed in the stationary section, and in the gland packing, where sealing is used in the sliding shaft section of the valve. In both cases, the type of fluid, temperature, and pressure must be considered in choosing the appropriate material.

(1) *Gasket* Asbestos and Teflon® materials are widely used. Metal gaskets are used for high-temperature and high-pressure applications. Asbestos is heat-resistant and is an excellent material because of its chemical stability. It is the most widely used of gasket materials, but there is a world-wide trend to regulate its use because of the harmful effect of asbestos fibers on the human body. As a substituted, packing made of pure black lead (Grafoil®) is employed because of its usefulness at high temperatures.

(2) *Gland packing* Most control valves, except the Saunders valve, use gland packing. Gland packing must keep the fluid from leaking out from the sliding section of the valve shaft. In must also minimize slide friction and minimize hysteresis. Because there is a conflict between these two functions, packing material must be carefully chosen. Figure 6.7 shows an example of gland packing installation.

(i) *Teflon® V packing*: Of the various types of packing, this sealing exhibits the least sliding resistance and excellent sealing properties, and also shows excellent resistance to chemicals. It is widely used as gland packing for control valves, but may not be used at high temperatures.

(ii) *Asbestos coil packing*: Asbestos coil packing is classified roughly into

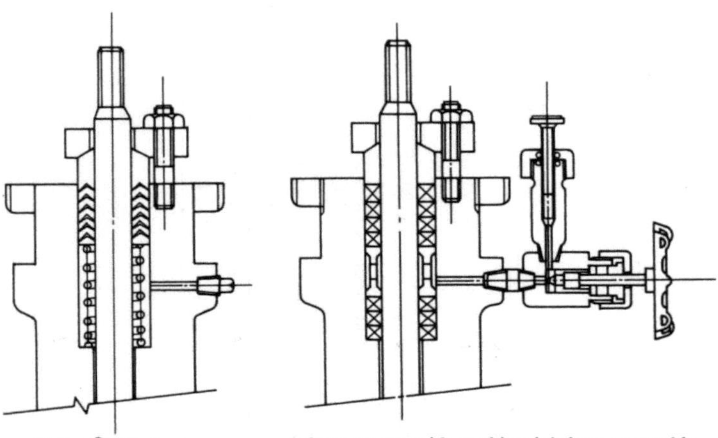

Teflon® V-rings Asbestos yarn with packing lubricator assembly

Fig. 6 · 7 Gland packing

types impregnated with Teflon®, which exhibit relatively little sliding resistance and have excellent sealing properties, and types with a special alloy additive, which are used in high-temperature/high-pressure applications. Both types require use of a lubricant grease, and application of a lubricant is usual for high-temperature use.

(*d*) *Ceramics*

Ceramics have been used recently as materials in control valves for their resistance to corrosion and wear as well as their resistance to heat. Ceramics are superior to metals in these areas. On the negative side, they are still rather expensive and are mechanically weak and susceptible to heat shock. Manufacture is problematic when complicated configurations are required, and it is difficult to use them jointly with metals. Thus there are many points to be considered, but nevertheless one can expect to see new uses for ceramics in the future.

6.3 Control Valve Bodies

Because the valve body comes into direct contact with the fluid, it is necessary to construct the valve body to fit the process conditions and purpose of use. In this section, we will show the various types of control valve configuration and their characteristics, the valve body ratings, and methods of connecting the valve to the piping.

6.3.1 Characteristics of various types of valves

Control valves are classified as linear types or rotary types according to the movement of the main valve shaft. The various relevant valve body types are listed in Table 6.2.

Table 6.2 Motion types and valve bodies

Motion types of valve stem	Types valve bodies
Linear motion type	Globe valve, Angle valve, Three way valve, Diaphragm valve, Gate valve
Rotary motion type	Butterfly valve, Ball valve, Eccentric rotating plug valve

(*a*) *Globe valves*

The globe valve has a spherical valve body, and the center line between the intake and outlet ports is a straight line. The fluid flow forms an S-shape within the valve. This is called a spherical valve.

According to the construction, globe valves are classified as single-seated, double-seated, or cage valves. These are the most commonly used among the control valves. The construction of this group

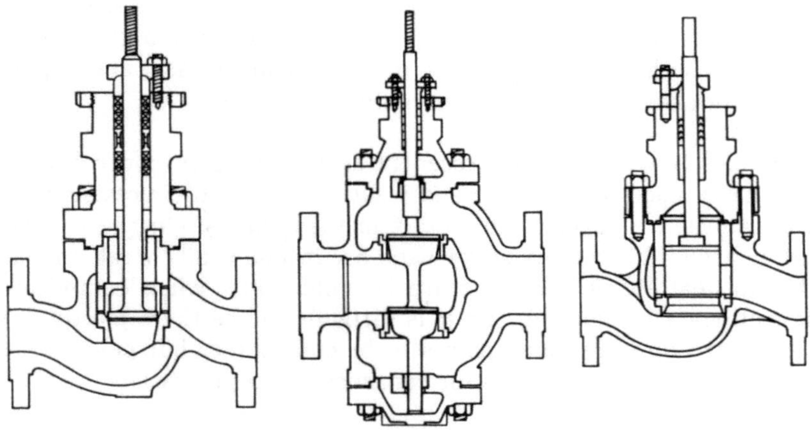

(a) Simgle seated valve (b) Double seated valve (c) Cage valve

Fig. 6・8 Construction of globe valves

of valves is shown in Fig. 6.8.

Because of the shape of the globe valve, it is possible to increase the pressure resistance of the valve body. The valve can be used at either high or low temperatures, depending on the quality of material chosen and the shape of the bonnet. By choice of plug shape, the flow can be given any desired characteristics. The rangeability differs according to the shape of the plug, but is typically limited to a value between 30 : 1 and 50 : 1

(1) *Single-seated valves* These are the most basic type of control valve; they have excellent valve shut-off ability. The quantity of leakage at the valve seat is rated with a C_v value of less than 0.01%, and with seat-face treatment, this can be reduce to less than 0.0001%. if a soft seat is used, it is possible to obtain tight shut-off. Because the full differential pressure is received by one plug, the required actuative power is greater than for a double-seated valve under the same conditions of use. It is generally suitable to use a control valve with a small intake diameter. In addition, a cut-off valve or an on-off valve may be used to obtain excellent shut-off.

(2) *Double-seated valves* These control valves have two ports, and quantity of leakage at the seat is rated at a C_v value of less than about 0.5%. In this respect, they are inferior to the single-seated valves. However, the required actuative power is much smaller than for the single-seated valve under the same conditions of use, because the power imbalance is eliminated by the upper and lower plugs. Generally, a control valve with at least a medium-size intake diameter is desirable.

(3) *Cage valves* Cage valves are equipped with a plug that slides within a cage. The cage guides the plug directly, and because the guiding area is large, vibration of the plug is suppressed and movement is stabilized when there is a high differential pressure. On the other hand, this valve is not suitable for use with fluids that contain admixtures. Regarding the inherent flow characteristics, this valve can be specified with a variety of shapes of the openings or windows in the cylindrical cage. There are both balanced and unbalanced shapes; the unbalanced shape is most frequently a single-seated valve, and the balanced shape is usually a doble-seated valve.

(*b*) *Angle valves*

The center line running from the intake of the valve body to and the outlet forms a right angle, and the direction of the flow also changes by 90° as it passes through the valve. Usually, the fluid moves from the side, and then downwards. It is used to ensure that slurries and viscous fluids flow easily, and is also employed when a drain may not be used on the outlet side of the valve. The inner construction is roughly the same as the globe valve, and there are both single-seated and cage configurations (Fig. 6.9 (a)).

(*c*) *Three-way valves*

This valve has three directional openings. There is both a mixing type that is used to mix two different fluids, and a diverting type that channels one fluid into two directions. As the purpose of these valves is to mix or divide fluid flows, they normally have linear characteristics and are constructed such that the total amount of flow is constant regardless of the degree of opening (Figs. 6.9 (b) and (c)).

(*d*) *Diaphragm valves* (*Saunders valves*)

The shape of these valves is such that there is a weir in the center of the valve body and the course of the fluid is pinched off by this weir, which is opened or closed by means of a diaphragm. The course of the fluid is simple, so the lining of the interior surface of the body is simple and easily made resistant to corrosion. These valves are also appropriate for use with slurries and viscous fluids. The inherent flow characteristic is nearly linear. Control is not good when the opening is small and the rangeability is about 10 : 1 or 15 : 1(Fig. 6.9 (d)).

(*e*) *Gate valves*

The opening and closing of these valves is performed by a vertically aligned disk that partitions the flow of fluid. Because diameter of the valve inlet is the same as the inner diameter of the pipe when the valve is fully opened, pressure loss is slight. Also, it has a high shut-off function, so it is often used as a cut-off valve.

When used as a control valve, a seat ring with a V-shaped orifice is chosen, and desings to improve the inherent characteristic and range-

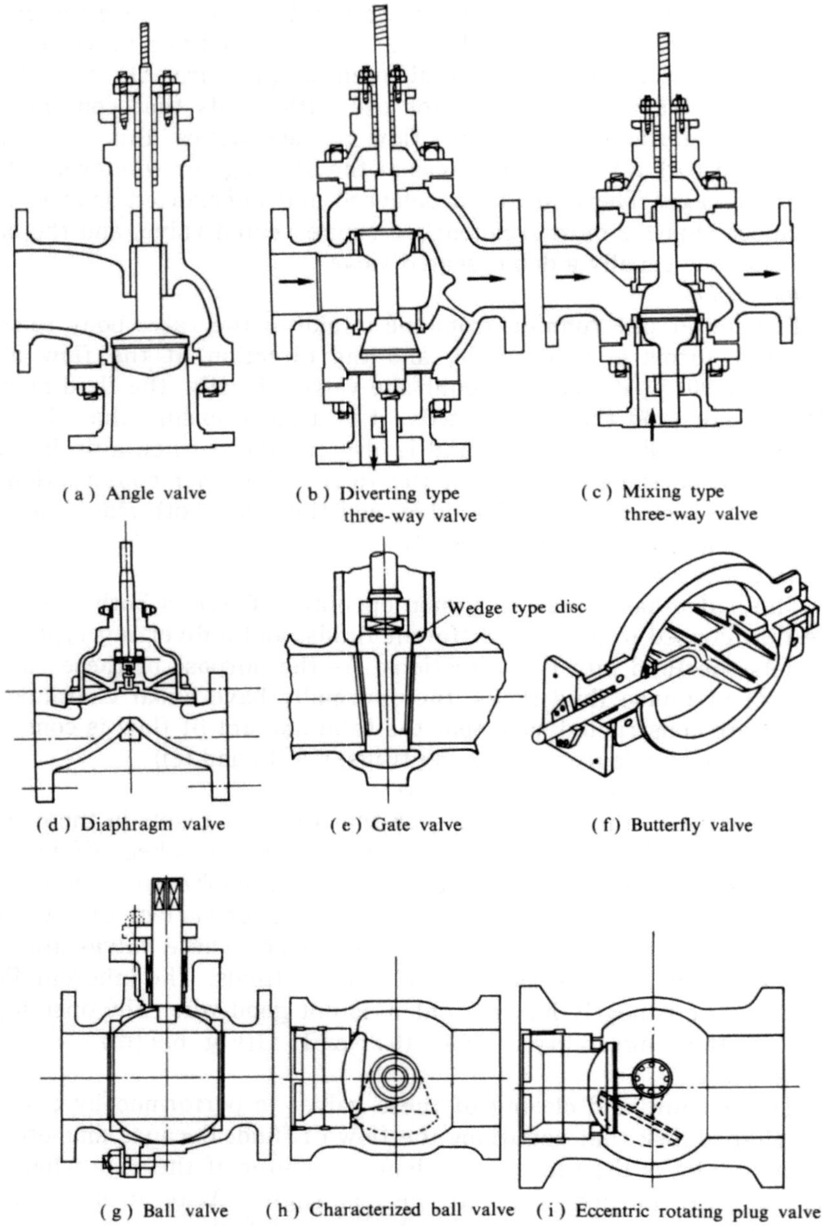

(a) Angle valve (b) Diverting type three-way valve (c) Mixing type three-way valve

(d) Diaphragm valve (e) Gate valve (f) Butterfly valve

(g) Ball valve (h) Characterized ball valve (i) Eccentric rotating plug valve

Fig. 6 · 9 Construction of vasious valves

ability have been developed (Fig. 6.9 (e)).

(*f*) *Butterfly valves*

In these valves, a disk rotates inside the valve body to open and close the valve. This construction is relatively inexpensive, and these valves are appropriate in cases where the diameter of the opening is large and when the differential pressure is low. According to the pressure of the fluid, maximum torque occurs when the valve is opend to about 70° and when this level is exceeded, the torque decreases abruptly. Thus when this valve is employed it is most commonly used at 0° to 60°. The inherent charcteristic of the valve is close to the equal-percentage characteristic. When a 0° to 60° opening is used, rangeability is sbout 20 : 1 or 30 : 1 (Fig. 6.9 (f)). Often, the butterfly valve dose not have a seat, and at full shut-off the amount of leakage is about 1% to 3%. To decrease the amount of leakage, various methods of sealing using soft materials such as rubber and Teflon® have been devised.

(*g*) *Ball valves*

In these valves, the valve body contains a ball which has an opening in it. The ball is rotated to open and close the valve. When a soft seat is used, the shut-off function is extremely good and it is possible to achieve tight shut-off. A metal seat has superior heat resistance, but the shut-off function is inferior to a soft seat. This type of valve is often used as a cut-off valve or as an on-off valve because of its good shut-off and low resistance when completely open (Fig. 6.9 (g)). Inherent characteristic improvements have been devised for when the ball valve is used for control. Such improvements include using a partial sphere with a V-shaped notch. With this configuration, the inherent characterisitic of the valve is close to equal-percentage and the rangeability is large (100 : 1 to 300 : 1).

When used as a control valve, the ball valve is similar to the globe valve, but is used under a gentle range of temperature and pressure conditions. It is particularly appropriate for use with slurries, including fiber slurries such as pulp liquors, because there is no obstruction in the fluid course and there is shear between the disk and the seat when the valve is shut off (Fig. 6.9 (h)).

(*h*) *Eccentric rotating plug valves*

The plug of this valve opens and closes the valve as it rotates about an eccentric axis. Structurally, it is similar to the ball valve when used for control purposes as listed above, but its use of an eccentric axis is its distinctive feature. Because of the eccentric axis, the seat and plug do not come into contact except when the valve is shut off.

Improvements in the shut-off function and reduction of torque are being developed. The inherent characteristic of the valve is close to linear, so a cam positioner is sometimes used to produce an equal-percentage characteristic. The rangeability is relatively large at about 100 : 1 (Fig. 6.9 (i)).

(*i*) *High-temperature/high-pressure valves*

Generally, in the design of valves, high temperature is defined as 350°C or higher, and high pressure is defined a pressures above JIS 63 kg/cm² class (ANSI class 900). Under these conditions, cavitation and noise become problematic, and proper choice of valve structure and materials is important. Countermeasures for cavitation and noise will be touched on here.

(1) *Cavitation countermeasures* One countermeasure is to choose materials to prevent damage from cavitation. Hard materials such as Stellite® facing or SUS 440C are used. Another countermeasure is to employ a valve structure that make it difficult for cavitaion to occur. There are many cage valves suitable for this purpose from a structural perspective.

(2) *Noise countermeasures* Noise produced by a cotrol valve can be classifided into the following three types.

(1) Noise arising from mechanical vibration of the structural parts.
(2) Noise from the flow of liquid.
(3) Noise from the flow of gas.

To eliminate the causes of the first type of noise, manufacturers study the natural vibration of the parts (mainly the plug), and then design the parts so as to shift the resonance point. The use of a cage valve with a sturdy guide lessens the likelihood that such a phenomenon will occur.

In the second type of noise, caused by fluid flow, almost all the noise results from cavitation. It is necessary to consider the fluid conditions and the structure of the valve to avoid cavitation.

The third type of noise, from gas flow, is the most difficult to predict, and dependig on the situation, the noise generated can be very great. Noise is produced as gas passes through a contraction. High-speed flows create turbulence in the gas. Also, a shearing force can develop when a high-speed flow and a neighboring low-speed flow mix, producing shock waves. However, the mechanism of this effect is complex and difficult to analyze.

One way to lessen then noise produced is to first break down the high-speed flows and average the speed distribution, at the same time shifting the noise into a high frequency region which is more easily decreased. The pressure differential that occurs in the control valve is dispersed. See Fig. 6.10 for an example of this effect.

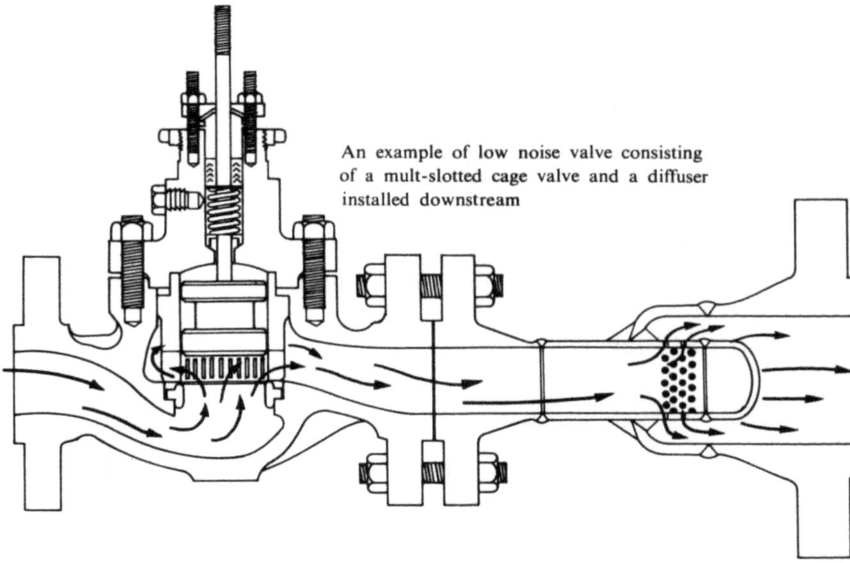

Fig. 6 · 10 Low noise valve for gas flow (Fisher controls Inc.)

(*j*) *Cryogenic valves*

These are for use with very low temperature fluids such as LNG, liquid nitrogen, and liquid oxygen. Characteristics that differ from other valves are note below.

(1) *Materials* Austenite stainless steel is used. Copper alloy or aluminum may also be used. Teflon® and Kel–F® are used as seal materials, but care must be taken with Teflon®, because of its high heat contraction rate.

(2) *Construction* The bonnet must be sufficiently long to maintain the gland packing at 0°C or above. The gap between the valve shaft and the bore is made extremely small so that the upper part of the bonnet is filled with gas. The valve body must be designed to have as small a heat capacity as possible, and for this purpose a welded connection between the valve and the piping is desirable.

(3) The interior of the valve should be carefull washed and all traces of water completely remove.

6.3.2 *Rating*

The valve body is treated as a pressure container. Consequently, given safety considerations, the allowable amount of stress at actual use temperatures is determined by the quality of the materials. However, making strength calculations and valve body designs from the pressure of the fluid each time becomes complicated and uneconomical. Moreover, compatibility with available valves is lost. For this reason, valves are given ANSI standard pressure ratings. The ratings are divided into classes ranging from 150 to 4 500. These are standardized (ANSI B 16.34) and are called basic rating pressures. Similarly, there are JIS ratings, such as temperature/pressure rating of pipe flanges (JIS B 2201). These are classified into nominal pressures from 2 K to 63 K.

6.3.3 Connection to piping

Among the various ways of installing control valves on piping and machinery are the flange method, wafer method (flange–less), welding method, and screw method. The most commonly used connection methods are listed in Table 6.3.

6.4 Control Valve Actuators

6.4.1 Conditions under which an actuator should be installed

The actuator opens and closes the valve. It receives the control signal from the controller and generates the driving power to open or close the valve with certainty. An actuator must provide the following.

(1) A driving power that is sufficient to overcome the friction in the valve shaft and thrust generated in the valve by the liquid pressure, and maintain the normal position of the valve shaft.

(2) A rigidity that overcomes the change in thrust on the valve shaft between full shut–off and full opening and maintains the stability of the valve shaft.

(3) A responsiveness that allows it to follow changes in the control signal swiftly and precisely.

6.4.2 Power sources

Actuators operate on hydraulic, pneumatic and electric power. Among these, pneumatic actuators are the most widely used for the following reasons.

(1) Compared with other types, the construction is simple and breakdowns are rare. Also, a large actuating force can be obtained.

(2) It possesses intrinsic anti–explosive properties and is easily serviced.

(3) It can be installed to operate in conjunction with long–distance electrical transmission signals by joint use with a current–to–pneumatic positioner, current–to–pneumatic converter, or electromagnetic valve.

(4) It is less expensive than other types.

However, air is compressible, so its use in high–response and

Table 6.3 Types of piping connections and use

Description		Type	Use
Flanged end	Raised face type	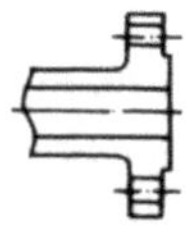	Most widely served. For below the class 600, this type is generally used.
	Flat face type	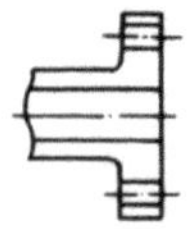	Used for cast iron valve. For low pressure such as JIS 10 K and less than the class 125.
	Tongue and groove type	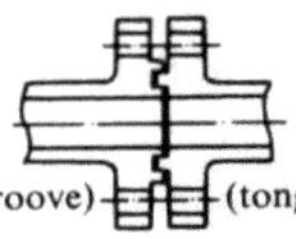 (groove) (tongue)	This type used cases of hazardous fluids and high pressure. Usually, valve mounts on groove side.
	Ring joint type	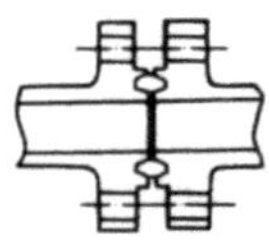	In many cases, this type is used for high pressure such as the class 900 or higher.
	Wafer type	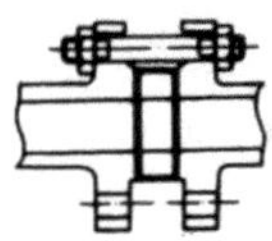	This type is often used for rotary type valves such as butterfly valves.
Welding end	Socket welding end type	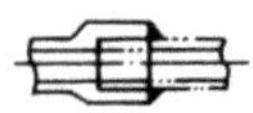	For high temperature and ihgh pressure. Used for 1½B or up to 2 B.
	Butt welding end type	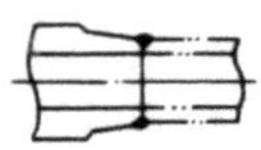	For high temperature and high pressure. Used for more than 2 B.
	Threaded end type	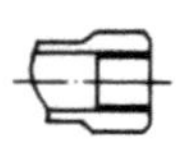	Thig type used for non-hazardous and low pressure fluids. Adopted for 1½B or up to 2B.

high–accuracy applications is limited. Thus the use of servo–motor and current–to–hydraulic models continues to increase, despite cost considerations.

6.4.3 Types of actuators and their characteristics

There are many types of actuators. These are classified according to type of valve and input signal in Table 6.4. In recent years, the rotary valve has come into wide use. There are various actuators suitable for use with this type of valve.

Table 6.4 Classification of actuator

Pneumatically-operated actuator	Linear motion Rotary motion	Diaphragm actuator Cylinder actuator Air motor actuator
Electro-mechanical actuator	Linear motion Rotary motion	Electro-mechanical actuator Servo actuator (electronic type)
Electro-hydraulic actuator	Linear motion Rotary motion	Electro-hydraulic actuator Servo actuator (hydraulic type)
Hydraulic actuator	Linear motion Rotary motion	Cylinder actuator
Pneumatic-hydraulic actuator	Linear motion Rotary motion	Servo actuator (hydraulic type)

(Note) Sometimes, spring type is called single-action type as well as springless type is called double-action type.

Representative examples of the construction of actuators and their characteristics are listed in Table 6.5 and explained below.

(a) Pneumatically–driven type

(1) *Diaphragm actuators* These are the most widely used actuators. They have simple construction and excellent movement accuracy and response. They are also highly dependable. Diaphragm actuators are divided into linear types and rotary types. Both types can be installed to provide direct action and reverse action. The constructions of these types of diaphragm actuators are shown in Figs. 6.11 and 6.12. In the linear diaphragm actuator, direct action occurs when the actuator shaft is pushed out as a result of increased air pressure; reverse action results when the actuator shaft is drawn in.

The most important parts of this actuator are diaphragm and the spring. Neoprene® rubber and ethylene–propylene rubber reinforced with fabric, such as nylon, are used in the diaphragm to heighten pressure resistant properties. A spring with stability and a high degree of

Table 6.5 Schematic diagrams of actuator and its features

Discription of operation	Output stem motion	Type	Schematic diagram	Output	Stroke	Power source	Size	Positioner	Action on power fail
Pneumatically operated actuator	Linear motion	Diaphragm actuator Spring type	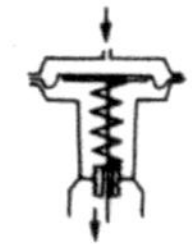	Small	Small	Pneumatic 1.2 to 4.0 kg/cm^2	Large	Required or not required	Spring controls
		Cylinder actuator Springless type	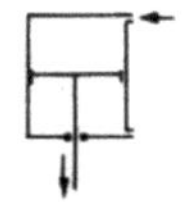	Large	Large	Pneumatic 3 to 7 kg/cm^2	Large	Required	No control
	Rotary motion (combination of linear motion valve)	Air motor actuator Springless type	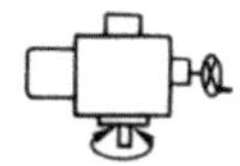	Large	Large	Pneumatic or gas 4 to 70 kg/cm^2	Large	Required	Locking up the stem position
	Rotary motion	Diaphragm actuator Spring type	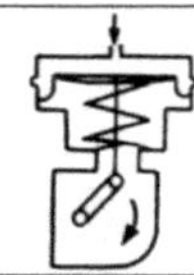	Small	60° or 90°	Pneumatic 1.2 to 4.0 kg/cm^2	Small	Required	Spring Controls
Electro-mechanical actuator	Rotary motion (Combination of linear motion valve)	Electro-mechanical actuator Springless type	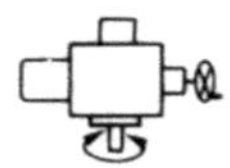	Large	Large	AC 200 V 220 V 440 V	Middle	Required	Locking up the stem position
	Linear motion	Electro-mechanical actuator Springless type		Small	Small	AC 100 V 200 V	Small	Required (built in)	Locking up the stem position

Discription of operation	Output stem motion	Type	Schematic diagram	Output	Stroke	Power source	Size	Positioner	Action on power fail
Electro-mechanical actuator	Linear motion	Servo actuator (electronic type) Springless type		Small	Small	AC 100 V	Small	Built in	Locking up the stem position
	Rotary motion	Servo actuator (electronic type) Springless type		Small	60° or 90°	AC 100 V	Small	Built in	Locking up the stem position
Electro-hydraulic actuator	Linear motion	Electro-hydraulic actuator Springless type	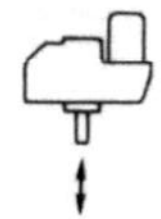	Middle	Small	AC 100 V 200 V	Middle	Required	No control (fixed)
	Linear motion	Electro-hydraulic actuator Spring type	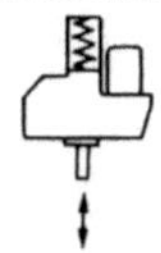	Small	Small	AC 100 V 200 V	Middle	Required	Spring controls
Hydraulic actuator	Linear motion	Cylinder actuator Springless type	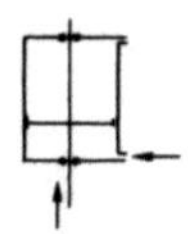	Large	Large	Hydraulic 10 to 210 kg/cm²	small	Required	No control
	Rotary motion	Cylinder actuator Spring type	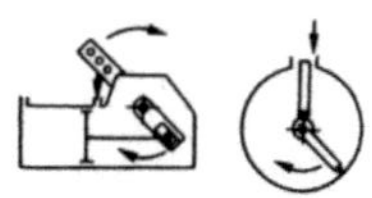	Large	60° or 90°	Hydraulic 10 to 210 kg/cm²	Middle	Required	No control

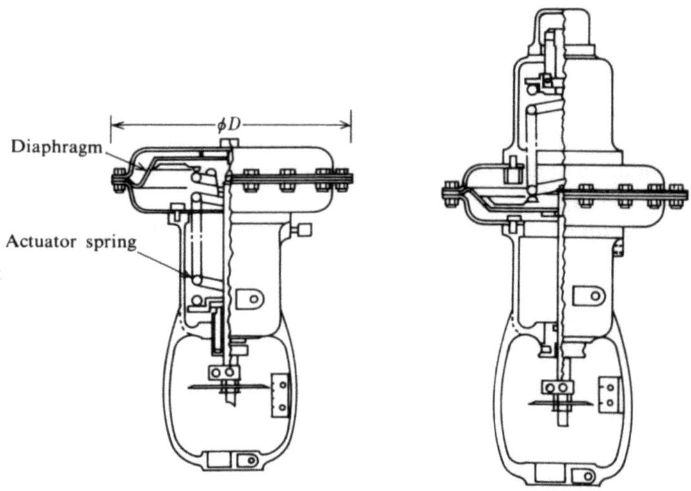

Direct action type (DA) Reverse action type (RA)

Fig. 6 · 11 Linear motion type diaphragm actuator

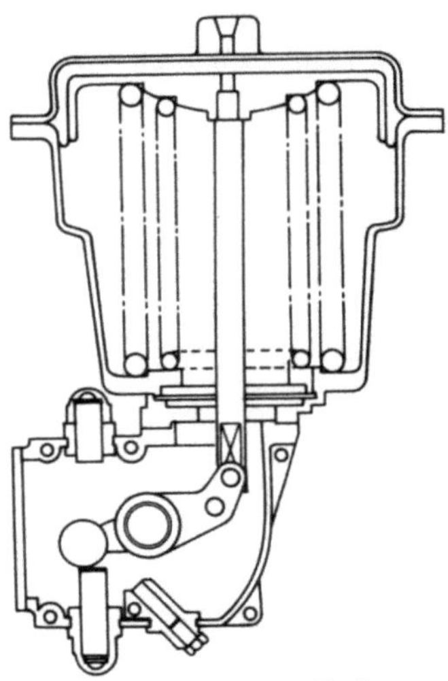

Fig. 6 · 12 Rotary motion type diaphragm actuator

accuracy is needed.

The output power of the actuator is determined by the effective area of the diaphragm and the operative pressure as well as by the spring. The range of the spring is usually 0.2 kg/cm^{2}G to 1.0 kg/cm^{2}G or 0.4 kg/cm^{2}G to 2.0 kg/cm^{2}G, but lately actuators with higher output power in special ranges such as 0.8 kg/cm^{2}G to 2.4 kg/cm^{2}G, and

1.6kg/cm^{2}G to 3.5 kg/cm^{2}G are being produced. Actuators with outer diameters (D) from about 150 mmϕ to 650 mmϕ are available.

When the signal pressure (operative pneumatic pressure) enters the diaphragm case, the air pressure causes the diaphragm to move and compresses the spring. As a result the output power shaft is held in place by a balance between the pneumatic pressure and the opposing pressure of the spring.

(2) *Cylinder actuators* These actuators have poorer response than the diaphragm actuators. However, they have many advantages such as being usable with high operating pressures and achieving high output power with a small actuator. They are used particularly often with rotary valves, on-off valves or large-port valves. The output power shaft is either a linear motion type which moves up and down, or a rotating type. Both types are available as spring or springless models. An example of a spring type is shown in Fig. 6.13. The operating pressure of cylinder actuators is generally above 4 kg/cm^{2}G, and a positioner is required to achieve proportional movement.

The output power of the springless type is far greater than the spring type. It is also more responsive to large strokes. However, if loss of air occurs it looses orientation. Thus, when used with cut-off valves and so on, a volume tank and lock valve should be provided, and

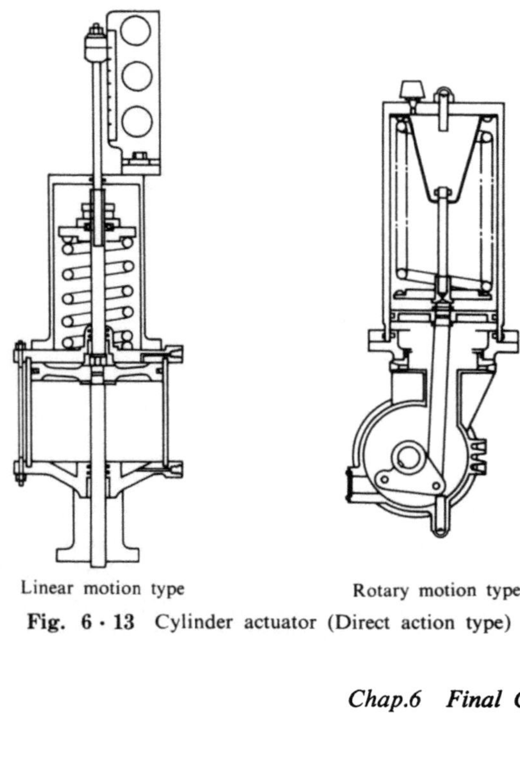

Linear motion type   Rotary motion type

Fig. 6 · 13 Cylinder actuator (Direct action type)

the valve should operate in the safe direction or should be locked in the present position.

(*b*) *Electric actuators*

The merits of the electric actuator include procurement of a power supply is easy, a strong operating force is obtained, and there is no lag in signal transmission. On the other hand, their construction is more complex, and explosion-proofing is need in environments where explosion can occur. They are also more expensive than the pneumatic types. Making the most of their merits, electric actuators are widely used for valves with large ports such as oil tank source valves and water cut-off valves for waterworks and sewer systems, or when no pneumatic power source is available. A current-to-current power source is built in, and devices that use a servo-motor to improve the control accuracy are being developed. The range of uses for electric actuators is increasing.

(1) *Electric motor type* These actuators are further divided according to their construction into those for use with valves having small-bore intakes and those for use with large-bore intake valves. The smaller actuators are driven by commercial 100V AC power supplies. The rotation of the motor is transmitted to the output shaft through gears. As with globe valves, linear valves use an added linking mechanism to convert the rotational motion of the output shaft into a linear motion. An exterior view of a small motor-drive valve is shown in Fig. 6.14. There is also a built-in current-to-current positioner type that operates by a 4 mA to 20 mA DC input signal. The larger models are driven by a 200V AC or 400V AC three-phase power supply. An example of a large motor-drive valve is shown in Fig. 6.15.

Fig. 6 · 14 Small type motor-drive valve

Fig. 6 • 15 Large type motor-drive valve

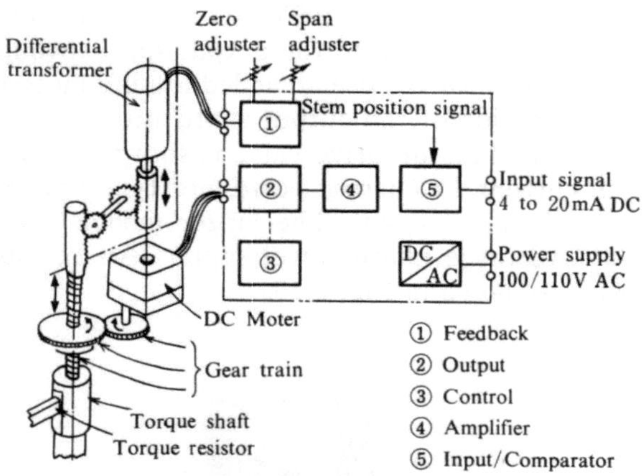

Fig. 6 • 16 Construction of servomotor type

(2) *Servo-actuators* This type is used with relatively small bore valves. Although it does not have a very large output, it is highly accurate and has good response as it most useful merit. Figure 6.16 shows an example of its construction.

A DC input signal of 4 mA to 20 mA drives the motor in a direction such that the difference between the input signal and the degree of valve opening is eliminated. The rotation of the motor moves the trapezoidal thread of the gear train up and down. This action provides feedback to a differential transformer that detects the degree of valve opening. The result is that the output shaft position is proportional to the input signal. In cases where rotating motion is required, rotation is accomplished by a worm-to-worm gear assembly.

(*c*) *Electro-hydraulic actuators*

The electro-hydraulic type combines the hydraulic pump and hydraulic equipment into a single unit. It has the following merits.

(1) It has high rigidity and operating accuracy due to use of a non-compressible fluid.

(2) The hydraulic part is completely packaged, and the electric power supply is the only input, thus it is easily handled.

On the other hand, construction is complex, and as with the electric motor types, an explosion-proof construction is required in environments where explosion is a possibility. A linear motion spring model is shown in Fig. 6.17, which illustrates both the hydraulic pump and each hydraulic device.

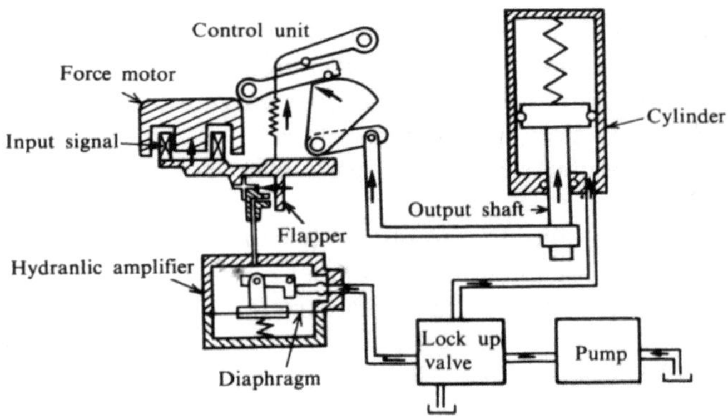

Fig. 6 · 17 Direct action spring type (Single-motion, reverse-action type)

Each hydraulic circuit is connected inside the container, which is constructed to prevent oil from leaking out. In this example, a nozzle flapper mechanism is used with the positioner, and excellent response is obtained.

(*d*) *Hydraulic actuators*

This type is used at high hydraulic fluid pressures of 10 kg/cm^2G to 210 kg/cm^2G. Consequently, a large output can be obtained from a relatively small unit. However, unlike the electro-hydraulic actuators, a hydraulic power source and high-pressure piping are necessary. The piping circuit is constructed to prevent oil from leaking out. After movement, the hydraulic fluid is returned to the hydraulic source.

6.5 *Positioners and Accessories*

6.5.1 *Positioner functions*

Positioners are used to allow the final control element to fully perform its function in the control loop. They maintain a precise equality between the signal from the cotroller and the position of the final control element drive shaft, resisting external forces. As they are installed in a direct control valve, they can be influenced by vibration in the piping, temperature, environmental gases, etc. Thus they must have a solid construction. They are used in the following cases.

(1) The differential pressure of the valve is great and the effects of change in fluid pressure are strong.

(2) Friction with the gland packing is great and hysteresis and linearity become bad.

(3) The controller signal and actuator signal differ.

(4) The distance between the controller and control valve is large and response is slow (for pneumatic signals).

(5) The intake diameter of the control valve exceeds 100 mm, and the response is slow due to heavy load.

(6) Two or more control valve operate in a split range in reponse to one signal.

(7) The operation signal is amplified because a large operating power is required (for pneumatic signals).

(8) The pressure of the fluid has a strong effect, as with butterfly valves.

(9) As with Saunders valves, it is necessary to improve the inherent characteristics of the valve to obtain the required characteristics.

6.5.2 *Pneumatic pressure positioners*

These positioners move upon receiving a pneumatic signal such as from a pneumatic controller. They are used particularly in high-temperature applications and as double-acting positioners in combination with double-acting cylinders that required a large operating pressure. The operating principle of pneumatic positioners is illustrated in Fig. 6.18

6.5.3 *Current-to-pneumatic positioners*

Most currently-used control loops consist of combinations of electronic controllers and pneumatic control valves. Current-to-pneumatic positioners are used to receive the electric current signal and move the pneumatic final control element. A torque motor or force motor that employs a permanent magnet and coil is used to receive the signal in

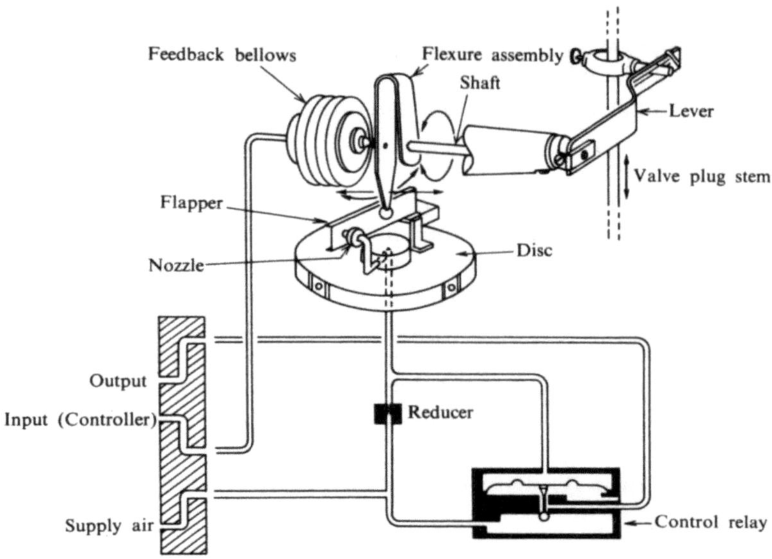

Fig. 6 · 18 Operational diagram of pneumatic positioner

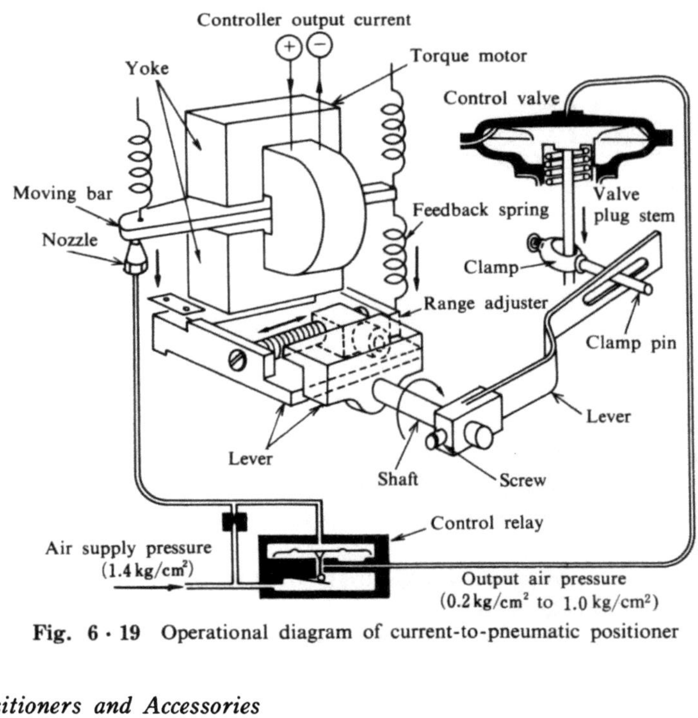

Fig. 6 · 19 Operational diagram of current-to-pneumatic positioner place of the metal bellows used to receive a pneumatic signal. Fig. 6.19 illustrates the operating principle of a force–balance positioner with a torque motor. This positioner is equipped with a manual–automatic mode switch, which allows manual operation in case of failure in the signal channel.

6.5.4 Current–to–current positioners

These are used to improve the control of motor–drive valves. The operating principle is illustrated in Fig. 6.20. The positioner receives feedback signal x_l that is proportional to the degree of valve opening from the potentiometer that works with the motor–drive valve. It compares x_l and the valve control signal x_S with the two independent comparators A_1 and A_2. The valve drive motor then rotates one way or the other according to the #1 and #2 relay contact signals that are driven by A_1 and A_2. The positioner controls the system so that x_l and x_S coincide within the limits of the setpoint neutral zone.

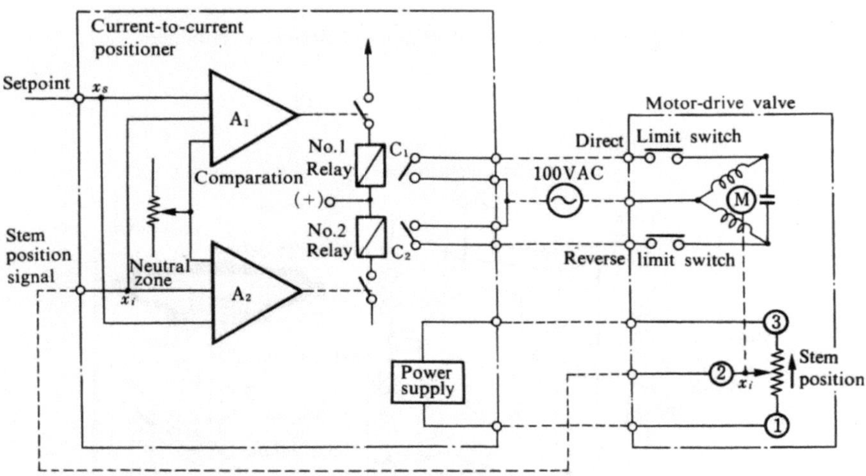

Fig. 6 · 20 Operational diagram of current-to-current positioner

6.5.5 Accessories

Aside from positioners, there are accessories such as volume boosters and speed controllers that improve the controllability of a control valve. The main types of accessories are described here.

(a) Volume boosters

These act to increase the quantity of air supplied and improve response. When the capacity of the control valve actuator is large, and when the signal piping is long, the volume booster is used to increase the operating speed of the control valve. However, because this alone cannot improve the static qualities of the control valve, volume boosters are often used together with a positioner.

(*b*) *speed controllers*

These constrict the flow of air and so control the speed of movement. The structure comprises a check valve and a variable restriction, and works by constricting the flow in one direction only.

(*c*) *Lock valves*

Lock valves are used to stop or divert the pneumatic circuit immediately in front of the actuator, thus forcing the control valve to move in the safe direction in the event of air supply failure. When air pressure is restored, the lock valve automatically returns the system to its original state.

(*d*) *Electro–magnetic valves*

These are used when it is necessary to switch the pneumatic circuit by means of an electrical signal. They are often used for the switching of on–off valves.

In addition to the accessories listed above, there are others such as air sets, pneumatic switching valves, volume tanks, and limit switches that are used as needed. Examples of the various types of accessories combined with actuators are shown in Fig. 6.21

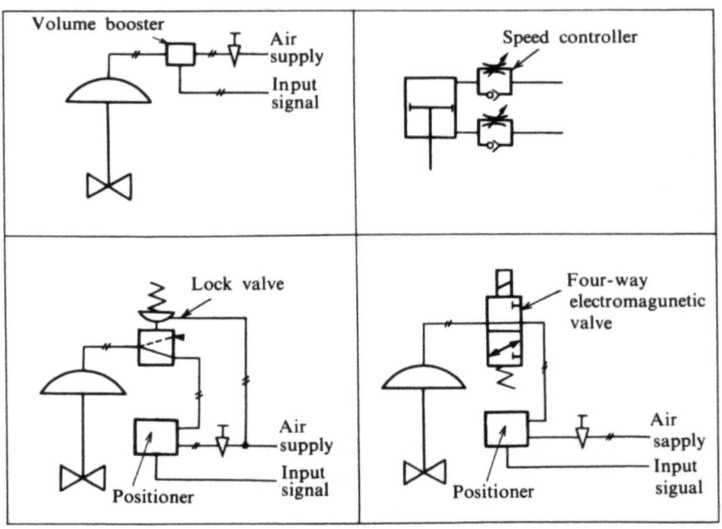

Fig. 6 · 21 Examples for using of control valve accesseries

6.6 *Self-powered Valves*

Mainstream process instrumentation systems are centralized, with the information center located in a central control room. Final control elements perform their functions according to commands from this control room. However, to reduce instrumentation costs, construction time, and maintenance costs, self-powered valves are still used in installations where supplementary power is unnecessary. The self-powered valve receives its energy directly from the controlled system, so the amount of control is limited. However, the structure is simple and economical, and when used appropriately the results are very good.

6.6.1 *Pressure-regulating valves*

These are generally called pressure-reducing valves, and are normally used to maintain stable secondary pressure without affecting the primary pressure. There are both direct-action types and pilot types (Fig. 6.22). In the direct-action type, a diaphragm lies between a spring and the opposing secondary pressure. Because the valve opens and closes according to the differential pressure, these valves are used with liquids or smallbore applications.

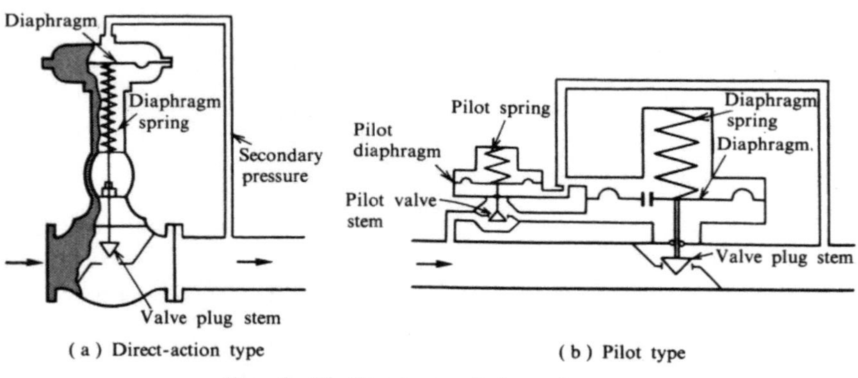

(a) Direct-action type (b) Pilot type

Fig. 6 · 22 Pressure regulating valve

In the pilot type valve, the pilot valve, which is actuated by a diaphragm, adjusts the primary pressure applied to a piston. Thus although the offset is small compared to a direct-action valve, this type of valve has the demerit of not operating when the differential pressure falls below a certain level.

In addition to these two types, there is also a fixed differential pressure control valve that maintains a fixed differential pressure between two flows. These valves are used for the atomizing of fuels.

6.6.2 Temperature control valves

This is a self-powered valve that controls the flow of steam or cold water to maintain a fixed process temperature. Although there are many ways of obtaining operating power from the heat, such as the steam pressure method, liquid expansion method, and gas expansion method, the steam pressure method is the most widely used.

In the vapor pressure method, the valve is controlled by the generated vapor pressure of the liquid that is contained in the temperature sensing section. The temperature is set by adjusting a spring in the same manner as the pressure-reducing valve.

As with the pressure-reducing valve, there are both direct-action and pilot types. The liquid in the temperature sensing section is ethylene chloride, acetone, alcohol, or some other liquid, chosen according to the temperature range. It is necessary to confirm that problems will not arise if the liquid leaks into the process fluid.

6.6.3 Flow control valves

The flow control valve combines a constriction such as an orifice with a fixed differential pressure valve such that the differential pressure created as the fluid that passes through the constriciton is maintained at a fixed level.

The flow rate is set by turning a handle which varies the aperture of the constriction. These valves are used mainly for heavy oil and light oil burners.

6.6.4 Float valves

These widely used valves are operated by the buoyancy of a float resting on the surface of the liquid. The buoyancy is amplified and transmitted to the valve by levers (Fig. 6.23). The size of the float and the leverage of the mechanism must be adjusted according to the valve size and the differential pressure.

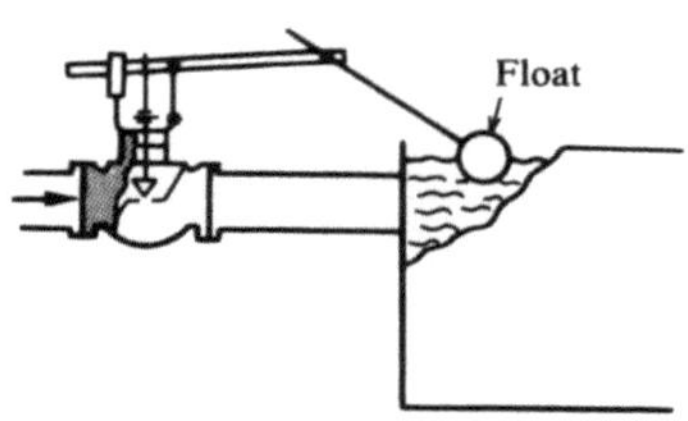

Fig. 6 · 23 Float valve

PRACTICE QUESTIONS

6.1 Choose the most appropriate valve size for the following application from the three possibilities listed below.
This valve regulates the flow of water. The maximum flow rate is 55 m^3/h, at which time the pressure conditions are $P_1=7$ kg/cm²G and $P_2=5$ kg/cm²G. The inherent characteristic of the valve is equal-percentage.

1. 50A rating $C_v=48$
2. 65A rating $C_v=72$
3. 80A rating $C_v=100$

6.2 Determine the required value for the following conditions.
Fluid: air
Flow rate: 2 000 Nm^3/h
Pressure: $P_1=5$ kg/cm²G, $P_2=3$ kg/cm²G
Temperature: 20°C

ANSWERS TO QUESTIOSNS

6.1 The required value of C_v is 45.5(from Ep. (6.1)). Normally, the control valve size is selected such that the degree of valve opening at maximum flow is 80% or less, allowing for various types of error and ease of use as a control valve. here, from Fig. 6.5, taking the vertical axis as flow rate = C_v, the approximate valve opening is obtained.

50A	C_v = 95(%),	valve opening = 95 to 100(%)
65A	C_v = 63(%),	valve opening = 80 to 90(%)
80A	C_v = 45.5(%),	valve opening = 75(%) (approx.)

Accordingly, in this example, item 3, 80A, the valve for which C_v = 100(%) is selected.

6.2 As $\Delta P<0.5P_1$ the calculation is performed using Eq. (6.2).

$$C_v=\frac{2\,000}{273}\sqrt{\frac{1\times(273+20)}{2(6.033+4.033)}}=27.9$$

REFERENCES

1) JIP Standard: JIP-7B-64-82 (in Japanese)
2) J. W. Hutchison: ISA handbook of contnal valve, 2nd Editon, ISA (1976)
3) T. Ishii: Process control Instrumentation Systems, Denki Shoin (1973) (in Japanese)
4) C. S. Beard: Final control elements, Vol.2, ISA (1973)

Chapter 7

SYSTEM ENGINEERING

In a wide range of fields today, the words "system" and "engineering" have become so popular that they are now felt by many to be overly used and ambiguous terms. In particular, on hearing the expression "system engineering," it would be difficult even to guess the actual work or process in question. This lack of clarity, however, can be somewhat alleviated if we take a look at the basic meanings of these two words. The word "system" originates from the Greek word meaning "to place together," and came to mean the coexistence of a variety of diverse things (*e.g.*, men, objects, money, information) utilized in a holistic way so as to achieve a new function independent of the composite parts. In addition, the word "engineering," apart from the modern connotations associated with Mechanical Engineering, Civil Engineering, and the like, also has the fundamental meaning of careful planning and efficient management.

In the light of the above, we define instrumentation engineering as design technology applied to the systemization of instrumentation facilities in order to achieve the objectives of a certain industrial plant or process. In this regard, if we view the instrumentation manufacturer as a specialist providing measurement and control technology and the user (customer) and plant builder as the source of process and plant design know-how, all objectives and functions must be clearly presented so that all parties concerned can together complete the project in question. This chapter describes, from the viewpoint of the instrumentation manufacturer, the elements of instrumentation engineering from the original planning stage up to installation and adjustment.

To begin with, in relation to basic system engineering, we describe an outline for plant construction as well as how project tasks should be divided amongst the user, plant builder and instrumentation manufacturer. We then explain the work involved at the "instrumentation system design" stage in which specification clarification meetings are held with the customer in preparation of actual instrumentation system construction. Also described is the work involved from this stage up to final test runs and turnover from the viewpoint of basic design and engineering. Items related to common technology such as instrumentation power sources and safety measures are covered as well.

Following the above is a description of "human engineering" factors as related to the design of a control room (operations room), which is continuing to increase in importance as the focal point of plant operations. Next, the planning and design of instrumentation work such as related to instrumentation installation, piping and wiring, and system startup (pre-commissioning) are described. Finally, the methodology behind a "design review" is introduced as an effective means of insuring product quality.

7.1 System Engineering Basics

7.1.1 Plant construction overview

In the process of constructing a plant, not only is the cooperation of the end user necessary but that of specialists as well, such as those in the chemical, mechanical, metal, civil, electrical and instrumentation engineering fields. Moreover, in addition to the primary engineering problems involved with building a plant, extensive consideration must also be given to other influential factors such as local environmental conditions and required manpower and living provisions during plant construction. A flow chart for plant construction is shown in Fig. 7.1.

(*a*) *Feasibility study* (*specifications evaluation*)

Cost studies: All factors contributing to the overall cost of the project are investigated, such as product marketability, raw material acquisition, site conditions, transport facilities, general construction costs, general production and sales costs, manpower costs and money flow planning.

Technological studies: Investigations are performed as to whether or not current technology may be used, and if not, whether development of new technology is feasible. In addition, other technologically-related factors such as licensing and the advantages and disadvantages of particular processes are reviewed.

Environmental studies: From the viewpoint of public nuisance, the possible affects of liquid waste, noise pollution, offensive odors and radiation are investigated, and in general, the possibility of any negative influences on society during both normal and abnormal plant operations is evaluated.

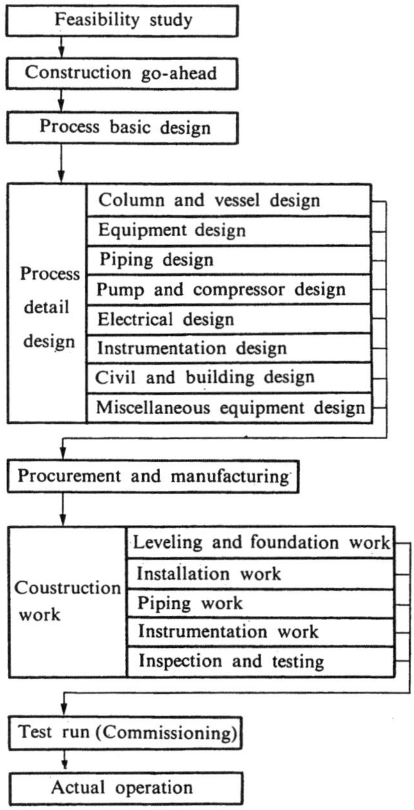

Fig. 7·1 Plant construction flow chart

(*b*) *Construction go-ahead*

On the basis of the above feasibility study, a decision is made as to whether construction planning shall proceed or not.

(*c*) *Process basis design*

At this stage, basic plant configuration and process data needed for plant construction are determined, as listed below.

(1) Verification of design conditions
(2) Material input and output, heat input and output
(3) Plant operation method, safety control measures
(4) Floor plan and basic structural specifications
(5) Process flow diagram
(6) P & I (Piping and Instruments) diagram
(7) Process layout chart
(8) Utility plans (electric power, water, heat, etc.)
(9) Basic specifications for major equipment (pumps,compressors,

blowers, heaters, heat exchangers, columns, etc.)

(10) Basic piping specifications (materials, conduit paths, etc.)

(11) Basic specifications for instrumentation and electrical systems

(*d*) *Detailed design*

In detailed design, which is also called manufacturing design, specification documents, drawings, etc. are prepared in a detailed manner on the basis of the above basic design. This will involve equipment design (*e.g.*, columns, tanks, pumps, blowers, etc.), piping design, electrical design, instrumentation design, and civil/floor plan design, for example.

Table 7.1 shows the allocation of engineering work from the feasibility study up to detailed design, as seen from the viewpoint of the instrumentation manufacturer.

(*e*) *Device procurement/manufacturing*

At this stage, each device specified in the above detailed design is either procured or manufactured. In addition, each device unit must eventually undergo inspection to verify that it meets specifications and achieves required performance.

(*f*) *Plant construction*

Actual plant construction is divided into the following steps: leveling and foundation work, major equipment installation, structural construction, piping, electrical wiring, instrumentation, heat insulation work, and painting.

Although the need for overall management of the plant construction process is obvious, note that due to the complicated relationships between the various construction steps in a modern plant as well as the detailed planning involved in each step, a powerful management technique should be employed such as a network analysis method or PERT (Program Evaluation and Review Technique).

(*g*) *Test run/system delivery*

After all construction work is completed, system evaluation begins on a partial basis by first checking the performance of individual devices. Next, as devices are adjusted, the range of evaluation broadens, and the performance between multiple devices is evaluated. Eventually, the entire process comes under scrutinization. In the final stage of testing, a so called "water–run" is performed in which overall performance is evaluated by substituting water or air for actual materials. Finally, an actual test run (commissioning) is performed before handing over the system to the end user. The above tests and evaluations should be performed with representatives from the end–user, plant builder, and the various electrical and instrumentation equipment manufacturers present.

7.1.2 System design considerations

(a) Clarifying required product quality

In order to facilitate the design and engineering of the plant in question, probably the most important consideration is product quality. In particular, the product quality required by the customer (plant user) must be accurately and fully understood so that design work can proceed in the right direction. Although the required quality is initially conveyed in the form of specification sheets and technical meetings, the descriptions or explanations may turn out to be insufficient or unclear and thus will need to be redefined. Accordingly, actual requirements must be thoroughly researched and specified as accurately and realistically as possible. In this regard, it is convenient to make use of quality development techniques such as that depicted by the diagram in Fig. 7.2. As shown, product quality passes through several stages of development where the quality in each stage is subject to problem analysis in relation to design and technical capabilities. In other words, quality development starts from required quality development, breaking down each stage step by step as quality function development and design function development. It eventually will be developed to a level at which target product quality can be realized in products.

(b) Incorporating new technology

Once the go–ahead has been given on a particular project, it is rare that all aspects of the technology to be employed (*e.g.*, theory, design, manufacturing methods) are fully understood. There are usually some elements of a project which involve new mechanisms or procedures (unknown technology), or at the least, known technology applied to new, untried conditions. Known and unknown elements in the technology to be applied must be thoroughly investigated and evaluated in a stratified fashion.

In particular, development of stratified unknown technology is carried out once the new technology required has been clarified by the quality development procedure explained above. However, in the case of projects such as plant building, *i.e.*, one–shot projects, which must be met within a limited cost range and time period, some trade–offs must be made between functionality, expense and delivery date. Thus, in the same way as described above, depending on the progress in design work, items requiring new development must be closely scrutinized in an organized manner and efforts made to minimize unknown technology by applying or combining with known techniques. Despite such efforts, however, verification and modification under actual plant conditions are still required. This development concept is illustrated in Fig. 7.3.

Table 7.1 Allocation of engineering work

	Work item	A	B	C	D	E	F	Description
End user	Proposal of basic specifications and operation techniques							A; Working scope desided by end user independently or by discussion with plant builder
	1. Production facility data	○						B; Working scope decided by plant builder independently or by discussion with end user
	2. Operation technique	○						C; Working scope decided by plant builder based on consultation with instrument manufacturer
	3. Initial investment and running cost	○						D; Working scope decided by plant builder depending on the proposal from instrument manufacturer
	4. Schedule	○						E; Working scope decided by instrument manufacturer based on discussion with plant builder or end user
	5. Site conditions	○						F; Working scope decided by instrument manufacturers independently
	6. Miscellaneous	○						
Plant builder	Basic design		○					
	1. Material and energy balance investigation							
	2. Operation technique and safety management investigation		○					Proposal for CRT displays and mathematical model
	3. P&I and schedule preparation		○					
	4. Equipment and machine layout preparation		○					
	5. Instrumentation system design			○				Proposal for instrument selection (computers, micro-computers, devices, etc.)
	6. Establishment of management data processing system and process control computer systems			○				Examination and proposal for optimal system based on items 1 and 2.

	7. Instrument specifications and data sheets		○			Preparation of specifications based on the basic data of items 2, 3, and 4 etc.
	8. Control panels and sequence logic specifications		○			Preparation of specifications based on the basic data of items 2, 3, and 4 etc.
	9. Process control computer system specifications		○			Preparation of specifications based on the basic data of items 1, 2, and 6 etc.
	10. Instrumentation works specifications		○			Preparation of specifications based on the documents of items 3, 4, 5, 7, and 8 etc.
	11. Manuals for over–all adjustment and delivery	○		◌		Drafting of overview manual related to the instrumentation system
	12. Manuals for operation and maintenance	○		◌		Drafting of overview manual related to the instrumentation system
	13. Manuals for routine inspections and emergency maintenance	○		◌		Drafting of overview manual related to the instrumentation system
	14. Miscellaneous	○		◌		
Instrument manufacturer	Instruments manufacturing design					
	1. Flow sheets (loop drawings of instrumentation flow)			○	○	
	2. Determination of instrument functional specifications			○	○	Preparation of loop diagrams, work sheets and data sheets
	3. Control panel design			○	○	CRT display and graphic display drawings
	4. Sequence logic design			○	○	Preparation of sequence tables
	5. Control panel internal power system design			○	○	
	6. Control panel internal connection diagram			○	○	
	7. Design of instrumentation work drawing			○	○	
	8. Miscellaneous			○	○	

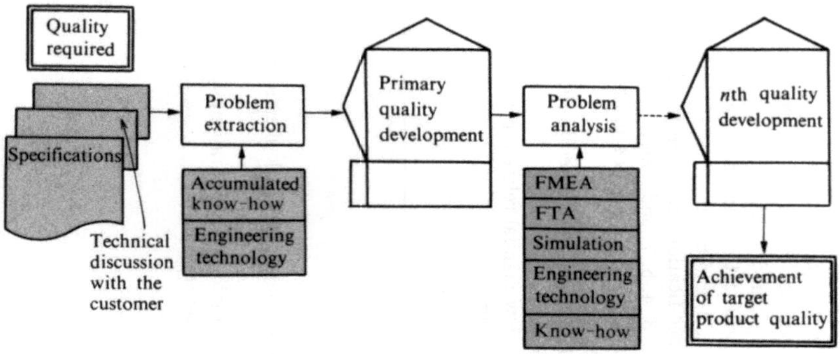

Fig. 7・2 Outline of quality development

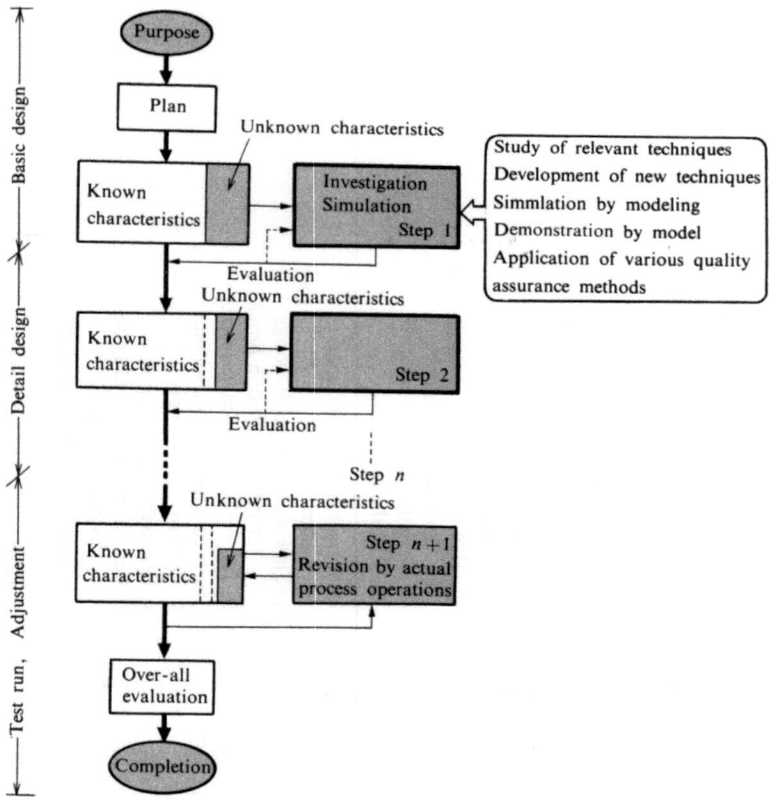

Fig. 7・3 System techniques for advancing the project

7.2 Instrumentation System Design

7.2.1 Job planning

In this stage the outline of the job is first clarified, after which a job organization is set up, a basic schedule established, and specifications narrowed down. During this time, all unclear points should be clarified to avoid any confusion in the job, keeping in mind at all times the slogan QCD (Quality, Cost, Delivery). The work flow is shown in Fig. 7.4. The relationship between basic design and the work overseen by the project leader is shown in Fig. 7.5.

(*a*) *Work preparations*

(1) *Job clarification* All relevant specification sheets and documents are thoroughly studied to ascertain the scale of the job, the time required, and the level of difficulty, and to help management setup methods for dealing with the job, the target quality, probable cost (man-hours), etc.

(2) *Documentation exchange* Specification sheets and documents to be supplied from the customer and those to be supplied from the instrumentation manufacturer to the customer are identified and listed. Examples of such documentation are described below.

(i) *Documentation from the customer*

(1) Purchase order (copy)
(2) Specification sheets (instrument list; instrument specifications; process data sheets; functional specifications; specifications of procured items, etc.)
(3) System specification sheets (system outline; control panel specifications; sequence specifications; P & I diagram; process flow diagram; utility specifications, etc.)
(4) Computer specifications
(5) Instrumentation work specifications (factory layout diagram; facility layout; control room plans; piping diagram; duct/pit diagram; drive system documentation, etc.)
(6) Miscellaneous (customer established standards; plant building schedule; safety standards; security standards; job organization chart, etc.)

(ii) *Documentation for the customer*

(1) Project quotation
(2) Specification sheets (instrument list; instrument specifications; instrument external view diagrams, etc.)
(3) System specifications (system configuration diagram; comprehensive flow chart; panel diagrams; functional description manuals; technical description manuals, etc.)

	Work flow	Description
1.	Narrowing down of work preparations	Understanding job outline, Establishment of a work organization, Preparation for narrowing down specifications
2.	Narrowing down of specifications (Over-all)	Items to be discussed with the customer
2-1).	Narrowing down of specifications (Basic items)	Project organization (Customer and manufacturer), Scope of work and schedule
2-2).	Narrowing down of specifications (Instrumentation in general)	Instrumentation system, Factory layout, Utilities and process data
2-3).	Narrowing down of specifications (Instruments and functions)	Determination instrument and function details (Types, Ranges and flow sheets)
2-4).	Narrowing down of specifications (Instrumentation work)	Instrumentation work (Materials, Wiring or piping paths, Scope of work, Organization, Schedule, etc.)
3.	Coordination	Manufacturing instructions and communication of specification information
4.	Approval drawings	Approval drawing (Final confirmation of specifications arranged with customer)
5.	Modification	Modifications and corrections
	Manufacturing	
6.	Inspection	Inspection manual, Company inspection and inspection with witness
7.	Delivery	Delivery inspections
8.	Instrumentation work	Installation and testing
9.	Start up and test-runs (Commissioning)	Test-run manual, Training for operators and handling complaints
10.	Completion activities	As-built drawings and job summary

Fig. 7・4 Flow chart of basic design work and job management

(4) Computer specifications

(5) Miscellaneous (safety organization chart; job organization chart; minutes of technical meetings, etc.)

(3) *Overseas jobs* The following items must be kept in mind while narrowing down the specifications in the case of an overseas job.

(1) Applicable laws and regulations

(2) Environmental conditions (climate; utility concerns such as air conditioning and power source, etc.)

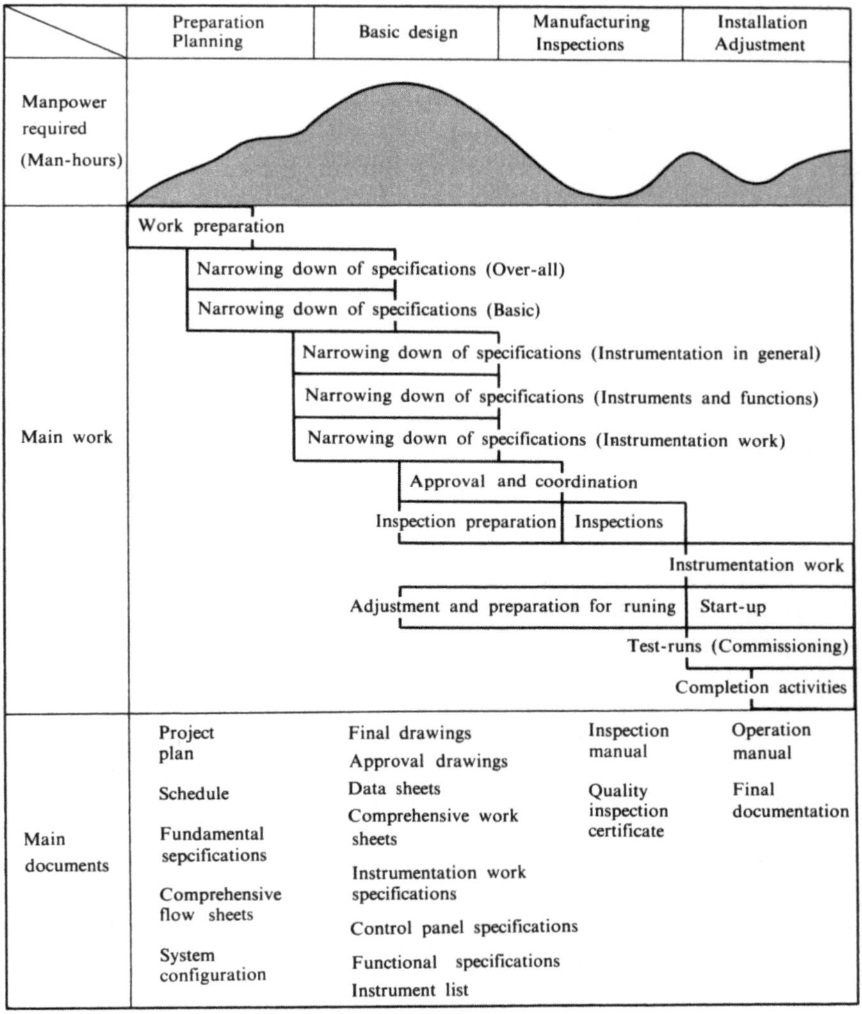

Fig. 7 · 5 Relationship between instrumentation design and job management

(3) Shipping, transport and storage conditions
(4) Documentation lists, language used and format requirements
(5) Training curriculum, language used, training place requirements
(6) Acceptance conditions, test methods

(4) *Scope of services provided* (*contract clarification*) In order to facilitate initial design work in common areas between the instrument designer/manufacturer, the end–user, and any other third party, mutual

consultation must be diligently held to clearly determine the scope of work to be performed by each party. Experience has shown that without close contact between the different groups, problems related to inadequate materials and insufficient facility tie-in at the time of instrumentation installation can easily occur at the site. Examples of areas for which the scope of responsibility must be clarified are piping interfaces, power interfaces, major equipment location, cable purchasing/layout, document preparation, and equipment transport and off-loading, etc.

(*b*) *Job organization*

The job organization required within the company (the instrument manufacturer) should be considered in relation to the scale of the job in question. In this regard, it may be necessary to contact relevant personnel or department heads to brief them on the job and obtain their feedback on organizing the job. At the same time, the organization agreed upon should be compared with the job organization of the customer and related engineering companies or manufacturers.

(*c*) *Job schedule*

Schedule charts must be prepared reflecting both the schedule of the entire project for the customer as well as that related to instrumentation design and installation. These schedules are provided to ensure that engineering work and instrument manufacturing proceed at an appropriate pace. They may be prepared with a master plan, intermediate plan, or plans of various degrees of detail as necessary using, for example, the PERT project management techniques mentioned above. For design work, in particular, it is important to include documentation scheduling in addition to that for instrumentation and installation.

It should be noted that although these schedule charts usually represent the schedule of the various departments in a time-series fashion, their possible use as proposal documents to the customer should be kept in mind when preparing them.

(*d*) *Proceedings of technical meetings*

The following describes important items which must be considered for facilitating engineering work.

Since general specifications usually exist only for the job at the time of initial negotiations, frequent meetings are required to arrive at more specific technical descriptions. As can be expected, many modifications and additions will be made to these specifications at this time. However, as technical matters usually come to overwhelm the discussions, there is a tendency to forget cost and schedule factors which are also essential to the project. All three of these factors should be kept in mind by all parties at these meetings. In this regard, efforts should be made to plan meetings carefully, *e.g.*, decide on the topics of discussion and exchange relevant documentation beforehand, so that all parties can arrive at the meetings well prepared. Minutes of each meeting should be carefully taken to provide an accurate record of what has been or has not been decided or considered to date. In this way, final specifications should be more accurate, *i.e.*, more inclusive of the three important factors: technology, cost and schedule, and should also be arrived at sooner.

7.2.2 System specifications

In this stage, specifications for the overall instrumentation system are finalized.

(a) Comprehensive flow sheet preparation

A flow sheet shows the relationship or interconnection between the various elements making up the measurement and control loops in the system. These elements are sensors, converters, receivers, controllers, final control elements, etc. Preparation of the instrumentation flow sheet concentrates on representing the system as seen from an instrumentation point of view and is based on various sources of information such as process, plant layout, and P & I diagrams. This flow sheet becomes a basis for further design work. An example of a comprehensive flow sheet is shown in Fig. 7.6. Major symbols used in flow sheets of this type are shown in Table 7.2.

(b) System configuration

A system configuration diagram (Fig. 7.7) shows the basic concept of the instrumentation system in terms of concrete system equipment such as transmitters, control panels, digital instrumentation devices and computer equipment. Accompanying this diagram will usually be a complementary manual called, for example, "System Overview," which helps to explain in more detail system concepts and makeup. This overview usually consists of the following items.

(1) *System outline description*

(2) *Operation* The man–machine interface as it relates to the operator. Information transfer and operations are described for control room operations as well as on–site instrument operations, both under normal and abnormal conditions.

(3) *Safety measures* Countermeasures such as redundant and analog backup systems are described for such problems as power and air supply failures and computer and DDC system failures.

(4) *Information management* Methods used to output and transmit operation data and to manage process data are described.

(c) Overall layout verification

Based on plant layout diagrams, the layout related to control

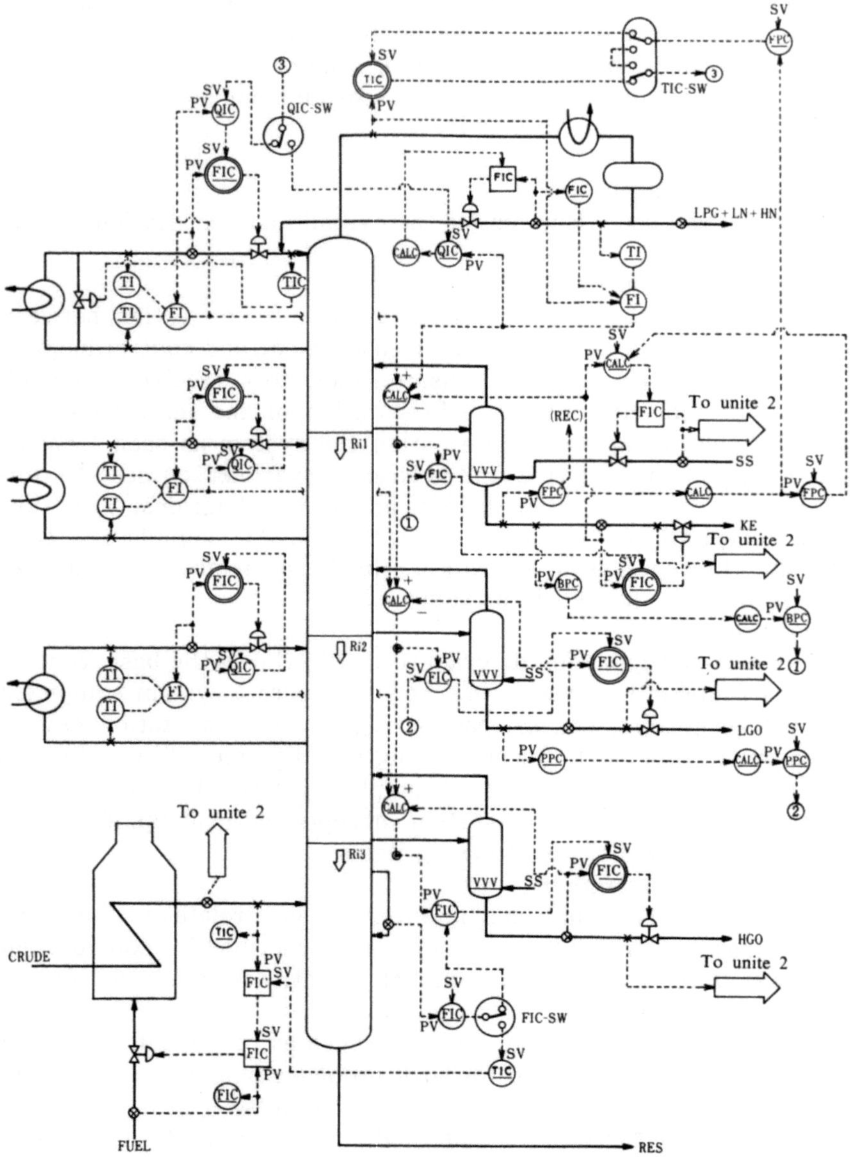

Fig. 7 · 6 Example of an instrumentation flow sheet for topping

Table 7・2 Main instrumentation symbols (Selected from JIS Z 8204)

Classfication	Graphic symbol
Electric signal	
Pneumatic-pressure signal	
Oil-hydraulic press signal	
Fine tube	
Orifice	or
Venturi tube	
Nozzle	
Area type flow meter	
Turbine type flow meter	
Volume type flow meter	
Electro magnetic flow meter	

Classfication	Graphic symbol
Valve (general)	
Angle valve	
Three-way valve	
Butterfly valve, damper louver	
Ball valve	
Valve motor-diaphragm type	
Valve motor-diaphragm type (Pressure balance type)	
Valve motor-electric motor type	
Valve actuator electromagnetic type	
Valve actuator piston type	

Classfication		Graphic symbol
Transmitter		
①		
Where the distinction of monitor and control location is required	Site	
	Control room	
	Rear panel installation	
	Local panel	
Change-over switch		

①: General case where the distinction of monitor and control location is not required

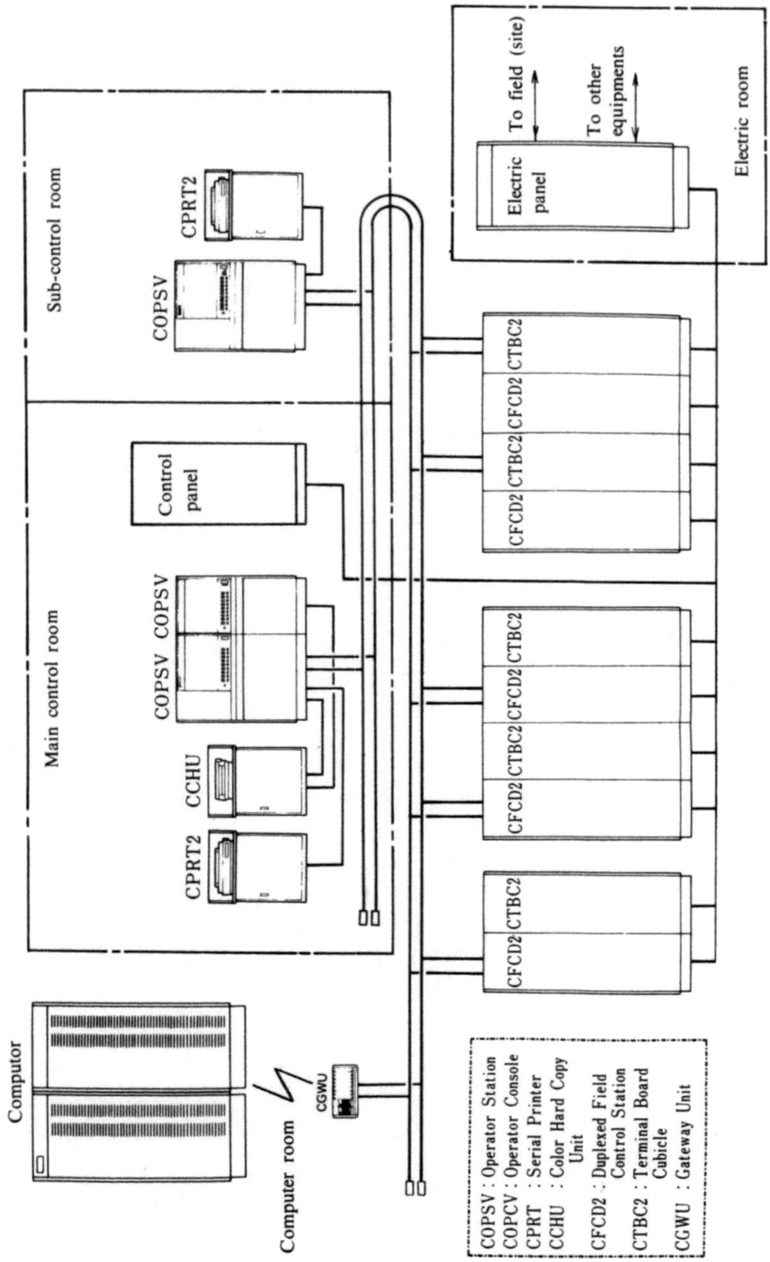

Fig. 7・7 System configuration example

rooms and field-mounted instruments and accompanying facilities, operation rooms, computer rooms, power supply rooms, etc. must be clarified. At the same time, signal transmission routes and transmission techniques should also be checked out.

(*d*) *Instrumentation layout in control rooms*

Based on the general plans for control room, computer room, power supply room and other specialized rooms, layout of control panels and other system devices is determined. This must be performed keeping in mind the functional layout for operation monitoring, as well as the need for maintenance space. Refer to Sec. 7.3 for details on the man-machine interface in control rooms.

(*e*) *Utility interconnection*

Verification of the amount, characteristics and location of the various equipment providing system utilities must be made with reference to the power system diagram, air-conditioning plans and other utility documentation. Such equipment usually includes power devices (*e.g.*, commercial power devices, uninterruptible power devices, batteries and power distribution boards, etc.), air supply devices for both factory and instrument use, air-conditioning devices, and miscellaneous devices such as for pressurized oil and N_2 gas. Refer to Subsec. 7.2.6 for details on power and air supply equipment used to drive instrumentation systems.

(*f*) *Process data*

The characteristics, data and desired control values related to process objects from the raw material stage up to the final product stage must be understood and described on a process data sheet. In addition, as these data will be used as a basis for the design of individual control loops and their driving means, it must be as accurate as possible, incorporating any changes made to date. Examples of specific process data items are pipe size, piping standards, type of fluid, temperature, pressure, flow rate, density, viscosity specifications, construction dimensions, etc.

7.2.3 *Device and function specifications*

In this stage, individual loop flow sheets are drawn up, control systems designed, and concrete specifications for devices and system functions determined.

(*a*) *Preparation of loop flow sheets and control system design*

Once the characteristics of a particular process are understood, an appropriate control system must be decided upon. In recent years, the use of microprocessors in distributed DDC (Direct Digital Control) and in small-size controllers (single loop controllers) has enabled the selection of a wide variety of control functions. These range from simple PI and PID control up to high-level "advanced" control. Refer to the appropriate chapter in this book for more details on this topic.

Given that a control system has been determined for a particular control loop, a loop flow sheet can then be prepared detailing the control and interlock system as well as the interface with other equipment. In addition, since the loop flow sheet will be used as a basic reference for instrumentation design, other information relevant to the control loop should be prepared as well at this time. Examples of such information are equipment list, equipment manufacturers, models, measurement range, custom specifications, process data (physical properties, characteristics, pipe size, other mechanical data), control calculation format, inter-loop interface, interlock conditions, and man-machine interface (*e.g.*, CRT, digital display). An example of a loop flow sheet is shown in Fig. 7.8.

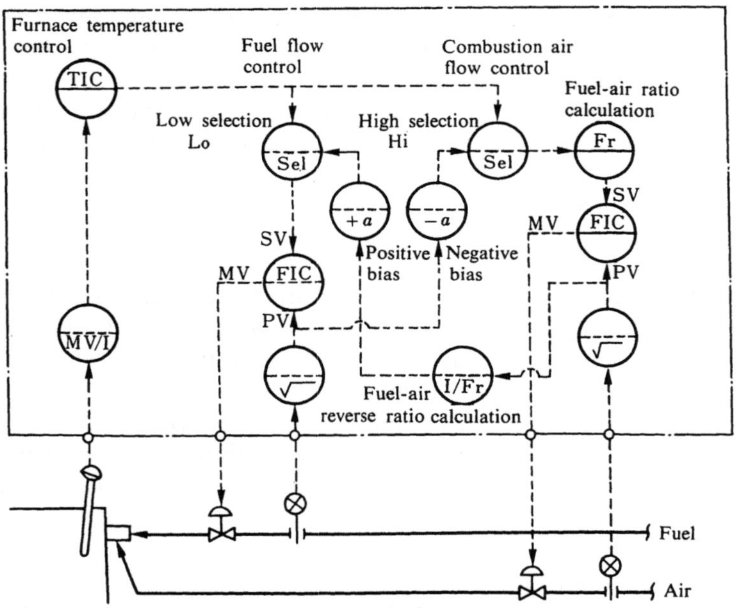

Fig. 7・8 Example of a flow sheet for anti-black smoke combustion control

(*b*) *Signal interface list*

This document lists signal characteristics, *e.g.*, destination, source, and signal type, for all signals transmitted between instruments, devices or other equipment making up the system. Signal interface is divided as follows:

(1) Signals between centralized instrumentation equipment such as control panels, digital instrumentation devices and computers, and

field devices such as transmitters and converters;

(2) Signals between various centralized instrumentation equipment; and

(3) Signals between centralized instrumentation equipment and electric, pneumatic, or other facilities.

Note that "interface list" as used in this book refers to the same type of documentation which may be called "input list," "input/output list," etc. depending on the work involved. An example of an interface list is shown in Fig. 7.9.

(*c*) *Instrument list*

Detailed information pertaining to all instruments employed is listed, such as instrument name, tag number, model, manufacturer and quantity. This document is used as process data for the design and manufacture of transmitters or as simple sizing data. An example of an instrument list is shown in Fig. 7.10.

(*d*) *Instrument sizing*

Sizing based on process data for such elements as orifices and valves must be documented. Since such elements are directly installed on process equipment and piping and may come in contact with high-temperature or corrosive process liquids, the materials used in their construction must be carefully selected. Examples of orifice and valve data sheets are shown in Figs. 7.11 and 7.12, respectively.

(*e*) *Specifications of procured instrumentation devices*

In the process of configuring an instrumentation system, it is frequently necessary to procure instrumentation devices from outside one's own company. Accordingly, purchase specifications for items to be resold must be prepared, as must prices, specification sheets and function descriptions. The necessity of additional services such as quality guarantees, user manuals, training, and maintenance must also be considered.

(*f*) *Sequence control*

In the case of implementing PA (Process Automation) or FA (Factory Automation), sequence control functions have become indispensable. Moreover, the demand for such functions has particularly been felt along with the increased automation of production facilities in recent years. For the definition and basic concepts of sequence control, refer to Sec. 2.5. This section concentrates on how to determine sequence specifications from an engineering point of view.

(1) *Overview of sequence control equipment* In order to perform sequence control, a variety of hardware is possible, such as relays, semiconductors, PCs (Programmable Controllers), distributed DDC, and computers. PCs in particular have become very popular in recent years due to the advances in microprocessors. An example of a PC configura-

Interface list

SHEET NO. 7

Signal No.	Signal name	Signal specification		Scope (plant builder side)				scope (End user side)				Inter face point	Remark
		Type	Signal form	Location	Panel name	Terminal No.	Wiring	Spec.	Terminal No.	Panel name	Location		
	Combustion stop (2)	$\frac{M}{D}$		Control room	CUNB(1)					Operation panel	LOCAL	CUNB(1)	6 points
	COMP/LOCAL Change over (3 notchs)	$\frac{S}{D}$		Control room	CUNB(1)					UD	main operation room	CUNB(1)	6 points×3
	Fuel oil shut down valve close (East)	$\frac{S}{D}$		Control room	CUNB(1)					——	LOCAL	CUNB(1)	6 points
	Fuel oil shut down valve close (West)	$\frac{S}{D}$		Control room	CUNB(1)					——	LOCAL	CUNB(1)	6 points
	Air control valve close	$\frac{S}{D}$		Control room	CUNB(1)					——	LOCAL	CUNB(1)	6 points
	Air valve Fully open	$\frac{S}{D}$		Control room	CUNB(1)					Operation desk	main operation room	CUNB(1)	6 points
	Air valve Fully closed	$\frac{S}{D}$		Control room	CUNB(1)					Operation desk	main operation room	CUNB(1)	6 points
	Auto mode dumper operation switch	$\frac{S}{D}$		Control room	CUNB(1)					Operation desk	main operation room	CUNB(1)	6 points

Fig. 7 · 9 Interface list

Instrument list | SHEET NO.

ITEM	SERVICE	TAG NO.	INSTRUMENT		MFGR	MODEL NUMBER	RANGE		SPECIFICATIONS	Q'TY	LOCA-TION	Documentation	REMARKS
2													
2-1	LD TD Molten steel temperature		(1)	mV generator	YOKOGAWA	PDC-16-1	0~10mV			1	RP		
			(2)	Cold-junction compensator	KOMATSU ELECTRONICS	ZC-001-6			Double elements	1	RP		
			(3)	mV·Temp. /voltage converter	YOKOGAWA	STED-210-TR*B	0~10mV			1	RP		
			(4)	Alarm setter	〃	SNYD-200*A			1 Input 2 Set	1	RP		
			(5)	Recorder	〃	ER182-G95SV-N-N-SV-N*B/2CSST/BAL1/BAL1-2			with terminals for chart start Chart speed 240mm/min 2 Pen	1	R		
2-2	Measurement		(1)	mV generator	〃	PDC-16-1	0~10mV			1	RP		
			(2)	mV·Temp./voltage converter	〃	STED-210-MV*B	0~10mV			1	RP		
			(3)	Alarm setter	〃	SKYD-200*A			1 Input 2 Set	1	RP		

Fig. 7·10 Insturment list

Orifice data sheet

No.	Loop	Tag No.	Q'ty	Fluid specifications									Pipe specifications			Allowable pressure loss (mmH$_2$O)	Differential pressure across the orifice (mmH$_2$O)	Orifice specifications			Remarks
				Fluid	Flow rate			Pressure Nor. (mmH$_2$O)	Temp. Nor. (°C)	Specific gravity (kg/Nm^3)	visco-sity (CP)	Humidity (%)	Nominal diameter/Connection	Material	Inner diameter (mm)			Type of taps	Material	Type	
					Scale (Nm^3/h)	Nor.	Min.														
	Fuel flow	F0100-1	1	COG	0~1200	950	200	420	20	0.5	0.01	100	150A/ JIS10K	SGP	155.2	200	200	Corner	SUS304 (SUS304)	F2DIC-150JIO-UDA-BUA	
1																					
2																					
3																					
4																					
5																					
6																					
7																					
8																					
9																					
10																					
11																					
12																					
13																					
14																					

Fig. 7 · 11 Orifice data sheet

Control valve data sheet

No.	Loop	Tag No.	Q'ty	Fluid specifications									Pipe specifications		Valve specifications								Remarks
				Fluid	Flow rate		Pressure				Tempe-rature Nor. (℃)	Specific gravity (kg/Nm^3)	Nominal diameter Connection	Mate-rial	Valve				Actuator			Accesso-ries	
					Max. (Nm^3/h)	Min.	Primary press.								Type	Body size × Port size	Mate-rial (Body/trim)	Port characte-ristics	Type	Action	Air supply pressure (kg/cm^2)		
							at Max. flow (mm H_2O)	at Min. flow	at Max. flow	at Min. flow													
ex.	Air blow	FCV-301	1	Air	6 000	1 000	500	550	450	400	25	1.293	450A/ JIS5KFF	STPY -41	Butter-fly valve	350A	SCPH-2 SUS304	—	Single action cylinder	DA	2.8	PP.AS. NH	with companion flang B.N.P. with reducer (450×350 A)×2
1																							
2																							
3																							
4																							
5																							
6																							
7																							
8																							
9																							
10																							
11																							
12																							
13																							
14																							

Fig. 7 · 12 Control valve data sheet

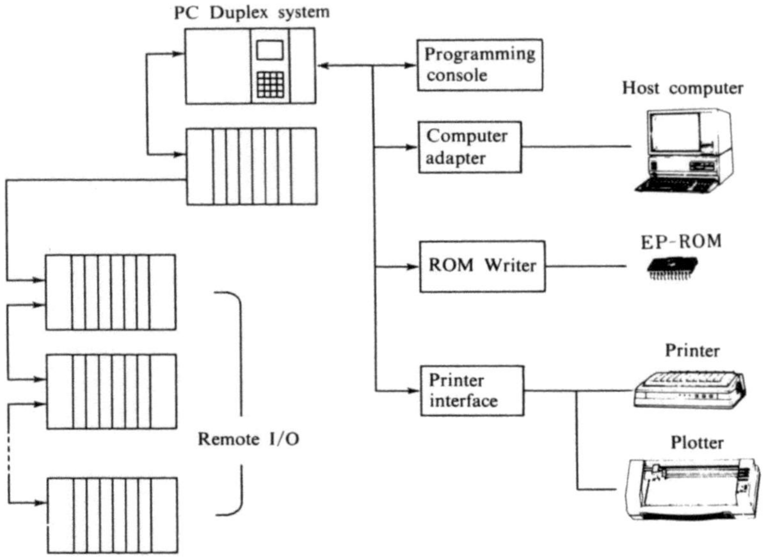

Fig. 7 · 13 A PC system configuration

tion is shown in Fig. 7.13. PCs feature easy creation and modification of sequences as well as highly-reliable, compact and lightweight equipment. They also feature communication functions which, due to their ease of implementation, facilitate system construction and expansion. Moreover, PCs have recently come to incorporate more sophisticated arithmetic/calculation functions for achieving even more powerful control. Examples of PC functions and commands are listed in Tables 7.3 and 7.4, respectively, and an example program is shown in Fig. 7.14.

Distributed DDC, apart from the features described above for the PC, is also characterized by being able to execute both feedback control functions and sequence control functions within the same piece of equipment. Specifically, it can easily receive and transmit feedback control and sequence control information. For example, if a tank level rises above its specified value, a drain pump can then be activated, or if the concentration of a mixing bath reaches its specified value, the stirrer can be stopped and an output pump activated. Thus, since measurement control and sequence control are very closely related, the ability to realize both feedback control functions and sequence control functions in the same controller is extremely convenient and efficient when designing an instrumentation system.

(2) *Sequence specifications* When defining sequence specifications, it is important to first determine the operation specifications for the machines and equipment in question. Major specifications in this regard

Table 7.3 Typical PC functions

Items	Contents
Input/output points	From approximately 64 to 2048 points depending on unit size
Arithmetic functions	Relay, Timer, Counter Four basic arithmetic operations Square-root extraction and exponential computation PID control computation Numerical comparison (>, <, =) FOR···NEXT etc.
Communication functions	Remote input and output CPU link Computer interface Data high way (Data communication cable)
Intelligent input and output	Numerical input and output Analog input and output High speed pulse counter Pre-set counter Registration PID parameters ASCII

relate to devices and drive methods used, operation and display equipment, and input and output signals. Moreover, personnel safety and equipment and plant protection during both normal and abnormal operations must also be incorporated into design plans.

(3) *Specification description* The main methods used today for expressing sequence actions are relay ladders (development diagram), logic circuits, flow charts, time charts and decision tables, as described earlier in Sec. 2.5. Each of these methods has its advantages and disadvantages in regards to a particular application. Moreover, the format of any one method will vary somewhat depending on the engineering field, the type of process, or the equipment involved. Nevertheless, several of the above should normally be used for any particular sequence description, and all sequence documentation should be clearly annotated and as error free as possible in the fashion of engineering blueprints. Examples of a time chart and a sequence signal chart as commonly used in instrumentation design are shown in Figs. 7.15 and 7.16, respectively. Symbols generally used in flow charts are listed in Table 7.5.

(4) *Safety measures* In addition to drive methods and operation pro-

Table 7.4 Typical PC instructions

Instruction	Symbol	Function	Instruction	Symbol	Function
Load	LD	Logical start operation.	Out	OUT	Outputs the result of a logical operation to the specified output relay, internal auxiliary relay, latching relay or shift register.
Load not	LD NOT	Logical NOT start operation.	Out not	OUT NOT	Inverts the results of a logical operation and then outputs them to the specified output relay, internal auxiliary relay, latching relay or shift register.
And	AND	Logical AND operation.	Timer	TIM	On-delay timer operation.
And not	AND NOT	Logical AND NOT operation.	Counter	CNT	Down counter operation.
Or	OR	Logical OR operation.	Shift register	SR	Shift register operation.
Or not	OR NOT	Logical OR NOT operation.	Latching relay	KR	Latching relay operation.
And load	AND LD	Logical AND operation with the previous condition.	Temporary memory relay	TR	Temporary memory relay operation.
Or load	OR LD	Logical OR operation with the previous condition.	Interlock	FUNC IL	Causes all the relay coils between this instruction and the LC instruction to be reset or not reset according to the result immediately before this instruction.

Instruction	Symbol	Function	Instruction	Symbol	Function
Interlock clear	[FUNC] [ILC]	Clears the IL instruction.	Compare (CMP)	[FUNC] [2 / CMP]	Compares data between I/Os.
Jump	[FUNC] [JMP]	Causes all the contents of a program between this instruction and the JME instruction to be ignored or executed according to the result immediately before this instruction.	Add (ADD)	[FUNC] [8 / SET +]	Adds data between I/Os.
Jump end	[FUNC] [JME]	Clears the JMP instruction.	Subtract (SUB)	[FUNC] [9 / RESET -]	Subtracts data between I/Os.
Move (MOV)	[FUNC] [5 / MOV]	Transfers data between I/Os.	Diagnostic (FAL)	[FUNC] [FAL]	Indicates the type of failure or abnormal mode.
Move not (MOV, NOT)	[FUNC] [5 / MOV]	Inverts and transfers data between I/Os.	End	[FUNC] [END]	The end of a program.

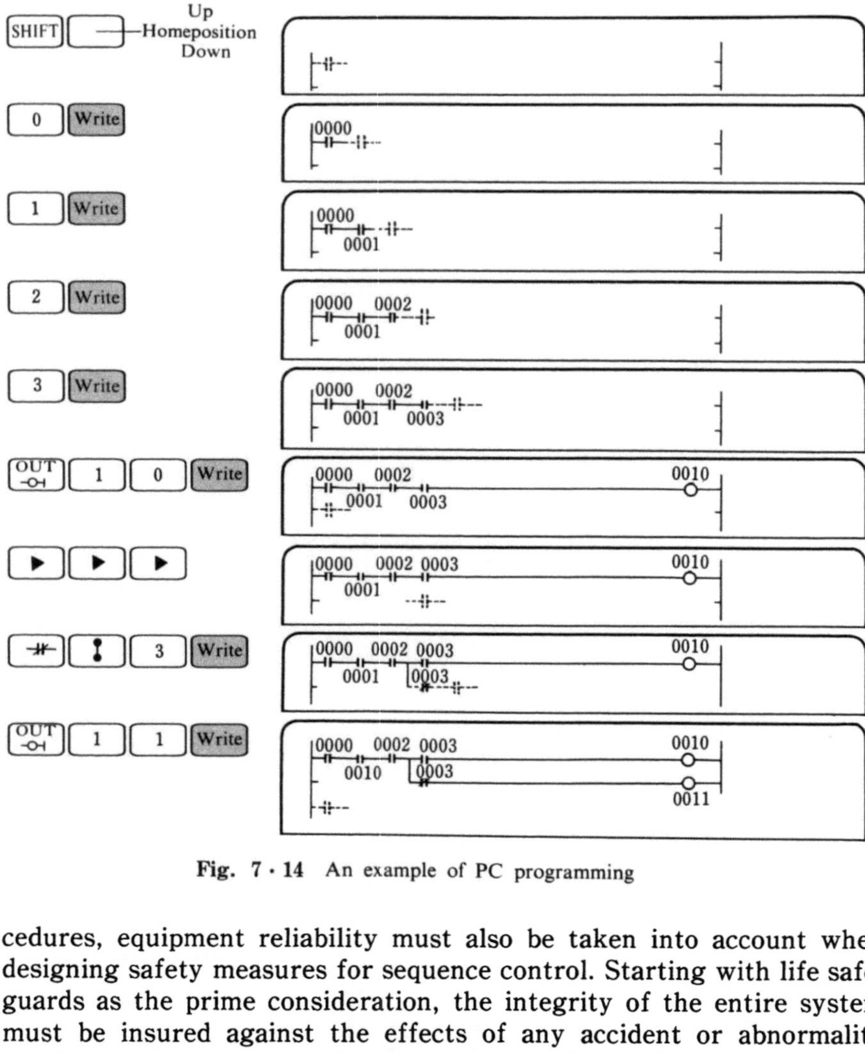

Fig. 7·14 An example of PC programming

cedures, equipment reliability must also be taken into account when designing safety measures for sequence control. Starting with life safeguards as the prime consideration, the integrity of the entire system must be insured against the effects of any accident or abnormality which may possibly occur. Needless to say, this requires a thorough investigation into the relationship between control objects, drive conditions, and hardware reliability. Specifically, the following items should be investigated when designing in safety measures:

(1) Power supply abnormalities (dividing the power supply system into subsystems should be considered as a countermeasure)
(2) Power loss/release measures (affects of power loss on safety; restarting measures after release)
(3) Reliability of sequence control equipment (backups, redundancy

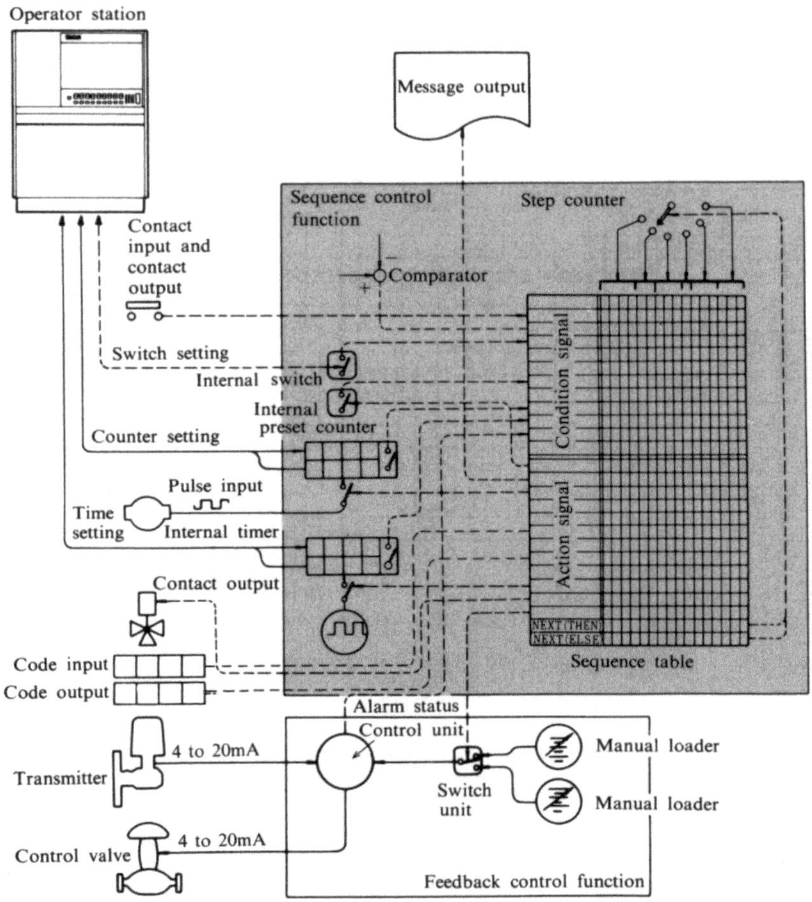

Fig. 7 · 15 Components of sequence control functions and their signaling diagram

as countermeasures)

(4) Emergency halt conditions (halt conditions should be considered within the range of central operations, on-location operations, and device operations, respectively)

(5) Erroneous operation countermeasures (wrong switch operation, interlock conditions)

(6) Information transfer (lamps, buzzers, sirens, CRT displays)

(7) Accessibility to maintenance check-points

(5) *Sequence specification documentation* Sequence specification sheets, which are used as a basis for specifications approval and detailed design work, must be prepared based on customer specifications

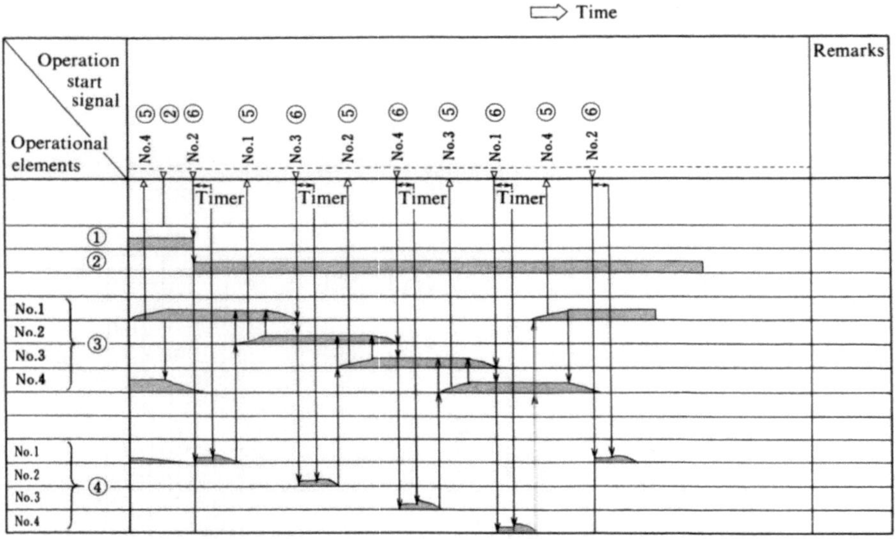

① H-Parallel blast
② S-Parallel blast
③ Cold-blast control valve
④ Mixing-cold-blast control valve
⑤ Mixing cold blast valve close
⑥ HS Switching signal

Fig. 7 · 16 Time chart example

Table 7.5 Flowchart symbols

No.	Name		Symbol	Description
1	Process	Automatic		Represents the sort of any processing function. Name of the function is written inside the symbol. This symbol is also used for representing logic element AND. Refer to No. 7.
		Manual	On line 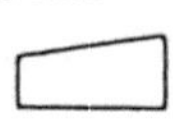	Represents manual processing functions related to sequence control.
			Off line 	Represents manual processing functions not related to sequence control.
2	Decision			Represents a decision to separate the process flow into two or more branches depending on a condition. Branches are indicated by yes, no, high, low, etc. adjacent to this symbol.

Table 7.5 (continued)

3	Predefined process		Used when a predefined flowchart indicated in the flow.
4	Lamp		Usage 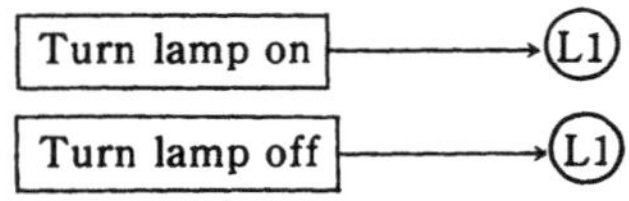
5	Display other than lamps		Represents data displayed an any common information medium.
6	Buzzer, bell, siren,		Usage 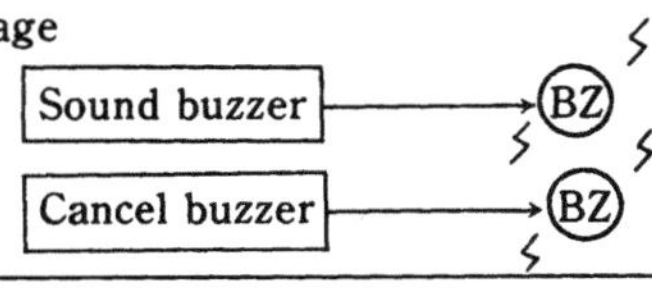
7	AND	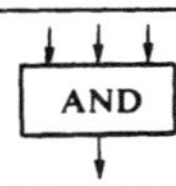	Example of symbol No. 1.
8	OR	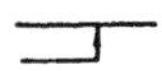	
9	NOT		
10	Timer	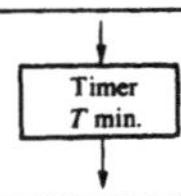	Example of symbol No. 1. Indicates both "timer start" and "*T* minutes passed" function.
11	Switch	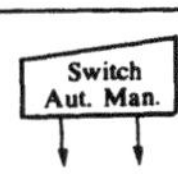	For instance, for changing over from automatic operation to manual operation.
12	Connector		Used to show which process on the same page should be connected with a conditional branch, or to show the relation between different sections so as not to complicate the flow chart. Connection No. is indicated in this symbol.
13	Connector		Same as No.12 above, but indicates connection between different pages. Connection No. is indicated inside the symbol and page references may also be given beside the symbol.
14	Terminal interrupt		Represents flowchart boundaries. For example, startpoint, endpoint, abort, stop, etc.

Table 7.5 (continued)

15	Comment Annotation	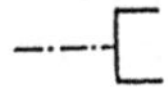	Used for descriptive comments. The dashed line is drawn from left or right to connect comment annotation.
16	Flow line		Represents connection of symbols and direction of flow. Arrows are not needed for directions from top to bottom or left to right. But, arrows must be used for directions opposite to those. ——— ; sequence flow —//— ; process flow (Note 1) ------ ; flow needs human intervention (Note 2)
17	Cross	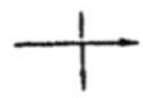	In two flow lines cross, they should do so at right angles.
18	Branch		Used to branch flow line into two or more lines in which branched lines have the same flow meaning as before branch. Branch point is indicated by a black dot.

Note: 1. For example, this flow line can be used for the following process.

Valve open ⟶ Level reaches upper limit ⟶ After valve is open and tank level reaches upper limit, the sequence advances to the next step.

Valve open ⟶ Limit switch turns on ⟶ After valve is opened by openning signal, the valve limit switch is closed. The sequence advances to the next step.

2. This line includes the case where the sequence cannot advance to the next step without human operation.

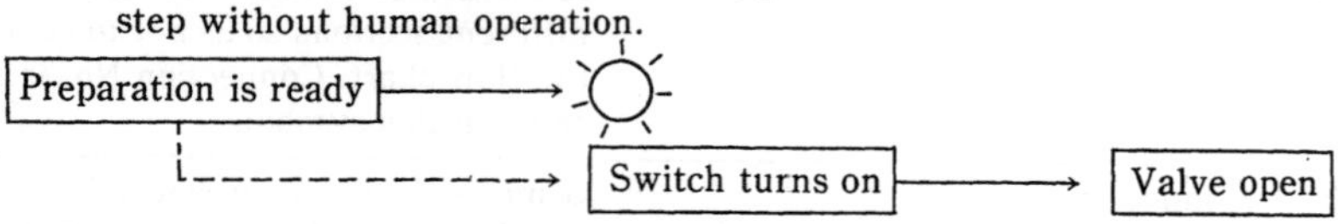

3. Location and tag No. may be indicated inside of symbol so that the flowchart can be used as an instruction manual for operators.

and on the results of technical meetings. Such documentation usually includes the following:

(1) Cover sheet (table of contents, statement of objectives)

(2) Process overview (describing sequences covered)
(3) Process flow sheet (illustrating entire process)
(4) Power supply diagrams (total power supply system, power supply diagrams, dividing diagrams for sequence systems, etc.)
(5) Interface list (type, conditions, specifications, etc. of input/output signals, etc.)
(6) Safety measures (redundancy, fail-safe measures, etc.)
(7) Operation description (sequence operations, operator procedures, flow charts, time charts, etc.)

(g) Control panel design

The main role of a control panel is to provide centralized management of the instrumentation system installed throughout the plant. A control panel normally consists of controllers, recorders, display units, rack mounted instruments, and operation switches.

Control panels are usually located in an environmentally-controlled room from which various system parameters such as flow rate, pressure, temperature and tank levels can be measured and controlled as well as from which control valves can be manipulated, process status monitored, and warning and halt signals output.

The configuration of a control panel depends on a variety of factors such as its main function, installation location, size and instrumentation system. Moreover, in the case of instrumentation systems employing one major type of instrumentation, conditions such as the amount of operator information, operator's actions, display readability, emergency actions, and fail-safe operations also influence the configuration of a panel. On the other hand, in instrumentation systems combining various instrumentation equipment such as the increasingly popular computerized and distributed DDC devices, information display, monitoring and operation functions are becoming centralized on CRTs. As a consequence, the role of control panels is changing. It should also be mentioned that as additional factors are considered such as providing an efficient and pleasant control room environment, *e.g.*, instrument/furniture layout, spacing, colors and lighting, control panels complementing the overall work environment have come to be demanded.

Thus, in the design of control panels, a comprehensive engineering approach must be taken which includes human engineering considerations. Refer to Sec. 7.3 for more details on control rooms and man-machine interfaces.

(1) *Common control panel configurations* Some common panel configurations used today are the upright type, bench type, console type, cabinet type, outdoor type and pipestand type, as illustrated in Fig. 7.17.

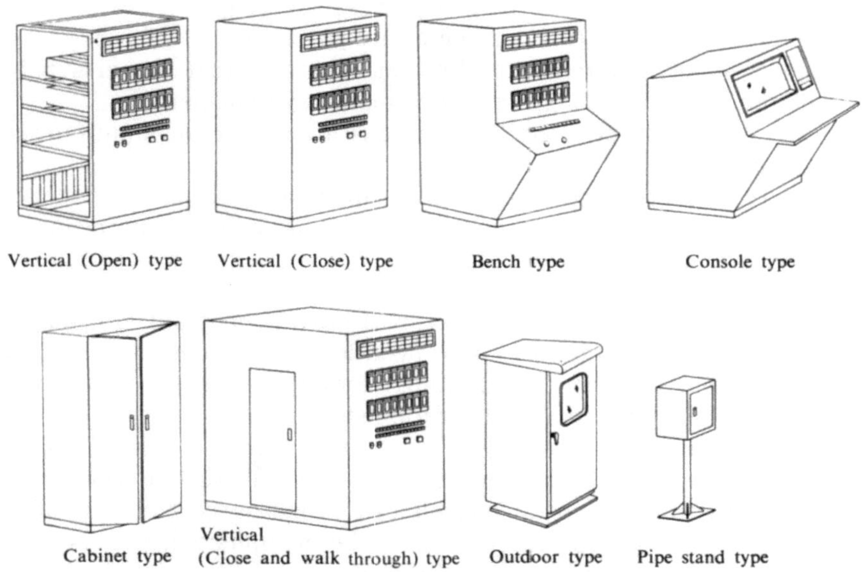

Fig. 7 · 17 Types of control panels

(2) *Control panel instrument layout* Although operability and maintainability is first taken into account in deciding on the general configuration of a control panel, also of prime importance here is the layout of instruments, *i.e.*, positioning of switches, displays, meters, knobs, etc., in particular as they effect the operator. Determining a final layout requires thorough consultation with the customer. An example of instrument layout for an upright type of control panel is shown in Fig. 7.18.

(3) *Design materials* (*engineering data*) For detailed design of control panels, the following list of materials are required. During the engineering phase, as control panel plans are being drawn up, these materials should be studied and used for establishing the various levels of design, as well as to help narrow down specifications at technical meetings.

(1) *General specification sheets*: General design objectives, requirements standards, regulations, etc.

(2) *Instrument list*: instrument name, model, manufacturer, tag number, etc.

(3) *Instrument configuration diagram*: diagrams showing instrument loop configuration

(4) *Instrumentation system input/output list*: number of input/output points, tag number, terminal number, etc.

(5) *Control room floor plan*: entrance dimensions, space allocation,

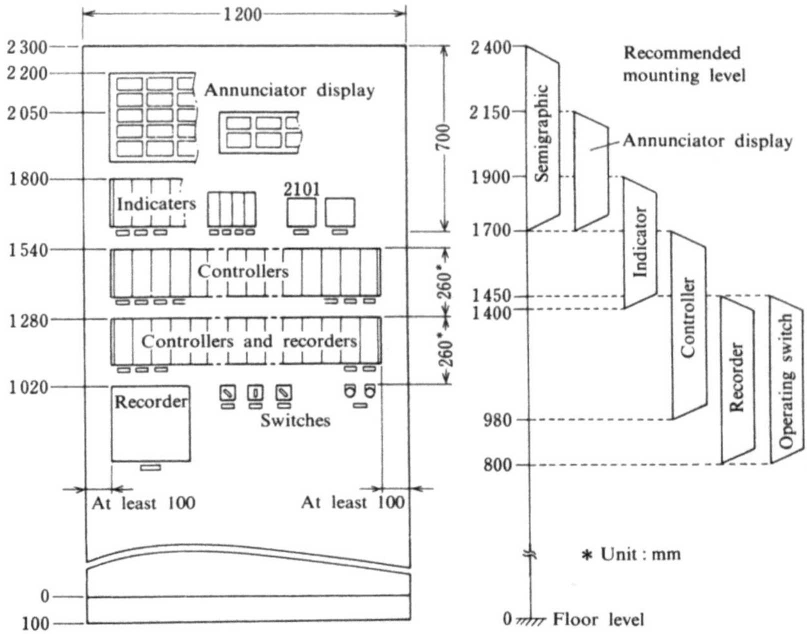

Fig. 7 · 18 Example of an instrument layout on a vertical panel

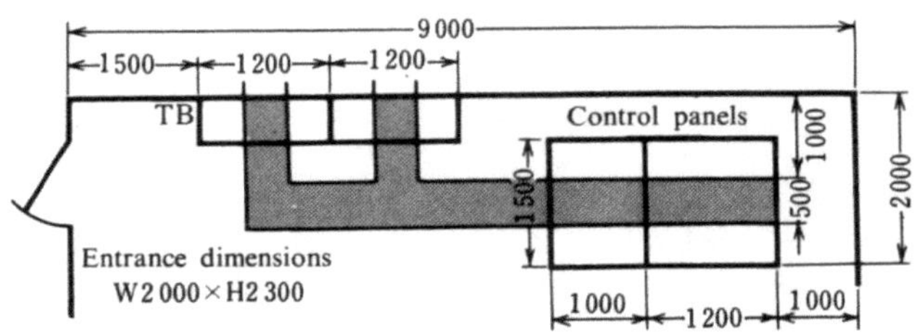

Fig. 7 · 19 Control room and control panel layout

pits, etc. (Fig.7.19)

(6) *Control panel external views, instrument layout*

(7) *Draft of graphic displays*: P&I diagrams, plots, etc.

(8) *Power supply, grounding diagrams*: control panel and power distribution board tie-in

(9) *Sequence action description*: logic diagrams, flow charts, sequence tables, time charts, etc.

(10) *Annunciator display lamp layout and inscriptions*

(11) *Name plate inscriptions*

(4) *Preparation of control panel specification document* This document is made up of the various detailed specification sheets for the con-

trol panel. It should be brought along to technical meetings during the engineering phase as a reference for modification and approval. These specifications usually include the following items:

(1) *General specifications*: installation location, air-conditioning requirements, weight limits, size limits, structural format, number of panels, dimensions, paint, etc.

(2) *Structure*: materials, thickness, etc. of front/back panel, side panels, top panel, doors; channel base, anchor bolts, hanging bolts, etc.

(3) *Wiring*: Wiring format, power supply, incorporated power supply equipment and power distribution board, power distribution format, wiring materials and colors, wiring standards/regulations, etc.

(4) *Annunciators*: Type of display lamps and warning bells and buzzers, layout, functions (non-lock-in, lock-in, double lock-in, etc.)

(5) *Sequence circuit diagram*: format, power, sequencer, input contact points, wiring format, etc.

(6) *Piping*: air supply, air supply devices, air header, bulkhead, piping materials, etc. (Refer to Fig. 7.20 for a conceptual diagram of an air supply device.)

(7) *Auxiliary devices, parts*: internal lighting, fans, maintenance power plug, replacement parts, etc.

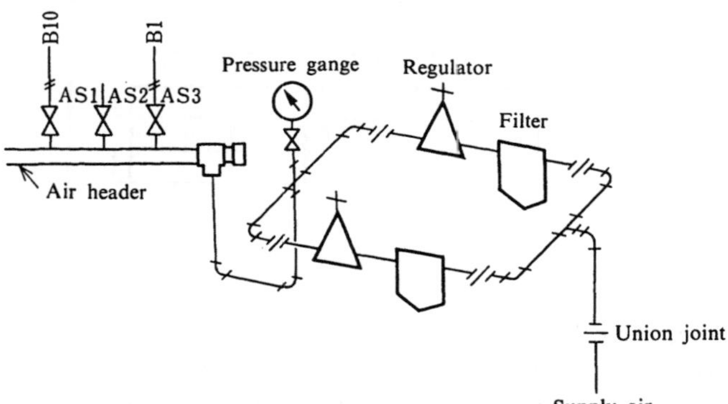

Fig. 7・20 Outline of an air suppling device

(5) *Control panel design and CAD/CAM tools* Up to the point that the final configuration and instrument layout of a control panel is finally decided upon, a good number of meetings will have taken place during which engineering plans and drawings undergo repeated correction and modification. As a result, CAD (Computer Aided Design) sys-

tems have been increasingly used as an aid in preparing control panel proposals and approval drawings since corrections and modifications as well as databasing can be easily and quickly performed. Usage of CAD tools, which not only saves on labor costs and provides neat and professional output, has increased in all areas of control panel design and documentation to the point that hand-written documents are all but obsolete. In addition, on the manufacturing side, CAM (Computer Aided Manufacturing) systems have long been in use for sheet metal and panel cutout processes. CAD/CAM systems have already proven themselves as effective in a wide range of functions from control panel price estimation to its manufacture. From now on, attention will be paid to increasing labor savings and raising quality levels.

(6) *Submittal of approval drawings* Once detailed design work has been completed based on engineering data received from the customer, the following approval drawings and related documentation must be submitted. Control panel manufacturing normally starts as soon as drawing approval is received.

Control panel specifications, Channel base drawing, External view drawings, Instrument layout drawing, Instrument list, Internal device layout drawing, Instrument loop drawing, Sequence circuit diagram, Power supply diagram, Power switch allocation, External terminal designation, Annunciator list, Name plate list, Graphic display drawings, Miscellaneous drawings, etc.

(*h*) *Digital instrumentation*

Although distributed DDC engineering is not fundamentally different from general instrumentation engineering, it must incorporate a variety of functions which conventionally have been designed and constructed separately (*e.g.*, control panels, graphic displays, sequencers, records, loggers, operator instructions, communications). This is in addition to DDC control functions. As a result, engineering has become more concentrated than in other forms of instrumentation engineering. Since the requirements, development methodology and functions of distributed DDC are explained in detail in Chap. 5, an example configuration and main functions only are summarized here in Fig. 7.21.

(1) *Digital instrumentation engineering* The main engineering steps here are specification confirmation, basic design, detailed design, work sheet description, hardware specification determination and customer software design. The engineering work flow is shown in Fig. 7.22. As can be seen, in addition to strictly engineering processes, other processes such as hardware manufacturing, system generation (software preparation), debugging and acceptance inspection are included as well.

(i) *Confirming customer's specifications*: The scope within distributed DDC can be realized must be defined based on customer specification

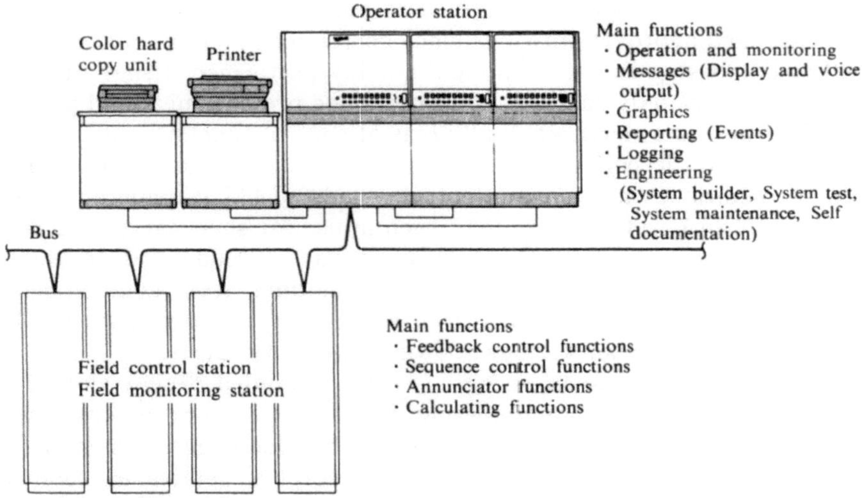

Fig. 7 · 21 Example of a distributed DDC configuration and main functions

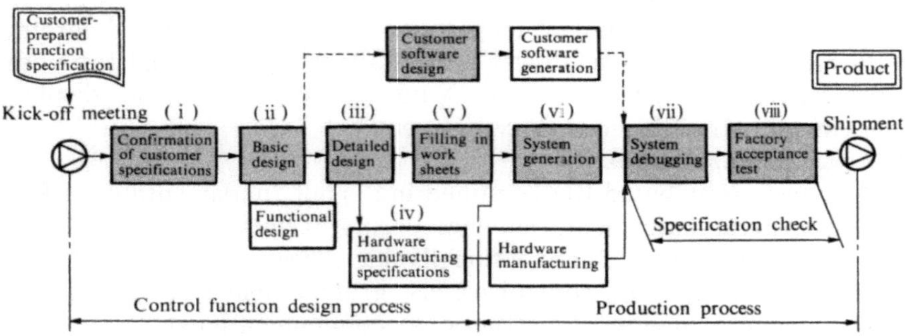

Fig. 7 · 22 Flow chart of distributed DDC engineering work

sheets and the results of technical meetings. Along with these initial specifications, which may include P&I diagrams, input/output signal lists, desired functions, and operation scope and methods, other factors such as operability, controllability, reliability, maintainability and safety must also be considered when determining this scope.

(ii) *Basic design*: Basic design is performed based on the specification sheets for the various functions desired by the customer. These include feedback control, sequence control, CRT display functions (*e.g.*, operations, monitoring, trends, graphics, operator instructions, etc.), logging, communications (*e.g.*, items, frequency of use, processing) and system configuration (*e.g.*, grouping, redundancy, additions, modifica-

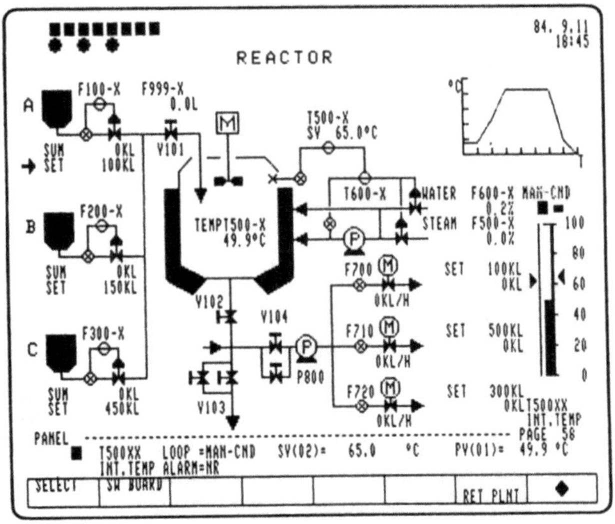

(a) Graphic display screen

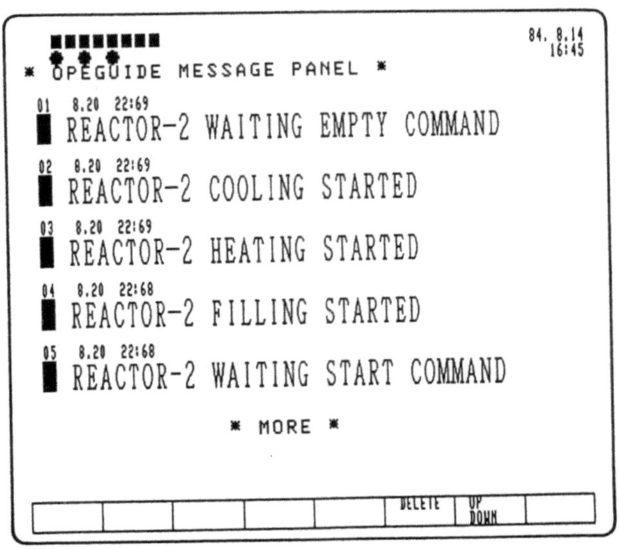

(b) Operator guide message screen

Fig. 7 · 23 Example of a distributed DDC and CRT display tions). Examples of CRT displays for distributed DDC are shown in Fig. 7.23.

(iii) *Detailed design*: Based on the basic design performed above, detailed design is carried out on feedback control loop diagrams, sequence control flow charts, logic diagrams, time charts, and display-design drawings.

(iv) *Hardware manufacturing specification sheets*: Manufacturing specification sheets related to hardware factors are prepared from such documents as system configuration drawings, man-machine function descriptions, detailed input/output signal lists, and equipment layout diagrams.

(v) *Work sheet description*: This document is prepared based on the results of detailed design work.

(vi) *System generation*: Based on the contents of work sheets, a control system for the user is constructed through use of a system generator, and typically, a system floppy disk is prepared for customer use.

(vii) *System debugging*: In the system debugging phase, individual system components (*e.g.*, operator console, field control station) are first checked out individually, followed by a comprehensive system-wide test. These tests should be performed according to previously prepared test procedure manuals and the test results recorded on prepared forms.

(viii) *Acceptance inspection*: Final approval of all system functions should be performed by carrying out a customer-attended final acceptance test (FAT) based on a previously prepared final-acceptance inspection manual. The results of the final acceptance test should be recorded on a prepared form. Figure 7.24 shows the engineering documentation system for distributed DDC. An example of the guidelines for document transfer between the customer and manufacturer is shown in Table 7.6.

7.2.4 Instrumentation work specifications

In this stage, the information listed below is collected, organized and verified as needed for carrying out instrumentation work.

Plant layout diagram, Facility layout diagram, Equipment placement diagram, Equipment external views, Plant piping diagram, Drive system documentation, Safety standards, Facility interface list, Plant wiring diagram, Duct/pit diagram, Materials list, Work schedule, etc.

In particular, during the course of technical meetings with the customer, the above information can be used as a basis for arriving at detailed specifications regarding the scope of instrumentation work, process and facility interfaces, instrumentation layout, wiring, piping, etc.

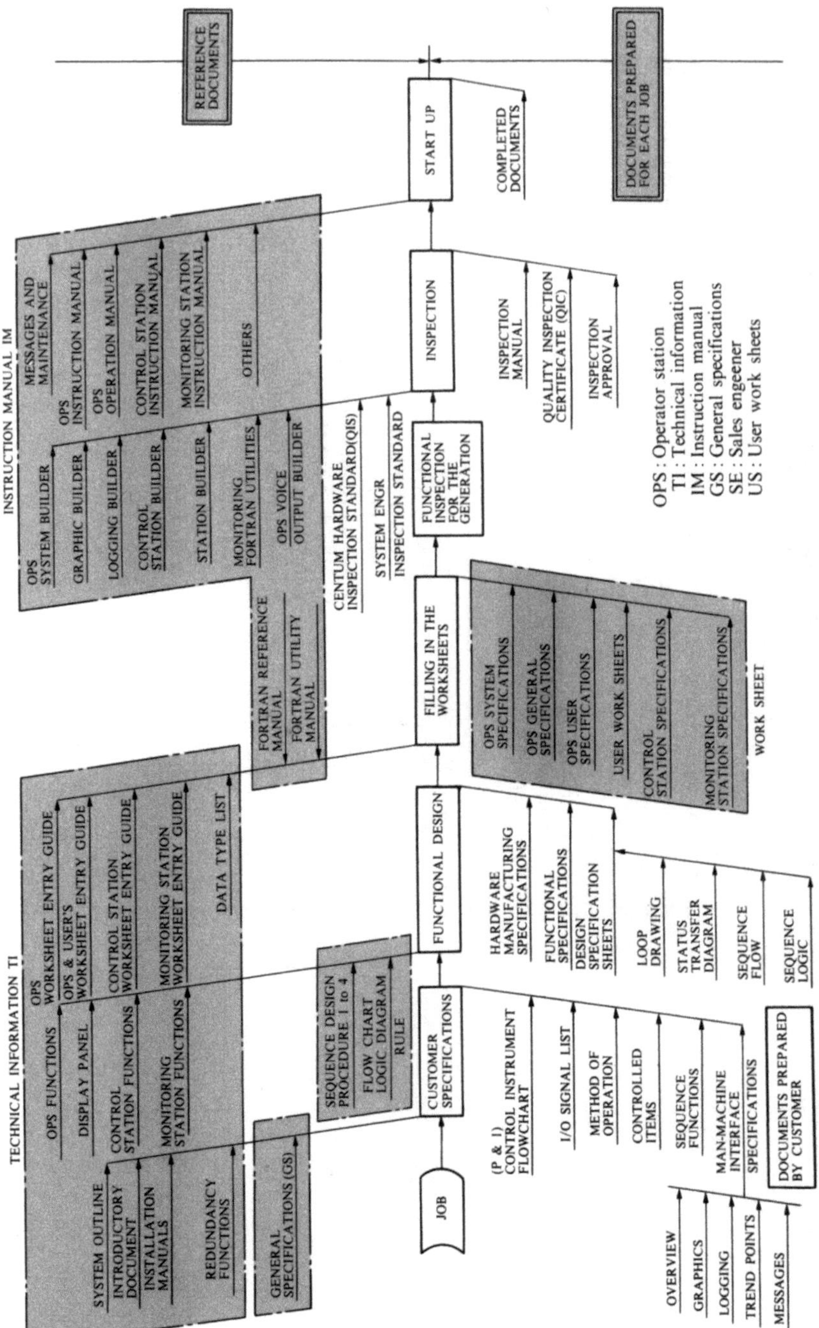

Fig. 7·24 Engineering documentation system for distributed DDC

Table 7.6 Guidelines for document transfer between the customer and manufacturer

Steps	Purpose	Item	Documents that the customer supplies to the manufacturer	Documents that the manufacturer supplies to the customer
1	System configuration and job overview	1. Confirmation of whole process from meetings up to shipment and start-up 2. system configuration 3. Interfaces with other equipment 4. Reasons for wanting new system 5. Confirmation of other rules	○ Control function flow diagram (PID) ○ List of I/O signals ○ Requested specification sheets ○ Proposed method of operation	○ Factory progress schedule ○ Organization chart ○ Hardware manufacturing specs. (for discussion) ○ List of documents requested
2	Confirmation of customer specifications	1. Explanation of customer specifications (by customer) 2. Explanation of method of operation (by customer) 3. Man-machine interface · Display panels (proposed)	○ Sequence control data · Flow charts · Logic diagrams · Time chart · Verbal description ○ Feedback control · Verbal description ○ Man-machine interface specifications	○ Approved hardware manufacturing specifications and drawings ○ Original user work sheets
3	Presentation of basic design specifications	1. Explanation of functional specs. (by manufacturer) 2. Explanation of any deviations from customer specs. (by manufacturer) 3. Request for approval	○ Graphic panel (proposed) ○ Overview panel configuration (proposed) ○ Completed user worksheets	○ Functional specifications · Basic design specifications · Feedback control composite loops · Sequence framework design

Steps	Purpose	Item	Documents that the customer supplies to the manufacturer	Documents that the manufacturer supplies to the customer
4	Confirmation of design specifications	1. Confirmation of detailed functions based on design specifications 2. Final request for approval 3. Documents sent to customer before factory acceptance test Confirmation of schedule		○ Design specifications · Feedback control loop diagrams · Sequence control flow chart · Logic diagram · Graphic panel specifications · Logging specifications

In regards to wiring, particular attention should be paid so that the wiring path or wiring method does not cause noise or other adverse affects in other plant facilities, and the maintainability of such wiring should be seriously considered. In addition, plans for actual work operations must be drawn up specifying work management, labor requirements, schedule, safety management, emergency communication system, etc.

7.2.5 Related work

In addition to the engineering work described above, the following processes or inter-department support will also be required as a part of system engineering or project engineering work.

(*a*) *Coordination*

In this stage, which precedes detail design, technical information gathered as a result of basic design work (engineering) is passed on to the manufacturing department (order entry). Specifically, the material listed below is prepared reflecting overall job planning. Moreover, an initial kick-off meeting should be held for presenting a job overview and describing system specifications to detailed-design groups and related departments.

System configuration diagram, Comprehensive flow sheet, Loop flow sheet, Power supply system diagram, Equipment/device list, Function specifications, Procurement specifications, Instrumentation work specifications, Control panel specifications, Sequence specifications, Work sheets, Job overview, Detailed schedule, Work organization chart, etc.

(*b*) *Approval documentation*

After a sufficient number of technical meetings have been held with the customer, the various specifications decided upon should be collected together and organized in the form of an approval document. On the basis of this document, the customer and instrumentation manufacturer should then mutually arrive at final approval. In regards to the form of this document, although one overall project approval document is ideal, there are also cases in which, due to manufacturing schedules and deadline considerations, multiple volumes are prepared, such as a system volume, device/function volume, control panel volume, instrumentation work volume, etc.

After checking approval documentation, the customer returns it to the instrumentation manufacturer. At this time, a certain amount of corrections and modifications is normal, and in some occasions major modifications may be required. In any case, such modifications should be made as quickly as possible and specification revision procedures instigated within manufacturing. Moreover, since specification changes may significantly effect current functions and cost estimates, the manufacturer should meet with the customer again to go over the contents of these changes. Discussions will include negotiations as to how the cost of such changes will be shared and how they will affect the manufacturing schedule.

(*c*) *Inspections*

In general, inspections are divided up into in-house inspections and customer inspections.

(1) *In-house inspections* The individual in charge of the basic design for a particular project task must verify that the target quality has been achieved.

(2) *Customer inspections* A similar inspection to the one described above is also carried out by an individual on the customer's side involved with design work or quality control. A quality inspection certificate must be prepared at this time. In addition, in the event that specification changes occur during inspection, these must be closely verified by reviewing specification change sheets, correction directive sheets, and the minutes of technical meetings.

(*d*) *Shipping*

Shipping and acceptance conditions are often indicated in separate documents or by contractual provisions with the customer.

(*e*) *Instrumentation work*

Support must be provided for instrumentation installation specifications and items to be negotiated with the customer from the standpoint of a job designer.

(*f*) *Start up/test run*

At start-up time, a job designer will perform device adjustment, device running, etc. through mutual consultation with the customer, and if necessary, witness an actual test run.

(*g*) *Final processing*

Final design documents must be prepared which reflect all changes made after the submittal of approval documents and which conform to the system as actually delivered to the customer. These documents should be compiled in the same way as that of the approval documents for consistency sake. In addition, since the customer will mainly be using final design documents for maintenance purposes, instruction manuals and maintenance manuals should be included as well. The type of binding should also be selected so as to prevent excessive wear due to frequent use or long-term storage. Needless to say, customer feedback and agreement as to the form and contents of final design documents should be obtained in advance of preparation.

(h) Job summary

A job summary document is useful for recording the specialized know-how obtained during the course of a job, and as such can be an invaluable asset. This document should concentrate on describing new technology appearing throughout the job from basic design to system operation. Moreover, it should serve as a useful reference for future instrumentation jobs.

7.2.6 Instrumentation drive system design

(a) Instrumentation power supply

The instrumentation power supply provides power to instrumentation equipment enabling it to collect information from throughout the plant and to take appropriate system actions based on this information. Thus, power to such equipment, which can be looked upon as the plant's brain and nerve system, is of supreme importance. Moreover, in the event of power reductions or total power loss throughout the plant, plans must be made to ensure safety in all plant processes. Common items requiring attention in instrumentation power supply design are listed below:

- (1) Countermeasures to power outages
  - Momentary outage
  - Extended outage
- (2) Countermeasures to power fluctuations
  - Voltage fluctuation
  - Wave form distortion
  - Frequency fluctuation
- (3) Load characteristics and levels
  - Priority devices
  - General devices
  - Sequencers
  - Computers
  - Miscellaneous
- (4) Power capacity selection
  - Plans for increase, etc.

An example of a power supply system diagram for an instrumentation system is shown in Fig. 7.25. Note that the private power-supply generator shown here is intended for use during extended power-down periods and moreover takes several minutes to start-up. Consequently, plant operations must be stopped for a short time to change over to the power supply system.

Due to the widespread popularity of microprocessors in instrumentation equipment, it has become difficult to configure a system with equipment not having them. This situation has demanded that power supplies for modern instrumentation be of even higher quality than traditional systems. The following paragraphs describe power sup-

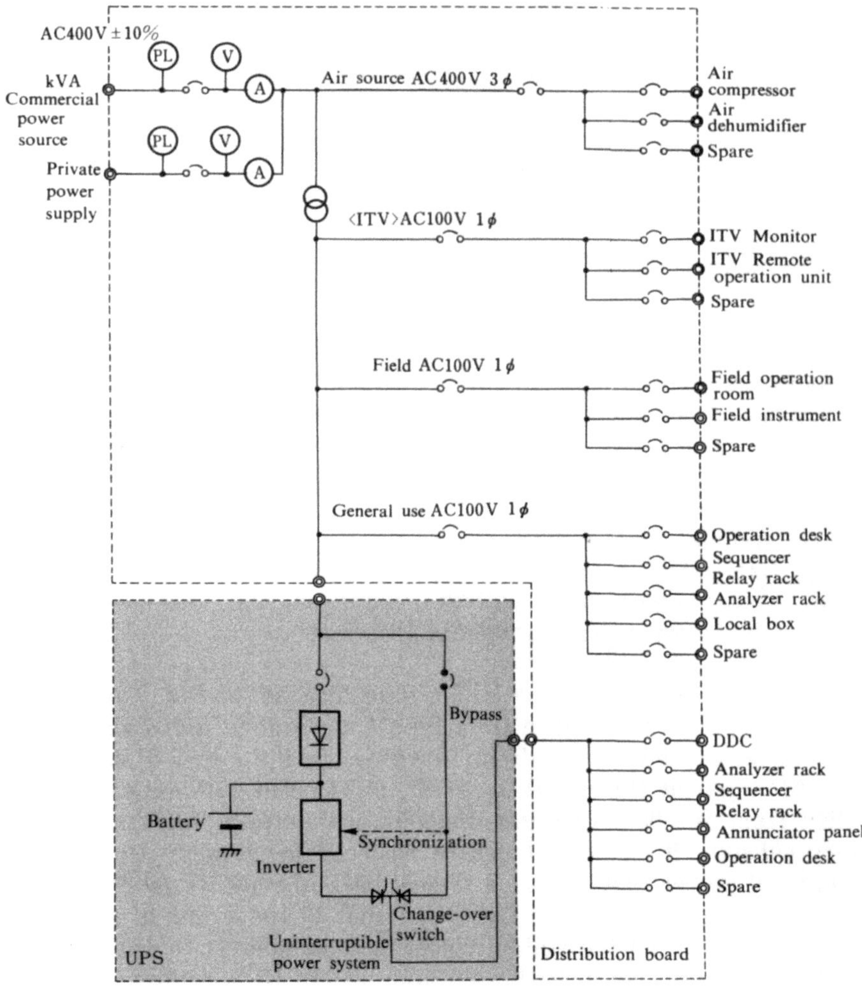

Fig. 7 · 25 Power supply system diagram for instrumentation

plies for modern instrumentation use and countermeasures for momentary power loss.

(1) *Uninterruptible power supply* Since instrumentation equipment plays the role of the plant's brain and nerve system, power supply problems that can halt operations can cause major damage to plant business. In addition to power outages, since microprocessor memory protection processing is immediately initiated if power voltage fluctuation

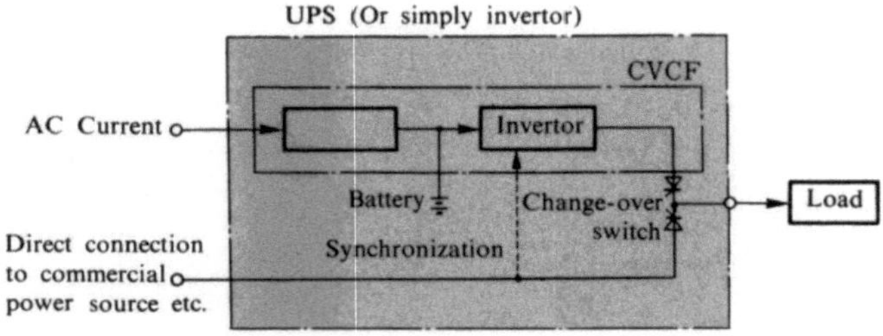

Note: CVCF (Constant Voltage Constant Frequency)

Fig. 7·26 Uninterruptible power system for instrumentation

or wave form distortion is detected, even within one cycle, small amounts of fluctuation or distortion cannot be tolerated. Consequently, it has become common to supply power to instrumentation in the form of an uninterruptible power supply (UPS). A typical UPS configuration is shown in Fig. 7.26, and standard requirements for a UPS are described below:

(1) Minimal maintenance, continuous operation
(2) Extended operation during power outages
(3) Power quality (*e.g.*, voltage, frequency, wave form) as required for instrumentation equipment functions
(4) High-reliability

(2) *UPS operation* In the UPS circuit diagram of Fig. 7.26, power is usually supplied to the load from the AC current terminal via the rectifier, inverter and SCR switch. However, in the event of problems in the AC input or the inverter circuit, power will automatically be supplied from the battery (instantaneous switchover). Although the battery capacity in such a structure varies depending on the estimated length of power-down time, a typical backup capacity is 30 to 40 minutes. Note also in this configuration that in the event of a faulty inverter, the SCR switch will automatically changeover to direct commercial power input. In selecting an appropriate UPS, points which must be considered are rush current, peak current, and wave form distortion, as described below.

(3) *Rush current* The inverter possesses drooping, or cut-off, characteristics for times of overload as a self-protection mechanism. As a result, rush current in the load can have negative effects. Although all electrical devices possess some form of rush current characteristics, devices requiring particular care are those having rectifier input forms (capacitor input forms). In particular, computers, instrumentation equipment and communication devices usually have internal single-phase all-wave rectifier circuits in which, as shown in Fig. 7.27,

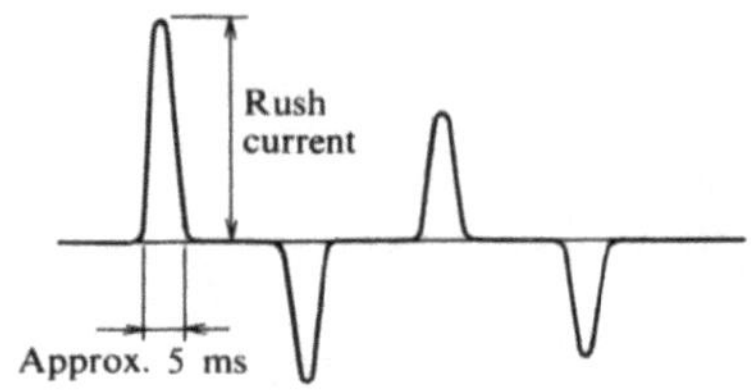

Fig. 7 · 27 Rush current waveform

rush currents are generated due to capacitor charging when current is applied. The amount of this current can reach ten times or more that of the stationary load current.

(4) *Peak current* A load in which a sinusoidal current flows on application of a sinusoidal voltage is called linear load, and a non-linear load otherwise. A typical example of a non-linear load is that having a capacitor input form. In this case, the peak value of the current wave form can become 2.5 times the effective value, and the corresponding voltage wave form will exhibit significant distortion, as shown in Fig. 7.28.

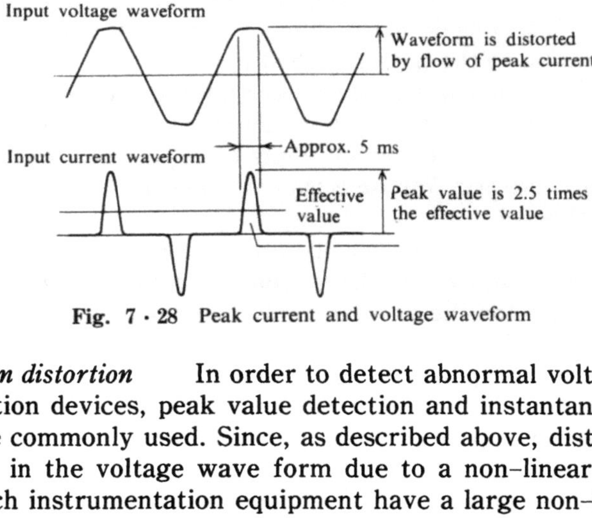

Fig. 7 · 28 Peak current and voltage waveform

(5) *Wave form distortion* In order to detect abnormal voltage within instrumentation devices, peak value detection and instantaneous value sampling are commonly used. Since, as described above, distortion will be exhibited in the voltage wave form due to a non-linear load, systems in which instrumentation equipment have a large non-linear load will require the placement of a rectifier on the UPS side as a load countermeasure.

In addition, even if the effective value of an input voltage wave form having distortion is the same as that of one not having distortion, the terminal voltage of its smoothing capacitor will be lower, and the instrumentation equipment may detect a power-down situation. For ex-

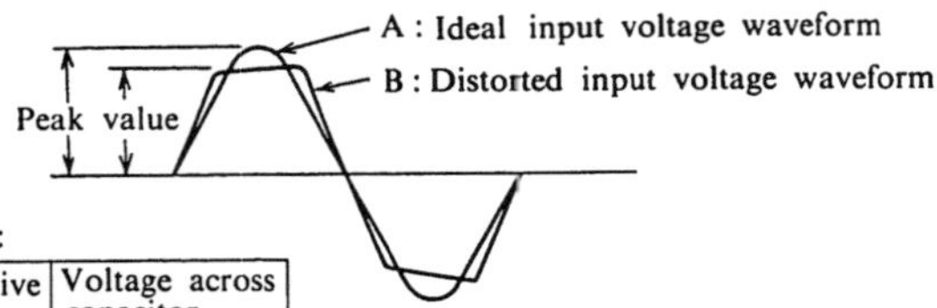

Example :

	Effective value	Voltage across capacitor
A	100V AC	138V DC
B	100V AC	120V DC

Fig. 7・29 Distorted input voltage waveform

ample, as shown in Fig. 7.29, even though input voltage wave forms A and B both have an effective value of 100 V AC, the terminal voltage of the smoothing capacitor for the B wave form will be lower, or in other words, the voltage supplied will be equivalent to an 85 V AC wave form without distortion.

(6) *Power–supply equipment output capacity* Assuming a linear load, the output capacity of power–supply equipment (UPS) will be characterized by a sinusoidal current in which voltage and phase agree. For a non–linear load, however, the required output capacity must be determined with due consideration given to the current wave form and phase in addition to the power needed by the load.

Also to be considered when determining the output capacity of power–supply equipment are such factors as equipment type, format, and the manufacturer. Based on years of experience and actual equipment results, criteria for selection of power–supply equipment are shown in Table 7.7.

Note that the output capacity obtained from Table 7.7 is greater than the power required by the system. This factor does not represent simply a margin for the power–supply equipment but is the power capacity necessary for normal operation in the case of non–linear loads.

In general, the power–supply equipment manufacturer should also be consulted to help determine output capacity requirements, or to assist when power–supply equipment already procured do not satisfy output capacity selection criteria.

In addition, power–supply equipment on the host side should be given the same consideration as described above. This also holds for UPS when determining its output capacity as host–side power–supply equipment, since it possesses rectifier load characteristics.

(*b*) *Instrumentation air supply*

Similar to the instrumentation electric power–supply described above, the instrumentation air supply also plays a major role as a drive

Table 7.7 Criteria for the selection of power facilities.

Type of power facility	Excessive variation in commercial voltage	Excessive variation in commercial frequency	High susceptibility of commercial line to high-frequency noise	High liability of commercial power to momentary failure (or where the system equipment requires a protection against a momentary power failure within 0.5 sec.)	Output capacity factor (service factor) (Note).	Price
Isolation transformer	×	×	○	×	2	Low
AVR (automatic voltage regulator)	○	×	○	×	3.5	↑
MG (motor generator set)	○	×	◎	○	1.5	↓
UPS (Uninterruptible power system)	○	○	◎	○	2	High

(×: not good; ○: good; ◎: best)

Note: The installed capacity of the power facility is calculated by multiplying the total power demand of all the system equipment by the output capacity factor. This factor serves as a measure to determine the output rating of a specific power facility. If the output capacity factor is set lower than the value specified, it is usually necessary to provide some protection against high-frequency currents.

source for the instrumentation system. Although the use of pneumatic instruments has been decreasing in recent years, they still find use in some areas such as processes susceptible to explosions or local control devices. In addition, the majority of control valves are still driven pneumatically. As for quality requirements, an air supply must be clean and stable, and measures must be taken for pulsation absorption, momentary power loss, and compressor breakdown. In the first two cases, a typical solution is an air supply tank, while for compressor breakdown, multiple compressors or switchover to a factory air supply are often used. Another example would be the temporary use of N_2 gas, or for that matter any technique which is both safer and less expensive. An example of air-supply equipment for instrumentation use is shown in Fig. 7.30.

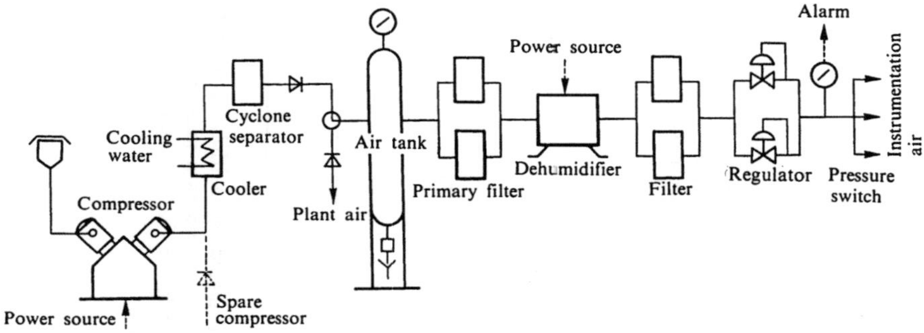

Fig. 7 · 30 Instrumentation air source system diagram

(1) *Air supply quality* Air purity for instrumentation use must in general meet the values listed below. Compressed air can be purified by such devices as coolers, multi-stage filters, drain separators and dehumidifiers.

Foreign particle diameter:	10 μm or less
Foreign particle density:	0.1 mg/m or less
Humidity (dew point under pressure):	10°C or less
Oil droplets:	1 ppm or less
Harmful gas:	20 ppm or less

Even if the air used for instrumentation eventually reaches atmospheric pressure, dry conditions without condensation must be insured. Figure 7.31 shows the relationship between atmospheric pressure dew point and dew point under pressure. If we set the dew point under pressure to 10°C, the atmospheric pressure dew point for a pressure of 7 kgf/cm^2 becomes −17°C, and dry air with no condensation can be obtained anywhere in the air-supply system (the dew point may be lower depending on the environment). Water content in the air can be elimi-

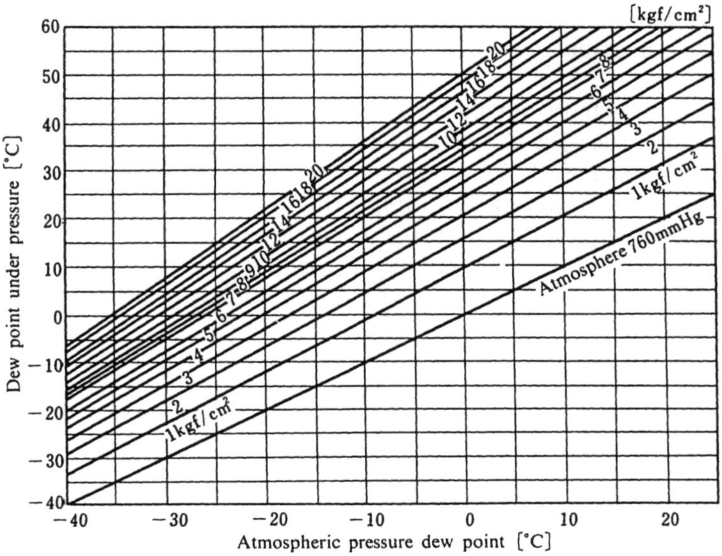

Using this chart :
This chart is used for converting the dew point under pressure to the atmospheric pressure dew point and conversely.
(ex.) To determine the atmospheric pressure dew point under a pressure of 7kgf/cm^2, draw a horizontal line from 10°C point on the vertical axis over to the 7kgf/cm^2 line, and then draw a vertical line from this point of intersection to the horizontal axis. The atmospheric pressure dew point is found to be −17°C as shown in the figure to the right.

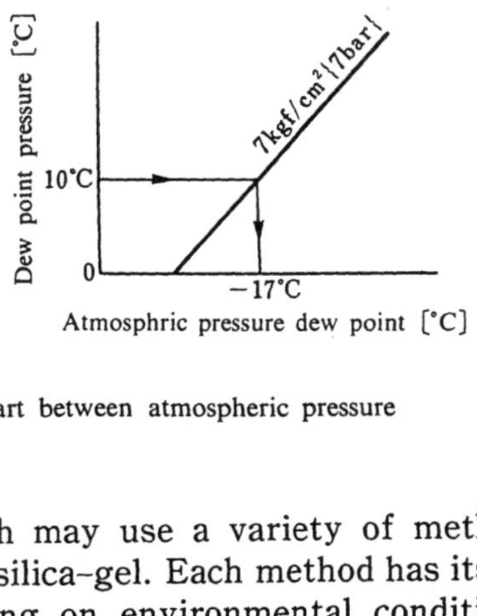

Fig. 7 · 31 Dew point conversion chart between atmospheric pressure and indicated pressure

nated by dehumidifier devices, which may use a variety of methods such as mechanical refrigeration and silica-gel. Each method has its advantages and disadvantages depending on environmental conditions, and should be selected accordingly.

(2) *Determination of air-compressor capacity* When determining air-compressor capacity, it is common to treat it as a function of total instrumentation system air consumption. The capacity of air-supply equipment is expressed by Eq. (7.1):

$$Q_0 = K_1 q \ [\mathrm{Nm^3/h}] \tag{7.1}$$

Here, Q_0 is the air-supply equipment capacity rating (Nm^3/h), q is the total air consumption for all devices (Nm^3/h), and K_1 is a safety factor (2.0 to 2.5).

In order to clarify device air consumption, we quote its definition from JIS standards (JIS C 1801): "Given that a device is supplied with its rated air pressure, if the amount of air flow is measured at various points throughout its operation range, the maximum air flow value obtained is taken to be the amount of air consumption for that device."

The amount of air consumption for devices operated intermittently such as shut off valves and cylinder valves is given by Eq. (7.2):

$$Q_n = \frac{V}{1\,000}\left(\frac{P_1}{1.033}+1\right)n \text{ [Nm}^3\text{/h]} \tag{7.2}$$

Here, Q_n is the amount of air consumption (Nm^3/h), V is the cylinder or diaphragm volume (l), P_1 is the supplied air pressure (kgf/cm^2), and n is the number of operation cycles per hour (cycles/h).

The air reservoir (backup tank) is installed to smooth the pressure pulsations from the air-compressor and also serves to temporarily support control valves and other equipment if air-compressor operation is halted by a breakdown or power loss. Accordingly, air system parameters such as air consumption, backup time, and the lower limit of the supplied air pressure must be investigated taking into account operability, maintenance and safety of the entire instrumentation system. A nomogram for selection of air reservoir capacity is shown in Fig. 7.32.

The air reservoir capacity can be calculated by Eq. (7.3) below. A backup time of 30 to 40 minutes is common.

$$V = \frac{qT}{P_1 - P_2} \text{ [Nm}^3\text{]} \tag{7.3}$$

Here, V is the tank capacity (m^3), q is the total device air consumption (Nm^3/h), T is the backup time (h), P_1 is the normal pressure = supplied pressure (kgf/cm_2), and P_2 is the minimum operating pressure (kgf/cm^2).

7.2.7 Other system functions (safety, failsafe and redundancy measures)

Any discussion of safety measures for an instrumentation system must necessarily involve the reliability of instrumentation devices making up the system. If this reliability cannot be grasped quantitatively, then it is impossible for system reliability to be predicted qualitatively by such predictive analysis methods as FMEA (Failure Mode Effect Analysis) and FTA (Fault Tree Analysis). A positive approach to improving the safety of each individual instrumentation device must be adopted.

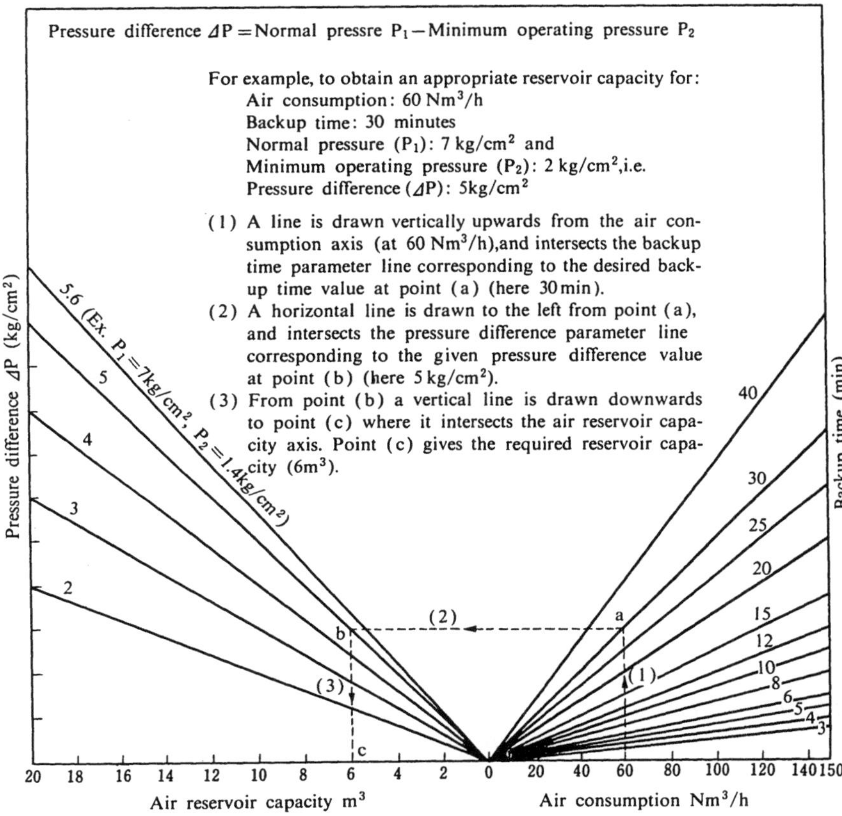

Fig. 7 · 32 Air reservoir capacity selection nomogram

Some examples of methods for improving device and system reliability such as explosion-proof and earthquake-proof techniques, redundancy and failsafe systems are described below.

(*a*) *Instrumentation device safety measures*

(1) *Explosion-proof construction* Instrumentation devices used in areas containing explosive gas or dust must not be the source of sparks or heat which may set off an explosion. Methods used as a countermeasure to explosive gases include explosion-protected, pressurized, increased-safety, and intrinsic-safety techniques. Countermeasures to flammable dust consist of both special and standard dust-proof configurations, while special dust-proof configurations are used for explosive dust. These are explained in more detail below.

(i) *Explosion-proof technique*: In order to achieve an explosion-proof construction, an enclosure must be strong enough to withstand the force of an explosion caused by the chance ignition of an explosive atmosphere which has somehow penetrated the enclosure from the outside. In addition, gaps in enclosure components such as the spindle and composition plane must be appropriately designed so that flames generated by an explosion are cooled as they escape to the outside, resulting in a temperature below the point at which explosive gases in the immediate area can ignite.

(ii) *Pressurized technique*: In this technique, the internal pressure of an enclosure storing air or an inactive gas is maintained at 5 mmH_2O above the pressure outside the enclosure. As a result, an explosive atmosphere is prevented from penetrating the enclosure.

(iii) *Increased-safety technique*: This is a commonly used technique for ensuring the absence of any factors which may act as a source of ignition. Specifically, a safety factor is added to the creepage distances and clearances and to the temperature-rise value associated with winding so as to lower the possibility of any form of ignition.

(iv) *Intrinsic-safety technique*: Here, a system is configured so that it intrinsically does not have the capability of igniting an explosive atmosphere not only during normal operation, but even in the event of arcs, sparks, or overheated components due to accidental shorts, groundings, or open wires.

(v) *Special dust-proof technique*: In this technique, a sealed enclosure is made dust-tight so that no outside dust can penetrate it.

(vi) *Standard dust-proof technique*: In this case, a sealed enclosure is configured so that outside dust cannot easily penetrate it.

(2) *Earthquake-proof construction* For plants in which abnormal reactions to earthquakes may have a significant impact on society (*e.g.*, nuclear power plants), earthquake-proof construction must be applied to instrumentation devices, in particular, to related safety assurance mechanisms. Earthquake-proof construction should also be considered for control panels in control rooms and outdoor cubicles.

(*b*) *Instrumentation system safety measures*

(1) *Redundant systems* Due to the fact that the reliability of instrumentation devices can only be raised so far, and since devices which are totally fault-free do not actually exist, system reliability can only be raised further by incorporating duplicate, triplicate, or higher redundant systems.

Basic types of redundant systems and their corresponding MTSF (Mean Time to System Failure) are shown in Table 7.8. As can be seen, incorporation of a maintenance system (repair work is performed at the time of a breakdown) has a significant effect on the MTSF of a redundant system. These formulas can be easily applied to instrumentation devices. For example, assuming a device with a MTBF (Mean

Table 7 · 8 Redundant systems and MTSF

Redundant system	MTSF	
	Non-maintenance system	Maintenance system
Parallel redundancy 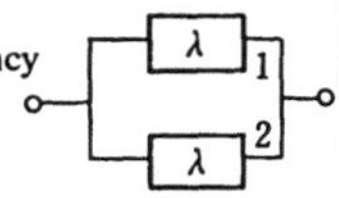	$\mathrm{MTSF}_0 = \frac{3}{2\lambda}$	$\mathrm{MTSF} = \frac{3}{2\lambda} + \frac{\mu}{2\lambda^2}$
Stand-by redundancy 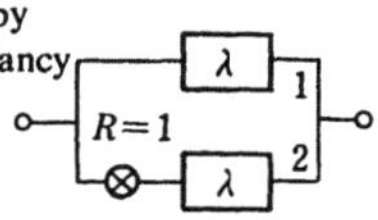	$\mathrm{MTSF}_0 = \frac{2}{\lambda}$	$\mathrm{MTSF} = \frac{2}{\lambda} + \frac{\mu}{\lambda^2}$
2-out-of-n redundancy 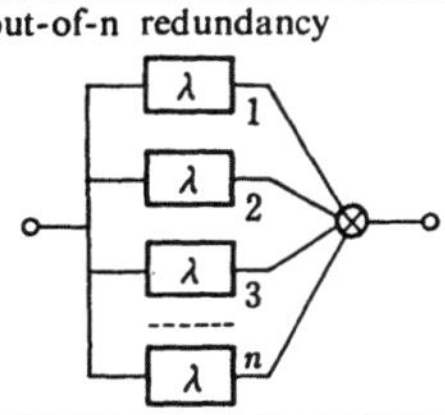	$\mathrm{MTSF}_0 = \frac{1}{n\lambda} + \frac{1}{(n-1)\lambda}$	$\mathrm{MTSF} = M \times \mathrm{MTSF}_0$ $M \approx \dfrac{\frac{1}{n-1} + \frac{\mu}{n(n-1)\lambda}}{\frac{1}{n-1} + \frac{1}{n}}$

Time Between Failure) of 10^5 hours (breakdown rate $\lambda = 10^{-5}$/h), the improved reliability in the case of a parallel redundant system would be given as:

Non-maintenance system:

An improvement of about 1.5 times can be obtained from the result of $\mathrm{MTSF}_0 \approx 1.5 \times 10^5$ (h)

With a maintenance system:

An improvement of about 500 times can be obtained from the result of $\mathrm{MTSF} \approx 5 \times 10^7$ (h)

(assuming a repair time of 100 hours: repair rate $\mu = 10^{-2}$/h)

The parallel redundant system of Table 7.8 is often used in the case of duplicate detection probes or relays, in which a normal condition from at least one of the two is acceptable. The stand-by redundant system is commonly employed in battery backup systems for DC current. The 2-out-of-n redundant system consists of n devices and is based on the principle that if 2 devices suffer breakdown at the same time, a system breakdown has occurred. Such a system is commonly used for parallel operation of a DC power supply box in an instrumentation system. Moreover, the configuration of $n = 3$ is especially appropriate for 2-out-of-3 majority selection formats commonly used for shutdown (emergency halt) sequences. An example of a 2-out-of-3 selection format as used in a nuclear reactor neutron measurement system

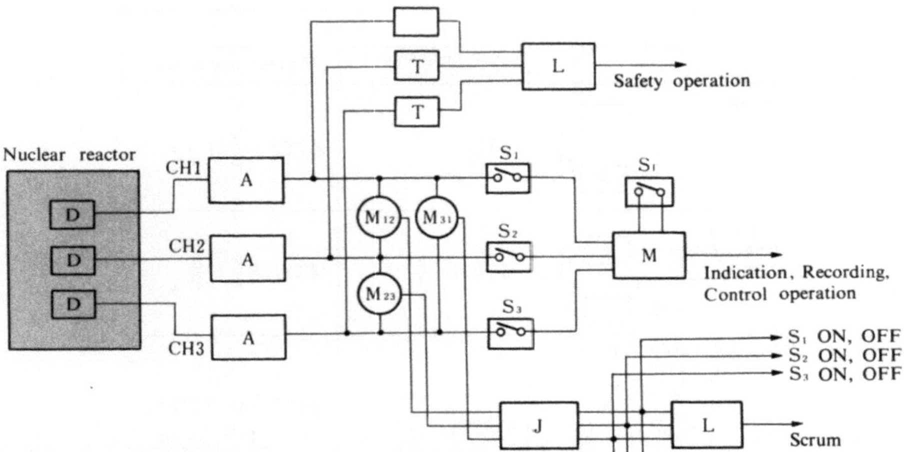

D : Detecting element
A : Amplifier system
J : Failure judgment system
M : Mean circuit
L : 2-out-of-3 logic circuit
M_{12} to M_{31} : Deviation detector
OR : OR circuit
S_1 to S_3 : Exclusion switch for failure channels
S_4 : 3-channel to 2-channel changeover switch
T : Trip circuit

Fig. 7 · 33 2-out-of-3 measuring system with failure judgment system

is shown in Fig. 7.33.

(2) *Failsafe measures* In a failsafe device, the occurrence of a fault will result in measures which force equipment conditions to move in a safe direction. Two examples of failsafe systems are described below.

(i) *Alarm system*: The configuration for a failsafe alarm system is shown in Fig. 7.34. The main point of consideration here is whether the output relay should be normally energized or energized only during abnormal occurrences. If energized only during abnormal occurrences, then if by some chance the power to the alarm system is shut off, the relay will remain in an deenergized state, thus indicating a normal system condition regardless of the input signal. This configuration has the danger that power failure cannot be noticed.

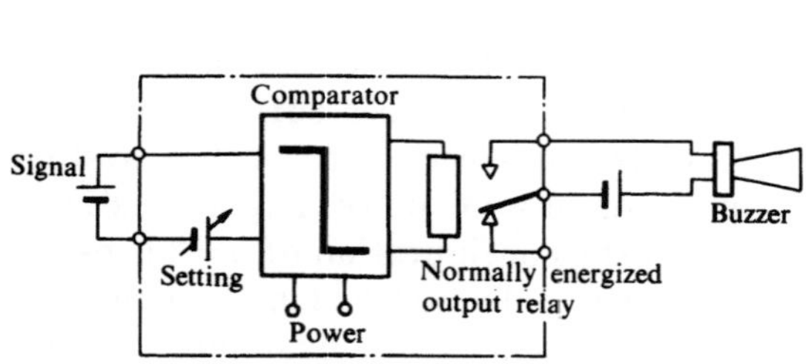

Fig. 7 · 34 Failsafe alarm system

If, on the other hand, the relay is normally energized, then if the power is cut-off, the alarm will sound regardless of any input signal, advising of an abnormal condition. Thus, an alarm system in which its output relay is normally energized is considered to be a failsafe system.

(ii) *Reverse-signal transmitter*: In the temperature control loop of Fig. 7.35, if the output signal from a transmitter employing positive characteristics goes off, the controller will interpret this as a low-temperature signal, and will begin to adjust the system to higher temperatures. Since this configuration does not provide for any safety measures in the event of an erroneous off signal, it is not a failsafe system.

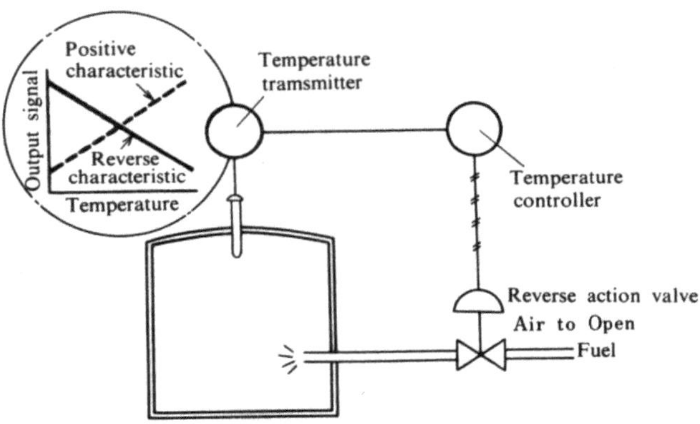

Fig. 7 · 35 Temperature control loop and reverse characteristic transmitter

If this control system is based on reverse characteristics, however, an off transmitter signal will signify a high-temperature condition to the controller, resulting in adjustment to a lower temperature state, *i.e.*, a safer or failsafe condition.

Thus, for systems in which high values of measured variables such as temperature and pressure can be dangerous, failsafe measures can be implemented by applying reverse characteristics to the early stage of a loop, *i.e.*, detection probes or transmitters. Conversely, if low values of measured variables are dangerous such as in the case of a tank level control loop, application of sensors or transmitters having positive characteristics will produce a failsafe system.

(3) *Distributed DDC redundant configurations* A one-unit distributed DDC configuration usually has about 30 to 40 loops in its control domain, and about 100 total points if monitoring functions are taken into account. Such a system obviously requires high reliability, and a common method for raising the reliability of distributed DDC employing microprocessors is duplicate systems. Moreover, in addition to this pri-

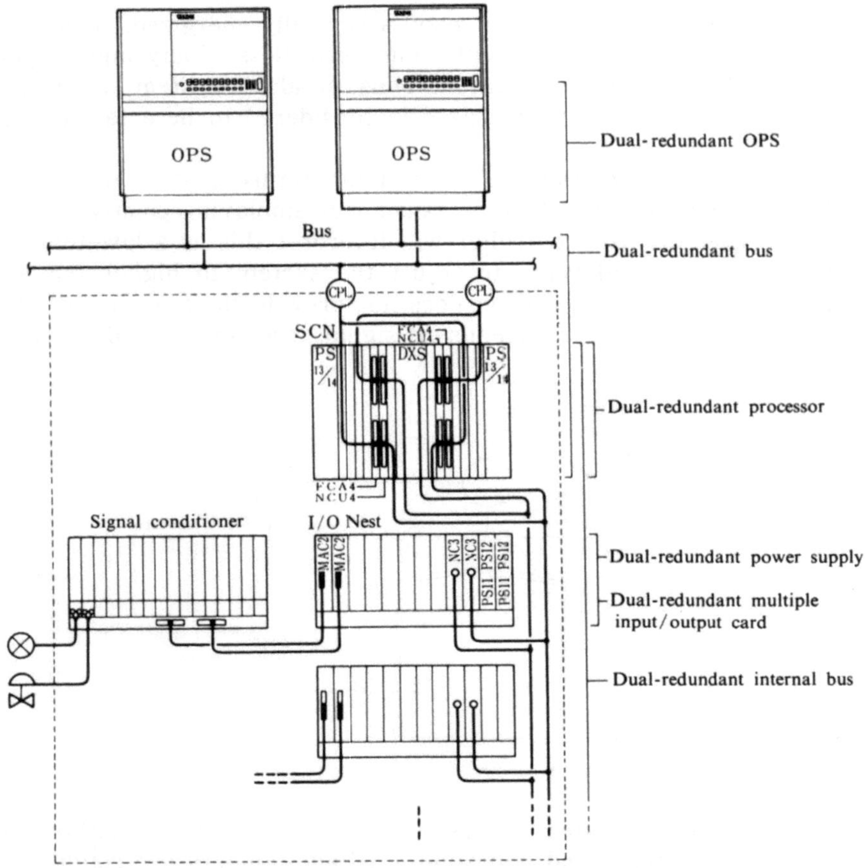

Fig. 7 · 36 Example of a redundant configuration for distributed DDC

mary objective of improving the reliability of a single unit system, a duplicate system can achieve system objectives in a far more cost-efficient manner. An example of a redundant configuration for a distributed DDC system is shown in Fig. 7.36. This configuration features the following redundancy techniques.

(i) *Duplicate OPS* (*Operator station*): By duplicating the OPS, a system characterized by mutual backup can be configured.

(ii) *Duplicate processors*: With two processors, one processor can automatically take over process control in the event that the other processor breaks down.

(iii) *Duplicate internal buses*: By duplicating the internal bus which connects the processor and I/O nest, data transfer can be continued in a normal fashion by one bus if the other one becomes inoperable.

(iv) *Duplicate power supplies*: By duplicating the I/O nest power supply card, one power supply card can continue to supply power to the I/O nest if the other one breaks down.

(v) *Duplicate multipoint input/output cards*: By duplicating the analog input/output cards used for multipoint control, process control can be transferred automatically to one card without influencing the process if the other card suddenly breaks down.

(4) *Instrumentation system interlock* In addition to the need of determining how to effectively implement a plant "trip" as a safety measure for process instrumentation, of more importance is how to plan for safe intermittent plant operation in response to instrument faults. The following describes two methods for safe operation based on an analog instrumentation system having internal decision-making functions.

(i) *Temporary locking of a final control element*: Although use of this technique may be naturally expected, it is still an emergency measure. However, in the case where the problem is determined to be a general one and not partial or localized, this technique is obviously not suitable. For major system faults, step-wise action for bringing about a plant trip is necessary.

Situations appropriate for locking of final control elements can be classified into instrument faults (abnormal signals) and loss of electric or air supplies.

In the case of instrument faults, the type of fault can be determined by detecting abnormalities in signal absolute value, signal rate of change, relevant process variable ratios and process variable deviation (exceeding tolerance values). It should be noted here that except for particular transmitters and clear-cut faults such as thermocouple open wires, it is difficult to definitely distinguish between instrument faults and actual process abnormalities with the above type of abnormal signals. Such a situation is the basis for adopting a redundant control system. For example, in the case of duplicate transmitters where excessive deviation between their two output signals has been detected, both transmitters can be checked during locking of the final control element and the faulty transmitter disconnected. The control loop can then be switched over to the operable transmitter and the interlock cancelled.

In the case of partial electric power loss or air supply loss in pneumatically operated final control elements, locking of final control elements will naturally take place as an emergency measure; direct move to a failsafe condition is not allowed here.

Depending on the importance of the process or control loop, some power-loss countermeasures are separate power supplies for each loop (*e.g.*, battery backup), dedicated power supply for final control elements only (remote manual operation), and local manual operation.

As an example of interlocking during power loss, consider a pneumatic drive mechanism in which interlocking is performed by cutting off the air pressure circuit of the diaphragm or cylinder section. In order to effect a "bumpless" return to automatic mode when the normal air supply is restored, the position of the final control element must be detected during interlock. In addition, controller output which is performing reset windup due to feedback must be pulled back. In the particular example of Fig. 7.37 which incorporates a pulse/pneumatic pressure converter along with a pneumatic pressure/current converter, the current is substituted by the position signal of the final control element and is feed back. The PMD (Pulse Motor Drive) in the figure performs tracking for the controller up until the output signal agrees with the feedback signal from the operating air pressure. Following this, automatic mode is restored, and a bumpless transfer performed.

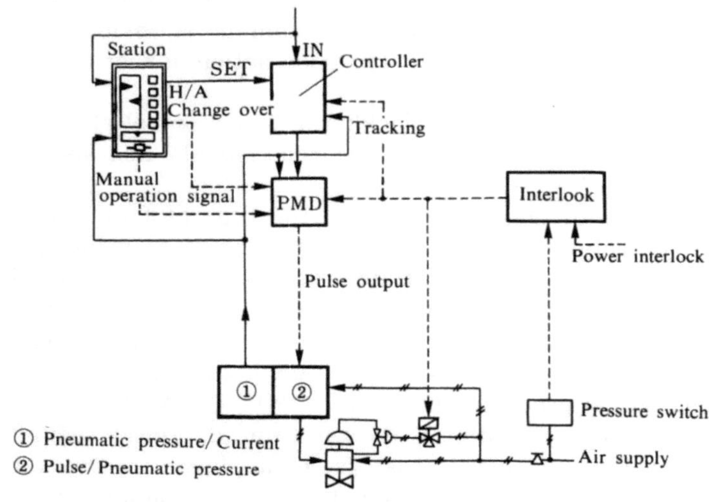

Fig. 7 · 37 Interlock for air-supply loss

(ii) *Running at reduced operating levels*: In the event that one side of a group of auxiliary plant equipment running in parallel is tripped, the flow rate or combustion rate of the associated system must be dropped to a load corresponding to the capacity of the remaining equipment. For example, in the case of boilers in which the rate of combustion is determined by a boiler master signal, if a trip occurs in the water supply pump or the forced draft fan of boiler number 1, the boiler master signal is reduced (runback) in coordination with the turbine governor to a value corresponding to the capacity of the remaining boilers.

The massive buildup in industrial plants and the improvements in efficiency in recent years has also been accompanied by major changes in instrumentation systems, as reflected by the move from analog equipment to distributed DDC, and the rapid adoption of CRT-based operations as a man-machine interface. Moreover, as a result of the high-performance and high-capacity systems made possible by recent advances in electronics and industrial materials, system monitoring and equipment operations has also undergone major transitions, leading to an increased burden on the operator. This situation has required designers to reevaluate the objectives and operability of control panels as centered about the operator, and at the same time has forced them to look at the problem from a total-engineering perspective combining both human engineering and instrumentation engineering elements.

7.3.1 *Human engineering and control panel design*

(a) *The role of the control panel*

As the brain-center of the entire system, the control panel plays the very important role of operating the plant in a safe and efficient manner. To do so, however, it must possess control and monitoring functions to act as in intermediary between man (operator) and machine. Thus, control panel design must consider operability, maintainability and safety in a way that conforms to the type of plant and its objectives, while at the same time easing the burden on the operator and providing for overall safe operation of the entire plant. The role of the control panel is outlined in Fig. 7.38.

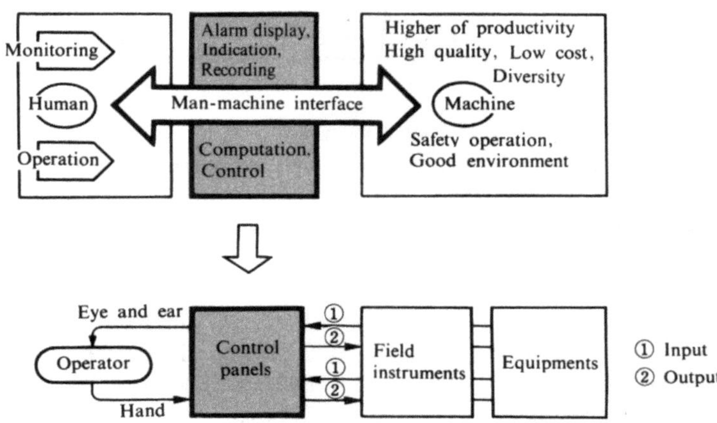

Fig. 7 · 38 Role of a control panel

(*b*) *Body measurements and control panel design*

Although in Japan various studies have been made on the physical dimensions of the human body, data expressly for control panel design has not yet been published. Here, basic body dimensions as taken from human engineering studies are established as a basis for control panel design, and standard configurations for control panel monitoring and operations are proposed based on the results of extensive experimenting with these values.

(1) *Operations and monitoring in a seated position* (Fig. 7.39)

Seated height	880 to 911mm
Chair to eye level	790mm
Chair to shoulder position	590mm
Chair height	410mm
Eye level	1 200mm (1 183mm to 1 213mm)

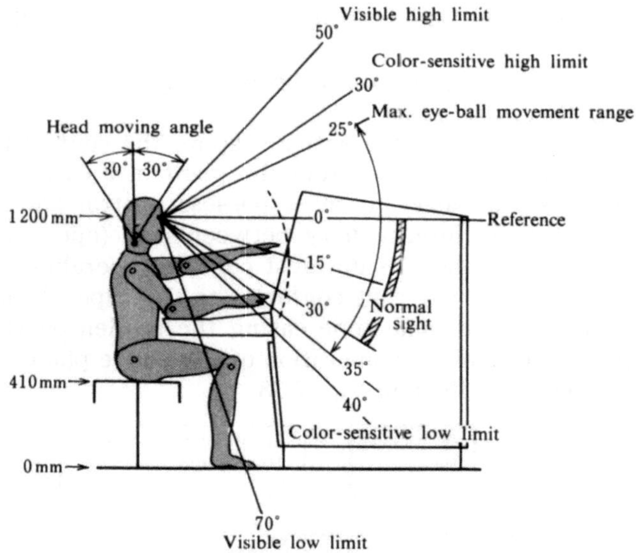

Fig. 7 · 39 Reference dimensions for a sitting-type console

(2) *Operations and monitoring in a standing position* (*basic body dimension + shoe height of 25mm*) (Fig. 7.40)

Body height	1 675mm (1 650mm + 25mm)
Eye level	1 565mm (1 540mm + 25mm)
Shoulder height	1 355mm (1 330mm + 25mm)
Waist height	845mm (820mm + 25mm)

(3) *Operation reach radius* 650mm

(4) *Optimal field of view* 30 degrees from eye level downward, 30 de-

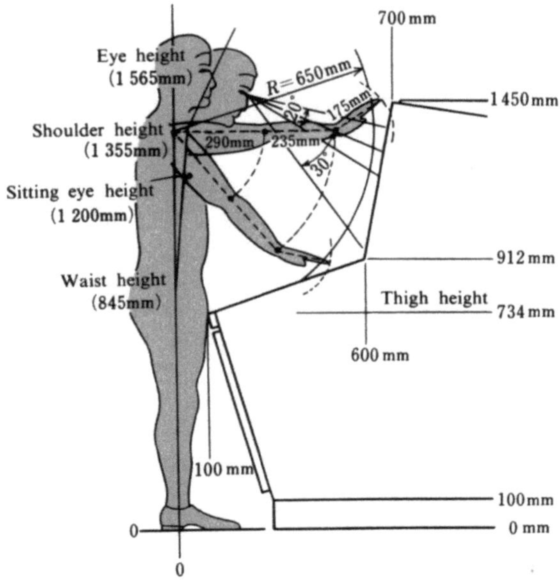

Fig. 7・40 Relationship between monitoring and operation at a console

grees to the left and right of body center.

(*c*) *Monitoring and operations considerations*

We now apply the above body dimensions to some actual examples involving control panel design.

(1) *Vertical type* In the case of a small-scale plant operation, occasions for performing control panel operations are generally few, and such operations are simply done in front of the panel in a standing position. For a vertical control panel incorporating semi-graphics for monitoring, the operator will usually place himself at a position where he can observe the entire panel, and for actual operations, will move right up to the panel. This monitoring position is usually about 2 to 3 meters, depending on the size of the graphics. The relationship between monitoring and operating for a vertical control panel is shown in Fig. 7.41.

(2) *Bench type* In this type of control panel, although controllers, indicators and displays are installed in the vertical section, the operation of various devices such as switches and displays in the bench section is of primary concern. Under usual conditions, the operator will be located in a position from which he can observe the entire panel, and for actual operations, will walk right up to the bench. The control panel should not be overly wide—at the most about 3 meters; some designs include a semi-graphic section on the upper panel.

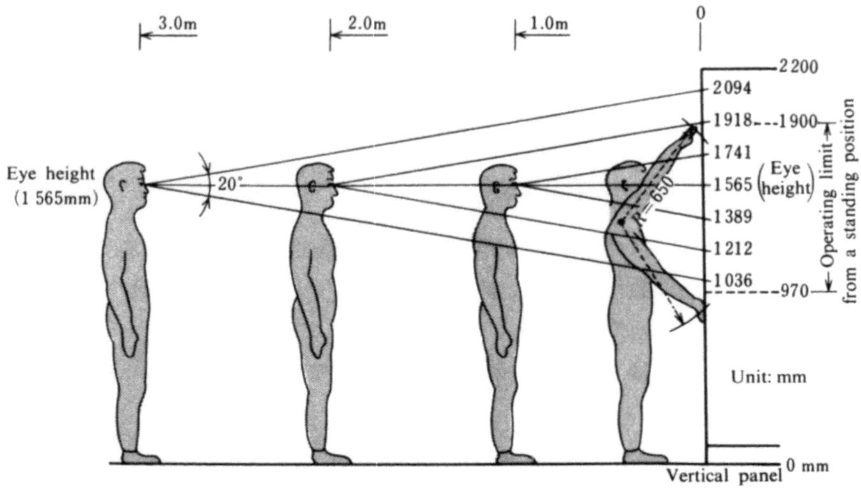

Fig. 7 · 41 Relationship between monitoring and operation at a vertical panel

(3) *Console type* Since the console type usually acts as a main control panel, plant equipment or other instruments (*e.g.*, large-size displays) are often visible behind the console type control panel. Moreover, due to the importance of this type of a control panel, it is desirable to have all operations related to controllers, indicators, recorders and displays within easy reach. In addition, although the operator will stand directly in front of the console during operations, he will normally be situated at a slight distance from the panel. The relationship between operating and monitoring ranges is shown in Fig. 7.42.

(*d*) *Operator and instrument layout*

Design of a control panel must take into consideration both the type of plant operations involved and the operator's responsibilities. For example, typical plant operations will usually include both close monitoring of instrumentation system automatic functions as well as active participation in system control based on operator decisions. Control panel configuration will be different in these two cases. In the former situation, a control panel will consist mainly of graphic displays for monitoring of status and warning information. The latter situation, however, is characterized mainly by bench type and console type control panels incorporating operation instruments such as controllers and switches. The following points must be considered for instrument layout on a control panel.

(1) If possible, the layout of instruments on the control panel should agree with that of corresponding equipment in the plant work area.

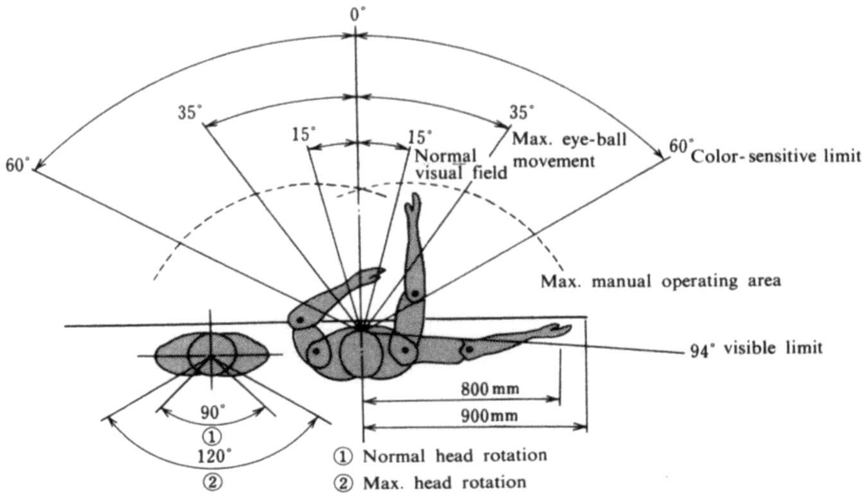

Fig. 7 · 42 Visible range and operation range

(2) Instruments used for abnormal occurrences should be physically separated from general operation ones, or alternatively protected by guard structures or distinguished by color. In addition, such instruments should be located with the optimal field of view as described earlier (within 30° on either side of the normal line of sight).
(3) In the case of two or more instruments making up a sequential operation, they should be arranged so as facilitate such sequential actions (*e.g.*, left to right, arc formation).
(4) Instruments should be laid out with due consideration given to the effectiveness of right-handed operation; controls requiring additional physical strength should also be installed on the right.

7.3.2 Control room engineering

The control room must be a safe and pleasant place to work in; the operator must be able to perform control system operations in an efficient and accurate manner without becoming over fatigued. The following describes the factors that help contribute to optimum design of a control room.

(a) Space and workability

During control room internal-layout design in which the minimum space required must be determined, it is necessary to take into account not only physical factors such as room size and shape, door and window location, number of control panels, pit and duct placement, and maintenance space, but other relevant factors as well such as system objectives, number of operators, and emergency maintenance. Specifically, the following basic items should be considered.

(1) There should be enough space to enable operators and maintenance personnel to perform their work in a smooth and efficient manner.

(2) In the case of several operators performing monitoring and operations, the visibility and field of vision of each operator must be considered, and the layout should be arranged to enable easy communication. There should also be sufficient table or bench area for document handling.

(3) Adequate maintenance space must be provided.

(4) Control panel shape, placement, color, etc., should be determined with due consideration given to their psychological and aesthetic influence on the operator.

(*b*) *Livable environment*

A livable environment is one in which people can comfortably perform their daily activities whether they be work or rest related. For a control room, this would require not only adequate air conditioning, low noise and low dust, but ideally a room with a relaxing color scheme as well. In other words, proper selection of ceiling, wall and floor color, control panel paint, graphic display background color, etc., can contribute significantly to achieving an optimal environment. The following points should be considered when deciding on a control room color scheme (see Fig. 7.43).

(1) *Ceiling color* The color of the ceiling is important since it effects the overall lighting of the control room. Colors such as white, ivory or creme which offer good reflectivity (80 to 90%) and do not tire the eyes are recommended.

(2) *Wall color* Wall colors, having some dependency on ceiling color and windows, are not in general fixed. A common selection, however, are non-glossy colors which prevent glare such as light green or ocher, having a reflectivity of 50 to 60%.

(3) *Floor color* The floor color should provide a reflectivity in the range from 15 to 30%. Extreme bright or dark colors will not only tire the eyes but will bring on general fatigue as well.

(4) *Graphic displays* Since graphic displays have the function of transmitting crucial information to the operator, they must be of a color different from background walls to enable quick and easy recognition. In addition, in order to clearly indicate symbols such as pumps and tanks and their interconnecting pipelines, a low chromatic color is preferable.

(5) *CRT console* As the center of operations, the CRT console

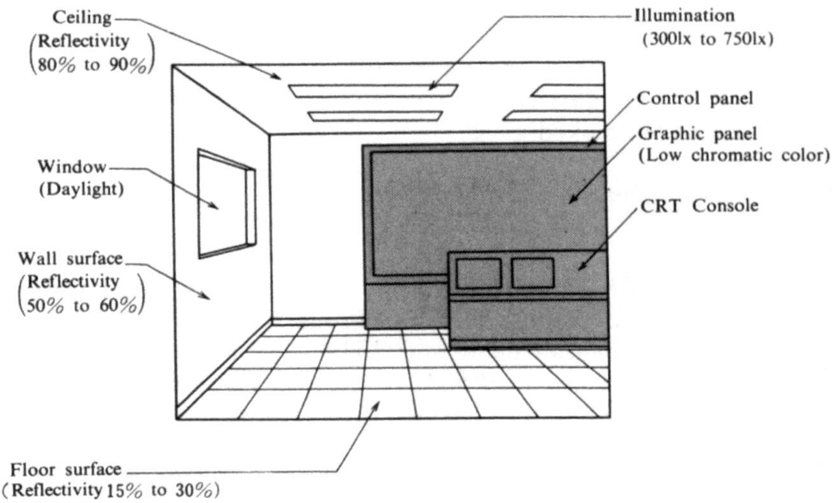

Fig. 7 · 43 Lighting and color arrangement for a control room

should be of a color which is both harmonious and conspicuous with respect to the rest of the control room. Moreover, the CRT console should be placed in a position reflecting its role as the nucleus of the man–machine interface.

(*c*) *Room lighting*

By maintaining appropriate illumination within the control room, operator fatigue can be lessened and efficiency and safety levels improved. Conversely, immoderate lighting and light from outside the control room can place an unnecessary burden on the operator in the form of eye fatigue or chronic visual hindrance. Measures must be taken to eliminate such negative effects.

Control room lighting in general should be in the range from 300 to 750 1x, and should definitely be as uniform as possible within one room. Moreover, lighting fixtures should be selected and installed so that CRT displays and other indicators are not difficult to read due to reflected light. The following points should be carefully considered during room lighting planning.

(1) In the case that CRT and graphic monitoring displays are combined on one control panel, the graphic panel should be relatively bright, and the CRT selection dim.

(2) Lighting in the peripheral area about a CRT screen of about 300 lx is recommended.

(3) As a countermeasure to light reflected on CRT screens from lamp sources, louver–style lighting fixtures should be employed,

and their installation location and angle selected accordingly.

(4) Windows should not be positioned directly in front or behind CRT screens. If for some reason this is not possible, some means of controlling undesired brightness must be employed such as window blinds or smoked glass.

7.4 Instrumentation Work and Startup

7.4.1 Overview

In the design of an instrumentation system, the traditional course taken is as follows: from among numerous process variables, those which best reflect overall plant objectives are first selected, and the loop configurations which measure, monitor and control these variables are then constructed. This is generally true today, but the instruments (*e.g.*, sensors, transmitters, manipulators), monitoring devices and controllers which make up these loops have undergone significant advances in performance and digitization due to the recent progress in electronics and industrial materials. Moreover, their handling and operation have also undergone major changes compared to conventional devices.

For a particular instrumentation system, verification that the above devices are performing satisfactorily is undertaken during instrumentation work and startup. Specifically, if we think of instruments and control devices as the brains, senses and appendages of the instrumentation system, then the role of instrumentation work is to construct a central and peripheral "nervous system," and that of startup is to verify that all functions of the system are operating normally. Instrumentation work and startup has the special responsibility of guaranteeing safe operations, optimizing performance, conserving energy, and preventing harmful environmental effects. Needless to say, this can only be achieved by careful, detailed planning and thorough implementation.

Just as construction of a plant involves a variety of specialized areas dealing with oil, chemicals, steel, power, gas, plumbing, paper pulp, food, first-aid, etc., instrumentation work and startup operations likewise consist of various components each characterized by specialized know-how. In this section, however, we will discuss only those elements common to the overall plant; particular know-how required at each stage of instrumentation work and startup will be taken up in a different publication.

7.4.2 Instrumentation work planning

"Instrumentation" is defined as "the outfitting of measurement, control and other related equipment." Instrumentation work performs the installation, wiring and piping of this equipment so as to enable mutual interaction and holistic functions.

Instrumentation work is distinguished from usual electric and plumbing work for a variety of reasons. Measurement and control equipment require more emphasis on accuracy, and they must function using relatively weak signals. In addition, the actual form of instrumentation implementation may vary depending on actual usage objectives. The following describes the main elements making up instrumentation work planning.

(*a*) *Instrumentation work estimation*

One of the most important factors in carrying out instrumentation work is projecting expenses, *i.e.*, providing an estimate. In this regard, it is common to add up the expenses of all sub-components when determining the total cost of instrumentation work.

Instrumentation work estimation is roughly divided into budget estimation and detailed estimation. In budget estimation, which is performed during plant construction planning, an overall budget structure must be drawn up at a time when many points are naturally still unclear. As a result, total cost calculations must be performed along the way as various conditions are established. Detailed estimation, however, is performed after the instrumentation work order has been set (at least unofficially) and at a time when the engineering phase has already began and the specifications and drawings for instrumentation work are essentially complete. Factors which must be considered when performing estimation are as follows.

(1) *Conditions affecting estimation*

(i) *General conditions*: Work location, payment conditions, time period, source of materials, special work conditions (*e.g.*, night work, hazardous work), temporary utilities (*e.g.*, site electric power, water).

(ii) *Extent of instrumentation work*: Extent of instrumentation installation; wiring, piping, painting and insulation.

(2) *Estimation breakdown*

(i) *Materials cost*: The materials required for each area of instrumentation work are summed up and multiplied by unit costs.

(ii) *Labor cost*: The man-hours required for each area of instrumentation work are summed up and multiplied by unit costs.

(iii) *Other expenses*: These include design expenses, on-site expenses, transport expenses, supervisory expenses, safety measures expenses, on-site out-of-pocket expenses, general management expenses.

(*b*) *Design planning*

Instrumentation work is only one of the various types of construction work (*e.g.*, civil structural construction work, equipment installation, piping, wiring) going on more or less at the same time during plant construction. Accordingly, it is quite important that basic design plans for instrumentation work are submitted to the customer as soon as possible, and that they be included from an early stage in design plans of related construction work (structural construction, electrical work, etc.) as well.

Specifically, instrumentation work specifications which should be determined by this stage are categorized as follows.

(1) *Basic specifications* These include work items, facility interfacing, materials required, extent and type of instrumentation work design, and work schedule.

(2) *Instrumentation room plans* Instrumentation rooms generally consist of operation rooms, control rooms and relevant instrument and device rooms in which plant processes are controlled, monitored and operated. The size and layout of each of these rooms must be determined based on the number, shape, operability and maintenance of the instruments and devices to be installed within each room.

(3) *Cable layout plans* Main cable paths must be determined employing either of two methods: Overhead method which must take into account the process piping framework as well as the plant ceiling, pillar, and wall configuration; and underground method which involves pit layout and construction. The former is considered more appropriate from a cost point of view. In order to negotiate the scope of work here, the planning drawing, including structural size and cable routing, should be submitted to each department, *e.g.*, for the overhead method, to the structural department, and for the underground method, to the earth works department.

(c) On-site instrumentation work control

Control at the work site is divided into schedule control, safety control, and materials control, as described below. Each of these are essential to the smooth execution of instrumentation work.

(1) *Schedule control* Since instrumentation work is only one of several types of construction work going on at the same time as described above, it of course cannot advance at its own pace ignoring the status of other closely related construction tasks. Schedule supervision will thus see that instrumentation work progresses in step with other areas by taking part in the following activities.

(i) *Plant-schedule meetings*: A basic plant schedule, called the "Plant Construction Schedule," is prepared based on a master schedule provided by the customer. This schedule summarizes the beginning and completion dates of each type of construction work required for plant completion.

In connection with the Plant Construction Schedule, representatives from each construction area must hold a plant-schedule meeting regularly to ascertain work progress and to discuss means of resolving any problems, *i.e.*, work behind schedule. The instrumentation work manager should always be in attendance at these meetings to discuss points of particular concern to instrumentation work, and he should obtain the minutes of the meetings for future reference. An example of a Plant Construction Schedule is shown in Table 7.9.

(ii)*Schedule preparation*: The particular schedule for instrumentation work must follow that of the Plant Construction Schedule and must be based on the master schedule mentioned above. Moreover, any changes in the work schedule must be clearly reflected on the appropriate schedule charts after mutual discussion at plant-schedule meetings.

(2) *Safety control* Out of all types of industrial accidents, it is generally felt that the highest rate of occurrence is at construction sites, as is the severity of injuries. Nevertheless, there is also agreement that accidents can be prevented by effective application of engineering, education, and enforcement in a combined, thoroughgoing manner. All parties involved with plant construction must take an active part in implementing safety education and making safety inspection tours, and should strive to maintain an organized, clean and orderly work environment with the attitude of preventing industrial accidents at all cost.

The most common forms of industrial accidents in order of frequency is as follows:

(1) Falls and falling objects
(2) Collapsing structures
(3) Electrical shocks
(4) Heavy machinery (cranes, construction equipment) accidents
(5) Traffic accidents
(6) Fires

(3) *Materials control* Construction materials used in instrumentation work should be kept in on-site storage lockers or in the operations office, and inventory accurately recorded. A specialist in materials supervision should be chosen to oversee such operations and to insure that no losses due to fire or theft occur. In addition, even instrumentation equipment and devices delivered to the customer must be stored and carefully recorded for later instrumentation work. It should always be kept in mind that since much equipment and materials are stored for extended periods of time, unexpected shortages or losses that are suddenly discovered can seriously impede the progress of instrumentation work.

7.4.3 Instrumentation work design

In instrumentation work design, detailed specifications are deter-

Table 7.9 Plant construction schedule

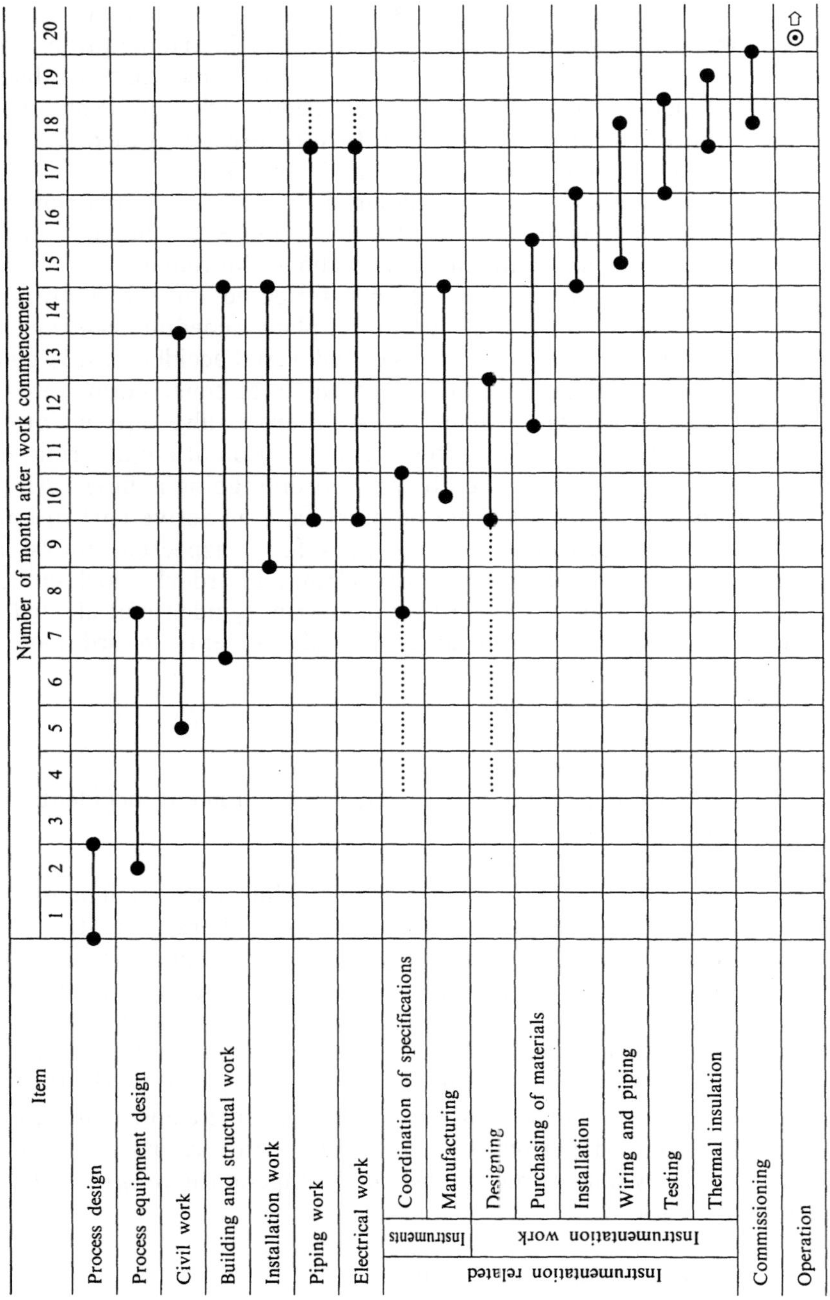

mined and drawings prepared so as to enable cost-efficient installation and hookup, easy maintenance, and high-functionality of the instrumentation system. In order to achieve these goals, it is necessary to prepare for construction work after investigating plant peculiarities and the properties of materials to be measured while following related standards and regulations. In addition, the interlap between the various types of work such as civil structural construction work, installation, wiring, piping, etc., must also be clearly specified.

(*a*) *Design procedure and method*

Instrumentation work design usually begins once drawings and documentation related to plant configuration and processes have been completed, as summarized below.

(1) *Plant documentation* Includes process flow sheets, general plant layout, plant facility layout, process piping diagram and piping standards, process devices and transmitter installation diagrams, structural construction drawings.

(2) *Instrumentation equipment documentation* Includes instrumentation equipment specifications, control-device system configuration diagram, external views of instrumentation and control devices and instruction manuals, control panel drawings, power supply system drawings.

(*b*) *Standards, regulations and guidelines*

In addition to various standards, instrumentation work must also consider a variety of regulations and guidelines. The following are examples of standards, regulations or guidelines which may have to be reflected in design drawings.

(1) *Standards* Industrial, construction, electric, measurement, lighting, painting, pressurized gas, non-destructive inspection, welding

(2) *Regulations* Building, electric, electric components, pressurized gas, anti-pollution, fire prevention

(3) *Guidelines* Electric facilities technology, electric component technology, gas and dust explosion prevention, static electricity safety measures

(*c*) *Design drawings and documentation*

The following types of drawings and documentation may be required in order to carry out instrumentation work (actual drawings required depend on the situation).

(1) *Equipment layout drawing* Shows the layout of control and monitoring equipment used in plant processes in a manner which facilitates operation and maintenance.

(2) *Field-mounted instruments location drawing* Used for verifying the location or installation of instrumentation equipment; this drawing becomes the basis of the instrumentation wiring and piping drawings.

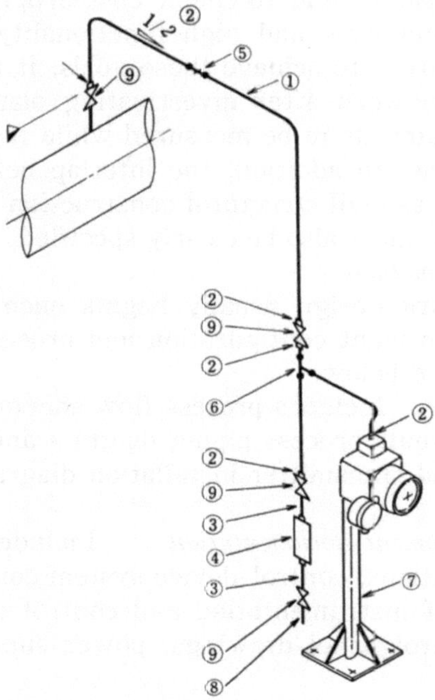

NO.	Description	Material	Dimension	Q'ty	Technical standard
①	Pipe	SUS316	12φ(OD)	12m	
②	Connecter	SUS316	12φ×1/2NPT	5	
③	Nipple	SUS316,SCH40	1/2NPT	2	
④	Drain pot	SUS316,SCH40		1	
⑤	Union joint	SUS316	12φ×12φ	1	
⑥	Union tee	SUS316	12φ×12φ×12φ	1	
⑦	Stanchion	SGP	50φ	1	
⑧	Half nipple	SUS316,SCH40	1/2NPT×50	1	
⑨	Valve	SUS316	1/2NPT	4	JIS10K

Fig. 7・44 Hookup drawing of a pressure transmitter installation

(3) *Hookup drawing* A basic drawing indicating the layout of connection pipes used for conveying pressures and differential pressures from process to field instrumentation devices; materials used are also indicated on the same drawing. An example of a hookup drawing for a pressure conveyor is shown in Fig. 7.44.

(4) *Loop configuration drawing* A system-wide drawing showing the various loops between field instrumentation devices and control and monitoring equipment.

(5) *Instrumentation wiring system drawing* Drawings covering the en-

tire instrumentation wiring system, including signal system drawings, power supply system drawings, and ground connection system drawings.

(6) *Instrumentation piping system drawing* Drawings covering the entire instrumentation piping system, including air supply system drawings, pneumatic signal system drawings, utility piping system drawings, and hydraulic piping system drawings.

(7) *Instrumentation main trunk line drawing* Shows the main trunk lines along which instrumentation wiring, piping and cables are run; also indicates the size and installation method of ducts, racks, pits, etc.

(8) *Detailed instrumentation wiring and piping drawings* These drawings provide detailed information for carrying out actual instrumentation wiring and piping; they are prepared on the basis of the above drawings.

(9) *Auxiliary hardware drawings* Indicates hardware in addition to instrumentation equipment which must be constructed during the instrumentation work phase, such as field control panels, terminal boxes, connection boxes, etc.

(10) *Wiring and piping connection drawings and tables* Provides information on equipment and device input/output terminals for performing instrumentation wiring and piping.

(11) *Work planning documents* These documents provide overall guidelines to help instrumentation work progress smoothly, *e.g.*, work schedules, work organization, and materials management.

7.4.4 Startup execution

Startup operations involve checking, adjustment and commissioning with witnesses of all instrumentation devices and equipment installed during plant construction, *i.e.*, verification that the instrumentation system exhibits the functions as originally planned.

In general, all delivered instrumentation devices and equipment will have been manufactured and delivered under strict quality control procedures. However, even if all devices and equipment are individually in top condition, and installation, wiring and piping work has been performed satisfactorily, it still remains to be seen as to whether the above have been interfaced with other facilities and utilities (*e.g.*, electric and air supplies) to achieve a unified, smooth running system. This type of overall functional verification is referred to as startup operations, or "pre-commissioning."

In recent years, instrumentation systems have been becoming more complex and highly-functional as the scale of industrial plants increases. As a result, startup operations for a new plant have become indispensable, and a high-level of technology in these operations have

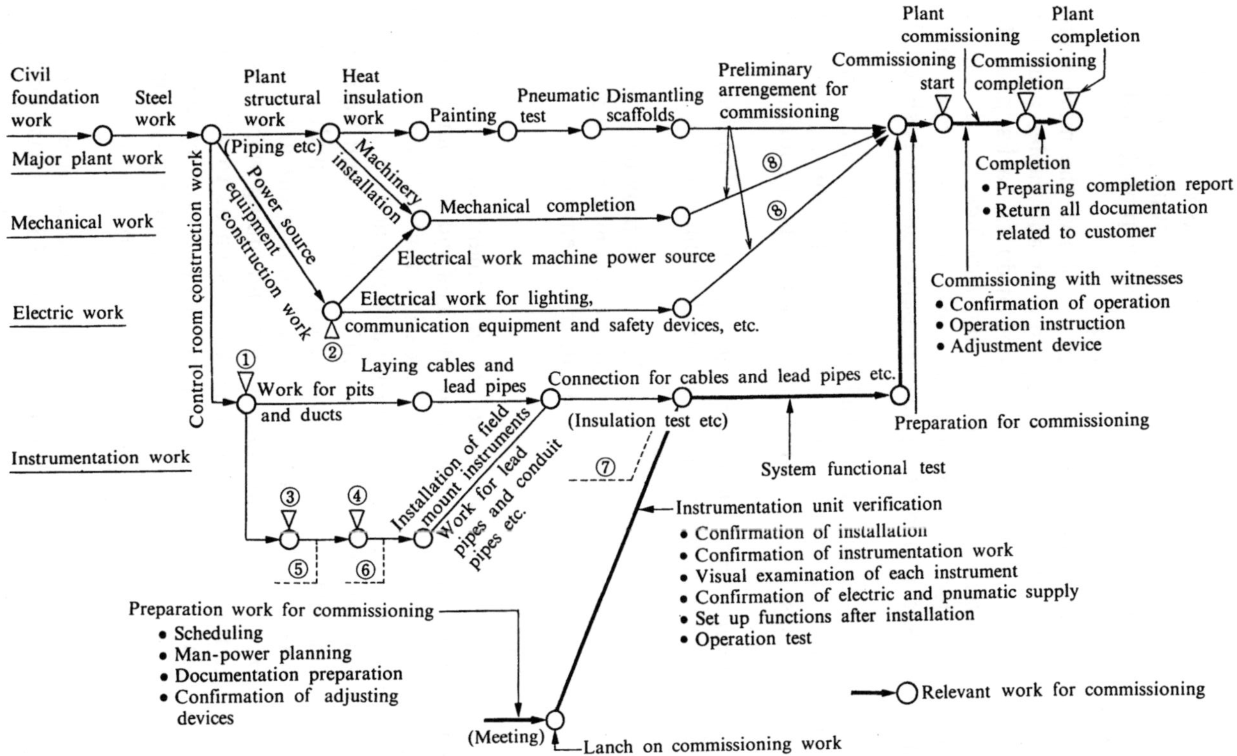

① Control room building completion
② Power source equipment completion
③ Transport of control panels
④ Installation of control panels
⑤ Unpacking and inspection with witnesses
⑥ Acceptance inspection for each instrument
⑦ Test on completion of instrumentation work, Confirmation of instrument installation status

Fig. 7·45 Plant construction process and placement of commissioning work also come to be required.

The schedule and placement of startup operations within plant construction work is shown in the diagram of Fig. 7.45.

7.4.5 Startup operations

(a) Startup preparation

Some preparation work is required to enable on-site startup operations to run smoothly, such as schedule preparation, personnel planning, and general interface meetings, as described below.

(1) *Schedule preparation and verification* The schedule for startup operations must be prepared in conjunction with the overall schedule for plant construction.

(2) *Personnel planning* The number of people required for executing startup operations must be determined in conjunction with the above schedule.

(3) *Documentation/tool preparation* All documentation and references required for carrying out startup operations and those devices and tools required for adjusting and verifying the operation of instrumentation equipment must be organized in advance. A checklist should be used to insure that all necessary items (*e.g.*, documents, procured devices, carried-in devices, calibration equipment) are included.

(b) Individual operation verification

After installation of a particular instrumentation device, its individual operation must be verified, as follows.

(1) *Installation verification* Installation must be performed in a manner that does not hinder device or equipment functions and maintenance.

(2) *Hookup verification* Wiring and piping must likewise be performed in a manner that does not hinder device or equipment functions and maintenance.

(3) *Visual inspection* Equipment must be visually checked for any damages that may have occurred during shipment or for any conditions not within specifications.

(4) *Power/air supply verification* The electric and air supply systems must be checked for proper voltage and pressure levels.

(5) *Equipment setup* After device or equipment installation, some setup procedures may be necessary before actual operation, such as removing internal "stoppers" used to prevent damage during shipment, adding ink to recording devices, and installing radiation sources for radiation analysis devices.

(6) *Individual operation inspection* The accuracy of individual equipment as well as the operation of graphic displays and sequences must be verified by inputting test data.

(*c*) *Loop verification procedures*

The operation of individual loops between field devices and panel instruments must be verified as preparation for commissioning with witnesses.

(1) *Loop operation verification* Various types of loop configurations and operation must be verified with simulation input including computer interfaces, system functions, sequence operations, and analyzer loops.

(2) *Test run preparations* Based on previously prepared materials and documentation, various system constants and transmitters must be readied for operation (so that measurements can be taken).

(*d*) *Test runs*

Apart from verifying with the customer that the instrumentation system is up and running, operation descriptions and adjustment advice are also provided during test runs (commissioning).

(1) *Operation verification* Whether or not the instrumentation system is demonstrating the functions desired is verified.

(2) *Operation description* Operation of instrumentation devices and equipment are explained especially as it concerns operators.

(3) *Adjustment advice* Also intended for the operators is advice relating, for example, to PID parameters tuning and setting of various constants.

(4) *Condition setting of sampling equipment* Various conditions for analyzer sampling equipment are set at this time.

7.5 Quality Assurance

7.5.1 Engineering quality

System engineering performs the role of collecting, integrating, and unifying numerous and diverse forms of technology. Consequently, the task of determining quality standards in such a complex situation can be quite difficult. In many cases, quality will be judged satisfactory or unsatisfactory merely by verbal agreement on a qualitative basis. As a result, quality standards often come to depend on the ability and temperament of the lead engineer and project team members. Also coming into play here is the fact that both current technology and its applicability as well as undeveloped technology are involved, the latter in particular posing new and unsolved problems. Thus, in the light of the above, the question can be asked, "What in engineering has the most sway over engineering quality?" The following describes one approach to this problem.

Engineering capability can be defined as the amount of current technology that can be applied, the extent to which undeveloped technology can be explored, and the level of problem solving skills. These can be broken down as follows.

(1) *Current technology*

The quality and number of engineers

The quality and amount of accumulated know-how

(2) *Undeveloped technology*

Application know-how development capability

(3) *Standardized operations*

Extent to which operation procedures are documented and applied (manual preparation)

Extent to which know-how reuse is systematized and employed

Effectiveness of productivity improvement measures

As can be seen, the quality of engineering and the capability of developing new application know-how still heavily depends on the ability and resourcefulness of individual system engineers and the project team as a whole. Thus, overall quality can hardly be improved without education and some form of self-improvement, and the importance of organized instructional programs along these lines should be evident.

In regards to the reuse of previously obtained know-how, of importance here is the extent to which know-how is systematically collected and made available as a common resource to engineers. A poorly organized system will inevitably result in individual know-how scattered amongst engineers, eventually becoming lost with time. In addition, productivity improvement measures such as know-how standardization and automation should be adopted to promote efficient engineering processes.

The contribution made by each of the elements above to engineering capability is illustrated in Fig. 7.46.

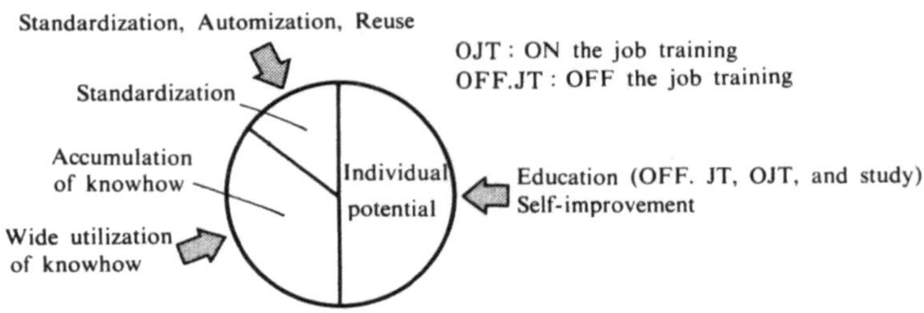

Fig. 7 · 46 Contribution to a company's engineering capability

7.5.2 *Design review (DR)*

Since system engineering has the role of collecting, integrating, and unifying various forms of technology as described above, the amount of knowledge and ability employed has become enormous in proportion to the increased complexity and scale of industrial systems. It is thus not uncommon for an engineer or even the whole project team to lose some control over the situation and make mistakes. This situation has lead to the introduction of "design review" as an effective method of reducing such problems.

Design review involves an organized, detailed review by specialists of each process stage from the beginning of sales operations to the delivery, test run and turnover of the system. Any predicted problems which are uncovered during the review are studied in advance, and appropriate countermeasures taken. Figure 7.47 shows various types of design reviews as applied to the different stages of a job from inquiry to turnover, in particular from a system engineering point of view. Note how for each design review, various methods of quality management such as quality development, FTA (Fault Tree Analysis) and simulation are employed for resolving unclear points.

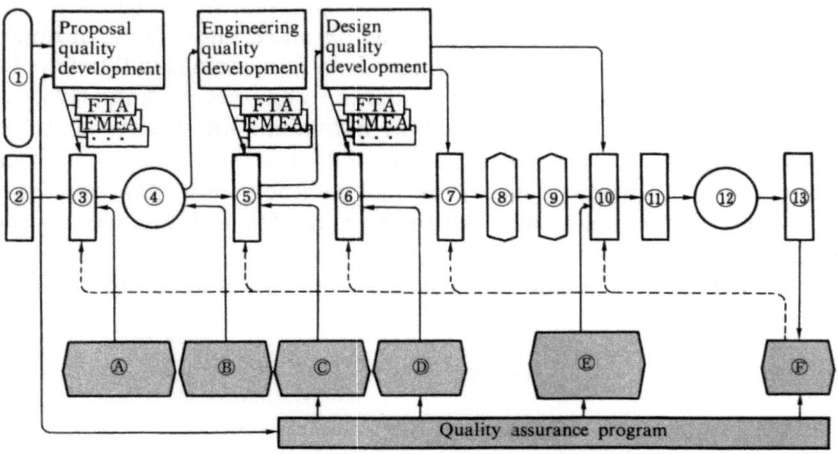

① Customer required quality
② Inquiry
③ Proposal
④ Award
⑤ Basic design
⑥ Detail design
⑦ Manufacturing
⑧ Company inspection
⑨ Witnessed inspection
⑩ Instrumentation work and precommissioning
⑪ Commissioning
⑫ Turn over
⑬ Operation

Ⓐ Proposal review
Ⓑ Proposal final review
Ⓒ Basic design review
Ⓓ Detail design review
Ⓔ Startup preparation review
Ⓕ Overall evaluation

Fig. 7 · 47 System engineering quality assurance

(*a*) *Proposal review*

After acquiring plant planning information in this design review, strategies and tactics employed for accepting a particular order are checked, and countermeasures to problems occurring both before and after order acceptance are reviewed. Moreover, in regards to estimation, the contents of the estimation document are examined to see whether customer needs can be realistically achieved, *i.e.*, whether the quality, cost and schedule requirements of the job are balanced in such a way as to be mutually beneficial to both the instrumentation manufacturer and the customer.

(*b*) *Proposal final review*

Once the order has been finally accepted (at least unofficially) and ready for execution, contract conditions and estimations must be reverified, and target quality, cost and schedule (basic design conditions) clarified.

(*c*) *Basic design review*

In order to obtain concrete estimations of the cost and time required to meet the quality requirements of the customer, basic design work must be checked in so far as it correctly takes into account all relevant factors such as process data confirmation, equipment selection, general specifications, equipment interface, and total system reliability and safety.

(*d*) *Detailed design review*

At this point, the contents of detailed design (also called manufacturing design) are checked to see whether the principles and ideas established by basic design have been faithfully carried over. In particular, since it is common for various departments to be placed in charge of different types of instruments and devices, much emphasis is placed on whether the interface between this equipment has been designed in a coherent, logical manner. Also evaluated is the amount of consideration given to maintenance and flexibility (system additions, modifications, etc.).

(*e*) *Startup preparation review*

Construction of the instrumentation system usually involves final on-site procedures (installation, adjustment and test run). Indeed, it is no exaggeration to say that the installation and adjustment procedures give the finishing touch to the system. Consequently, those engineers or the project team in charge of these areas must become thoroughly familiar with all aspects of the system, and the preparations undertaken for system startup must be perfected as far as possible. The design review in this case thus checks the extent of system comprehension, documentation organization, task descriptions, work scheduling, on-site organization. work conditions, and the like.

(f) Overall evaluation

This evaluation covers the entire length of the job from order acceptance through specifications meetings, manufacturing, inspection, delivery, installation, adjustment, and test runs. Examples of items checked are the level of quality achieved, *i.e.*, does it meet customer specifications, the effectiveness of joint meetings, and the cause of schedule holdups. Note that in order to prevent a self-pleasing review from being prepared, the customer should be requested to participate either directly or indirectly (*e.g.*, questionnaire).

(g) Objects of review and reviewers

Since applying the same depth of investigation to all job tasks is not necessarily the most efficient, the level of review should be decided depending on the difficulty and scale of the task in question. An example of allocating design reviews is shown in Table 7.10.

(h) Design review breakdown

Table 7.11 lists the types of documentation normally required for design reviews, Table 7.12 shows a section of a typical checklist, and Figure 7.48 shows a typical format for recording a design review.

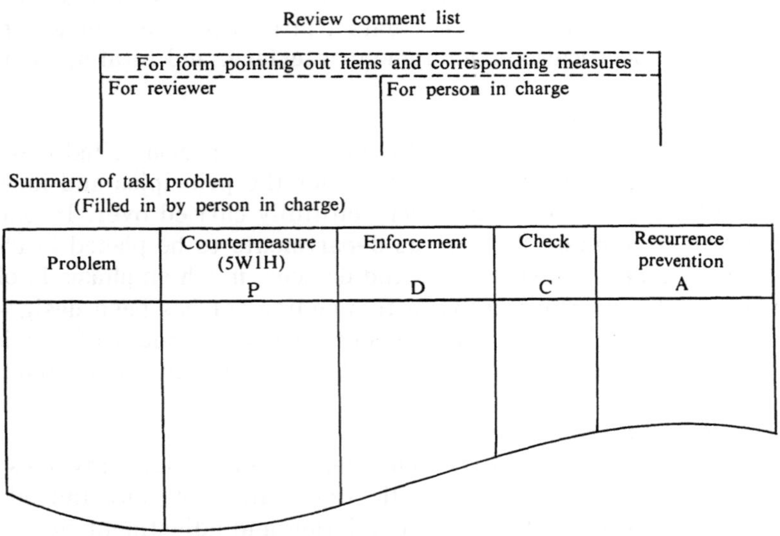

Fig. 7·48 Example of a design review record

Table 7.10 Object of design review and review number

Review unit	Sales headquarters		Department or branch office		Section	
Review subject	(1) Highly-difficult job (2) Large scale job (3) Division manager guaranteed job		(1) Highly-difficult job (2) Medium scale job (3) General manager or branch manager guaranteed job		All small scale jobs	
Review member	Proposal review	Proposal final review	Proposal review	Proposal final review	Proposal review	Proposal final review
Reviewer						
Sales headquarters	[Division manager]	[Division manager]	–	–	–	–
Sales coordination department	General manager	General manager	Manager	Manager	–	–
Sales planning department	General manager	–	Manager	–	–	–
Sales department or branch office	General manager*	General manager*	[General manager*]	General manager*	[Manager]	Manager
System engineering devision	General manager	General manager	Manager	[Manager]	Manager	[Manager]
System order processing devision	–	General manager	–	Manager	–	–
System design division	–	General manager	Manager	–	–	–
Quality control devison	–	General manager	Manager	–	–	–
Review respondent (including salesman in charge)	General manager*	General manager*	Manager	Manager	Section member	Section member
Assistant (including system engineer in charge)	Manager	Manager	Section chief	Section member	Section member	Section member

Note: * : Branch manager, in case of branch office.

[ ] : Chief reviewer

Table 7.11 Typical documents to be prepared for design review

No.	Name of document	Proposal review	Proposal final review	Remarks
1	Job information management table	○	○	From 1. This also serves as a cover sheet.
2	Review record	○	○	From 2
3	Proposal document list	○	○	From 3
4	Review check sheet	○	○	From 4
5	Comment list	○	○	From 5
6	Problems summary table	○	○	From 6
7	Job situation file (with job execution plan)	○	○	From 7
8	Bid analysis	○		
9	Job request sheet		○	In case of an overseas job the overseas check sheet should be attached.
10	Approvals for major terms	○	○	
11	Customer's specifications	○	○	
12	Quotation	○	○	
13	Request for quotation	○	○	
14	Basic data file for quotation	○	○	
15	Competition comparison table	○	○	From 8
16	Approval for award	○	○	
17	Contract document		○	
18	Award analysis		○	
19	Conference notes file	○	○	

Table 7.12 Checklist

Basic design review items			
Evaluation item	Description	Content	Evaluated score
Quality (Q)	Design	What item was most important in choosing the control system? What item was most important in choosing the operation system? What item was most important in choosing the operation method? How can non-standard processing such as start up/shut down etc. be handled? Was there any trade-offs between cost and reliability, maintainability, safety and product performance? (Trade-off is to make compromises between the three essential product elements of quality, cost and delivery date.) How did you arrive at the optimum answer for the trade-off? Is the scope of service defined? Are the specifications for resale products defined? Are the specifications for instrumentation work defined? Are the specifications for service defined? Have past problems been investigated and countermeasures taken? If new devices have been designed, was a check made for any infringements of patents or utility models? Did you apply for patents or utility models? Are extensions or revisions considered?	
	Reliability	Is the definition for reliability of the system clear? Did the customer agree to them? Are the requisitions from the customer clear and does this system satisfy them? How is the reliability evaluated (quality function development, FMEA and FTA etc.)? FMEA(failure mode and effects analysis) FTA(fault tree analysis)	

Table 7.12 (continued)

Evaluation item	Description	Content	Evaluated score
	Reliability	How was the reliability certified (redundancy, failsafe and foolproof etc.)? Redundancy(providing the alternative measures in the case of failure) Failsafe (safe is insured if the function is lost) Foolproof (prevention of human errors or careless mistakes) Are environment specifications clear and does this system safety them? Are the load factors and allowances sufficient? Is the derating sufficient?	
	General	Are you familiar with the background for the award of this job? Are you familiar with the production plan of this plant? Are you familiar with the object and expected use of this system? On realizing them, what are the important points of the system? Are points of disagreement and alternatives confirmed with the customer and the agreement reached documented? Are detailed discussion items confirmed and agreement reached? Are the problems resolved for the proposal review and the proposal final review? And if there are any problems remaining, are the countermeasures for the remaining problems clear? Are all the specifications of the customer's requisition described on the drafts of approval drawings? How is the satisfaction of the customer confirmed for the requisition specifications? Are there any problems to be resolved regarding the requisition specifications? If problems exist: Are the causes understood? Have countermeasures been considered already? Have prevention procedures for the recurrence of these problems been implemented? Are there any undetermined items in the specifications? If so: Are the schedules for determination of them established? How did the prevention for specifications changes perform?	

Design	Did you check all applicable laws and regulations for compliance? Are the countermeasures adequate for preventing explosions? Is there consistency between system configuration and instruments? In the case that the control system includes new products to be developed, does the agreement include each section concerned (such as development, manufacturing, documentation, and postservices)? Are the alternatives to nonstandard components investigated using standard components? Are the specifications, functions and characteristics of nonstandard components defined? Are mill certificates necessary?
Reliability	Is the countermeasure for excess heating complete? Is the countermeasure for excess noise complete? Are the measures for dust-proof and drop-proof complete?
Maintainability	Is space for maintenance secured? Is failure investigation difficult? Is installation or dismantle easy? Is inspection easy? How much time is forecasted for maintenance? How much is forecast for availability (reliability or production ratio)? To what extent does it safety customer requests? Are the measures for preventive maintenance defined? Are the PL and PS requested? What kind of measures are taken? PL (product liability; product responsibility) PS (product safety; safety for products) Is the requisition for maintenance personnel defined? Does the customer agree to them? Is the selection criteria for spare parts and tools appropriate?

Table 7.12 (continued)

Evaluation item	Description	Content	Evaluated score
	Safety	How does the system react to emergencies in the process and how does its behavior reflect on system design? How does the system react to emergencies in utilities and how does its behavior reflect on system design? How does the system react to emergencies in the system and how does its behavior reflect on system design? Has the affect of momentary and long-term failures of the systems been considered? How was hazard-analysis (safety evaluation) carried out? Are there any problems? Are the countermeasures sufficient?	
Cost (C)		Is the building cost at the time of the construction award confirmed? Was the cost verified at the start of design work? What is the cost ratio (actual cost/forecasted cost) at the end of design work? Are there any improvements in cost by comparison with the cost verified at the start of design work? Are any improvements developed for this job applicable to other projects? Are there any problems? Are countermeasures defined after looking for the causes of any problems? Are preventive procedures for these problems implemented? Are all problems on proposal review and proposal final review resolved? Are countermeasures for unresolved problems sufficient? Were there any changes in specifications? Did you estimate these changes? Is the cost to be forecasted for future work increasing? Are countermeasures against this matter ensured? Is there any possibility for improvement? If so, what measures are effective?	

Delivery date (D)	Was the master schedule issued? Are the any schedules unwritten in the master schedule? Was the master schedule agreed to by the customer? Was the master schedule agreed to by the department and section in charge? What is the critical path? Are there any problems that will affect the process? Are there any special comments for this job? What kind of response is made to problems that arise? Is there any plan? Are all the problems regarding proposal review and proposal final review resolved? Are the countermeasures for unresolved problems sufficient?
Quality (Q)	Did the customer agree with the specifications for witnessing inspection? Are there any items for special inspection and their preparation ready? Are the style, content, language, type, and number of approval drawings and as-built drawings defined? Are the persons in charge of preparing inspection manuals with witnesses defined? Is the curriculum of education, text books, language, instructors, place, date, and number of trainees defined? Are the points to be considered regarding transportation defined? Are the points to be considered regarding storage defined? Is the packing manual defined? Is the language to be used in documents defined?

REFERENCES

1) *Automation*, **28**, 1 (1983) (in Japanese).
2) *Automation*, **30**, 4 (1985) (in Japanese).
3) A View of Engineering Industry, Toyo Keizai Shinpo Sha (1977) (in Japanese).
4) *Yokogawa Tech. Rep.*, **19**, 3 (1975) (in Japanese).
5) The Association of Japan Instrumentation Industry: Instrumentation Work Manual (1984) (in Japanese).
6) T. Fuzita: Process Instrumentation, The Association of Japan Instrumentation Industry (1984) (in Japanese).
7) A. Kobayashi, *et al.*: Instrumentation Work Planning and Execution, Tokyo Denki University Press (1984) (in Japanese).
8) Editorial Committee of Anthropometry for Ergonomics: The Measure of Man, Ningen to Gijutsu Sha (1979) (in Japanese).
9) Study for Terminology of Project Management: ENAA–1984–MANAG–3 The Japan Machinery Federation and Engineering Advancement Association of Japan (1985) (in Japanese).

Chapter 8

ADVANCED CONTROL

Although feedback control in distributed-type instrumentation systems is still centered about PID control, "advanced control" techniques which make use of the special features of digital equipment are also coming into wide use. Many functions which have been difficult to realize in analog devices can now be implemented relatively easy in a digital format. For example, dead-time units and also complicated loops such as non-interacting control can be easily accommodated in digital systems. This chapter describes advanced control as a control format on a somewhat higher level than PID control having relatively high applicability.

8.1 Control Theory Considerations

In process control which started out with an emphasis on PID control, the control system is analyzed and designed based mainly on frequency response. This was a relatively familiar form of control even amongst control technicians. In the 1960s, however, the state-space method was introduced in which a system is described by simultaneous differential equations. This method was used as a basis for the development of various optimal control theories. Following this, design methods related to the frequency domain were reconsidered, resulting in the appearance of INA and other methods applicable to multivariable control system design. This led to the realization that modern design theory cannot be separated into that for the frequency domain and that for the time domain, but rather must be developed on the basis of a uniform theory.

An overview of the applicability of process control theories is shown in Table 8.1. This table shows typical techniques employed for the different categories of system expression, control system design, control system configuration, and identification method. Although the

Table 8.1 Over view of process control theory

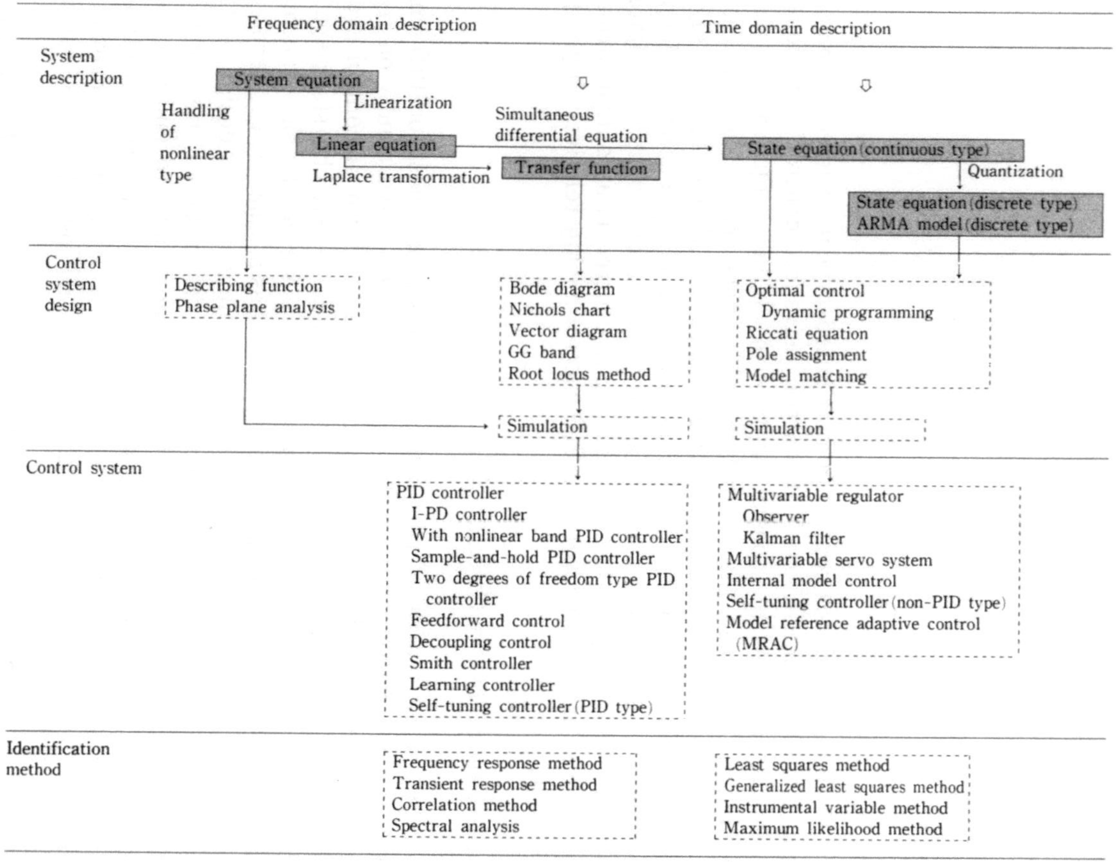

table is generally divided into frequency domain and time domain applications for convenience sake, it must be kept in mind that, as mentioned above, this is not absolute, and that a very definite mutual relationship exists between the two.

A question that might now be asked is: where within the overall system are process control techniques applied? Figure 8.1 shows how process control techniques can extend over a rather wide range of system related components. Although for matters of control, concentration on only control characteristics is relatively straightforward, design work must also take into account the adjustment and maintenance of the control system. As shown in the figure, various tech-

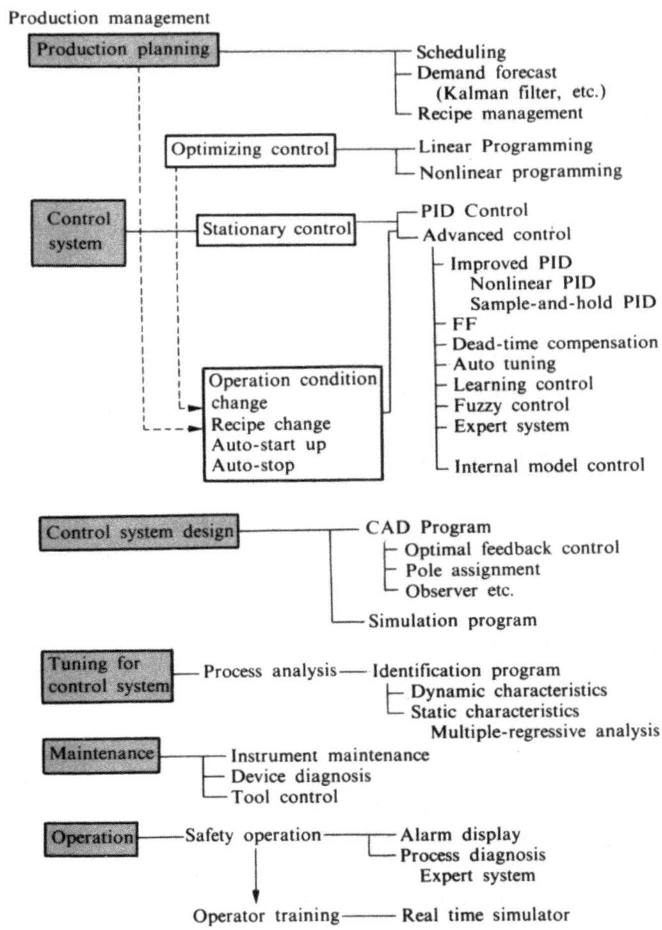

Fig. 8·1 Applications of process control techniques

niques are being developed for improving overall maintenance and system operation in addition to direct system control. However, this chapter is not intended to cover such a broad range of topics, and we will concentrate on the area of advanced control as delineated in the figure.

Advanced control can be divided into control which simply adds new functions to a traditional PID controller and that having structures different than those of PID, as summarized in Table 8.2.

8.2 Feedforward Control

Feedback control, as described in Chapter 2, is characterized by the detection of any deviation arising between the desired variable and the controlled variable, followed by attempts to eliminate this deviation. Thus, if no deviation is detected, no control actions are taken. If, on the other hand, disturbance in the system can be measured and its influence predicted, countermeasures to prevent the appearance of any deviation can be taken beforehand. This type of control is called "feedforward control."

8.2.1 Feedforward control in a heat exchanger

We consider the example of a heat exchanger to help explain the fundamentals of feedforward control. In the traditional feedback control format shown in Fig. 8.2, the temperature T_2 of the outgoing fluid is the object of control. If the amount of incoming fluid is suddenly increased, the temperature T_2 naturally decreases, and the temperature controller TC 1 is activated. As a result, the amount of steam is increased, and a new equilibrium state is eventually reached. The problem here, however, is that the temperature goes through a period of fluctuation before setting down, as shown in Fig. 8.3. If, however, the amount of incoming fluid is measured and the amount of steam increased in a corresponding manner, a temperature variance should not appear.

Heat equilibrium in the system of Fig. 8.2 can be expressed by the following formula:

$$W_S H_S = W_P C_P (T_2 - T_1) \tag{8.1}$$

Here, W_S is the rate of steam flow, H_S is the latent heat of the steam, W_P is the rate of incoming fluid flow, C_P is the specific heat of the fluid, T_2 is the output temperature, and T_1 is the input temperature. Solving Eq. (8.1) for the amount of steam required, we get

Table 8.2 Various measure of process control

	Type of control	Main purpose of control system	Example
	PID control		
PID+Advanced control	With nonlinear band PID controller	Countermeasure for non linearity	PH control
		Countermeasure for random noise	Level control
	Sample-and-hold PI controller	Countermeasure for dead time	Component control
	Smith controller	Countermeasure for dead time	Component control
	Feedforward controller	Countermeasure for disturbance (load change etc.)	Heat exchanger
	Non-interacting control	Countermeasure for interaction	Distillation column
	Override control	Constraint control	Compressor
	Two degrees of freedom PID control	Optimal response both disturbance and setpoint change	
	Self-tuning controller	Countermeasure for parametiric change	
Non-PID control	Pulse transfer function control	Countermeasure for dead time	
	Optimal feedback control	Advancement of controlability	
	Sliding mode control	Advancement of tracking robustness for setpoint change	
	Adaptive control	Countermeasure for parameter change	
	Fuzzy control	Applied to the difficult process in making process model	
	Expert method		

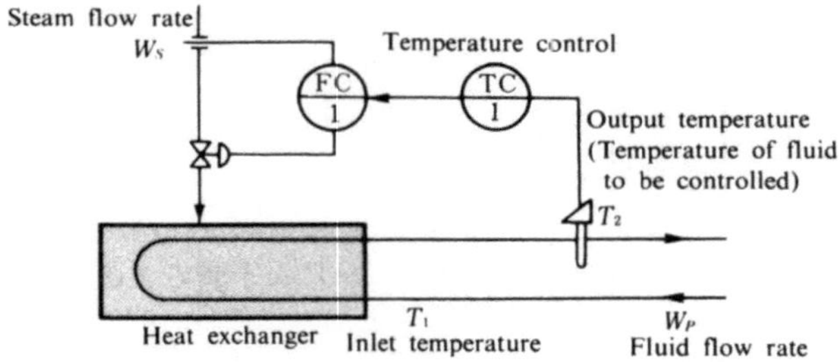

Fig. 8・2 Control of a heart exchanger

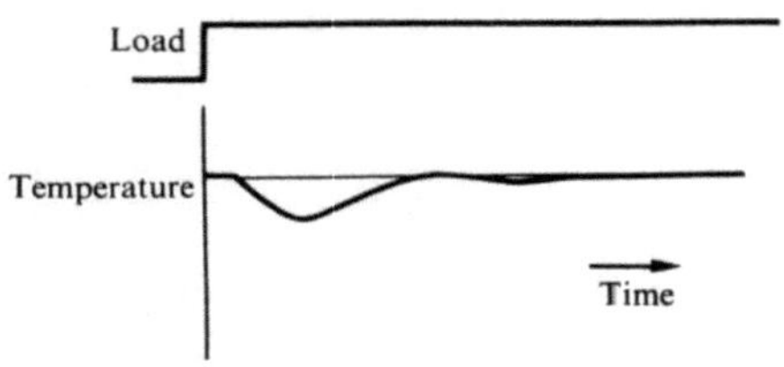

Fig. 8・3 Temperature change due to the load step change

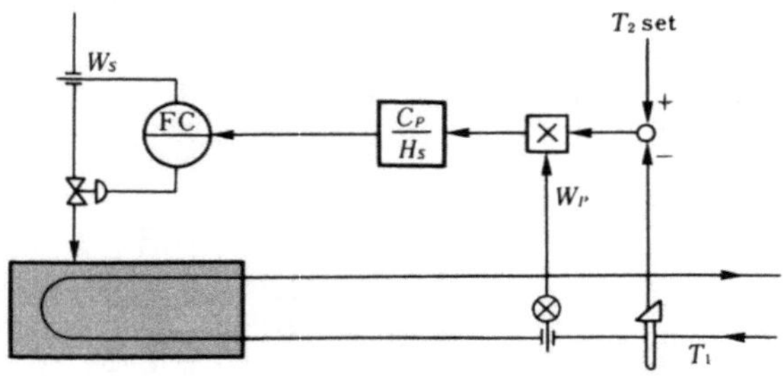

Fig. 8・4 Feedforward control of a heat exchanger

$$W_S = \frac{C_P}{H_S} W_P (T_2 - T_1) \tag{8.2}$$

Thus, in order to supply this amount of steam to the system, we can employ the feedforward configuration shown in Fig. 8.4.

8.2.2 Combining feedforward control and feedback control

Although feedforward control is highly effective, it is difficult to completely match its gain with the process. Consequently, the use of only feedforward control is seldom employed, and combined use with feedback control is more common. An example of such a combination is illustrated in Fig. 8.5. In the figure, TC 1 output α is used to negate

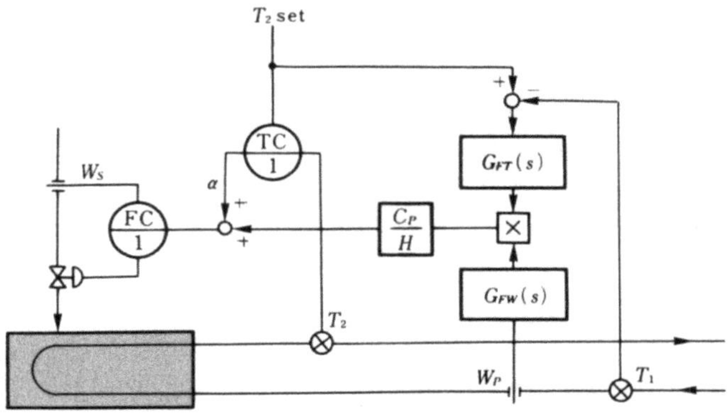

Fig. 8・5 A combination of feedforward and feedback control

the error in the feedforward control, and transfer functions $G_{FT}(s)$ and $G_{FW}(s)$ compensate for dynamic characteristics. These elements are explained in more detail below.

8.2.3 Determination of feedforward elements

The above example considers equilibrium characterized by a steady-state situation between the fluid load and equalizing steam. In actuality, however, the time delay following changes in the amount of load and steam up until a change in temperature appears must be taken into account. To better comprehend the nature of this delay, we make use of the generalized feedforward control system shown in Fig. 8.6. Disturbance D having characteristics $G_D(s)$ produces a change in process variable PV, and a change in manipulated variable MV appears after passing through $G_P(s)$. Note that manipulated variable MV is orig-

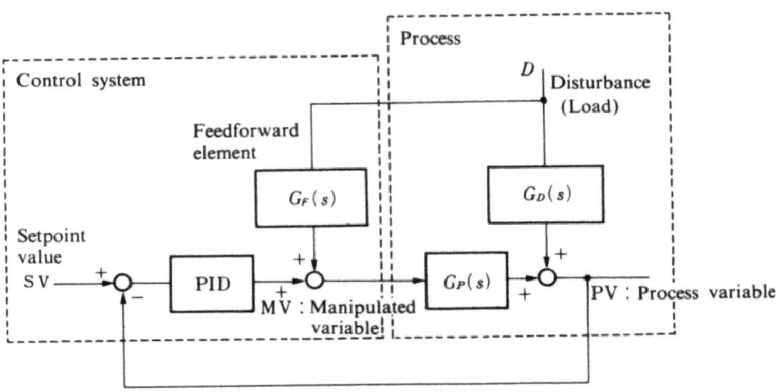

Fig. 8・6 Feedforward control system

inally the sum of the output of feedforward element $G_F(s)$ and that of the feedback-oriented PID controller.

Now, if we assume that process characteristics $G_P(s)$ and $G_D(s)$ are already known, how should feedforward element $G_F(s)$ be determined? Since the objective is to prevent disturbance D from affecting process variable PV, one method is to have the sum of the two paths by which D can reach PV, *i.e.*, the sum of $G_D(s)$ and $G_F(s)G_P(s)$, be zero. In other words,

$$G_D(s)+G_F(s)G_P(s)=0 \tag{8.3}$$

Solving for $G_F(s)$, we get

$$G_F(s)=-\frac{G_D(s)}{G_P(s)} \tag{8.4}$$

If, for example, the process is characterized by first-order lag and dead time, the denominator and numerator of the above would become

$$G_P(s)=\frac{K_P e^{-L_P s}}{1+T_P s} \tag{8.5}$$

$$G_D(s)=\frac{K_D e^{-L_D s}}{1+T_D s} \tag{8.6}$$

and the feedforward element $G_F(s)$ on substitution of the above would then be

$$G_F(s)=-\frac{K_D}{K_P}\cdot\frac{1+T_P(s)}{1+T_D(s)}e^{-(L_D-L_P)s} \tag{8.7}$$

However, a point of caution here. If $L_D \geq L_P$, there are no problems, but if $L_D < L_P$, it is impossible to realize, and a situation unsuitable for feedforward control appears. If feedforward control is attempted, then, excluding the dead-time term in Eq. (8.7), the lead in the lead/lag element will be somewhat overly used. The characteristics expressed by Eqs. (8.5) and (8.6) are appropriate for measurement of step responses in an actual plant environment.

8.2.4 *Feedforward control application examples*

(1)*Feedforward control in a furnace* Figure 8.7 shows an example of temperature control in a furnace. Here, feedforward control is performed by manipulating the amount of fuel in response to the amount of raw materials input into the furnace.

(2)*Feedforward control in a distillation column* Figure 8.8 shows an example of feedforward control in a distillation column, in which the amount of reflux and reboiler steam is adjusted corresponding to the

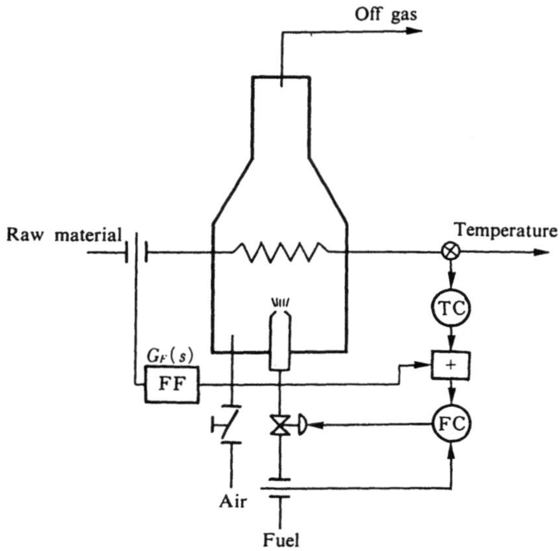

Fig. 8・7 Feedforeward control of furnace

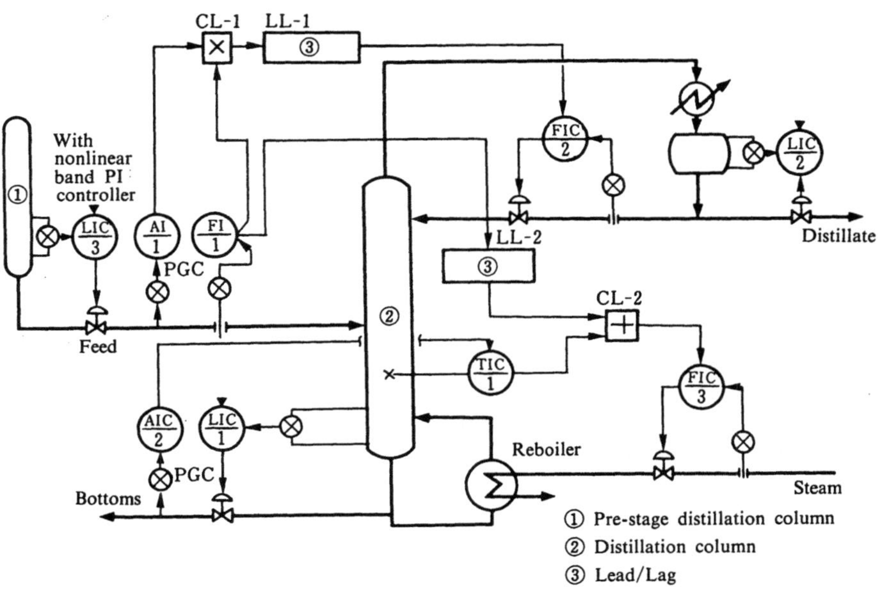

Fig. 8・8 Control of distillation column

rate of feed flow and its composition. Also employed here is feedback controller TIC–1 which treats the amount of steam as a manipulated variable. In addition, lead/lag elements LL–1 and LL–2 are used for compensation of dynamic characteristics.

8.3 Control of Dead–Time Processes

8.3.1 Dead–time processes

The period during which the effects of input applied to a process or system have not yet appeared is referred to as "dead time." Dead time may take on a variety of forms within a process control loop, ranging from pure dead time as in the delay in conveying materials or in performing analysis in an analyzer, to apparent dead time in higher–order lag systems. In any case, if dead time exists within a closed loop, control will usually be difficult. The above can be easily understood by the following example. There are probably many people who have experienced the unexpected difficulty of trying to adjust the water temperature of a shower (Fig. 8.9). To raise the temperature of the water slightly, the first step is to open the hot–water faucet in what is hoped is a corresponding amount. If, however, no change is felt after a short period, it is usually thought that the faucet was not opened far enough, so that it is then quickly opened even further. The end result may be a sudden onslaught of scalding hot water. The desired temperature will never be reached in this way, and the best alternative, as most people have learned, is to manipulate the faucet slowly in small increments.

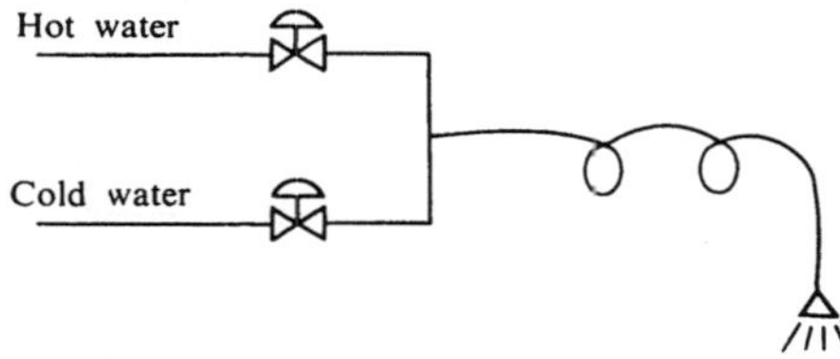

Fig. 8・9 Temperature control of shower

This situation also holds true for automatic control in a controlling device. Using Fig. 8.10 as the basis for a slightly theoretical explanation, the following conditions related to the gain and phase delay must be satisfied for oscillation of the closed loop to occur:

$$|G_c(s)G_p(s)|=1 \tag{8.8}$$

$$\angle G_c(s)G_p(s)=180^\circ \tag{8.9}$$

where $G_c(s)$ and $G_p(s)$ are the controller and process transfer functions,

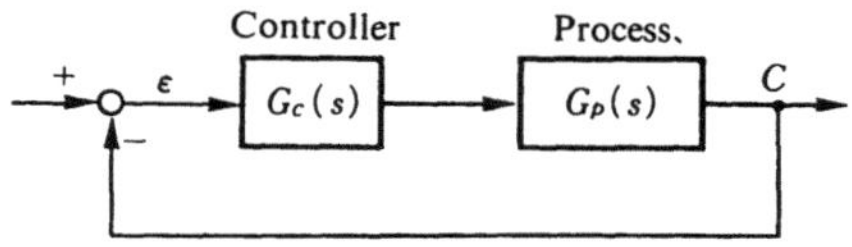

Fig. 8 · 10 Block diagram of control system

respectively.

Case 1: $G_p(s)=1/(1+T_s)$ For a process having a first-order lag, if we set $s=j\omega$, the phase delay is given by

$$\angle G_p(s) = -\tan^{-1}\omega T \tag{8.10}$$

As can be seen, even if it is assumed that $\omega=\infty$; the most the phase delay can be is 90°. Accordingly, for control by a P controller or a PI controller, stability, at least theoretically, can be achieved whatever the gain.

Case 2: $G_p(s)=e^{-Ls}$ In this case, in which the process includes dead time, we obtain the following using Euler's theorem:

$$\begin{aligned}\angle G_p(s) &= \angle e^{-j\omega L} = \angle(\cos\omega L - j\sin\omega L)\\ &= \tan^{-1}\left(-\frac{\sin\omega L}{\cos\omega L}\right) = -\omega L\end{aligned} \tag{8.11}$$

Thus, as ω increases, the phase delay becomes infinitely large, and if the controller's gain becomes sufficiently high, instability will occur. Since an actual process will be characterized as superimposed by both first-order lag and dead time, its transfer function can be approximated as

$$G_p(s) = \frac{e^{-Ls}}{1+Ts} \tag{8.12}$$

Thus, the ratio L/T can be used as indicators of the difficulty of control. For example, the larger the value for L/T, the more difficult is the control. Figure 8.12 shows examples of step responses to step disturbances applied to the control system of Fig. 8.11. As can be seen, if T is held constant and L varied, longer L times result in longer settling-down periods.

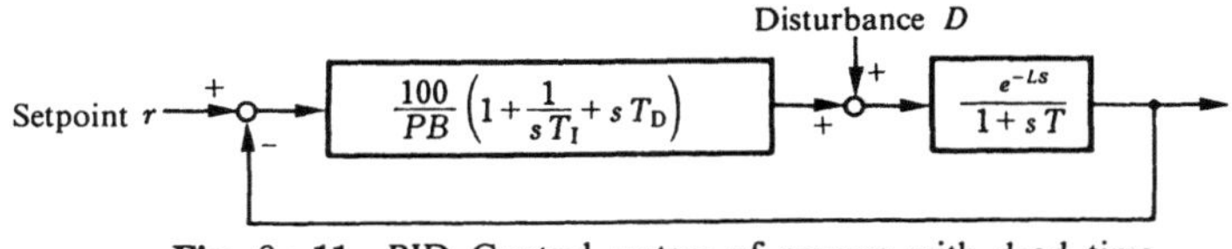

Fig. 8 · 11 PID Control system of process with dead time

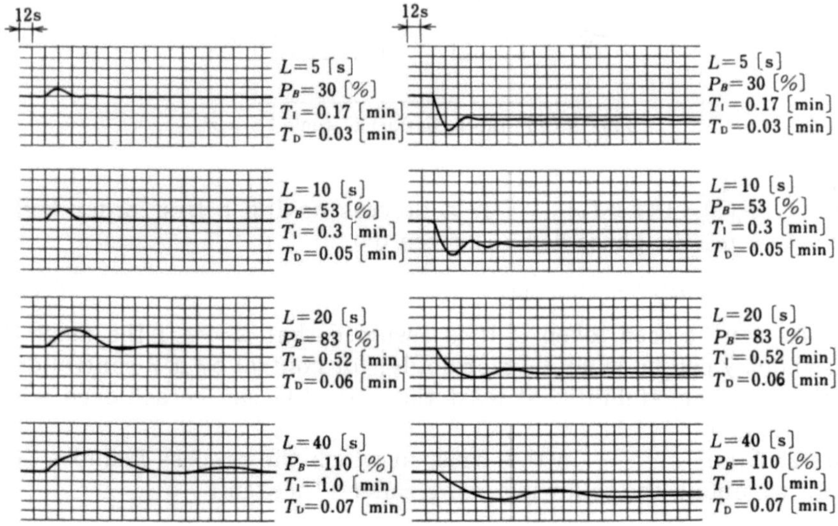

Fig. 8・12 Step response to setpoint change in control system using the PID controller (T=20[s])

8.3.2 Smith controllers

(a) Principles of Smith controllers

Two common techniques for improving the controllability of processes with large dead times are the Smith controller and sampling controller. The following will first discuss the Smith controller.

The Smith controller technique[1)] was first proposed by Otto Smith. As shown in Fig. 8.13, a process model within the control system is included, and if we exclude the dead-time, the process is expressed simply as $G(s)$. In order to better understand the actions of the control system here, we employ the equivalent transformations shown in Figs. 8.14 (a),(b) and (c), disregarding disturbance D. Comparing the equivalent diagrams in (a) and (c), we can see that the dead-time factor has been moved outside the loop. Thus, since no dead-time is present with-

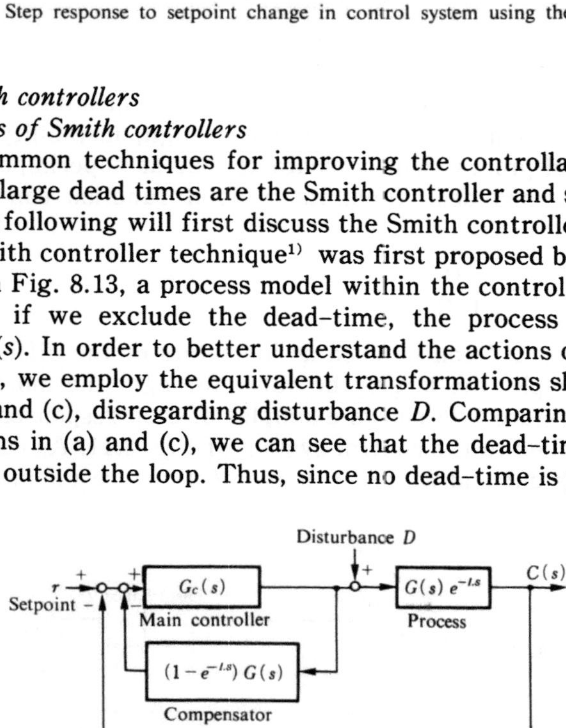

Fig. 8・13 Control system using Smith controller

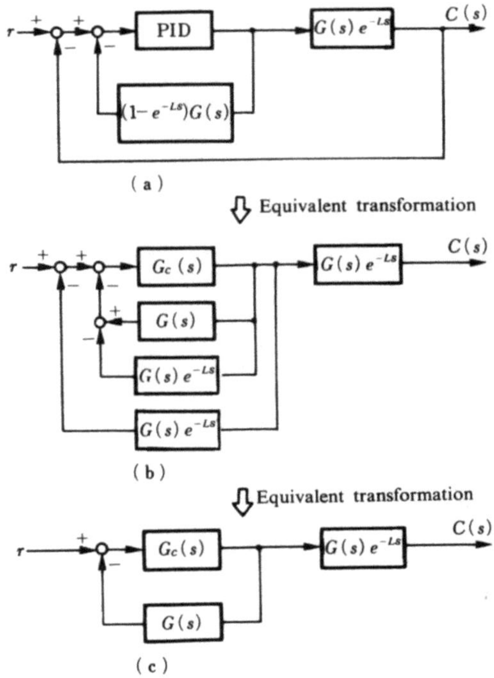

Fig. 8・14 Equivalent transformation in Smith controller system

in the closed loop, it is possible to construct a controller with high gain and fast response characteristics.

Step responses for a variety of L/T values are shown in Fig. 8.15. The improvement over the responses in Fig. 8.12 are readily apparent. An application example of the Smith controller is shown in Fig. 8.16.

(*b*) *Process model mismatch*

Although good control can be realized if the process model within the Smith controller faithfully matches the actual process, in actual systems this is nearly impossible to achieve. Expressing the process by dead time and first-order lag in itself includes some error, as does the values of L and T. Accordingly, the effects of mismatch must always be considered. The effects of dead-time mismatch are illustrated by simulation in Fig. 8.17, and those of time constants mismatch in Fig. 8.18. It should be noted here, however, that the extent of mismatch influence depends on the strength of main-controller tuning. In other words, if the main controller experiences strong tuning, the mismatching margin becomes small, while if it experiences weak tuning, the mismatching margin becomes large. Thus, as a compromise between the two must be reached, the extent of mismatch must be predicted and

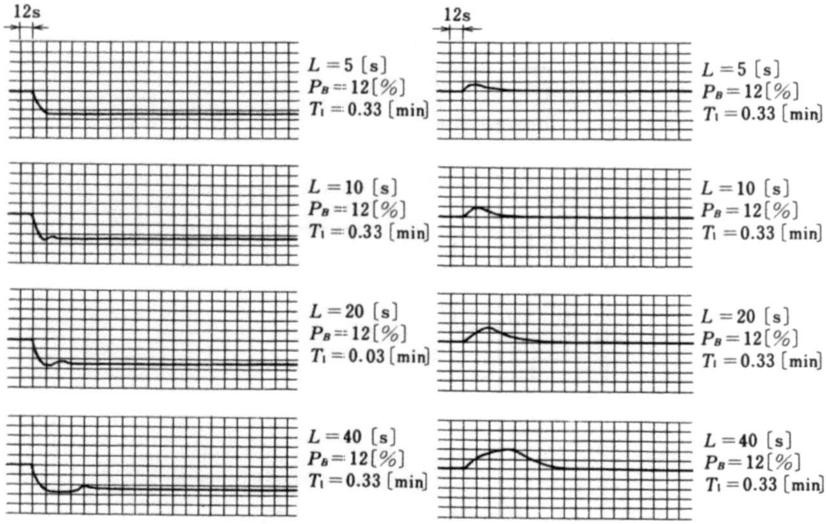

Fig. 8・15 Step response of setpoint in control system using Smith controller (T=20[s])

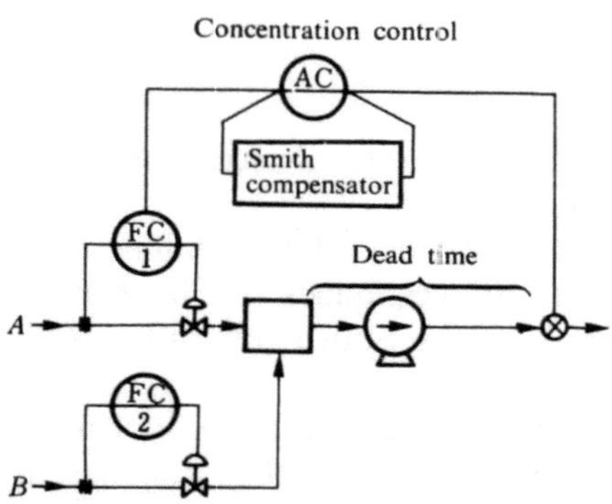

Fig. 8・16 An application of Smith method in concentration control system

main controller tuning performed within this range so that instability does not occur.

(c) Caution on applying to integrative processes

Although the Smith controller can provide good response even with respect to disturbance, special consideration is necessary in the case of integrative processes. In other words, even if integral action is incorporated in a controller, offset with respect to disturbance will arise. Referring again to Fig. 8. 13, if we solve for the output due to a

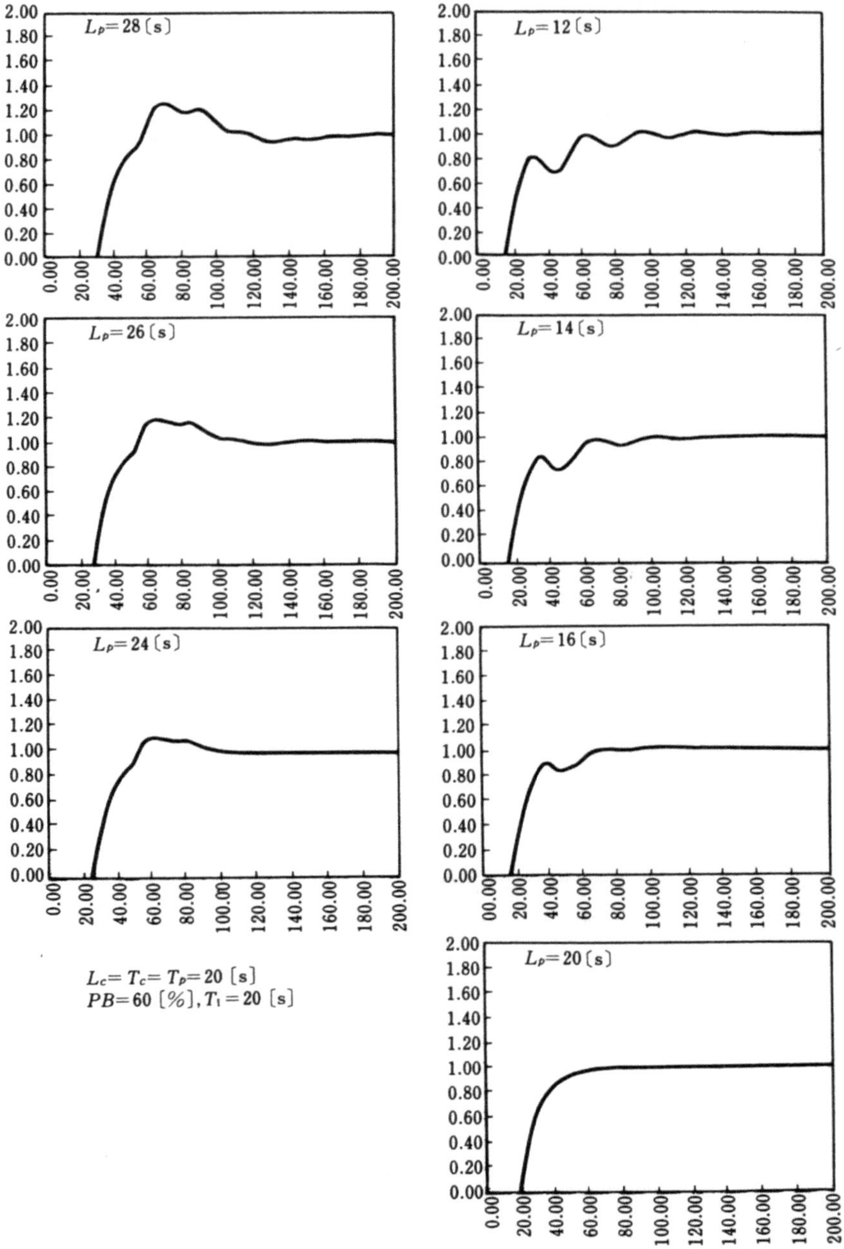

Fig. 8・17 Response of Smith controller system due to the process dead time changes

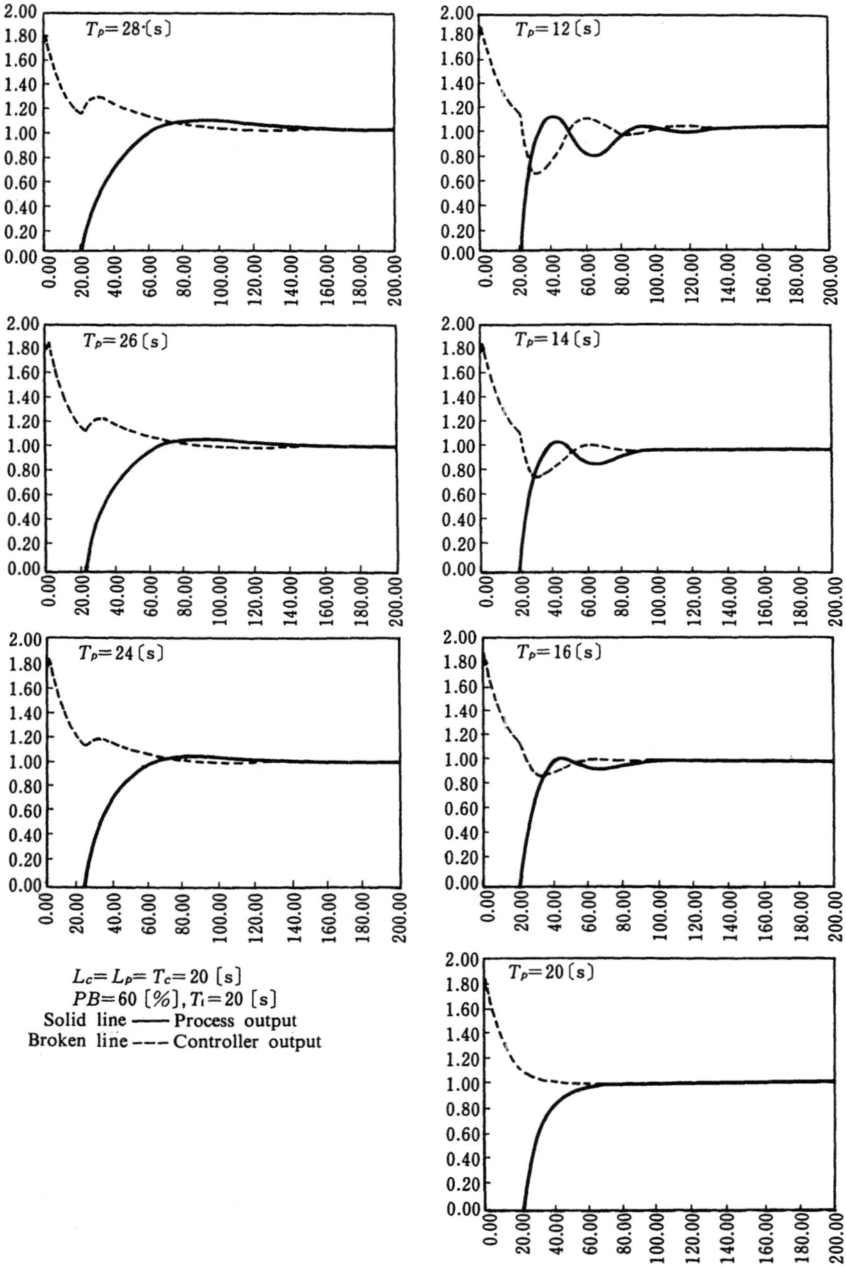

Fig. 8・18 Response of Smith controller system due to the process time constant changes step disturbance, we obtain the following expression:

$$C(s)=\frac{G(s)e^{-Ls}}{1+G_c(s)G(s)}\cdot\frac{D}{s}+\frac{G_c(s)G(s)e^{-Ls}}{1+G_c(s)G(s)}$$
$$\times\{G(s)-G(s)e^{-Ls}\}\frac{D}{s} \tag{8.13}$$

where D is the size of the step disturbance.

Specifying the size of the output when $t\to\infty$ as C_∞, then from the final-value theorem for the Laplace transform, we obtain

$$C_\infty=\lim_{s\to 0} sC(s)=\lim_{s\to 0}\left[\frac{G(s)e^{-Ls}}{1+G_c(s)G(s)}D+\frac{G_c(s)G(s)e^{-Ls}}{1+G_c(s)G(s)}\right.$$
$$\left.\times\{G(s)-G(s)e^{-Ls}\}D\right] \tag{8.14}$$

Here, if $G_c(s)$ indicates a PI controller and $G(s)$ first-order lag, then

$$G_c(s)=K_p\left(1+\frac{1}{T_I s}\right)=\frac{1+T_I s}{T_I s} \tag{8.15}$$

$$G(s)=\frac{1}{1+Ts} \tag{8.16}$$

In these equations, K_P and T_1 are the controller proportional gain and reset time, respectively, and T is the first-order-lag time constant for the process. We now get

$$\lim_{s\to 0} G(s)=\lim_{s\to 0}\frac{1}{1+Ts}=1 \tag{8.17}$$

$$\lim_{s\to 0} G_c(s)G(s)=\lim_{s\to 0}\frac{1+T_I s}{T_I s}\cdot\frac{1}{1+Ts}=\infty \tag{8.18}$$

$$\lim_{s\to 0} e^{-Ls}=1 \tag{8.19}$$

$$\lim_{s\to 0}\{G(s)-G(s)e^{-Ls}\}=\lim_{s\to 0}\left\{\frac{1}{1+Ts}-\frac{1}{1+Ts}e^{-Ls}\right\}=0 \tag{8.20}$$

and consequently, from Eq. (8.7):

$$C_\infty=0$$

or in other words, no offset remains. However, since for an integrative process:

$$G(s)=\frac{1}{s} \tag{8.21}$$

then

$$\lim_{s\to 0}\{G(s)-G(s)e^{-Ls}\}=\lim_{s\to 0}\left(\frac{1-e^{-Ls}}{s}\right)$$

$$=\lim_{s\to 0}\left(L-\frac{L^2s}{2!}+\frac{L^3s^3}{3!}-\cdots\right)=L \quad (8.22)$$

Thus, even if a controller possesses integral action, some offset will remain. Several techniques have been proposed to eliminate this negative point. In the Sawano method[2)], even for a K_P/sT_P process, dynamic characteristics of the Smith compensator can be treated in first-order lag form as $K/(1+Ts)$. In addition, in the Watanabe and Ito method[3)] , a disturbance compensator such $H(s)$ as shown in Fig. 8. 19 is introduced. (Refer to the references at the end of the chapter for more details.)

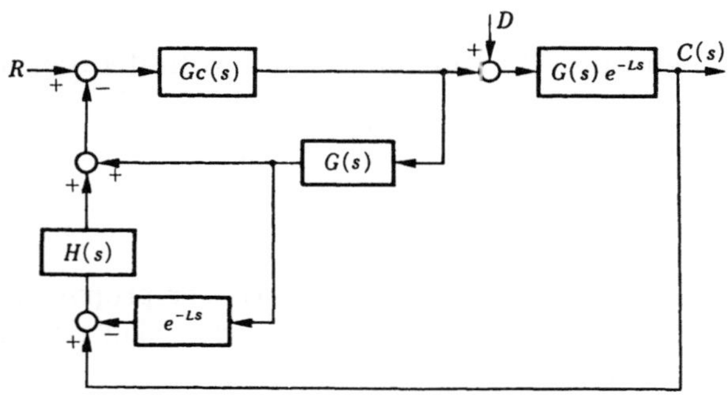

Fig. 8・19 Disturbance compensation in Smith method

8.3.3 Sampling PI controller

(a) Principles of operation

The basic principle behind the sampling PI controller is that after a certain operation has been performed, the controller waits a sufficient amount of time for the result of that operation to appear before performing the next operation (Fig. 8. 20). Switch Sw is closed periodically for only a fixed time τ, and open otherwise. If a deviation ε continues to appear, controller output will be incremented in a step-wise

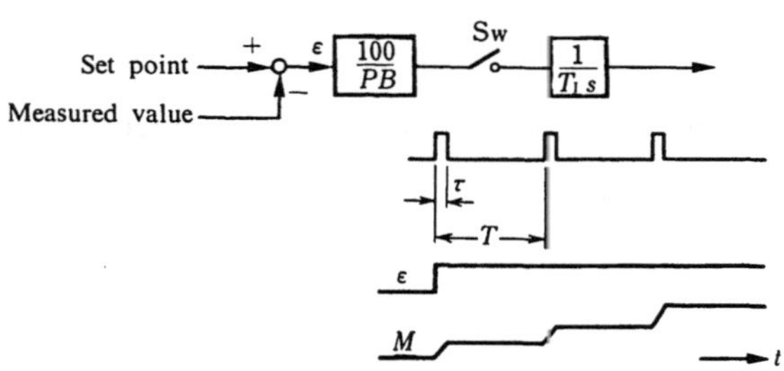

Fig. 8・20 Principle of the sample-and-hold PI controller

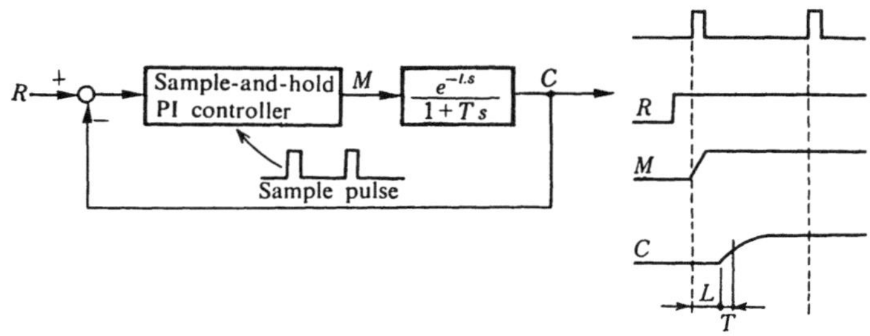

Fig. 8 • 21 Control system using the sample-and-hold PI controller

fashion, as shown. Thus, as this is an integral controller for the most part, one might ask: why is such a controller effective for processes having a large dead time? To help answer this question, Figure 8.21 shows as example in which, although a step change R in the setpoint value is input, the system nevertheless settles down in a minimum amount of time. Given that the controller's gain is correctly adjusted, the controlled variable will match the setpoint value after only one sampling cycle, and the system will settle down with no corrective action necessary in the next sampling cycle.

(*b*) *Effects of parameter changes*

Although the basic principle of the sampling controller is relatively simple, if controller gain or process characteristics are changed, stabilization cannot be actived after one sampling. The effects of changes in the controller gain on response curves are shown in Fig. 8.22. Figure 8.22 (a) shows a proper amount of gain, (b) a gain which is somewhat too large, and (c) a gain which is too small. For these last two cases in which the gain is not properly set, it will be necessary to adjust it while observing the response during actual plant operations. However, since this controller operates on a simple principle, such adjustment is relatively easy to perform, reflecting another feature of the sampling PI controller.

(*c*) *Application considerations*

As described above, although the sampling PI controller for the most part acts as an integral controller, it cannot be applied to an integrative-type process. The reason for this is the 90°phase delay of such a process. Even for exceedingly low frequencies, care must be taken since the system can be oscillatory as a 90° phase delay will easily arise in both the controller and process.

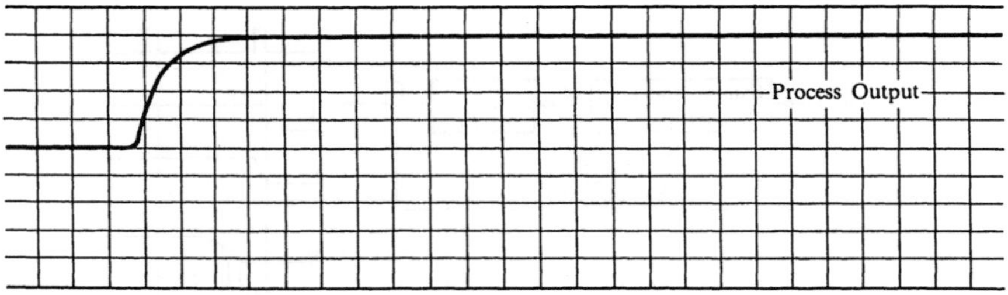

$T_s = 20$ [s] $PB = 24$ [%] 5 s/div
$\tau = 1.8$ [s] $T_I = 8$ [s]

(a) Optimal gain

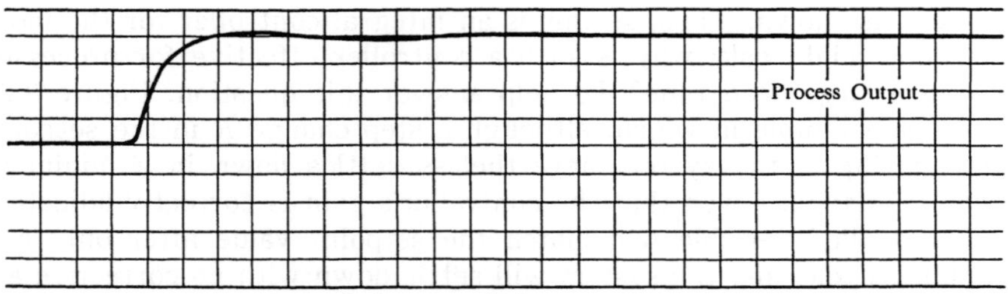

$T_s = 20$ [s] $PB = 20$ [%] 5 s/div
$\tau = 1.8$ [s] $T_I = 8$ [s]

(b) Case of slightly high gain

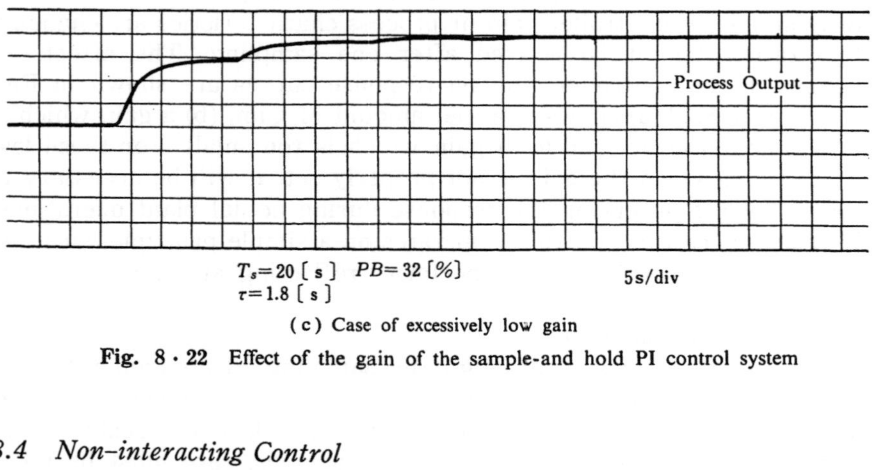

$T_s = 20$ [s] $PB = 32$ [%] 5 s/div
$\tau = 1.8$ [s]

(c) Case of excessively low gain

Fig. 8 · 22 Effect of the gain of the sample-and hold PI control system

8.4 Non-interacting Control

8.4.1 Interaction between process variables

For processes in which two or more variables must be controlled, mutual interaction between the variables will occur. For example, Figure 8.23 shows a process in which both pressure P and flow rate F are being controlled concurrently. If control valve V_a is manipulated to con-

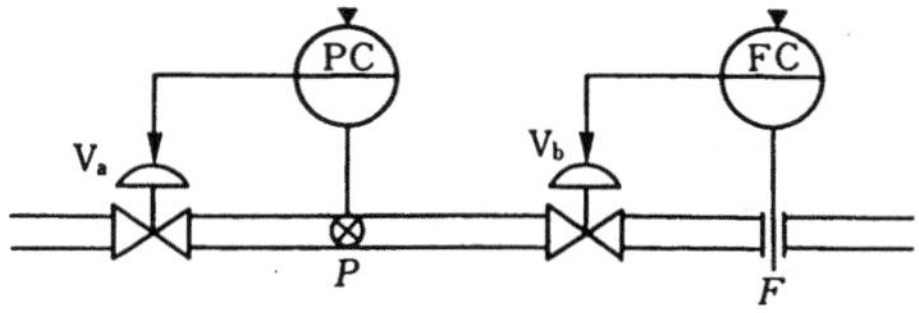

Fig. 8 · 23 Flow and pressure control loop

trol the pressure, the flow also will be affected. Likewise, if control valve V_b is manipulated to control the flow, the pressure will be affected. Thus, if F happens to be at a desired value but P is not, manipulation of V_a to control P will adversely affect the flow rate, changing what was up to now a correct value. Moreover, if we now manipulate V_b to readjust the flow, P will be changed from its correct value.

In addition to the process described above, there are many other kinds of processes in which interaction exisits. In the two following examples, interaction amongst all the variables exist. Figure 8.24 shows an example of temperature control in a distillation column. Here, the top and bottom temperatures cannot be independently altered; if either the amount of reboiler steam or reflux flow is manipulated, both temperatures will be affected. In the example of Fig. 8.25, the flow of two types of composites are mixed, and control of either one naturally leads to mutual interaction.

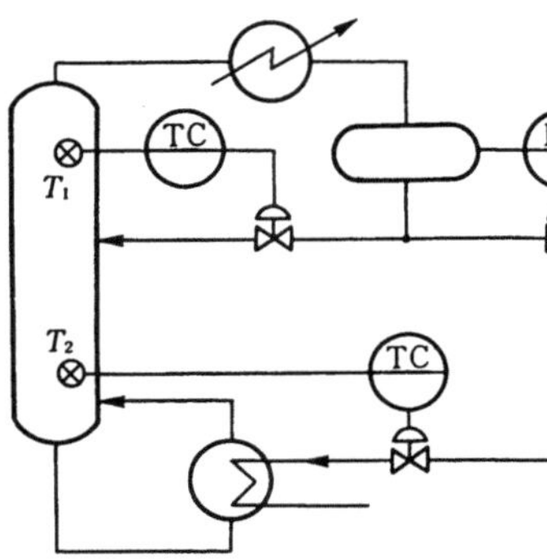

Fig. 8 · 24 Top and bottom temperature control of distillation column

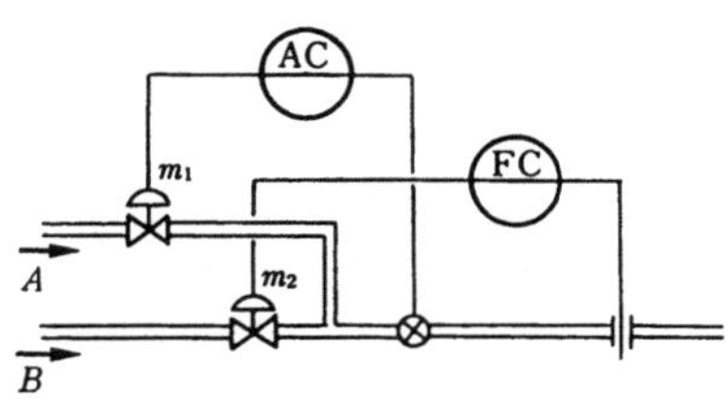

Fig. 8 · 25 Flow component control loop

8.4.2 Influence exerted by mutual interaction

In order to determine the actual extent of mutual interaction, we investigate the problem using simulation. In the control system shown in Fig. 8.26, process elements $G_3(s)$ and $G_4(s)$ both exhibit mutual interaction. The first chart in Fig. 8.27 shows a case with absolutely no interaction when a change in the setpoint value is applied to the first loop only. The dashed line in the chart represents the second loop, which in this case is totally unaffected. If, however, we hold controller parameters constant and raise the gain K_3 and K_4 of the interacting elements, the response gets steadily worse as shown in the following charts. (The parameters for the controllers and process model are as indicated.)

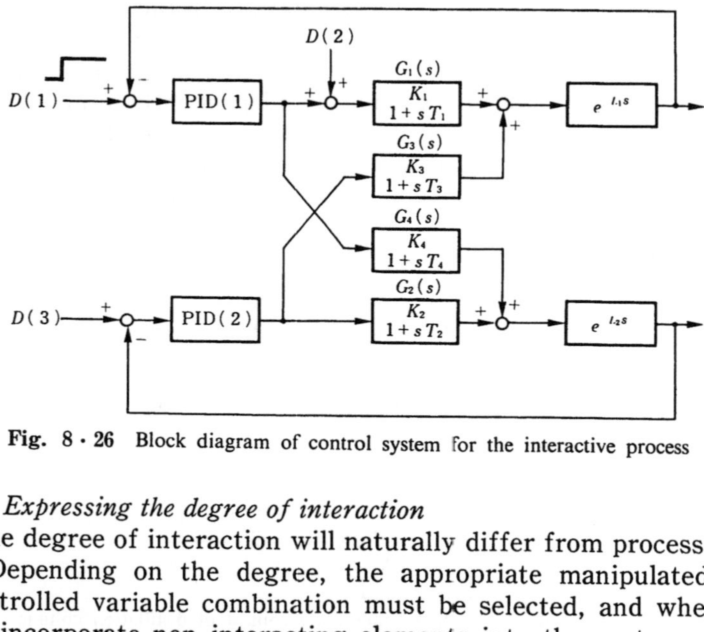

Fig. 8・26 Block diagram of control system for the interactive process

8.4.3 Expressing the degree of interaction

The degree of interaction will naturally differ from process to process. Depending on the degree, the appropriate manipulated variable/controlled variable combination must be selected, and whether or not to incorporate non-interacting elements into the system must be decided, as discussed below. The following considers methods for estimating the degree of interaction.

(*a*) *Relative gain*

The two-variable processes shown in Figs. 8.23 and 8.24 can be expressed as shown in Fig. 8.28, where g_{11}, g_{12}, etc. indicate transfer functions. To begin with, let's focus our attention on the relationship between manipulated variable u_1 and controlled variable y_1. For the case in which the gains of controllers 1 and 2 are both zero, *i.e.*, the loops

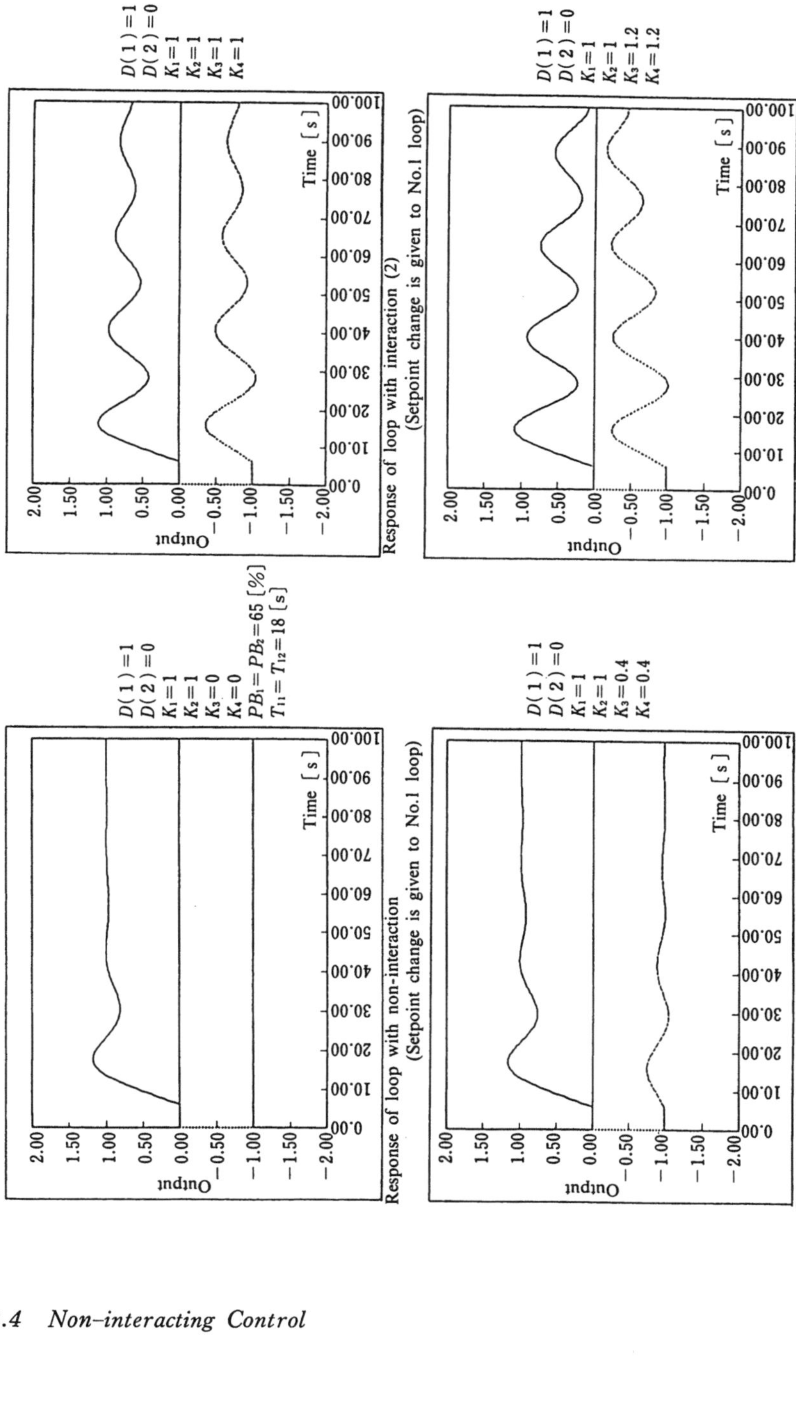

Fig. 8・27 Simulation for the loop with interaction

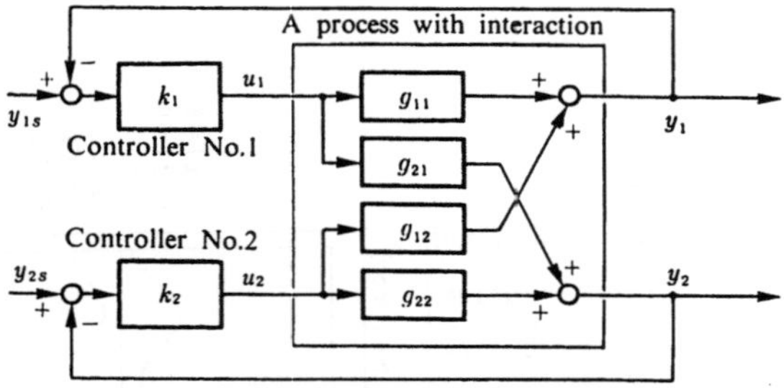

$$\begin{pmatrix} y_1 \\ y_2 \end{pmatrix} = \begin{pmatrix} g_{11} & g_{12} \\ g_{21} & g_{22} \end{pmatrix} \begin{pmatrix} u_1 \\ u_2 \end{pmatrix}$$

Fig. 8 • 28 Block diagram of two-variable control loop

are open, if we denote the relationship between u_1 and y_1 as $y_1/u_1|_{\text{open}}$, then clearly

$$\left.\frac{y_1}{u_1}\right|_{\text{open}} = g_{11} \tag{8.23}$$

Next, if we close only the second loop and denote the transfer function between u_1 and y_1 as $y_1/u_1|_{\text{close}}$, we obtain the configuration in Fig. 8.29. Consequently, this transfer function becomes

$$\left.\frac{y_1}{u_1}\right|_{\text{close}} = g_{11} - \frac{g_{21}k_2g_{12}}{1+k_2g_{22}} \tag{8.24}$$

Examining this equation, the second term on the right arises due to interaction, and the larger this term becomes relative to the first term, the greater the interaction. If we now take the ratio between the above expressions for the open loops and closed loops, we can define the relative gain λ_{11} as follows:

$$\lambda_{11} = \frac{\left.\frac{y_1}{u_1}\right|_{\text{open}}}{\left.\frac{y_1}{u_1}\right|_{\text{close}}} = \frac{g_{11}}{g_{11} - \frac{g_{21}k_2g_{12}}{1+k_2g_{22}}} \tag{8.25}$$

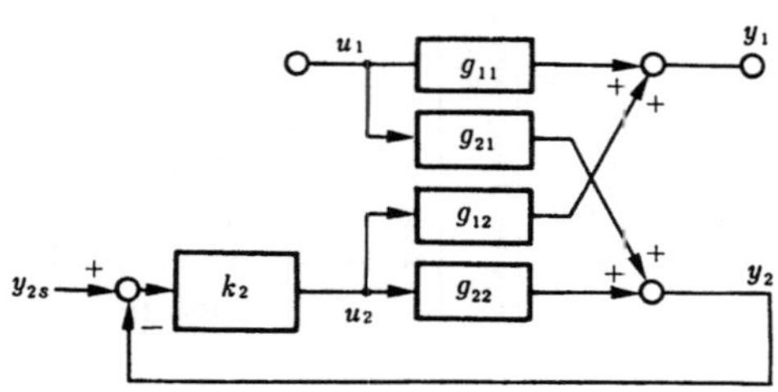

Fig. 8 • 29 In case of No.2 loop is closed in two-variable control loop

In other words, as this value indicates the degree of interaction as received from the closed loop, a situation with absolutely no interaction will correspond to $\lambda_{11}=1$, and the degree of interaction will be indicated by the amount that λ_{11} differs from 1.

In the same way, the relative gain for the other combinations of $u_1 \rightarrow y_2$, $u_2 \rightarrow y_1$ and $u_2 \rightarrow y_2$ can also be defined as λ_{21}, λ_{12} and λ_{22} respectively.

$$\lambda_{12}=\frac{g_{12}}{g_{12}-\dfrac{g_{11}k_2 g_{22}}{1+k_2 g_{21}}} \tag{8.26}$$

$$\lambda_{21}=\frac{g_{21}}{g_{21}-\dfrac{g_{11}k_1 g_{22}}{1+k_1 g_{12}}} \tag{8.27}$$

$$\lambda_{22}=\frac{g_{22}}{g_{22}-\dfrac{g_{12}k_1 g_{21}}{1+k_1 g_{11}}} \tag{8.28}$$

If we now consider the gain of the two controllers to be sufficiently high so that k_1 and $k_2 \rightarrow \infty$, Equations (8.25) to (8.28) can be rewritten as follows (assuming g_{11} to g_{22} represent static gain only):

$$\lambda_{11}=\frac{g_{11}g_{22}}{g_{11}g_{22}-g_{12}g_{21}} \tag{8.29}$$

$$\lambda_{12}=-\frac{g_{12}g_{21}}{g_{11}g_{22}-g_{12}g_{21}} \tag{8.30}$$

$$\lambda_{21}=-\frac{g_{12}g_{21}}{g_{11}g_{22}-g_{12}g_{21}} \tag{8.31}$$

$$\lambda_{22}=\frac{g_{11}g_{22}}{g_{11}g_{22}-g_{12}g_{21}} \tag{8.32}$$

On examining the above equations, the following relationships between the four interaction coefficients appear:

$$\lambda_{11}=\lambda_{22} \tag{8.33}$$

$$\lambda_{12}=\lambda_{21} \tag{8.34}$$

$$\lambda_{11}+\lambda_{12}=1 \tag{8.35}$$

$$\lambda_{21}+\lambda_{22}=1 \tag{8.36}$$

These relationships are useful when solving for interaction coefficients in an actual process. Specifically, in the case of two inputs and two outputs, once λ_{11} is determined, the other three follow from the above relationships. In addition, the relationship between process variables and interaction coefficients can be expressed as follows:

	u_1	u_2
y_1	λ_{11}	λ_{12}
y_2	λ_{21}	λ_{22}

(8.37)

This is referred to as the relative gain matrix. For a process with no interaction, the above becomes

	u_1	u_2
y_1	1	0
y_2	0	1

On the other hand, a relative gain matrix of

	u_1	u_2
y_1	0.2	0.8
y_2	0.8	0.2

indicates that y_2 should be controlled by u_1 and y_1 by u_2. The relative gain matrix is thus useful for determining the proper combination of controlled variable and manipulated variable. This will be described in more detail later.

(*b*) *Interaction coefficients*

Relative gain as described above has been used to express the degree of interaction based on static characteristics. This is not the only method, however, for defining the degree of interaction. The following equation, which includes dynamic characteristics, defines the interaction coefficient λ. Here, a higher value of λ indicates a higher degree of interaction.

$$\lambda(s)=\sqrt{\frac{g_{12}(s)g_{21}(s)}{g_{11}(s)g_{22}(s)}} \tag{8.38}$$

This equation can be used for graphically determining system stability.

8.4.4. *Controlled variable and manipulated variable combination*

As described above, the relative gain is a useful device for determining the proper combination of controlled variable and manipulated variable. Ignoring for the time being dynamic characteristics and considering only static gain, the relative gain for a process in which $g_{11}=0.8$, $g_{12}=2$, $g_{21}=1$, and $g_{22}=0.5$ is given by

$$\lambda_{11}=\lambda_{22}=\frac{0.8\times0.5}{0.8\times0.5+2\times1}=0.17$$

Thus, since $\lambda_{12} = \lambda_{21} = 1 - \lambda_{11} = 0.83$, the relative gain matrix becomes

	u_1	u_2
y_1	0.17	0.83
y_2	0.83	0.17

The closer the relative gain is to one, the smaller the amount of interaction from the other loop. Thus, in this example, y_2 should be controlled by u_1, and y_1 by u_2. However, it must be noted here that the relative gain expresses only static gain without taking into account dynamic characteristics at all. Thus, the stability of a system cannot be immediately judged based solely on the size of the relative gain, which should be treated as only one of several possible criteria. As described above, the interaction coefficient equation (Eq. (8.38)) considers dynamic characteristics and can be used to draw a Gershgorin band to enable a graphical determination of process stability. Although this method will not be described in detail here, the reader is referred to the references at the end of the chapter.[4)]

8.4.5 *Non-interacting control*

As described in the sections above, processes having mutual interaction are in general difficult to control. To control such processes, one technique is to fine-tune controller parameters, and another is to add elements to a system which eliminate interaction. The latter technique is described here. In the example of Fig. 8.30, by adding elements c_1 and c_2 to the system, interaction can be effectively removed.

The following relationships are present in the system of Fig. 8.30:

$$\begin{bmatrix} u_1 \\ u_2 \end{bmatrix} = \begin{bmatrix} 1 & c_1 \\ c_2 & 1 \end{bmatrix} \begin{bmatrix} m_1 \\ m_2 \end{bmatrix} \tag{8.39}$$

$$\begin{bmatrix} y_1 \\ y_2 \end{bmatrix} = \begin{bmatrix} g_{11} & g_{12} \\ g_{21} & g_{22} \end{bmatrix} \begin{bmatrix} u_1 \\ u_2 \end{bmatrix} \tag{8.40}$$

Substituting Eq. (8.39) into Eq. (8.40), we get

$$\begin{aligned} \begin{bmatrix} y_1 \\ y_2 \end{bmatrix} &= \begin{bmatrix} g_{11} & g_{12} \\ g_{21} & g_{22} \end{bmatrix} \begin{bmatrix} 1 & c_1 \\ c_2 & 1 \end{bmatrix} \begin{bmatrix} m_1 \\ m_2 \end{bmatrix} \\ &= \begin{bmatrix} g_{11} + c_2 g_{12} & g_{11} c_1 + g_{12} \\ g_{21} + c_2 g_{22} & g_{21} c_1 + g_{22} \end{bmatrix} \begin{bmatrix} m_1 \\ m_2 \end{bmatrix} \end{aligned} \tag{8.41}$$

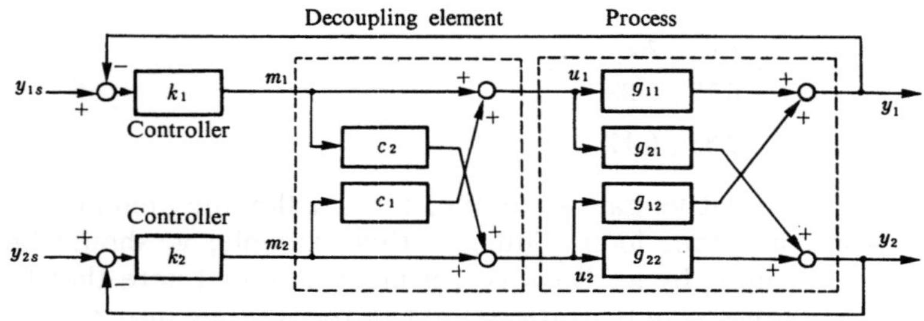

Fig. 8・30 Non-interacted control loop

Here, if we let

$$g_{11}c_1+g_{12}=0 \tag{8.42}$$

$$g_{21}+c_2g_{22}=0 \tag{8.43}$$

then the matrix in Eq. (8.41) is made a diagonal matrix. In other words, by letting

$$c_1=-\frac{g_{12}}{g_{11}} \tag{8.44}$$

$$c_2=-\frac{g_{21}}{g_{22}} \tag{8.45}$$

the above matrix becomes

$$\begin{bmatrix} y_1 \\ y_2 \end{bmatrix}=\begin{bmatrix} g_{11}-\dfrac{g_{12}g_{21}}{g_{22}} & 0 \\ 0 & g_{22}-\dfrac{g_{12}g_{21}}{g_{11}} \end{bmatrix}\begin{bmatrix} m_1 \\ m_2 \end{bmatrix} \tag{8.46}$$

Accordingly, for mutual interaction to be eliminated, the control system must be equivalent to two independent loops, as shown in Fig. 8.31.

When performing non-interaction control, however, there are some situations which must be given special attention. For example, consider the situation in which

$$g_{11}g_{22}-g_{12}g_{21}=0 \tag{8.47}$$

In this case, the gains $g_{11}-g_{12}g_{21}/g_{22}$ and $g_{22}-g_{12}g_{21}/g_{11}$ in Eq. (8.46) both become zero with the result that control becomes impossible. Thus, in such a situation, this method cannot be used to create a non-interact-

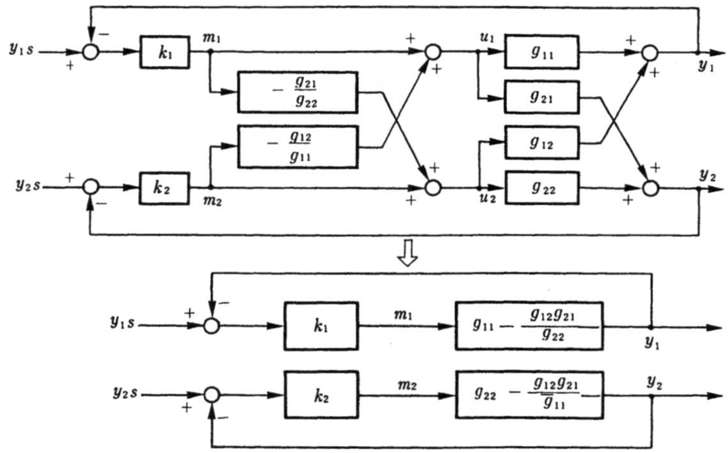

Fig. 8 · 31 Non-interacted control loop and its equivalent loop

ing system.

Another situation to be considered is one in which g_{11} and g_{22} are transfer functions as the denominator of the non-interacting terms of Eqs. (8.44) and (8.45). Therefore, if these transfer functions are higher than the order of the numerator g_{12} and g_{21}, non-interacting control cannot be realized.

As for methods of actually implementing c_1 and c_2, dynamic characteristics can be ignored and static gain only compensated, or assuming dynamic characteristics, a first-order lead/lag element at the most can be applied, after which the time constant can be adjusted.

8.4.6 An example of non-interacting control

Figure 8.32 shows an example of non-interacting control in the case of a distillation column. As can be seen, a non-interacting element has been added between TIC at the bottom and the reflux line. Of note here is the fact that interaction elimination elements do not necessarily have to be applied in both directions. In the distillation example here, addition of a non-interacting element on only one side is sufficient to obtain satisfactory results.

8.5 Self-tuning Controller

8.5.1 Overview

A PID controller in which each parameter does not have to be set, or a controller in which parameters automatically adjust to changes in

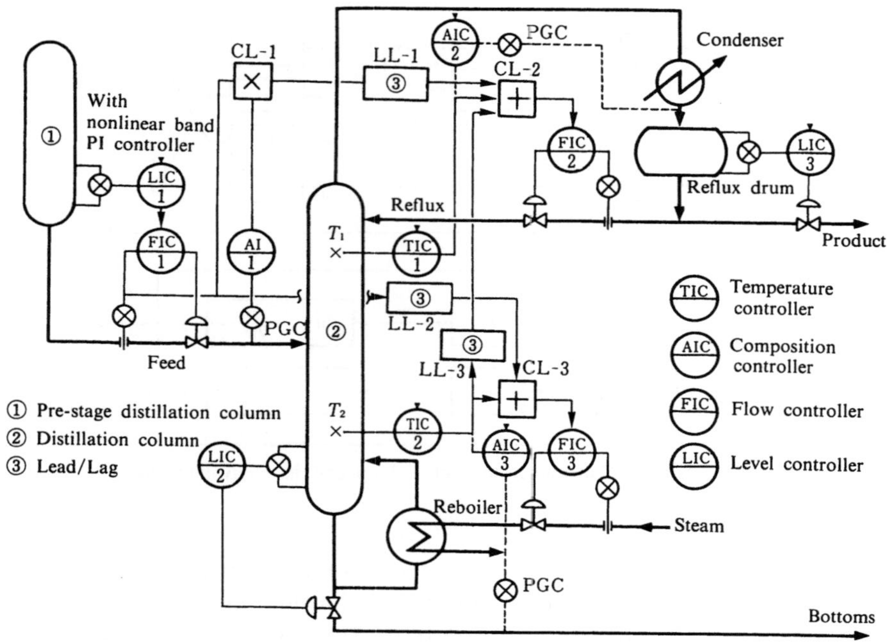

Fig. 8 · 32 Advanced control of distillation column

Table 8.3 Types of adaptive control

Changes in process characteristics	System	Example
Predictable	Open loop adaptive control	Gain scheduling adaptive control
Unpredictable	Closed loop adaptive control	Model reference adaptive control (MRAC)
		Self-tuning adaptive control (STC) · PID type · Non-PID type

process characteristics, provides extremely convenient features. Such a controller is called a self-tuning controller, or STC for short. STC is one type of adaptive controller. Table 8.3 classifies adaptive control formats.

If for a certain process, the functions relating process variables and change in process characteristics are understood, then controller parameters can be changed accordingly. This type of format is called gain-scheduling adaptive control and is characterized by open-loop operations. In contrast, if process characteristics cannot be predicted, they must be continuously re-estimated until parameters are adjusted satisfactorily. This format is generally divided into model reference adaptive control (MRAC) and self-tuning control (STC).

Figure 8.33 shows the structure of gain-scheduling adaptive control. Here, the influence exerted by the environment on plant characteristics are measured, and controller characteristics are changed so that the response of the closed loop will agree with desired characteristics. On the other hand, in the MRAC system shown in Fig. 8.34, a closed-loop response model is incorporated as a reference, and controller parameters are changed so that closed-loop response characteristics will agree with those of the reference model. The type of controller used here is not necessarily a PID controller, but may be of a much more complex configuration. Figure 8.35 shows the structure of STC. The main feature here is an on-line identification device which estimates plant characteristics based on plant input and output. The type of controller here is PID or controllers of higher order.

At this point, let's consider where STC can be applied. In addition to its original intention of responding to changes in process characteris-

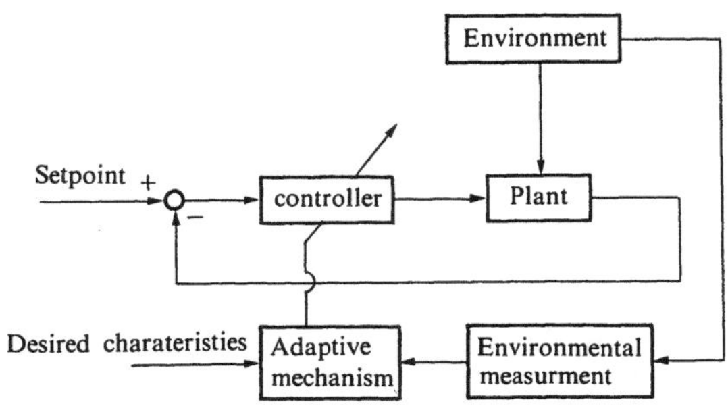

Fig. 8 • 33 Gain scheduling adaptive cnotrol

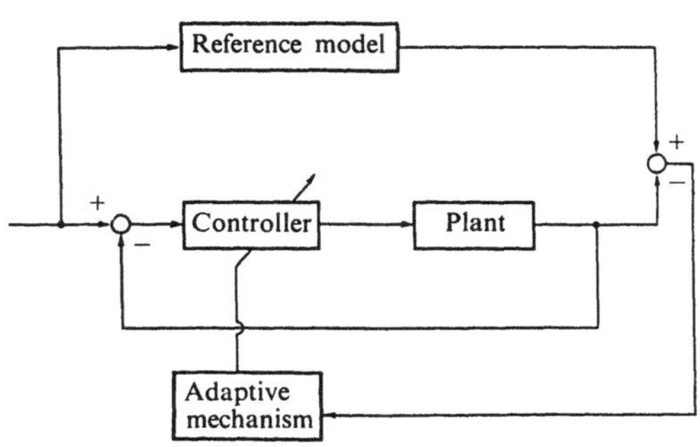

Fig. 8 • 34 Model reference adaptive control system

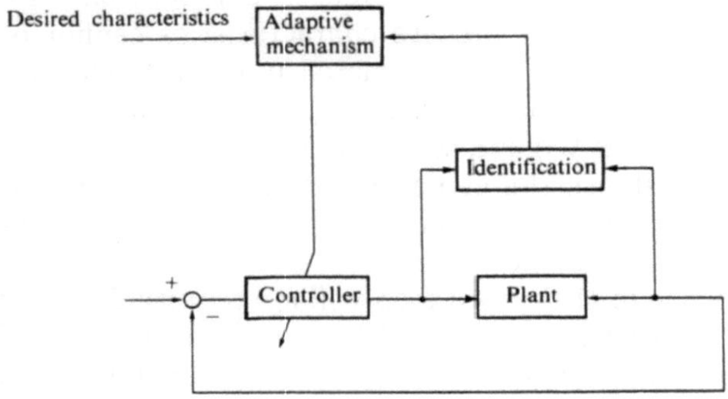

Fig. 8 · 35 Self-tuning controller

tics, it is also employed to make tuning operations more simple. These two general applications are summarized below.

(1)*Responding to changes in process characteristics*

Startup time / shutdown time
Load changes
Internal–state changes (*e.g.*, batch reactor)
Non–linear characteristics in final control elements
Elapsed–time changes

(2)*Tuning simplification*

Elimination of operator idiosyncrasies
Reduction of plant startup time

8.5.2 Gain–scheduling control

As described above, gain–scheduling control is used when the functions relating process variables and parameter changes are know. Use of the gain–scheduling technique for temperature control of a reactor is shown in Fig. 8.36. Given that reactor temperature and process gain are related as shown in (b), then if the controller gain is changed as shown in (c), the closed–loop gain can always be held constant.

Figure 8.37 shows an example of calorie control for a mixed gas. The system here mixes three types of gases B,C and L having different calorie ratings to obtain a mixed gas with a certain desired calorie rating. In particular, C gas or L gas is mixed in as a ratio with respect to B gas. Thus, in order to perform calorie control, the C/B ratio undergoes feedback control in which a Smith controller is applied. However, as dead–time varies depending on flow velocity, the dead–time incorporated in the Smith controller will have to be likewise varied. Accordingly, gas flow must be measured after the gases are mixed to enable dead–time calculation and appropriate compensation. In addition, the gain in calorie variation with respect to C/B ratio operations is given

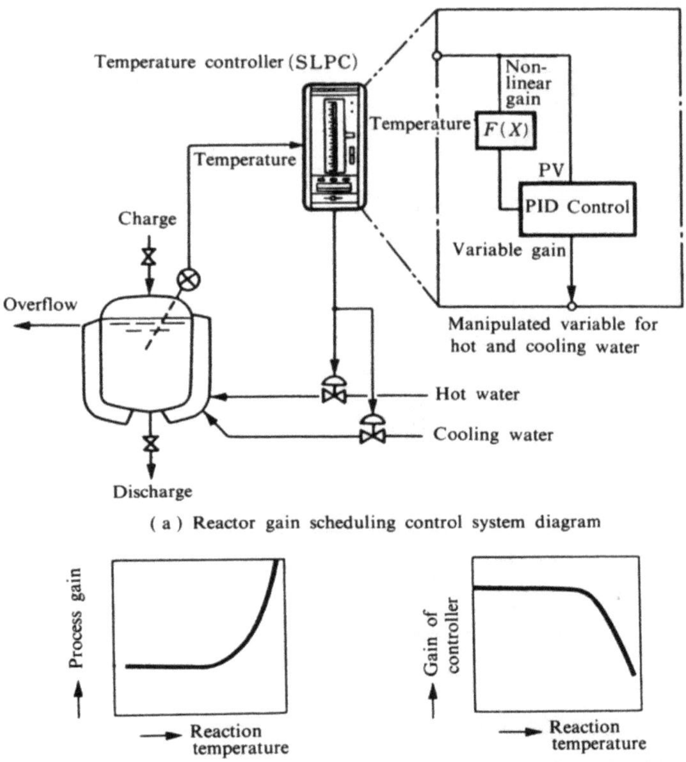

(b) Process gain characteristic (c) Characteristic of F(X)

Relationship between process gain characteristic and function generator F(X)

Fig. 8 · **36** Gain scheduling control of reactor

by the following formula:

$$K=\frac{dQ_M}{d(C/B)}=\frac{Q_C-Q_B+(Q_C-Q_L)L/B}{(1+C/B+L/B)^2} \tag{8.48}$$

Here, Q_C,Q_B,Q_L and Q_M represent the calorie rating of the C,B,L and mixed gas, respectively, and C,B and L the respective gas flows. By calculating this gain by computer, the proportional gain of the controller can be changed.

The above examples of gain-scheduling adaptive control, though not particularly interesting from a theoretical point of view, find wide practical application.

8.5.3 Self-tuning controller (STC)

To date, various types of STC formats have been conceived, with

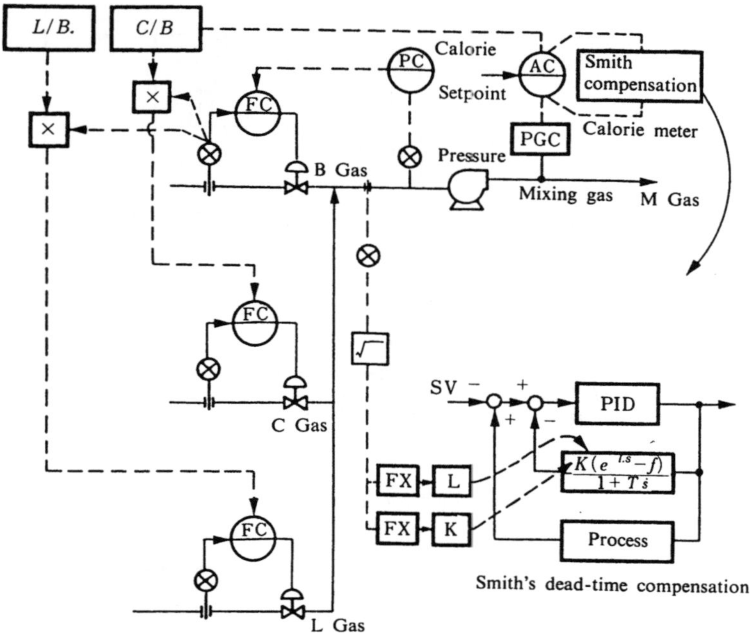

$$K=\frac{dQ_M}{d(C/B)}=\frac{(Q_C-Q_B)+L/B(Q_C-Q_L)}{(1+C/B+L/B)^2}$$

Fig. 8 · 37 Gain scheduling of calorie control

four major ones listed below.

(1) Ultimate-sensitivity method[5)]

(2) Limit cycle method[6)]

(3) ARMA model (successive identification) method[7)]

(4) Expert database method

In the ultimate-sensitivity method as shown in Fig. 8.38, an auxiliary controller is incorporated in parallel with the main controller. As the proportional gain of the auxiliary controller is gradually raised, a critical oscillation having small amplitude is generated. PID parameters can then be set from this critical oscillation using the Ziegler–Nichols method.

For the limit cycle method, on the other hand, an on–off controller is incorporated in parallel with the main controller, as shown in Fig. 8.39. Here, a small-amplitude limit cycle is generated whose period and amplitude are used to estimate the process dead time and time constant. Based on these estimations, PID parameters are then determined using the Ziegler–Nichols method.

The ARMA model method is not described here. The following sec-

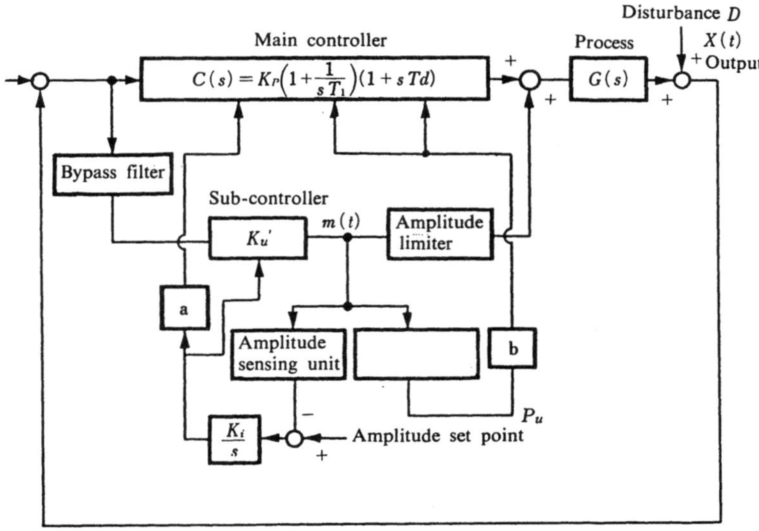

$K_u = K_u' + K_P$

Where, K_u': Gain of sub-controller; K_P: Gain of main controller

$K_P = aK_u$

$\therefore\ K_P = \frac{a}{1-a} K_u'$

Auto-tuning system applying the ultimate-sensitivity method

Fig. 8・38 System utilizing for the ultimate-sensitivity method

tion takes up the expert database method.

8.5.4 STC based on the expert method [8)]

The expert method is characterized by observation of the closed-loop response to disturbance or a change in the setpoint value, followed by tuning of PID parameters based on previously established rules. As shown in Fig. 8.40, the configuration for this method is divided into response-curve analyzer and PID paramter tuning sections. In the PID parameter tuning section, the observed waveform is classified into a set of basic response curves and compared with a collection of similar curves stored in the knowledge base. Based on the results of this search, a fixed rule is then applied to modify PID parameters.

(a) *Control target*

As STC based on the expert method is still not a firmly established technique, we continue our description using as an example the design incorporated in the Yokogawa Electric Corporation YS-80 series single-loop controller. Targets for closed-loop responses are divided into four types as shown in Table 8.4. Determining a control target

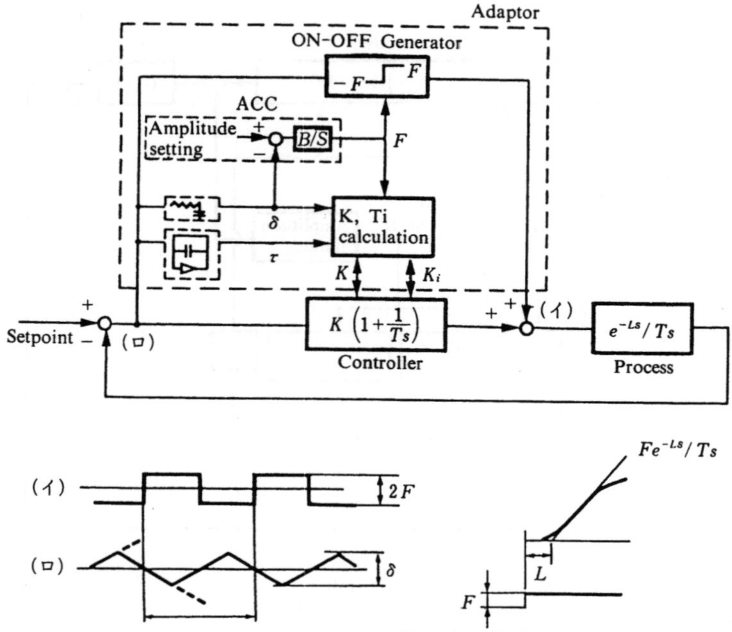

$$\begin{cases} T=\frac{\tau}{2\delta}\left(F+\frac{k\delta}{\pi}\right) \\ L=\frac{\tau}{4}\left(1-\frac{k\delta/\pi^3}{F+k\delta/\pi}\cdot\frac{\tau}{T_i}\right) \end{cases}$$

τ : Period of limit cycle
δ : Amplitude (p-p) of limit cycle
F : Amplitude of ON-OFF generator output
k : Proportional gain of main controller
T_i : Integral gain of main controller

Fig. 8 · 39 Limit cycle method[4]

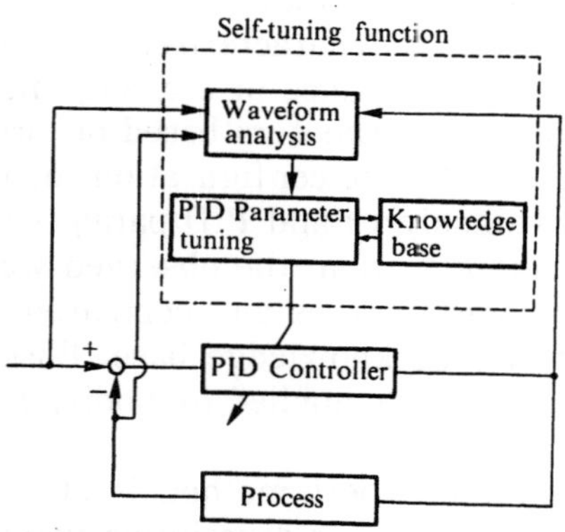

Fig. 8 · 40 Functional brock diagram of self-tuning controller

Table 8.4 Control target type

	Target	Performance criteria
0	Overshoot: Non	Overshoot: 0
1	Overshoot: Small (5%)	$\min\int \lvert e \rvert t dt$ (ITAE)
2	Overshoot: Medium (10%)	$\min\int \lvert e \rvert dt$ (IAE)
3	Overshoot: Large (15%)	$\min\int e^2 dt$ (ISE)

e ; control deviation

based on the amount of overshoot has been found to be a very convenient method in actual use.

(b) *Tuning rules*

Figure 8.41 shows typical closed-loop response curves. Figure 8.41(E) shows an optimal response curve, while (A) shows one having no oscillation at all; since the latter implies a delayed response, both P and I parameters are made small. In (B), since the oscillation of the response curve has a long period and thus slow stabilization, P is made small and I large. In (C), however, the high-amplitude damped oscillation calls for a large P and a large I. Finally, for the response curve in (D) in which the damped oscillation has a short period, P is made large and I small. Although these examples show the direction of parameter change, the response curves must be closely analyzed and their difference with target response curves determined to obtain the actual amount of tuning needed.

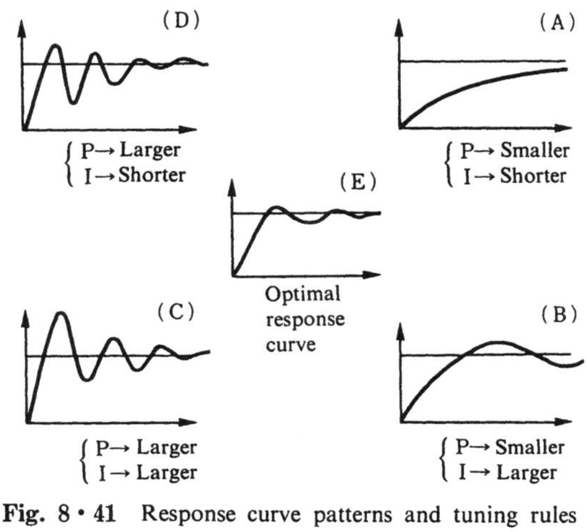

Fig. 8 • 41 Response curve patterns and tuning rules

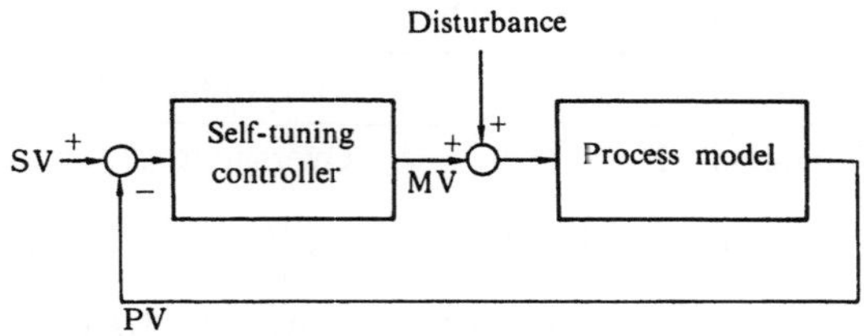

Process model : $e^{10s}/(1+10s)$

Fig. 8・42 Simulation test configuration diagram

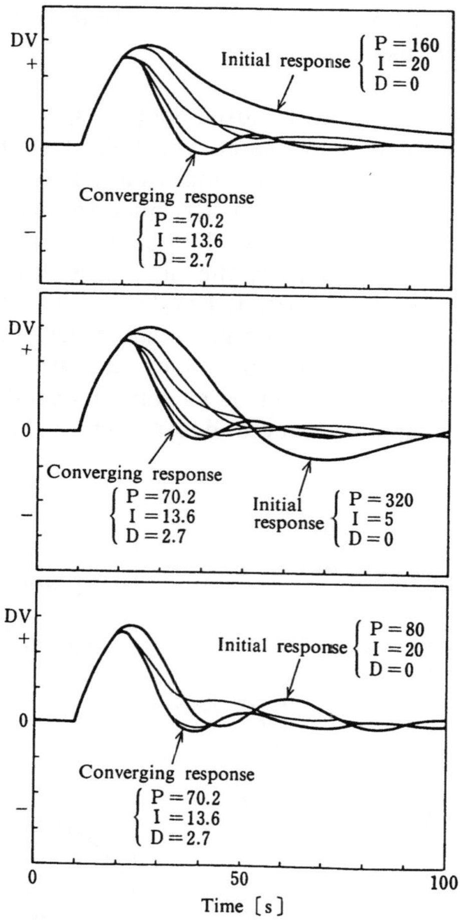

Fig. 8・43 Results of simulation

(c) *Simulation test*

A control system in which disturbance can be repeatedly applied is shown in Fig. 8.42, and observations of convergence behavior in this system are shown in Fig. 8.43. Starting with three different initial response curves and PID parameters, it can be seen that the PID parameters converge to the same values after the several tuning actions in each case.

8.5.5 *STC application considerations*

STC, though an effective means of control, is not a cure–all for process control; control objectives must be thoroughly considered and the most appropriate method selected. In particular, the following points should be considered at this time.

(1) Has the reasons for introducing STC been clarified?
(2) Can gain–scheduling control be used?
(3) Is on–line identification necessary? Can off–line identification be used?
(4) Is identification signal marking feasible? Size, type?
(5) Can STC catch up with the speed of process change?

8.6 *Optimal Control*

8.6.1 *The meaning of "state"*

This section introduces the field of optimal control based on the concept of "state feedback." First of all, however, the meaning of "state" must be clarified, and although a thorough study of the subject would fill many pages, a short review will be covered here.

Assume a process defined by the following transfer functions:

$$\frac{Y(s)}{U(s)}=\frac{1}{(1+T_1s)(1+T_2s)} \tag{8.49}$$

Rearranging this equation, we get:

$$\frac{Y(s)}{U(s)}=\frac{1}{T_1}\cdot\frac{1/s}{1+1/T_1s}\cdot\frac{1}{T_2}\cdot\frac{1/s}{1+1/T_2s} \tag{8.50}$$

The above two equations correspond to the block diagrams of Fig. 8.44. In the figure, the $1/s$ elements are integrators, situated between input signals $\dot{x}_1$ and $\dot{x}_2$ and output variables x_1 and x_2, respectively. The relationship between the signals in the figure can be given as follows:

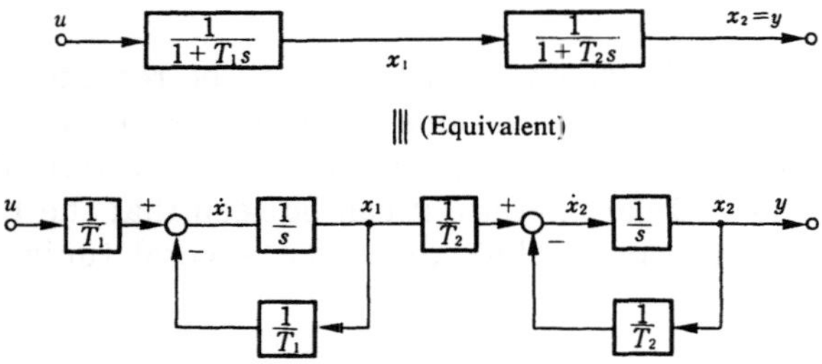

Fig. 8・44 Brock diagram of $1/(1+T_1s)(1+T_2s)$

$$\begin{cases} \dot{x}_1 = -\dfrac{1}{T_1}x_1 + \dfrac{1}{T_1}u & (8.51) \\ \dot{x}_2 = \dfrac{1}{T_2}x_1 - \dfrac{1}{T_2}x_2 & (8.52) \end{cases}$$

$$y = x_2$$

At this point we define the following:

$$X = [x_1, x_2]^T, \qquad Y = [y], \qquad U = [u]$$

$$A = \begin{bmatrix} -\dfrac{1}{T_1} & 0 \\ \dfrac{1}{T_2} & -\dfrac{1}{T_2} \end{bmatrix}, \qquad B = \left[\frac{1}{T_1}, 0\right]^T, \qquad C = [1, 0]$$

where the symbol "T" indicates matrix or vector transposition. Based on these definitions, Equations (8.51) and (8.52) can be expressed as

$$\begin{cases} \dot{X} = AX + BU & (8.53) \\ Y = CX & (8.54) \end{cases}$$

Equation (8.53) is called the state equation, Equation (8.54) the output equation, and X simply the state. Note that in contrast to the transfer functions of Eq. (8.49) which only express the relationship between process input and output, the above state equation expresses the state of the process as well.

8.6.2 Integral optimal regulator

Introduction of the concept of state as described above has brought internal process variables into the picture as well. If the values of these variables are used in configuring feedback control, even better control characteristics can be expected, as discussed below.

Given the system in Fig. 8.45 with setpoint value y_r, the deviation

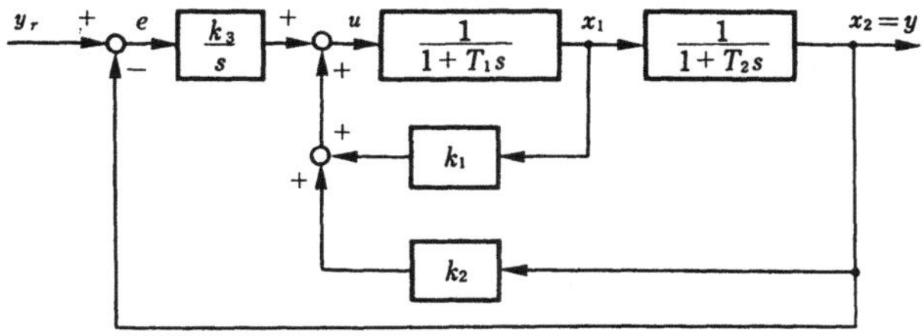

Fig. 8・45 Integral feedback control

and corresponding integral are then defined as follows.

$$e=y-y_r \tag{8.55}$$

$$q=\int e dt \tag{8.56}$$

Substituting the above into Eqs. (8.53) and (8.54) we obtain

$$\frac{d}{dt}\begin{bmatrix} \boldsymbol{X} \\ q \end{bmatrix}=\begin{bmatrix} \boldsymbol{A} & \boldsymbol{0} \\ \boldsymbol{C} & \boldsymbol{0} \end{bmatrix}\begin{bmatrix} \boldsymbol{X} \\ q \end{bmatrix}+\begin{bmatrix} \boldsymbol{B} \\ \boldsymbol{0} \end{bmatrix}\boldsymbol{U}+\begin{bmatrix} 0 \\ -1 \end{bmatrix}y_r \tag{8.57}$$

Here, if we let

$$\boldsymbol{A}_e=\begin{bmatrix} \boldsymbol{A} & \boldsymbol{0} \\ \boldsymbol{C} & \boldsymbol{0} \end{bmatrix}, \quad \boldsymbol{B}_e=\begin{bmatrix} \boldsymbol{B} \\ \boldsymbol{0} \end{bmatrix}$$

then

$$\frac{d}{dt}\begin{bmatrix} x_1 \\ x_2 \\ q \end{bmatrix}=\boldsymbol{A}_e\begin{bmatrix} x_1 \\ x_2 \\ q \end{bmatrix}+\boldsymbol{B}_e\boldsymbol{U}+\begin{bmatrix} 0 \\ 0 \\ -1 \end{bmatrix}y_r \tag{8.58}$$

The control objective here is to determine the control input $\boldsymbol{U}$ which minimizes the performance function expressed by the following 2nd-order function at $y_r=0$

$$J[u(t)]=\int_0^\infty\{[x_1, x_2, q]\boldsymbol{Q}\begin{bmatrix} x_1 \\ x_2 \\ q \end{bmatrix}+\boldsymbol{P}u\}dt \to \min \tag{8.59}$$

Here, $\boldsymbol{Q}$ and $\boldsymbol{P}$ are weighting functions. The above is referred to as a LQ (Linear Quadratic) problem, and is generally solved using the Riccati equation with respect to $\boldsymbol{K}$ as follows:

$$\boldsymbol{KBeP}^{-1}\boldsymbol{B}^T\boldsymbol{eK}-\boldsymbol{A}^T\boldsymbol{eK}-\boldsymbol{KAe}-\boldsymbol{Q}=0 \tag{8.60}$$

Using this $\boldsymbol{K}$, control is determined as

$$U = -P^{-1}Be^{T}KX \tag{8.61}$$

For the system in Fig. 8.45, we thus get

$$\begin{aligned} u(t) &= k_1x_1 + k_2x_2 + k_3q \\ &= k_1x_1 + k_2x_2 + k_3\int(y - y_r)dt \end{aligned} \tag{8.62}$$

which is called an integral optimal regulator. As for a means of solving Eq. (8.60), the following equation can be solved numerically.

$$\frac{dK}{dt} = K(t)BeP^{-1}B^{T}eK(t) - A^{T}eK(t) - K(t)Ae - Q, \quad K(t) = 0 \tag{8.63}$$

8.7 Kalman Filter

8.7.1 Kalman filter formula

The Kalman filter is used in a wide variety of fields. It is expressed as a set of iterations estimating the current state of the system, and it is suitable for on-line computerized calculations. The formula for this filter is presented below in a discrete data system.

For a system described by continuous state equations such as Eqs. (8.53) and (8.54), such equations can be converted to a discrete data system as shown below (derivation omitted). In particular, if we let

$$P = e^{AT} \tag{8.64}$$

$$Q = (e^{AT} - I)A^{-1}B \tag{8.65}$$

we obtain

$$\begin{cases} X_{k+1} = PX_k + QU_k & (8.66) \\ Y_k = CX_k & (8.67) \end{cases}$$

Given that input noise W_k and observation noise V_k is applied to the system (Fig. 8.46), their variance can be denoted by Gaussian white noise W and V, respectively. The problem for the Kalman filter is to

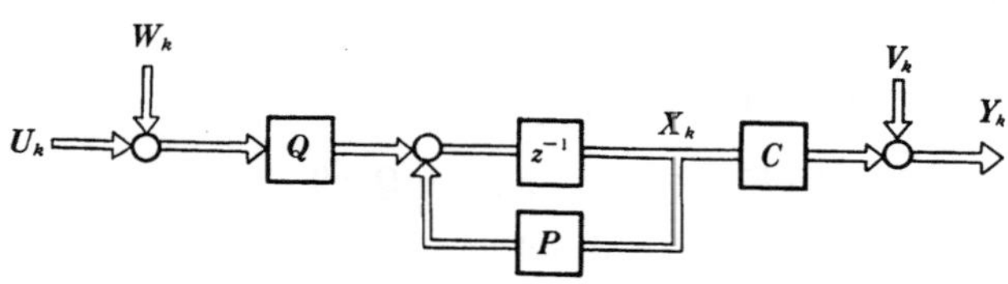

Fig. 8・46 Block diagram of system

measure input/output values Y_0, $Y_1, \cdots$, Y_{k+1}, U_0, $U_1, \cdots$, U_k and then perform an optimal estimation of X_{k+1}. The formula for doing so is given below without proof or derivation.

$$\hat{X}_{k+1}=(P\hat{X}_k+QU_k)+K_{k+1}\{\tilde{Y}_{k+1}-C(P\hat{X}_k+QU_k)\} \tag{8.68}$$

Here, $\hat{X}_{k+1}$ and $\hat{X}_k$ denote the estimate of X at sampling times $k+1$ and k, respectively, and $\tilde{Y}_{k+1}$ the measured value of Y at sampling time $k+1$. In addition, gain K_{k+1} can be determined from the following formulas.

$$M_{k+1}=P(I-K_kC)M_kP^T+QWQ \tag{8.69}$$

$$K_{k+1}=M_{k+1}C^T(CM_{k+1}C^T+V)^{-1} \tag{8.70}$$

The Kalman filter can be implemented on a computer by starting off with initialized values and then calculating the above three equations at every sampling time, *i.e.*, by performing iterations. The term $C(P\hat{X}_k + QU_k)$ in the brackets on the right side of Eq. (8.68) represents the estimate of Y at sampling time $k+1$ based on the estimate at sampling time k; this follows from Eqs. (8.66) and (8.67). As a result, the entire expression in brackets is the difference between the actually measured value and the estimate of Y. Thus, Equation (8.68), the estimate of X_{k+1} at time $k+1$, can be interpreted as consisting of a corrective term obtained by multiplying this difference by gain K_{k+1}. Gain K_{k+1} is given by Eq. (8.70), in which M_{k+1} denotes the estimation error and V the variance of the observation error. Accordingly, if the observation error is large compared with the estimation error, K_{k+1} is made small. In other words, since for a large obserbation error the reliability of measured variable $\tilde{Y}_{k+1}$ is low, strong corrective action is not taken in Eq. (8.68). Estimation error M_{k+1} is given by Eq. (8.69). Since the first term on the right of the equation takes on a form in which the expression $(I-K_kC)$ is multiplied by the previous estimation errorM_k, it can be seen how the larger K_k becomes, the smaller $(I-K_kC)$ and hence M_{k+1} become. This can be interpreted as follows: the greater the previous correction, the closer we are to the true value, which means a smaller estimation error.

8.7.2 Application to the parameter estimation problem[9)]

A certain water management office has come to use the Kalman filter to predict the daily amount of water demand. Water usage changes in a complex way depending on a variety of factors such as weather, season, and day of the week. Specifically, a different formula is used for each of the four seasons, as follows:

$$\text{Spring:}\quad Q_p = a_{11}t^2 + a_{21}t + a_{31}r + a_{41} \tag{8.71}$$
$$\text{Summer:}\quad Q_p = a_{12}t^2 + a_{22}t + a_{32}r + a_{42} \tag{8.72}$$
$$\text{Fall:}\quad Q_p = a_{13}t^2 + a_{23}t + a_{33}r + a_{43} \tag{8.73}$$
$$\text{Winter:}\quad Q_p = a_{14}s_S + a_{24}s_{yf} + a_{34}t + a_{44}r + a_{54}Q_y + a_{64} \tag{8.74}$$

In these equations, Q_p denotes the 24-hour total demand increment, t the temperature at the time of prediction, r the weather factor increment ($r=1$ during rain and 0 otherwise), s_{yf} the amount of snowfall during the previous 24-hour period (mm), s_S the accumulated amount of snow (mm), and Q_y the actual amount of demand for the previous 24-hour period.

Setting up of the above equations probably included some elements due to experience. Nevertheless, to the extent that we assume them to be correct, the Kalman filter can be applied for determining their coefficients. For example, if we rewrite the equation for Spring as

$$\boldsymbol{Y} = [Q_p] \tag{8.75}$$
$$\boldsymbol{X} = [a_{11}\ a_{21}\ a_{31}\ a_{41}]^T \tag{8.76}$$
$$\boldsymbol{C} = [t^2\ t\ r\ 1] \tag{8.77}$$

then the system equations become

$$\begin{cases} \boldsymbol{X}_{k+1} = \boldsymbol{X}_k & (8.78) \\ \boldsymbol{Y}_{k+1} = \boldsymbol{C}\boldsymbol{X}_{k+1} & (8.79) \end{cases}$$

and the problem is to estimate $\boldsymbol{X}_{k+1}$ To apply the Kalman filter formula, we substitute

$$\boldsymbol{P} = \boldsymbol{I} \text{ (unit matrix)}$$
$$\boldsymbol{U}_k = \boldsymbol{0}$$
$$\boldsymbol{Q}_w = \boldsymbol{0}$$

to arrive at the following three equations:

$$\begin{cases} \boldsymbol{M}_{k+1} = (\boldsymbol{I} - \boldsymbol{K}_k\boldsymbol{C})\boldsymbol{M}_k & (8.80) \\ \boldsymbol{K}_{k+1} = \boldsymbol{M}_{k+1}\boldsymbol{C}(\boldsymbol{C}\boldsymbol{M}_{k+1}\boldsymbol{C}^T + \boldsymbol{V})^{-1} & (8.81) \\ \hat{\boldsymbol{X}}_{k+1} = \hat{\boldsymbol{X}}_k + (\tilde{\boldsymbol{Y}}_{k+1} - \boldsymbol{C}\hat{\boldsymbol{X}}_k) & (8.82) \end{cases}$$

Calculation of these three equations once every day enables the coefficients to approach more correct values. Use of this method for estimating water demand showed effective results, as shown in Fig. 8.47.[9]

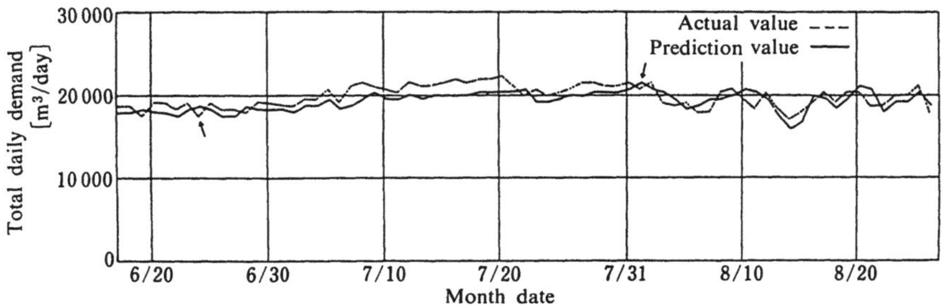

Fig. 8・47 Result of total daily demand Prediction for a distributing reservoir

8.8 Other Forms of Advanced Control

For cases in which disturbance is repeatedly applied, or in which changes in the setpoint value are performed in a periodic fashion, the results of previous control actions can be stored and used as a basis for parameter correction. Such a method is referred to as "learning control." Moreover, parameter correction based on simple exponential filters are gradually coming into use.

Testing is also being performed on fuzzy control, a form of control which applies fuzzy set theory. The object here is to incorporate within the control system the experience possessed by a skilled operator, experience which naturally includes elements of fuzziness. The extent of ambiguity is expressed through use of "membership functions." Fuzzy control is appropriate for systems in which preparation of a process model is overly difficult. In all, it is expected that the development work in this field will lead to even more advanced forms of control based on expert systems applied artificial intelligence.

REFERENCES

1) O. J. M. Smith : "A controller to overcome dead time," *ISA J*, **6**, 2 (1959) 28–33.

2) Nozawa : "Control for astatic control object including dead time," *Automatic Control*, **7**, 5 (1960) (in Japanese).

3) K. Watanabe, *et al.*: "Disturbance rejection of Smith predictor control system," *Trans. Soc. Instrument & Control Eng.*, **19**, (1983) 187–192 (in Japanese).

4) M. Araki : "Computer-aided-design of multivariable control systems: Further developments of the inverse Nyguist array method," *Systems & Control*, **26**, (1982) 218–227, 353–362, 489–497 (in Japanese).

5) T. Kitamori : "Study on an adaptive control system based on the ultimate sensitivity method," *Trans. Soc. Instrument & Control Eng.*, **6**, 6 (1970) 537–545 (in Japanese).

6) A. Sumi : "Adaptive control system applied with limit cycle," *Trans. meeting for Study of Analog Tech.*, **12**, 8, (1972) 39–47 (in Japanese).

7) K. J. Åström, *et al.*: "Theory and applications of self-tuning regulators," *Automatica*, **13**, (1977) 475–476.
8) Y. Sakai, *et al.*: "Enhanced YEWSERIES 80," *Yokogawa Tech. Rep.*, **30**, 3 (1986) 31–33 (in Japanese).
9) H. Kamei, *et al.*: "On-line water demand predictions with Kalman filter," *Yokogawa Tech. Rep.*, **25**, 4 (1981) 36–40 (in Japanese).

Chapter 9

CONTROL OF PROCESS UNITS (Application I)

The purpose of this chapter is to describe the basic elements of representative process units used in the process industry. This material should serve as a basic reference when planning control systems for such process units.

9.1 Overview

Analog instrumentation is usually characterized by independent application to each control loop, with integrated configurations rate. The introduction of distributed control systems, however, brought about the widespread use of coordinated control systems as typified by feedforward control in continuous processes and totally automatic systems in batch processes. The application of selective control for protecting equipment and ensuring safe operations has also come to be widely adopted.

Thus, in order to effectively apply these new forms of control devices, it is felt that a thorough understanding of the various factors affecting process units is now more than ever indispensable. Examples of such factors with relevant control techniques indicated in parenthesis are; principles of process operations and control objectives, adjustment of equipment load (throughput control)*, relationship between disturbance and manipulated and controlled variables (feedforward control), load variation and change in process gain (PID optimal setting), particular process-dynamic characteristics (application of advanced control methods), equipment constraints (selective control), operation economy (energy conservation, optimal cost control) and startup and shut down operations (labor-saving control).

At present, knowledge related to the above is mostly obtained by

* The contents in the parenthesis show the relevant control techniques.

experience, namely by plant operation technicians. This knowledge, however, is not enough and a quantitative understanding is also considered to be essential. It is in this regard that the further development of chemical engineering geared to process operation and control (plant operation engineering)[1] is eagerly awaited.

Accordingly, the application examples in this chapter, although by no means exhaustive, aim to provide the beginner with a deeper understanding of process unit control in the light of the above considerations.

9.2 Control of Fluid Transport Processes

9.2.1 Pump control

(a) Centrifugal pump capacity control

Small motor-driven pumps often employ a control method in which a control valve incorporated on the discharge side can be manipulated (Fig. 9.1). Capacity control of this pump can be illustrated by the curves shown in Fig. 9.2(a) in which one type of curve indicates pump characteristics (flow versus discharge pressure) while the other indicates total pipe-pressure loss including that of the control valve. Thus, if we consider the flow rates at two points where these curves intersect, we can see that as we increase the pressure loss in the control

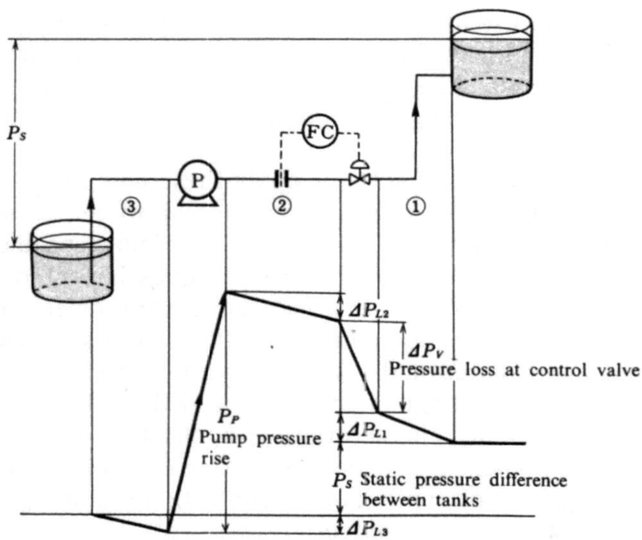

ΔP_{L1} to ΔP_{L3}······ Pressure loss in each piping section

Fig. 9·1 Pressure balance in centrifugal pump line and its capacity control

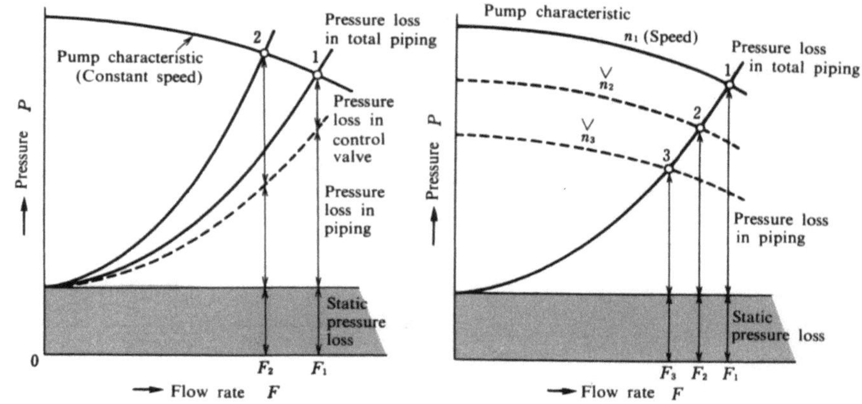

(a) Discharge valve operation　　(b) Pump speed operation

Fig. 9・2 Capacity control of centrifugal pump

valve from points 1 to 2, the flow rate decreases.

Pump characteristics are in general given by the following expression:

$$P=K_1n^2-K_2F^2 \tag{9.1}$$

where P is the discharge pressure, n the pump speed, F the flow rate, and K a constant.

Based on the relationship above, an energy-conservation control technique can be devised for a large-type pump. In this devise, as shown in Fig. 9.2(b), flow rate is controlled by changing pump characteristics, which is achieved by altering the pump speed without a control valve. In order to employ this technique as effectively as possible, the ratio of dynamic pressure loss* to total pipe-pressure loss must be large and the time during which low-load operations are performed must be sufficiently long with respect to the total operation period. This can best be clarified by performing calculations for a dimensionless model as shown in Fig. 9.3(a). The curve for pump characteristics in this figure represents a typical example of Eq. (9.1) and is given by:

$$P=1.2-0.2F^2 \tag{9.2}$$

With $F=1$ and the ratio of dynamic pressure loss to total pipe-pressure loss as f, the total pipe-pressure loss P_L is then

* The pressure loss changing in proportion to the square of the flow rate.

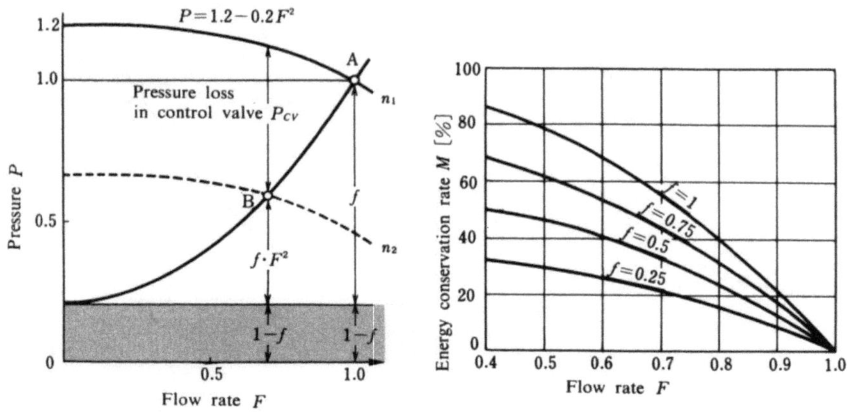

(a) Dimensionless model for comparison

(b) Relationship between flow rate and energy conservation rate
(*f* is partial rate of dynamic pressure loss in total loss)

Fig. 9 · 3 Comparison in energy consumption between valve control and speed control

$$P_L = (1-f) + fF^2$$

in which $(1-f)$ is static pressure loss.

In the same figure, if the flow is decreased from a state of total load (1) to a flow F, we can then compare the power W_1 required when using control valve regulation with the power W_2 required when adopting pump speed control. Expressing pump power as (discharge pressure) × (flow rate), W_1 and W_2 are then:

$$W_1 = (1.2 - 0.2F^2)F$$
$$W_2 = \{(1-f) + fF^2\}F$$

If we now define the energy–conservation rate M as

$$M = 100\,(W_1 - W_2)/W_1$$

then from the two expressions above:

$$M = \frac{(0.2+f)(1-F^2)\times 100}{1.2-0.2F^2} \tag{9.3}$$

Figure 9.3(b) shows the results of calculations based on this equation. The affect that changes in F and f have on the energy–conservation rate M can be quantitatively understood by these curves.

(b) Capacity control of a constant–volume pump

Capacity control of rotary pumps using gears or vanes can be per-

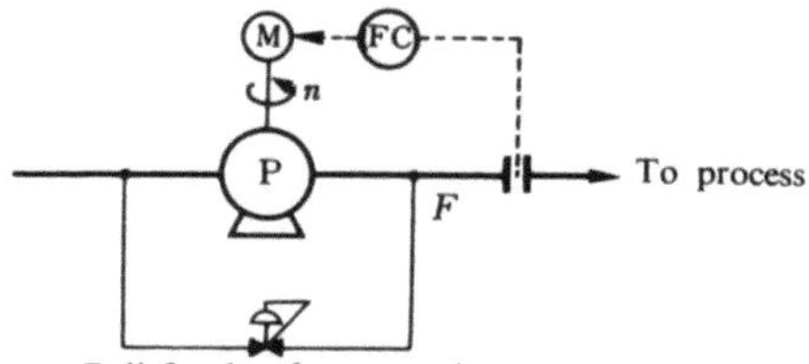

(a) Capacity control by speed adjustment

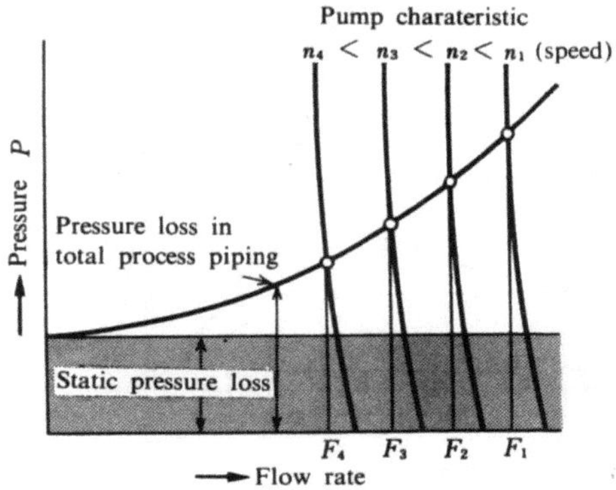

(b) Change in pump charateristic

Fig. 9 · 4 Capacity control of constant volume pump (1)

formed by the speed adjustment technique shown in Fig. 9.4 or the by-pass–valve adjustment technique shown in Fig. 9.5.

9.2.2 Compressor control

(a) Capacity control of a compressor and pressure loss in piping

Although capacity control of a compressor within a process is in principal not very different than that of a pump, some differences do exist as shown in Fig. 9.6. Here, pressure controller (PC)* is often used instead of flow controller (FC) and a change in capacity can be instigated from either the feed side or demand side via selection of the respective controlled variables; suction pressure and discharge pressure.

The dynamic pressure loss of compressed gas in the piping can be obtained by the following equations. Here, the symbols V, F and M correspond to those in the figure and represent volumetric flow rate, mass flow rate and molecular weight, respectively.

---

* Equivalent to LC (level controller) in the case of fluids.

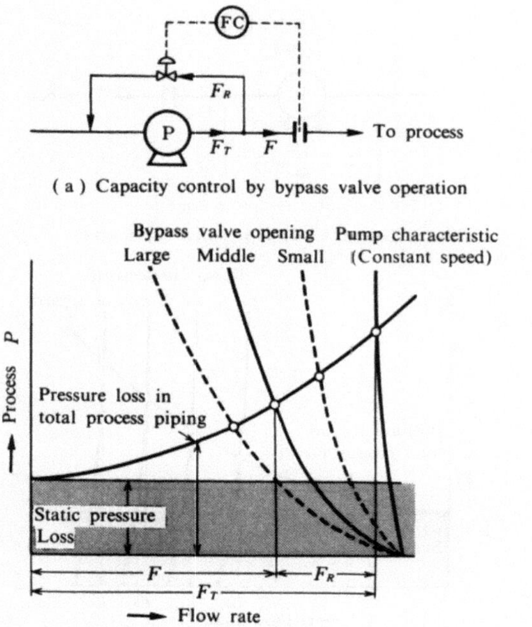

(a) Capacity control by bypass valve operation

(b) Bypass valve opening and flow rate in each part

Fig. 9 · 5 Capacity control of constant volume pump (2)

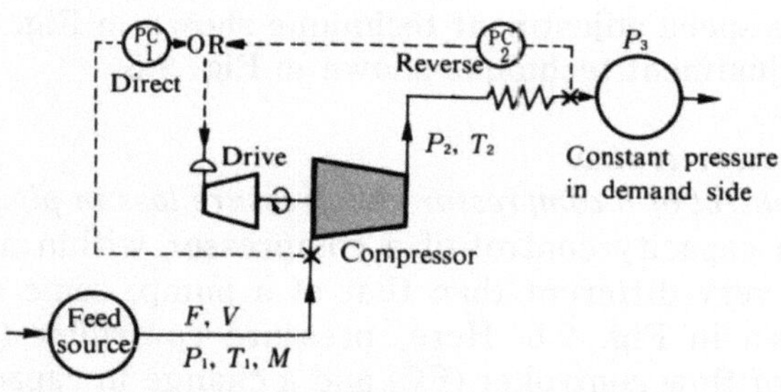

Fig. 9 · 6 Capacity control of compressor in process

$$V = \frac{R_g T_1}{M P_1} F \qquad (R_g\text{: gas constant}) \tag{9.4}$$

$$F = c\sqrt{(P_2 - P_3)\frac{MP_2}{T_2}} \qquad (c\text{: flow coefficient of piping}) \tag{9.5}$$

From these two equations, we arrive at:[2)]

$$V = \frac{cR_g T_1}{\sqrt{MT_2}}\sqrt{\left(\frac{P_2 - P_1}{P_1} + 1\right)\left(\frac{P_2 - P_1}{P_1} - \frac{P_3 - P_1}{P_1}\right)} \tag{9.6}$$

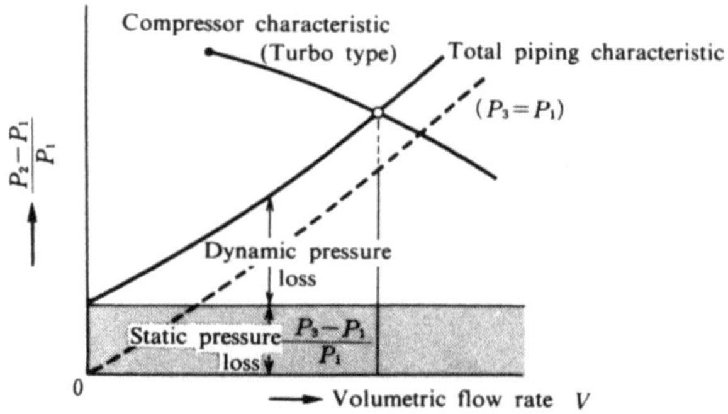

Fig. 9 · 7 Compressor characteristic and total piping characteristic

The above equations are used to obtain the relationship, as shown in Fig. 9.7, between the curves for total pipe-pressure loss and compressor characteristics (turbo type). Here, the flow rate V can be controlled by changing compressor characteristics (*e.g.*, altering speed) as in the case of pump control.

(*b*) *Turbo compressor characteristics*

The following rules (fan rules*) are established for turbo compressors having a relatively low compression ratio and handling a gas with a low molecular weight (such compressors are also referred to as fans or blowers depending on the compression ratio value):

(1) The flow rate is proportional to compressor speed

(2) The discharge pressure is proportional to the square of the compressor speed

(3) The shaft power is proportional to the cube of the compressor speed

Note, however, that characteristics different from these will appear as the compression ratio becomes large.

Changing of turbo compressor characteristics with the objective of capacity control can be achieved by changing either compressor speed, suction vane opening or suction valve opening. Figure 9.8(a) shows a case in which characteristics change by varying compressor speed. Also shown in the figure is a surge limit line which defines the operable part of turbo compressor characteristics. Within this limit, discharge pressure falls along with a decrease in flow rate, exhibiting so-called negative resistance characteristics. This gives rise to cyclic fluctuation in flow rate and discharge pressure, which cross over the limit line. This pulsation is referred to as "surging," and its negative affects can be countermeasured by implementing appropriate opera-

* These rules also hold for pumps.

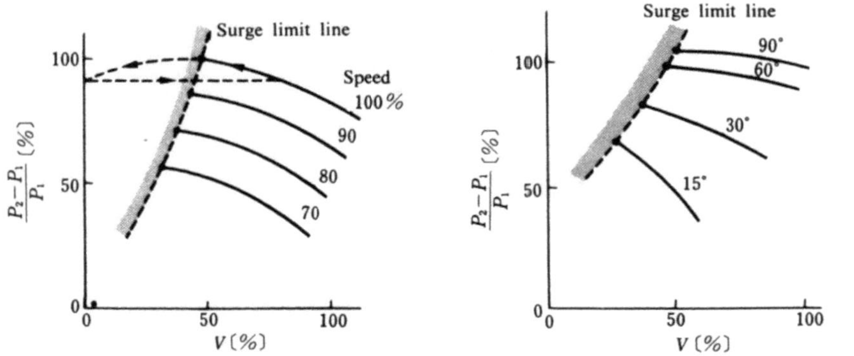

(a) Characteristic change by speed variation (b) Characteristic change by various opening of suction vane

Fig. 9・8 Characteristic of turbo type compressor

tions such as 1) increasing the flow rate as the pressure loss on the discharge side decreases, or 2) decreasing compressor speed, so that operation continues only up to the point where crossover would occur. In particular, a large compression ratio can result in immediate damage to a compressor due to the shock on entering the cycle. The anti-surge control method described in following paragraphs is indispensable in such a case.

Figure 9.8 (b), on the other hand, shows a case in which characteristics are changed by varying suction vane* opening under conditions of fixed compressor speed.

Although the varying of suction valve opening also changes characteristics in a very similar way, the range of adjustment is narrow compared to the former techniques, and moreover since energy loss is somewhat large, this technique is usually adopted only for compressors having relatively small volume.

(c) Anti-surge control for turbo compressors

The position and form of a surge limit line will vary depending on gas composition and suction temperature. Assuming, however, a compressor for which these conditions do not change and moreover in which speed is fixed, then by incorporating a bypass flow operation as shown in Fig. 9.9(a), flow rate will not drop below the minimum set value and surging can be prevented. On the other hand, for adjustable-speed compressors in which the minimum set value is not fixed, if we lower the set value along the surge control line** corresponding to a reduction in speed (Fig. 9.9(b)), then if the process flow rate falls be-

* Installed on the compressor body and manipulated by a control signal.

** A line drawn parallel to the surge limit line at a certain allowable distance.

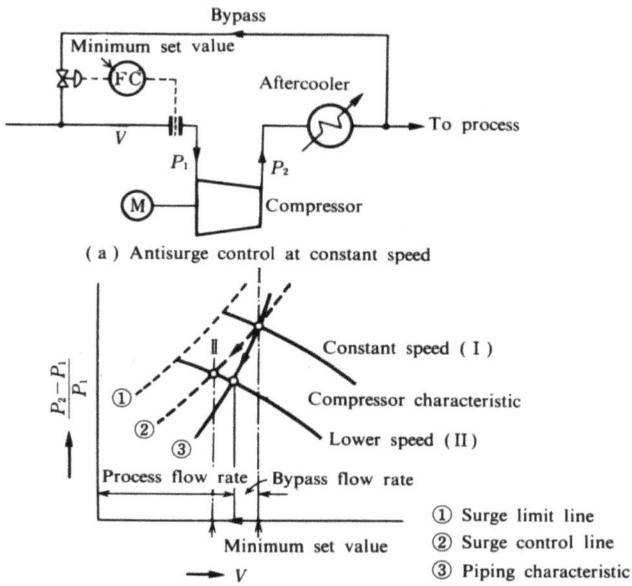

(a) Antisurge control at constant speed

(b) Modification in minimum set value at variable speed

Fig. 9・9 Antisurge control on unchangeable condition in suction

low the fixed minimum set value, useless power can be reduced without bypass flow.

In the above control systems, it is assumed that flow controllers incorporate an anti-reset-windup mechanism and control valves (positive plug*) possess linear characteristics and rapid opening/closing capability so that control actions can be performed as fast as possible.

Considering now the case in which gas composition and suction temperature vary, if the compression ratio P_2/P_1 is relatively small (under 3), then the surge limit line can be approximately given as:[2)]

$$\frac{P_2-P_1}{P_1} \approx c_1 \frac{MV^2}{T_1} \tag{9.7}$$

where c_1 is a machine-dependent constant.
In addition, the following equation holds for an orifice incorporated on the suction side:

$$V = c_2 \sqrt{\frac{hT_1}{MP_1}} \tag{9.8}$$

* The valve opens (safety side) in the event of less air.

where c_2 is an orifice constant.
From the above two equations, we get:

$$h=\frac{P_2-P_1}{c_1c_2^2}=C(P_2-P_1) \tag{9.9}$$

Thus, as shown in Fig. 9.10, by selecting a coefficient* C_c having a value larger than C as determined by the above equation and by controlling the bypass valve on the basis of $h=C_c(P_2-P_1)$ surging can be prevented.

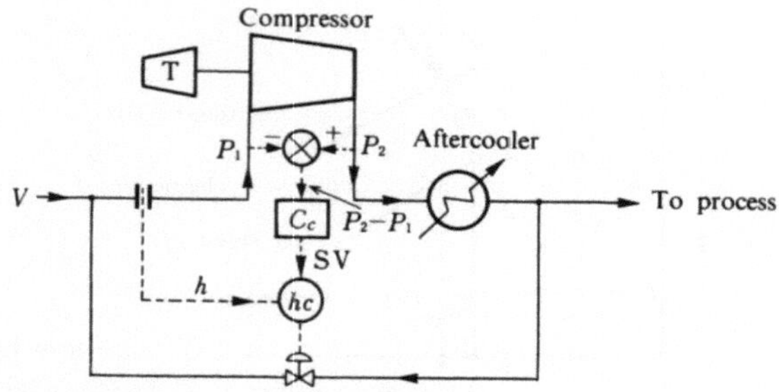

Fig. 9 · 10 Antisurge control in case of change in suction condition

(d) Capacity control of a rotating positive-displacement compressor

Compressors of this type, although employing a variety of techniques such as Roots, vane, Nash, or screw, all possess flow and discharge pressure characteristics similar to positive-displacement pumps. Thus, the methods employed for capacity control of pumps may also be applied here.

(e) Capacity control of a reciprocating compressor

This type of compressor usually contains several pistons and cylinders in which the compressor capacity of each cylinder is controlled through use of a clearance pocket actuator and unloader actuator (Fig. 9.11). The unloader, which manipulates the suction valve, can set compressibility to zero by keeping the valve in a open position. The clearance pocket, on the other hand, can increase or decrease the volume of compressed space by completing opening or closing its operation valve, *i.e.*, the amount of compressed gas delivered within the full piston stroke can be controlled in a two-stage fashion. Referring to the figure, if the clearance pocket valve is completely closed, the piston will move up to position A and the pressure within the cylinder will rise up to P_2. The volume of gas corresponding to residual piston stroke 1 will then be delivered. If the clearance pocket valve is now completely open-

* To provide a safety margin.

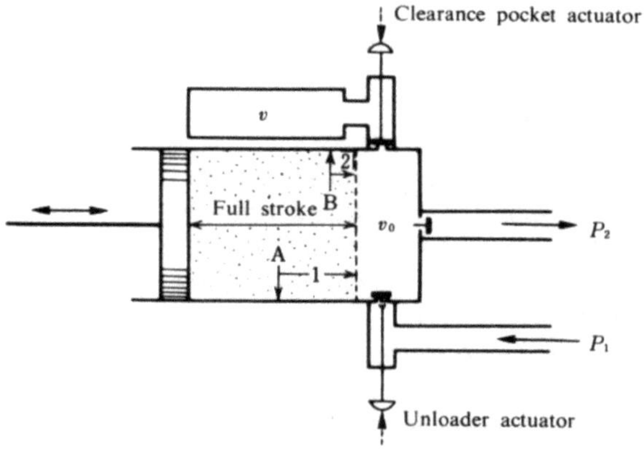

Fig. 9 · 11 Operation principle of unloader and clearance pocket

ed, the piston will move up to position B with the internal pressure again reaching P_2 and the volume of gas corresponding to residual piston stroke 2 delivered.

The difference in delivered volume (at suction conditions) for these two situations can be determined by the following expression, where v specifies the volume of the clearance pocket:[2)]

$$v\left\{\left(\frac{P_2}{P_1}\right)^{1/\gamma}-1\right\},\quad \gamma=\frac{\text{specific heat at constant pressure}}{\text{specific heat at constant volume}}$$

Figure 9.12[4)] illustrates how the capacity of a fixed-seed reciprocating compressor can be adjusted in 5 stages when both unloader and clearance pocket actuators are used.

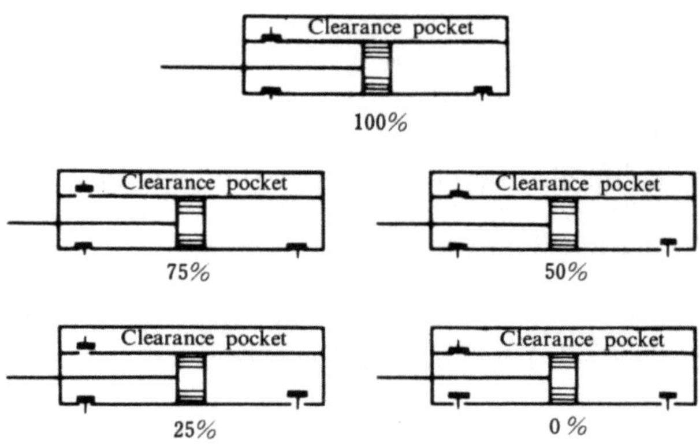

Fig. 9 · 12 Multi-stage capacity control of reciprocating compressor

9.3 *Control of Heat Transfer Processes*

9.3.1 *Control of heat exchangers*

(*a*) *Heater control by steam valve adjustment*

Figure 9.13(a) shows a temperature control process in which steam flow G decreases proportionally with a reduction in flow F in response to a high-load situation. As shown in (b) of the figure, the pressure P inside the vessel is reduced at this time since the heat load becomes small.* Thus, if the steam feed pressure P_S is kept constant, the pressure drop across the valve increases. As a consequence, if the control valve possesses linear characteristics, the open-loop gain for temperature control increases considerably. To compensate for this, a valve having equal-percentage characteristics should be adopted instead of one with linear characteristics in practice.

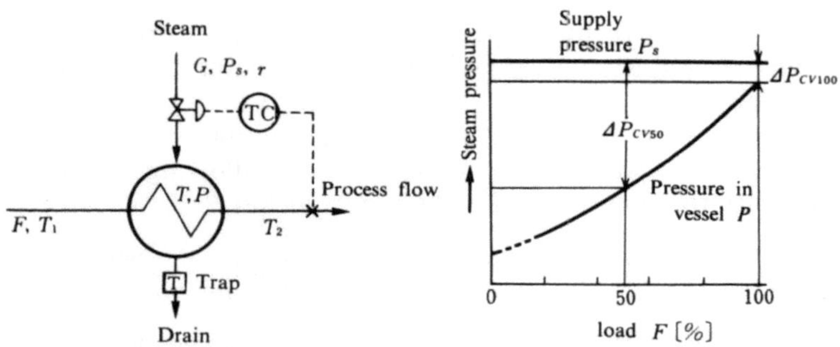

(a) Temperature control process (b) Change in control valve pressure difference by load change

Fig. 9 · 13 Heater control by steam valve

In such a process, although the value of pressure drop across the valve at maximum load often presents a problem at the design stage, 7 to 10% of the feed pressure (absolute) is considered a sufficient pressure drop. It should also be noted that if the load decreases remarkably, the valve will move to nearly complete closure, and since the C_V value becomes discontinuous in this state, an on–off line operation will occur. In addition, in the case of a steam heater with the set temperature around 100°C, internal pressure P will approach atmospheric pressure with a decrease in load and the steam trap will cease to function.

---

* The temperature difference between the steam within the vessel and the process fluid may be small.

(b) Heater control by drain valve adjustment

Heating capacity can also be controlled by manipulating drain flow,* as illustrated by the system in Fig. 9.14 (a). With this technique, there is no interference with drain discharge due to load reduction as described above, and a wide range of load variation can be accommodated. In addition, since valve size here is considerably smaller than that in steam control, a definite savings in equipment cost can be realized. However, control response in this technique is somewhat inferior to that of control by a steam valve. Referring to the block diagram of this process in Fig. 9.14 (b), the following expressions are obtained for determining the transfer function expressing the heat transfer rate corresponding to drain flow manipulation.

$$\frac{Q(s)}{G_c(s)}=\frac{r}{Ts+1}, \qquad T=\frac{r}{(\Delta A/\Delta W_h)U\overline{\Delta T}} \tag{9.10}$$

It can be seen that this process has a first-order capacity delay.

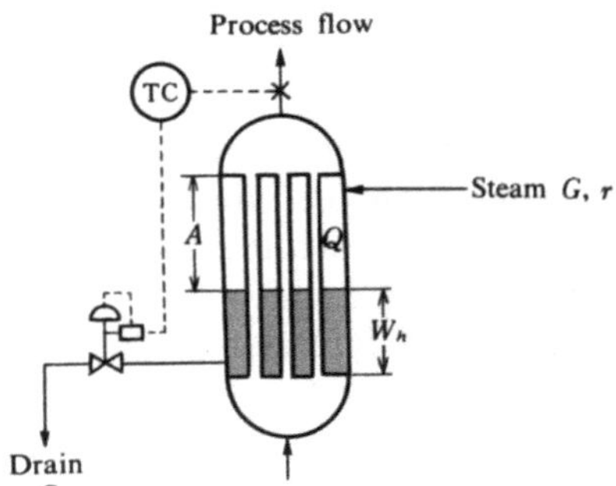

(a) Temperature control process

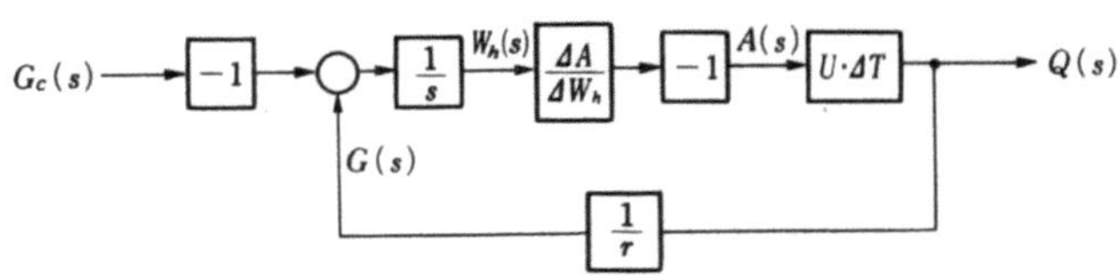

$G_c(s)$: Drain flow rate

$Q(s)$: Heat transfer rate for condensation

$G(s)$: Steam flow-in rate

$W_h(s)$: Amount of drain in vessel

$A(s)$: Effective heat transfer area for condensation

U : Over-all heat transfer coefficient on condensing surface

$\overline{\Delta T}$: Average temperature difference between steam and process fluid

r : Latent heat of steam for condensation

(b) Heat rate response for drain flow operation

Fig. 9・14 Heater control by drain control valve

* The value of the overall heat transfer coefficient of the condensation surface is much larger than that of the drain soaking surface.

The time constant in the above expressions is usually a few minutes in most equipment (refer to practice question 9.3). Therefore, a controller having a narrow proportional band (20 to 30 %) with an integral time of about one minute and having a control valve with a positioner shall be used in this technique. Although the closed-loop natural oscillation cycle here is quite long compared to control by steam valve adjustment, it can be used with almost no interference in heat control for equipment such as distillation column reboilers.

(*c*) *Control of fluid-to-fluid heat exchangers*

We first consider a case of heat exchange in which the flow of a high-temperature fluid is controlled and a low-temperature fluid is heated to a set temperature (Fig. 9.15). It is assumed in the following description that the only disturbance is the flow rate and temperature of the low-temperature fluid, and that the inlet temperature of the high-temperature fluid is invariable. In this process, if the inlet temperature of the low-temperature fluid drops or if its flow rate increases, temperature controller TC will act to increase the flow rate of the high-temperature fluid. This results in a large average temperature differential between the two fluids which in turn increases the amount of heat-transfer (although the increase in flow rate brings about an increase in the heat-transfer coefficient which aids control, this is exceedingly small compared to the effect exerted by the change in average temperature differential).

However, since the effect of increasing the average temperature differential by increasing the flow rate of the high-temperature fluid is limited when the exit temperature of the high-temperature fluid ap-

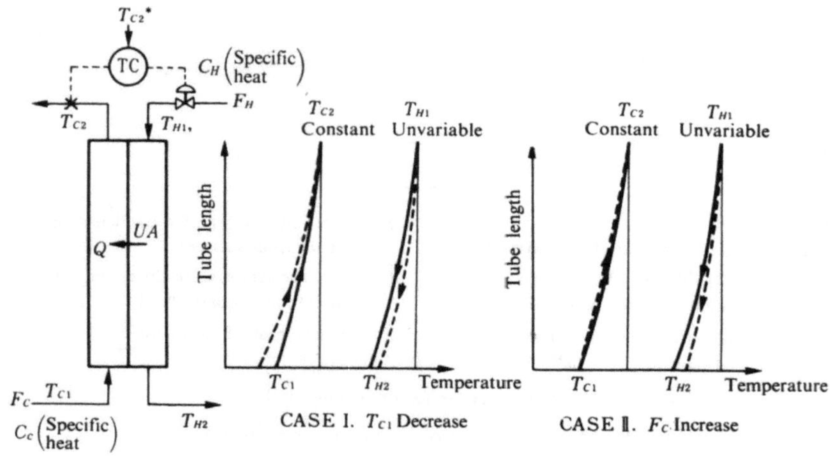

Fig. 9・15 Temperature control for fluid-to-fluid heat exchanger proaches its inlet temperature, rapid saturation tends to occur. Thus, in heat processes similar to this one*, gain characteristics for the change in heat-transfer rate due to flow manipulation become extremely non-linear, presenting a problem to control in not a few cases.

The following relationship holds for the variables shown in Fig. 9.15:[3)]

$$T_{C_2} = T_{C_1} + \frac{T_{H_1} - T_{C_1}}{(F_C C_C/UA) + (1/2)\{1 + (F_C C_C/F_H C_H)\}} \tag{9.11}$$

We consider a particular process in which the variables in the above equation are as follows: $T_{C1} = 92°C$, $T_{C2} = 200°C$, $T_{H1} = 260°C$, $T_{H2} = 230°C$, $F_C = 50$ kg/min, $F_H = 180$ kg/min, and $C_H = C_C = 1$. If either T_{C1} or F_C varies, this equation can now be used to calculate the desired change in F_H to maintain T_{C2} at 200°C (Fig. 9.16).

In addition, the sensitivity (gain) of the change in T_{C2} due to adjustment of F_H can be obtained by differentiating Eq. (9.11) as follows:

$$\frac{\partial T_{C_2}}{\partial F_H} = \frac{T_{H_1} - T_{C_1}}{[(F_C C_C/UA) + (1/2)\{1 + (F_C C_C/F_H C_H)\}]^2} \cdot \frac{F_C C_C}{2F_H^2 C_H}$$

Using this equation, Table 9.1 shows the results of calculating the change in gain for changes in F_C in the above example. Just how remarkable this change is can be seen by the values listed. Although this tendency can be somewhat relaxed by adopting an equal-percentage

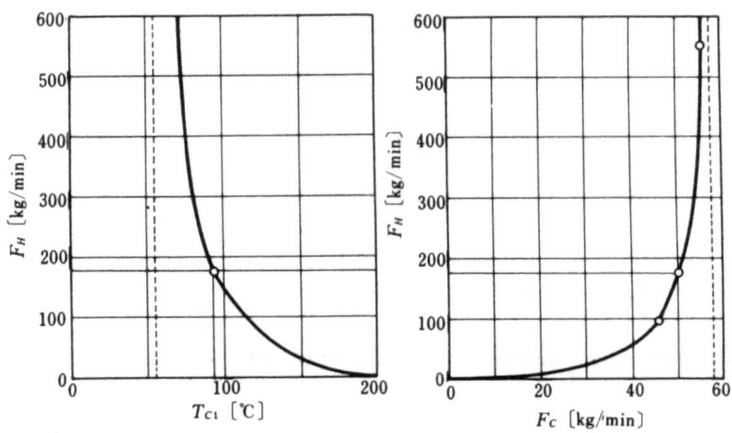

(a) Disturbance T_{C1} (b) Disturbance F_C

Fig. 9·16 Change in heating flow rate for process side disturbances

* Such as in condenser adjustment by cooling water, or stirred-tank temperature adjustment by jacket-coil cooling water.

Table 9.1 Process gain change of fluid-fluid heater for heated flow disturbance

F_C (kg/min)	F_H (kg/min)	$\frac{\partial T_{C_2}}{\partial F_H}$ (°C/kg/min)	$\partial T_{C_2} \Big/ \left(\frac{\partial F_H}{F_H}\right) \times 100$(°C%)
55	555	0.006	3.33×10^{-2}
50	180	0.055	9.90×10^{-2}
45	97	0.166	16.1×10^{-2}

control valve, examining the last column in the table reveals that such compensation is not sufficient.

(d) Temperature control by heat exchanger bypass

Temperature control as performed by the bypass configuration to the heat exchanger shown in Fig. 9.17 exhibits excellent response compared to the control techniques so far described. As a result, this technique has come to be frequently used. However, if the bypass valve opening changes due to variation in process flow rate or the temperature controller (TC) setpoint value, control gain will also change resulting in degraded control. In addition, since the parallel-piping flow resistance of the heat exchanger and bypass valve changes, disturbance will be applied to process flow F. These undesired affects can be alleviated by using a three-way valve instead of a two-way one, as shown in Fig. 9.18.

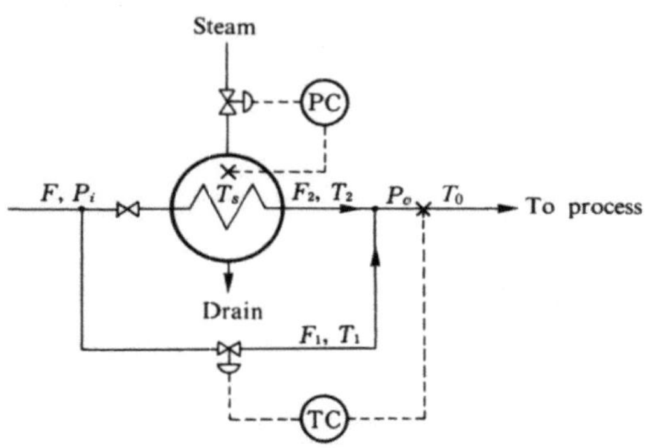

Fig. 9 · 17 Temperature control of heater by process bypass valve

The following equation holds for the two-way valve of Fig. 9.17.

$$T_0F = T_1F_1 + T_2F_2$$

If we now designate the bypass valve C_V value as c_1 and the C_V value equivalent to the flow resistance on the heat-exchanger side as c_2, the following equations also hold.

$$F_1 = ac_1\sqrt{P_i - P_o}$$

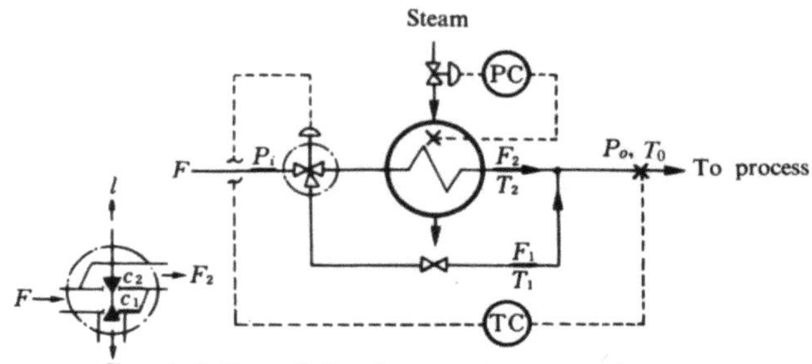

(a) Case of diverting type three-way valve

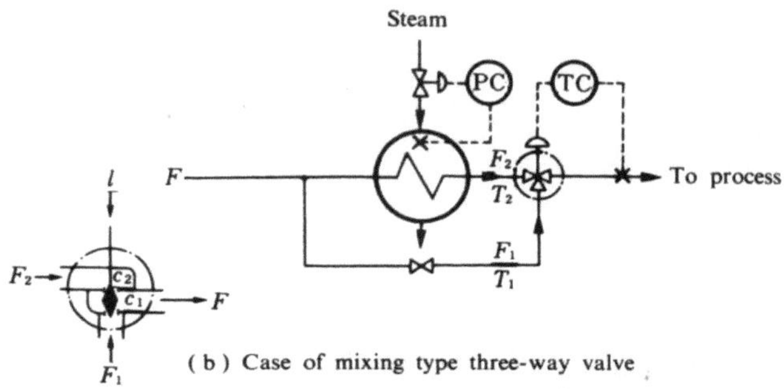

(b) Case of mixing type three-way valve

Fig. 9·18 Temperature control of heater by three-way valve

$$F_2 = \alpha c_2 \sqrt{P_i - P_o}$$

$$F = F_1 + F_2 = \alpha (c_1 + c_2) \sqrt{P_i - P_o}$$

where α is a constant. From these equations, we arrive at the following:

$$T_0 = \frac{T_1\{1 + (T_2/T_1)(c_2/c_1)\}}{1 + c_2/c_1} \tag{9.12}$$

It can be seen from this equation that if T_1 is invariable, then T_0 will not change as long as T_2 remains constant. In regards to T_2, since it will approach steam temperature T_S if heating capacity of the heat exchanger has a sufficient margin, its amount of change is small. However, it can also be seen that c_1 becomes small if T_1 decreases with T_0 fixed, or alternatively, if set value T_0 is raised while T_1 is fixed.

The affect that change in c_1 exerts on T_0 can be obtained from Eq. (9.12) as follows:

$$\frac{\partial T_0}{\partial c_1} = -\frac{c_2}{(c_1 + c_2)^2}(T_2 - T_1) \tag{9.13}$$

This equation shows how process gain will vary due to change in c_1

even if T_2-T_1 remains constant.

We next investigate cases in which a three-way valve is used instead of the two-way bypass valve above. For ease of description, we will consider cases in which flow resistance of both the heat exchanger and bypass can be ignored.

Since the sum of the C_V values for both ports (linear characteristics) of the three-way valve is constant, we can obtain equations for the valve characteristics of each port as follows:

$$\left.\begin{aligned} c_1 &= (1-l)c_{VT} \\ c_2 &= lc_{VT} \end{aligned}\right\} \tag{9.14}$$

where c_{VT} is the above sum and l is the valve lift (full stroke = 1).

Since the relationship in Eq. (9.12) still holds with the above conditions, we substitute Eq. (9.14) into that equation to get:

$$T_0 = T_1 + (T_2 - T_1)l$$

$$\therefore \quad \frac{\partial T_0}{\partial l} = T_2 - T_1$$

From these two equations we can see that if T_1 and T_2 are fixed, T_0 is determined only by l and process gain is constant.* In addition, the piping flow resistance does not change but takes on the constant value c_{VT} according to l.

In actuality, however, bypass piping resistance and heat exchanger flow resistance do exist to an extent and adjustment of the manual bypass valve can be performed to make the two equivalent. Although this situation is somewhat different than the ideal one described above, the same results are generally obtained and control is greatly improved over the two-way valve bypass.

Although the above descriptions have dealt with bypass temperature control of heating processes, the results obtained completely apply to cooling processes as well.

9.3.2 Heating furnace control

(a) Control system configuration

Furnace control, as depicted by the system shown in Fig. 9.19, is performed by integrating various inter-related control functions. The following describes each of these control functions.

(b) Total heat control

Primary disturbances to the furnace process are the amount and quality of the feed fuel. Such disturbances, however, can be suppressed by keeping the total amount of heat input into the process con-

---

* In actuality, F does cause a change in T_2 to some extent.

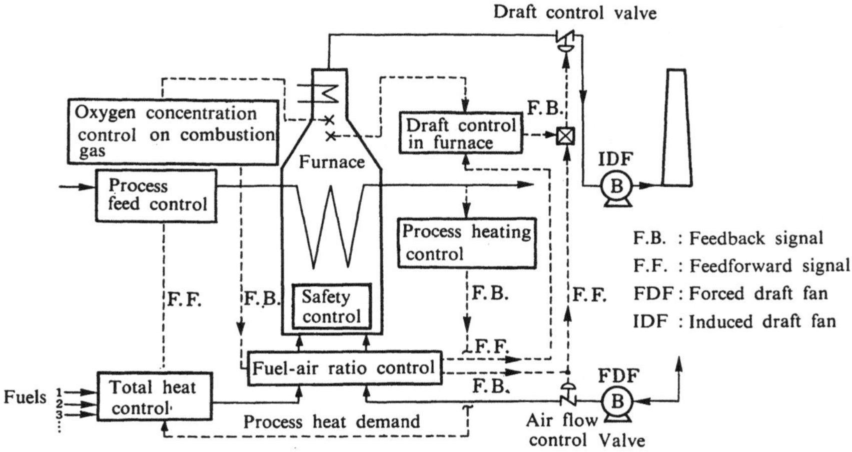

Fig. 9・19 Configuration of furnace control system

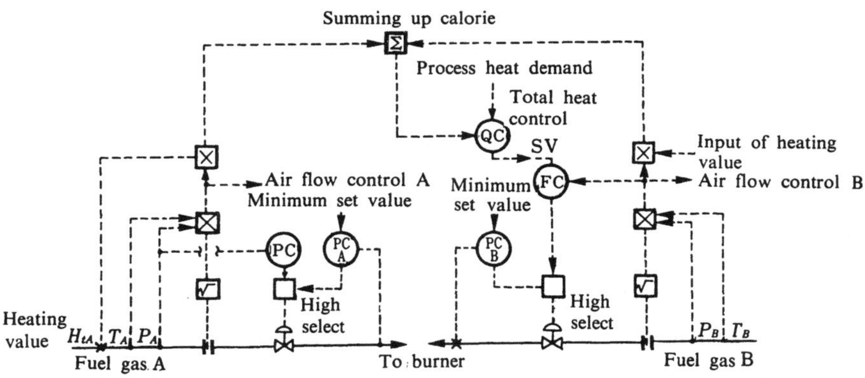

Fig. 9・20 Total heat control system at fuel gas mixing

stant, as shown by the system in Fig. 9.20. Here, a process residual gas A whose amount and quality vary and a stable supplementary gas B are combined as fuel to the furnace. Gas heating value H_t (kcal/Nm^3) is either calculated from the value detected by an analyzer (*e.g.*, process gas chromatograph, density meter) or obtained from a gas calorie meter. Pressure controller A (PCA) and pressure controller B (PCB) in the figure maintains the minimum value of the burner inlet pressure in all situations and prevents dangerous conditions from arising by flame–out (Reber to item (f) below).

(c) Fuel and air flow control

The flow rate of fuel supplied to the furnace is determined by the amount of heat required by the process. In order that the fuel is burned as completely as possible, however, an amount of air A slightly more than the theoretical amount of air A_0 determined by stoichiometry must be supplied. This ratio, A/A_0, is called the excess air ratio and is usually denoted by m. An optimal value for m is determined by the type of fuel and combustion equipment and combustion within the furnace is controlled with m as a desired value.

In addition, the relationship between the concentration of O_2 in the combustion gas and m is given by the following well-known equation:

$$O_2\ \text{vol}\ \% = \frac{0.21(m-1)}{0.21a+m} \times 100 \approx \frac{0.21(m-1)}{m} \times 100 \quad [\%]$$

Here, a is a constant dependent on the type of fuel, *e.g.*, 0.5 for methane, 0.33 for light oil, 0.2 for heavy oil, and 0 for carbon.

Thus, based on this equation, the value of m can be controlled through use of an oxygen analyzer. In general, if the furnace heat load is above 50% of the total load, the optimal value of m can be considered constant. On the other hand, for that below 50%, m tends to increase rapidly. Accordingly, the concentration of O_2 in the combustion gas (a desired value) may also have to be increased depending on the load (Fig. 9.21).

A widely used fuel-air ratio control system[5] is shown in Fig. 9.22. The basic function of this system is to set the fuel flow rate based on the process heat-demand signal while at the same time setting the amount of combustion air by multiplying this signal by an appropriate ratio g_1.

This ratio, which must be a value promoting optimal combustion, receives feedback correction based on the output from the controller governing O_2 concentration in the combustion gas. The setpoint value for O_2%, moreover, is automatically set according to the heat load (fuel flow rate) for the reason described above. In addition, during transient periods when the load varies, the occurrence of black smoke due

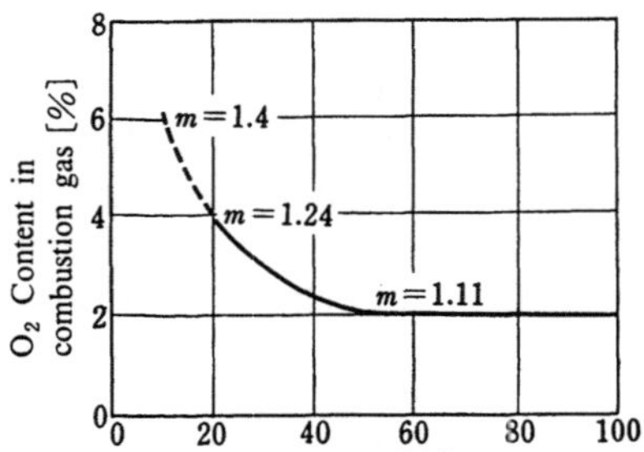

Fig. 9・21 Relationship between heat load and optimum oxygen concentration in combustion gas

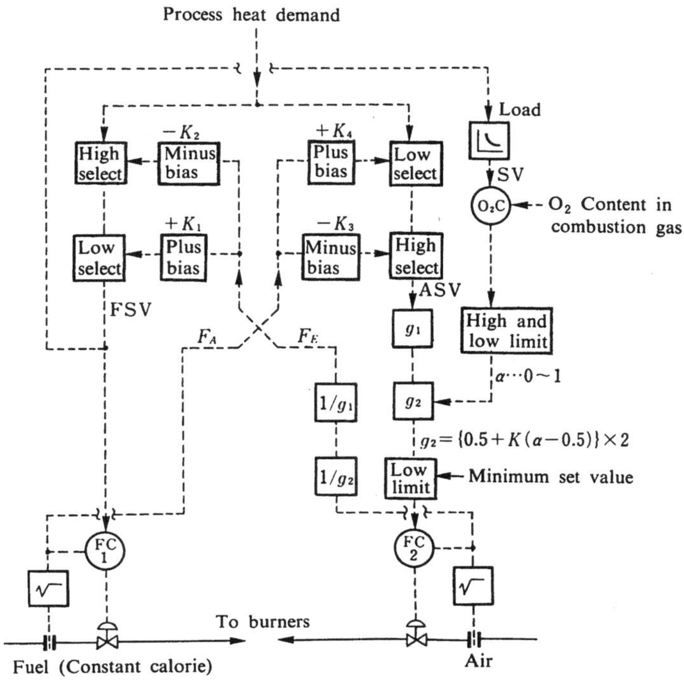

Fig. 9・22 Fuel-air ratio control system

to inadequate combustion air and heat loss caused by excess air are prevented by adding limiting functions (dash lines in the center of the figure) to the flow controller (FC) setpoint values for fuel (FSV) and air (ASV). The relationship between the various values in the figure is given by the following expressions in which F_E is the calculated fuel flow rate corresponding to the air flow:

$$(\text{excess air limit } F_E - K_2) < (\text{FSV}) < (\text{black-smoke generation limit } F_E + K_1)$$

$$(\text{black-smoke generation limit } F_A - K_3) < (\text{ASV}) < (\text{excess air limit } F_A + K_4)$$

(d) Furnace draft control

Draft within the kind of furnace shown in Fig. 9.19, which commonly employs a forced draft fan, is controlled by a draft control valve. In such a configuration, draft control can be improved by adding the manipulation signal of the air flow control valve to that of the draft control valve via a time-delay factor.*

* Feedforward control.

(e) Heating tube control

Piping within a furnace is formed by many branched parallel pipes. As liquid within the pipes are heated and then evaporated, it becomes necessary to control the flow in each pipe in order to uniformly distribute the feed flow.*

Although in most cases the furnace piping outlet temperature can be kept constant by controlling the amount of fuel, this is not necessarily easy to achieve because of the large time delay in feedback control. However, feedforward control from process flow to fuel flow can be effective when the process flow varies frequently. In addition, if disturbance is present on the fuel side, it should be suppressed by employing the countermeasure such as described in item (b) above.

(f) Safety control

In order to prevent problems such as burn-out of heating tubes, damage to the furnace structure, or explosions in the combustion chamber, an emergency operation system can be installed as described in Table 9.2. Sequence control is also appropriate for automating startup operations and for securing safety in general.

9.4 Control of Distillation Processes

9.4.1 Binary-component distillation column control

(a) Process and related equipment

The following describes the process and functions of the continuous distillation equipment shown in Fig. 9.23.

(1) *Functions* The feed liquid consists of two components, one having a low boiling point and the other a high boiling point. After passing through a preheater, it enters the distillation column at a middle stage where it becomes mixed with liquid falling from the upper column section (enriching section). The resulting liquid then itself falls to stages in the lower column section (stripping section) coming in contact with vapor rising through the column. At each stage, residual liquid in upper stages are heated by vapor from the lower stages and boiled. In particular, since the low-boiling point component can be evaporated from the liquid by this vapor, the concentration of this component in the liquid will decrease for stages closer to the bottom of the column. The major liquid component at the bottom of the column thus becomes the high-boiling point one, and part of this becomes a bottom

* In case of fluid flowing uniformly at constant rate in each pipe without the flow controller (FC), then if the flow should happen to decrease in one pipe due to disturbance, vaporization will begin increasing the flow resistance and decreasing the flow rate even more; flow rate in other pipes will thus increase by a corresponding amount.

Table 9.2 Emergency operation for furnace protection

Factors of operation	Reason for operation
Flow in heating tube decrease	Buring out or coking of heating tube
Outlet temperature of heating fluid rise	Buring out or coking of heating tube
Supply pressure of fuel low	Flame-out (trigger on explosion in furnace) Back fire in fuel gas
Differential pressure between fuel oil and atomizing fluid	
Increase	Flame-out
Decrease under the limit	Incomplete combustion
Draft pressure rise (depends on stopping of fan etc.)	Internal gas flow out of furnace
Temperature of gas after combustion rise (depends on bursting out of heating tubes etc.)	Furnace gets damaged
Emergency stop button	Operater make decision depending on circumstances

Emergency operation	Note
Fuel cut	For some kind of fuel, down flow piping of shut off valve should be purged by N_2 or steam
Feed cut for heating tube	In order to prevent heating tubes from over heat, steam etc. should be flowed
Steam purge for stack	This operation is necessary if the draft pressure rise on IDF stop and exhausting of combustion gas is still required

product while the remainder is reheated by a reboiler to generate rising vapor.

On the other hand, while the vapor rising through the enriching section comes in contact with falling liquid (internal reflux), the concentration of the low-boiling point component gradually increases in the same way as above. As a result, the vapor which eventually reaches the top of the column (overhead vapor) consists mainly of the low-boiling point component and is cooled within a condenser to become liquid again. A part of this liquid then becomes distillate product and is removed, while the remainder is returned to the column as reflux, and falls down inside the column.

(2) *External material balance* The following equations are obtained

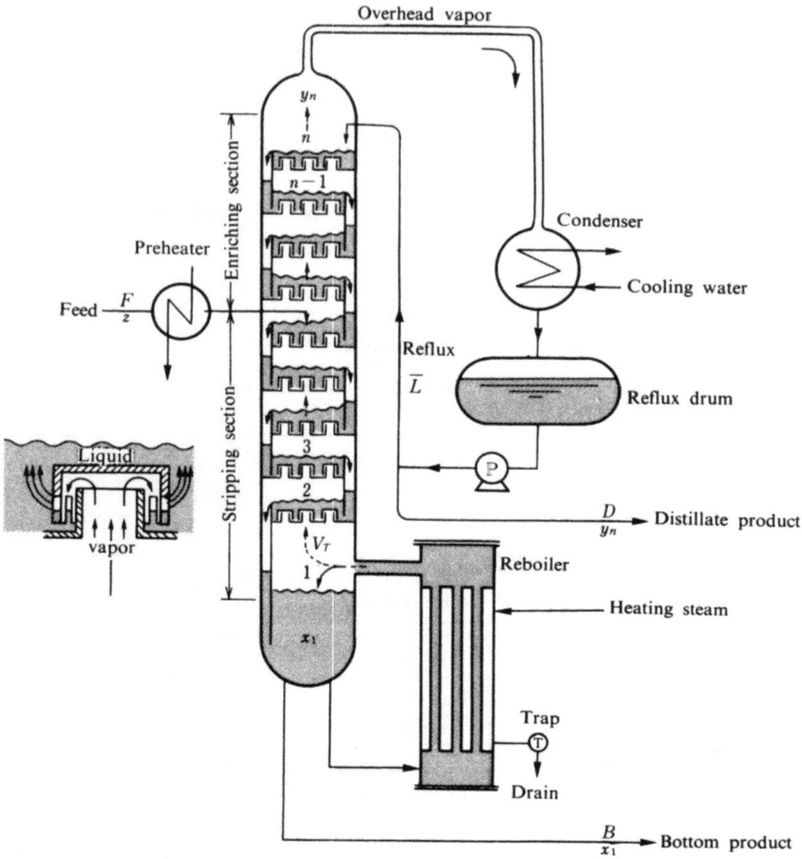

Fig. 9・23 Outline of continuons distillation equipment

for inflow and outflow of the system in Fig. 9.23 during steady-state conditions.

$$F = D + B \qquad \left(\begin{array}{l}\text{material balance of low-boiling}\\ \text{point component}\end{array}\right) \tag{9.15}$$

$$Fz = Dy_n + Bx_1 \tag{9.16}$$

from which we obtain:

$$\frac{D}{F} = \frac{z - x_1}{y_n - x_1} \tag{9.17}$$

Here, z is the feed liquid concentration (low-boiling point, the same as follows), y_n the liquid phase concentration at the nth stage (distillate product concentration), x_1 the liquid phase concentration at the 1st stage (bottom product concentration), F the feed liquid flow rate, D the distillate product flow rate, and B the bottom product flow rate.

(3) *Control variables and disturbances in continuous distillation processes* The object of distillation process operation is to separate the supplied liquid in the most economical way and to produce a distillate or bottom product having the desired concentration (specific value). Product concentration thus becomes the main controlled variable and the flow rate and concentration of feed liquid becomes the main disturbances which must be suppressed. In general, since distillate and bottom product concentration change quite slowly in responding to these disturbances or other manipulated inputs, and in particular, when load changes occur frequently, the adoption of feedforward control for handling such disturbances should be considered.

The suppression of thermal disturbances to the distillation column is also important for stable operations. To begin with, if the enthalpy of the feed either increases or decreases, the amount of vapor rising from the feed plate will correspondingly increase or decrease and the amount of liquid falling from this plate will change conversely. Accordingly, control is carried out to stabilize preheating conditions of feed liquid (if the temperature of the feed liquid is the same as that at the feed plate, vapor condensation or generation will not occur). Likewise, if the reflux temperature increases or decreases, the amount of overhead vapor will increase or decrease and conversely the amount of liquid falling through the column (internal reflux). Thus, in cases when such temperature change is large and needs to be stabilized,* an internal reflux flow control system can be adopted in which reflux flow is either increased or decreased to compensate for changes in its temperature.

(*b*) *Vapor–liquid equilibrium and column separation*

(1) *Vapor–liquid equilibrium of a binary-component solution* We consider an equilibrium situation in which a mixed solution consisting of two components having a high and low boiling point is brought to a boil under constant pressure. In such a system, a fixed relationship holds between the concentration of the two components in their vapor and liquid phases depending on the type of components. This relationship can be measured by the equipment shown in Fig.9.24 (a) under atmospheric pressure conditions. The results of measurement are shown in Fig.9.24 (b) for a benzene–toluene system. For most mixed solutions in such an equilibrium state, the concentration of the low-boiling point component in vapor phase (y) is higher than that for the same component in liquid phase (x), which is the basis of separation as used in distillation.

* For example, in the case of an air-cooled condenser.

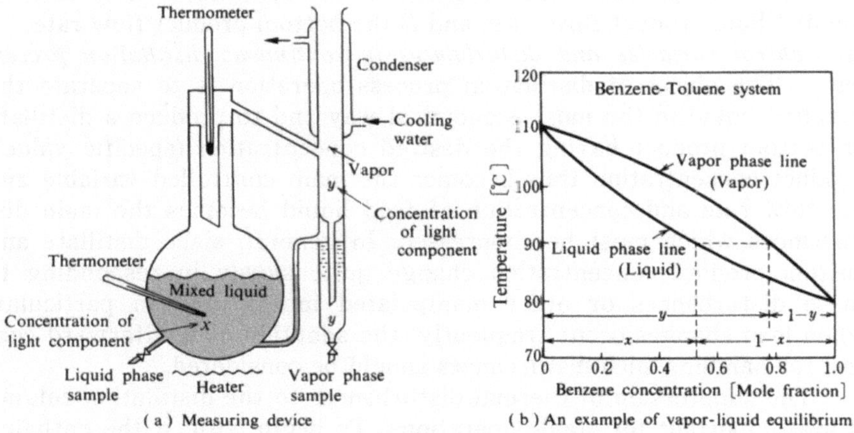

(a) Measuring device (b) An example of vapor-liquid equilibrium

Fig. 9 · 24 Vapor-liquid equilibrium under atmospheric pressure

(2) *Relative volatility* The ease or difficulty of separation in the distillation process can be understood by the size of relative volatility as described below. To begin with, the relative volatility α_{AB} of a mixed solution composed of a low-boiling point component A and a high-boiling point component B is defined by the following equation:

$$\alpha_{AB}=\frac{y_A/y_B}{x_A/x_B}=\frac{y/(1-y)}{x/(1-x)} \tag{9.18}$$

Relative volatility α_{AB} is usually greater than 1 and separation by distillation becomes easier as this value increases. Moreover, as shown by Table 9.3, relative volatility tends to increase generally as the pressure decreases for a fixed-components solution. Thus, the energy required for constant separation decreases for distillation performed under lower pressures.

(3) *Total reflux operation for binary-component distillation* A continuous distillation process operation in which the respective flow rates

Table 9.3 Change in relative volatility of benzene-toluene system depends on pressure

Total pressure (mmHg)	Benzene boiling point (°C)	Relative volatility α
860	83.9	2.58
760	80.1	2.60
660	75.4	2.66
560	70.3	2.72

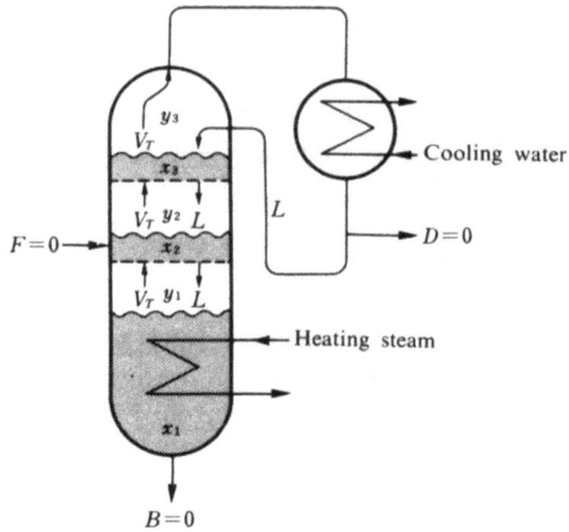

Fig. 9 · 25 Total reflux operation

for the feed liquid, distillate product, and bottom product are zero and in which rising vapor V_T generated by heating is completely condensed into reflux L is called a total reflux operation. The following investigates such a system using the simple example shown in Fig. 9.25.

In this example, since each plate is in a vapor–liquid equilibrium state, the following equations can be obtained where $\bar{\alpha}$ indicates the average relative volatility* for a stage.

$$\bar{\alpha}=\frac{y_1/(1-y_1)}{x_1/(1-x_1)},\quad \bar{\alpha}=\frac{y_2/(1-y_2)}{x_2/(1-x_2)},\quad \bar{\alpha}=\frac{y_3/(1-y_3)}{x_3/(1-x_3)}$$

Multiplying these three equations by each other, we get:

$$\bar{\alpha}^3=\frac{\cancel{y_1}/\cancel{(1-y_1)}}{x_1/(1-x_1)}\cdot\frac{\cancel{y_2}/\cancel{(1-y_2)}}{\cancel{y_2}/\cancel{(1-x_2)}}\cdot\frac{y_3/(1-y_3)}{\cancel{x_3}/\cancel{(1-x_3)}} \tag{9.19}$$

The following equations are also obtained from the material inflow and outflow of the low-boiling point component at the top plate and bottom:

$$V_T y_2+L y_3=V_T y_3+L x_3,\quad V_T y_1+L x_3=V_T y_2+L x_2$$

Since in total reflux operation $V_T=L$, it can be seen from the first equation above that $y_2=x_3$ and from the second equation that $y_1=x_2$. Applying these relationships to Eg. (9.19), cancellation of some of the numerators and denominators can be performed giving us:

---

* Since in general the relative volatility differs somewhat between the top and bottom of the column, an average value is taken.

$$\bar{\alpha}^3 = \frac{y_3/(1-y_3)}{x_1/(1-x_1)}$$

This relationship can be extended to a nth plate configuration in the same manner. Specifically, if we designate y_{n0} and x_{10} as the concentration of overhead vapor and bottom liquid, respectively, we obtain:

$$\bar{\alpha}^n = \frac{y_{n0}/(1-y_{n0})}{x_{10}/(1-x_{10})} \tag{9.20*}$$

(4) *Separation of distillation* In order to express the degree of separation for continuous distillation equipment under operation, the column separation S is patterned after Eq. (9.20) and defined as follows:

$$S = \frac{y_n/(1-y_n)}{x_1/(1-x_1)} \tag{9.21}$$

For example, under operational conditions represented by $y_n = 0.95$ and $x_1 = 0.05$, column separability S becomes 361. Although a number of equations are known for which S can be calculated from related variables, the following Shinski[3] equation is the most concise and easy to use:

$$S = \left[\frac{\bar{\alpha}}{\sqrt{1+1/\{(V_T/D-1)z\}}}\right]^n \tag{9.22}$$

where V_T is the flow rate of rising vapor.

Examining this equation, it can be seen that in the case of total reflux operation for which D is 0, it reduces to Eq. (9.20). Note also that the value n in Eq. (9.22) is not the actual number of column plates, N, but is the effective number of plates ($n = NE$) obtained by multiplying N times the plate efficiency** E.

(*c*) *Control system for binary-component continuous distillation*

The control system for a particular piece of distillation equipment will in general be determined by target and disturbance conditions. Since it would be impossible to cover all of the various cases here, a few representative examples are described.

(1) *Control of distillate concentration during changes in feed flow* We consider a system in which the concentration of the distillate product is controlled by manipulating its flow under conditions of changing feed flow rate and constant heating steam flow rate.

Referring to Eq. (9.22), if V_T is constant and z, $\bar{\alpha}$, and n all invariable, then S will clearly depend only on D. Moreover, by rearranging

* Fenske equation.

** As vapor-liquid separation in each plate is not performed under ideal conditions, correction is necessary based on efficiency.

Eq. (9.21), we get:

$$x_1 = \frac{y_n}{S - y_n(S-1)} \tag{9.23}$$

If we now eliminate x_1 between this equation and Eq. (9.17), then if y_n is kept constant, we can see how D/F depends on S, or in other words, on D. A fixed relationship thus holds between F and D. Figure 9.26 shows several curves representing this relationship based on calculations of successive approximation by computer with the parameters shown.

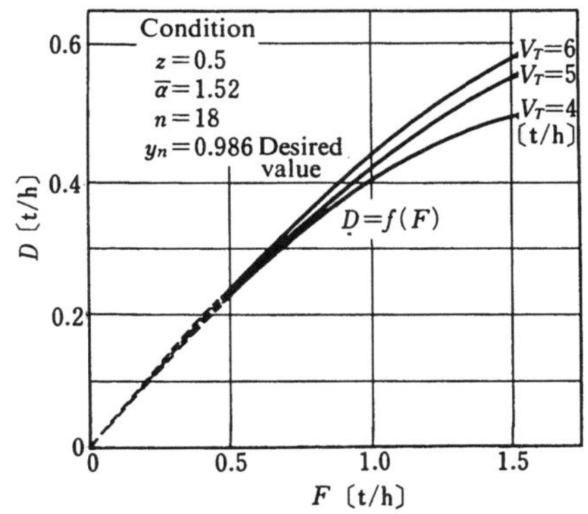

Fig. 9 · 26 Relationship between each flow rate in feed and distillate when concentration of distillate is controlled

Based on the above description, it can now be seen that the distillate concentration y_n in relation to disturbance F can be maintained at a fixed value by setting distillate flow D according to the relationship $D=f(F)$ as shown in the figure. Figure 9.27 shows two types of control systems based on this conclusion employing (a) feedback control, and (b) feedforward control. Note that in the latter case, a dynamic compensator element L/L (Lead/Lag) and feedback corrective operation must be included.

(2) *Control of bottoms concentration during changes in feed flow* In this system, the concentration of the bottom product is controlled by manipulating its flow under conditions of changing feed flow rate and constant reflux flow rate. Approaching this system in the same manner as with the previous description, we rewrite Eq. (9.22) under steady-state conditions of $V_T = L + D$ and $D = F - B$:

$$S = \left[\frac{\bar{\alpha}}{\sqrt{1 + 1/\left(\frac{L}{F-B} z\right)}} \right]^n \tag{9.24}$$

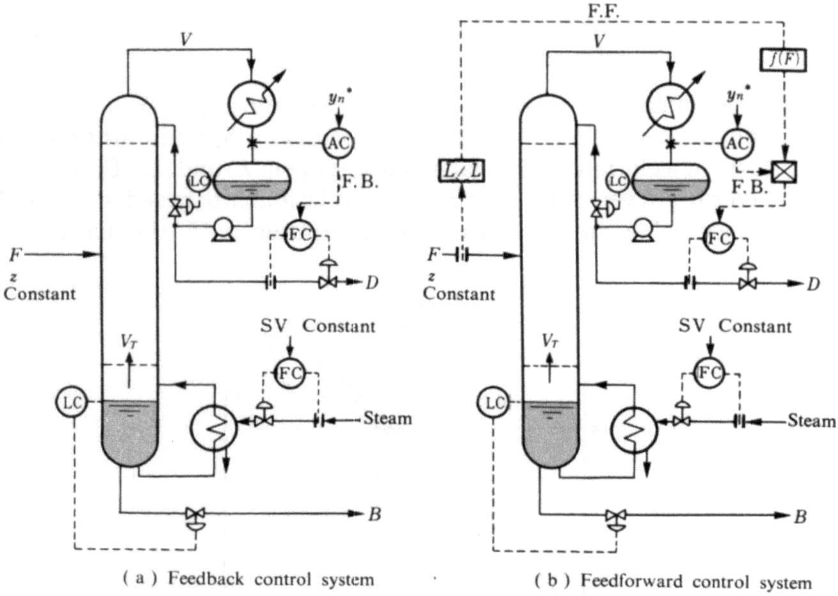

Fig. 9・27 Concentration control of distillate for feed flow change

From this equation it can be seen that S depends only on $F-B$.

If we now eliminate D between Eqs. (9.15) and (9.16), we get:

$$\frac{B}{F}=\frac{y_n-z}{y_n-x_1} \tag{9.25}$$

In addition, from Eq. (9.21):

$$y_n=\frac{Sx_1}{1+x_1(S-1)} \tag{9.26}$$

If x_1 is held constant in these two equations, then B/F depends on S, or in other words, on $F-B$, and a fixed relationship between F and B must hold as $B=g(F)$.

Figure 9.28 shows two control systems based on this result.

(3) *Control of distillate or bottoms concentration during changes in feed concentration* We first consider a system in which the concentration of distillate product is controlled by manipulating its flow rate under conditions of changing feed concentration and fixed feed and heating steam flow rate. With these conditions in mind, we first eliminate x_1 from Eqs. (9.17) and (9.23) and see that D is dependent on S and z. In addition, since from Eq. (9.22) S depends on z and D, we can now eliminate S from these last two equations to obtain a relation between z and

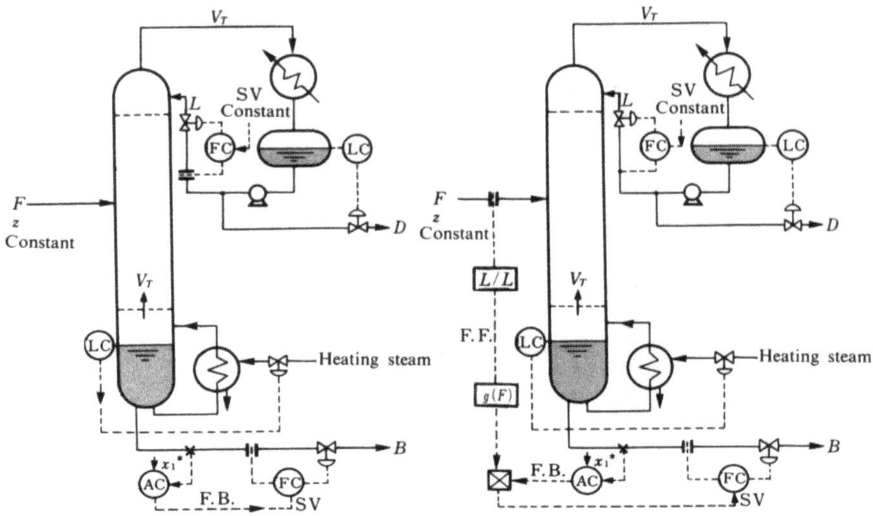

(a) Feedback control system (b) Feedforward control system

Fig. 9・28 Concentration control of bottoms for feed flow change

D: $D=\varphi(z)$. With this result, a system exactly like the feedback control one in Fig. 9.27(a) can be realized. In order to realize a feedforward control system, however, the feed flow rate in Fig. 9.27(b) must be substituted by a concentration signal, and the function $f(F)$ by the relation $D=\varphi(z)$ described above.

Considering next a system in which the concentration of bottom product is controlled under conditions of changing feed concentration, we can draw an analogy from Fig. 9.28 in the same way as above (Refer to practice question 9.4).

(4) *Control of distillate and bottoms concentration during changes in feed flow* Assuming that feed concentration z is invariable, then in order to keep distillate and bottom product concentration y_n and x_1 at fixed value, the ratio D/F as expressed by Eq. (9.17) must be held constant, as must the value S from Eq. (9.21). In addition, if S is constant, then from Eq. (9.22) the ratio V_T/D must also be constant, which in turn means that V/F is constant as well.

Based on the above discussion, we examine the control system[6] shown in Fig. 9.29. Although by employing a concentration controller for both the distillate and bottom product, feedback control for both the distillate flow rate and heating-steam flow rate is possible, controllability can be improved even further by adding two feedforward paths from the feed flow rate. Note also that due to the mutual interference

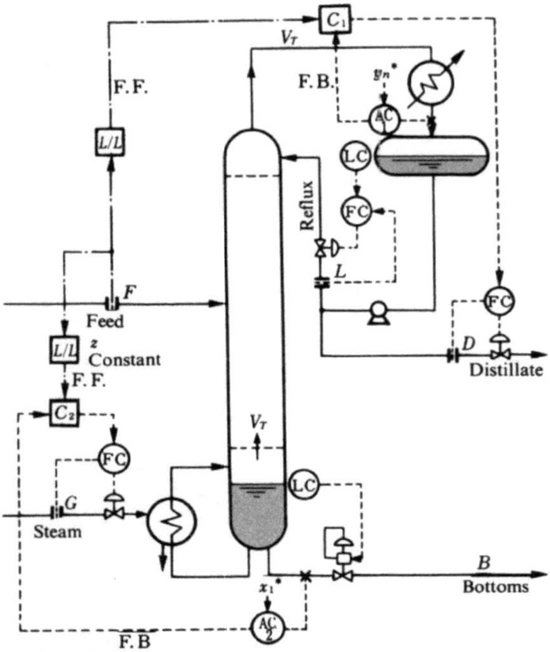

Fig. 9・29 Concentration control of both distillate and bottoms for feed flow change (Case of high purity in bottoms)

between the two concentration controller loops in this system, it is recommended that the bottom product exhibit high purity (*e.g.*, 0.999) while that of the distillate product be relatively low (0.8 to 0.9). If the opposite case is needed, a system can be configured in which the distillate concentration is controlled based on reflux and the bottom concentration is controlled by manipulating the bottom flow rate.

In either of the above two systems, mutual interference is low, so that no problems should arise in practical use.[3)]

(*d*) *Control of distillation column pressure*

Stable operations in distillation equipment require column pressure to be controlled in a steady-state fashion. Although various types of control systems for achieving this have been processed,[7)] this section presents only several examples which have come into wide use, all characterized by the condenser using cooling water.

(1) *Inert gas injection and purge technique* Referring to the system in Fig. 9.30(a), if the pressure in the column should happen to drop, inert gas is injected to maintain pressure, while if the pressure rises, gas in the column is purged to the outside, thus lowering the pressure. Two

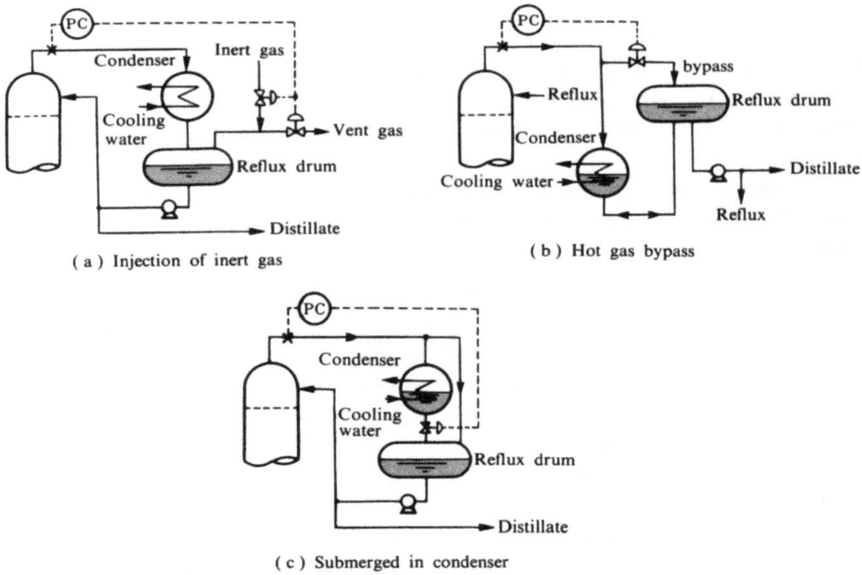

Fig. 9·30 Pressure control at distillation column

control valves are employed here operated on a split-range basis. Despite some negative features such as increase in vent-gas flow rate and solubility of inert gas in distillate, the high responsiveness of this technique has come to be widely used in pressurized columns operating near atmospheric pressure.

(2) *Hot-gas bypass technique* In the example shown in Fig. 9.30(b), if the amount of vapor from the top of the column increases causing the pressure to rise above its setpoint value, the pressure control valve will be moved in the closed direction (in the opposite direction if pressure drops below setpoint value) decreasing the amount of bypass vapor. Consequently, the amount of condensation on the surface of the liquid in the reflux drum decreases, which in turn causes the surface temperature to drop and the pressure in the drum to fall. As a result, liquid in the condenser tends to flow into the reflux drum increasing the condensation heat-transfer area in the condenser and raising the speed of vapor condensation. This finally causes the pressure in the column to drop back towards its setpoint value. In an actual situation, if the condensed liquid at the condenser outlet is sub-cooled* to about 4°C, the control valve pressure differential will become about 0.2 to

* Cooling to a temperature even lower than saturation temperature; bypass vapor flow rate is governed by this temperature.

0.3 kg/cm. Because of the excellent response of this technique, it has come to be applied in many pressurized columns.

(3) *Submersion of condenser heat transfer area by condensed liquid* As shown in Fig. 9.30(c), column pressure is controlled in this technique by manipulating the amount of condensed liquid flowing out of the condenser thereby increasing or decreasing the condensation heat-transfer area.

(4) *Condenser cooling water adjustment technique* This technique should be easy to understand without the aid of an illustration. In particular, referring to the earlier description of fluid-to-fluid heat exchangers, it should be clear that if the heat load drastically increases, the sensitivity of pressure control by cooling water adjustment will decrease significantly. Another disadvantage here is that as the cooling water flow drops during times of low load, scale deposits from the heat transfer surface are produced by low flowing rate. Thus, this technique requires careful consideration on application.

(5) *Floating pressure control*[2] During times of fixed load, it is possible to set the value of the distillation column pressure even lower if the temperature of condenser water drops. Since a lower column pressure is directly related to energy conservation as described in (2) of (b) (about 2% reduction in energy for every 1°C drop in cooling water temperature[3]), maintaining a column pressure as low as possible by making the condenser always operate fully corresponding to the cooling water temperature in the condenser can be quite effective (floating pressure control). On the other hand, as previously described, the column pressure must be kept as close as possible to a steady-state value. An example of a control system satisfying these conditions is shown in Fig. 9.31.

In the figure, the function of the pressure controller PC is funda-

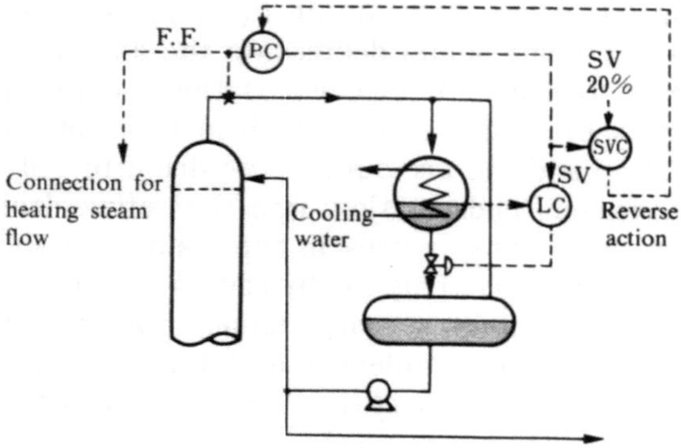

Fig. 9 · 31 An example applied for floating pressure control mentally the same as that in Fig. 9.30 (c). If cooling water temperature drops, then if the column load is fixed, the cascaded setpoint value of the condenser's level controller (LC) becomes large. However, if the pressure controller (PC) setpoint value is lowered via setpoint controller (SVC) output, the liquid level can be kept around normal low levels (20%). In addition, since a radical change in column pressure disrupts column stability, slow integral action* is adopted for SVC. Such operation is considered adequate since changes in water temperature themselves occur slowly. Moreover, since the amount of heating vapor must be decreased corresponding to the drop in column pressure, the column pressure signal is applied to the heating steam flow setpoint value as a feedforward signal.

(*e*) *Loop enhancement in a distillation column control system*

Up to this point, we have only presented basic configurations for distillation column control systems. The following discussion examines ways to enhance system control performance in an efficient manner.

(1) *Enhancing the distillate concentration control system*[8] Concentration control for the systems shown in Figs. 9.27 and 9.29 can be greatly improved by adding local feedforward control as shown in Fig. 9.32. Here if $K=0$, we have simply level controller/flow controller (LC/FC) cascade control. If $K=1$, however, a change in distillate flow D is compensated by an equivalent change in reflux flow leaving the liquid level unchanged. The effects of distillate flow manipulation thus reach the column as changes in reflux without a time delay. Moreover, if $K>1$, a lead element is employed to compensate for the time delay required for reflux manipulation to effect changes in concentration, thus improving controllability even further. On the basis of the above, it has been reported that the oscillation period of the concentration control loop can be reduced from 5 hours to 30 minutes.

(2) *Distillation equipment constraints and control* The operation range

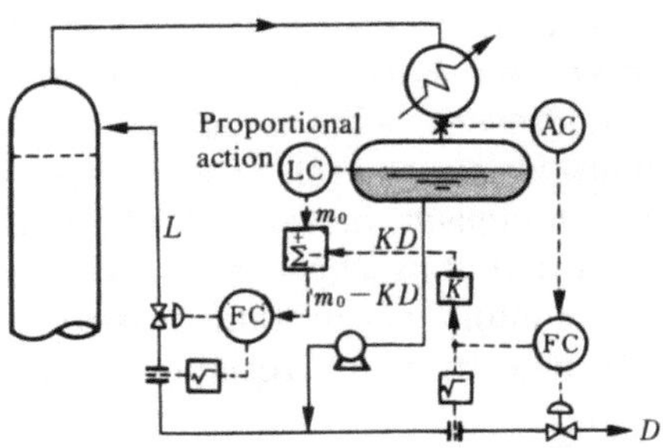

Fig. 9・32 Improvement in concentration control of distillate[8]

* An integral time of about one hour is selected.

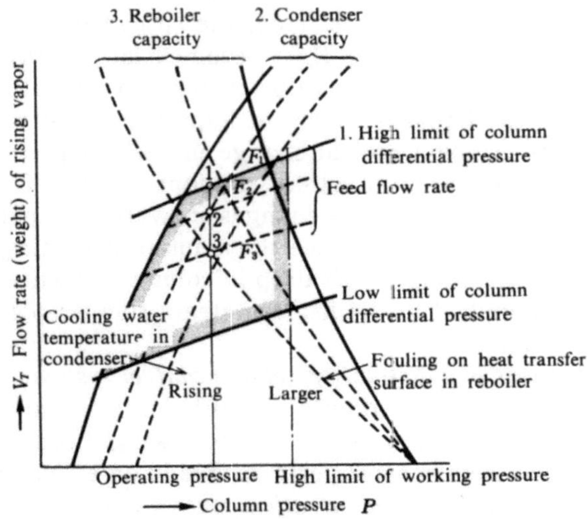

Fig. 9 · 33 Constrained condition for load of distillation column

of distillation equipment is constrained by limits in the column pressure and in the handling capacities of various parts with respect to rising vapor flow as shown in Fig. 9. 33[8]. To begin with, if the flow velocity of rising vapor in the column becomes excessive, liquid will likewise rise with vapor (flooding), creating problems for distillation. However, since the column pressure differential is proportional to the square of the vapor flow velocity, the flow velocity limit can be determined from this relationship. For example, in the event that a concentration control system like the one in Fig. 9.29 requests an excessive amount of heating steam, the control configuration shown in Fig. 9.34(a) can halt the request if the column pressure differential goes above its limit.

Next, referring again to Fig. 9.33, we see that with a rise in column pressure, condenser handling capacity increases while reboiler handling capacity decreases. This is due to the fact that as the pressure rises, the temperature in each vessel rises, with the temperature differential of the condenser increasing and that of the reboiler decreasing. In addition, handling capacity in either case will drop remarkably with a rise in cooling water temperature in the condenser, and with an increase in heat-transfer resistance due to surface fouling in the reboiler.

Points 1, 2, and 3 in the same figure show the limits of rising vapor flow determined from the handling capacity of the elements under a particular operating pressure. In this example, the allowable upper

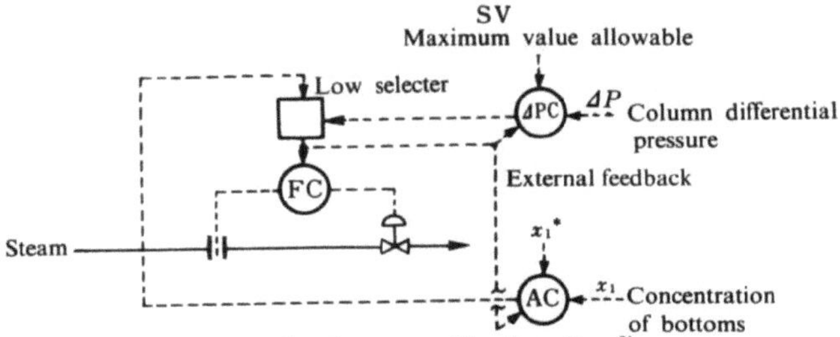

(a) High limit control for flow rate of heating steam[8)]

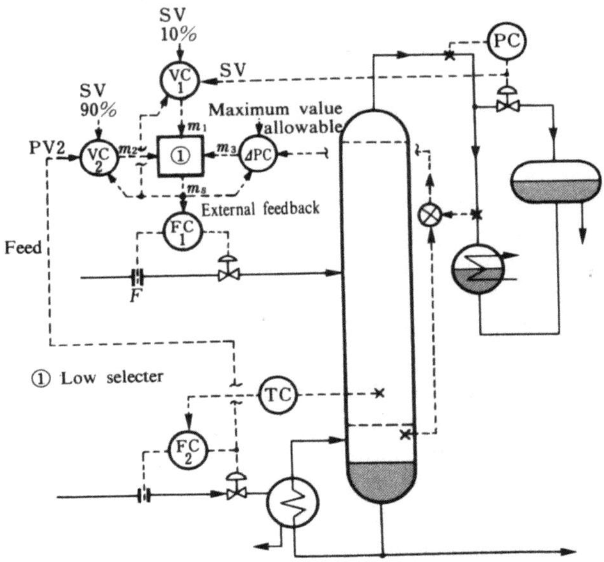

(b) Capacity-limit control of distillation column

Fig. 9 · 34 Constrained control of distillation column

limit of rising vapor flow will be governed by reboiler performance 3. Moreover, since feed flow of the column corresponds to rising vapor flow, the control system shown in Fig. 9.34(b) will set the feed flow to the governing value determined by the current handling capacity of each section. In this way, distillating capacity will always be kept at its maximum attainable limit.

In the same figure, the upper limit of column pressure is obtained from the output of differential pressure controller ΔPC, and the capacity limit of the condenser and reboiler from the output of respective control valve controllers VC1 and VC2. Thus, by setting the feed flow

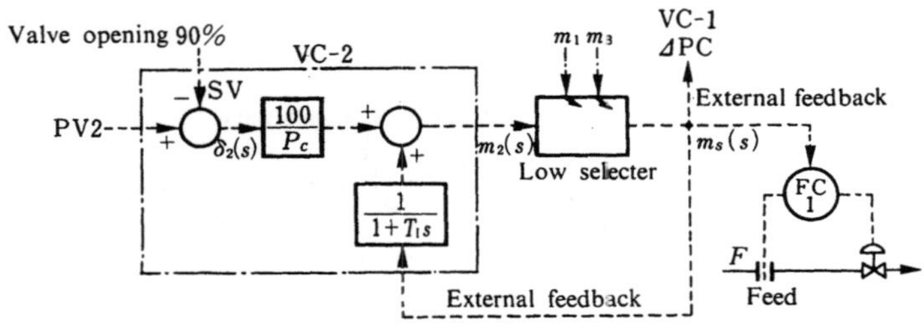

If m_2 is selected, the following equation is obtained

$$\frac{m_s(s)}{\delta_2(s)}=\frac{100}{P_c}\cdot\frac{1}{1-\dfrac{1}{1+T_I s}}=\frac{100}{P_c}\left(1+\frac{1}{T_I s}\right)$$

P_c= Proportional band, T_I= Integral time

Fig. 9 · 35 Controller with external feedback used in selecter control

controller with the minimum value from the output of each controller, the above objective can be realized.

Each controller in Fig. 9.34(b) having external feedback is configured in the manner shown in Fig. 9.35.[9] As can be seen from the inductive equation in the box, only the controller chosen by the selector takes on P+I action, while the remaining controllers apply bias to selector output by P action only. Since at the time of signal switchover by the selector the signal must cross over the setpoint, the control deviation of the P action controller becomes zero and switchover is performed in a smooth and reversible manner.

9.4.2 Multi-component distillation column control

(*a*) *Two-product distillation column control*

The most common form of distillation involves feed liquid which, in addition, to heavy and light "key" components, includes a relatively small amount of heaviest and lightest components (non-key components), all of which are separated in one multi-component distillation column. Such a column is also used as a basis for forming a multi-column train system.

(1) *Material balance and product composition* Concentration breakdown in each section of a two-product distillation column and concentration distribution throughout the column are shown in Figs. 9.36(a) and (b). As shown, the lightest (heaviest) component is ordinarily completely separated into distillate (bottom) liquid. We thus obtain the following equation:

$$z_{ll}F=y_{ll}D \qquad (9.27)$$

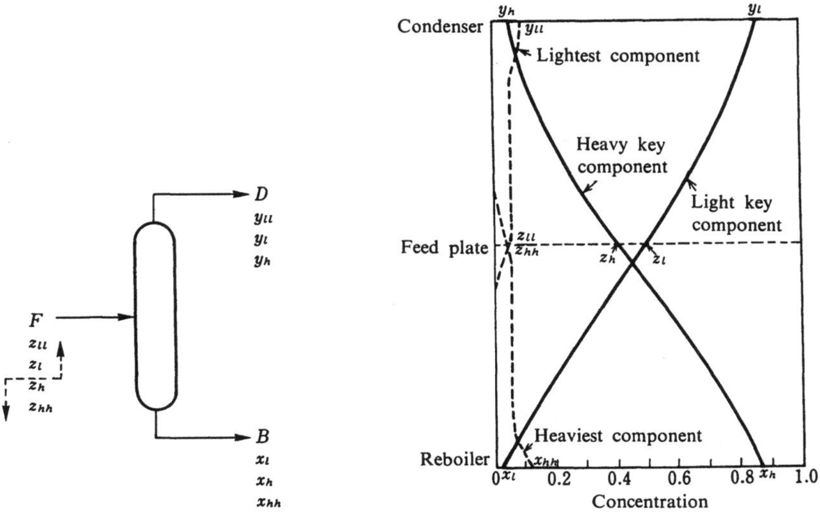

(a) Concentration of each part in column

(b) Concentration distribution of each component in column

Fig. 9 · 36 Two products distillation column for multi-component system

In addition, since $y_{ll}+y_l+y_h=1$, we get:

$$y_h=1-y_l-\frac{F}{D}z_{ll} \tag{9.28}$$

Moreover, since the relation $z_lF=y_lD+x_lB$ holds for the light key component (l) after material balance, we use it along with the simple relation $B=F-D$ to obtain:

$$\frac{F}{D}=\frac{y_l-x_l}{z_l-x_l} \tag{9.29}$$

Substituting this equation into Eq. (9.28), we arrive at

$$y_h=\frac{z_l-x_l(1-z_{ll})-y_l(z_{ll}+z_l-x_l)}{z_l-x_l} \tag{9.30}$$

Equation (9.30) expresses the relationship between x_l and y_h (key-component separation) given feed compositions z_{ll} and z_l and provided that distillate key-component concentration y_l is a stipulated value. Figure 9.37 shows the results of applying this equation to a debutanizer column. It can be seen how, if heating steam is increased to lower bottom butane concentration x_l, distillate pentane concentration y_h will increase. Accordingly, since y_h+y_{ll} is fixed, y_{ll} will decrease, and from Eq. (9.27), it follows that a greater amount of butane product D will be obtained from a specific composition of feed liquid. However, the ex-

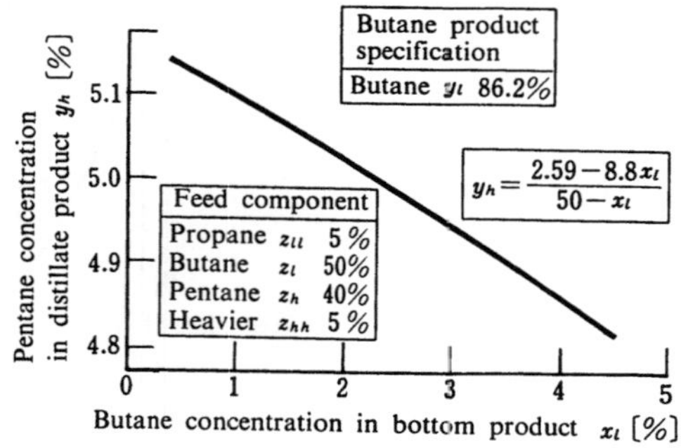

Fig. 9·37 Relationship between x_l and y_h at debutanizer

tent to which x_l can be lowered is determined by considering the trade-off between the cost of increased heating steam and more (distillate) product.

In the same manner, a relationship similar to Eq. (9.30) can be derived for bottom liquid. We omit the derivation here and present only the equation:

$$x_l = \frac{(1-y_h)(z_{ll}+z_l+z_h-x_h)-(z_{ll}+z_l)(1-x_h)}{z_h-y_h} \tag{9.31}$$

In this case, the relationship between y_h and x_l is obtained given z_{ll}, z_l and z_h and the stipulated bottoms key-component concentration x_h.

(2) *Relationship between light key-component concentration and light non-key component concentration* Eliminating the ratio F/D from Eqs. (9.27) and (9.29), we obtain:

$$\frac{z_{ll}}{z_l} = \frac{y_{ll}}{y_l} \cdot \frac{1-(x_l/z_l)}{1-(x_l/y_l)}$$

Ordinarily, since x_l (bottom liquid impurity) $\ll z_l$ and $x_l \ll y_l$, the following approximation holds:[8)]

$$\frac{z_{ll}}{z_l} \approx \frac{y_{ll}}{y_l} \tag{9.32}$$

From this, it can be seen that since both components change while maintaining a proportional relationship, control of multi-component distillation can be approximately treated as that for a two-component system.

(*b*) *Concentration control in distillation column train*

A process in which feed liquid consisting of N components is sequentially separated by $(N-1)$* distillation columns has found wide

* By using more than $(N-1)$ columns, intermediate products may also be separated.

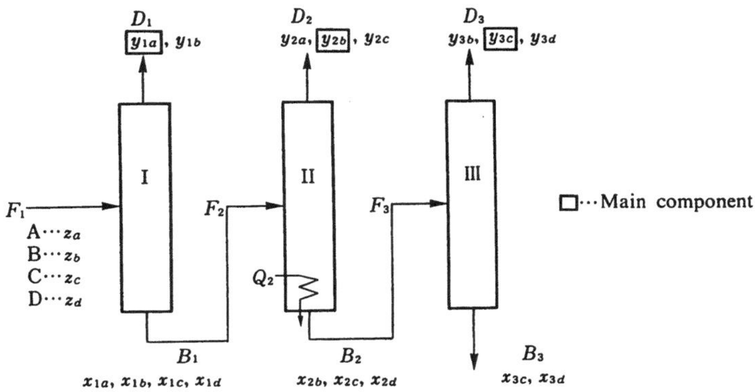

Fig. 9·38 Separation for four components system using three columns

application. An example of economic optimal control of such a process is described below.[10)]

(1) *Process description* Figure 9.38 shows a system in which three columns are used to separate an initial feed liquid consisting of four components: A, B, C and D (from light to heavy components in that order). Here, the concentration of each feed component is represented by z_a, z_b, z_c and z_d, the distillate and bottoms flow of each column by D and B, and their respective concentrations by y and x, with subscripts 1, 2, and 3 corresponding to columns I, II, and III, respectively. The following conditions are assumed for the distillate product of each column: y_{1a} must be held constant, $y_{2b} \geq 0.955$, $y_{2c} \leq 0.015$, y_{3c} must be held constant and y_{3b}/y_{3c} ($\approx x_{2b}/x_{2c}$)* $= 0.03$.

In addition, the relationship between the economic value V of each component unit amount is assumed to be $V_a < V_b < V_c$.

Since the distillate compositions of columns I and III are already set, only column II becomes the object of economic operation. As shown in Fig. 9.39(a), y_{2b} of column II's distillate concentration is fixed to 95.5 %, the minimum value satisfying specifications. Of the remaining 4.5 %, however, if the amount occupied by least-expensive component y_{2a} is extended, the economic value V_2 for D_2 unit amount increases, with the amount of change given by

$$\Delta V_2 = (y_{2a} V_a + y_{2c} V_c) - (0.03\, V_a + 0.015\, V_c)$$

But, since an increase in y_{2a} causes a decrease in y_{2c}, the amount of steam, which is required to separate B and C in column II, increases, resulting in an increase in cost C_2 per unit amount of D_2.

On the other hand, an increase in y_{2a} allows an increase in x_{1a} (re-

* Refer to Eq. (9.32).

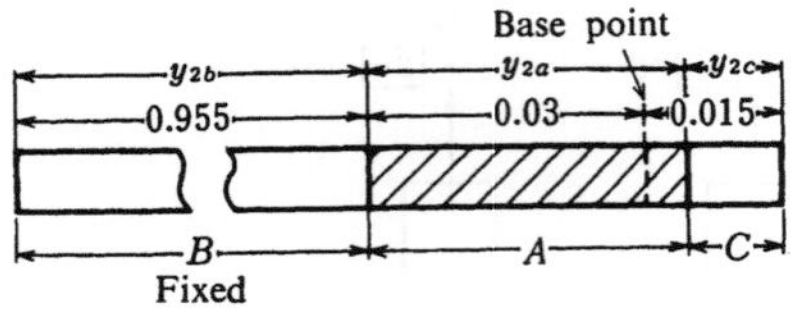

(a) Composition of distillate product

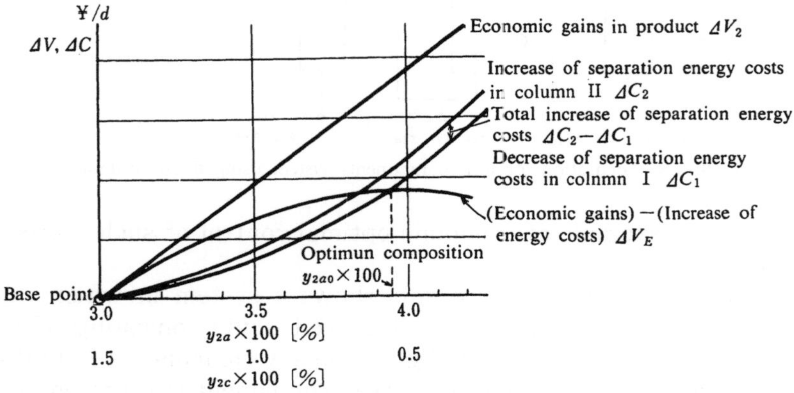

(b) Change in economic gains

Fig. 9・39 Relationship between bottoms composition of column II and economic gains

fer to Eq. (9.32)), so that the heating steam in column I can be decreased, thus lowering steam cost C_1 slightly.

In general, the following approximation holds between the amount of rising steam and the concentration in a distillation column.[11)]

$$\frac{V_T}{F} \approx a \log \frac{1}{x_l y_h}$$

We can now obtain the increment C_1 and C_2 from the above equation.

$$\Delta C_1 \approx c\beta \log \frac{x_{1a}}{x_{1aB}} \approx c\beta \log \frac{y_{2a}}{0.03}$$

$$\Delta C_2 \approx c\alpha \log \frac{0.015}{0.045 - y_{2a}}$$

Here, c is unit cost of steam, α and β are constants, and x_{1aB} is the reference.

From the above considerations, we can obtain the net economic value increment ΔV_E per unit amount of D_2 as the following equation.

$$\Delta V_E = \Delta V_2 - (\Delta C_2 - \Delta C_1)$$

Figure 9.39(b) illustrates the various relationships described above.

The value of y_{2a} for attaining economic optimal operation (referred to as y_{2a0}) can be obtained from these curves.

(2) *Economic optimal control system* The logical configuration of a system for economic optimal control of the above three-column process is shown in Fig. 9.40. The basic loop controls section and product quality controls section perform concentration control by feedforward mechanisms such as those in Fig. 9.29. The economic optimum control section, on the other hand, calculates the setpoint values of each concentration satisfying various given conditions using relevant values. In the case of using floating-pressure control, since the change in column pressure here affects the amount of required heating steam flow, column pressure P_2 must also be taken into account. (Refer to Subsec. 9.4.1(d)(5).)

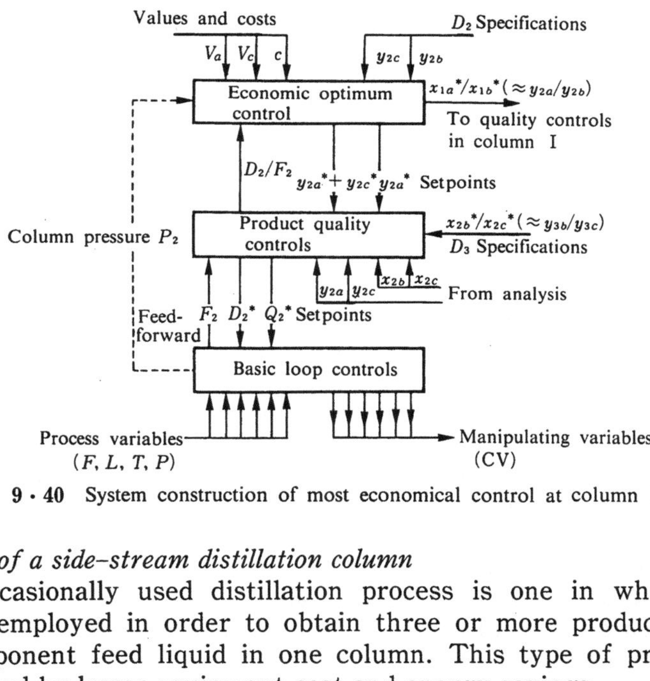

Fig. 9・40 System construction of most economical control at column II

(*c*) *Control of a side-stream distillation column*

An occasionally used distillation process is one in which side stream is employed in order to obtain three or more products from multi-component feed liquid in one column. This type of process is characterized by lower equipment cost and energy savings.

(1) *Side-stream concentration control* In most cases, side stream implemented at a plate above the feed plate draws liquid in order to reduce as much as possible the concentration of light components. On the other hand, side stream below the feed plate draws vapor with the object of reducing the concentration of heavy components. In Fig. 9.41(a), if we increase the flow of side-stream liquid, internal reflux below the drawing plate decreases, and as a result heavy-component con-

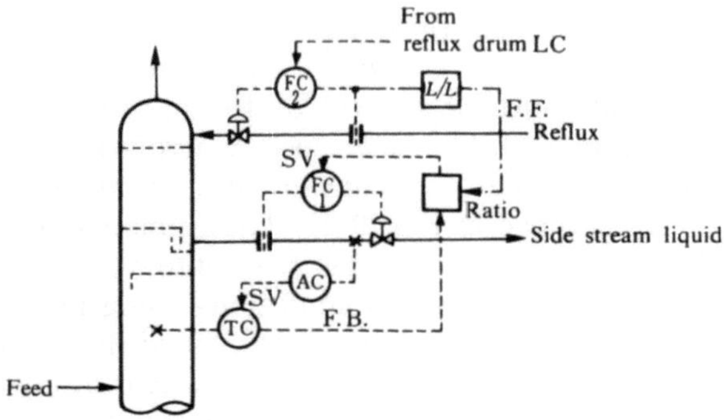

(a) Side stream liquid control

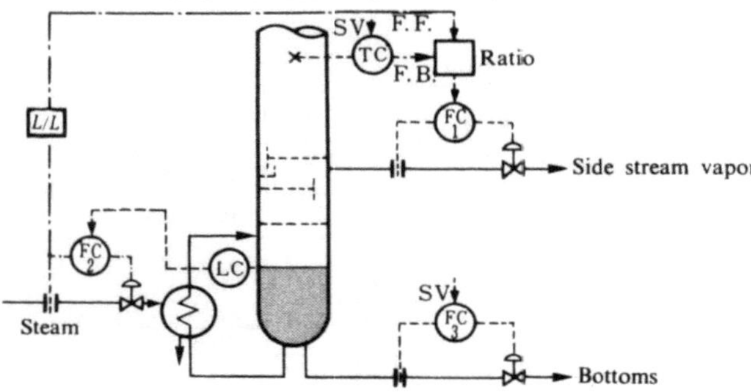

(b) Side stream vapor control

Fig. 9 · 41 Side stream control of distillation column

centration within rising vapor increases. Consequently, by reducing side-stream flow when the temperature controller (TC) measured value becomes higher than the setpoint value, heavy-component concentration in the side stream can be controlled. Moreover, in order to accurately maintain concentration at a certain level, the temperature controller (TC) setpoint value can be corrected by the output from the concentration controller. Finally, column conditions can be stabilized by incorporating feedforward control in the side steam based on variations in reflux.

Considering now the process in Fig. 9.41(b), if the flow of side-stream vapor is increased, the amount of rising vapor above the drawing plate decreases and the concentration of light components in falling liquid increases. Thus, by reducing the flow of side-stream vapor when the temperature controller (TC) measured value becomes less than the setpoint value, light-component concentration in the side stream can be controlled. In addition, feedforward control based on the amount of heating steam is useful for maintaining column stability as in the case of reflux above.

(2) *Distillation column control with little side stream* In a multi-component distillation column, the concentration distribution of certain components in the lightest or heaviest component does not change as simply as that shown in Fig. 9.36; maximum values may be exhibited at an intermediate plate in the column. In order to lower these concentrations in distillate or bottom liquid, a small amount of side stream can be drawn at an intermediate plate having such a maximum concentration. In this case, side-stream flow is controlled by keeping it at a fixed ratio with the feed flow.

9.5 Control of Reaction Processes

Reaction equipment, in which feed material (reactant) is subjected to chemical reactions in order to obtain a product of higher value, plays a central role in chemical plants. Although there are various types of reactors depending on such factors as phase of feed materials or products, heat-transfer method and operational mode, this section will take up the essentials of controlling a stirred-tank polymerization reactor and a gas-phase catalytic reactor. These two types of reactors are widely used and are good examples to understand reaction process control.

9.5.1 Control of a stirred-tank polymerization reactor

(*a*) *Continuous polymerization process and temperature control*

(1) *Process overview* Although the main material fed to the reactor is a monomer,* a solvent may also be supplied simultaneously in the case that the reaction is to be performed within a solvent. In addition, a small amount of catalyst may be supplied continuously to promote the reaction, a process which can greatly effect reaction speed and polymer quality. An additive may even be provided in order to control the molecular weight of the polymer product.

Monomer polymerization is a heat-generating reaction in which reaction temperature exerts a great influence on product molecular weight. Consequently, reaction temperature must be accurately controlled (*e.g.*, to within $\pm 0.5°C$) employing any appropriate method which removes the generated heat (see Fig. 9.42). Moreover, since it is necessary to polymerize as much monomers as possible under conditions of limited reaction speed, it is common to set retention time**of reaction liquid to a large value (several minutes to several hours).

---

* If producing a copolymer, two or more types of monomer may be fed.

* (reaction system volume)/(liquid volumetric flow)

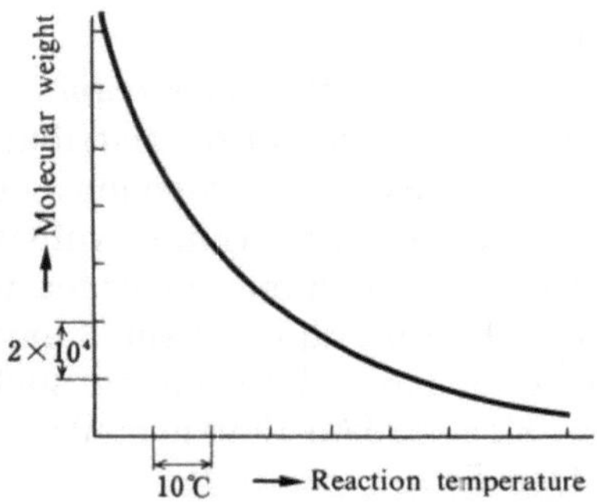

Fig. 9 · 42 An example for influence of reaction temperature to polymer molecular weight

(2) *Control of reaction temperature* A variety of heat-removal techniques are available for temperature control, depending on actual circumstances. For example, a sensible heat technique may be used in which cooling water flows through a reactor jacket or internal coil, or in which reactor substances are circulated externally through a cooler. Latent heat techniques may also be used, such as cooling vapor-generating monomers and solvents in a condenser.

Figure 9.43 shows an example of such a reactor in which both of the above techniques are used. Here, reactor temperature is adjusted by varying the temperature of the water circulating in the reactor jacket. Specifically, water temperature is controlled only by a proportional action controller (proportional band about 20%), with the desired value set by the output from the reactor temperature controller (PID action) as a cascade control. In addition, since the amount of heat removed by

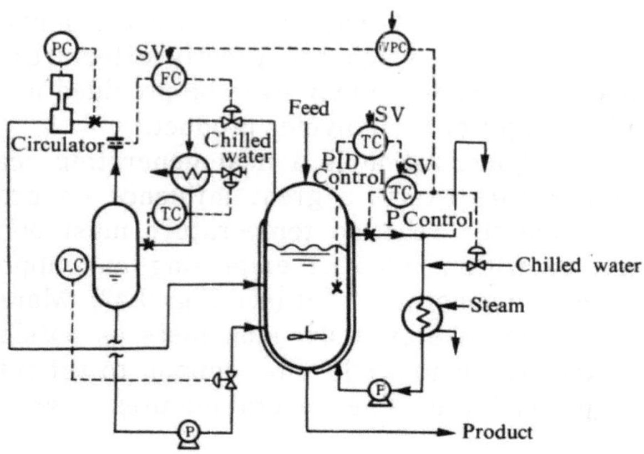

Fig. 9 · 43 Temperature control system in polymerization process

jacket cooling is not totally sufficient, the amount of circulating gas is adjusted based on the output from the valve controller while keeping the cooling-water valve at an appropriate opening; this enables the amount of heat removal from condensation to be increased or decreased. Also incorporated here is a steam heater in the circulating water system which has the role of bringing reactor substances up to reaction temperature at startup.

(b) Quality control in continuous polymerization

(1) *Problems in control* The polymerization reaction process is controlled with the object of producing a variety of polymers each requiring a certain amount and level of quality as stipulated by the end user. Since the number of various polymers in demand is usually large, product specifications switchover must take place quite frequently (*e.g.*, every day or every several days), at which time control conditions must be changed accordingly. However, since on-stream quality measurement is usually impossible, lab analysis is often employed. Form a control point of view, this can be the source of a large dead time (several hours). Moreover, since the large retention time described above provides a significant time delay in operation results, quality feedback control becomes exceedingly difficult, resulting in long product-specifications switchover times. In addition, as odd products formed during switchover are mostly off-specifications, the reduction of switchover time is a prime economic consideration, and adequate control measures for alleviating this problem are eagerly desired.

(2) *Steady-state quality control*[12] Taking up the problem of quality-switchover response, we first consider quality control in a steady-state situation. In the process shown in Fig. 9.44, product quality p_1 and p_2 can be controlled by adjusting the concentrations of addi-

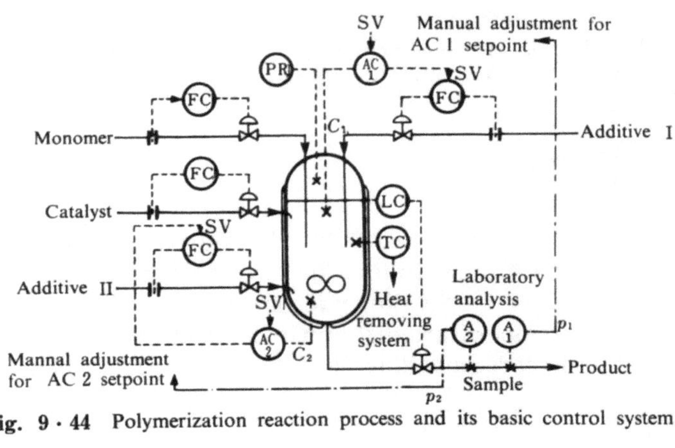

Fig. 9 · 44 Polymerization reaction process and its basic control system.

tives I and II respectively within the reactor liquid. Thus, if the following formulas can be determined, a means for performing quality control under steady-state conditions can be obtained.

$$\left.\begin{aligned} p_1 &= f_1(K_1, T, C_1, C_2) \\ p_2 &= f_2(K_2, T, C_1, C_2) \end{aligned}\right\} \tag{9.33}$$

Here, T is the reaction temperature, C_1 and C_2 are the mole concentrations of additives I and II, respectively, and K_1 and K_2 are calibration coefficients based on the results of lab analysis.

In more detail, values C_1 and C_2 in the figure are controlled by adjusting additive flow by respective concentration controllers. Moreover, as C_1 and C_2 are desired values corresponding to specified values p_1 and p_2, they can be calculated by functions (9.33), and by making them setpoint values for each controller, the desired quality can be achieved.

(3) *Predicting the quality value*[12] For the sake of clarity, the following description considers only the change in product quality p_1 resulting from varying the feed flow of additive I (the description applies in the same way to p_2). Referring to Fig. 9.45, the reaction is in steady-state before changing the flow of additive I, and only the polymer product corresponding to product quality p_{10} (black dots in the figure) determined by function (9.33) is generated at a speed R_p. If feed flow of additive I now undergoes a step increase at time 0, reaction conditions change in a short time, and from this time on, only the polymer product corresponding to new product quality p_{1f} (white dots in the figure) is generated at the same speed. Thus, at any arbitrary time, outlet product quality p_{1t} (the same as that of substances within the reactor if mixing is complete) will reflect an average value due to the mixture of generated products corresponding to product qualities p_{10} and p_{1f}. If we designate the polymerization holdup volume within the vessel as V_p, p_{1t} can be determined by the following mixed-process equation,* by which we see that it varies by the first-order capacity lag of time constant V_p/R_p.

$$\frac{V_p}{R_p}\cdot\frac{dp_{1t}}{dt}+p_{1t}=p_{1f} \tag{9.34}$$

In this equation, V_p is obtained from the value resulting from multiplying the reactor substance volume by corresponding liquid density. R_p is the value obtained by dividing the rate of heat removal (obtained by multiplying the temperature difference between the cooling water inlet and outlet by the corresponding flow rate and specific heat) by the po-

* A similar equation holds for p_{2f}, p_{2t}.

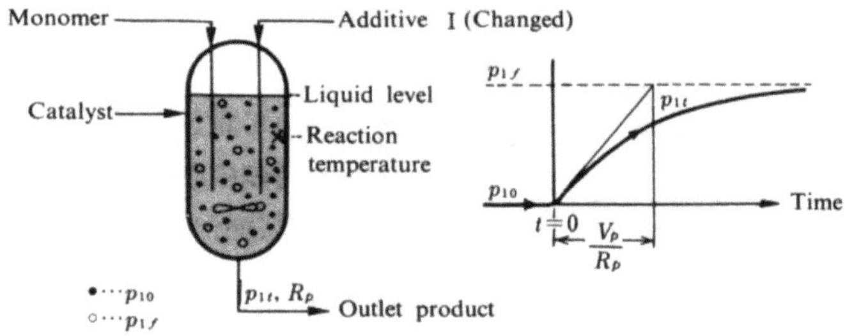

Fig. 9 · 45 Change in product quality by time

lymerization heat (kcal/monomer kg). Product quality p_{1f}, being that of polymer substances generated in the reactor at the current time, can be determined by substituting actual measured values into function (9.33). Thus, predicted product quality p_{1t} for the outlet product at the current time can in turn be obtained from Eq. (9.34). An example of a system for performing quality prediction is shown in Fig. 9.46. Here, constants K_1 and K_2 are constantly being calibrated so that the predicted product quality agrees with the actual value as close as possible. Specifically, every time lab analysis results are obtained, process variables corresponding to the time that data for the analysis was sampled are searched for in a historical memory file. These values are then consolidated and values which best conform to the actual situation are calculated for K_1 and K_2. In this way, if the current value of the outlet product quality can be predicted, feedback control can be performed without difficulty and applied to the reduction of product-specifications switchover time as described next.

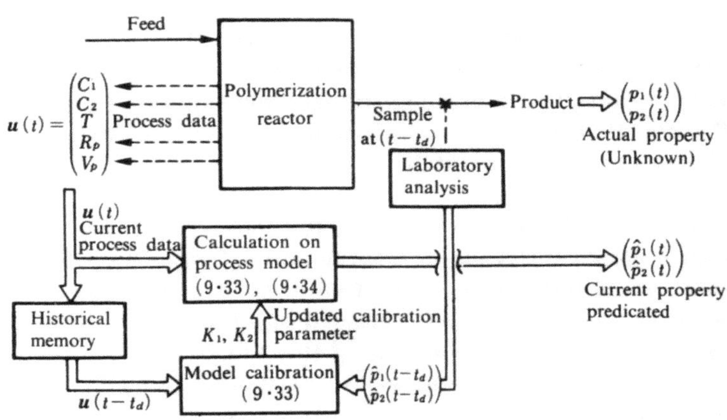

Fig. 9 · 46 Prediction system of polymerization product property

(4) *Product specifications switchover control* For a reactor in which polymer substances having quality p_{10} and p_{20} are generated, we consider operations in which a change in process conditions results in a switchover to product qualities p_{1f} and p_{2f}. As described above, if functions (9.33) can be determined, by setting values C_1 and C_2 to C_{1f} and C_{2f} corresponding to new quality values, the generated substances will achieve new product quality after an elapsed time as shown in Fig. 9.45. However, in order to complete switchover in as short a time as possible, a control scheme as illustrated in Fig. 9.47[13] is highly effective. Here, when switching outlet product quality from p_{10} to p_{1f}, initial value C_1, rather than being set to value C_{1f} corresponding to product quality p_{1f}, is set to a value C_{1m} corresponding to p_{1m}, which is sufficiently greater than p_{1f}. As a result, the product quality p_{1t} of the outlet product (a mixture of polymers corresponding to product qualities p_{10} and p_{1m}) rises quickly. If this value p_{1t} can be predicted momentarily by the method of Fig. 9.46, then if C_{1m} is switched over to C_{1f} at the time B where p_{1t} reaches p_{1f}, subsequent outlet product quality will remain constant at p_{1f}.

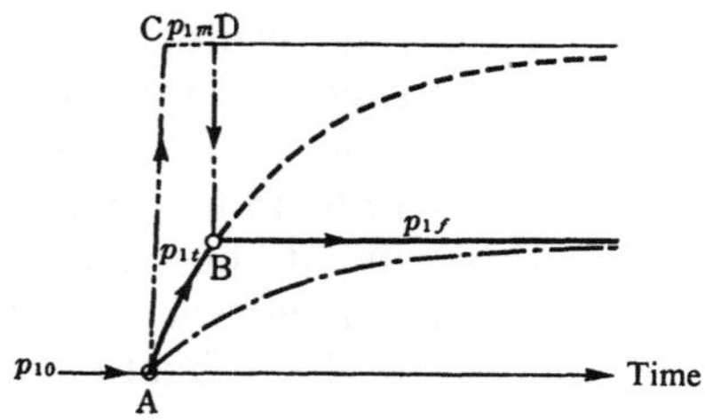

Fig. 9・47 Switching control of product-specifications

As can be seen from Fig. 9.47, the larger the value p_{1m} is made, the more switchover time can be reduced. However, the step width here cannot ordinarily be made extremely large due to a separate quality limit associated with the p_{10}, p_{1m} and p_{1f} mixed substance produced shortly after switchover point *B*.

(c) Control of a batch polymerization reactor

(1) *Process overview* A batch reaction system consists of various processes such as initial charge, pre-treatment, reaction, post-treatment and discharge, and is repeatedly started and halted in a high-frequency, cyclic fashion. Total automation of this kind of system is being rapidly adopted due to the application of distributed digital control, which has come to be characterized by a functional synthesis of feedback control and process sequence control.

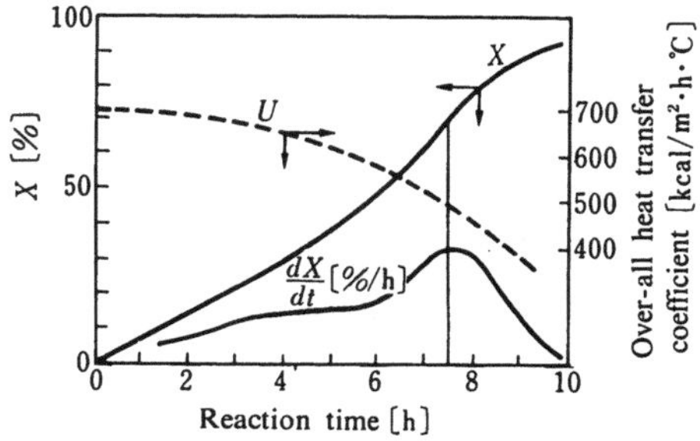

Fig. 9 · 48 An example for reaction of batch polymerization process

Figure 9.48 shows an example of the reaction process for suspension vinyl chloride polymerization. Although the percent X of monomer converted to polymer is increasing with time here, its rate of change (reaction speed) dX/dt exhibits a peak at a certain point. In addition, this rate of change is proportional to the rate of generated reaction heat. Furthermore, the overall heat-transfer coefficient U between jacket cooling water and reaction liquid decreases* as polymerization advances. Accordingly, since in this kind of batch polymerization reaction a wide range of heat disturbance usually exists, temperature control planning must carefully take this into account.

(2) *Temperature control* Temperature control of batch reactions is basically the same as that of continuous reactions described before. However, as the reduction of total cycle time in batch reactions is closely coupled with raising productivity, those particular processes involved with initial temperature raising must be performed as rapidly as possible. A point of concern here, though, is the danger of runaway reaction due to rapid acceleration of the reaction if the upper limit of internal reactor temperature is exceeded; the rate of generated heat surpasses heat removal capability resulting in a significant rise in temperature.

If the initial temperature raising process is controlled by standard PID action, the described danger may occur by excessive temperature overshoot caused by reset-windup. The methods shown in Fig. 9.49 can be used to prevent this problem. In Fig. 9.49(a), the heating steam manipulated variable is kept constant until the deviation reaches a specified value. On reaching the switchover point, fixed bias is decreased and after returning to an appropriate manipulated variable, the process is switched over to PID control. In Fig. 9.49(b), on the other hand, the manipulated variable is reduced stepwise along with the decrease in

* According to the rise in produced liquid viscosity.

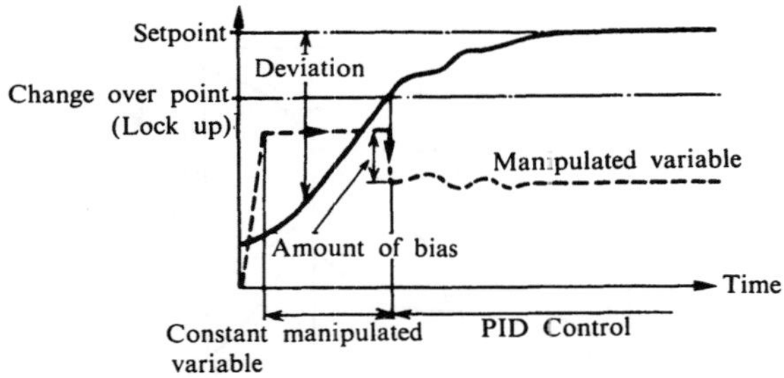

(a) Manipulated variable is held in constant until changing over PID control

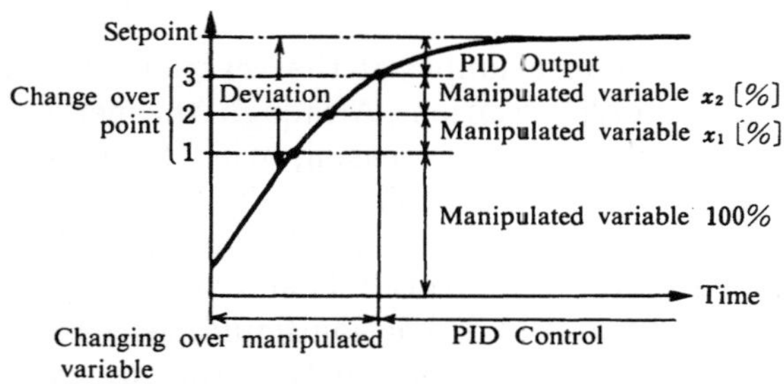

(b) Manipulated variable is reduced gradually until changing over PID control

Fig. 9 · 49 Control without overshoot in rising temperature at batch reaction

deviation.

Although control after attaining reaction temperature is essentially the same as that for continuous polymerization, heat removal over a wide range is necessary due to extensive heat disturbance as described above. It should also be noted that since variation in the rate of generated heat and heat-transfer coefficient changes process dynamics, control parameters are automatically updated with advance in polymerization so as to maintain optimal control settings over the entire process.

9.5.2 Control of a gas-phase solid-catalytic reactor

(a) Multi-stage fixed-bed reactor control

(1) *Process overview* A reactor made up of multiple catalyst beds is considered as shown in Fig. 9.50. Here, feed gas reacts adiabatically in a catalytic bed and turns hot gas,* which after mixing with cooling gas, experiences a drop in temperature. This gas then passes to the next bed where a similar reaction is repeated. With this type of reactor, it is common to control temperature, concentration and pressure in each stage within desired values with the object of maximizing the

* In the case of an exothermic reaction.

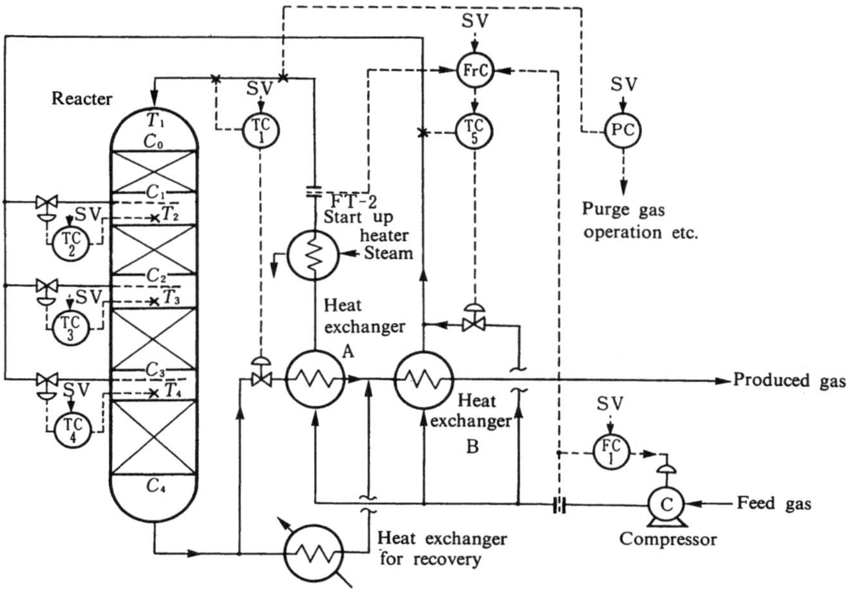

Fig. 9·50 Control system of multi-stage reactor

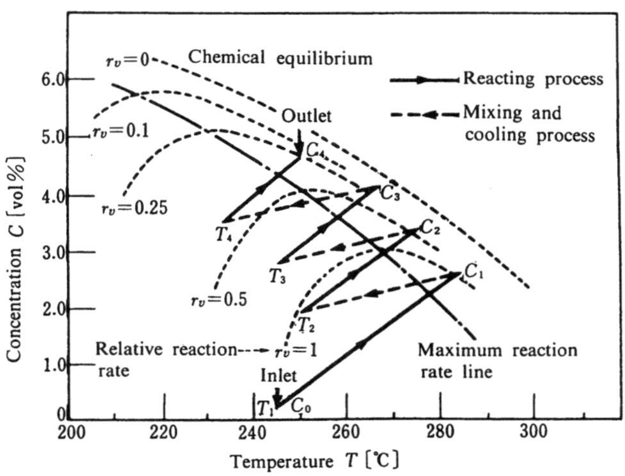

Fig. 9·51 Change in temperature and concentration at multi-stage reactor[14]

reaction rate. An example of a methanol synthesis reactor in which temperature and concentration in each stage vary is shown in Fig. 9.51.[14]

In this process, reactor feed gas having methanol concentration*

* Since feed gas includes recycled gas (see item (e)), a small amount of methanol exists.

C_0 enters the reactor with a temperature of T_1, where it undergoes reaction in the first catalyst bed. The temperature of the gas subsequently rises as does the concentration, which becomes C_1. After mixing with cooling feed gas, however, both temperature and concentration fall, which marks the beginning of another reaction in the next catalyst bed. This process is repeated until the end of the reactor where the concentration at the outlet is C_4.

In general, the reaction speed in a reversible exothermic catalyst reaction under constant pressure (amount of generated product per unit time and catalyst unit volume) will at first increase along with the rise in temperature. In time, however, the rate of increase decreases so that reaction speed eventually reaches a peak, after which it then begins to decrease. In addition, if the concentration of the generated product is high, reverse reactions will come to predominate and reaction speed will decrease. This tendency is illustrated in Fig. 9.51 by the equi-reaction-velocity curves (broken lines). If the temperature of each catalyst bed is set so that reaction speed approximates the maximum reaction velocity curve connected with the points of maximum reaction velocity as close as possible, the process can approach maximized reaction rate. Reaction operations in this figure nearly satisfy this condition.

(2) *Control system* In order to achieve stable optimal operations in the reactor of Fig. 9.50, the control system will set each temperature controller (TC) to the corresponding temperatures shown in Fig. 9.51. In addition, in order to suppress as much as possible transitions in reaction temperature and concentration patterns due to system changes such as the amount of feed gas, the ratio between feed gas flow and cooling gas flow is kept constant by manipulating cooling gas temperature. Furthermore, reaction pressure is kept at a specified optimal value by pressure controller (PC). (Refer to item (e)).

(b) Multi-tubular reactor control

(1) *Process overview* In multi-tubular heat-exchanger type reactors as shown in Fig. 9.52, each tube is filled with catalyst and heat generated in a reaction tube is removed through a heat-transfer medium, by which reaction temperature can be controlled. Such reactors are often used for reactions characterized by large reaction heat and narrow optimum reaction temperature range, and by catalytic time which must be adjusted as needed. In regards to the tubes employed here, temperature distribution in the radial direction of each tubes must be kept uniform, and consequently small apertures are usually selected. For temperature distribution in the axial direction, however, a peak appears about halfway down the tube, as shown in the same figure. This distribution is affected by reactant gas flow, composition and temperature,

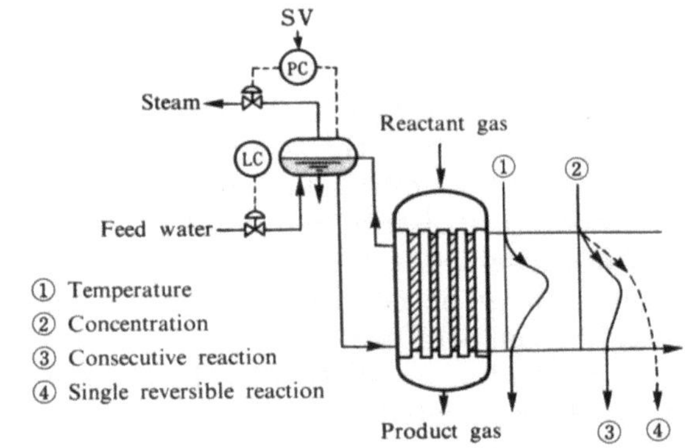

Fig. 9・52 Distribution of axial temperature and concentration in catalytic tube of multi-tubular reactor

and if catalytic activity deteriorates with time, the maximum temperature value will be less, flattening out the curve. As a countermeasure here, temperature of the heat-transfer medium can be raised (by raising the pressure controller (PC) setpoint value), thus raising the temperature level of all reactor tubes and maintaining reaction speed.

Also of interest here is the concentration distribution of product components within a pipe, which varies according to the type of reaction. For example, for a single reversible reaction, concentration will increase smoothly as shown by the dashed curve in Fig. 9.52. For a consecutive reaction as shown in Fig. 9.53, however, the concentration of object product B will peak out as the chemical reaction which changes B to C begins to advance. In the former type of reaction, temperature will be controlled so as to maximize yield per unit time of the object product. In the latter type of reaction, on the other hand, both yield and selectivity (the percent of reactant A converted into object product B) must be raised and temperature controlled appropriately with these

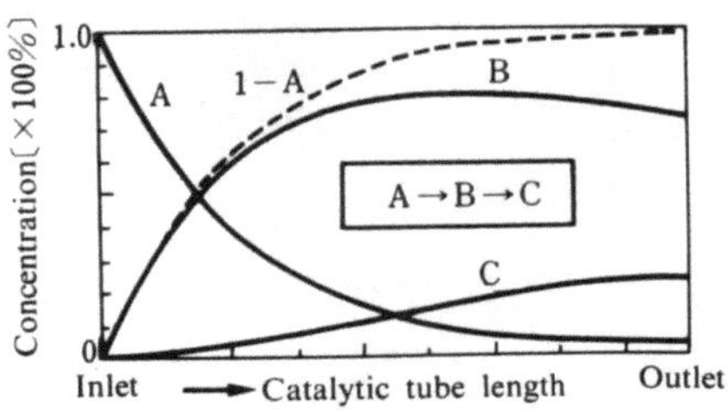

Fig. 9・53 Change in concentration of reactant and products at consecutive reaction

two requirements in mind. In general, if yield is improved, the selectivity tends to become lower.
(2) *Temperature control* The number of catalyst tubes in a multi-tubular reactor is proportional to the size of production and is thus generally quite high (several thousand in large-scale reactors), therefore, it is very difficult to make all of the tubes in uniform state. In addition, since temperature distribution in tube's axial direction varies due to a variety of causes as described above, temperature within a catalyst tube usually cannot be controlled directly. Instead, indirect methods for keeping the temperature of the heat-transfer medium constant is adopted. In particular, in addition to the example of Fig. 9.52 in which natural circulating heated water is used as the heat-transfer medium, a forced-circulation process is also possible. Moreover, the process shown in Fig. 9.54 is often adopted, in which heat-transfer mediums having high-boiling points such as oil, dowtherm and molten salt are forcibly circulated to remove generated heat by the sensible heat technique. In any of the above processes, however, it is important to suppress reaction runaway. This can be achieved, for example, by making circulation flow as large as possible and the temperature differential between inlet and outlet small, by employing a heat-transfer medium with a large overall heat capacity, or by making the temperature differential between the heat-transfer medium and catalyst as small as possible. Each of these countermeasures improves the degree of temperature self regulation in the face of disturbance, thus contributing to reaction stability.

Finally, because of the danger of reaction runaway or explosion associated with stopping circulation flow, reactant gas must be immediately shut-off according to a lower limit of circulation flow or an up-

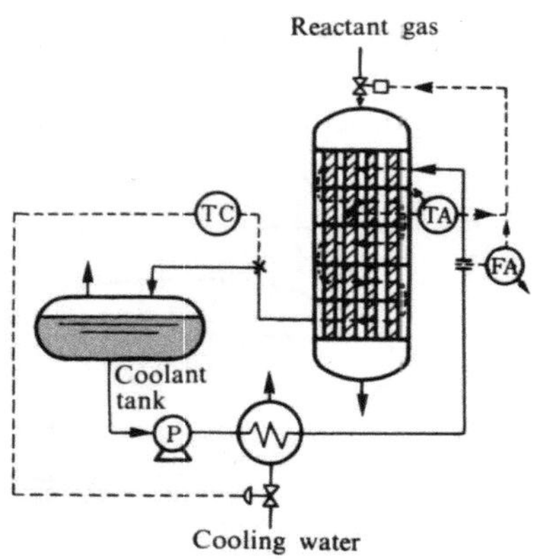

Fig. 9 · 54 Temperature control and safty control of multi-tubular reactor per limit of maximum catalyst temperature in some representative tubes.

(c) *Fluidized-bed reactor control*

(1) *Process and temperature control* In the vertical-cylindrical type of reactor shown in Fig. 9.55, a granular or powdery type of solid catalyst is employed with reactant gas injected from the bottom, and the catalyst is made to react while a fluidization state is maintained. Since the catalyst section of the reactor undergoes severe agitation here, there is hardly any temperature distribution in the radial or axial directions, which makes it possible to maintain a uniform reaction temperature across the entire catalyst. Moreover, reaction heat can be removed by incorporating multiple cooling tubes within the catalyst fluidized bed. Here, the amount of heat so removed can be sensitively changed by raising or lowering the temperature of the cooling fluid because of a large heat-transfer coefficient.* Referring again to the figure, pressurized water is passed through the cooling pipes and heat is removed via steam generated from reaction heat; reaction (catalyst) temperature can thus be controlled by manipulating the setpoint value for saturated steam pressure (corresponding to heated-water temperature) according to temperature controller output.

(2) *Catalyst fluidized-bed adjustment* One special feature of this type of reactor is that catalyst material may be added or removed as needed along with gas flow during actual operation. Consequently, catalyst filling, catalyst activity and height or condition of the fluidized bed can be adjusted so as to maintain optimal conditions for reaction. As an aid for achieving this, the following relations can be used which include the measure values ΔP_1 and ΔP_2 shown in Fig. 9.55:

$$W_C = \Delta P_2 S_A, \qquad \rho = \frac{\Delta P_1}{h_1}, \qquad h_2 = \frac{\Delta P_2}{\Delta P_1} h_1$$

Here, W_C is the amount of catalyst filling, S_A is the cross-sectional area of the fluidized bed, and ρ is the density of the fluidized bed.

In addition, since with experience we can guess the abnormal state of the fluidized bed (such as gas channeling or slagging) from the behavior of the ΔP recording pens, such information is effective for stable operation.

(d) *Preparation of reactant-gas mixture*

(1) *Mixed reactant gas* Reactant gas in gas-phase catalytic reactors as described above is, in most cases, prepared by mixing several components. The mixing ratio, although set to an optimal value for reactor operations, can present problems if varied over a wide range. In partic-

* By direct contact between solid catalyst and heat-transfer tube.

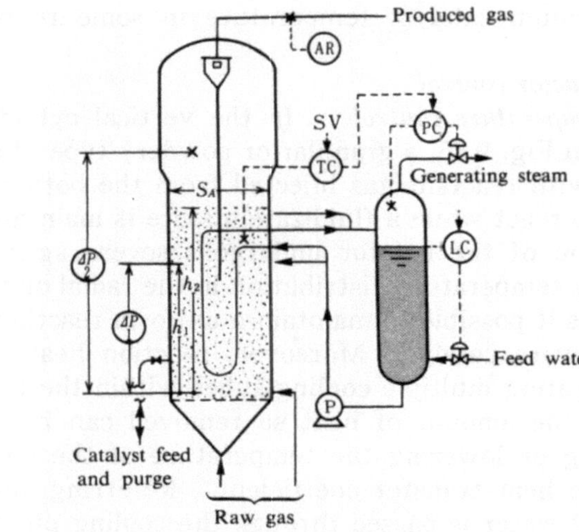

Fig. 9・55 Control of fluidized-bed reactor

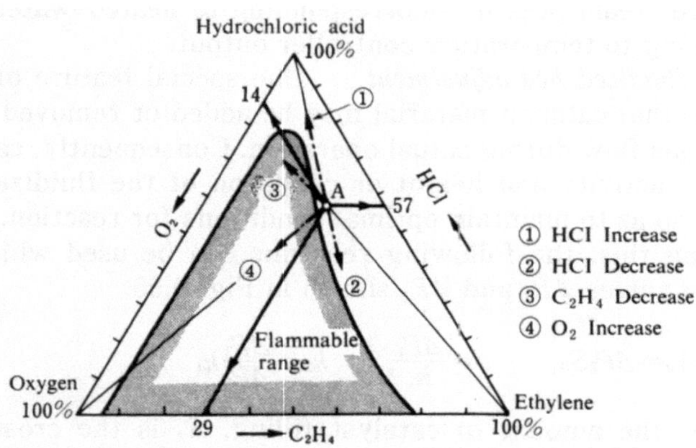

Fig. 9・56 Composition of mixed gas and flammable range

ular, dangerous situations such as catalyst deterioration, overheating and explosions can occur. Accordingly, gas mixing ratio should be reliably controlled during process operations.

The triangular coordinate system shown in Fig. 9.56 can be used to represent the respective concentrations in a three-component gas. The example shows the point A indicating the composition of feed gas required for the following reaction in which ethylene dichloride is formed from ethylene, hydrochloric acid gas and oxygen. Also shown is the associated flammable zone.

$$2C_2H_4+O_2+4HCl \rightleftharpoons 2C_2H_4Cl_2+H_2O$$

Concentrations are read from this system as follows. Taking hydrochloric acid as an example, we draw a line through point A parallel to the axis opposite the triangle vertex corresponding to hydrochloric acid 100%. The intersection of this line with the HC1 scale gives us 57%, as shown. Concentrations of the other components can be determined in the same manner.

(2) *Mixture preparation control for a reactant gas*[15] A control system for preparing the above mixed gas (Fig. 9.56) is shown in Fig. 9.57. Here, load adjustment is first performed by changing controller settings for hydrochloric acid gas flow. Following this, however, the flow rates of oxygen and ethylene are set so as to obtain an optimal gas ratio using the output from respective hydrochloric-acid ratio units. Although change of ratio settings are actually performed by an operator, a system can be adopted in which settings which mistakenly move the system in a dangerous direction can be refused and an alarm sounded. Reliability of the system can be improved as follows. As show in Fig. 9.56, as increase in the flow rate of oxygen moves mixed-gas composition towards the oxygen 100% point, the dangerous flammable zone is entered. Thus, in order to prevent a sudden increase in oxygen flow due to an erroneous drop in output from a faulty flow detector, a redundant detector is incorporated so that the highest value from the two is used as controller input. In addition, if oxygen flow should at-

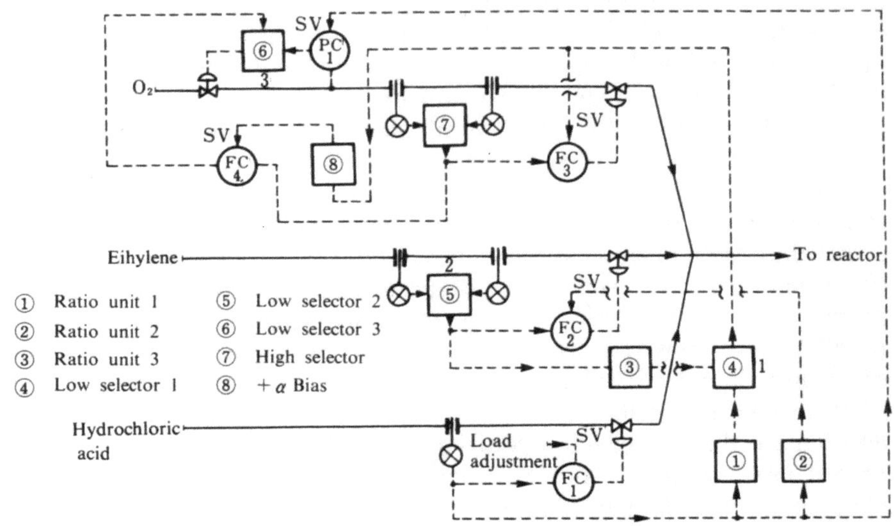

Fig. 9 · 57 Mixing and safety control system of raw gases tain its setpoint value $+\alpha$ due to a defect in the flow controller FC3 of the main control loop, auxiliary flow controller FC4 will take over manipulation of the pressure controller PC1 control valve in order to maintain this value. Moreover, if the system experiences a sudden jump in load from a low-load situation so that the FC3 control valve quickly opens, the PC1 setpoint value will be lowered immediately in the load change and prevent a momentary massive flow of oxygen.

We next examine the case in which ethylene flow rate is decreased. Here, if the composition of the mixed gas moves away from the ethylene 100% point, the danger zone is entered (Fig. 9. 56). As in the case of oxygen above, a redundant detector is employed to prevent a drop in ethylene flow due to an erroneous rise in output from a faulty flow detector. In addition to this countermeasure, oxygen flow is set using the lowest value output from ratio units 1 and 3. This is done to prevent the oxygen flow setting based on output from ratio unit 1 from exceeding its allowable upper limit in regards to ethylene flow rate.

Note here that increase or decrease* in hydrochloric acid flow presents little danger, as shown in Fig. 9.56.

Although the control system of Fig. 9.57 provides sufficient protection of the reactor, incorporation of emergency shut down equipment for each feed gas will assure reactor safety. In this case, it should be clear from the above discussion that in such an interlock circuit, oxygen must be the first to be cut-off, followed by ethylene and then hydrochloric acid provided that confirmation of oxygen cut-off is received. In this regard, although duplicate shutoff valves for oxygen may be the first to be considered as a means for implementing redundancy, the designer must be careful of "common" faults (*e.g.*, drive power failure) which can just as easily degrade system reliability.

(*e*) *Control of a recycle reaction system*

(1) *Materials balance* Since in many reaction processes, passing materials through a reactor only once does not totally complete a reaction, a recycling type of process has come to be often used. Here, unreacted gas remaining after separation of produced materials is mixed at a proper concentration with reactant gas and returned to the reactor. A problem here, however, is the presence of inert gases in the reactant gas which do not contribute to the reaction, or inert gases which may be generated by a side reaction. In either case, inert gases will tend to accumulate due to recycle operations resulting in a drop in react gas partial pressure and a slower reaction. As a countermeasure to this, some of the recycle gas can be purged so as to keep the concentration

* However, a drastic decrease in hydrochloric acid is dangerous.

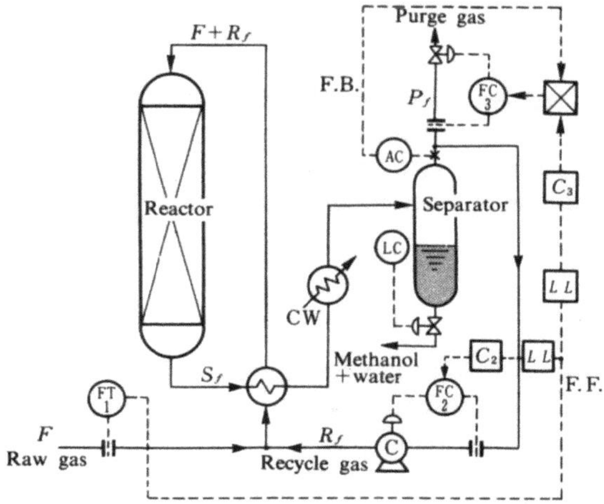

$CO_2 + 3H_2 \rightarrow CH_3OH + H_2O$ (Unit: mol)

Component	F	$F+R_f$	S_f	P_f	R_f
CO_2	0.665	1.000	0.400	0.065	0.330
H_2	1.993	3.000	1.200	0.194	1.010
Inert gas	0.012	0.082	0.082	0.012	0.070
CH_3OH	0	0	0.600	0	0
H_2O	0	0	0.600	0	0
Total	2.670	4.082	2.882	0.271	1.410
Flow rate ratio	1	—	—	0.101	0.528

Fig. 9・58 Material balance and control in recycle reaction system

of inert gases within a fixed limit.

Figure 9.58 shows the flow of materials in each section of a recycle system in which methanol is formed from reactant gas composed mainly of hydrogen and carbon dioxide (mole ratio 3) and raw gas contains inert gases (0.012 mole). In this system, 60 % of the feed gas reacts after one pass through the reactor, and since a certain amount of inert gases brought into the system is now expelled as a component of purged gas, concentration of inert gases in the system is held constant.

(2) *Control system* Given that the composition of reactor feed gas has been determined, the reaction ratio (conversion ratio) for one pass through the reactor can be controlled in a stable fashion by the method described in item (a) above. The problem of interest here is to maintain the composition ratios shown in Fig. 9.58 even during a change in raw gas flow by controlling the flow of circulation gas and purge gas. In particular, for a high-pressure reaction system having a large gas hol-

dup, controlling each flow rate by feedforward control based on the ratios listed can be quite effective against fluctuations in raw gas flow. In regards to purge flow, feedback correction can be employed using output from a concentration controller in order to keep gas composition in the system fixed. Alternatively, since an accumulation of inert gases brings about a rise in system pressure, a pressure controller can be used instead of a concentration controller with the same objective in mind.

9.6 Other Process Control

This section describes typical examples of three types of processes, refrigeration, evaporating and drying, which are frequently used in a variety of applications.

9.6.1 Control of refrigeration equipment

(a) Control of refrigerant compression Examples of equipment used for cooling process fluid and two control configurations for keeping the fluid at a constant temperature are shown in Fig. 9.59. In configuration (a), refrigerant evaporation pressure is held constant by controlling the amount of compressor suction and as a result evaporation temperature becomes constant due to a fixed functional relationship with pressure. Consequently, by setting the level controller (LC) via cascaded output from the temperature controller (TC), the area for heat transfer by evaporation can be increased or decreased as needed and process fluid temperature controlled. This technique can accommo-

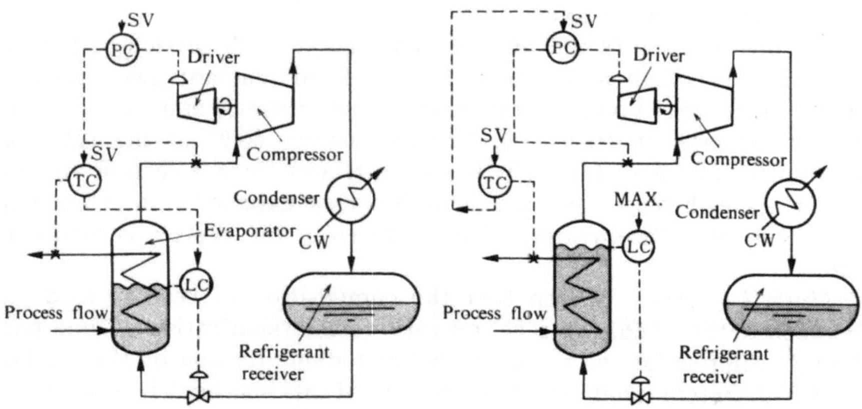

(a) Constant-press · changed-level type (b) Changed-pressure · constant-level type

Fig. 9·59 Control of refrigerant compression cycle

date a wide change in heat load (process flow, temperature) and has relatively good control response. Form the standpoint of energy efficiency, however, configuration (b) is definitely superior.

In configuration (b), the evaporator level is always kept at its highest position in order to make use of the maximum evaporation heat-transfer area. Here, process outlet temperature is controlled by setting evaporation pressure controller (PC) accordingly using cascaded output from temperature controller (TC).

Comparing the above two configurations, if we assume equal heat loads, configuration (b) is more suitable for holding higher evaporation temperature (pressure) than (a). Accordingly, the discharge pressure/suction pressure ratio of the compressor is always smaller in configuration (b), which means less power required by the compressor. However, the larger the amount of refrigerant stored in the evaporator, the worse control response becomes.

(*b*) *Feedforward control in refrigeration equipment*

In configuration (b) above, if a more accurate control system is required with the object, for example, or preventing process fluid freeze-up, feedforward control can be used.

(1) *Process description* Symbols for designating process variables are those shown in Fig. 9.60. Since the heat removed from the process fluid is equivalent to that supplied to the refrigerant via the evaporation heat-transfer surface, the following approximation holds:

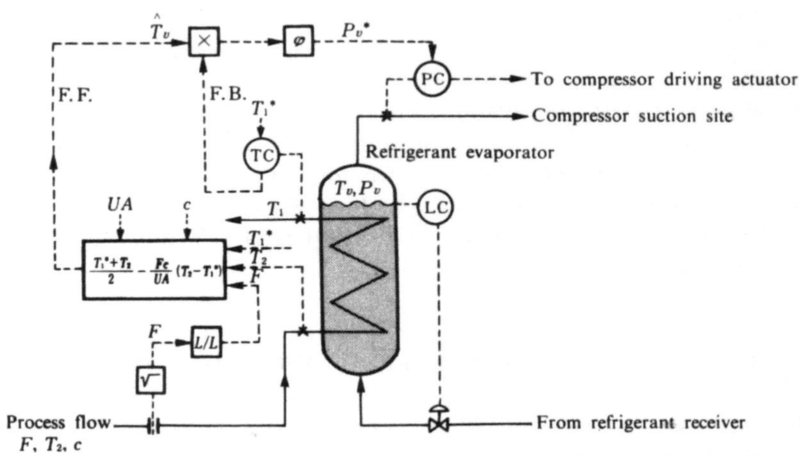

Fig. 9·60 Feedforward control in changed-pressure/Constant-level type refrigerating process

$$Fc(T_2 - T_1) \approx UA\frac{(T_1 - T_v) + (T_2 - T_v)}{2}$$

$$\therefore \quad T_v \approx \frac{1}{2}(T_1 + T_2) - \frac{Fc}{UA}(T_2 - T_1) \qquad (9.35)$$

Here, c is the process fluid specific heat and UA is the product (overall heat-transfer coefficient) × (area) for the evaporation surface.

In addition, the fixed relationship $P_v = \varphi(T_v)$ generally holds between evaporation temperature and pressure.

(2) *Feedforward control system* Referring to Fig. 9.60, given values UA and c and using input values T_1, T_2 and F, value T_v can be estimated using relationship (9.35). Consequently, a value P_v corresponding to this T_v can then be provided as a setpoint value to the pressure controller (PC). In addition, the gain along the feedforward path requires feedback correction from the output of TC, the process outlet temperature controller. Note also that the dynamic compensation element L/L consists of a delay element for synchronizing variation in F with the time delay associated with change in outlet temperature, and a lead element for negating the time delay from P_v manipulation until its effect on outlet temperature. In general, since variation of inlet temperature with time is similar to flow rate in that it does not change rapidly, dynamic compensation for T_2 can be omitted.

9.6.2 Evaporator control

(*a*) *Process description*

Figure 9.61(a) shows the most basic configuration of an evaporator in which feed liquid, which includes non-volatile components, is heated to separate volatile components. In order to cut down on the amount of heating steam, however, the multiple-effect evaporator in Fig. 9.61(b) reuses vaporized steam from the previous stage. The kind of configuration is commonly used in large scale operations.

Using the symbols shown in the figure, the following relationship holds for the process variables:

$$F_i = F_o + F_V \qquad (9.36)$$

In the case of the multiple-effect evaporator, this becomes $F_V = F_{V1} + F_{V2} + F_{V3}$. In addition, if we designate the non-volatile concentration of the feed liquid and concentrated liquid as C_i and C_o, respectively, we obtain from the balance of materials:

$$F_i C_i = F_o C_o \qquad (9.37)$$

The following approximation holds due to the fact that the amount of heat supplied by heating steam is nearly equal to the amount of evapo-

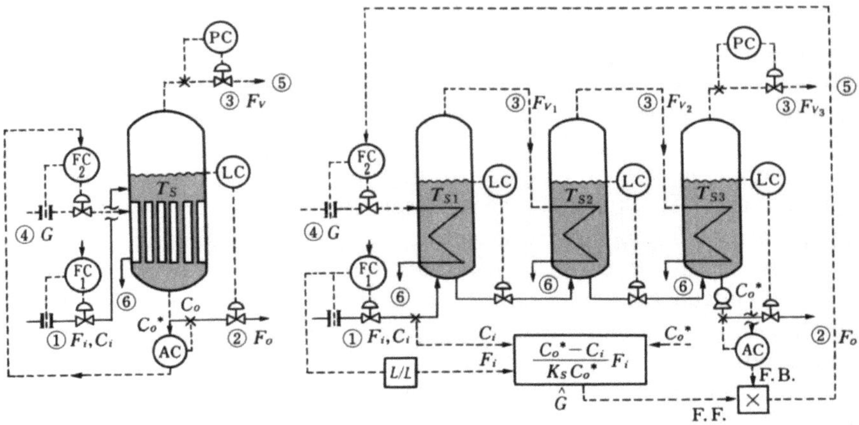

① Feed liquid ③ Vaporized steam ⑤ To vacuum system
② Concentrated liquid ④ Steam ⑥ Drain

(a) Single-effect evaporator (b) Multiple-effect evaporator (Forward feed type)

Fig. 9 · 61 Control system in case of changes in feed flow rate

rated heat accompanying separation of the solvent from the feed liquid:*

$$F_V \approx K_s G \tag{9.38}$$

Here, K_s is a constant dependent on the process; it becomes large as the number of stages in the multiple-effect evaporator is increased.

If we now eliminate F_o and F_V from the above three equations, we obtain:

$$G \approx \frac{(C_o - C_i)F_i}{K_s C_o} \tag{9.39}$$

Or by eliminating F_i and F_V:

$$F_o \approx \frac{K_s C_i G}{C_o - C_i} \tag{9.40}$$

From the above discussion it should be clear that these equations represent general relationships regardless of the type of evaporator. The relationships serve as a basis for configuring an evaporator control system, as described next.

---

* Feed liquid is normally preheated close to the internal temperature.

(*b*) *Control system*

The object of evaporator control is to maintain the concentration of non-volatile components in the concentrated liquid constant. During this time, the process receives disturbance in the form of feed liquid flow variation (product amount variation) and change in the concentration of non-volatile components. A basic control system for such a situation is shown in Fig. 9.61(a), while the feedforward control system for the multiple-effect evaporator in Fig. 9.61(b) is used to improve controllability. In the latter case, since the time delay required for a change in feed liquid flow to affect concentrated liquid concentration is relatively small compared to that for heating steam to effect corresponding concentration (in particular if retention time of each vessel is short), a time lead element incorporated in the *L*/*L* dynamic compensator is used.

The multiple-effect evaporator in Fig. 9.61(b) is also characterized by parallel flow, *i.e.*, steam and liquid flow in the same direction (parallel-flow type). As a result, however, vessel pressure as well as temperature is lower than that of the previous vessel. This low temperature, in turn, results in increased viscosity of the concentrated liquid (*e.g.*, caustic soda aqueous solution). As a countermeasure to this, the counter-flow multiple-effect evaporator shown in Fig. 9.62 can be used to keep the temperature of the concentrated liquid high. Although control system configuration in this case is essentially the same as that for Fig. 9.61, since time delay relationships are opposite that of parallel flow, dynamic compensation for such time delays will be different.

Another type of system to consider is that in which the amount of available heating steam varies, which necessitates a corresponding change in feed liquid flow. Such a control system is shown in Fig. 9.63. Although a single-effect evaporator is shown here, a multiple-effect evaporation (parallel flow) can be applied just as well. The feedforward

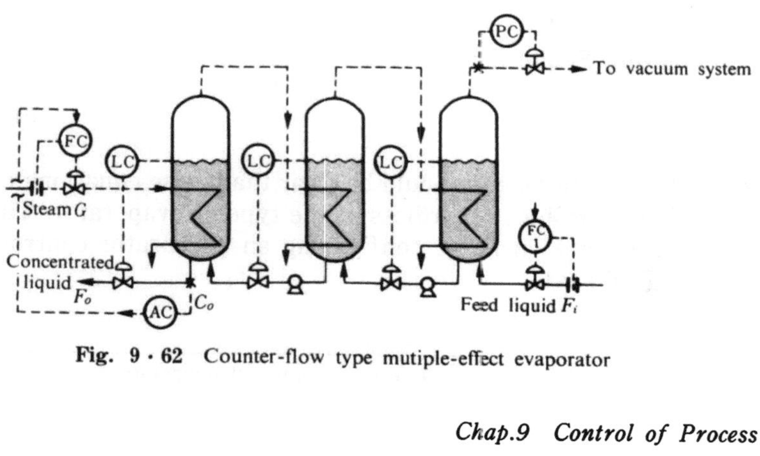

Fig. 9 · 62 Counter-flow type mutiple-effect evaporator

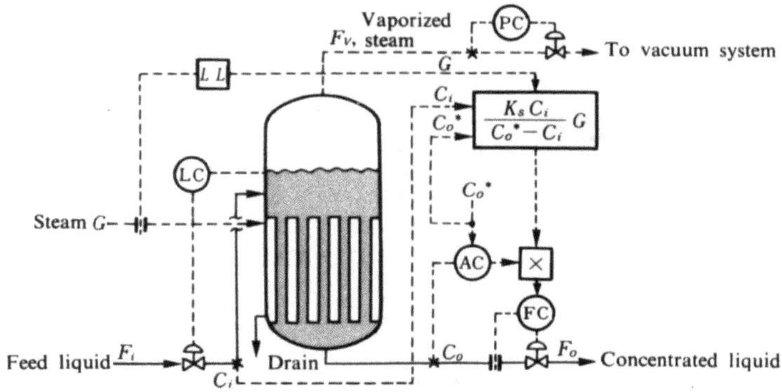

Fig. 9 · 63 Control system used with variable steam flow rates

mechanism here, which is based on relationship (9.40), becomes even more effective in a multiple-effect configuration. Note also that the *L/L* dynamic compensator employed here acts as a time delay element.

In each of the above control systems, the means of measuring non-volatile concentration may employ a variety of forms. For example, a density meter may be used in which the measured value corresponds to a concentration value, or by making use of the relationship between boiling point and non-volatile concentration, liquid temperature can be measured to give the concentration. However, in the latter case, since boiling point itself will vary depending on vapor pressure, and automatic pressure compensator becomes necessary which converts the boiling point to that under standard pressure. In addition, an alternative method which has come to be widely used is to measure the temperature differential between evaporation vapor and liquid to obtain a corresponding concentration value.

9.6.3 Drying process control

(*a*) *Control of a batch fluidized-bed dryer*

Although there are many types of batch dryer configurations, the fluidized-bed type has come to be adopted in a wide range of fields. Since this type is also demonstrative of drying processes in general, a typical example will be described here.

(1) *Drying process* In the equipment shown in Fig. 9.64, a granular or powdery substance for drying (referred to as material below) is placed in a vessel having a perforated bottom tray and is dried by having heated air pass through it for a certain length of time. During this time, the particle water content (H_2O kg/non-water material kg) decreases in the manner shown in Fig. 9.65(b). In addition, the particle water content rate of change (drying speed) shown in Fig. 9.65(c) bec-

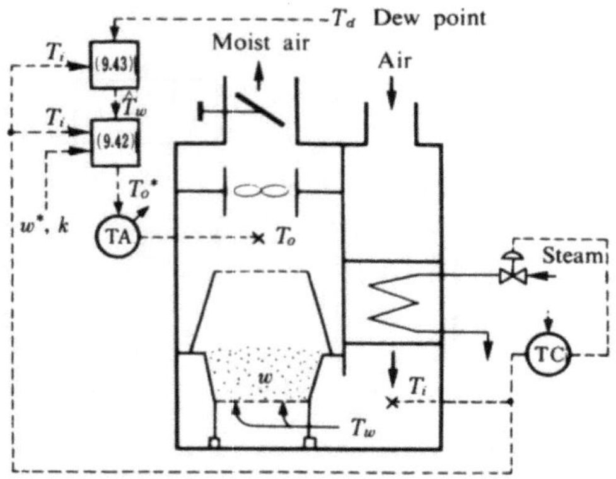

Fig. 9・64 Control of batch fluidized-bed dryer

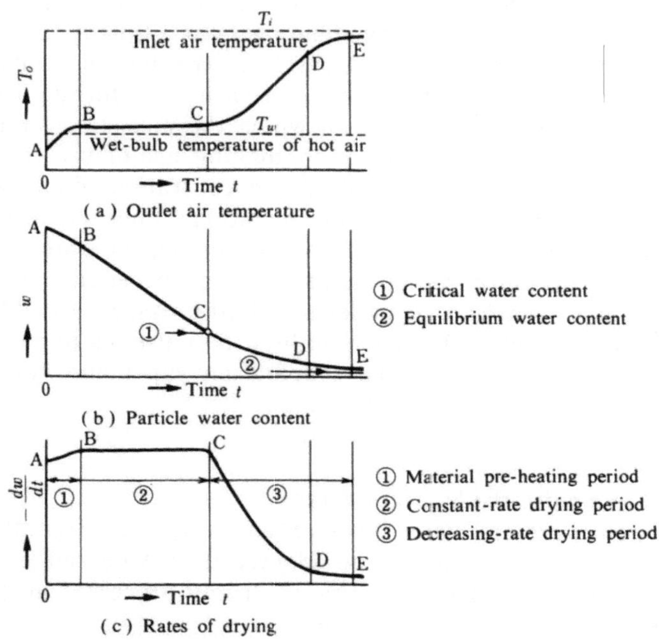

Fig. 9・65 Batch fluidized-bed drying process

omes constant shortly after material pre-heating (constant-rate drying period). The reason for this is as follows. As water comes to cover the surface of the material, material temperature reaches equilibrium with wet-bulb temperature T_w, since heat transferred from the passing air is used completely for water evaporation. This gives rise to constant evaporation, and as a consequence, air outlet temperature T_o during this period is constant but slightly higher than T_w, as shown in Fig. 9.65(a).

As the drying process advances, moisture transfer from the interior of the material falls behind evaporation so that some of the material surface dries, raising its temperature. Accompanying this is a small temperature differential arising between the hot air and material which decreases the amount of heat transfer and lowers the drying speed (decreasing-rate drying period). As a result, air temperature T_o at the dryer outlet gradually rises at this time and approaches inlet air temperature T_i (at completion of drying, outlet air temperature should, in principal, be equal to inlet air temperature). The material water content w is given by the following equation in the above situation:[2)]

$$w = k \log_e \frac{T_i - T_w}{T_o - T_w} \tag{9.41}$$

Here, k is a constant for a particular type of material and amount and constant air flow.

(2) *Control system*[2)] Given a process in which the value k in Eq. (9.41) above is obtained from actual results, if we now designate the desired final water content of the material as w^* and the corresponding outlet air temperature as T_o^*, the following relationship is obtained from this same equation:

$$\frac{T_i - T_w}{T_o^* - T_w} = e^{w^*/k} = \frac{1}{K} \quad \text{(constant value)}$$

$$\therefore \quad T_o^* = KT_i + (1-K)T_w \tag{9·42}$$

Thus, since T_o^* can be determined from measured values T_i and T_w, the current value of T_o can be checked to see whether or not the de-

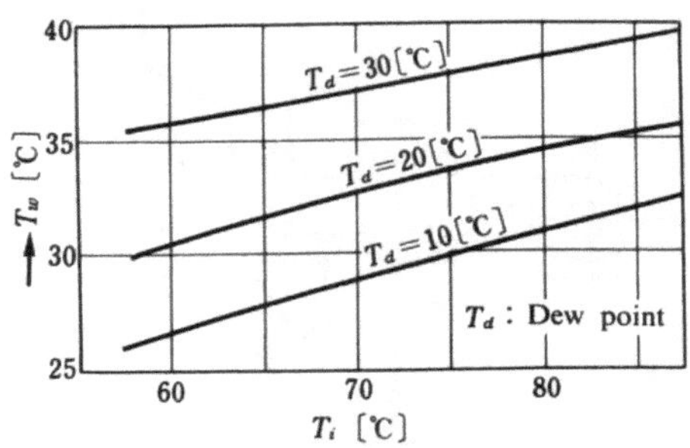

Fig. 9 · 66 Relationship between dry-bulb temperature and wet-bulb temperature of heated air sired material water content has been reached. However, as direct measurement of hot air wet-bulb temperature is technically difficult, the curves to get T_w can be drawn up instead based on easy-to-measure inlet air temperature T_i and atmospheric dew point T_d, using general humidity curves. From the relationships obtained, T_w can then be surmised, as shown by the example in Fig. 9.66. In addition, these relationships can be approximated by the following expression:

$$\hat{T}_w \approx \beta T_i + \alpha(T_d) \tag{9.43}$$

where $\hat{T}_w$ indicates the estimated value of T_w.

A control system based on the above ideas is shown in Fig. 9.64. Specifically, air temperature T_i is kept constant, and from this value and atmospheric dew point T_d, the value of T_w is determined from the above expression. In addition, using Eq. (9.42), desired value T_o^* corresponding to w^* is calculated. If the actual measured value of T_o should reach this value, an alarm is sounded announcing the end of drying operations.

(b) Control of a continuous fluidized-bed dryer

Although there are also many configurations for continuous dryers, we select the fluidized-bed type as an example for the same reason stated above for batch systems.

(1) *Process description* Referring to Fig. 9.67, a granular or powdery material having water content w_f is supplied to the dryer with a mass

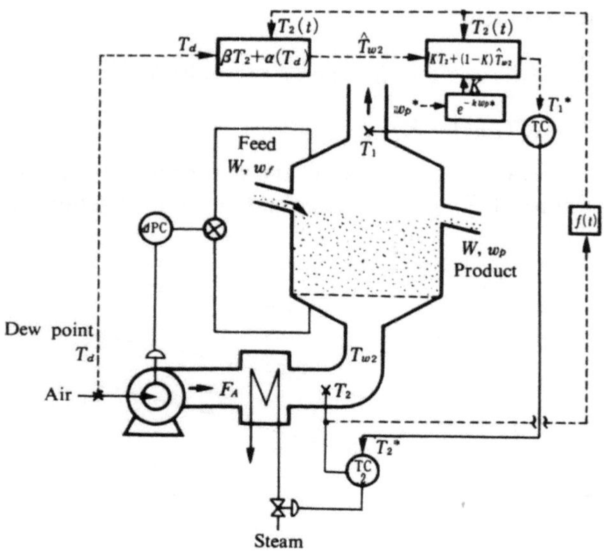

Fig. 9 • 67 Continuous fluidized-bed dryer and its control system

flow (non-water based) of W, and drying is performed through contact with hot air supplied to the fluidized bed with a mass flow (dry based) of F_A. Since the mass of material in the vessel here is kept constant through control of the fluidized-bed pressure differential (see Subsec. 9.5.2(c)), material whose water content decreases due to passage through the vessel is expelled to the outside at a flow rate equal to feed flow.

In this dryer, the heat required for water evaporation is completely supplied via heat transfer from hot air (adiabatic type).

Since mixing in the fluidized bed is completely closed in, water content w of material within the vessel is considered to be equal to water content w_p of expelled material. In addition, provided that material quality is invariable and due to the fact that hot air flow F_A is nearly kept constant, the following equation similar to Eq. (9.41) above can be written:

$$w = k \log_e \frac{T_2 - T_w}{T_1 - T_w} \tag{9.44}$$

(2) *Control system*[16] The object of control here is to maintain product water content w_p ($= w$) constant in the face of variation in feed values (W and w_f) or change in atmospheric moisture content (T_d). A problem here, however, is that from a control point of view, w_p is technically difficult to measure directly. However, if control is performed based on the following equation corresponding to Eqs. (9.42) and (9.44), objectives can be achieved without direct measurement.

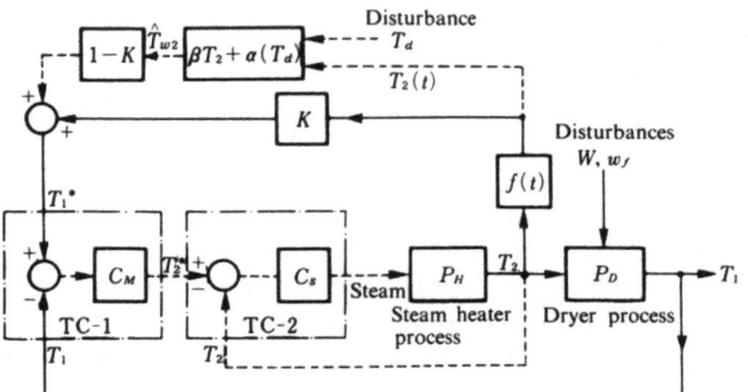

Fig. 9 · 68 Block diagram of continuous fluidized-bed dryer control system

$$T_1 = KT_2 + (1-K)T_{w2} \tag{9.45}$$

Here, T_{w2} can be determined by Eq. (9.43).

A control system satisfying the above conditions is configured as shown in Fig. 9.67. Here, desired value w_p^* for product water content is input into the system as the value K. Also of importance here is that the path of $f(t)$ in the figure makes up a positive feedback circuit, (see also Fig. 9.68). However, since it exists in parallel with another negative feedback circuit in the drying process, if the gain of the former becomes larger than that of the latter, the control system will become unstable. To stabilize this situation, an appropriate time-delay element $f(t)$ can be incorporated to bring the gain of the former lower than that of the latter at the operating frequency.

PRACTICE QUESTIONS

9.1 A centrifugal pump used for plant cooling water is operated in a piping system for which the ratio of dynamic pressure loss to total pressure loss at times of 100% load is 75%. If load percentage changes during the year depending on the season as shown below, determine the rate of energy savings when employing pump speed control.

Period	Load Percentage	Period	Load Percentage
December–February	50%	July–August	100%
March–June	85%	September–November	85%

9.2 Explain the reason for adopting a three-way control valve in bypass control of steam heater.

9.3 Some steam heaters may adjust the heating rate by a drain valve. Determine the response time-constant for the change in heating rate due to drain flow manipulation. Assume the following conditions in the heater:

Working steam	2.5kg/cm², 137°C
Condensation latent heat	522 kcal/kg
Total condensation heat-transfer surface	100 m²
Overall heat-transfer coefficient	488 kcal/m²·h·°C
Total heat-transfer area submersion drain amount	433 kg
Process-fluid average temperature	74°C

9.4 In a binary-component distillation column in which the feed flow is constant, but feed composition changes, illustrate a feedforward control system for keeping the concentration of bottom liquid constant under conditions of constant reflux flow.

9.5 Discuss the performance required for the temperature control system used in a stirred tank type batch polymerization reactor.

9.6 Given a gas-phase fluidized-catalyst reactor having a cylindrical shape of diameter 3 m, total pressure differential over the catalyst bed during operation of 1000 mm, and pressure differential at the pressure taps installed at a height of 2.5 m within the catalyst bed of 450 mm, calculate the total weight of catalyst filling the reactor and the height of the fluidized catalyst bed.

9.7 Given the case of compressor-refrigeration equipment using ammonia as a refrigerant in which process fluid is cooled from -3°C to -7°C; the level of ammonia liquid within the evaporator is about halfway, and its temperature when the effective evaporation heat-transfer area is half is -10°C. If the liquid level is now raised and the heat-transfer surface is completely submersed, what will the refrigerant temperature value become? Assume heat load is constant.

9.8 A triple-effect evaporator may be used to concentrate the sugar content of an aqueous solution from 25 wt% to 85 wt%. In this case, since the flow of heating steam will vary, it is desirable to implement feedforward concentration control to increase or decrease the flow of concentrated liquid accordingly. What would be a satisfactory value for the ration between the two flows? In addition, in regards to dynamic here, consider whether lead or lag elements should be incorporated. Assume in this process that for every 1 kg of heating steam, 2.7 kg of water is evaporated.

9.9 In an adiabatic continuous fluidized-bed dryer using heated air as a heating source, name the major disturbances involved with trying to keep the water in the product constant.

ANSWERS TO QUESTIONS

9.1 The rate of energy savings for each period can be determined from Fig. 9.3(b) (f= 0.75) and averaged as follows:

$$\bar{M}=(0.62)(3/12)+(0.25)(7/12)+(0)(2/12)\approx 0.3\cdots 30(\%)$$

9.2 Please see Subsec. 9.3.1(d).

9.3 Substituting the various values into Eq. (9.10), we get $T\approx 0.75$ h = 4 min.

9.4 Using Z instead of F in Fig. 9.28(b) results in a feedforward control system. Moreover, from Eqs. (9.24), (9.25), and (9.26), the relation $B = \varphi(Z)$ is now used in place of $g(F)$.

9.5 Please see Subsec. 9.5.1(c)(2).

9.6 From the equations in Subsec. 9.5.2(c)(2), W_c = 7069 kg, h_2 = 5.56 m.

9.7 Substituting the first condition (heat-transfer area 0.5 A) into Eq. (9.35), we obtain F_c/UA = 5/8. Using this value, T_v = -7.5°C.

9.8 From Eq. (9.38), K_s = 2.7, and using this value and the concentration values in Eq. (9.40), F_0/G = 1.125; delay element. (see Subsec. 9.6.2(b) regarding dynamic compensation.)

9.9 Feed flow and water content, and atmospheric humidity (dew point). (See Subsec. 9.6.3(b).)

REFERENCES

1) T. Takamatsu : "Roles of operator in chemical plants," *Kagaku Kogaku*, **49**, 11 (1985) 845–848 (in Japanese).
2) F.G. Shinskey : Energy Conservation through Control, Academic Press (1978).
3) F.G. Shinskey : Process-Control Systems, 2nd Edition, McGraw-Hill (1979).
4) B.G. Liptak : Instrument Engineers' Handbook, vol, II, Chilton Book (1970).
5) Ichioka, *et al.* : " Application status of microcomputer on combustion control," *The Energy Conservation*, **33**, 5 (1981) 61.(in Japanese).
6) F.G. Shinskey : "The material-balance concept in distillation control," *The Oil and Gas Journal* (July 14, 1969).
7) T.G. Chin : "Guide to distillation pressure control methods," *Hydrocarbon Processing* (Oct. 1979).
8) F.G. Shinskey : Distillation Control–For Productivity and Energy Conservation, McGraw-Hill (1977).
9) P.S. Buckley, W.L. Luyben, and J.P. Shunta : Design of Distillation Column Control Systems, ISA (1985).
10) G.F. Fauth and F.G. Shinskey : "Advanced control of distillation column," *Chem. Eng. Prog.* (June 1975).
11) F.G. Shinskey : Controlling Multi-Variable Process, ISA (1981).
12) G.C. Ardell and B. Gumowski : "Model prediction for reactor control," *Chem. Eng. Prog.* (June 1983).
13) A. Mori : Transition Operation for Preparation of Polyolefin Having Different Melt Index, Lade Open Gazette 142007 (1980). (in Japanese).
14) Osaki, *et al.* : "Development of new type reactor as well as energy conservation and scale-up on methanol plant," *Plant and Process*, 9 (1984) (in Japanese).
15) J.T. Illidge and J. Wolstenholme : "Hazard of oxyhydrochlorination," *Loss Prevention*, 12 (1979).
16) F.G. Shinskey : "How to control product dryness," *Instrumentation Tech.* (Sept. 1968).

Chapter 10

INSTRUMENTATION TO MANUFACTURING INDUSTRIES (Application II)

As was described in Chap. 9, the various kinds of processes used in manufacturing industries are generally comprised of processing units. Different industries, however, employ different operational methods. Thus, the instrumentation and control methods will also vary reflecting those differences.

Our thinking with respect to instrumentation has undergone much change in recent years; the concern used to be operational supervision and automatic control for the sole purpose of mass production. Today, however, such considerations as safety, quality, labor savings, and energy conservation are at least equally important in the development of instrumentation. Another significant development that has become increasingly noticeable with the introduction of digital devices and systems, is the rapid evolution toward so-called total factory automation (FA) with the capability for small production runs of a diverse range of products.

How extensively is instrumentation employed today? While this is not easy to determine precisely, the Japan Electric Measuring Instruments Manufacturers' Association (JEMIMA) has come up with an instrumentation ratio, which they define as follows:

$$\text{instrumentation ratio (IR)} = \frac{\text{investment in instrumentation}}{\text{investment in plant and eguipment}}$$

Investment in instrumentation includes investment in industrial instruments, process computer systems, and instrumentation work used in productive processes (analyzers used for experimental purposes in laboratories, for example, are excluded). *Investment in plant and equipment* includes all costs associated with equipment and facilities, but not including land.

Although the instrumentation ratio is only calculated triennially, it is nevertheless apparent that the ratio is increasing every year. In

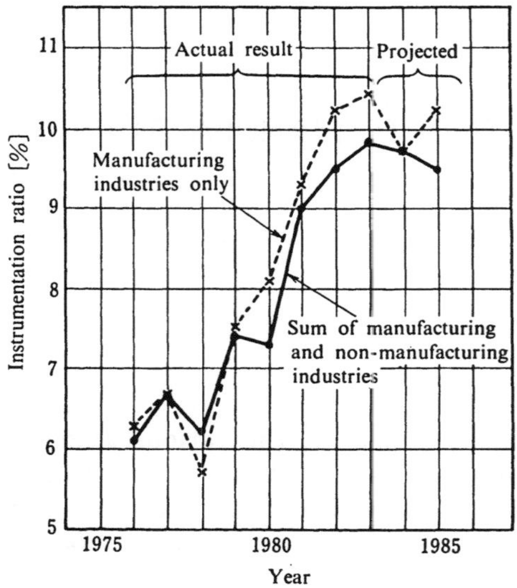

Fig. 10・1 Progress in instrumentation ratio

fact, as is shown in Fig. 10.1, the ratio jumped sharply from 6 to 10% between 1976 and 1983.

This chapter is intended to provide a broad overview of how instrumentation is applied to such diverse but representative industries as oil, steel, and electric power given their very different and unique operational and control requirements. With the limited number of pages we have available, this will be no more than a cursory introduction. The authors also acknowledge that they are affiliated with corporations that manufacture the types of instruments described. While a certain amount of bias may be reflected in the presentation, it is nevertheless hoped that the information presented in this chapter will serve as a useful reference.

10.1 Instrumentation Applications in the Petroleum Industry[1)–6)]

10.1.1 The petroleum industry and instrumentation

Among process industries, the petroleum industry serves to illustrate one of the most advanced in terms of process automation.

Although the existence of oil has been known since before Greek times, active exploration for underground deposits and widespread use didn't begin until the latter half of the nineteenth century. In the present century, control over oil resources was an important concern in the two world wars, and of course today the importance of oil is universally recognized. In the post-war years, the petroleum industry has continued to make enormous strides. It is no exaggeration to say that the rapid economic recovery of Japan and the West after the war, and the evolution of society as we know it today was made possible by the ready availability and use of oil.

However, the knowledge that oil is a limited natural resource and the abrupt price increases of the 1970s (the co-called oil shocks in 1973 and 1978) led to an industry-wide effort in Japan to conserve energy. This effort was highly successful and, in fact, has contributed to the huge oversupply of oil that prevails on the world market today. Diversification of energy sources has also made much progress. It is estimated that by the year 2000, the world's dependence on oil as an energy source will have fallen by about 30%.

From the standpoint of instrumentation, the petroleum industry is unique in that it is a constant-load continuous process and is basically fairly easy to control. As we leave the era of mass production behind, however, and enter the age of diverse feedstocks and value-added products, it will be necessary to re-evaluate our approach to instrumentation.

In the earliest days when oil first started to be used, distillation of crude oil was carried out in a single kiln. Successive separation of the various distillates proceeded while the operator kept a close watch on the temperature. Instrumentation of the time didn't go beyond simple measurement.

By the turn of the century, the demand for oil had increased tremendously, which led to the development of more advanced distillation units and continuous processing. Instrumentation systems were also improved so they could be applied to control as well as to measurement. Control logic also saw much improvement at this time thus enabling even more advanced process control devices. This in turn made it possible to increase the scale of processes to unprecedented proportions. Today, equipment economy is not the only consideration. To cope with the greater diversity of crude feedstocks and the more stringent requirements for product quality, instrumentation and equipment manufacturers today are working closely together to develop and improve systems. It goes without saying that a major support for this development is the state of contemporary instrumentation technology. Looking ahead, the challenge will be to achieve even more precise levels of control by optimizing process equipment through the use of microprocessors and process-application computers.

The petroleum industry includes many activities encompassing everything from extraction to refining and blending. In fact, if we regard the industry in its widest sense, shipment of petrochemical products should also be included. We will organize this chapter according to a major distinction made by the industry itself—the distinction between *on site and off site*. On site is used to designate equipment that is actually involved in the distillation, separation, and polymerization of petroleum products, whereas off site refers to processing and/or handling that is done away from the refining complex such as blending, shipping, and processing orders for petroleum products. Subsection 10.1.2 will consider the application of instrumentation to an on-site topping unit. Subsection 10.1.3 will then discuss off-site instrumentation, especially as pertaining to a distribution control system for petroleum products.

10.1.2 Topping unit instrumentation

The processing of crude oil into various finished petroleum products is referred to as refining. The starting raw material is crude oil which is subjected to various refining processes including separation, refining, and reforming (also called cracking) to produce a range of value-added products: gasoline, kerosene, gas oil, and fuel oil. Typical oil refining processes are shown as a block-flow diagram in Fig. 10.2. The

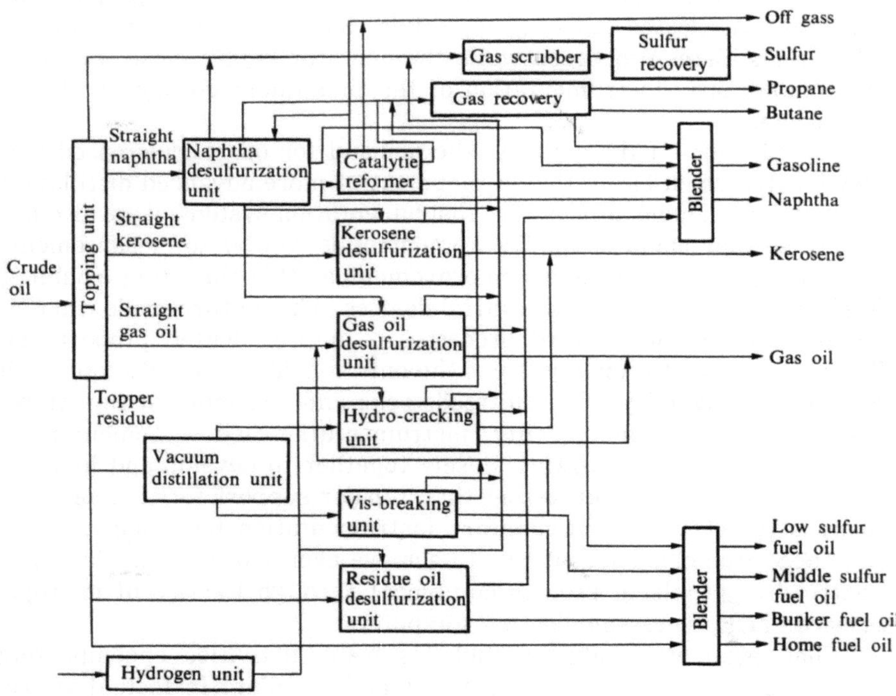

Fig. 10 · 2 Typical petroleum refining process

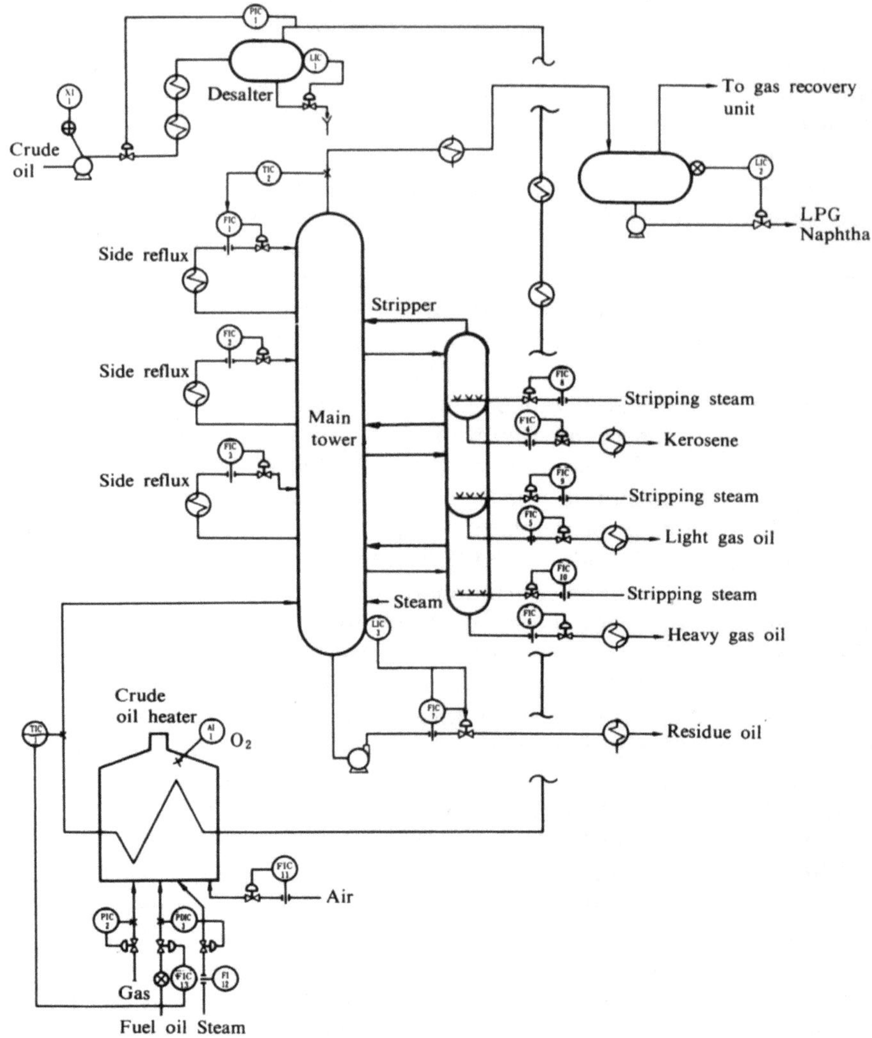

Fig. 10 · 3 Flow diagram of topping unit

initial processing of crude oil is carried out in a topping unit, such as shown in Fig. 10.3.

Crude feedstock is continuously supplied to the topping unit. It is routed through a heat exchanger where it picks up heat from side refluxes and product streams, and is then sent to a desalter. The desalted crude, after recovering heat from even higher temperature liquid, is brought up to a constant temperature (300 to 350°C) in a crude oil heater, and then charged into a topping unit. The topping unit functions to separate the crude oil by fractional distillation into components of differing boiling ranges. Typically, the crude will be distillated into naphtha, kerosene, gas oil, and topped residue. The lighter fractions in the main tower are still mixed together, so they are sent to a stripper for further separation.

In the past, the various topping unit parameters such as temperature, pressure, and level were separately controlled with discrete devices for each control loop. In recent years, however, this approach has generally been replaced by distributed control systems (DCS) using microprocessors. In the remainder of this section, with continued reference to Fig. 10.3, we will consider the process controls that are required regardless of whether a DCS or a non-DCS approach is adopted. Finally, an example of advanced control using a DCS will be described.

A steam-driven turbine pump is generally used to move the crude oil from the crude storage tank to the main distillation tower. A governor adjusts the output of the turbine in response to the load on the pump. The operating principle of the governor is to adjust the flow of steam with a throttle valve to control the speed of the pump.

The crude is pumped continuously to the desalter, where inorganic salts are removed from the crude by chemical or electrical desalting. The separated salts are then withdrawn from the desalting tank according to a tank level indicator control (LIC-1) called a displacer.

Since the displacement-type level gauge is described in detail elsewhere in this volume, here we will only note that this type of gauge has many applications at petroleum plants; most commonly they are used as fieldmount indicating controllers and transmitters.

Of particular importance in the desalter unit are temperature control to assure complete separation of water and oil and pressure control (PIC-1) to assure the light fractions in crude aren't vaporized. After desalting, the crude is heated in tubular heaters up to around 350°C. The fuel for the heaters is regulated by temperature controllers (TIC-1) that use thermocouples mounted at the outlets of the heaters for the detecting elements. Generally a mixture of gas (methane, ethane) and fuel oil is used for the fuel; the gas facilitates combustion and the fuel oil controls the temperature.

Control of the combustion system is essentially the same as that used for a boiler system. When fuel oil is used for the fuel, a displacement-type flowmeter is used. Fuel oil has to be atomized for optimal combustion, which means that a differential pressure controller (PdIC-3) is required to maintain constant pressure. Optimizing the air input in the fuel is also important. Today, this is often done by measuring the oxygen content in the exhaust gas emission using a zirconia oxide analyzer (AI–1), and making any adjustment necessary with an air-flow compensation controller. Finally, as a safety precaution, the combustion is monitored with a flame detector coupled to a mechanism for shutting off the fuel supply if the flame goes out.

Crude oils are extremely varied in terms of the hydrocarbons they are composed of. They can nevertheless be turned into a diverse range of petroleum products suited to different applications as long as the crude can be separated into blends of specific boiling ranges. Referring again to Fig. 10.3, it can be seen how constant temperature is maintained at the top of the main tower by regulating the volume of side reflux returned to the tower with a temperature controller (TIC–2). The lightest fractions in the crude are withdrawn from the top of the tower and passed through a condenser. The non-condensable component of the gas is sent to a gas recovery unit and the naphtha fraction is sent to a stabilizer section by means of a level controller (LIC–2), where it is further separated to produce LPG (liquefied petroleum gas).

In the topping unit, the fractions are condensed in the order of boiling ranges. In the middle section of the tower, a constant volume of the fractional streams is withdrawn as regulated by each flow controller (FIC–4, 5, 6). In the side stripper, stripping steam controlled by steam flow controllers (FIC–8, 9, 10) strips the low-boiling point fractions. To assure good resolution of the fractions, side reflux is returned to the column. At the same time heat is recovered to heat the crude, flow control is maintained (FIC–1, 2, 3) based on constant load conditions in order to achieve good fractional distillation of the various product streams.

Low-volatility residue oil (*topped residue*) accumulates at the bottom of the topping unit, but the lighter components can be extracted from this bottom residue by blowing steam through it. Then, in order for the topped residue to be withdrawn and charged to the vacuum distillation unit and desulfurization unit, a control loop to eliminate any external disturbance on the down streams must be implemented. The conventional approach has been to use a nonlinear controller (*dead band*) for this purpose; within a given range, equilibrium would be maintained within the tower and the down streams would not be adversely affected by any external disturbance. Figure 10.3 shows a cascade control loop configuration linking a nonlinear controller (LIC–3) and a flow controller (FIC–7) for eliminating external disturbance.

Now we will consider some of the special control features that are available with modern digital control systems (DCSs).

(*a*) *Crude oil feedpump control*

Feedpump control has traditionally been maintained not by adjusting the pump rpm, which was held constant, but by controlling the load flow of the crude with a pressure control valve at the outlet of the pump. With this method, only the pump discharge pressure and consumption of steam has to be increased to compensate for the pressure loss of the pressure control valve. An alternative scheme that conserves energy is to leave the pressure control valve in the constant opening alone and regulate output by controlling pump rpm.

Figure 10.4 shows a design that enables both direct pressure control and rpm control. When the pump is first started and the load characteristic of the turbine is still poor, the pressure control valve is directly controlled by a loop consisting of the discharge pressure controller (PIC) and pressure control valve. Then, when the pump gets up sufficient speed for rpm control, the system switches over to rpm control through a steam flow control valve regulated by a cascade controller linking discharge pressure controller and rpm controller. This figure shows a loop for application to a continuous control system. In actual implementations, however, to prevent any external disturbance from being introduced by transfer lag on the control, it is combined with sequence and sampling control.

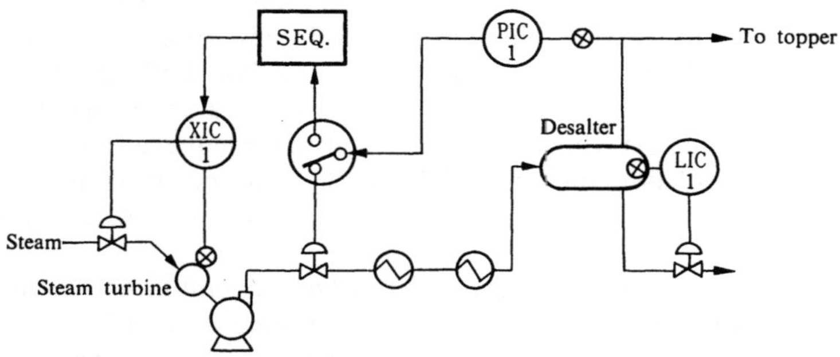

Fig. 10・4 Control diagram around crude oil feed pump

(*b*) *Heat control of the topping unit*

To bring about an enrichment of the vapor in the more volatile components than could otherwise be accomplished with a single distillation, the typical distillation tower is designed to return reflux to the tower at two to four points to countercurrently contact the rising vapors.

The heat load of the side reflux is the difference between the heat quantity of the feedstock entering the process and the heat quantity of the products leaving the process, and is used for preheating the incoming crude. In order to control the side reflux heat exchange, it is necessary to indirectly determine this exact heat quantity. Conventional analogue controllers are incapably of handling this type of task. With a distributed control system (DCS), however, the heat quantity can be easily calculated according a computational function, simulating the reaction taking place in the tower. It is not surprising, therefore, that DCSs are being increasingly applied to control panels for monitoring and adjusting the heat volume of distillation tower reboilers and heat exchangers.

Heat quantity control is discussed in detail in another contribution to this volume. Here we will be concerned only with the safety function that is implicit in such control.

A heat quantity control loop for a distillation tower is shown in Fig. 10.5. Here, TI represents temperature at the inlet and outlet of a heat exchanger, ΔTI is temperature differential, and QIC is a calorie controller. The difference in heat between the inlet and outlet of the heat exchanger is transferred to the crude on the shell side of the exchanger, thus maintaining constant reflux load in the tower. In this system, the heat quantity Q is given by

$$Q=\rho F\{H(T_1) - H(T_2)\}$$

where $H(T_1)$ is the enthalpy at the inlet, $H(T_2)$ is the enthalpy at the outlet, F is the flow rate, and ρ is the density.

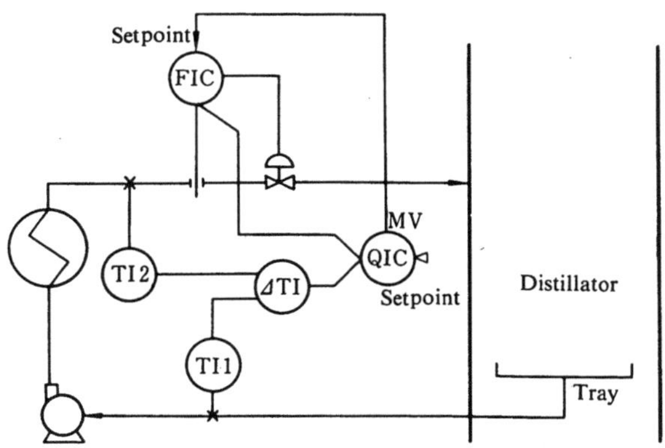

Fig. 10 · 5 Heat control diagram for side reflux of distillation tower

A calorie controller calculates the heat quantity based on signals detected at three points, and outputs compensation signals adjusting the setpoint value of the secondary reflux flow controller to control the side reflux load.

Even with a system as uncomplicated as this, if any one of the detection points ceases to function normally, control capability is lost. The DCS thus provides monitoring over the several detection points and safety interlock circuits that can be switched to maintain viable control in the event a detection point malfunctions. Figure 10.6 illustrates interlock for heat quantity control.

The interlock monitors input signal wires breaks (IOP) from two temperature gauges. If a break occurs on one of the input wires, the secondary side flow controller to reflux changes from cascade mode to automatic mode.

One additional interlock is provided for use in the event of operational error. Referring again to Fig. 10.6, when the calorie controller is set to manual mode, in this case too the secondary side flow controller changes from cascade mode to automatic mode. To avoid any adverse affect on the secondary side as a result of operational error, this interlock is applied at the controller for the final output terminal. These safety interlock features are more or less mandatory to assure the safety of the equipment and the reliability of the system as a whole. The control loop, therefore, should be given very careful consideration regardless of how complicated it becomes.

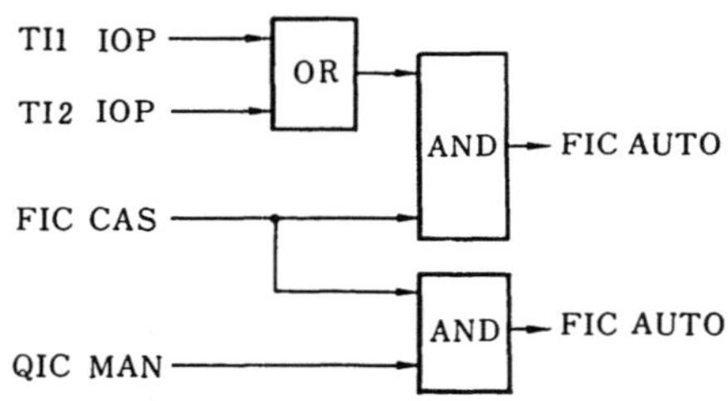

IOP : Breaking of input signal wire
CAS : Cascade mode
AUTO : Automatic mode
MAN : Manual mode

Fig. 10・6 Safety interlock logic diagram

(*c*) *Topping unit control incorporating analyzers*

In Subsec. 10.1.1, we described how the rapid progress in energy conservation has helped create the oil surplus we have today. Nevertheless demand for intermediate distillates such as gasoline, kerosene, and gas oil has continued to grow. This has led refineries around the world to convert their facilities to increase the output of those intermediate distillates for which there is such a strong demand.

One measure, for example, with the objective of boosting the recovery rate of designated topping unit fractions is to mount on-line analyzers to side stream outlets. To illustrate the principal, we will consider an analyzer mounted on a kerosene distillate line (refer to Fig. 10.7). The figure illustrates flash point control for kerosene. The flash

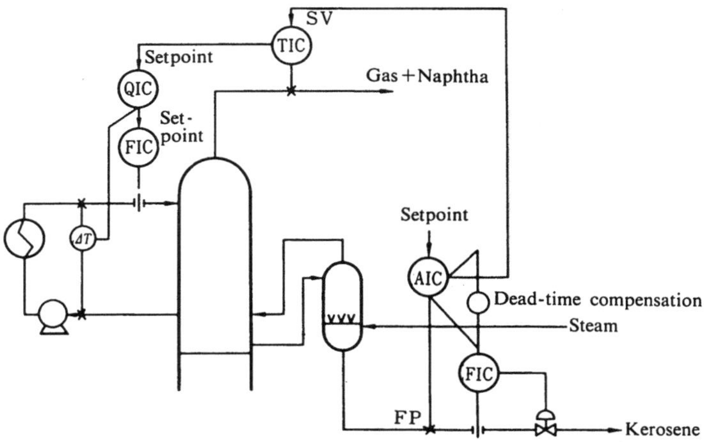

Fig. 10 · 7 Control diagram of kerosene on-line analyzer

point is the lowest temperature at which combustion begins and continues in a substance; the safe handling of petroleum products is obviously contingent on the constancy of this temperature value for a given product.

The conventional method for determining the properties of petroleum products is to take a sample of the product back to the laboratory and analyze it. Since the analysis had to be repeated several times over a day's time to assure the reliability of the results, a great expenditure of time was required to stabilize the process after any alteration of the running conditions or the kind of crude oil. On-line analyzers are extremely important for maintaining a continuous operation, and, in fact, they are even mandatory when modifying running conditions.

Recently, the durability and reliability of these analyzers have been enhanced to the point they can be used on-line. An example will serve to clarify how control is effected. When the kerosene flash point raises above the setpoint, this results in insufficient lighter fractions in the kerosene which in turn has the effect of reducing the setpoint of the temperature controller at the top of the tower. As the temperature controller at the top of the tower lowers the distillate temperature, this signals the need for more reflux to be returned to the kerosene. This raises the setpoint of the temperature controller, and the volume of reflux (FIC) is increased. Thus, as the temperature in the upper section of the tower is reduced, the temperature gradient within the tower becomes large. Since the temperature of the extracted kerosene distillate is lower, the distillate contains more lighter fractions.

Since the standard for kerosene involves limits on the heavy-side as well, about 95% of distillation point analyzers can also be employed for heavy-side control. The increasingly widespread use of on-line instruments, including pour-point analyzers and gas chromatograph analyzers in addition to the instruments already described, have made refinery facilities more efficient and have contributed to improved product quality.

10.1.3 Off-site instrumentation

In contrast to on-site, direct-production processes carried out at oil refineries (distillation, cracking, polymerization, etc.), supporting facilities such as handling and the physical distribution system of petroleum products is referred to as off-site.

Given the concerns of off-site instrumentation, it mainly involves processing and control of data accompanying the physical distribution. Systems configured out of panel instruments and process computers have been available in the past, but with the appearance of distributed control systems and high-performance mainframes, the trend is toward much more comprehensive management of data.

The production activity at refineries generally follows a monthly production plan from headquarters. It is only natural, however, that the actual production management structure of each individual refinery proceed in accordance with the local flow of production activity. The various activities that fall under off-site—the presence or absence of a marketing network, the blending of half-finished or finished products, land and maritime shipping, etc.—will vary greatly depending on distribution facilities and/or mechanisms that are available. Instrumentation methods are also quite different depending on the type of operation. Nevertheless, control of distribution data is the primary concern; increasingly the recent trend is toward consolidated systems that incorporate production management.

(*a*) *Operations control system*

The functions of this system are to control; production plan data, receiving and shipping plan data, quality control data, incoming/outgoing shipping capacity, tank storage capacity, etc. Tank storage capacity is especially critical and must be checked every two hours.

(*b*) *Operational control system* (*maritime blending*)

The main function of this system is to control blending operations aboard tankers while the product is being shipped. Five subsystems of the operational control system will be briefly described.

(1) *Order processing subsystem* This subsystem receives and processes information from external sources that is necessary for the operational control system. After the data is verified, it is compiled as a data file. The information contained in the data file is necessary for the smooth operation of all the other subsystems. A summary of the or-

Table 10.1 Order processing information

	Input data item
List for assignment of ships and handling of oil	Date, Product, and Quantity
Sailing schedule	Date, Customer, Retailer, Delivery area, Ship name, Product, and Quantity
Received oil properties	Quality control standard value
Production schedule	Load of equipment, Rundown, Target properties and schedule of bonded oil
Distillation schedule	Feed tank No., Rundown tank No., Feed amount and change-over time
Crude oil tank handling schedule	Crude oil tank operational plan
Maintenance work	Date, Work items and work period
Operational data	Feed amount to each equipment, Distillate amount and product
Others	

der processing information is presented in Table 10.1.

(2) *Handling schedule subsystem* The task of this subsystem is to determine the optimum tank configuration for distribution based forecasts in one-week units. The following factors are especially relevant: data received from the order processing subsystem, production plans, equipment operation plans, maintenance work plans, the inventory situation for different products, quality control data, etc.

The tank capacity projections for different products are derived from production volume data available on a per-day basis (volume of product produced from all operations, volume produced for each product). The weekly forecast is then calculated based on the one-day units.

Figure 10.8 shows the logic flow that is used to determine the optimum handling tank. Note that in the handling tank, or handling berth, logic there are a number of factors such as individual refinery rules and conditions, external conditions (*e.g.*, maritime conditions), and so on that vary tremendously and make it extremely difficult to drive an optimized calculation. Thus, human experience and judgment have been incorporated into an expert-system program (the so-called tank, berth selection support program) that is now available.

(3) *Operational schedule subsystem* Once the determination has been

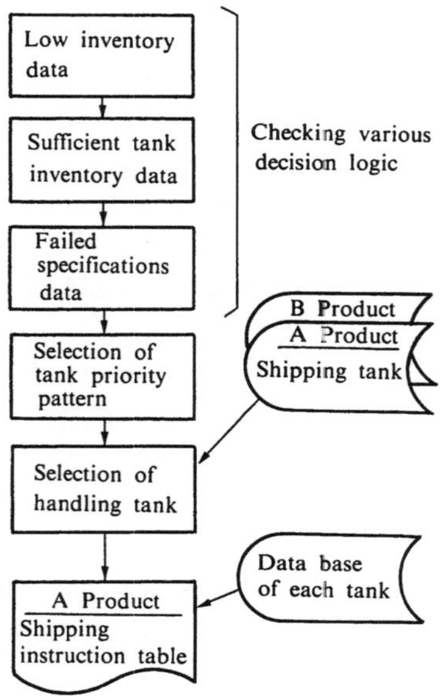

Fig. 10 · 8 Flow chart of handling tank

made in the handling schedule phase that the oil of a particular refinery is to be transported, the object of this system is to formulate a schedule either on the same day or by the next day, of oil moving work units. The two objectives of the logic—both realized by scheduling the minimum time needed to perform the various tasks involved—are to reduce the work time. This is done not only to conserve labor, but also to improve product quality by minimizing the time the product is exposed to contamination.

Once the work is actually under way, external elements often intervene making it necessary to depart from the original schedule. To cope with these inevitable alterations in the work schedule, the schedule is kept on an interactive basis through a CRT.

Specific scheduling tasks that fall under the purview of the operational schedule subsystem include loading, unloading, blending, transfer productivity, mixing, and scheduling documentation.

(4) *Tank-inventory control subsystem* This subsystem manages all data pertaining to tank inventories.

The system monitors the following areas; tank inventory calculations (current and revised inventory calculations), confirmed, cyclic, and effective tank inventories. This subsystem also maintains a tank capacity chart.

(5) *Quality control subsystem* The main function of this subsystem is

to handle sample testing (from the initial test request to the issuing of the test report) to establish the properties of each lot. Some of the areas covered include loading and unloading operations, oil moving between equipments and tank productions operations. This subsystem also performs inferential calculations of oil properties and ratio calculations to assist blending operations.

Figure 10.9 shows how the quality control subsystem is related to the other subsystems.

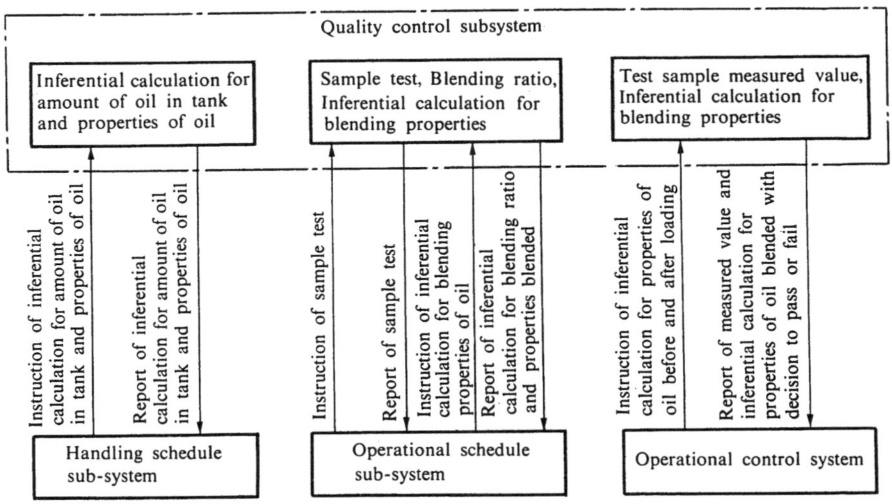

Fig. 10 · 9 Relationship between quality control subsystems and each subsystem at terminal

(c) Operational control system

The task of this system make sure the various operations are conducted safely and go smoothly based on operational booking information obtained from the upper-level operational management system.

Here we will describe two components of the system, automatic lineup and job control.

(1) *Automatic lineup* Lineup is the task of determining the optimum route to move the product from a point of origin to a point of destination.

Automatic lineup represents the off-site system as a network of pipelines and devices. As can be seen from the outline of automatic lineup shown in Fig. 10.10, the pipeline network is shown with device numbers and node numbers allocated to pipeline branching points. From a given starting point, a line is found to a node number. Assuming, for example, that our objective is to get from Tank 1 (the starting point)

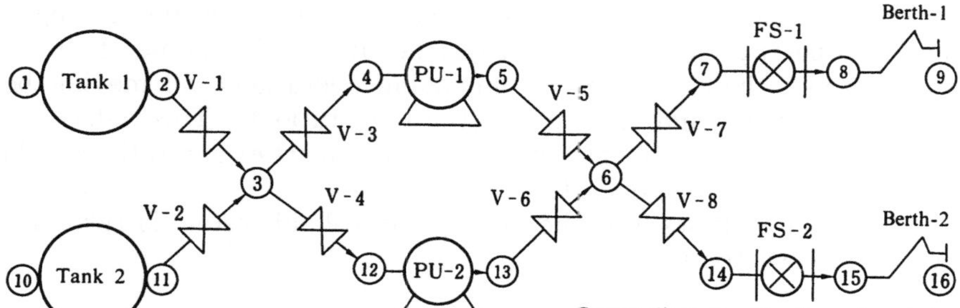

System diagram with allocated node Nos.

No, of paths = 2
Examples of selected lines

① → ② → ③ → ④ → ⑤ → ⑥ → ⑭ → ⑮ → ⑯

① → ② → ③ → ⑫ → ⑬ → ⑥ → ⑭ → ⑮ → ⑯

Fig. 10 · 10 Outline of automatic lineup

to Berth 2 (the destination point), one of two lines could be selected.

Then, the lineup conditions of the two possible routes are evaluated: to what extent are the pumping equipment and flowmeters already being used along the two lines? Is there a difference in capacity between the two lines? And so on. Finally, one line is selected based on the results of the evaluation.

(2) *JOB control* Another important off-site concern is moving oil around: shipment out of a tank, blending and moving the product to a holding tank, shifting oil from one tank to another. This type of operation that involves movement of oil from one place to another is referred to as an *oil handling operation.* The unit of oil handling operations is the *JOB*, which is specified by a sequentially assigned number (called a *JOB No.*). Every movement of oil is thus designated and handled according to a JOB No.

JOB control thus consists of centralized control over the string of yard devices (flowmeters, pumps, valves, etc.) that correspond to a selected automatic lineup covered by a given JOB No. JOB control also involves such functions as start and stop sequences, alarm monitoring and emergency shut-down procedures, and collection of data. A system diagram of JOB control is shown in Fig. 10.11.

Blending operations while en route aboard ocean-going tankers (maritime blending) calls for especially tight control. Today, even sulfur-content control (a feedback control process) and viscosity control, tasks that used to require conventional box-type meters, have become routine with the DDC and sequential functions that are available with

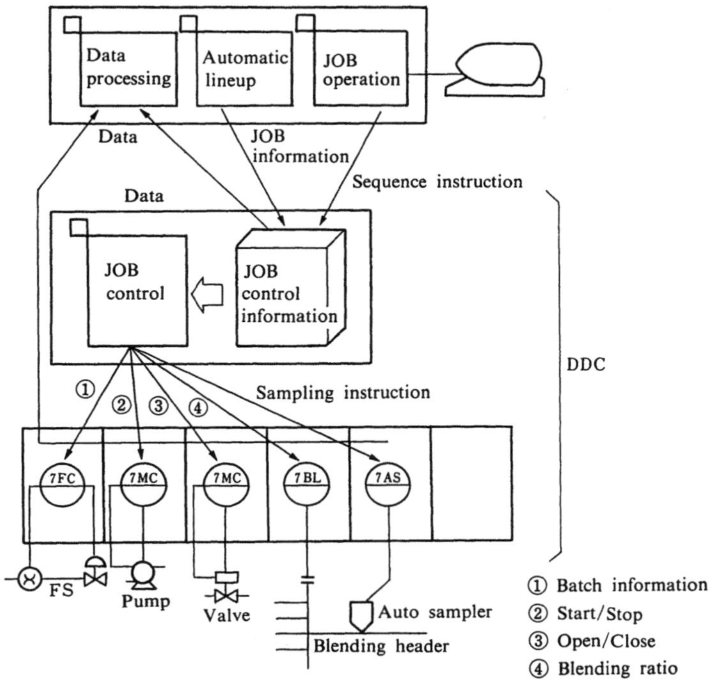

Fig. 10 · 11 System diagram of JOB control

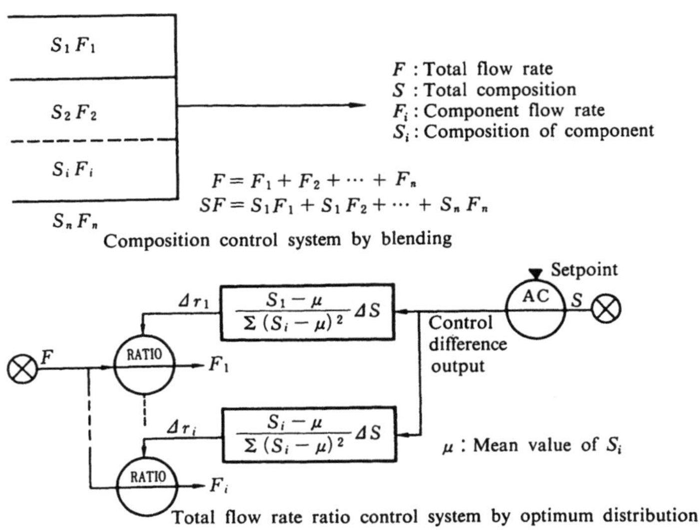

Fig. 10 · 12 Composition control by blending

distributed control systems.

Merely maintaining the sulfur content of the fuel oil at a fixed level is not enough; some version of the blend composition control scheme shown in Fig. 10.12 is necessary to obtain a stable, standard product.

In maritime blending, the DDC algorithm can be set not only for fixed–volume batch control but also, by adjusting the setpoint appropriately, for flow–rate control. Also there are two control algorithms; which algorithm is selected depends on the blending requirements of the particular lot under shipment.

The maritime blending control algorithm has a number of addition functions:

(1) temperature compensation (ASTM compensation, and primary and secondary compensation);
(2) non–reset integrator (uncompensated value, compensated value);
(3) emergency shut–down function in response to anomalies; and
(4) a simulation function.

The determination of specific gravity when making temperature compensation adjustments has especially been improved by making this a software function. The older semi–fixed method produced a mean specific gravity, but now it is possible to determine the specific gravity of every load at the time of shipment so the weight can be precisely regulated.

(*d*) *Land shipping control system*

This system maintains operational control of land shipping in fuel oil and LPG tanker trailers and rail tank cars.

Generally, there is a higher level order processing control system that passes contract information to the overland shipping control system. The specific functions of the overland shipping control system are then to issue dispatch slips, verify the orders, pass control information to the lower level control system, and compile data on the shipments. In land shipping, especially when long hauls are involved, rail tank cars are preferred because volume shipments are more economical. Oil companies generally have swap agreements in effect so they can exchange oil to save transportation expenses. The trend is away from shipment of oil in large–scale (20 k*l* capacity) truck trailers because the transportation costs of this mode are comparatively high.

Fuel–oil and LPG tank trailers are dispatched early in the morning and can accommodate only 4 k*l* (formerly, only 1 or 2 k*l*) compared to 20 k*l* for the large–scale tanker trailers. A fast, large–capacity computer is essential for this system to keep up with the frequent trips, the numerous control items involved, and the rapid accumulation of control data.

Since operating conditions change quite frequently, the operators must be left in charge of collecting and issuing dispatch slips. A general upgrade in system performance is strongly desired that would enable immediate execution of dispatch slips and a hatch control function for safety. In the overland shipping control system, as in the other systems described, safety measures must continue to be developed that are effective with fewer personnel. Basing operational information on feedback from the operators and interlock control are particularly important. Interlock control confines and minimizes the danger to shipping outlets and field depots in the event of a system failure.

In LPG tanker shipping systems, commodity weight has generally been determined by subtracting the tare weight from the actual weight. This method is being rapidly replaced by DDC weighing techniques, which are far more accurate.

10.2 Instrumentation Applications in the Iron- and Steel-Industry[7)–15)]

10.2.1 Overview of instrumentation in the iron- and steelmaking process

In periods of high growth, the steel industry has generally sought to achieve economic growth through production expansion by increasing the scale and speed of equipment, and through the introduction of continuous processes. The first oil crisis in 1973, however, caused a fundamental re-evaluation of this conventional wisdom. New demands emerged for conservation of resources, energy, and labor; for reduced costs and improved quality. These demands have served as a powerful impetus for the development of equipment that consumes less energy; for processes that are simpler, faster, and continuous; and for equipment that is direct coupled, faster, and that permits continuous processing.

Instrumentation has also been responsive to the changed environment and the new demands; as a result, a diverse range of instruments, sensors, and system devices has been developed. The point has been reached where the post-oil crisis objectives of lower costs and improved quality are all but unrealizable without the benefit of the instrumentation that is available today. In fact, it is no exaggeration to say that the reliability of instrumentation exerts a major influence on productivity. Two developments in particular have contributed to the elevation of instrumentation equipment to this important role: the emergence of digital control systems through the development of microcomputer technology and improved sensor technology. For example, digital instrumentation systems have been developed that satisfy operational needs by providing centralization of a diverse array of functions together with manipulation and supervision. It is not surprising that in recent years these systems have been adopted to most processes.

(*a*) *Overview of iron– and steelmaking*

Figure 10.13 shows the production flowchart for a typical iron– and steelmaking process. As is shown in the figure, the main process divisions are extraction of iron from iron ore in the ironmaking plant, production of steel and steel slab in the steelmaking plant, and production of various semifinished and finished steel products in the rolling plant.

(1) *Ironmaking plant* While the blast furnace is the central unit of an ironmaking plant, other equipment is also necessary because the ore must undergo various preliminary processing to make it more desirable feed for the blast furnace. For example, to equalize the quality of the feed ore and achieve stable operation of the blast furnace, a sintering plant is necessary to fuse or frit pulverized iron ore, limestone, and coke breeze together into uniform-size agglomerate called sintered ore or sinter. Also requisite is a coke oven. Coke produced in this unit serves not only as a thermal agent to promote fluxing of the iron ore and limestone but also is an indispensable reducing agent. Sintering and coke play a further important role in creating the optimum permeable burden which permits a high rate and also a uniform gas flow within

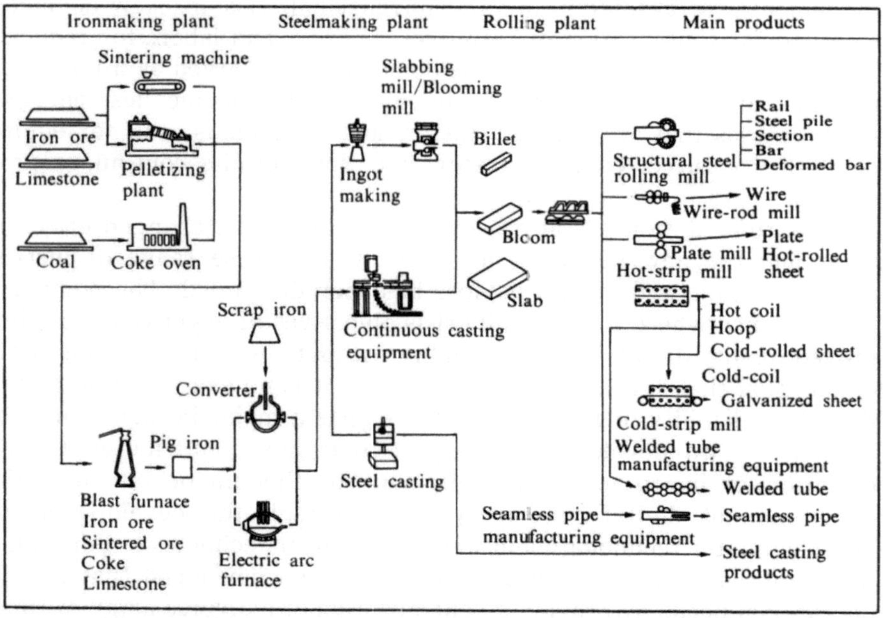

Fig. 10 · 13 An example for production flowchart of a typical iron– and steelmaking process

the blast furnace. In the blast furnace proper, the pre-processed sintered ore and coke are charged at the top of the furnace along with such auxiliary material as limestone. Hot-air blast introduced from the tuyeres burns the coke to heat, reduce, and melt the charge as it descends to produce pig iron.

With a high carbon and impurities content, pig iron has a hard, brittle characteristic. For this reason, most pig iron is sent on to the steelmaking plant to be processed as steel. Finally it should be noted that the blast furnace gas (BFG) and coke oven gas (COG) are valuable heat sources that are reused in the plant.

(2) *Steelmaking plant* In the steelmaking process, the carbon content is reduced and the impurities are removed from the pig iron to produce steel, a material that is malleable yet strong and can be readily worked. The center unit in the steelmaking process is the converter. Traditionally, top-blowing oxygen was generally used in the converter, but more recently bottom-air-blown and combination blowing (top and bottom) converters have been developed. The combination-blowing converter produces a homogenous-quality steel very quickly with a low level of impurities, and has thus come to be widely employed in the industry. Gas generated in the converter during blowing is recovered and is employed as a valuable heat source elsewhere in the plant.

The liquid steel produced in the previous process step then undergoes secondary refining to achieve the precise qualitative requirements of the finished product and also to reduce the amounts of hydrogen, oxygen, and nitrogen gases which have an adverse effect on product quality. The two principal secondary refining methods are vacuum and injection; which method is used normally depends on how the finished steel is to be used.

In the next step, the liquid steel is teemed into molds as ingots or continuously cast to form semifinished shapes (*e.g.*, slabs, blooms, or billets). In recent years, more and more use is being made of continuous casting because of its advantages over conventional casting into ingots. These advantages include simplified processing, reduced investment in equipment (the need for heavy rolling-mill equipment is eliminated), conservation of energy, and increased yield. In Japan, for instance, the ratio of steel tonnage processed by the continuous slab-cast method has reached 90%.

The most recent advance in process technology is a method to bypass the reheating furnace altogether so that semifinished slabs can be sent directly from the continuous casting machine to the rolling plant.

(3) *Rolling plant* Rolling mills can be divided into two fundamental types: hot-strip mills in which the hot-roll processing occurs at a high temperature and under pressure, and cold-reduction and temper mills

in which the cold-roll processing is carried out at normal temperatures (structural changes occur at temperatures below about 720°C). In this area too, rapid advances have been made in the development of rolling-mill processes that are direct-coupled, faster, and continuous. The procedure of processing one coil at a time has been largely superseded by continuous rolling and continuous annealing equipment, and this has had a very beneficial effect on quality and yield.

(*b*) *The trend in instrumentation equipment*

Iron-and steelmaking processes are fundamentally batch or semi-continuous processes. Research nevertheless continues in an effort to simplify and speed up processes in order to conserve energy and labor. Especially in recent years, in response to the demand for small-lot production of diverse products, not only the enhancement of equipment and operation technologies but also the development of advanced instrumentation equipment such as process computers, digital instrumentation systems, and sensors has come to assume an extremely important role in the iron and steel industry. Also, as communications technologies have become more sophisticated, this has spurred the trend toward increased integration and centralization of discrete equipment units. Integrated control systems based on optical communications and other advanced technologies have also continued to grow in importance in last few years.

10.2.2 Blast furnace instrumentation

A blast furnace is a tall refractory-lined furnace in which iron ore is indirectly reduced by coke to produce pig iron. Iron ore, sinter, coke, and other auxiliary materials are charged at the top of the furnace, and hot-air blast introduced through tuyeres burns the coke. CO gas given off by the burned coke reduces iron oxides (*e.g.*, FeO) to produce pig iron, which is removed through a taphole. Impurities contained in the ore are removed as slag through slag holes.

The blast furnace is often regarded as a symbol of iron- and steelworks. The trend has been toward larger furnaces, and in recent years, a number of giant blast furnaces have been constructed with capacities in excess of 5 000 cubic meters. Figure 10.14 shows the system layout of a blast furnace and auxiliary equipment.

As in other industries, the principal roles of instrumentation in the iron and steel industry are equipment maintenance, stabilization and supervision of operations, and data collection.

The overall instrumentation system of a modern steelworks can comprise as many as 700 to 800 loops, but, of these, at least 80% are control loops for supervisory control. In the case of conventional analogue instrumentation, the instrument panel must be capable of super-

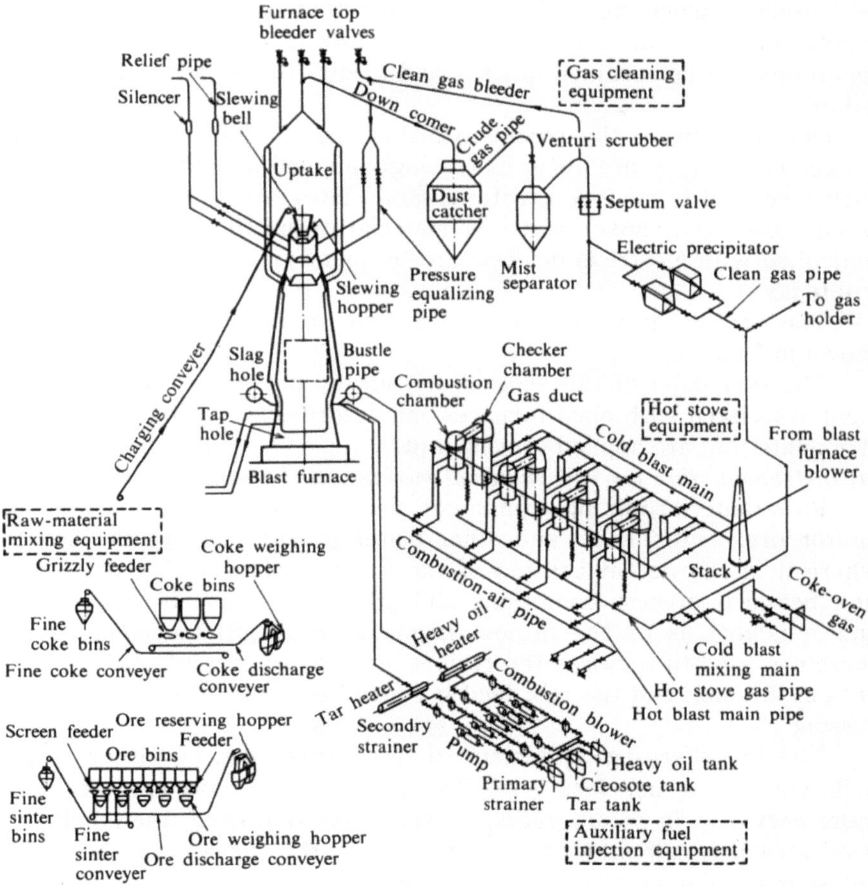

Fig. 10 · 14 System layout of blast furnace equipment

vising the entire loop in order for the operator to maintain a comprehensive grasp of the work situation. As a result, the instrument panel for such systems generally ranges from 10 to 15 meters in width which must be accommodated in a fairly large control room and attended by numerous personnel. With the rapid development of digital control instruments, operator–system interaction is carried on through a CRT; essential data respecting the work can be quickly and easily brought up on a CRT and acted upon. This type of arrangement needs fewer operators and thus the control room can be reduced in size.

Once a problem or accident has occurred within a blast furnace, major repercussions are felt by all other processes. Thus, the importance of maintaining a safe and stable blast furnace operation cannot be overemphasized. Redundant implementation of all critical loops is mandatory. An uninterruptible power supply is also essential so that operations can be resumed quickly and safely in the event of a power failure.

One outcome of the recent availability of digital instrumentation devices is that they are being increasingly applied to continuous and sequence control to achieve greater organic integration. Sequences such as hot-stove switchover and raw-material outputting, conventionally controlled with electrical devices, are frequently put under digital control today.

The configuration of a typical digital instrumentation system is shown in Fig. 10.15.

The remainder of this section will survey the main types of equipment associated with blast furnaces and describe the various control instruments that are applied to each unit.

(*a*) *Instrumentation for raw-material mixing and charging equipment*

Raw-material mixing equipment controls the mixing and feeding of iron ore, sinter, coke, and other materials and conveys the mix to where it is charged at the top of the furnace. Considering the direct link between process conditions and product quality, high precision and reliability is of the utmost importance. A process computer is therefore used to regulate the recipes and proportions of the constituents in the mix, and the mix feed is controlled through sequential processing.

The top-charging equipment of the furnace can be either of the bell type or the bell-less type. The bell-less type has become much more pervasive in recent years, because with that type, the stock-line level at which raw material is charged into the furnace can be freely set at any point or thickness. The bell-less type is also considerably lighter than the bell type. In terms of instrumentation, however, the requirements of the two types are almost identical. One difference is that in the case of the bell-less type the seal valve is susceptible to damage from the raw material if the valve isn't closed after all the material is discharged from the hopper. With this design, therefore, a sensor is mounted on the hopper to detect load cutoff.

(*b*) *Blast furnace instrumentation*

Blast furnaces are basically cylindrical vessels with a slight pear-like bulge about one-third up from the bottom. Thermal expansion of the material causes it to be pushed out against the walls of the interior of the furnace with considerable force; this, coupled with the bulge in the lower central section of the furnace, prevents the material from sliding down inside the furnace. As iron is formed, the material shrinks greatly in volume. For this reason, the bottom section of the

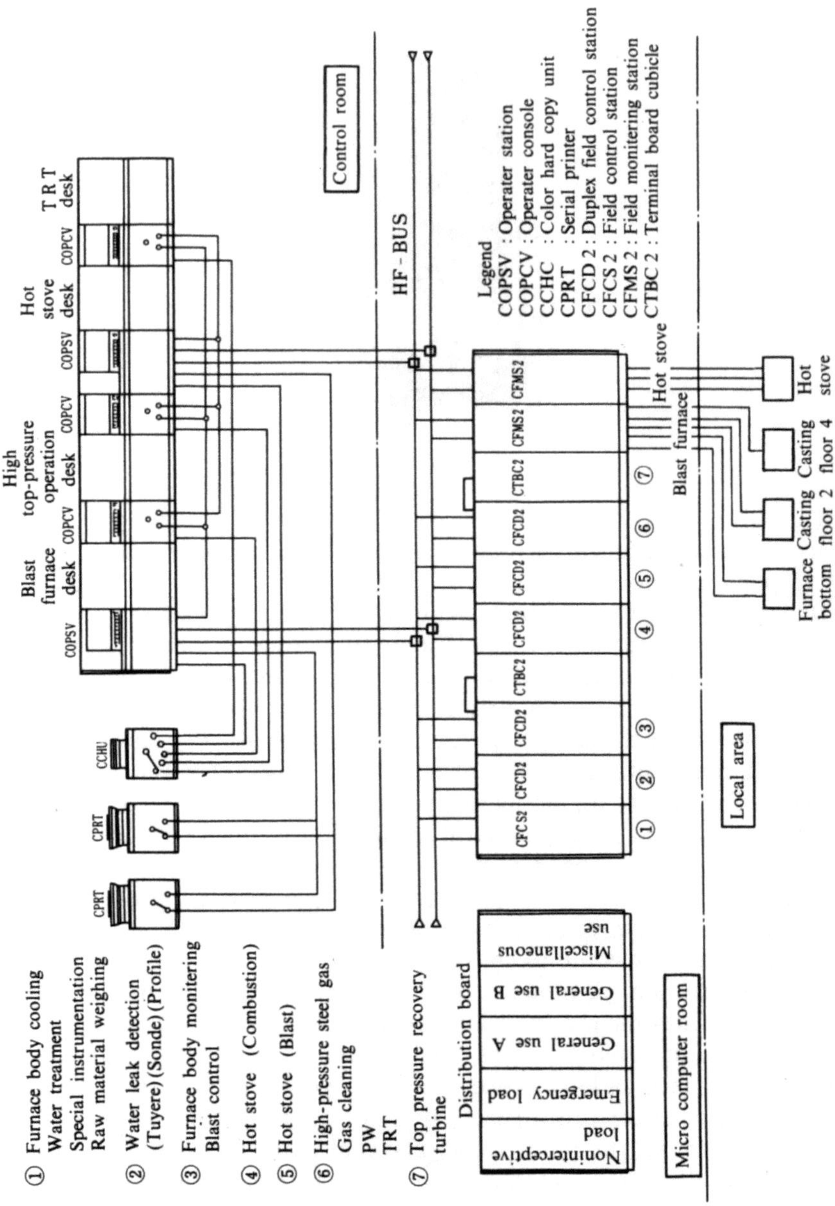

Fig. 10·15 Typical configration of blast furnace instrumentation system

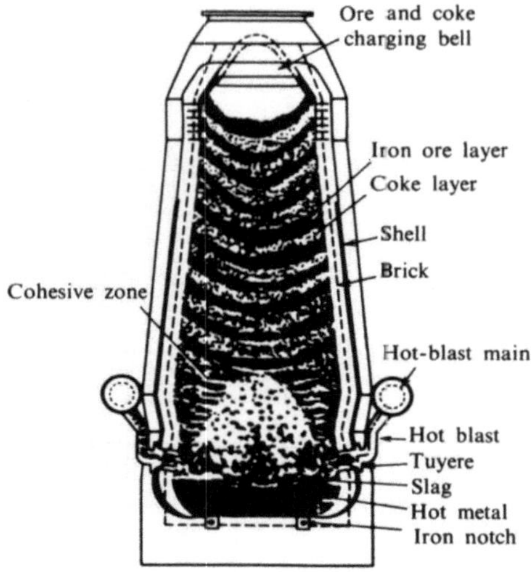

Fig. 10 · 16 Shape of blast furnace and reaction status inside

furnace is made narrower. Figure 10.16 shows a cutaway diagram of a blast furnace in reaction.

Due to the extremely high temperatures and pressures that are reached inside the furnace, it is difficult to control the reaction directly. Thus, indirect monitoring is used by regulating certain parameters and variables. The more important of these variables would include: (a) composition and weight of the raw material; (b) temperature, weight and composition of the hot metal and slag; (c) volume of blast; (d) pressure of blast; (e) temperature of the hot blast; (f) pressure at the top of the furnace; (g) temperature inside the furnace and temperature distribution; (h) upper furnace gas flow and composition. While this describes the situation up to present, we would note that new instrumentation techniques and special sensors are rapidly being developed that make it possible to control directly more and more aspects of the furnace reaction. A modern blast furnace is depicted in Fig. 10.17 showing the locations and functions of the most important sensors.

Recall that hot blast enters the furnace through openings at the base of the furnace called tuyeres. If the tuyeres malfunction allowing cooling water to enter the furnace, it is not only the furnace that is adversely affected. In the worst case, a steam explosion could result endangering the lives of personnel. To make sure such a scenario does not actually occur, it is necessary to detect for tuyere cooling water leaks. A number of conventional methods of detecting for tuyere leaks are available including:

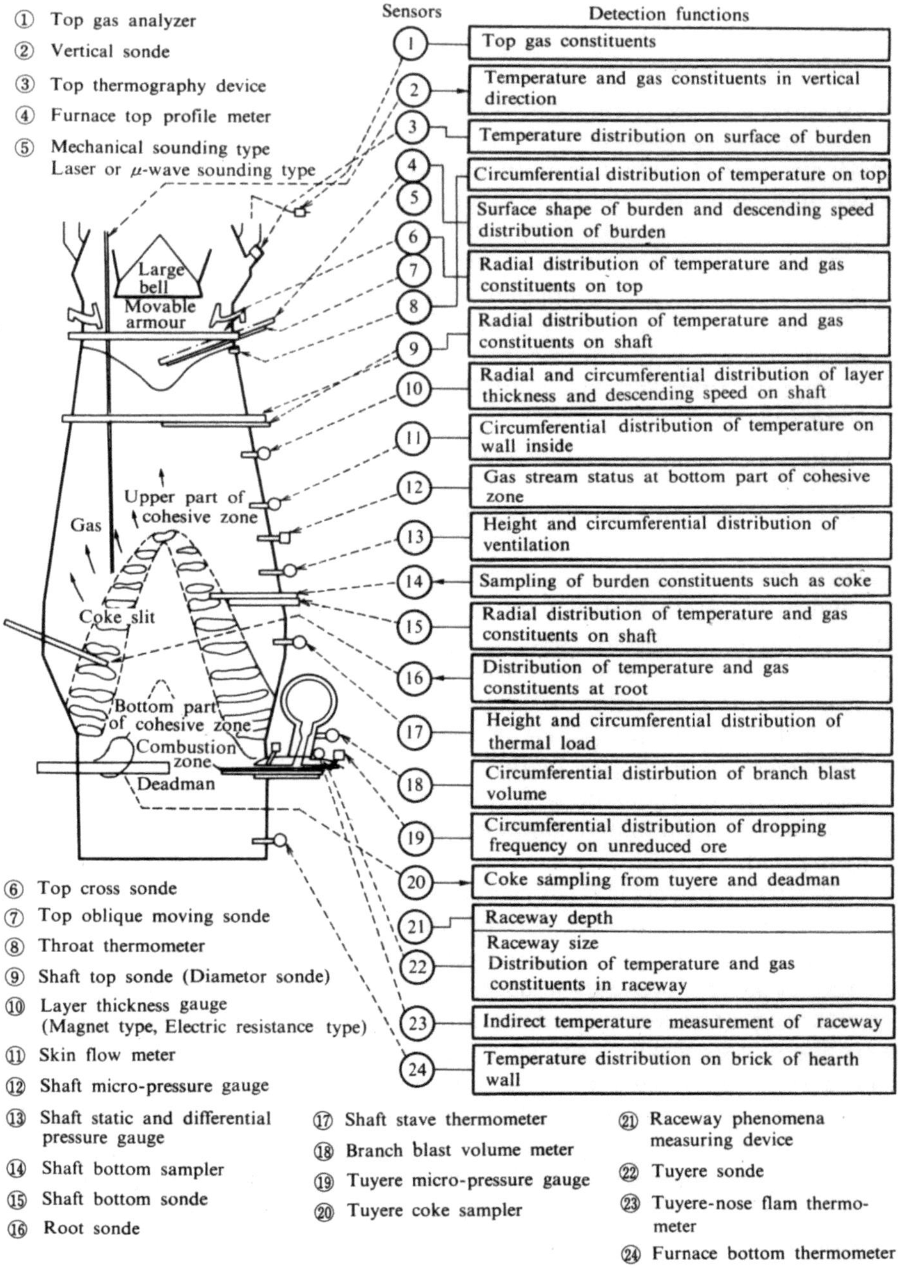

Fig. 10 · 17 Functions and locations of blast furnace sensors

(1) monitoring the temperature of the tuyere cooling water outlet;
(2) monitoring H_2 with a gas analyzer at the top of the furnace;
(3) monitoring CO and CO_2 gas in the tuyere cooling water outlet; and
(4) relying on the judgment of an experienced operator.

The basic problems that are shared by all of these conventional methods is that they are indirect, they involve a certain amount of dead time, and they yield less than satisfactory detection accuracy. An alternative direct method is to continuously compare the volume of supply water against the volume of discharge water. An advantage of this method is that it can be applied to larger furnaces with more tuyeres that require a faster cooling–water flow rate to satisfy more stringent cooling efficiency requirements. Using an electromagnetic flowmeter for the detecting element, digital instrumentation can provide the following functions:

(1) automatic compensation and control of the flowmeter;
(2) alarm supervision of actual and differential flows;
(3) collection and recording of actual and differential flow data.

To assure a safe operation, the temperature distribution inside the furnace must be closely monitored. This is done with numerous thermocouples mounted on the inner walls of the furnace. The large–scale blast furnaces of recent years are equipped with anywhere from 300 to 500 sensors which are supervised by a remote scanner.

(c) Gas–cleaning equipment instrumentation

Blast furnace gas (BFG) is captured so it can be used as a heat source elsewhere in the facility. However, BFG is loaded with dust which has to be removed before it can be used. Here, a brief description of some of the more common kinds of gas–cleaning equipment will be provided.

A dust catcher is a device that separates out the courser–sized dust particles by reducing the gas flow rate. Venturi scrubbers, usually configured in pairs, are gas–cleaning devices in which liquid injected at the throat of a venturi is used to absorb the dust from the gas flowing through the venturi. A septum valve situated between the two scrubbers is generally used as a final control element for pressure control at the top of the furnace, but septum valves are extremely noisy when they open and close. Ring–slit scrubbers are being substituted for septum valves in many contexts because they are so much quieter. Ring–slit scrubbers are generally installed between two venturi scrubbers. Regardless of the configuration, a hydraulic drive system is used for the actuator.

For venturi scrubbers, the main instrumentation function is water–level control. This is usually accomplished in one of two ways: either with a gas-seal function or by monitoring scrubbing discharge. In the past, the latter method had been much more common due to the harsh environment a level-detecting element has to operate in. Recently, however, a differential pressure transmitter with diaphragm seal has become available that offers a method of direct detection. These elements are mounted in pairs for redundancy with the ability to switch over to the second element if the first element fails. The actuator has to be designed so it is immune from clogging; usually, a butterfly valve is used. In this environment, measures to enhance abrasion resistance are mandatory given the extreme abrasion from dust and slag. Here too, a hydraulic drive type actuator is used.

(*d*) *Instrumentation for high top-pressure operation equipment*

It is obvious that pressure must be controlled at the top of blast furnaces to maintain a safe, stable operation. Most modern blast furnaces operate at high pressure in order to realize two objectives:

(1) The volume of hot-air blast can be increased without increasing the gas flow rate. This reduces the volume of dust particles and helps to prevents *hanging* and *channeling* from occurring in the furnace. (*Hanging* describes a condition when the raw material in the upper part of the furnace becomes stuck and only the material in the bottom portion of the furnace descends. *Channeling* refers to a state in which the raw material collapses and the gas blows past it.)

(2) Operating at high pressure also makes it possible to slow the gas flow rate, thus prolonging the contact between gas and ore; this improves the overall reaction in the furnace and enables the ratio of coke to be reduced.

An essential instrumentation task is top pressure control. The loop that performs this function is the most important in blast furnace operation. Considering its importance, the top pressure detection element is usually mounted in duplicate or triplicate for redundancy with the ability to switch between elements. Moreover, since so much dust is present in the upper part of the furnace, back pressure is measured based on N_2 which is intermittently purged from the vicinity of the detecting element.

The control element for top pressure control consists of a septum valve and a ring-slit scrubber. More recently, in an effort to conserve energy, it has become quite common to install equipment to recover energy from the furnace gas. The equipment is configured in such a way that energy recovery and pressure control are accomplished at the same time.

Given the instrumentation available today, it is rare for abnormal pressure to develop in blast furnaces. In the event pressure does reach a critical level (*i.e.*, due to operator error, control equipment failure, and so on), it can be quickly dissipated via a bleeder valve to prevent the furnace from sustaining any damage.

Since the hopper and furnace top pressures must be equalized, raw material is charged to the hopper in batches.

(*e*) *Hot stove equipment instrumentation*

In order to achieve fast and effective combustion in blast furnaces, great amounts of hot gas are required. This requirement is met with hot stoves, a specially designed type of heat exchanger. Heat is stored in bricks by application of combustion gases. After the combustion is turned off, cold blast is brought into direct contact with the preheated brick, and heat is transferred to produce hot blast. A typical blast furnace is supported by 3 to 4 such stoves, which are rotated (*e.g.*, heated for a period of 3 hr and *on-wind*, or exhausting for 1 hr) to provide hot air to the furnace on a continuous basis.

Hot stove instrumentation has three main functions; switching system control to regulate the alternating heating and blast phases of the several stove units, combustion temperature control for the heating phase, and blast temperature control for the on–wind, or blast phase.

The hot stove instrumentation functions are shown schematically in Fig. 10.18.

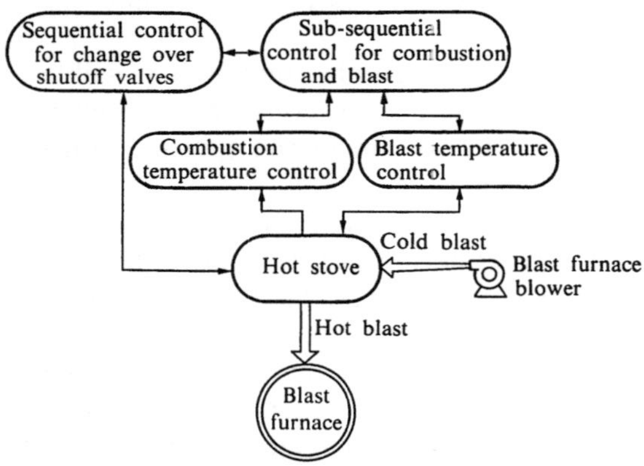

Fig. 10 · 18 Hot stove instrumentation functions

Combustion temperature control can be broken down into fuel gas calorie control and stove dome temperature control. Combustion control consists of (a) combustion gas flow control to regulate applied heat, and (b) air/fuel ratio control to regulate combustion temperature. The temperature of the stove dome can be controlled by regulating the above two factors. To optimize the combustion in order to conserve fuel, it is essential that the exact proportion of air is delivered to the fuel mixture for perfect combustion. This can be controlled by sensing the amount of oxygen (O_2) in the exhaust gas. Moreover, as stoves approach their target heat storage temperature, exhaust gas temperature increases, which is also a waste of fuel. Thus, exhaust gas temperature is monitored to maintain efficient combustion temperature control. For dome temperature detection elements, the most common arrangement is to mount two thermocouples (one is a spare) and a radiation pyrometer with switchover capacity between elements. Zirconia oxide analyzers are generally used as oxygen analyzers for the exhaust gas.

Blast temperature control is based on a signal from a temperature sensor mounted at the end of the hot-blast main pipe. Based on the signal output, hot blast is mixed with cold blast to produce the required blast temperature. There are different schemes for providing the blast to the furnace, but two common methods are: the single blast method where the blast is delivered by a single stove, and the staggered parallel method in which the blast is delivered by two stoves at a time on a cyclic or staggered basis. Controlling blast temperature is extremely important for blast furnace operation. A spare blast temperature sensor is therefore mounted along with the active sensor with switchover capability.

A change-over sequence controls a system of shutoff valves to regulate alternating steady-state periods of heating, blast, and rest. It also controls the periodic transitions between these steady states. Until recently, this control was maintained by electric equipment, but now instrumentation configurations are more common.

To monitor stove temperature, a typical stove is equipped with 40 to 50 sensors mounted on the shell and throughout the stove. The sensors are then supervised with a remote temperature scanner.

(f) Furnace cooling equipment instrumentation

This system is necessary to prolong the life of the refractory brick used to line the blast furnace. Two types of systems are encountered; the plate cooler and the stave cooler methods. With the migration toward higher pressure operation, the stave cooler method has come to predominate. With this method, steel tubes cast into cast iron are installed in the shell to cool the furnace. Either natural-flow or forced circulation evaporative cooling is piped through the tubes. Forced circulation systems are far more common today.

Instrumentation regulates a number of important variables in the connection between the deaerator storage tank and the stave drum. For example, monitoring the characteristics of the pure water used as coolant is particularly important; dissolved ion density and pH of the cooling water are monitored, alkalinity of the coolant is maintained to prevent tube and stave corrosion. Staves are vertical strips made of refractory brick that are placed edge to edge to form the lining of the furnace. A float switch is mounted near the top of each vertical stave system to detect for either damage to the staves or disrupted circulation of coolant.

(*g*) *Auxiliary fuel injection equipment instrumentation*

The principal fuel used in blast furnaces is coke. To reduce the proportion of coke, however, it is generally supplemented with other fuels. In the past, heavy oil and tar were commonly used as supplementary fuels, but since the oil crisis, every effort has been made to phase out petroleum-based fuels and substitute coal-based ones. Injection of pulverized coal has become the most common method in recent years.

Operation of pulverized coal injection (PCI) requires facilities to stock the coal, pulverized it, store the pulverized coal, and finally inject it. The pulverized coal enters the blast furnace by being injected through the tuyeres.

One thing that must be carefully considered in the instrumentation for the PCI is the prevention of explosions. Since the sensors must be in direct contact with the pulverized coal, they are mounted inside the pulverizing equipment. Control elements such as shut-off valves and control valves that come in contact with the medium carrying the pulverized coal must be designed to withstand extreme levels of abrasion. Moreover, while pulverized coal must be continuously injected into the furnace, other units in the overall operation are often run on a batch-cycle basis. Thus, the pulverized coal injection must be organically integrated with sequence control and continuous control.

(*h*) *Furnace blower equipment instrumentation*

Shutting down the blower is equivalent to shutting down the furnace, and therefore an auxiliary blower is always standing by. For a single-furnace operation, two blowers are configured, and for a two-furnace facility, three blowers are usually provided. Since blast furnaces require a stable supply of blast, turboblowers are equipped with instrumentation to maintain a constant volume of blast for normal operation. When switching between stoves, constant blast pressure control is applied to maintain the delivery pressure and assure a constant amount of blast is delivered to the furnace. Moreover, since the pressure and volume of delivered blast changes depending on the furnace load, a centrifugal compressor is used, which introduces the danger of surging. Surging prevention control has to be implemented to assure this will not occur.

To maintain a constant volume of blast, flow sensors are mounted on the gas main that delivers blast to the furnace. Flow control is effected by calculating a nominal flow value and then making any adjustments in temperature and pressure that are necessary to make the actual flow correspond to the nominal value. Final control is maintained by adjusting the angle of the stator blade on the turboblower. The controller employs an external feedback mechanism to regulate constant blast pressure.

Constant blast pressure control is automatically triggered by a signal that a particular stove is *on-wind*. The constant blast pressure control uses the discharge pressure value immediately prior to switchover as the setpoint value. Then, when a signal is received that a stove is about to go *off-wind*, control is switched back to constant blast volume control. The controller is based on external feedback (*i.e.*, when constant blast volume control is on).

Surging is another problem associated with blowers. Surging characteristics of each blower are determined so surge prevention can be designed into the lines. For control, a blow-off valve is opened to keep from entering the surging zone, which can be determined by the relation between measured discharge pressure and intake blast. If the controller indicates a large deviation from normal, the control output signal to close the blow-off valve overshoots. Since the control output signal must react quickly in the event the control zone is entered, the controller is equipped with an anti-reset function.

While this section has covered the fundamentals of blast furnace instrumentation, there are aspects of the reaction mechanism that are not fully understood even today. Blast furnace technology in the future will evolve toward a more competitive process that is attuned to the natural resources situation, environmental impact, and economic conditions. It goes without saying that the role of instrumentation in this development is extremely important. Even greater efforts must be applied toward the development of more sophisticated instrumentation systems and sensors.

10.2.3 Continuous casting equipment instrumentation

(*a*) *Overview of the equipment*

Referring again to Fig. 10.13, after the pig iron (hot-metal) is tapped from the blast furnace, refined in a converter and then conveyed by ladle to a continuous casting machine (an alternative route to the continuous casting equipment is via an electric furnace, as shown in the figure). The continuous casting equipment forms the molten steel into semi-finished shapes, *e.g.*, slabs, blooms, or billets. This is done, as the name says, on a continuous basis.

Figure 10.19 shows an overview of a continuous casting machine. Molten steel from the ladle is poured via a tundish into a water-cooled copper mold. Once the molten steel has been poured from ladle to tundish, it is teemed into the mold by means of an immersion nozzle. In a

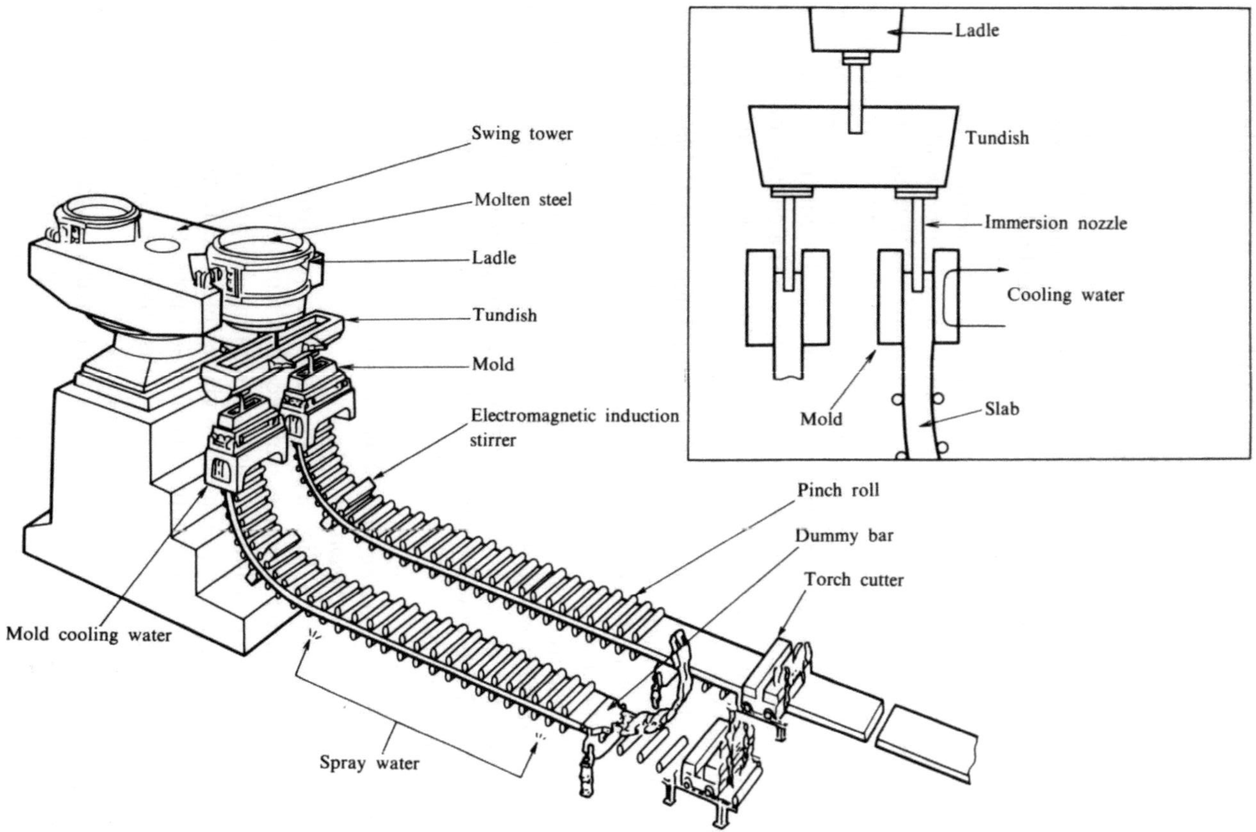

Fig. 10 · 19 Outline of continuous casting machine

continuous casting operation, the tundish serves a dual role; it serves as a buffer during cut-over to the next full ladle and also aligns precisely over the next mold before pouring. The inset in Figure 10.19 shows a tundish with two nozzles; this arrangement is for a two-strand continuous-casting machine. Pure copper is the material of choice for molds because of its good heat-conducting property. Water flows into the mold cooling the molten steel so that a solidified shell forms on the surface. Secondary cooling water then solidifies the rest of the steel. The solidified cast product is then continuously removed from the molds through a pinch-roll system, cut into appropriate lengths at a cutting station, and transported to the rolling mill.

In recent years, continuous casting has been applied to virtually all types and grades of steel. Even more recently—applying newer high-speed, high-temperature, defect-free continuous casting technologies—dramatic advances have been made in new continuous process technologies such as hot charge rolling (HCR), which makes efficient use of sensible heat and sends the high-temperature slab directly to the reheating furnace, and hot direct rolling (HDR), which does away with the need for reheating. Ultimately, these developments are expected to lead to a sheet casting method that greatly simplifies or eliminates intermediate rough-rolling processes.

(*b*) *Aspects of the instrumentation system*

Continuous casting control systems are basically well established, but new techniques have been developed in the areas of high-temperature slab technology for HCR and HDR (*e.g.*, high-speed casting and vapor cooling), and to realize better quality output (*e.g.*, electromagnetic induction stirrer and electromagnetic breaker). Also, while it is quite common to use process computers for continuous casting control systems, as supervisory functions have continued to multiply, the trend is to shift as many control functions to the instrumentation as possible. Instrumentation systems have also become increasingly complex. If a function such as spray cooling tracking (described in detail below) were given over to control by instrumentation, for example, such an implementation would require an operation data base and high-level general purpose language functions.

With respect to system configuration, strands are recognized as separate units with control stations set up for each strand. The most common arrangement is for shared control functions to be processed by the independent control stations. Moreover, although there are more billet strands than slab strands in operation, since the spray-cooling equipment for slabs is heavier than for billets, more control loops are required per strand. One final point must be noted. if instrumentation is going to be used to control such functions as cutting, mold width alteration, or equipment diagnosis (*e.g.*, break-out prediction, mold oscillation monitoring, etc.), so-called digital instrumentation control systems are inadequate by themselves because they lack sufficient calculating speed and data memory capacity. For such applications, an integrated computer system such as YEWMAC or YEWCOM developed by Yokogawa Electric Corporation is mandatory.

(*c*) *Instrumentation overview*

An overview of the main continuous casting process functions is shown in Fig. 10.20. Here we will be primarily concerned with mold molten steel level control and spray cooling water control.

(1) *Mold molten steel level control* The quality of a continuous casting product is largely determined at the mold; especially to reduce the amounts of harmful nonmetallic inclusions, it is extremely important that the molten steel be maintained at a constant level. Quite a variety of molten steel level meters have been developed for this purpose, as shown in Table 10.2. In recent years, however, the general trend has been to move away from radiation type meters toward electromagnetic (esp., detecting coil) type meters.

To control the level of the molten steel, while the volume of metal teemed into the mold is controlled by the stopper-rod nozzle or sliding-gate nozzle of the tundish, the cast slab has to be withdrawn from the mold at a varying rate. However, since the optimum withdrawal speed depends on the type of steel being cast and the dimensions, it is not desirable from the standpoint of level control that the withdrawal speed be frequently altered. This is, of course, assuming the withdrawal speed change is used in the level control function as the feedforward value.

Operator intervention has necessarily been involved at the beginning and end of a pour, but some modern implementations have even included autostart (flow-in) and autostop (flow-out) functions in the instrumentation system. Especially with respect to autostart, however, the detection range of the molten-steel level meters that are available today is still inadequate, and thus there is room for improvement in this area.

(2) *Spray cooling water control* Once the molten steel has been teemed into the mold, a solidified shell forms on the mold surface. By the time the slab has passed through the spray (secondary) cooling zone, it is completely solidified. The objective of mold cooling control therefore is to make sure sufficient layers, or shells, of molten metal have solidified and the slab formed by the time it reaches the pinch-roll. The task of the spray cooling water control is even more critical; assurance that the still partially molten slab is completely solidified. The setpoint of the volume of spray cooling water is set in pro-

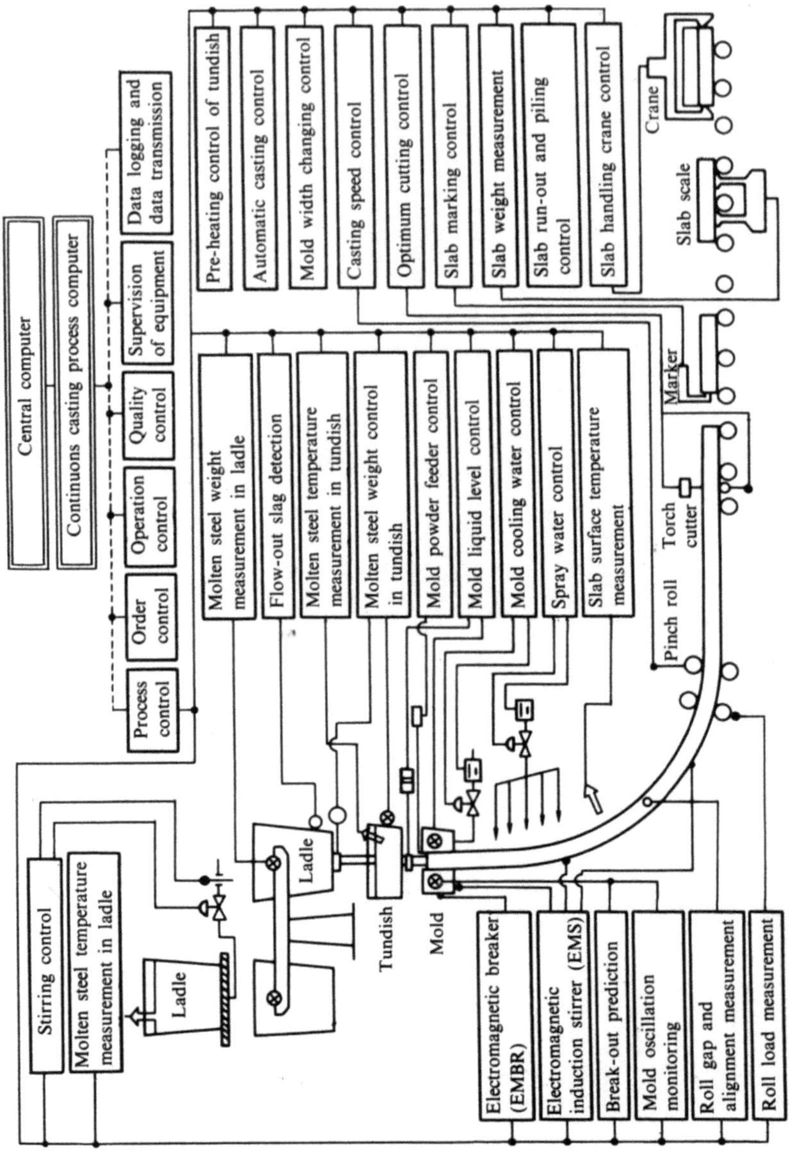

Fig. 10 · 20 Overview of continuous casting process functions
(Sato, et al, Kawasaki Steel Giho Vol.14, No.3, 1982)

Table 10.2 Comparison of different types of molten steel level meters

Electromagnetic type	Fixed detecting coil type	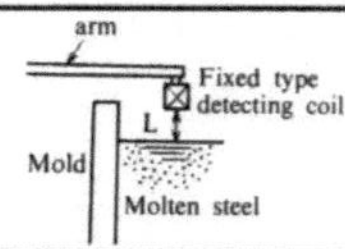	Level of molton steel can be measured by amplifying a detected change in impedance due to the change of distance between the coil and the surface of the molton steel.
	Attached detecting coil type	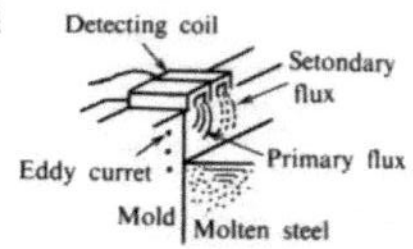	An eddy current is generated in the mold by primary flux from and oscillator. The level of molten steel can be measured by detecting the secondary flux caused by the eddy current because the eddy current reflects the level of molton steel.
	Liquid level tracking type	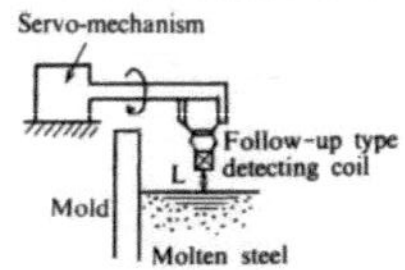	Link arm or telescopic mechanism is followed by servo device so that distance L, between the surface of molten steel and the detecting coil, is kept constant. The level of molten steel is shown by the amount of movement of the coil.
Radiation type		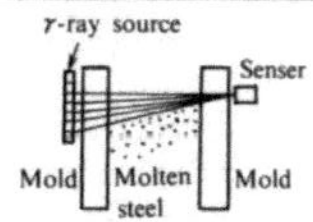	γ-ray source is installed in mold wall and sensor (scintillation counter) detects the radiation of γ-ray through molten steel to measure the level of molten steel.
Thermocouple type		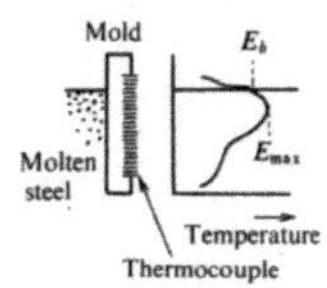	Level of molten steel is estimated from the temperature distribution pattern shown by the thermocouples installed in mold wall. $E_b = E_{max} \times 0.6$
Laser type		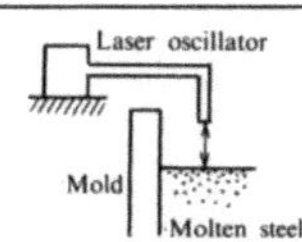	Laser light, directed to the surface of molten steel, produces a reflection from the surface. The level of molten steel can be scaled by measuring the reflection time.

portion to the casting speed, and the coefficient of proportional equation is determined by computer based on feedback from the surface temperature of the slab. While it is correct to say that the volume of spray is held proportional to the casting speed of the process, the proportional coefficient must be adjusted for various conditions. The three main variables that cause the proportional coefficient to be adjusted are discussed below. (Refer to Fig. 10.21)

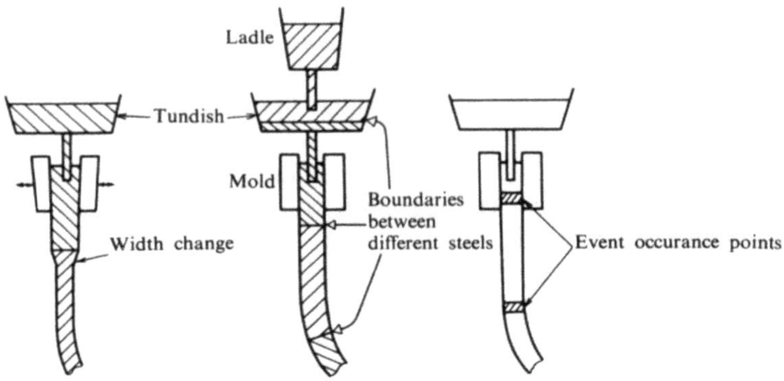

Fig. 10 · 21 Spray control tracking

(i) *Mold width (shape)*: Even with the same kind of molten steel, if the cross-sectional shape of the slab changes (mold change), the volume of spray also has to change. Since mold widths can be readily adjusted today, the flow coefficient is successively adjusted for the different width slab that passes through the spray cooling zone.
(ii) *Different kinds of steel*: When different kinds of molten steel are poured or even when the molten steel is poured from a new ladle, identical casting conditions must be maintained while the same kind of steel is being used. What appears to be a continuous slab, is actually made up of sections of different steels. Thus, at the boundary where different steels pass through the spray cooling zone, the flow coefficient must be successively adjusted.
(iii) *Miscellaneous events*: For example, when the immersion nozzle or tundish are being replaced, injection of molten metal from ladle to tundish must be stopped momentarily. When this occurs, withdrawal from downstream molds must also be temporarily stopped. After replacement is completed, pouring and slab withdrawal are resumed. The volume of cooling water is constricted during the interruption, but even so a section of slab receives excess chilling. Thus, the spray cooling control must make an adjustment recognizing the

points where such events have occurred when the operation starts up again (see Fig. 10.21). The essential point is that the slab must be subjected to the same processing even if the casting operation has to be temporarily halted to accommodate miscellaneous events.

This need to recognize different situations and adjust the quantity of spray as the affected slab passes through the spray cooling zone is called spray cooling tracking. The conventional method of handling this tracking function has been to use a process computer and to adjust the instrumentation setpoints as necessary. In recent years, however, as digital instrumentation systems have become generally available, plants where this function is handled exclusively by instrumentation have increased. In such cases, sufficient operational data must be incorporated in the system to manage and adjust the amount of spray for any event that might occur. At the same time, every spray control loop must be equipped with a setpoint control (SPC) function.

(d) Instrumentation considerations

(1) *Cooling water maintenance* If cooling water for either molds or spray is interrupted, this can mean that the whole continuous casting operation has to be shut down. The reliability of the water control system therefore must be given especially close attention in the design stage. It is equally important that provision is made for unlikely events that could occur in the control system; for example, the design should incorporate a system of values and water supply to insure an adequate flow to keep the operation up and running.

Electromagnetic flowmeters are generally selected for this purpose because they perform better in cooling water environments and operate with little pressure loss. To obtain precision cooling, gas-atomized cooling (a mixture of air and water) is applied to the slab as a secondary cooling. For the air source, a vortex flowmeter is used because it offers an especially wide range for this application.

(2) *Control cycle* As digital control systems have found more and more applications, the discrete-time control has commonly been based on a standard 0.1-to-1-second control period. For the mold molten-steel level control, however, which is a feedback control function, the control period has to be under 0.2 seconds to achieve satisfactory results. Also, to control mold width adjustment and cutting equipment, a control period of 0.05 to 0.1 second is required.

10.2.4 Instrumentation for an electrolytic galvanizing line

(a) Equipment overview

Slabs produced in a continuous casting machine are then processed by rolling into various shapes depending on how the finished product is going to be used. In this section, we will be primarily concerned with a line set up to apply zinc to the surface of cold-rolled strip (*i.e.*, surface treated steel sheet) to produce anticorrosion steel sheet. Such facilities are generally referred to as an electrolytic galvanizing line (EGL).

Figure 10.22 summarized the more common types of surface-treated steel sheet. Especially in recent years, numerous types of electrolytic alloy galvanized steel sheet have been developed that satisfy the anticorrosion standard commonly known as the Canada code.*

The basic equipment that goes to make up a typical EGL is shown in Fig. 10.23. The steel sheet enters the line in the form of coil, and then undergoes continuous galvanizing. Here we will provide an overview of the main EGL equipment functions (excluding the entry and de-

Surface treated steel sheet

- Galvanized steel sheet
  - Galvanized sheet (JIS G 3302)
  - Hot dip galvanized steel sheet (Non JIS specification)
  - Alloyed hot-dip galvanized steel sheet
  - Hot-dip zinc · aluminium alloy galvanized steel sheet
  - Electrogalvanized steel sheet (JIS G 3313)
  - Electrolytic alloy galvanized steel sheet
- Aluminized sheet (JIS G 3314)
- Terneplate
- Tinplate (JIS G 3303)
- Tin-free steel (JIS G 3315)
- Coppered steel sheet
- Pre-painted steel sheet
  - Pre-coated galvanized sheet
  - Welding pre-painted sheel sheet

Fig. 10 · 22 Classfication of surface-treated steel sheet

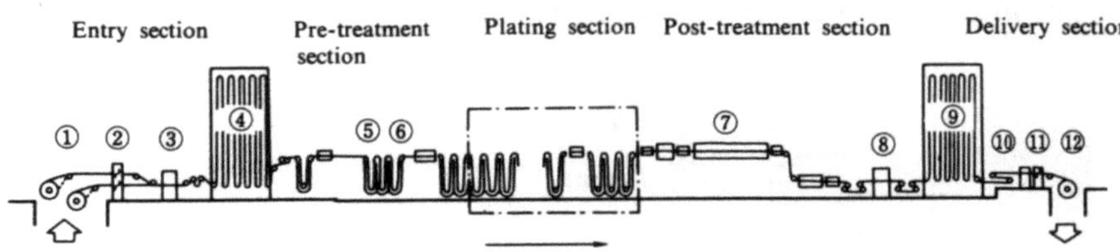

① Pay-off reels ⑤ Electrolytic cleaning ⑨ Delivery looper
② Entry shear ⑥ Pickling ⑩ Inspection table
③ Welder ⑦ Chemical treatment ⑪ Delivery shear
④ Entry looper ⑧ Tension leveller ⑫ Tension reel

Fig. 10 · 23 Typical EGL equipment

* Canada distributes a huge quantity of salt on its road during winter to keep them ice-free, therefore Canada's anticorrosion standard is especially rigorous.

livery sections).

(1) *Pre-treatment section*

Electrolytic cleaning: Oils adhering to the surface of the steel are removed in this section. A solution of sodium hydroxide is used for the electrolyte.

Pickling: In this section, rust that has accumulated on the surface of the steel is removed. A weak acid solution is the cleaning agent.

(2) *Plating section* This is the section where the actual electroplating takes place. Note that most lines are not limited to a single plating material; various types of metallic coatings can be applied to steel sheet by changing the plating solution.

(3) *Post-treatment section* After plating, the sheet metal undergoes further surface finishing in this section. For example, a phosphating treatment might be applied to the sheet as an undercoat prior to painting, or a special organic resin might be applied to surfaces that will not be painted.

The equipment for the plating and post-treatment sections varies considerably because the specific configuration largely depends on how the surface-treated steel sheet is going to be used.

A schematic of a typical plating section is shown in Fig. 10.24. Plating solution (an acid solution containing ions of the plating metal), circulates between the circulation tank and the plating bath. As the sheet is introduced to the plating bath by the contact roller, the sheet acts as a cathode and positive ions of the plating metal are deposited on the surface in a solid state. Essential equipment associated with the circulation tank includes: (a) ion-supply equipment to replenish ions of the plating metal in the circulation tank, (b) impurity removal equipment to eliminate contaminants that would have an adverse effect on the plating, and (c) plating solution exchange equipment that is needed

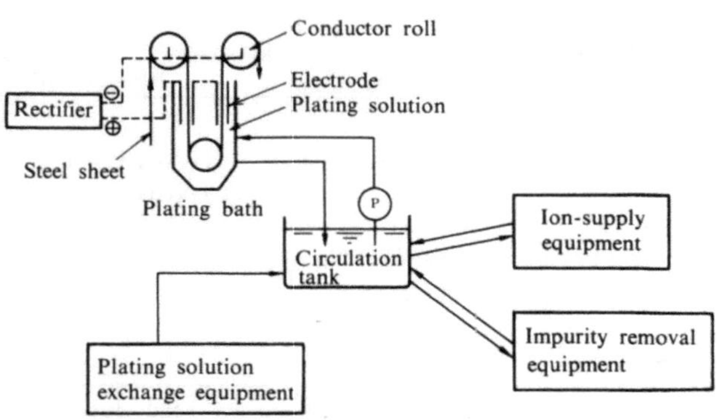

Fig. 10 · 24 Conceptual configuration of plating section

to clean the plating bath and circulation tank when changing over to a different plating solution.

(*b*) *Unique demands on EGL instrumentation*

Two characteristics of EGLs create special demands on the instrumentation. First, the plating equipment must accommodate various plating solutions (from strong base to strong acid), and second, a great number of valves and pumps are involved in switching between plating solutions. As far as the instrumentation is concerned, this means that

(1) special care should be taken in selecting types and materials for fieldmount devices that will hold up well in a harsh environment, and

(2) with so many pumps and valves, it is necessary to clearly define and compartmentalize which monitoring, operation, and interlock functions are to be handled by instrumentation and which are to be handled by electrical equipment.

This second point especially should be given very careful consideration because how the question is answered will have a major impact on the number of digital input and output points and also on the sequence volume. Note too that if the plating solution cutover sequence is automated, this will also require considerable sequence volume. This is assuming the present trend continues of concentrating the monitoring, operation, and interlock for all devices in the human-machine interface of instrumentation.

(*c*) *Instrumentation overview*

In contrast to ironworks equipment, it is difficult to standardize the instrumentation for application to EGL. This is particularly true for the plating and post-treatment sections, where each steel manufacturer has developed individual product lines that require unique facilities and instrumentation. In our treatment here, however, we will only be concerned with instrumentation that is common to different applications.

(1) *Electrolytic cleaning unit* With respect to the circulation tank, control must be maintained over temperature, concentration and solution level.

For detecting the concentration of sodium hydroxide solution, usually an electromagnetic concentration meter (specific conductivity meter) is employed. However, since it is not a pure solution, the objective is not so much to obtain an absolute concentration, as it is to derive a concentration range that can be used as a standard for the control limit. Since fast response is not required, the concentration can be adjusted by on/off control of the solution and water.

Quite a number of level-control methods have been devised. One typical technique is to charge a fixed proportion of water and solution once the low level limit is reached.

Concentration control and level control tend to counteract one another, but precision is not overly important in this context. The only essential thing is that the concentration control must be cut off as a safety measure when the upper level limit for the equipment is reached.

(2) *Pickling unit* Basically identical to electrolytic cleaning unit, the main control parameters for the circulation tank are temperature, concentration and solution level.

The conventional method of obtaining an approximate value of concentration c is by employing the first-order equation

$$c = K_1\sigma + K_2\rho + K_3$$

where σ is the reading from an electromagnetic concentration meter, ρ is the reading from a density meter, and K_1, K_2, and K_3 are constants. Here too, however, as was previously the case when we measured the concentration of sodium hydroxide solution in the electrolytic cleaning unit, one cannot expect to obtain absolute concentration. More recently, automatic titrators and a gamma-ray concentration meters have begun to be applied as on-line analyzers for concentration control. Level control too is essentially the same as described above for the electrolytic cleaning unit. One minor difference is that when automatic mixing is applied to the solution, the exothermic reaction of the sulfuric acid must be factored in when setting up the sequence.

(3) *Plating section* In addition to the continuous-control functions that are more or less the same as in the other sections, there are two control functions that are unique to the plating section: plating current control and contaminant removal control. Also, as was mentioned previously, relatively large-scale sequence control programs are required to switch back and forth between single- and double-side plating and to control plating solution changeover.

Control of zinc plating solution has conventionally involved regulation of the pH value and the density of solution, as shown in Fig. 10.25. As can be seen in the figure, the characteristics of the plating solution shift from the upper right in the figure toward the lower left as the plating proceeds (*i.e.*, from high density and pH to low density and pH). Control is thus necessary to restore the solution to its original condition by supply additional ions of the plating metal.

On-line analyzers such as automatic titrators and fluorescence x-ray analyzers have been widely applied to most electrolytic alloy galvanizing processes; they generally work by adding either acid or metal ions as required to restore the solution.

(*d*) *Instrumentation considerations*

(1) *Fieldmount devices* An obvious but important consideration is

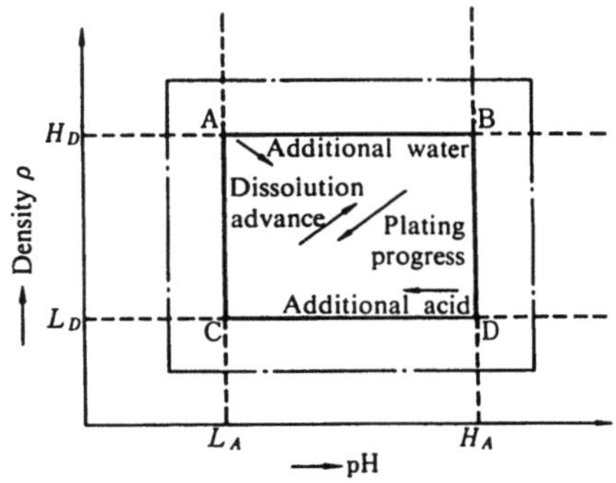

(Within solid-line border = control range)
(Within chain-line border = control limit)

Fig. 10 · 25 Concept of concentration control for plating solution

that fieldmount devices are constructed out of materials that are not going to be corroded by the strong solutions with which they come in contact. This corrosion factor is one reason purge-type detection elements have been so pervasively used for detecting level and density. When the purge tap (also known as a bubble tube) is made of resin, it is important to avoid long lengths (say, longer than 2 m) of unreinforced tube. A splash proof tube should also be installed to guard against rough agitation in the circulation tank. Finally, the bubble tube should not be subjected to heavy loads.

In using a purge-type density gage, if the change in density is so small that an elevation function alone doesn't produce sufficient accuracy above zero on the differential-pressure transmitter, one solution is to use a pressure adjusting tank (reference column), as illustrated in Fig. 10.26. With this kind of arrangement, the differential pressure detected by the gage is given by

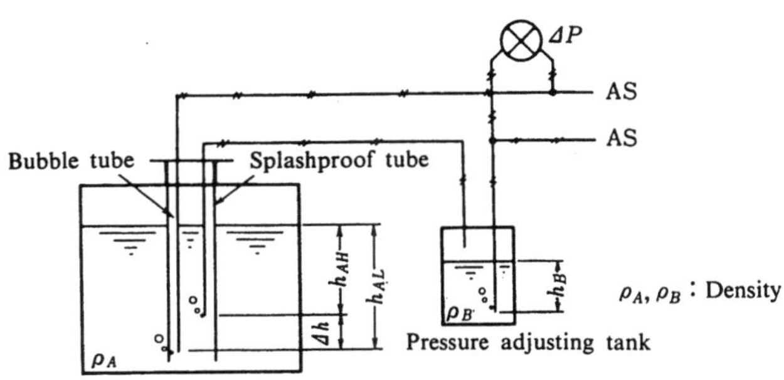

Fig. 10 · 26 Purge-type density measurement using pressure adjusting tank

$$\Delta P = \rho_A h_{AL} - (\rho_A h_{AH} + \rho_B h_B)$$
$$= \Delta h \rho_A - \rho_B h_B$$

and it is possible to eliminate the bias value $\rho_B h_B$.

One final consideration is that if bubbles or turbulence are present on the surface of the solution, ultrasonic level gages cannot be used.

(2) *Plating solution changeover* A sequence must be implemented to make sure the solutions are kept separate when changing over to a different plating solution. This involves verifying that valve A is fully closed before opening valve B.

(3) *dumping the plating bath solution* If sheet steel is left submerged in a strongly acidic solution for any length of time (*e.g.*, in the pickling and plating sections), there is a danger of the sheet being melted. To prevent this from occurring, an alarm should be implemented that is triggered a certain interval (say, 20 to 30 minutes) after the process is stopped with the steel sheet left in solution. Finally, as a precaution in case the instrumentation system malfunctions or fails, provision should be made so the plating bath can be easily dumped manually.

10.3 Instrumentation Applications in the Power Industry[16)]

10.3.1 Overview

The power industry consistutes a large-scale system encompassing hydroelectric, thermal, and nuclear power generation plants; transmission and substation facilities to distribute the power; and of course the factories and households that ultimately determine the power consumption load. A particular characteristic of the industry is that, while demand for power fluctuates, no means of storing electricity in commercial quantities has been devised, and as a consequence, generated power has to exactly respond to consumption demand. The implication for power plant instrumentation is that control results have to be extremely precise and responsive.

Today, hydroelectric power represents only about 6 percent of the power supplied in Japan and will therefore not be considered; our examples of power-plant instrumentation will be taken exclusively from thermal and nuclear power plants.

10.3.2 Thermal power plants

Conventional thermal power plants generate electricity through the combustion of coal, petroleum, and natural gas. Since these fuels are all derived from fossilized organic material laid down over the past 600 million years, this type of power plant is also commonly referred to as a fossil-fuel plant, a designation that clearly differentiates it from nuclear power plants*.

The working medium to effect the energy conversion from fuel to power in a thermal power plant is steam. A boiler produces steam, the steam turns a turbine, and the turbine turns an electric generator. After the steam has performed its work in the steam turbine, it is converted back to water in a converter and returned to the boiler. An overview of a typical thermal power plant is shown in Fig. 10.27.

① Coal yard
② Coal conveyer
③ Coal bunker
④ Boiler
⑤ Dust collector
⑥ Chimny
⑦ Ash handling system
⑧ Turbin-generator room
⑨ Transformer
⑩ Sub-station
⑪ Circulating water intake
⑫ Circulating water discharge

Fig. 10 · 27 Bird's-eye view of a thermal power plant (Courtesy Mitsubishi Heavy Industries, Ltd.)

10.3.3 Boiler control

Fuels burnt in the boiler impart thermal energy to the boiler water and the flue gas produced is discharged into the atmosphere through a chimney. Feedwater in the boiler is evaporated by the heat generated by the burnt fuel and leaves the boiler in the form of steam (Fig. 10.28).

The fundamental control system of the boiler is designed to maintain a mass balance of water and to maintain an energy balance by burning fuel in the boiler.

These two basic objectives are realized through three major control systems in addition to the steam temperature control system. These are:

(1) *Feedwater supply control* The quantity of working fluid is kept constant (mass balance). Drum water level is the control target for drum type boilers.

(2) *Combustion control* Steam evaporation is adjusted to meet the

* Some thermal power plants burn diesel oil for fuel, but since their output is small and their control systems relatively simple, they are also excluded from consideration here.

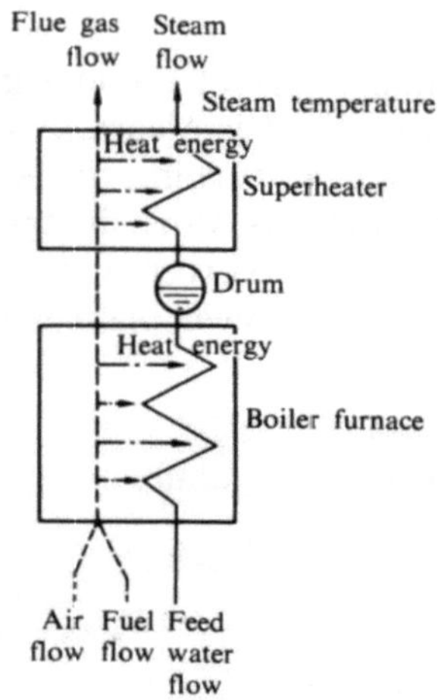

Fig. 10 · 28 Energy and mass flow in boiler

load (energy balance). Steam pressure is controlled at the boiler outlet (superheater outlet).

(3) *Steam temperature control* The steam temperature at the superheater (and reheater) outlet of the boiler is kept constant (emergency balance). Steam temperature at the superheater outlet is kept constant (this is also true of the reheater outlet in the case of reheat boilers).

However, since boilers exhibit complex nonlinear behavior, they are in fact quite difficult to control. To achieve good control results, boiler control systems have become extremely large and complicated. Application of adaptable estimate control methods based on advanced control theory has contributed to this development. The analog control systems of the past are no longer adequate for coping with present-day control problems.

Recent progress in microprocessor techniques, has led to the development of digital control systems (DCS) that greatly extend automation for thermal power plant control. Moreover, since the quantity and quality of information required by CDSs has increased markedly, computer terminals and associated software have also come to play and increasingly important role for the man-machine interface in thermal power plant control. Figure 10.29 shows a typical CRT operator console.

Boilers used in thermal power plants are classified into two types: the circulation type (also known as the drum type, because they employ drums) and the once-through type. The following explanation of boiler control systems is based on the circulation type.

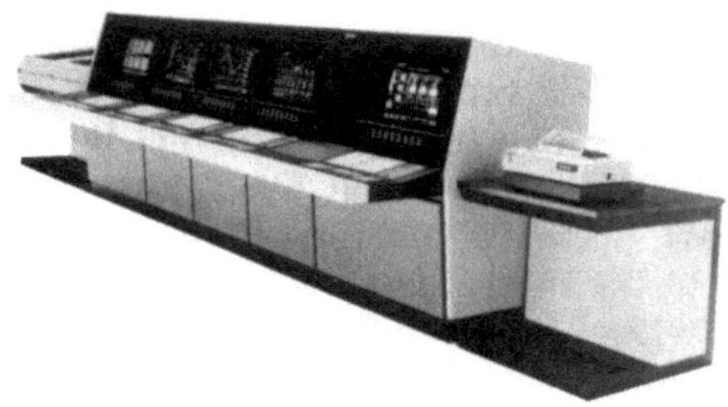

Fig. 10 · 29 CRT Operater console

(*a*) *Automatic combustion control* (*ACC*)

The purpose of boiler combustion control is to maintain constant steam pressure at the boiler outlet so that the fuel flow satisfies steam evaporation requirements. Fuel combustion in the boiler furnace requires that an adequate amount of air be supplied to the furnace. Combustion control can thus be broken down into fuel flow control and air flow control.

(1) *Fuel flow control* (*steam pressure control*) In fuel flow control, as shown in Fig. 10.30(a), steam pressure is detected and compared against a setpoint value, and the fuel flow is controlled so as to minimize the difference between the detected pressure and the set pressure. The steam pressure signal is called the master signal, and the control signal that is compared against the setpoint value is called the master control signal. Since boilers contain water of a large specific

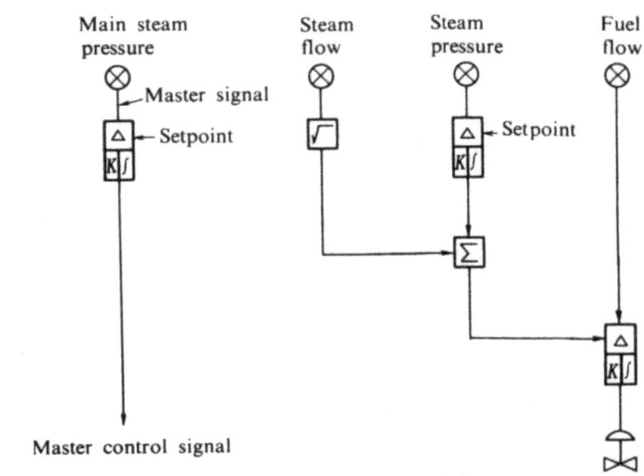

(a) Master control circuit (b) Fuel flow control circuit

Fig. 10 · 30 Fuel flow control system

heat and a large amount of steel, there is a time lag (called the time constant of boiler heat capacity, or boiler time constant) before changes in steam flow (or fuel flow) are reflected in steam pressure changes (see Fig. 10.31). This means that if fuel flow is only controlled by a master signal, control deviation tends to increase. As shown in Fig. 10.30(b), steam pressure control can be improved by adding a steam flow signal to the master control signal as a feedforward signal and by adjusting the fuel flow beforehand to meet the change in steam flow. The master control signal is transmitted to the fuel flow control (fuel flow control valve, coal feeder and so on) to control the fuel flow. The amount of fuel is measured as a result of control and compared with the master control signal as a feedback signal.

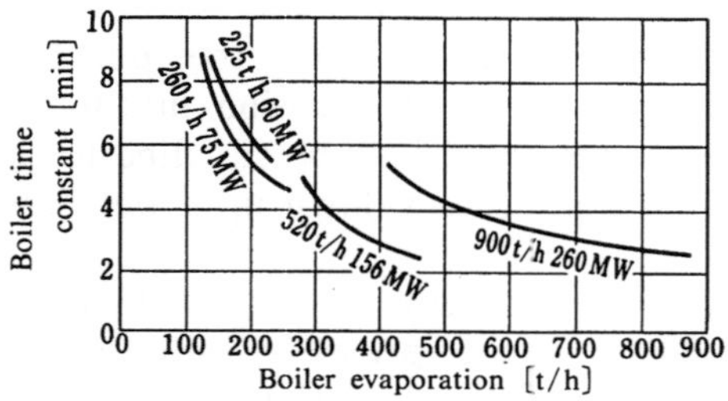

Fig. 10·31 Boiler time constant
(After Ito and Fukada et al. "Dynamic characteristics of boiler" in Mitsubishi Industries Technical Report, Vol.1, No.1.)

(2) *Air flow control (fuel/air ratio control)* Oxygen (air) is necessary to burn fuels, and combustion conditions are greatly affected by the air/fuel ratio. Combustion can be thought of as oxidation of fuels, and the air flow needed for theoretical oxidation (combustion) is called theoretical air. In actual combustion in boilers, however, complete combustion is not achieved unless slightly more than the theoretical air flow is supplied. The ratio of theoretical air flow to actually supplied air flow is called the excess air ratio and is expressed by μ. In actual boiler operation, the excess air ratio is found by the following equation, where the percentage of O_2 in the exhaust gas is measured.

$$\begin{aligned}\mu &= \text{actual air flow/theoretical air flow}\\ &= 21/(21 - O_2)\ (\%)\end{aligned}$$

where 21 is volumetric ratio of oxygen in air.

In boiler operation, when μ decreases, heat loss due to unburned fuel increases. When μ increases, on the other hand, heat is lost up the

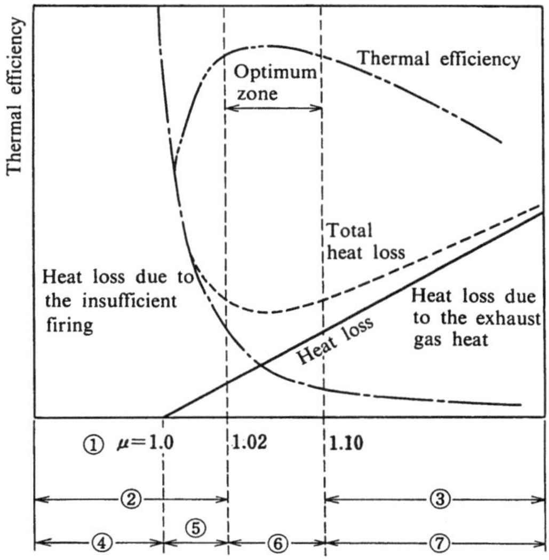

① Fuel/Air ratio
② Increase CO and smoke density
③ Increase NO_x, SO_x
④ Short air
⑤ Extreme low excess air
⑥ Low excess air
⑦ Over excess air

Fig. 10 · 32 Relationship between excess air ratio and combustion status

chimney, with a resulting decline in boiler thermal efficiency. This relation is shown in Fig. 10.32. From this figure it is seen that maximum boiler efficiency ranges from 1.02 to 1.10 of μ for oil-fired boilers (1.20 to 1.30 for coal-fired boilers).

The basic air flow control system is shown in Fig. 10.33(a). In actual boiler operation, the change of air flow supplied by forced draught fans (FDF) in the course of a load change is retarded than the change of fuel flow (due to the response speed lag of the control vane and air flow lag). Hence, when the boiler load increases, μ decreases in the form of a transient with the possibility of black smoke shooting up. Most cross-limit combustion control systems such as shown in Fig. 10.33(b) use a selective relay of low and high signals, to ensure a constant and sufficient supply of air even in the event of transient conditions during load changes.

The master combustion control signal is compared with the air flow signal by a low signal selective relay and whichever value is found to be smaller is transmitted to the fuel control valve. Thus, even though the boiler load increases and a signal indicating a fuel increase is received, the output of the low signal selective relay does not increase because the air flow still has not increased. In the case of a high signal selective relay, the rise of the master control signal together

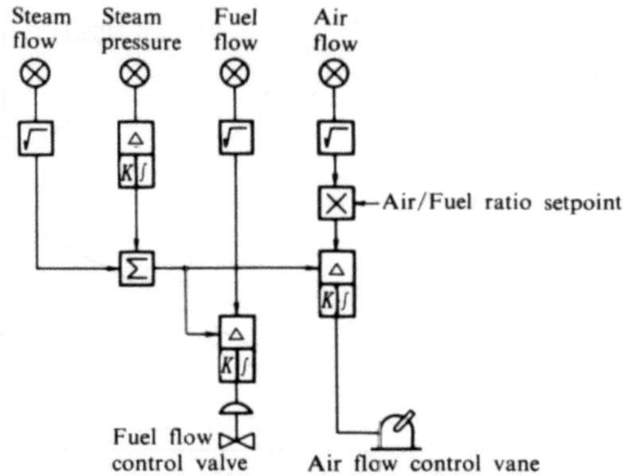

(a) Fundamental system of air flow control

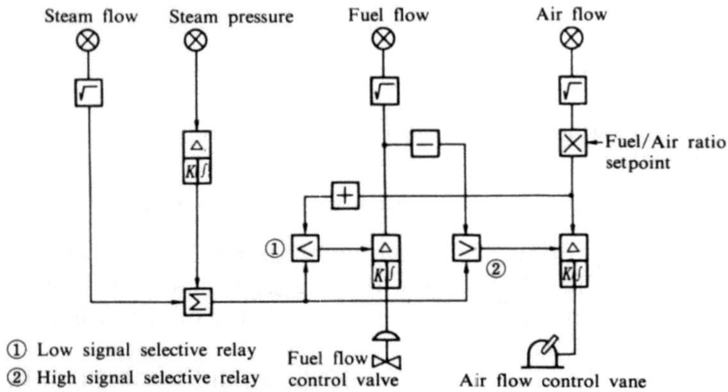

(b) Cross limit air flow control system

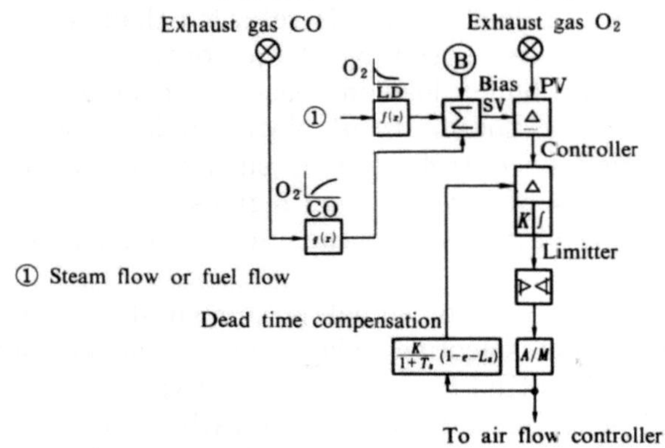

(c) Low O_2 air flow control system performed with set valve correction using CO value

Fig. 10 · 33 Air flow control system

with load increase is conveyed to the air flow control vane and the air flow increases first. Then, the output of the low signal selective relay increases to correspond to the increase in air flow, thus resulting in an increase in fuel flow. A bias signal is added to the air flow, and fuel flow signals of each selective relay define the μ fluctuations during the change. Contrary to the other operation described, air flow cannot be decreased with load decreases, until after the fuel has been decreased.

Recently, reliable combustion gas analyzers (O_2, CO) have been developed, so that air flow control employing O_2 values to perform the setting and CO values to correct the setpoint values is also used (Fig. 10.33(c)).

(*b*) *Feedwater control* (*FWC*)

Feedwater control is used to keep the quantity of water in the boilers constant. In drum type boilers, the water level in the drum is used as an index of the amount of water in the boiler. Feed water flow is generally controlled by feedwater control valves or by the speed of feedwater pumps. As shown in Fig. 10.34, however, there is a transient characteristic (termed the reverse response) of the water level in the drum* such that the water level rises transiently as steam flow increases. Feedwater control that only relies on detecting the drum water level, therefore, cannot produce good results. To eliminate the reverse response effect, steam flow is used as a disturbance compensation signal, and feedwater flow is used as a feedback signal. This type of feedwater control system is shown in Fig. 10.35. Because this con-

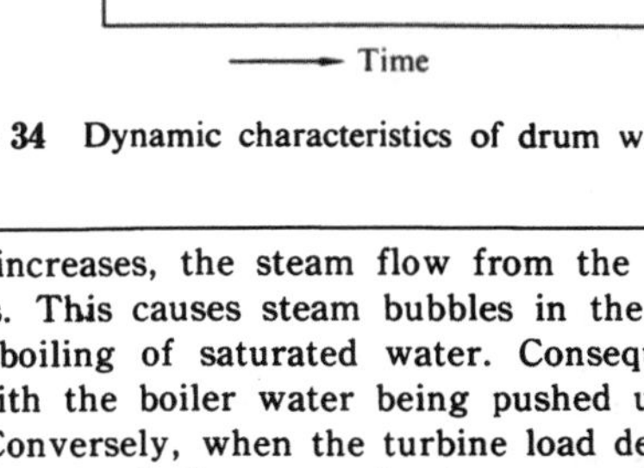

Fig. 10 · 34 Dynamic characteristics of drum water level

* As the turbine load increases, the steam flow from the boiler increases and the boiler pressure decreases. This causes steam bubbles in the generating tubes to increase due to the self boiling of saturated water. Consequently, the drum water level rises transiently with the boiler water being pushed up into the drum by increased steam bubbles. Conversely, when the turbine load decreases, the opposite occurs; feedwater flow increases, boiler water in the generating tubes cools, and the steam bubbles decrease. As a result, a transient drop in the drum water level occurs even if the feedwater flow increases.

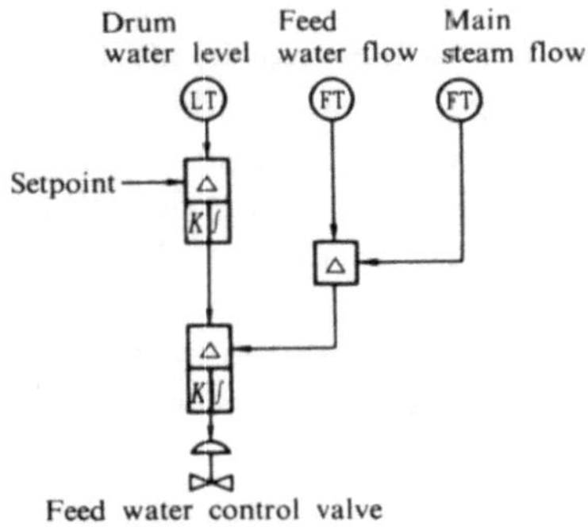

Fig. 10·35 Feed water control system

trol system uses drum water level, steam flow, and feedwater flow as control elements, it is termed the *three-element control system*. During boiler start-up, the flow measurement of both the feedwater and the steam flow are inaccurate, and therefore only the drum level signal is used for this drum level control system.

(*c*) *Steam temperature control* (*STC*)

Changes in steam temperature can have a great impact on turbine performance as well as on boiler and turbine service life.

(1) *Main steam temperature control* Main steam temperature control is, for the most part, performed by spray water injection into a desuperheater installed at an intermediate point in the superheater. In this case, to compensate for the response lag due to thermal inertia in the superheater, and the transfer lag due to steam flow, the temperature of the main steam before the spray water injection is detected for cascade control (Fig. 10.36(a)). In the case of large-capacity boilers that operate at high pressures and temperatures, the above lags become larger. This necessitates high-temperature resistant superheater tube materials and the installation of more than two spray point stages.

Since steam temperature exhibits a functional relation with boiler load (Fig. 10.37), steam flow is used as a lead signal to obtain better control results (Fig. 10.36(b)).*

(2) *Reheat steam temperature control* Temperature control by spray

---

* The control of steam temperature is extremely difficult for a number of reasons; (i) the time constant of a heating surface is large and incorporates a transportation lag, and (ii) the characteristics of heat transfer depends on the cleanliness of a heating surface. Therefore, a predictive control system has been developed that works quite well. The system performs advance control by predicting changes in steam temperature from boiler conditions (boiler load, cleanliness of heating surface and so on) using a mathematical model of the boiler in a computer.

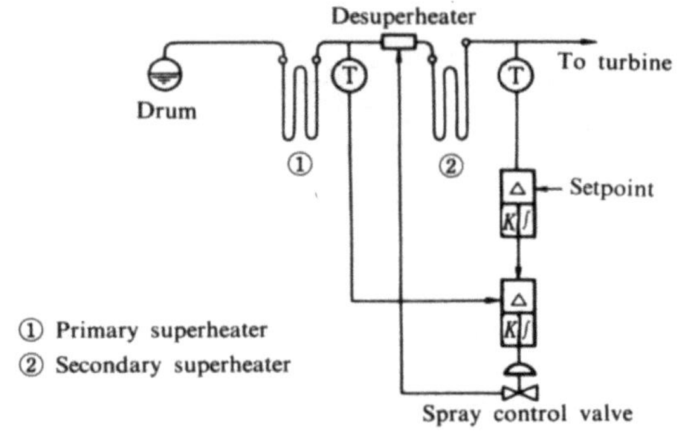

(a) Fundamental system of steam temperature control

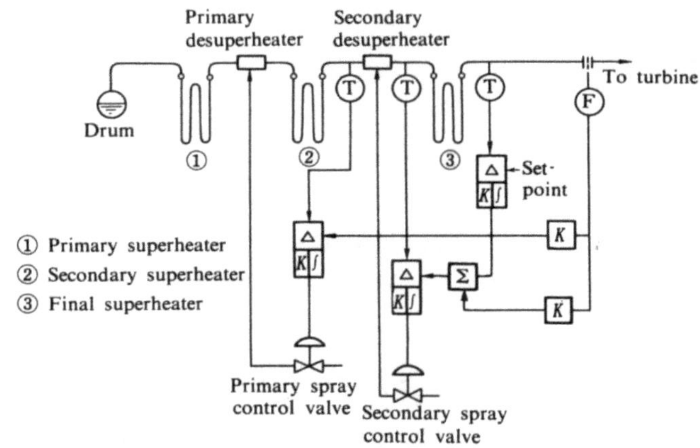

(b) Steam temperature control system (Cascade control)

Fig. 10 · 36 Steam temperature control system

water injection is not used* for reheat steam because this would degrade the efficiency of the Rankine cycle.

The control of reheat steam temperature, as shown in Fig. 10.38(a), (b), and (c), includes gas recirculation, burner tilting (the aim of both is to determine the heat absorption increase in the reheater) and gas distribution damper methods (by changing the heat distribution between the superheater and reheater). Here we will explain the gas recirculation control system (Figs. 10.38(a) and 10.39), since this is the most widely used method.

When a portion of the low temperature gas at the economizer out-

* A spray control system is installed for the reheat system, but it is only used in emergencies.

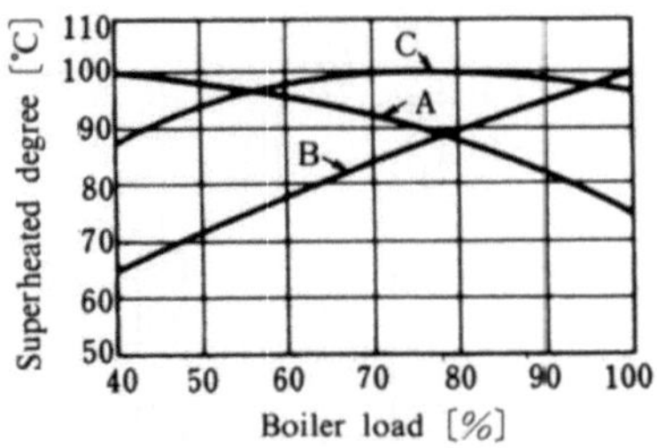

A : Radiation superheater, B : Convection superheater, C : Radiation/Convection superheater

Fig. 10 · 37 Temperature characteristics of superheater
(Reprinted from JSME mechanical engineers handbook : power plant engineering (1987))

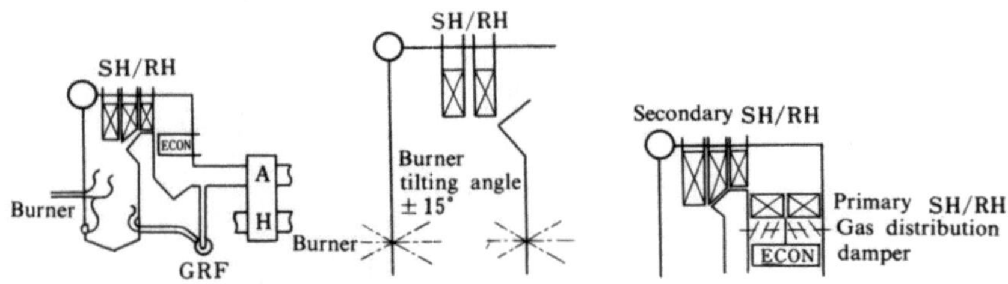

(a) Gas recirculation method (b) Burner tilting method (c) Gas distribution damper method

Fig. 10 · 38 Reheater steam temperature control methods

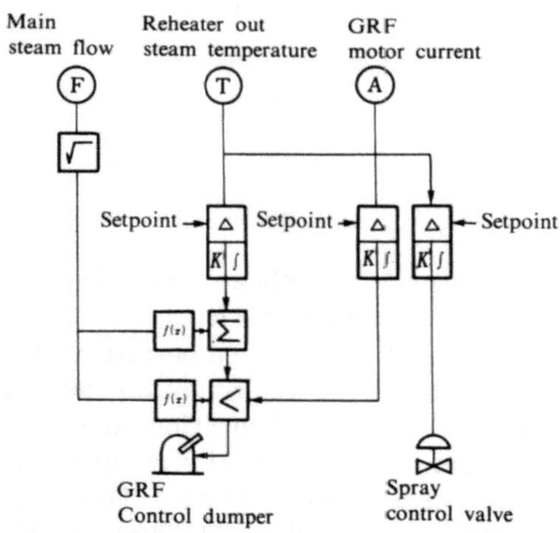

Fig. 10 · 39 Reheater steam temperature control system

let is blown into the furnace bottom by a gas recirculation fan (GRF), the gas temperature at the furnace outlet rises due to the reduction of radiation heat absorption in the furnace. By coupling this with the heat absorption increase due to gas flow speed increase while passing through the reheater, the steam temperature at the reheater outlet can be increased.

Because the steam temperature at the reheater inlet depends on turbine load, a steam flow signal is used as feedforward signal as an index of the lead. As gas recirculation (GR) flow increases at low load and the ratio of GR to combustion air flow becomes excessive, an unstable combustion condition occurs. Therefore, programming control system to decrease GR at low load is used in conjunction with a low selection relay to prevent overload of the GRF motor due to drops in GR gas temperature.

(*d*) *Automatic burner management system* (*ABMS*) *and boiler protection system*

Because the operation of burners have such a crucial role in overall boiler operations, and because misoperation of burners accounts for such a large proportion of serious boiler mishaps, most boilers are equipped with an ABMS to improve safety and reduce labor requirements. This system, including the furnace purge system before light-off, is so important for boiler operation control that it is designated the furnace safeguard supervisory system (FSSS).

(1) *Automatic burner management system* (*ABMS*) The quantity of fuel atomized at the burner is proportional to the pressure of the fuel entering the burner. If the fuel pressure at the burner inlet reaches the limit defined by the delivery press of the pump, the fuel pressure required during the fuel flow increase can be reduced by increasing the number of burners. If, on the other hand, in the case of a reduction in fuel flow, the fuel pressure at the burner inlet drops below the limit due to throttling by the fuel control valve, this adversely affects fuel atomization and combustion efficiency deteriorates. The fuel pressure therefore should be increased to the desired level by reducing the number of burners (Fig. 10.40).

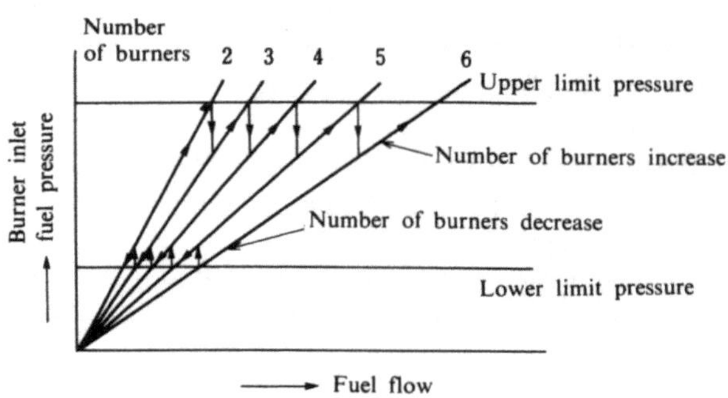

Fig. 10・40 Relationship between number of burners and fuel pressure

In the ABMS, burner inlet oil pressure is detected from the relationship of boiler load (fuel flow) and the number of burners to determine the necessary number of burners. Burners are sequentially* fired according to a command signal from the ABMS (Fig. 10.41).

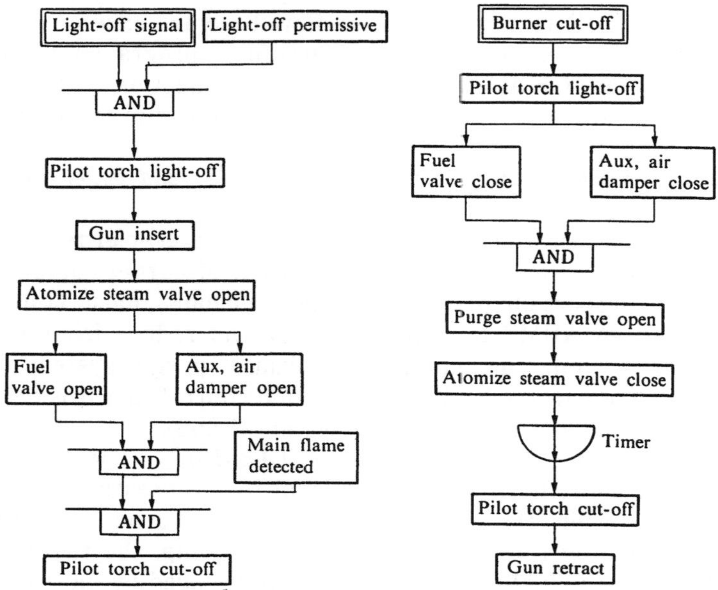

Fig. 10 · 41 Burner operation block diagram

(2) *Boiler protection system* Boiler protection systems safeguard against a number of dangerous operating conditions, and include a master fuel trip (MFT) system to rapidly cut off an entire fuel source (Fig. 10.42), a furnace purge system to prevent a detonation of unburnt gases or fuels remaining in the furnace or flue gas passage when igniting the boiler, (Fig. 10.43) and a fuel leakage testing system to detect for fuel leaks in the shut-off valves (Fig. 10.44).

(*e*) *Other controls*

Boiler plants employ numerous other control systems, but here we will mention only two; furnace pressure control and steam air preheater temperature control.

(1) *Furnace pressure control* With low-pressure fuels such as gaseous fuel or pulverized coal, the pressure in boiler furnaces is controlled

* ABMS and FSSS are open loop controls that are controlled by sequential logic, whereas the automatic control mentioned before is a closed loop control.

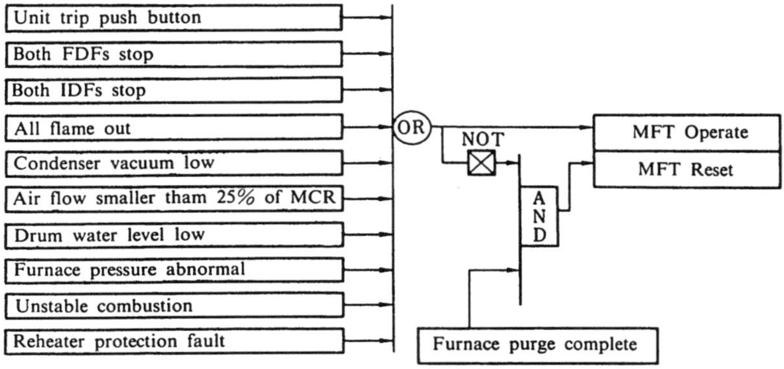

Fig. 10 · 42 Block diagram of master fuel trip system

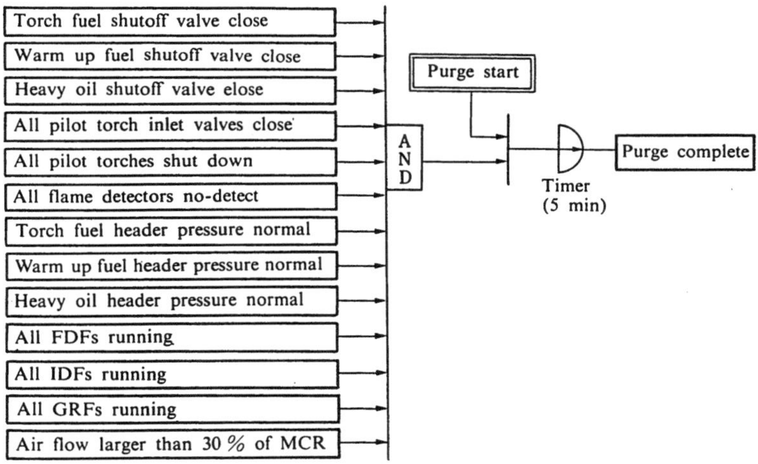

Fig. 10 · 43 Block diagram of furnace purge system

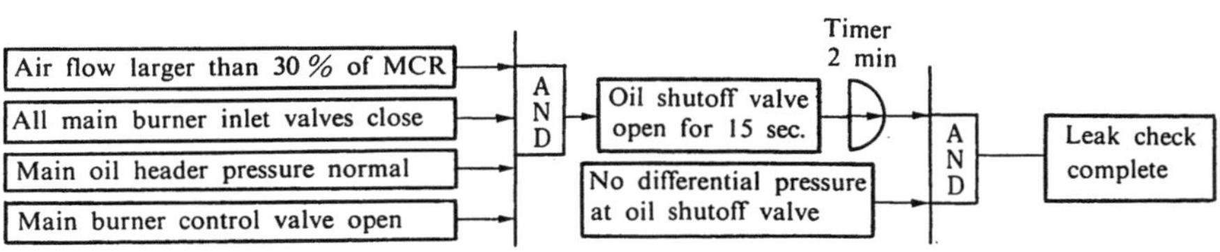

Fig. 10 · 44 Block diagram of fuel leakage testing system

at a level somewhat lower (*i.e.*, 5 to 10 mmH_2O lower) than atmospheric pressure by induced draught fans (IDF). A master control signal is added to the furnace pressure control signal as a feedforward signal (Fig. 10.45).

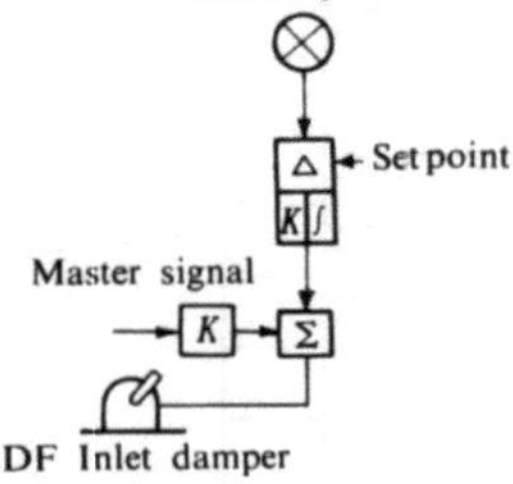

Fig. 10 · 45 Furnace pressure control system

(2) *Steam air preheater (SAH) temperature control* Air preheaters (AH) are used to improve combustion and to increase boiler efficiency by heating combustion air using boiler flue gas. As the boiler load decreases, however, flue gas temperature falls and SO_3 in the exhaust gas corrodes the AH. To prevent AH corrosion, a SAH is installed in the air duct at the AH inlet to prevent the AH temperature (represented by an average air temperature at the AH inlet and gas temperature at the AH outlet) from falling below the dew-point temperature of SO_3 (Fig. 10.46).

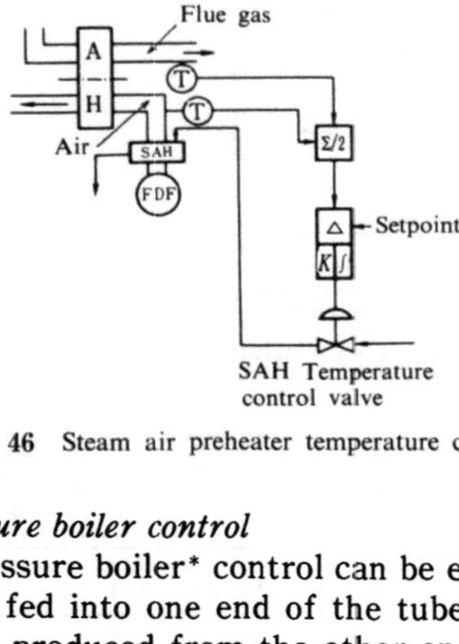

Fig. 10 · 46 Steam air preheater temperature control system

(f) Supercritical pressure boiler control

Supercritical pressure boiler* control can be explained by envisioning a tube; water is fed into one end of the tube, where the water is heated, and steam is produced from the other end. In this simple system, the evaporation quantity (corresponding to steam pressure) depends on feedwater flow. When feed water flow is kept constant and the firing rate is increased, the end point of evaporation moves toward

* Critical condition of the steam is 225.65 kg/cm² and 374.15°C.

to the feedwater inlet, thus expanding the superheating area. The steam temperature is controlled by the firing rate, but since the response is slow, it is used in combination with temperature control by spray.

Figure 10.47 compares the above control systems with drum type boilers.

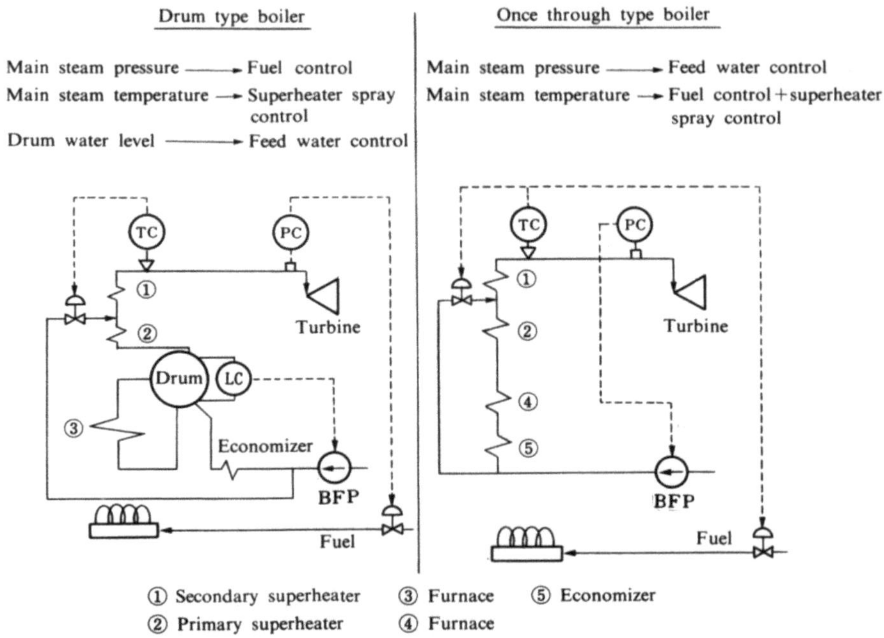

Fig. 10 · 47 Comparison of control system drum type boiler VS once through type boiler

10.3.4. Turbine control

The steam turbine plant consists of a turbine as the main machine, a condenser, feedwater heaters, a deaerator, boiler feed pumps, a circulating water intake system and a demineralized water treatment system. The controls for some of these systems will be considered here.

(*a*) *Turbine governor control*

The generator is driven by the turbine. If the supplied energy (the energy inherent in the steam) does not correspond to generator output, turbine shaft torque will be insufficient (or excessive) and rpm will fall off (or accelerate). This is illustrated in Fig. 10.48. If turbine load increases from P_1 to P_R and rpm decreases from N_R to N_S, and the aperture of the steam governing valve adjusts from A to B. This

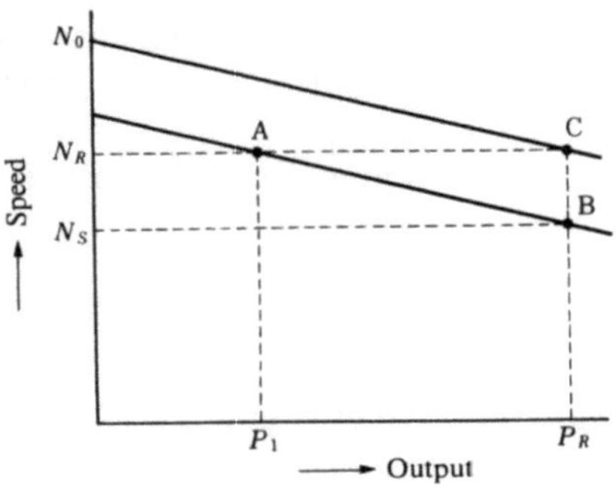

Fig. 10 · 48 Characteristics of turbine speed governor

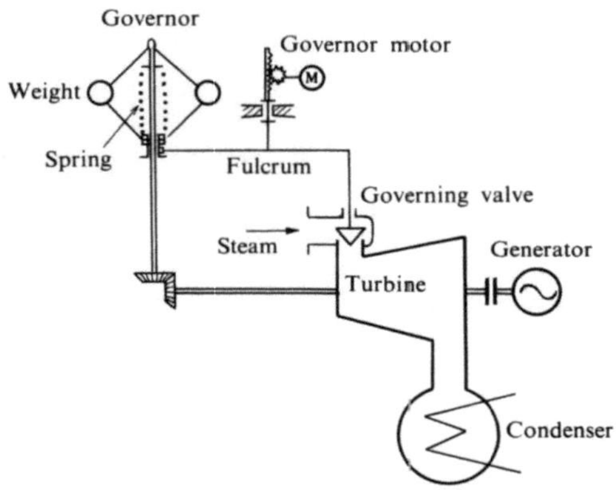

Fig. 10 · 49 Operation principle of speed governor

brings the output in line with load P_R, and equilibrium is restored. Then to bring the rpm back up to N_R, the governor motor controls the steam governing valve so it adjusts to point C (refer to Fig. 10.49). This relationship between turbine output and rpm is called speed regulation, and generally ranges from 4 to 5% for utility thermal power plants. Speed regulation, R_S, is given by

$$R_S=\frac{N_0-N_R}{N_R}\times 100 \quad [\%]$$

Here, N_0 is turbine rpm under a no-load condition, and N_R is turbine rpm under a rated load condition.

The above explanation applies to how the turbine main machine control device (turbine operating governor) regulates the functions of speed and load. Other essential control functions of the operating governor are speed control and emergency shutdown in the event it is necessary to dump the load.

There are two principal types of operating governors. Mechanical hydraulic (MH) operating governors, which have been in common use for a long time, and electron hydraulic (EH) operating governors, which have come into fairly wide-spread use recently and are better adapted to the increased scale and more diversified operation of modern turbines.

(*b*) *Condenser hot-well level control*

The condenser hot-well level is determined when exhaust steam from the turbine and feed water to the boiler are in balance. However, when the condenser hot-well level falls due to blow-off from the boiler or when extraction from the system disturbs the balance, the make-up water control valve (CV-1) operates to return the hot-well to its normal level. When the level becomes abnormally high, the spill-over control valve (CV-2) operates to return the condensate to the make-up water tank (Fig. 10.50).

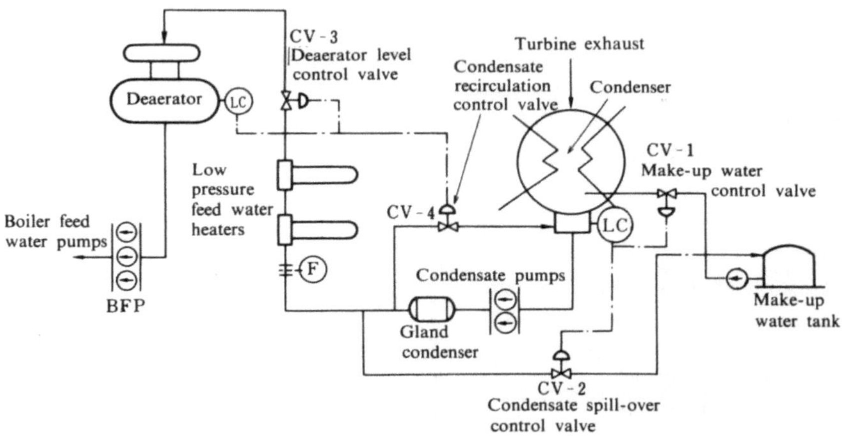

Fig. 10 · 50 Level control system around the condenser and the deaerator

(*c*) *Deaerator level control*

The deaerator holds approx. ten minutes of maximum continuous boiler evaporation feedwater and removes oxygen or carbonic acid gas. The deaerator also serves to cushion fluctuations in boiler feedwater during load changes.

Deaerator level control is performed by the deaerator level control valve (CV-3), but when the minimum water flow necessary for the air ejector or gland condenser (for cooling leak-off steam escaping from the shaft seal of turbine) is in short supply, the condensate recirculation control valve (CV-4) operates to maintain sufficient water flow. CV-4 is sometimes directly controlled by a condensate flow meter, but in the arrangement shown in Fig. 10.51, the deaerator level control signal is set at a split range to operate CV-3 and CV-4 in reverse.

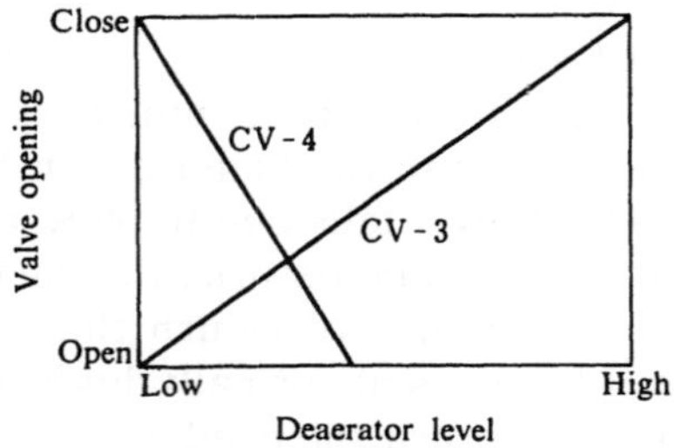

Fig. 10 · 51 Control characteristics of the deaerator level control and the condensate recirculation control

(d) Turbine and generator protection system

In the event the turbine or generator begin to operate abnormally, all steam valves supplying steam to the turbine are abruptly shut to stop the turbine and thereby protect it and the generators. A block diagram of the system is shown in Fig. 10.52.

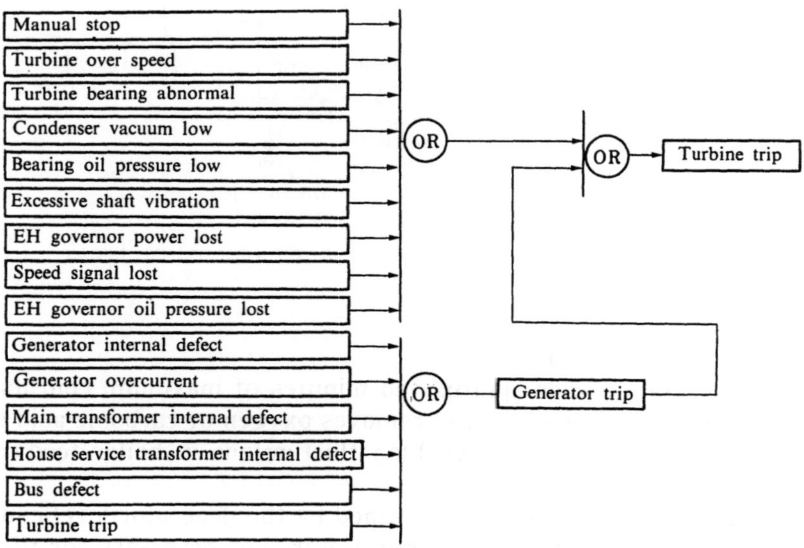

Fig. 10 · 52 Turbine-generator protection block diagram

10.3.5 *Power plant system control*

Electric power systems generally attempt to minimize frequency deviations in order to maintain the speed of motors and the accuracy of electrical equipment such as computers. Electric power generating systems in power plants, on the other hand, are directly connected to the power demand load. Such systems must accurately control the generation of electricity in accordance with load changes, as well as work to enhance the efficiency of the entire power plant. Various measures are implemented to facilitate such functions.

(*a*) *Coordinated boiler/turbine operation*

To deal with load changes, the power system controls the output of the generator in accordance with fluctuations in the load. In other words, the control adjusts steam flow into the turbine, and a change command is given to each power plant from a load dispatch control center which controls the overall electrical generation of the total network. In order for each individual power plant to respond efficiently to this change command, the following operational control methods are implemented.

(1) *Sliding pressure operation* In power plants, boilers are normally operated at a constant pressure, and the load is controlled by manipulating the turbine governing valves. Therefore, except for the case when the steam governing valves are wide open, a drop in turbine efficiency occurs due to the throttling action of the valves (Fig. 10.53). In the sliding pressure operation method, main steam pressure is set approximately in proportion to load (Fig. 10.54). This way, the governing valves can be almost completely opened regardless of load, and a drop in turbine efficiency due to the throttling action of the governing valves can be prevented. In a sliding pressure operation, the boiler/turbine combination that yields the optimum pressure condition is determined. That value is then used as the master signal setpoint in the program.

(2) *Coordination of boiler/turbine control* The control of boilers and

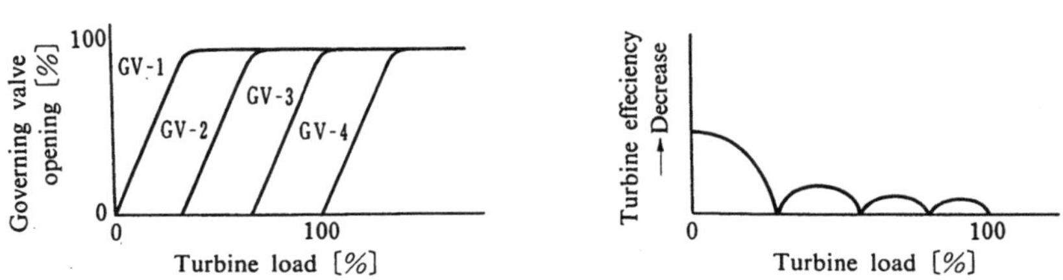

Fig. 10・53 Turbine governing valve opening and turbine effeciency (In case of four GV)

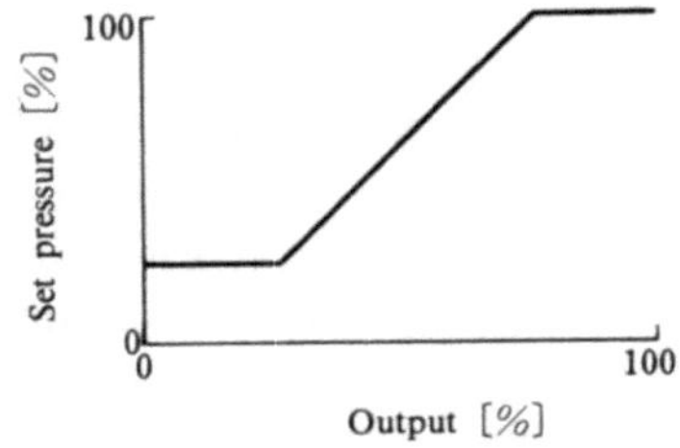

Fig. 10・54 Set pressure of main steam at the sliding pressure operation

Table 10.3 Control method of boiler/turbine block

Control method	Boiler-follow mode	Turbine-follow mode	Boiler/Turbine coordinate mode
Control system	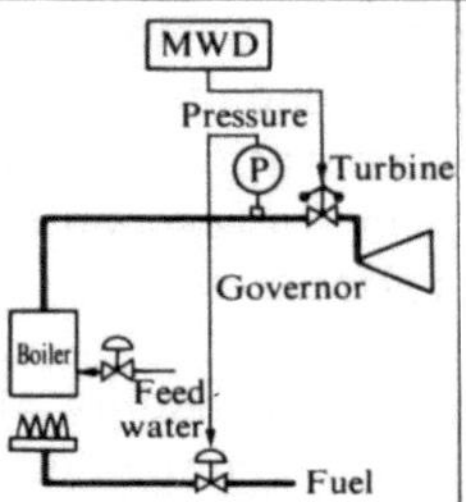	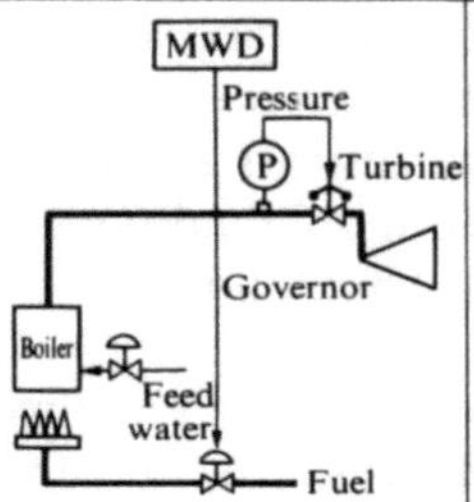	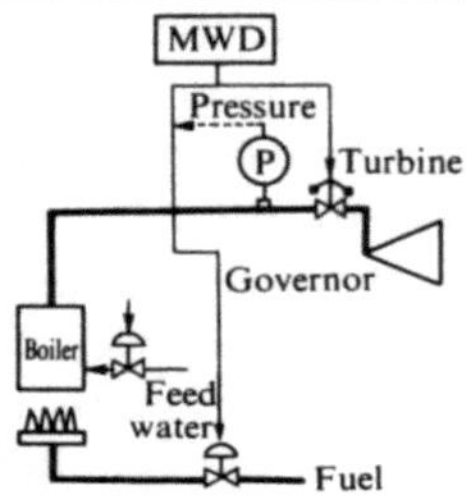
Fundamental control function	Load control···Turbine (Governor) Main steam pressure control ···Boiler (Fuel)	Load control···Boiler (Fuel) Main steam pressure control ···Turbine (Governor)	Load control···Turbine (Governor) Boiler (Feed water, Fuel) Main steam pressure control ···Boiler (Feed water, Fuel)

turbines in thermal power plants includes boiler–follow, turbine–follow, and boiler/turbine coordinate modes (Table 10.3).

In the boiler–follow mode, a megawatt demand signal (MWD) is transmitted to the turbine governor to control steam flow into the turbine, and steam pressure in the boiler is controlled by ACC. This mode is used in large–heat–capacity drum boilers and exhibits good power–output response during load changes. Operating conditions in this mode, however, are very hard on the boiler.

In the turbine–follow mode, the MWD is transmitted to the boiler master control signal, and the governor controls steam flow into the turbine to maintain constant steam pressure. In this mode, the turbine load is changed through the time constant of the boiler, and therefore the response is slow. An advantage of this mode is that the boiler can be operated in a stable condition. This mode is widely applied in Europe where once-through boilers have been widely adopted.

The boiler/turbine coordinate mode is intermediate between the other two modes already described and is employed with thermal power plants where large once-through boilers are used.

(*b*) *Load variation and its control*

The period and amplitude of load variation in an electric power system is shown in Fig. 10.55. Short period load variations of less than about two minutes are absorbed by the turbine governor control and load self-balance. (A good example of this principle is provided by an ordinary motor. As the cycle increases, the speed of the motor also increases until eventually the increased load and speed settle at a proper level). For slightly longer load variation periods of from two to twenty minutes, automatic frequency control (AFC) is applied to control the generation of electricity to maintain the frequency. For even longer load variations, which are predictable to some extent, economical load dispatching (ELD) is applied to dispatch demand to each power plant so as to minimize fuel costs and transmission losses.

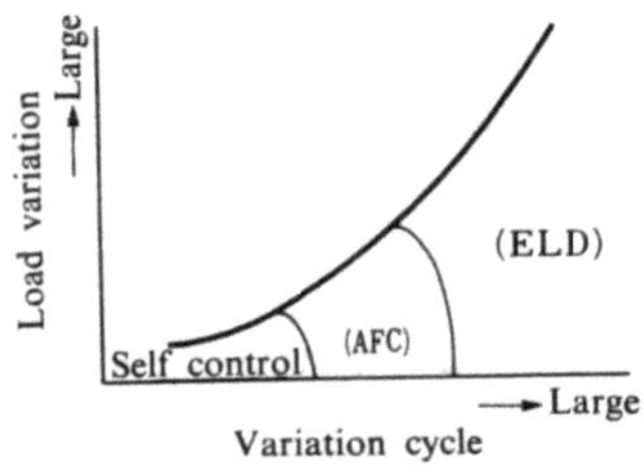

Fig. 10 · 55 Assignment of power generation control

10.3.6 Nuclear power plant overview

(*a*) *Nuclear reactor overview*

While the use of water power and combustion of fossil fuels to generate electricity has quite a long history, it was only in 1960 that the first nuclear power plant began operation in Japan. At present, there are more than 35 nuclear power plants in operation, and approximately 25% of the electricity consumed in Japan is generated by nuclear plants. A comparison of these three modes of power generation is shown in Table 10.4.

The operating efficiency of pressurized water (PWR) and boiling water reactors (BWR) in Japan is already extremely high at over 70%. In the coming years, the objectives are to boost this performance even higher and to move toward a more standardized type of light-water reactor. Also to be noted is the dramatic progress that has been made toward the development of a commercially viable advanced thermal

Table 10.4 Energy from various kinds of power generation

	Hydraulic power generation	Thermal power generation	Nuclear power generation
Energy source	Change in potential energy of water molecule (from high potential to low potential)	Oxidation reaction of carbon molecule. $C+O_2 \rightarrow CO_2$ + heat	Energy yield on fission of uranium etc.
Kind of force	Universal gravitation	Binding force between atoms	Nuclear binding force
Energy yield per molecule	2.9×10^{-23} (J) per 100m drop	6.4×10^{-19} (J)	3.2×10^{-11} (J)

Table 10.5 Comparison of various types of atomic reactors

	Neutron contributing to fission	Fuel	Moderator	Coolant	Conversion ratio
Light water reactor	Thermal neutron	Uranium 235 3% Uranium 238 97%	Light water (Ordinary water)	Light water	0.6
Advanced thermal reactor	Thermal neutron	Plutonium 239 1.5% Uranium 235 0.7% Uranium 238 97.8%	Heavy water	Light water	0.7
Fast breeder reactor	Fast neutron	Plutonium Approx.20% Depleted uranium (Blanket fuel Approx.80% Uranium 238)	–	Sodium	1.2

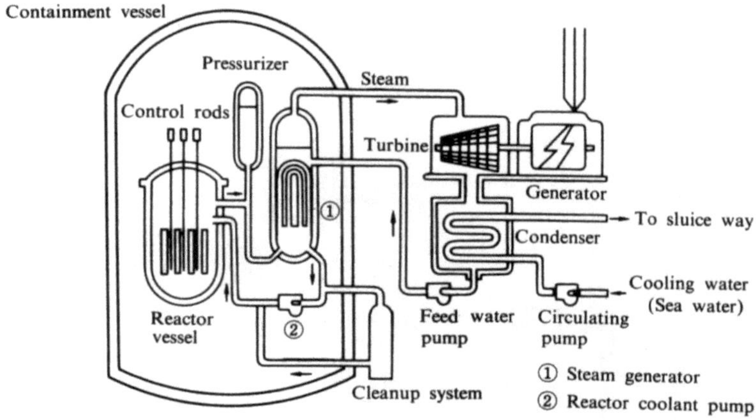

Fig. 10 · 56 Pressurized water reactor (PWR)

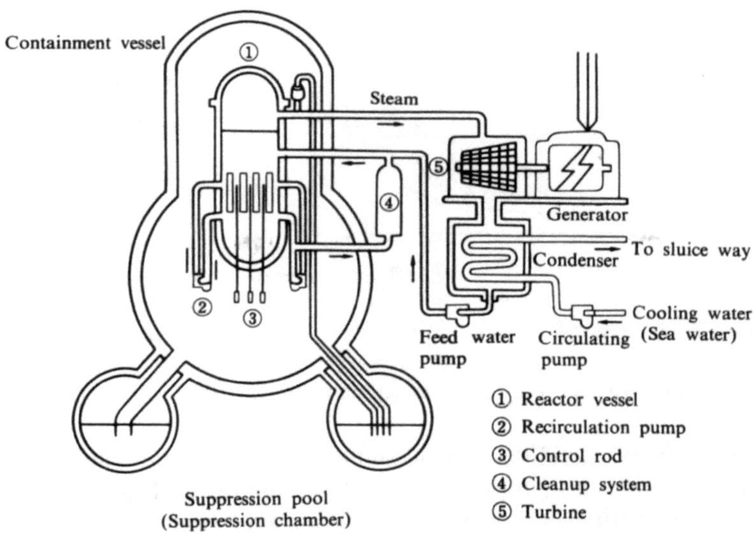

Fig. 10 · 57 Boiling water reactor (BWR)

reactor (ATR) and fast breeder reactor (FBR). These four types of reactors are shown schematically in Figs. from 10.56 to 10.59, and a comparison of some of their key features is summarized in Table 10.5.

Structurally, nuclear and fossil-fuel power plants have a great deal in common; both generate electricity using steam-driven turbogenerators, and condensers and water-supply heaters can likewise be found in both types of plants. Where they radically differ, of course, is in their heat source: in a nuclear plant, energy is generated from the

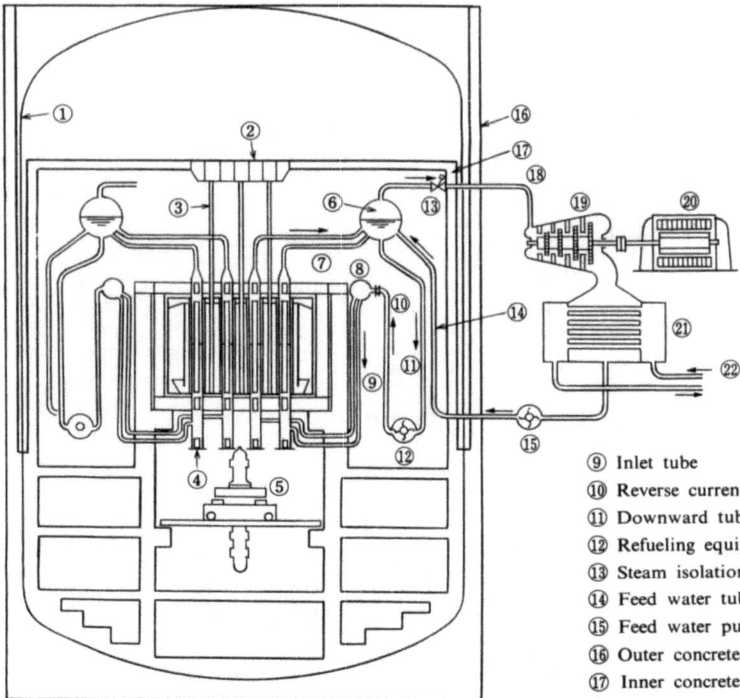

⑨ Inlet tube
⑩ Reverse current valve
⑪ Downward tube
⑫ Refueling equipment
⑬ Steam isolation valve
⑭ Feed water tube
⑮ Feed water pump
⑯ Outer concrete wall
⑰ Inner concrete wall
⑱ Main steam tube
⑲ Steam turbine
⑳ Generator
㉑ Condenser
㉒ Sea water for cooling

① Containment vessel
② Control rod driving mechanism
③ Control rod guide tube
④ Seal plug
⑤ Refueling equipment
⑥ Steam drum
⑦ Upward tube
⑧ Lower header

Fig. 10 · 58 Advanced thermal reactor (ATR) (Prototype "*Fugen*")

fission of heavy nuclides, and a conventional thermal plant uses the chemical combustion of fossil fuels such as bunker oil, gas, or coal.

The rest of this section will provide a broad overview of considerations pertinent to the control of nuclear reactors.

(*b*) *Nuclear power*

(1) *The atomic nucleus* The nucleus of an atom contains two kinds of primary particles, *protons* and *neutrons*. Both kinds of particles have roughly the same mass, but protons carry a positive electric charge and neutrons have no charge. The portion of the atom outside the nucleus contains only negatively charged particles called *electrons*. In a normal neutral atom, the number of electrons orbiting the nucleus is exactly the same as the number of protons in the nucleus. Thus, the opposite charges of protons and electrons cancel each other and the atom

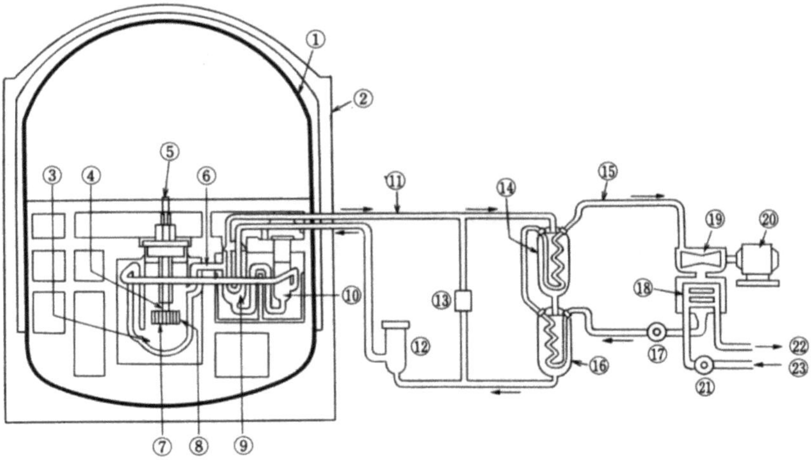

① Containment vessel
② Outer concrete wall
③ Reactor vessel
④ Control rod
⑤ Control rod driving mechanism
⑥ Primary sodium
⑦ Fuel
⑧ Blanket
⑨ Intermediate heat exchanger
⑩ Primary main circulating pump
⑪ Secondary sodium
⑫ Secondary main circulating pump
⑬ Air cooler
⑭ Super heater
⑮ Super heat steam
⑯ Steam generator
⑰ Feed water pump
⑱ Condenser
⑲ Turbine
⑳ Generator
㉑ Circulating water pump
㉒ To sluice way
㉓ Cooling water (Sea water)

Fig. 10 · 59 Fast breeder reactor (FBR) (Prototype "*Monjyu*")

is electrically neutral.

Hydrogen, the lightest element that occurs in nature, has only one proton and one electron. Chemically, the nucleus is the same as a deuterium atom with one proton and one neutron in the nucleus, orbited by one electron outside the nucleus. The number of protons, which is unique for each element, is called the *atomic number*, and the total number of protons and neutrons in a nucleus is called the *mass number* because it approximates the total mass of the nucleus.

The heaviest naturally occurring element is uranium with an atomic number (the number of protons) of 92. From this, we know that the number of electrons is also 92, but the number of neutrons varies. 0.7% of natural uranium has a mass number of 235, and 99.3% has a mass number of 238. Atoms of elements that have the same atomic number but a different mass number are called *isotopes*. The two uranium isotopes mentioned above are written $^{235}_{92}U$ and $^{238}_{92}U$. $^{235}_{92}U$ is inherently unstable and tends to break down to a more stable form, emitting radiation in the process.

(2) *Nuclear fission* If a neutron is absorbed by the nucleus of an

inherently unstable $^{235}_{92}U$ atom, this causes the nucleus to fission, giving raise to large amounts of energy and emitting two or three neutrons in the process. One condition for fission to occur is that the velocity of the neutron has to be slowed down so that its energy distribution is close to the Maxwellian distribution in the material in which it is found. Slowed-down neutrons are called *thermal neutrons*. If they are not slowed down, or *thermalized*, the neutrons tend to escape, or are dispersed, making them unavailable to fission.

There are many possible ways $^{235}_{92}U$ might fission, but a typical reaction is

$$^{235}_{92}U + ^{1}_{0}n \rightarrow ^{236}_{92}U \rightarrow ^{139}_{56}Ba + ^{94}_{36}Kr + 3^{1}_{0}n + \text{Energy}.$$

Ba and Kr designate *fission products*, and $^{1}_{0}n$ represents a neutron. Compared to $^{235}_{92}U$, $^{238}_{92}U$ almost never fissions.

It was mentioned that the percentage of fissionable $^{235}_{92}U$ constitutes only a very small fraction of natural uranium. Isotope separation is therefore used to boost the proportion of uranium-235, to produce so-called enriched uranium. Modern water reactors use uranium with a $^{235}_{92}U$ concentration of 2 to 3 percent.

(3) *Chain reaction* When excess neutrons produced by a fission reaction set off other fission reactions, this provides the basis for a self-sustaining chain reaction. The energy of a nuclear reaction is much greater than that of a chemical reaction, as is apparent from Table 10.4. For example, the fissioning of 1 kg of $^{235}_{92}U$ would yield approximately 2×10^{10} kcal of heat. To produce an equivalent amount of heat from the combustion of oil or coal would require 2 000 k*l* and 3 000 metric tons of these materials, respectively.

(4) *Criticality* To support a self-sustaining chain reaction, the uranium core of a reactor must be optimized so that the number of neutrons available for absorption and the number of neutrons that escape to the surrounding environment are precisely balanced to keep the chain reaction barely going. This optimal state is termed *criticality*. Criticality is not an absolute, but depends on the *multiplication factor k*, defined as the ratio of the present neutron population to that of the previous neutron generation. Thus, while the reactor operates at a steady power level, $k=1$. Criticality is characterized by a constant number of neutrons generated per second and a constant number of $^{235}_{92}U$ atoms that undergo fission per second.

(5) *Nuclear reactor* A nuclear reactor is a device where the chain reaction we have described can proceed safely under controlled conditions. To maintain a state of criticality in the reactor, the absolute number of thermal neutrons must be controlled, and the number of those that fission must be kept constant. This is accomplished using

rods made of cadmium or boron, materials that have high adsorption cross sections for thermal neutrons. These two materials are generally referred to as control materials.

To moderate or slow down neutrons form the high velocities at which they are created in the fission process, a moderator is introduced to the reactor. The most common moderators are light and heavy water.

A very obvious requirement for fission reactions to occur in the reactor is the nuclear fuel itself; uranium or plutonium are typical fuels.

To prevent neutrons from leaking out of the core, reflective material is applied around the reactor core.

Ultimately, the nuclear energy produced by nuclear reaction is converted to thermal energy in the nuclear fuel. The coolant must therefore lower the temperature of the fuel. Light or heavy water around the core commonly serves as both coolant and moderator.

(6) *Reactor control* The control rods are the main instruments of reactor control; they are gradually pulled out of the core until criticality is reached ($k=1$). In a typical reactor, the control rods can be manipulated over a multiplication-factor range from 0.96 to 1.30.

By way of illustration, suppose that in bringing the number of neutrons up in the reactor, the control rods are withdrawn a bit too far so the multiplication factor exceeds 1 and the number of neutrons begins to escalate exponentially. To restore the desired level of neutrons, the control rods are pushed inward until $k=1$. Of course, this procedure could also be applied in reverse.

We have stated that as the control rods are withdrawn, the number of neutrons and hence the number of fission reactions increases. Energy released from the fissions, however, raises the temperature of the coolant, and this causes a decline in reactivity. Hence, there is a homeostatic tendency toward equilibrium (reactivity is a measure of the extent of fission reactions).

10.3.7 Pressurized water reactor control system

(a) System configuration of a PWR

The reactor system of a PWR power plant consists of the reactor coolant system (see below) and various support equipment. In the reactor coolant system, heat generated in the core is transferred from the primary to the secondary system via steam generators (tube-and-shell type heat exchangers), thereby generating steam in the secondary system which is delivered to the turbogenerator. The support systems can be categorized as those involved in the routine starting, running, and shutting down of the reactor, and those implemented to deal with po-

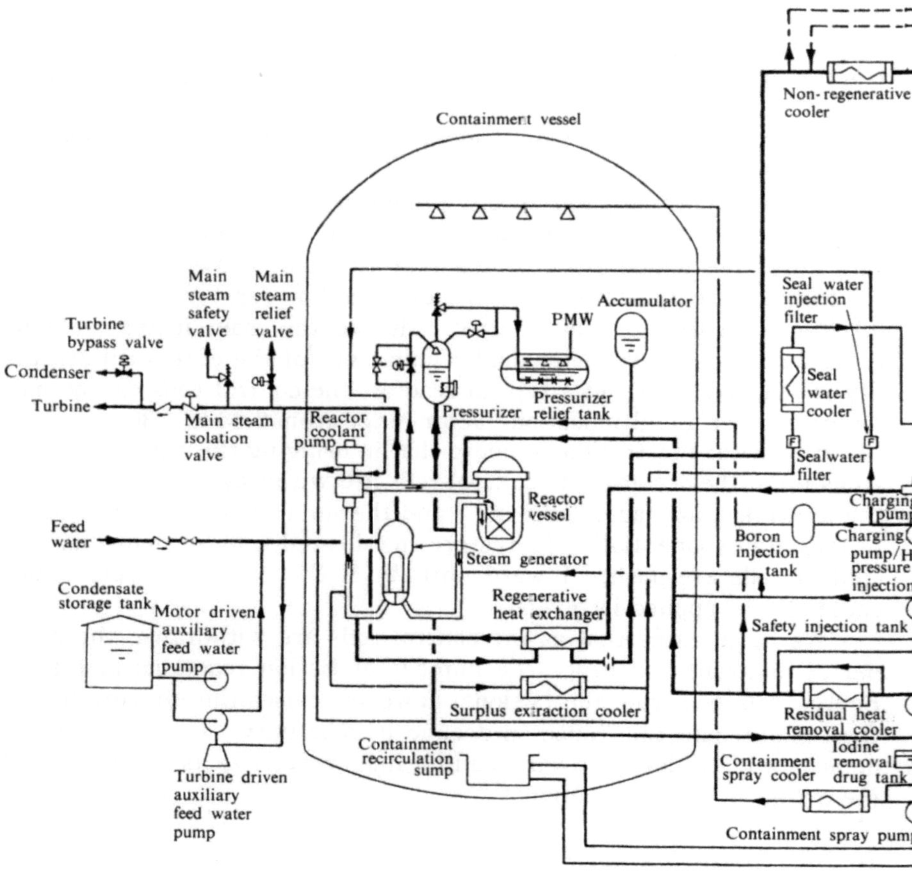

Fig. 10 · 60 System layout of a

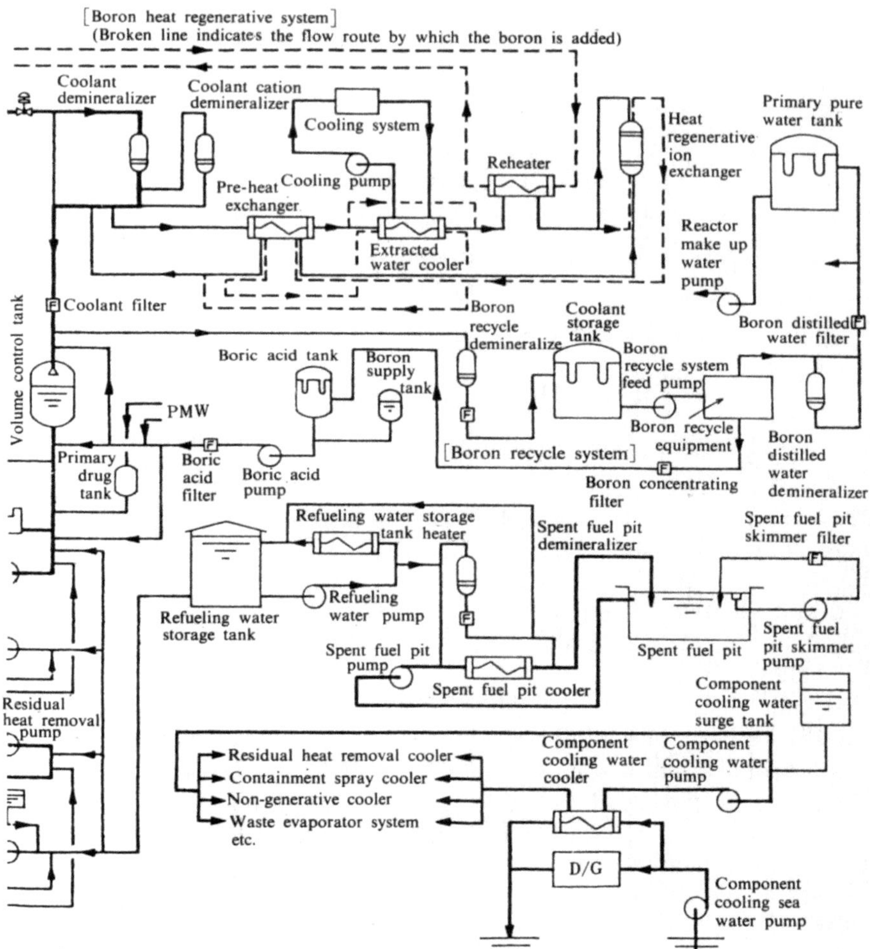

typical PWR Atomic reactor

tential emergencies.

The system layout of a typical PWR plant is shown in Fig. 10.60.

(1) *Reactor coolant system* (*RCS*) The principal function of the reactor coolant system is to remove heat from the reactor core and transfer it by means of steam generators to the secondary system where it is used to drive the turbines. The major components of the system are the reactor vessel, and 2 to 4 coolant units (called heat transfer loops), each containing a steam generator and a reactor coolant pump. A 1 100-MWe-class PWR is equipped with 4 such loops.

The coolant system is installed within the containment vessel and consists of (a) a number of heat transfer loops (each containing a steam generator and a reactor coolant pump), (b) the reactor vessel, (c) main coolant pipes connecting loops and reactor vessel, (d) pressurizer, (e) pressurizer safety valve, (f) pressurizer relief valve, (g) pressurizer spray valve, and (h) pressurizer relief tank.

Aside from the nuclear core, the reactor vessel also contains (a) a core-support assembly, (b) control-rod clusters, (c) thermal shielding, and (d) in-core assemblies. The steam generator is installed above the core so that even if the external power source supplying the reactor coolant pump were to fail, surplus heat from the core will continue to be removed through natural circulation of the coolant. For the coolant pump, a vertical-operation mixed-flow type pump with sealed shaft for leakage control is used. The typical pressurizer is a vertical vessel equipped with a spray nozzle located at the top where vapor is condensed and electric immersion heaters installed at the bottom of the vessel. The pressurizer is also equipped with a safety valve and pneumatic relief valve.

Pressure of the reactor coolant system is controlled by the pressurizer at 157 kg/cm^2. The immersion heaters in the pressurizer are divided into a proportional group and a back-up group. If there is a loss of pressure, the proportional-group heaters are activated first to restore thermal equilibrium. If equilibrium still cannot be restored with just the proportional heaters, the backup heaters are also turned on. If pressure raises above the set value, pressurizer spray volume is adjusted as necessary to maintain operating pressure. If the pressure variation is more than can be corrected with the spray valve, the pressure relief valve is opened as needed. A final defense against surplus pressure is provided by the pressurizer safety valve. The purpose of the pressure relief tank is to hold water to condense steam emitted by the pressurizer.

(2) *Chemical and volume control system* (*CVCS*) The primary functions of the CVCS are as follows:

(1) maintenance and control of the volume of coolant in the reactor coolant system;

(2) regulation of boric acid concentration in the coolant;
(3) removal of fission and corrosion products in the coolant;
(4) addition of corrosion-neutralizing chemicals to the coolant and control of water quality;
(5) extraction and recovery treatment to enable reuse of boric acid and coolant; and
(6) sealing water supply for the reactor coolant pump seal.

Chief components of the CVCS are: water purification and makeup system, boric acid recovery system, and boron thermal regeneration system.

(3) *Residual heat removal system* (*RHRS*) Principal functions of the RHRS are:

(1) Removal of residual decay and sensible heat that continue after the reactor has been shut down until the reactor cooling system is able to realize the prescribed temperature and pressure (177°C, 28 kg/cm^2).
(2) Functions as a component of the safety injection system. If there is a loss of primary coolant, the reservoir of boric acid water used during refuelings is injected into the core. After this water is injected, it is continuously withdrawn by the containment recirculation sump and recirculated in the core.
(3) Fills the fuel cavity with water during refueling. The cavity water is then pumped to a special tank designated for refueling water.

In a typical configuration, there are two trains, each containing one cooler and one pump.

(4) *Safety injection system* (*SIS*) In the event of an accidental loss of primary coolant, the SIS injects boric acid water into the core to cool it to prevent any major damage to the fuel or fuel cladding. Boric acid is also injected if the main steam line is ruptured or obstructed for some reason. By adding negative reactivity, the injected boric acid effectively quenches reactivity in the reactor.

Components of the safety injection system are the accumulator injection system, the high-pressure injection system, and the low-pressure injection system.

(5) *Component cooling water system* (*CCWS*) This system cools a diverse range of auxiliary equipment that is instrumental in the running of a nuclear reactor. For example, the pump and generator for the primary coolant, the non-regenerative cooler, and the seal-water cooler are all cooled by this system. The system also cools numerous other equipment during power operation, the residual heat removal system during shutdown, and this same system and the containment vessel spray system in the event of an accident.

(6) *Sea water system* (*SWS*) The sea water system is comprised of a sea water pump plus connecting pipeline and system valves. Auxiliary cooling equipment, diesel generators, and coolers for air conditioning purposes are commonly cooled by sea water.
(7) *Containment spray system* (*CSS*) The main components of this system are the containment spray pump, containment spray cooler, iodine removal drug tank, and refueling water storage tank. Primary functions are to reduce the heat within the containment vessel in the event of an accident involving the loss of coolant, and at the same time, remove iodine that is given off in the vapor part of the containment vessel.

The containment spray pump is activated by a signal that the containment spray has been triggered. It pumps water from the refueling water storage tank to the upper part of the containment vessel where the spray heads are mounted. Caustic soda stored in the iodine removal drug tank is added to the spray water. In recirculation mode, spray water is collected by the containment recirculation sump and returned to the spray system.

Pumps, coolers, and piping for the spray system are implemented in two independent trains.
(8) *Main steam & feedwater system* (*MSFWS*) The main steam system consists of the main steam line from the steam generator as far as the turbine's intake steam shut-off valve. Mounted on the main steam line are a number of valves including main steam safety valve, main steam relief valve, main steam isolation valve, check valve, and turbine bypass valve.

The turbine bypass system makes it possible for the operation to continue when there is a sudden decline from the rated load, without having to trip the reactor or emit a large quantity of steam, by diverting the surplus steam to the condenser.

Feedwater to the steam generator is generally supplied by the feedwater pump. The volume of feedwater supplied is regulated by the SG water-level control (the three elements governed are: water level, feedwater volume, and steam volume).
(9) *Other systems* In addition to the systems already covered, some of the other systems found in a typical PWR plant are:

- Waste disposal system (WDS)
- Sampling system (SS)
- Spent fuel pit cooling & clean-up system (SFPCS)
- Refueling water system (RWS)
- Primary makeup water system (PMWS)
- SG blowdown system (BDS)
- Instrument air system (IAS)
- Auxiliary steam system (ASS)

• Heating ventilation and air conditioning system (HVAC)

(*b*) *PWR instrumentation equipment*

(1) *Reactor instrumentation* Here we will be particularly concerned with instrumentation used to monitor and collect information regarding conditions in the reactor core.

The out-core nuclear instrumentation system is designed to supervise the level of neutron flux, which is proportional to the output of the reactor core. The system consists of neutron detectors mounted around the reactor vessel. Three ranges are detected: source, intermediate, and power. The system provides reliable supervision of core neutron flux levels from shutdown up to 120% of the reactor's rated output.

The in-core nuclear instrumentation system is designed to monitor the output distribution of the core, and consists of equipment to measure the temperature and neutron flux within the core. To supervise the output distribution of the core, the in-core temperature detection system uses chromel-alumel thermocouples to detect the temperature of the primary coolant at the outlet of the fuel assembles. The in-core neutron flux detection system uses flexible, small-scale neutron detectors that are inserted into the center of the fuel assemblies. This enables measurement of the neutron flux distribution parallel to the fuel assemblies.

The control-rod position indicator system is based on position detectors that are mounted on each control-rod cluster. The position detectors are configured out of 42 coils mounted on the pressure housing on the outside of the control-rod driving mechanism. Information indicating the positions of the control-rod clusters is converted to digital signals and displayed on the central control panel.

(2) *Process instrumentation* It is necessary to measure the volume flows of all essential processes. This applies not only to the primary coolant system, but to the processes of the various other support systems as well.

Process instrumentation consists of sensors and miscellaneous instruments installed in an instrument rack. Output from the instrument rack is sent to the (a) reactor protection system, which automatically shuts the reactor down in an emergency, (b) the engineered safety features operation system used to activate equipment essential in the event of an accident, and (c) the reactor control equipment for controlling reactor output. In addition, important parameters from the instrument rack output are displayed, recorded and used to activate alarms on the central control panel. To illustrate this type of instrumentation, we will examine the process instrumentation of the safety protection system, which provides essential parameters for the reactor protection system and the engineered safety features operation system.

The process instrumentation for the safety protection system is designed to be redundant. The channels are all implemented separately with their own sensors. The instrument rack is divided into four channels. In keeping with the functional segregation of the design, the power sources supplying the instrumentation of each channel are also independent. A portion of the safety protection system process instrumentation produces a control signal that is used by the reactor control system (described below). In this case, however, an isolation amplifier is used so that even if the reactor control system were to fail, no adverse effects could spread back to the safety protection system. Signals to the indicators and recorders are also mediated by isolation amplifiers. Finally, it should be noted that all process instrumentation can be tested to verify it is in good working order even while the reactor is being operated at full power.

(3) *Reactor control system* The control system of a nuclear reactor has two main functions, namely, to compensate for changes in operating conditions that occur while PWRs are operating at a steady power level and to regulate reactor power in response to load changes. Plant output control during operation is effected by regulating the volume of turbine steam and by controlling the reactivity of the core. Core reactivity is controlled through a combination of two basic methods; adjusting the positions of control-rod clusters and by controlling the concentration of boron in the coolant.

Control-rod clusters are adjusted according to need to compensate for short-term changes in plant operating conditions such as power and temperature, and to absorb excess reactivity in a high-temperature shutdown. Boron concentration control, on the other hand, is used to compensate for long-term reactivity changes involving fuel burnup and consumption of fission-product poisons, and to absorb excess reactivity in a low-temperature shutdown. Power control by adjusting control-rod clusters is done manually up to about 15% of the rated load of the reactor; beyond 15% of the rated load, control-rod adjustment is under automatic control. Within the automatic control range, load changes can be ramped at a rate of $\pm 5\%$/min or stepped at a $\pm 10\%$. Moreover, by making use of the turbine bypass system, extreme load changes equivalent to about 50% (or even 95%) of the rated load can be accommodated without have to shut down the reactor. In the rest of this section, we will discuss the main control systems involved in controlling nuclear reactors. A schematic overview of the main control systems is shown in Fig. 10.61.

The control-rod control system is under automatic control and consists of the control-rod clusters in the control group. As was noted ear-

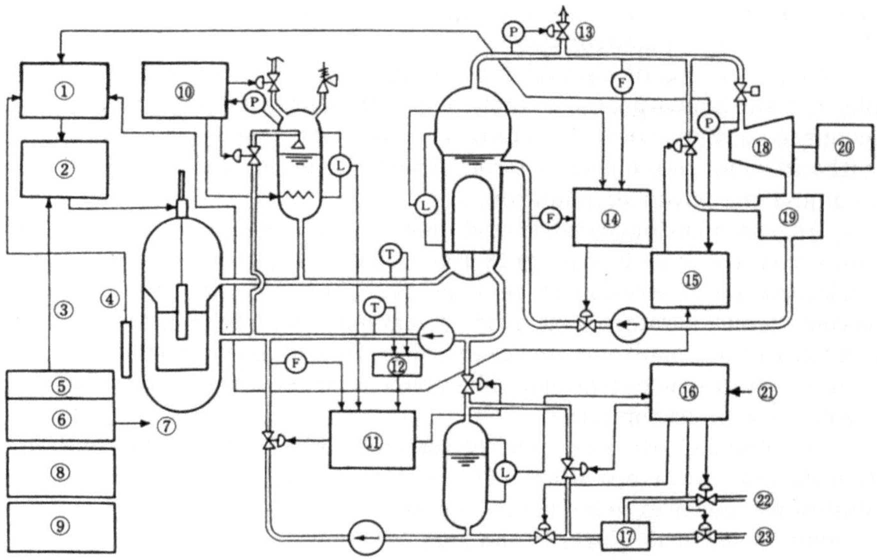

① Control rod control
② Control rod driving mechanism
③ Trip signal
④ Neutron flux
⑤ Reactor protection system
⑥ Engineered safety features operation system
⑦ Operation signal
⑧ Reactor instrumentation system
⑨ Other process instrumentation system
⑩ Pressurizer pressure control
⑪ Pressurizer water level control
⑫ Average
⑬ Main steam relief valve control
⑭ Feed water control
⑮ Turbine bypass control
⑯ Boron concentration control
⑰ Mixer
⑱ Turbine
⑲ Condensor
⑳ Generator
㉑ Control mode
㉒ Boric acid water
㉓ Pure water

Fig. 10 · 61 Systematic diagram of PWR Atomic reactor instrumentation equipment

lier, the system is generally used to compensate for shorter-term changes in plant operating conditions such as power and temperature changes. A primary coolant average temperature signal is compared against a program average temperature, which is proportional to turbine load, and a control signal is generated. This control signal plus a compensation signal that is generated when transients occur based on the rate of differential change between turbine load and neutron flux are used to govern the speed of the control-rod clusters of the control group and to maintain the primary coolant average temperature at the required value. The control-rod clusters of the control group are arranged in four banks, designated A, B, C, and D. When the rods are inserted or withdrawn, some overlap is deliberately provided to linearize the effects of the control rods.

Other groups of control rods perform other functions. For example, the shutdown group is kept in the full-out position, providing the means (together with the control group) for fast reactor shutdown without using automatic control. Another control-rod cluster is used to adjust the power distribution.

As was mentioned, boric acid dissolved in the primary coolant controls only the slowly changing reactivity effects. Adjustment of the boric acid concentration is done by the chemical and volume control system; concentrated boric acid is manually added to increase boron concentrations, and pure makeup water free of boric acid is added to reduce boron concentrations. The concentration level is supervised by means of regular sampling.

As much as possible, control rods are held in the withdrawn position during power operation. This not only provides a means of rapid shutdown, but also helps shape the power distribution within the core to make it uniform. Early in the core life, only bank D control-rod clusters in the control group are partially inserted (the so-called bite position) during power operation. If these control rods are either inserted or withdrawn beyond their position limits, the concentration of boron is adjusted so the rods can move back to their designated positions. The rod withdrawal limit is calculated to enable a return to full power.

The pressure of the primary coolant system is controlled by the pressurizer (pressurizer pressure control system). When pressure in the pressurizer increases, the coolant on the low-temperature side of the primary coolant system becomes proportional to the pressurizer pressure signal, steam in the pressurizer is condensed by a subcooled pressurizer spray, and this causes the pressure to drop. When pressure falls and the water level in the pressurizer raises above the setpoint, then heat is applied by proportional and backup immersion heaters causing the pressure to raise. Thus, the drop in temperature of the liquid vapor in the pressurizer is compensated. If excess pressure builds in the pressurizer beyond the capacity of the spray to correct, thermal equilibrium is restored by the relief valve.

Volume of the chemical and volume control system is maintained under automatic control by the pressurizer water-level control system based on a deviation signal reflecting the difference between programmed and actual water level in the pressurizer. If the water level drops to an abnormally low level, the extraction line valve is closed, thus halting any further decline in water level.

Feedwater control is configured separately for each steam generator. Three control elements determine the aperture of the feedwater control valve; feedwater volume, steam volume, and the steam-generator water-level signal. If the steam generator water level raises to an abnormal level, the feedwater control valve and bypass feedwater control valve are closed. In the opposite situation where the water level is too low, the auxiliary feedwater pump is automatically activated to ensure that the ability of the steam generator to eliminate heat is not impaired. If it is necessary to boost power, water-level control can be maintained either manually or automatically using the bypass feedwater control valve.

The turbine bypass control system diverts steam generated by the steam generator past the turbines, conveying it directly to the condenser. This makes it possible for the plant to adjust smoothly to large, step-shaped drops in load by eliminating heat transients in the primary coolant system. In the case of high-temperature standby or plant cooling, pressure control is maintained by main steam header pressure to either eliminate remaining heater or eliminate heat for cooling, as the case may be.

The main steam relief valve control system works as follows. In case of a large increase in the main steam pressure, due for example, to an abrupt drop in load, a main steam pressure signal is produced, opening the main steam relief valve. Recourse to the main steam safety valve is avoided if possible.

Immediately prior to reactor shutdown, the following interlocks are implemented to enable many of the automatic systems.

Interlock to prevent withdrawal of control-rod clusters is effected if intermediate range neutron flux, power range neutron flux, over temperature ΔT, and over power ΔT reach the excess high set point. This interlock prevents both automatic and manual withdrawal of the control rods.

Turbine runback interlock reduces turbine output in case over temperature ΔT and over power ΔT reach excessive high levels. Also, to restrict automatic control when reactor power is at a low level, automatic withdrawal of control-rod clusters is prevented when turbine output falls below the setpoint.

(4) *Reactor safety protection system* The system is designed to protect the system and keep it from failing in the event of abnormal transients or an accident. The system maintains the fuel assemblies, maintains the temperature and pressure of the containment vessel within permissible design parameters, and safeguards the coolant. Two subsystems are included in the reactor safety protection system: these are the reactor protection system, which activates the automatic reactor shutdown system, and the engineered safety features operation system, providing access to various equipment that would be necessary in

the event of a reactor failure.

The reactor protection system senses any abnormal condition within the reactor and begins shutdown before damage to the structure can occur by means of the out-of-core instrumentation and safety protection system process instrumentation described earlier. The protection system consists of two main elements; a logic circuit and a reactor trip breaker. The logic circuit outputs reactor-trip and interlock signals in response to a bistable signal (*i.e.*, alarm activation signal of the alarm setter) from the out-of-core and safety protection system process instrumentation. The reactor trip breaker is automatically opened when a reactor trip signal is received. For added safety, the reactor protection system is dispersed in either two or four independent trains. The principle is illustrated in Fig. 10.62 for a two-train configuration. All functions of the safety protection system (including the trip breaker) can be tested to verify they are in good working order without shutting down the reactor.

The operating principle of the reactor trip breaker is shown in Fig. 10.62. An M-G set, three-phase, alternating-current supply system is connected to the control-rod drive mechanism. If the trip breaker is opened, power to the control-rod clusters is cut off thus causing the control rods to drop into the core under the influence of gravity.

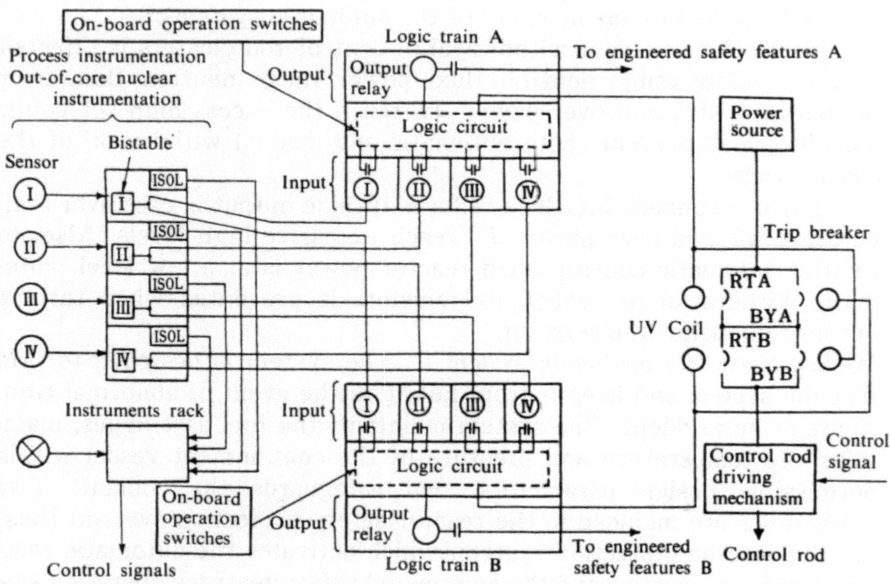

Fig. 10 · 62 PWR Atomic reactor control and protection system

The engineered safety features operation system is the means of activating the reactor protection system for automatic shutdown of the reactor in the event of failure. A number of equipment units are involved; a bistable signal from the safety protection system process instrumentation triggers the emergency core cooling system, the containment vessel separation valve, and the containment vessel spray equipment. Signals are also generated activating the various subsystems associated with these units. Note in Fig. 10.62 that here too the engineered safety features operation system is dispersed in either two or four trains. As with the reactor protection system, all system functions can be tested while the reactor is in full operation.

10.4 Instrumentation Applications in the Food Processing Industry[17)–19)]

10.4.1 Overview

Japan's food processing industry has evolved very rapidly as life-styles have become increasingly moderized and Westernized. Most food products are made in batches. Handling systems in the food-processing industry generally require automatic line cleaning systems and extra stringent quality assurance. These unique requirements call for unique instrumentation techniques compared with other process industries.

Since the food-processing industry by its very nature produces such a diversity of products, it is characterized more than other industries by a complex structure of many companies of all different sizes, some dealing with raw materials, others with semi-finished products, and still other companies with finished products. Nationwide, for example, there are some 85 000 establishments (employing approximately 1.15 million people) involved in the food-processing industry in Japan. Of these, 98% have fewer than 100 employees, leaving a scant 2% (about 1 700 firms) employing more than 100 people (1984 data).

Considering the great diversity of products, it comes as no surprise that almost as many unique automated techniques have been devised for their production.

In the following, we will describe some of the unique aspects of food-processing instrumentation and sanitary detectors that are an integral part of food-processing lines. For our specific applications, we will focus on two products that have achieved the highest degree of automation in their production: whiskey distilling and sugar refining.

(*a*) *Food-processing instrumentation*

When we speak of the food-processing industry, we are primarily referring to those processes shown in Fig. 10.63. In terms of instrumentation, the following attributes are unique:

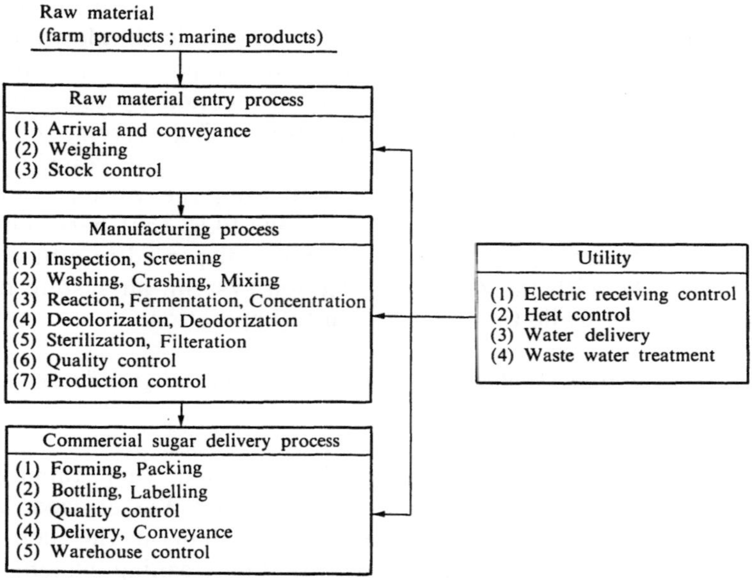

Fig. 10 · 63 From ran materials entry to delivery

(1) There are many processes where productive technology has not been adequately established; therefore, establishment of standardized operating methods for food–industry equipment is exceedingly difficult.
(2) Many situations call for sensory evaluation of flavor, color, and odor to determine product quality. Thus special sensors and analyzers are required.
(3) There is often considerable seasonal fluctuation in the operation of food–processing equipment. This tends to drag down the overall operating rate.
(4) Characteristics of the controlled objects can change quite readily with variations in temperature or over time. Also, since most food–industry materials are solids, material handling methods are required.
(5) The process scale is often small, and a diverse range of products are produced in small–lot batches; thus, sequence control is essential.
(6) Since the instruments come in contact with food products that are susceptible to spoilage, sanitation is a primary concern. To withstand the rigors cleaning and disinfecting, such devices must be heat resistant, chemical resistant, and of course capable of being sanitized. To meet these stringent requirements, the

food-processing industry uses sensors that are designed specifically for sanitary environments.

(*b*) *Sanitary detectors*

In contrast to sensors for general industrial applications, sensors for use in the food industry are subject to extra stringent sanitation requirements. This is only natural considering the direct link between food product quality and health.

In order to maintain the quality of foodstuffs, the following sanitation gidelines are observed:

(1) prevention of bacterial contamination;
(2) prevention of deterioration of food-product constituents;
(3) prevention of mixing contaminated foods; and
(4) proper use of food additives.

There are two international standards that stipulate in general terms how these are conditions are to be met. These are the 3-A Sanitary Standards and the international standards compiled by the International Dairy Federation (IDF). With respect to sensors used as instrumentation devices, these standards stipulate four requirements pertinent to sanitation that should be complied with; compatible with a high level of product safety, good cleanability, easy to disassemble, and easy to inspect. For cleaning, a particularly important concept is *cleaning in place* (CIP). All pipes and devices should be capable of being cleaned and disinfected using either base or acid cleansers and hot water (or, steam) without them being removed or disassembled. This CIP requirement also applies to sensors.

10.4.2 Whiskey distillery instrumentation

Whiskey can be broadly categorized into three types based on the mixture of spirits that goes to make it:

1. *Single-malt whiskey*: A single malt stock.
2. *Pure-malt whiskey*: A blend of malt stocks.
3. *Blended whiskey*: A blended product made from malt stock and grain whiskey.

Traditionally, the most commonly consumed whiskey has been blended whiskey, but the demand for pure-malt whiskey has increased in recent years. Malt and grain whiskeys are produced in exclusive distilleries, where the immature sprits are produced, allowed to age, and finally blended to make the final product. Here we will focus on the instrumentation in a malt-whiskey distillery and the instrumentation applied to lautering equipment.

(*a*) *Overview of a malt-whiskey distillery*

A malt-whiskey distillery consists of eleven basic process steps, as shown in Fig. 10.64, all of which are carried out as batch processes.

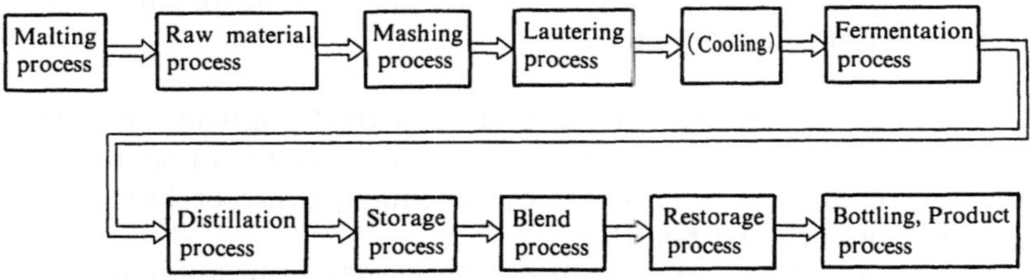

Fig. 10 · 64 Processes in a malt whiskey distillery

(1) *Malting*: Clean barley, the main material from which whiskey is made, is steeped, allowed to germinate, dried in a kiln, and screened to remove the culm, or dried sprouts.

(2) *Raw material processing*: The malt is stored, separated, and ground into a meal, or grist. Separation is to remove waste parts of the grain, and grinding stimulates the production of enzymes capable of saccharifying the raw grain. The grist is then stored as dictated by the production plan of the distillery.

(3) *Mashing*: The grist is mixed with hot water in a mash tub, and allowed to soak for 3 to 7 hours until the water has liquefied all of the starches, and the diastase has converted the starches into maltose and dextrin. The liquid now is known as wort.

(4) *Lautering*: The wort is transferred to a lauter tub, a tank with a perforated false bottom, where a filter press method is applied and the insoluble grain residue is separated from the wort.

(5) *Cooling*: The wort is cooled to 20°C using a plate coil or similar cooler, and then pumped into fermenting vats.

(6) *Fermentation*: Pure cultured yeast is added to lautered wort and fermentation continues for 3 to 4 days. When fermentation is completed, a liquid with an alcohol content of 5 to 8% has been produced known as mash.

(7) *Distillation*: The mash goes into a pot still for two distillations. Water is added to the now immature malt whiskey to reduce the proof to under 60%.

(8) *Storage*: The immature whiskey is then put up in oak barrels and aged for 3 to 8 years.

(9) *Blending*: After maturing, malt whiskeys are blended first and then grain whiskeys are blended in to achieve the proper aroma, flavor, etc. The blended whiskey is then rebarreled and allowed to marry for about a month before bottling.

(10) *Bottling and product processing*: After this final aging, the blended whiskey is reduced to a standard proof and bottled for market.

(*b*) *Lautering equipment instrumentation*

In this section, we will consider the instrumentation requirements for gravity–filter–type lautering equipment, a type of filtration equipment designed to separate the mash into wort and spent grain. The spent grain itself serves as a filter thus making the use of another filter unnecessary.

(1) *Lauter filter sequence control* As is shown in Fig. 10.65, the process begins by charging a fixed amount of water into the equipment (a process called spurgging). Then, after transferring the mash (*i.e.*, wort + spent grain) from the mashing process, the mash and water are circulated at a constant rate of flow for a fixed period of time (in some cases, the time is determined by the quantity of mash added). After the spent grain has been extracted, the wort is cooled as it is transferred to the fermentation process. Next, water is added at both top and bottom of the lauter tub, and husks and other grain residue are extracted by means of the awn–hacking unit. The water added at the top of the tub is called spurgging, and water added at the bottom is called backwash. After a fixed amount of water has been added, it is circulated at a uniform rate for a fixed interval. The second time, the liquid is extracted from the grain. The wort is then sent back to the fermentation process passing through the cooler on the way. This is called the No. 2 lautering.

By repeating the spurgging (backwash), circulation, and lautering

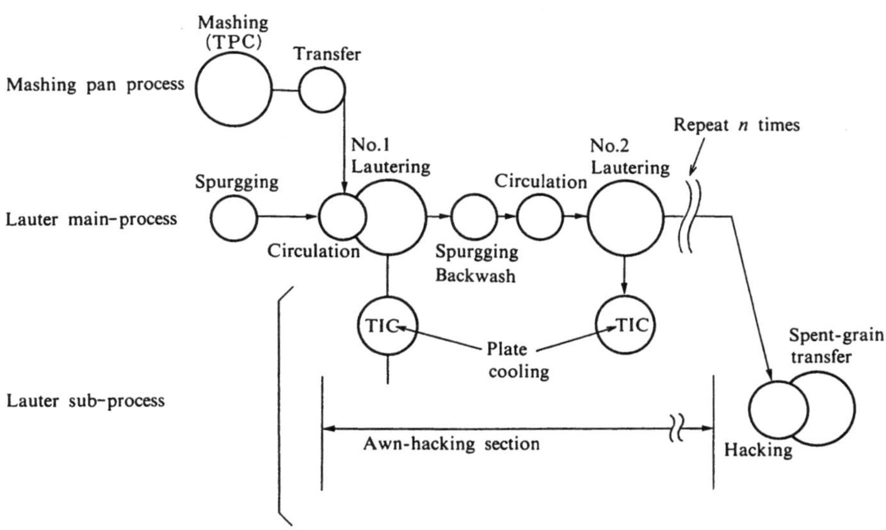

Fig. 10 · 65 Block diagram of Lauter filteration process

a number of times, the spent grain is completely removed, and one batch is completed.

(2) *Lauter equipment instrumentation and control* The instrumentation control involved in the operation of lauter equipment is shown schematically in Fig. 10.66.

(1) *Mash flow rate control* (*FQC–11*): Mash (wort + spent grain) from the mashing process is supplied to the main equipment in fixed amounts by means of batch–add control. Usually, an electromagnetic flowmeter is used to measure the flow rate.

(2) *Spurge and backwash flow control* (*FQC–8*): For this control, an electromagnetic flowmeter mounted on the mashing water supply

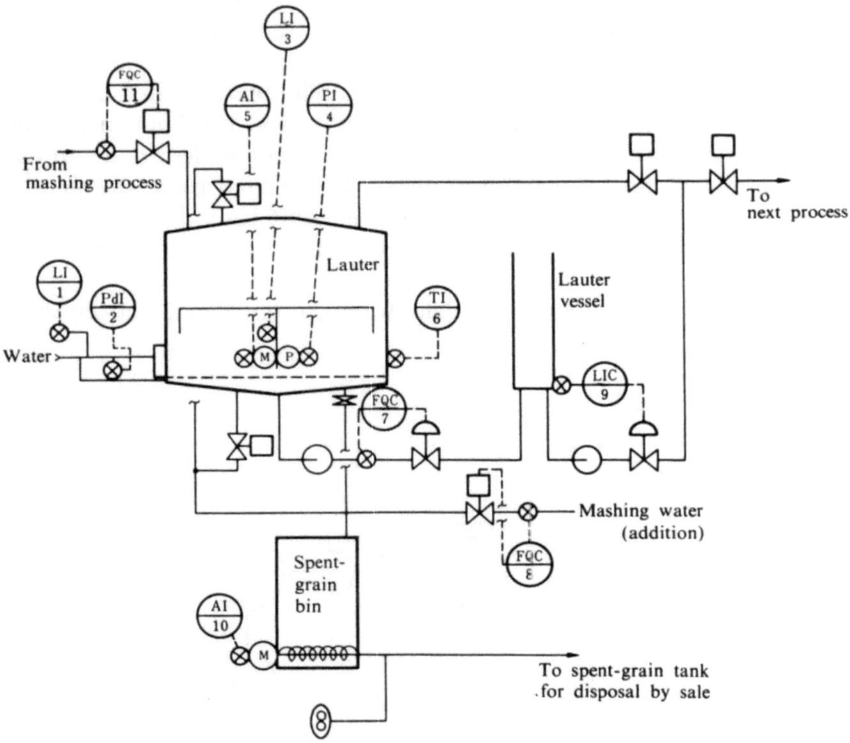

Wort level (LI-1)
Lauter differential pressure (Pdl-2)
Axial height (LI-3)
Axial oil pressure (PI-4)
Axial current (AI-5)
Wort temperature (TI-6)
Wort outlet flow rate (FQC-7)
Mashing water flow rate (FQC-8)
Lauter vessel level (LIC-9)
Spent-grain bin · Screw conveyer current (AI-10)
Mash flow rate (FQC-11)

Fig. 10 · 66 Overview of Lauter instrumentation

line is used to maintain batch-add control.

(3) *Circulation and lauter flow control* (*FQC-7*): Circulation flow rate and lauter flow rate to extract spent grain and lauter are carried out by fixed valve control using orifice, vortex, or electromagnetic flowmeters. Moreover, the valve-opening program control is governed by a control valve as well. Note that in cases where the wort extraction is based on lauter flow control, lauter differential pressure (PdI-2) is sometimes used to determine the timing.

(4) *In-lauter temperature* (*TI-6*): To monitor the quality of the wort, a recording device regularly records the temperature within the lauter.

(5) *Other measurements*: Other measurements besides those already mentioned include: axial height (LI-3), oil pressure (PI-4), current (AI-5), and spent-grain screw conveyer current (AI-2).

Another technique of regulating the circulation volume is to use level control (LIC-9) applied on the circulation line attached to the lauter vessel.

The smooth operation of the lauter equipment depends on the integration and synchronization of all the measurement controls and sequence controls described.

10.4.3 Sugar refinery instrumentation

Considered a daily necessity by people around the globe, sugar is an international commodity that is widely produced and widely consumed.

World-wide production is close to 100 million metric tons (in 1982), roughly 60% of which comes from cane sugar and 40% from beet sugar. There are three different kinds of sugar refineries specializing in the processing of raw sugar, beet sugar, and refined sugar. Here, we will focus on the typical manufacturing processes of a refined sugar processing plant. The various kinds of sugar products are shown in Fig. 10.67, and a glossary of technical terms used in the sugar industry is

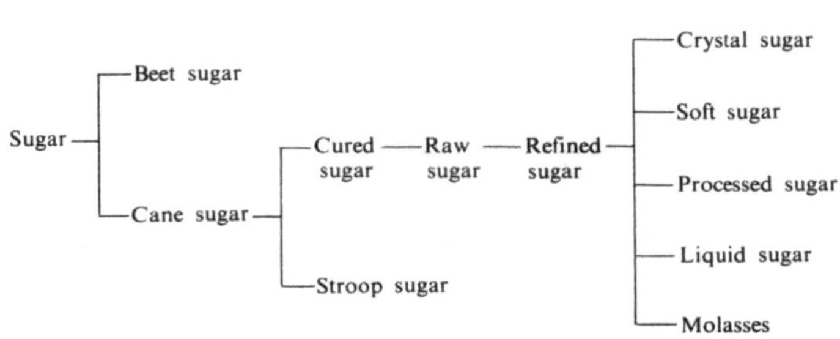

Fig. 10 · 67 Sugar products

Table 10.6 Technical sugar terms

Terms	Definition
Washed sugar or affinated sugar	Sugar which is washed and separated by centrifugal machine
Magma	Mixture of sugar and water, syrup or molasses
Liquor	Liquor consisting of sugar dissolved in water, so far not treated by crystalation process
Affination syrup	Syrup which is separated from magma of raw sugar and devided into two groups; green syrup and white syrup
Raw liquor	Liquor consisted of washed sugar dissolved in hot water or sweet water
Fine liquor	Final liquor of washing process, just before being supplied to crystalization pan
Sweet water	Thin liquor which is obtained after washing equipments
Syrup	At raw sugar plant; liquor which is concentrated in evaporator At sugar refinary; liquor which is separated by centrifugal machine
Juice	Liquor which is squeezed from raw materials (sugar cane, sugar beet)
Molasses	Syrup which is separated by curing machine (term used in raw sugar plant)
Visco	Thick invert sugar liquid, which is added in manufacturing of soft sugar and includes a mixture of more than 90% fructose and glucose
Massecuite	Mixture of crystal and mother liquor of sugar which is acquired from boiling at pan
Boiling	Operation which separates sugar cristals by heating in pan
Brix degree, Bx	Concentration unit of sugar shown by weight percentage of solid part Ex. 60 Bx (60 Brix degree) Liquor contains 60% sugar
Seed	Nucleus of crystal (seed)
Twin	Two single crystals stuck on a specified plane
Agglomerate	Crystal, based on agglomeration of multiple seeds other than twins
Smear	Crystal which grows spontaneously
MA (mean aperture)	Mean value of grain size in product
CV (coefficient of variation)	Dispersion of grain size in product
NCS (natural crystal score)	Ratio of monocrystal in product
Purity	Sugar ratio in solid content of liquor
Final molasses	Molasses, which has finished the recovery process

provided in Table 10.6.

(a) Overview of sugar refinery processes

(1) *Raw sugar feeding* Raw sugar transported by ship from overseas is unloaded and moved by conveyer belt to a raw warehouse. Then, in measured quantities, the raw sugar is transported by conveyer to a raw sugar storage bin.

(2) *Washing and melting* Many impurities adhere to the surfaces of sugar crystals. To remove these impurities, they are first loosened by mixing with juice in a mixer-type scroll conveyer called a mingler. The juice is then purged in a affination centrifugal, and dissolved in a melter to become coarse sugar juice.

(3) *Carbonization and filtration* Lime milk is added to the coarse sugar juice, which reacts with carbon dioxide gas in a carbonator to produce calcium carbonate. Impurities adhere to the calcium carbonate which is then passed through a primary filter to remove these impurities, yielding a clear amber-colored sugar juice.

(4) *Decolorization and concentration* The filtered sugar juice is passed through a char cistern and an ion exchange resin tower. These processes decolorize and refine the juice. Then, after passing through a check filter, the juice goes to a concentrator where it becomes more concentrated fine liquor.

(5) *Crystallization* The fine liquor output from the concentrator is then crystallized in vacuum crystallization pans. After the concentration of the syrup is increased even further, product-specific seed crystals are added and sugar crystals grown to uniform standard size as syrup continues to be added.

(6) *Purging, drying, and finishing* Granular sugar crystals are extracted, or purged, from the mixture of crystals and surrounding syrup (massecuite) in a product separator. Remaining syrup is returned to the crystallization process to reclaim further sugar. The final end product of the process is final molasses. The extracted sugar crystals are passed through a dryer and cooler, then stored in a humidity- and temperature-controlled sugar bin. Next, the sugar is processed through a multistage sieve to separate it into standard-grade products.

(7) *Packaging* How sugar is packaged depends largely on the type of sugar and how it is to be used. Sugar is usually either packed in bags and warehoused or shipped directly in loose bulk.

The entire process from receipt of the raw sugar to final packaging is shown schematically in Fig. 10.68.

(b) Key sugar refinery controls

(1) *Melter Brix (concentration) control in the sugar melting process*

Raw sugar, the raw material for this process, is mixed with heavy syrup in a mingler to form a magma. The magma is separated into wa-

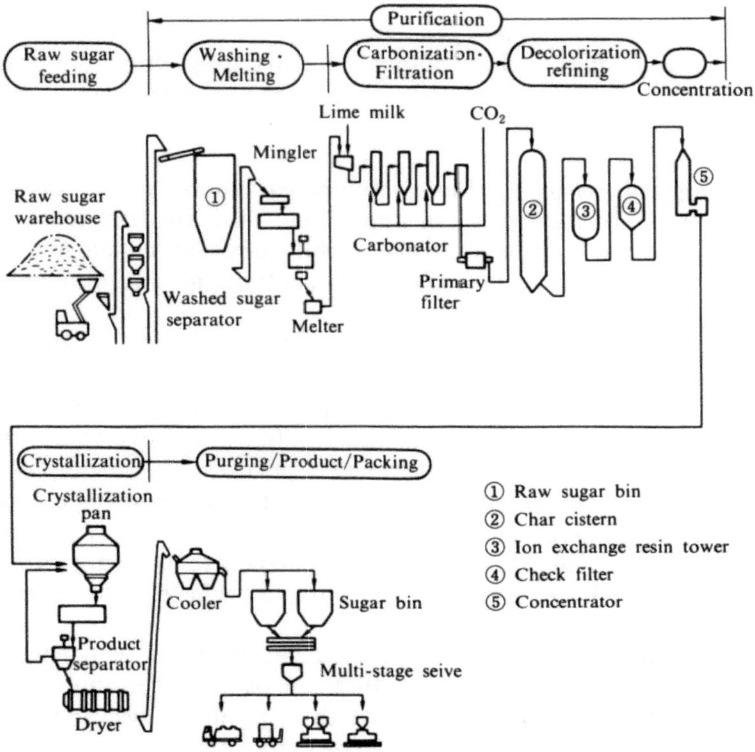

Fig. 10 · 68 Manufacturing process of sugar plant

shed sugar and green syrup in a centrifugal separator, and the washed sugar is then dissolved in sweet waters in the melter to become raw liquor, and sent to the carbonation process. An instrument called a Brix meter is used to determine the concentration of the raw liquor. Differential-pressure type Brix meters are especially widespread. the configuration of the control loop involved is shown in Fig. 10.69.

Key points respecting the instrumentation are as follows.

(i) Since washed sugar is intermittently withdrawn from the centrifugal separator, the proper amount of sweet water charged to the centrifugal is automatically adjusted and synchronized with the centrifugal's discharge signal.

(ii) Sufficient residence time is necessary to produce high-concentration raw liquor with a stable Brix (concentration) value. This is accomplished by adding most of the sweet water and approximately adjusting the Brix value in the pre-melter. The Brix value is then fine-tuned with only a small quantity of sweet water in the main

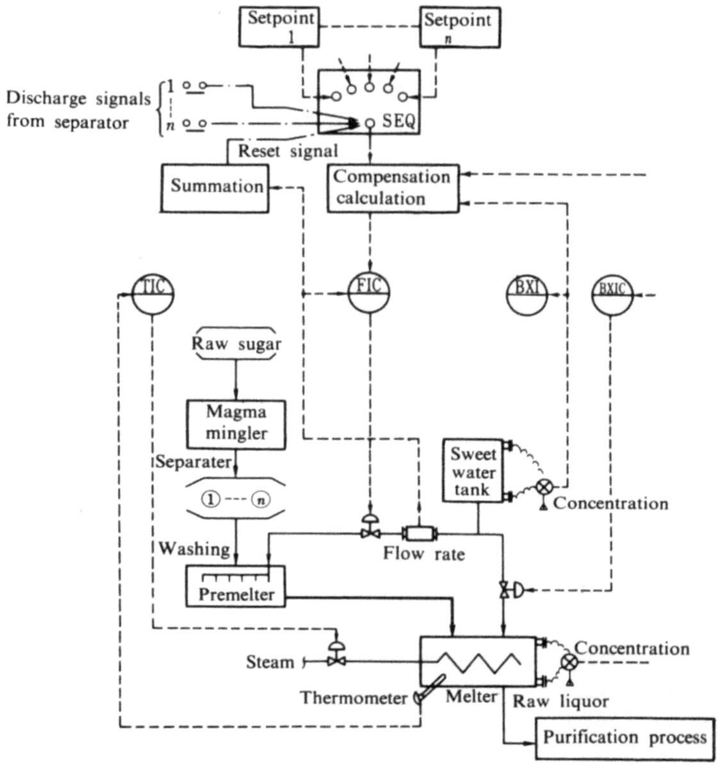

Fig. 10 · 69 Melter-Brix control system

melter.

(iii) By monitoring the Brix value of the sweet water, the quantity of sweet water going into the pre-melter is automatically compensated. This control makes it possible to stabilize the Brix value and obtain high-concentration raw liquor. The burden on down-stream processes is thus reduced and energy is conserved.

(2) *pH control in the carbonization process* In the carbonator, lime milk is added to raw liquor and it reacts with carbon dioxide gas to cohere impurities. Accurate pH control is extremely important in this process. The control loop configuration is shown in Fig. 10.70. The key points respecting the instrumentation are as follows.

(i) The flow rate ratio between the raw liquor and the lime milk in process must be controlled.

(ii) The lime milk flow ratio is corrected based on the concentration of the lime milk and the pH of the carbonator. The control not only stabilizes the pH but also improves the effectiveness of the

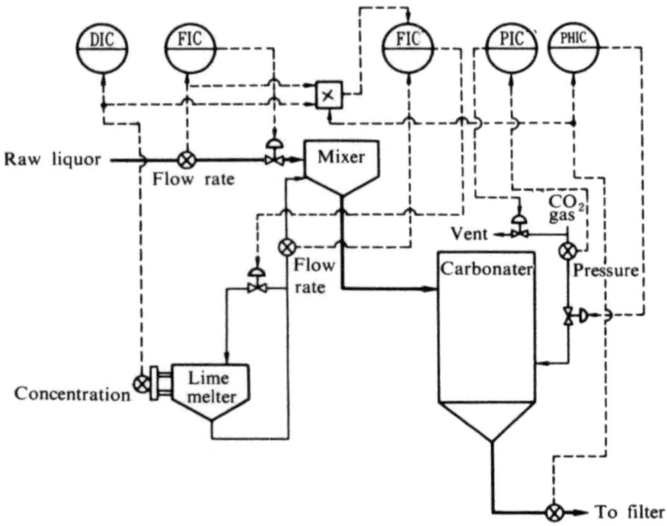

Fig. 10 · 70 PH Control in the carbonater process

following filtration process.

(3) *Flow control and sequence control for the filters* The primary filter removes impurities that cohere to calcium carbonate to yield a clear amber–colored sugar juice.

Sequence control is applied to this operation mainly to coordinate standby, pre–coat, filtration, residual blowing, desugaring, and sluicing processes. This filtration is carried out as a batch process, but sequence control is required because multiple filter lines are employed. Sequence control coordinates the different processes, and also synchronizes filter–line rotation.

The control loop configuration is shown in Fig. 10.71, and a time chart showing how the operation of the different filter lines are staggered is shown in Fig. 10.72. Key points respecting the instrumentation are as follows.

(i) Flow control cascaded by the carbonation tank level and pressure selector control at the filter are carried out.

(ii) The standby, sluicing, and multi–line filter system are under sequence control.

(iii) An operation should be implemented so as to maximize the efficiency of the filter cloth (*i.e.*, the quantity of liquid passed through the filters).

With these control measures, automatic operation can proceed on the basis of combined capacity of multiple filter lines with minimal waste, to achieve improved operational efficiency.

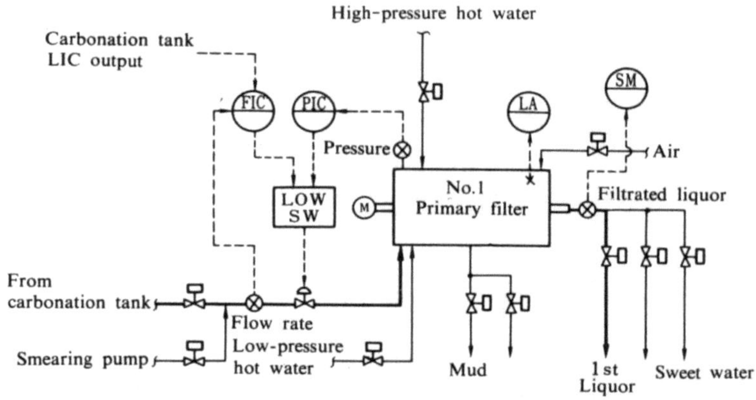

Fig. 10 · 71 Flow and pressure control for filter

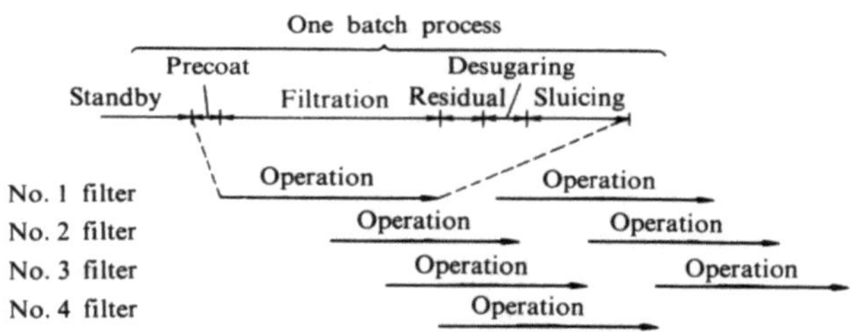

Since the precoat line is imprimented as a common line to the filtration operation, operation of the four filter lines should be staggered so that the precoat process is not duplicated

Fig. 10 · 72 Operation system of filters

(4) *concentrator control* In the concentrator process, decolorized, refined liquor is boiled down to the desired concentration, then sent to the crystallization process. The control loop configuration is shown in Fig. 10.73.

Key points respecting the instrumentation are as follows.

(i) Feedforward control is implemented with an operational element to calculate changes in feed concentration, the largest external disturbance. This measure reduces the adverse effects of lags in the outlet concentration control loop to a minimum.

(ii) Using (a) the difference between the desired outlet concentration and the measured concentration, and (b) the steam flow, the following calculation can be used to stabilize the set point of the desired inlet flow concentration without altering the steam flow:

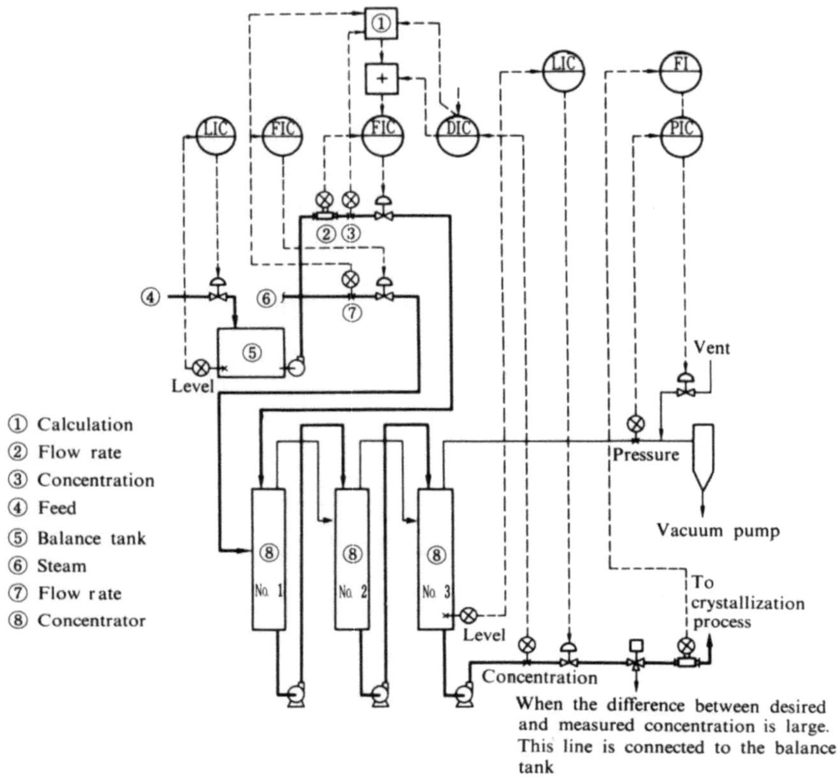

Fig. 10 · 73 Control system for the concentration process

$$F_i = \frac{B_q F_s (1 + k_2)}{(B_q - B_i)(1 + k_1 B_i)}$$

where F_i is the inlet flow (m^3/h), F_s is the steam flow (t/h), B_q is the desired outlet concentration (0 to 1.0), B_i is the inlet concentration (0 to 1.0), k_1 is the concentration conversion factor, and k_2 is the evaporation multiplying factor.

Homogenization of the syrup concentration is essential to deliver a stable volume of syrup to the crystallization pans in the next process.

(5) *Control between tanks in the purification process* The downstream crystallization process requires a stable supply of fine liquor from the purification process. Thus, the buffer tanks between equipment units must be monitored at all times by follow-up flow control of the amount of product (melt) making sure the tanks neither overflow nor run dry. The control loop configuration is shown in Fig. 10.74.

Key points respecting the instrumentation are as follows.

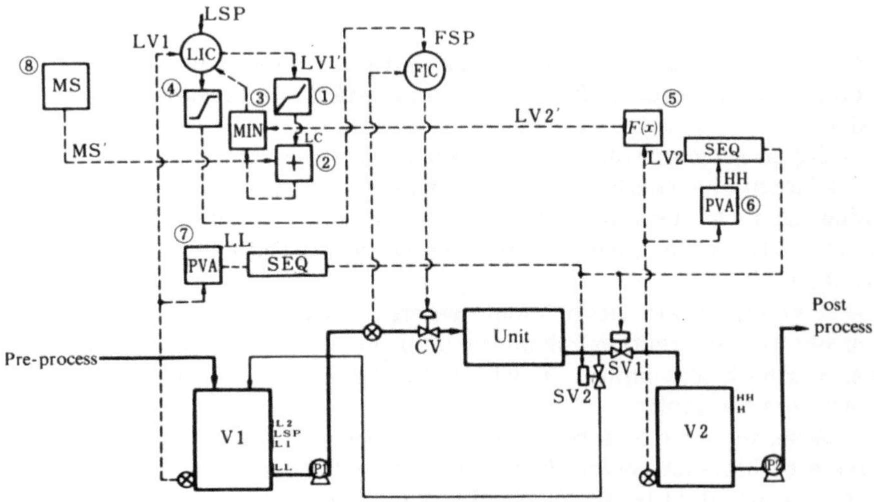

① Nonliner calculator
② Adder
③ Signal selector
④ High-low limiter
⑤ Function calculator
⑥ High-low alarm unit
⑦ High-low alarm unit
⑧ Constant value generator

Fig. 10 · 74 Loop configration of flow control between tanks

(i) The most basic control loop implements a cascade-control of the flow loop linked with the level. The flow and level setpoints for each process are calculated from the production (melt) volume.

(ii) The volume of liquor passed by each process is determined by supervising the level of buffer tanks both before and after the respective processes. The level signals are then used for the flow loop set points after correction.

(iii) The level high-low values of each buffer tank are monitored on a continuous basis. If the level deviates from the set range, a return valve is opened to redistribute liquor as required.

A number of points can be made with respect to the control loop configuration shown in Fig. 10.74. We will assume that the setpoint (LSP) of the level controller (LIC) for tank V1 has been pre-set to maximum capacity to make the best use of the tank. A nonlinear calculation of the deviation between the level setpoint (LSP) and the actual level (LV1) of the tank is calculated by calculator ① and input to calculator ②.

In the meantime, the processing volume setpoint (which includes the production volume) is output by the constant value generator (MS), and added in to the calculation of adder ② such that

$$F_k = M_s' + L_c$$

where F_k is the addition value, M_s' is the volume of processed liquor (determined by the production volume), and L_c is the level correction value.

Signal selector (the minimum value) ③ then selects the lesser value between the result obtained by adder ② (*i.e.*, the level correction value plus the processed liquor volume) and the level of tank V2 (LV2′). This value then becomes the setpoint (FSP) of flow loop controller FIC.

Two other calculators should be mentioned at this point; loops are configured for function calculator ⑤ used to correct the level of tank V2, and high-low limiter ④ whose output is used to assure minimum flow from the tank.

Also, in order to activate solenoid valves SV1 and SV2, the combination of high-low level alarm units ⑥ and ⑦ are used to detect the high-low limits of tanks V1 (sending) and V2 (receiving). Thus, if a situation develops where not enough flow is coming from tank V1, the overflow from downstream tank V2 is shut off.

(6) *Crystallizer control* Syrup close to saturation is delivered from the previous process and charged to the crystallizer (intake). Since the sucrose component would be decomposed at high temperature, evaporation is carried out in vacuum pans at low temperature (60°C) until supersaturation is reached (concentration). At the proper moment, seed crystals are added to form the grain (seeding). Then, as the boiling progresses, syrup and feedwater supply valves are intermittently controlled in response to hardness and level signals until the crystals reach their proper predetermined size (crystal-growing). The finished product is then discharged and sent to the next process (sugar discharge).

Key points respecting the instrumentation are as follows.

(i) Sensors (hardness meters, more properly, rheometers) are required to precisely monitor the progress of crystal growth in the crystallizer.

(ii) Crystallization is carried out based on a program control system that has been implemented based on the knowhow of a seasoned operator with extensive experience.

This control system, implemented with hardness meters and a microprocessor, is called an automatic intermittent syrup boiling system.

The control system configuration for the crystallizer is shown in Fig. 10.75, and a schematic overview of the control operation is shown in Fig. 10.76. A hardness meter, or rheometer, is shown in Fig. 10.77.

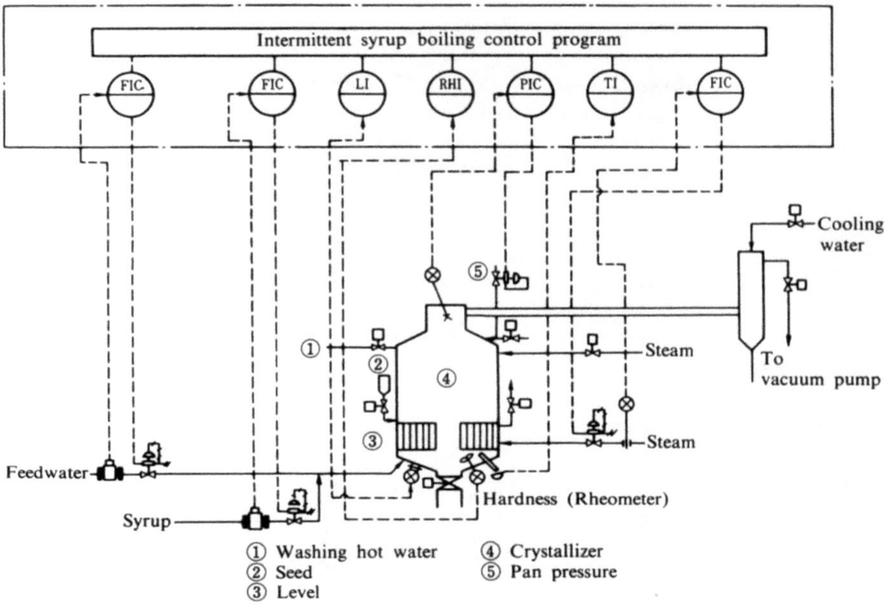

Fig. 10 · 75 Crystallizer control loop configration

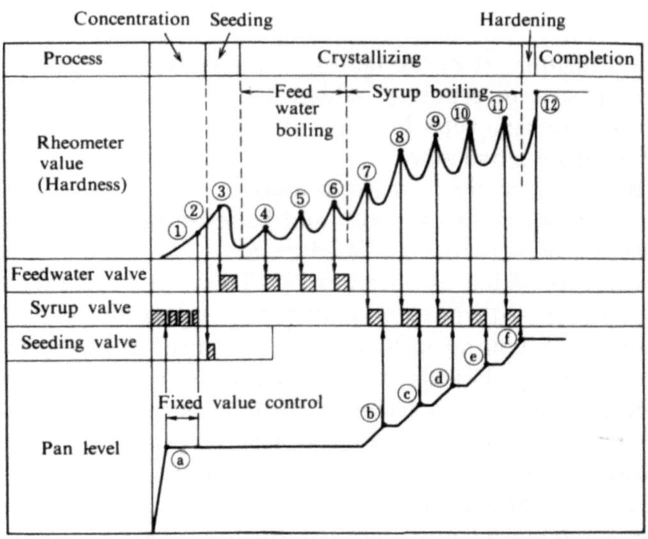

Fig. 10 · 76 Operation of intermittent syrup boiling system

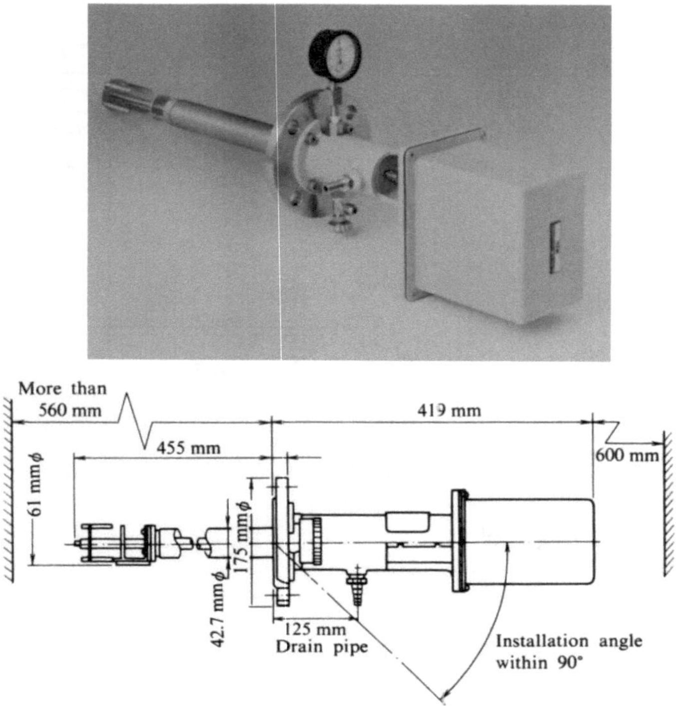

Fig. 10 · 77 Hardness meter (Rheometer)

In the intermittent syrup boiling system, fixed amounts of feedwater or syrup are alternately added to the pans as predetermined rheometer (hardness) setpoints are reached, until the sugar crystals achieve their prescribed size.

Briefly, the operation works as follows.

Syrup is charged into the pan and boiled rapidly until supersaturation is reached at setpoint ①. Seed crystals are added at setpoint ②, and feedwater is added at setpoint ③. Then, at setpoints from ④ to ⑥, feedwater is intermittently added (feedwater boiling). This is followed by syrup boiling in which syrup is intermittently added at setpoints from ⑦ to ⑪. Finally, the hardening is finished at setpoint ⑫.

It can be seen that various control actions are triggered in the crystallizer as rheometer values reach setpoints from ① to ⑪. Next, each addition of syrup during the syrup boiling phase is controlled by detecting a series of level in the pan from ⓐ to ⓕ.

(*c*) *Total production control system at a sugar refinery*

Today, sugar refineries, in common with many other kinds of plants, must find new ways to conserve energy, resources, and labor.

More than ever before, advances will be won by improving product quality and by catering to the more diversified tastes of consumers by implementing production systems capable of turning out a great range of products in small lots. Key to this development is the establishment of integrated production control systems based on computers.

(1) *Hierarchical production control system structure* A production control system is an information management structure for organizing the information required to produce a product of consistent quality and within in fixed time period by the most economical means possible. The task is complicated by circumstances, both within and external to the plant, that are easily subject to change.

Such a structure cannot be realized by merely mechanizing part of a single process. Rather, the entire productive system must be optimized so it achieves economic balance.

Integrated production control systems commonly assume a hierarchical configuration with purview over every step from receipt of the raw material to shipping the finished product.

At a sugar refinery, since the sugar process and the packing and delivery process have different functional requirements in terms of control, different control systems are applied in each case. Also, because the information used by the various systems making up the hierarchical structure is different, a centralized control system is necessary to combine and integrate the disparate information from the other systems.

The structure of a hierarchical integrated production control sys-

Fig. 10 · 78 Hierarchical system structure of sugar refinery tem for a sugar refinery is shown in Fig. 10.78.

(2) *Production control functions* Production control functoins can be separated into the optimal schedule level and the operation control level.

The optimal schedule level, starting with the production schedule based on the business plan, also includes the boiling schedule and the packing schedule. Once these basic schedules are in place, the production volume (melt volume) can be established for the process control system. This makes it possible to determine the operating schedules for the crystallizer, the product separator, and the packer.

The operation control level encompasses a great number of control items including boiling centrifuge control, product inventory control, and delivery control. Through interface with the various control systems which monitor conditions throughout the plant, it has become possible to implement on-line operation control in real time.

The hierarchical structure classified according to functions is shown in Fig. 10.79.

10.5 Instrumentation Applications in the Paper Manufacturing Industry[20)–23), 38)]

10.5.1 Overview of an integrated paper mill

Paper manufacturing plants can be broadly classified as either pulp plants or paper machine plants. Pulp is delivered to the latter type of plant as the main raw material for the manufacture of paper.

Pulp plants produce pulp from wood or other fibrous raw material. There are two general processes for producing wood pulps; mechanical and chemical. In this section we will be primarily concerned with *kraft* pulp processes, that fall under chemical processing.

A schematic overview of the kraft pump manufacturing process is shown in Fig. 10.80. The wood for raw material is first put through a chipping process where it is cut up into small, uniform-sized chips. The chips are then conveyed to the cooking process, where they are charged to digesters. A chemical solution, *i.e.*, cooking liquor, is added to the chips and the mixture is cooked with steam at a temperature of about 170°C at a pressure of several tens of kg/cm^2. After reacting over several hours, lignin and other interfiber bonding materials in chips are dissolved. When the cooked chips are discharged from the digester to the blow tank, they are reduced to a suspended fibrous slurry called brown stock that is already termed pulp. After washing and screening, the pulp is then sent to the bleaching process. Here the pulp is treated with the bleaching agents chlorine, sodium hypochlorite (hypo), chlorine dioxide solution, and sodium hydroxide. Oxygen is sometime also applied at this stage to remove residual lignin still adher-

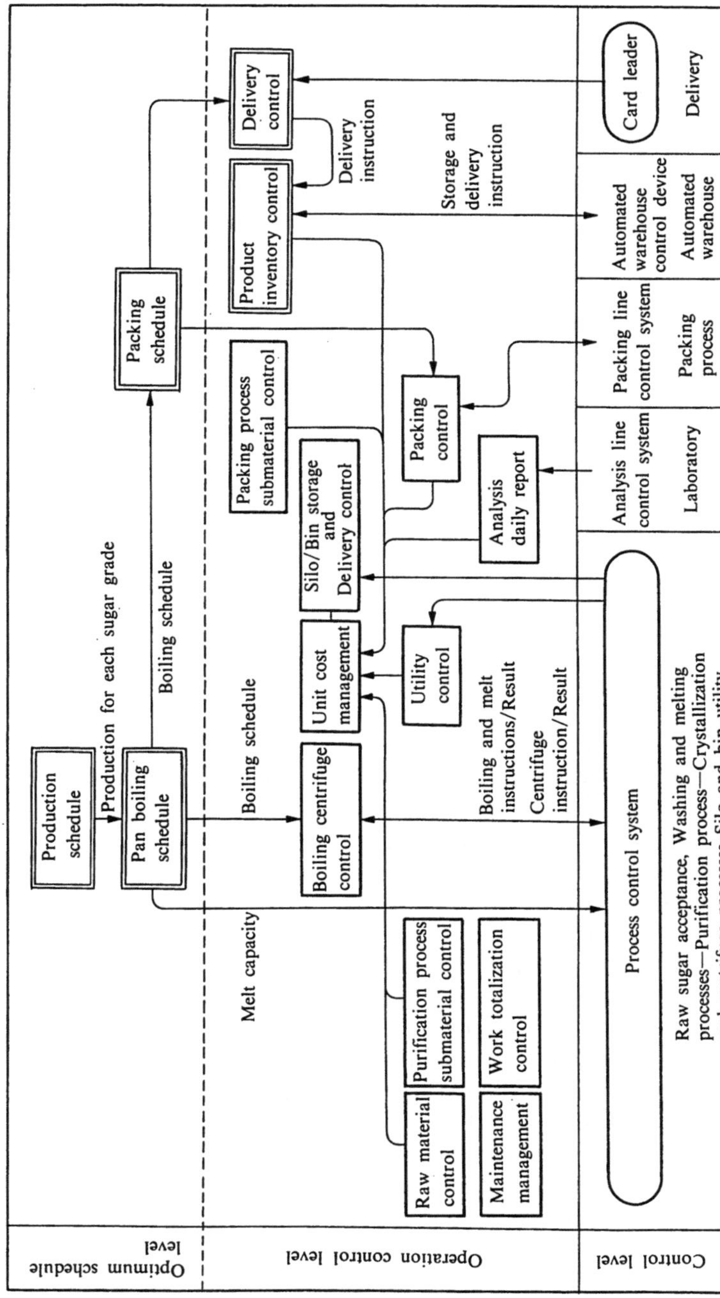

Fig. 10 · 79 Hierarchical structure classified by functions

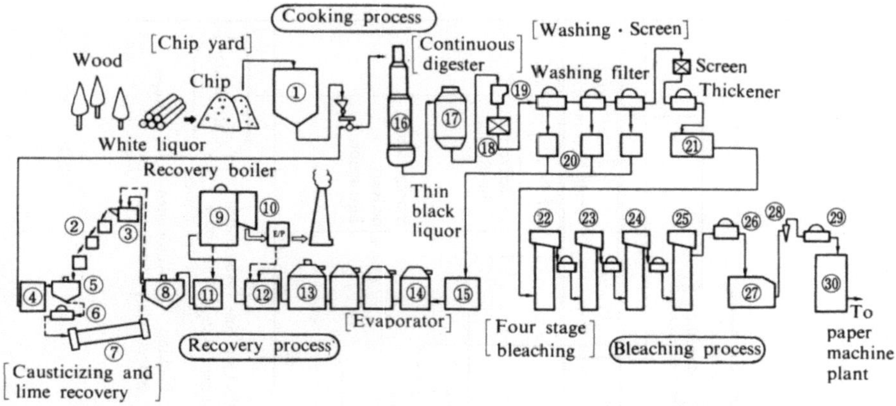

① Chip silo
② Causticizing tank
③ Lime slaker
④ White liquor tank
⑤ White liquor clarifier
⑥ Lime mud filter
⑦ Lime kiln
⑧ Green liquor clarifier
⑨ Recovery boiler
⑩ Economizer
⑪ Dissolving tank
⑫ Thick black liquor storage tank
⑬ High-concentration evaporator
⑭ Evaporator
⑮ Thin black liquor storage tank
⑯ Digester
⑰ Blow tank
⑱ Knotter
⑲ Screen
⑳ Filtrate tank
㉑ Unbleached pulp chest
㉒ Chlorination tower
㉓ Extraction tower
㉔ Hypo tower
㉕ Chlorine dioxide tower
㉖ Thickener
㉗ Bleaching chest
㉘ Cleaner
㉙ Thickener
㉚ Bleached pulp tower

Fig. 10 · 80 Outline of pulp plant

ing to the fibers. After passing through the bleaching process, the pulp is washed and screened again and sent as raw material to the paper machine plant. In the meantime, the spent cooking liquor, now called black liquor, is put through an elaborate recovery system. After being concentrated through a condensation process, the organic black-liquor solids are burned in recovery boilers and the inorganic compounds are subjected to reduction recovery. The inorganic compounds (green liquor) are put through an additional process where it is causticized with lime. The resulting clarified white liquor is returned to the digester as cooking liquor.

A typical paper plant is schematically shown in Fig. 10.81. In the initial stock preparation process, various kinds of pulps are mixed according to the type of paper that is to be produced. Also, fillers, dyestuffs, and other materials are added at this point depending on the surface smoothness, strength, color, and opacity desired for the final product. In the machine process, the prepared stock is delivered as a slurry (*web*) onto a moving endless finemesh wire screen (*wire*), where water is drained from the web. More water is forced out in the following press section by press rolls. Then, the web is dried in steam driers to

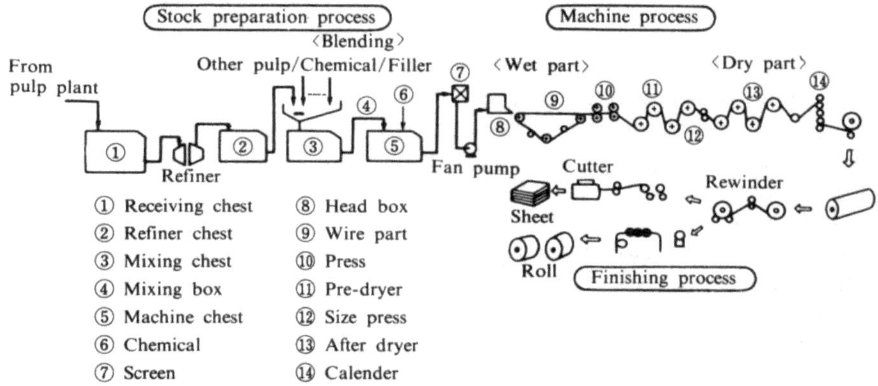

Fig. 10 · 81 Outline of paper machine plant

become paper. Throughout this processing, such qualitative factors as basis weight (weight per unit area), moisture content, and thickness are carefully controlled. The finished paper may then be sent on to a coating machine for application of sizing or coating to one or both surfaces.

Finally, the paper is sent to the finishing process where it is wound onto winders, cut, and/or trimmed to standard sizes, and wrapped for shipment.

10.5.2 Pulp plant instrumentation

[I] *Cooking-process instrumentation*

The disgestion process is carried out in a digester, of which there are two basic kinds; batch and continuous.

To illustrate instrument applications, we will describe a Kamyr-type continuous digester, the most common type of digester used in Japan.

(a) *Key instruments for continuous digesting*

A typical continuous digester is a cylindrical vessel 20 to 80 meters in height and 3 to 8 meters in diameter. As can be seen in Fig. 10.82, metered chips pass through steaming vessel where they are preheated and air is expelled. The steamed chips are then continuously fed concurrent with cooking liquor into the upper portion of the digester by a high-pressure rotary feeder. As the chips continuously move downward in the digester, the cooking reaction proceeds with additions of heat from the cooking liquor. More heat is provided by recirculating liquor that has been withdrawn and heated. By the time the chips have spent about half their retention time in the digester, the cooking reaction is more or less completed. As the cooked chips reach the bottom zone of the digester, they are plowed to a central well in the bottom of the digester while being counterflow-washed with washing liquor. The chips are then discharged from the bottom of the digester by means of a bottom scraper.

In the operation of a continuous digester, the ideal is to optimize two sets of values; the objective is to stay as close to the optimum chip-cooking values as possible while at the same time maintaining the various process parameters, which become primary factors of operation cost, at their optimum values. It is particularly important that the following are controlled:

(1) Chip/pulp balance,
(2) Cooking liquor addition ratio,
(3) Temperature at different zones in the digester, and
(4) Degree of washing.

(b) *Instrumentation methods*

(1) *Chip/pulp balance*

(i) *Chip feed control*: The amount of chips fed into the digester is determined by the rotational speed of the chip-meter. The chip-meter speed control is set based on the desired value of rotation, which is calculated from the target pulp production volume, the chip moisture content, the filling factor of the chip-meter, and the basic pulp yield.

(ii) *Chip-bin level*: To assure a uniform and stable chip-meter filling factor (*i.e.*, the quantity of chips delivered by one full turn of the chip-meter), the level of the chip bin must be controlled. The depth of chips in the bin is detected by means of either a ultrasonic level detector or a load cell, and control is maintained by regulating the feed volume from the chip silo.

(iii) *Chip level control in the digester*: To maintain the chip level inside the digester within a desired range, feedback data is used to adjust the chip feed volume. Chip level is detected with a gamma-ray level indicator.

(2) *Cooking liquor addition ratio* Referring to Fig. 10.82, it can be seen that the desired flow rate of cooking liquor is obtained by multiplying the ratio (addition ratio) times the chip feed volume. The basic addition ratio value is determined by the degree of cooking desired and the type of tree the chips are derived from. This basic value is then fine-tuned by factoring in cooking liquor alkalinity, moisture content of the chips, and so on.

(3) *Temperature at different zones in the digester*

(i) *Impregnation zone temperature control*: The upper portion of the digester above and around the cooking circulation strainer is called the impregnation zone, because it is here that the chips are impregnated with cooking liquor. The temperature in the impregnation zone is de-

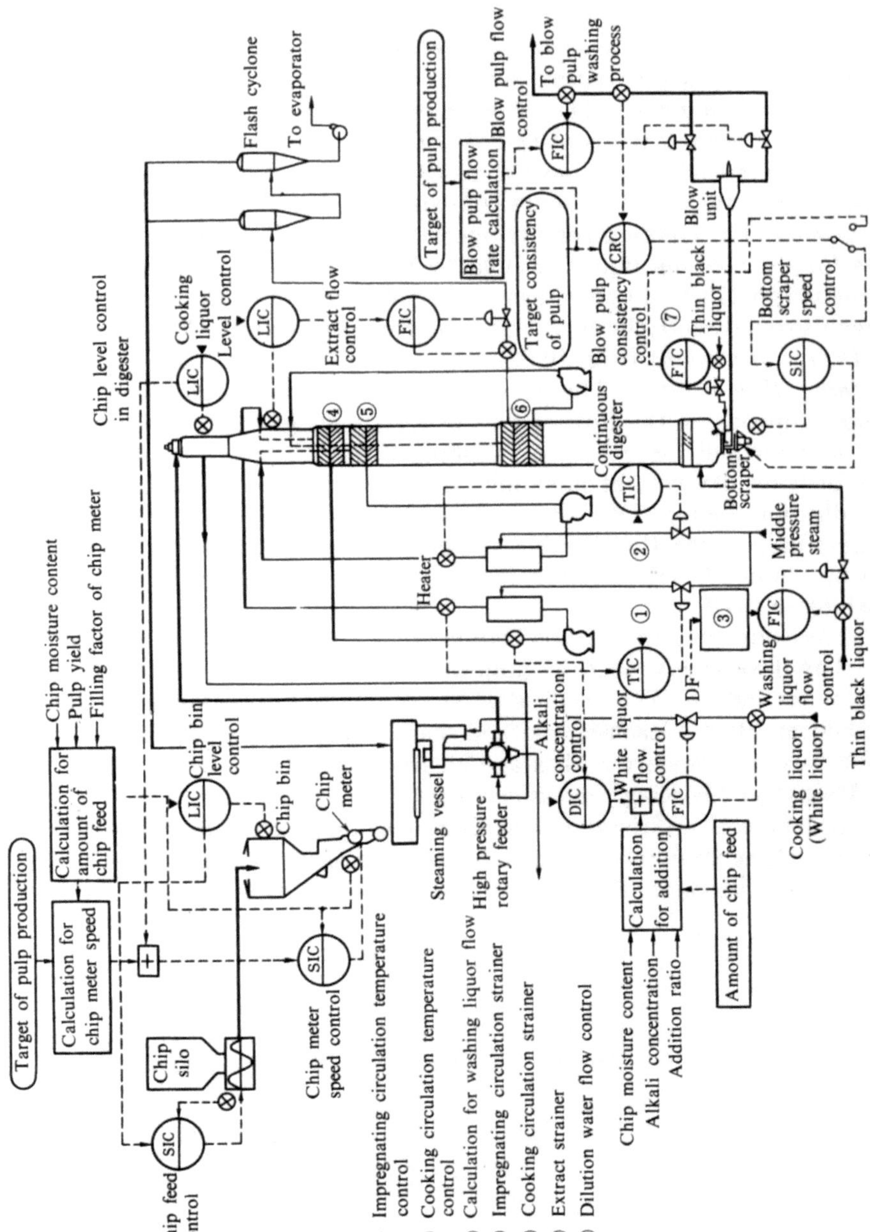

Fig. 10·82 Continuous digesting process and control

termined by the temperature of the cooking liquor that is circulated in the zone, which is controlled at about 145 to 150°C.

(ii) *Cooking circulation temperature control*: The section of the digester from the cooking circulation strainer to near the extract strainer is called the cooking zone. The cooking reaction, in which most of the delignification occurs, takes place here. Temperature as well as retention time in the cooling zone have a major influence on the cooking reaction; in fact, these are ultimate controlling determinants of the degree of cooking. The cooking zone temperature is regulated by the temperature for the cooking liquor circulating in the zone, and is controlled at about 165 to 170°C. In some cases the temperature of the cooking zone is monitored directly by inserting a thermal sensor into the digester.

(4) *Degree of washing* (Fig. 10.82)

Washing flow control: The flow rate of washing liquor (thin black liquor) depends on the flow rate of bone-dry weight pulp descending in the digester and the extent of washing desired. The ratio of the flow rate of bone-dry weight pulp and flow rate of washing liquor is called the dilution factor (DF). By multiplying the flow rate of bone-dry weight pulp times the dilution factor, the desired washing flow rate can be obtained. The dilution factor is set according to the degree of washing desired.

[II] *Bleaching-process instrumentation*

The bleaching process in pulp plants has undergone a true technological revolution. The traditional single-stage hypochlorite bleaching method or batch bleaching method using diverse bleaching agents has given way to a modern multistage bleaching process.

(a) *Multistage bleaching process overview*

In a typical multistage bleaching operation, separate reaction towers (bleaching towers) are implemented for each bleaching agent. The pulp passes through the towers one at a time where the successive

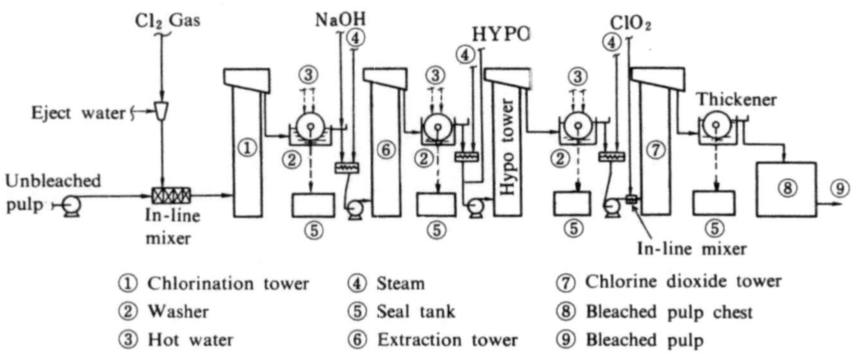

Fig. 10 · 83 Four-stage bleaching process flow

bleaching operations take place. Fig. 10.83 shows the flow of a typical four-stage bleaching process. The various bleaching stages have different functions as follows:

(1) *Chlorination stage* The unbleached pulp is subjected to chlorination treatment which facilitates extraction of the chlorinated lignin in the following process.

(2) *Caustic* soda stage In this stage, lignin and impurities that impart color are extracted.

(3) *Following the hypo stage* Remaining lignin is oxidized, extraction is repeated, and the desired degree of whiteness is obtained.

(b) *Instrumentation methods for each stage*

(1) *Chlorination stage* Figure 10.84 shows typical instrumentation for the chlorination stage. The final objective of this stage is to achieve a consistent *kappa* number (a measure of residual lignin) by the end of the caustic soda extraction. The amount of chlorine supplied is proportionate to the kappa number of the unbleached pulp.

(i) *Chlorine feed control*: The amount of chlorine injected is determined by first deriving the bone-dry weight which equals the volume flow of the pulp times pulp consistency. This product is then multiplied by the

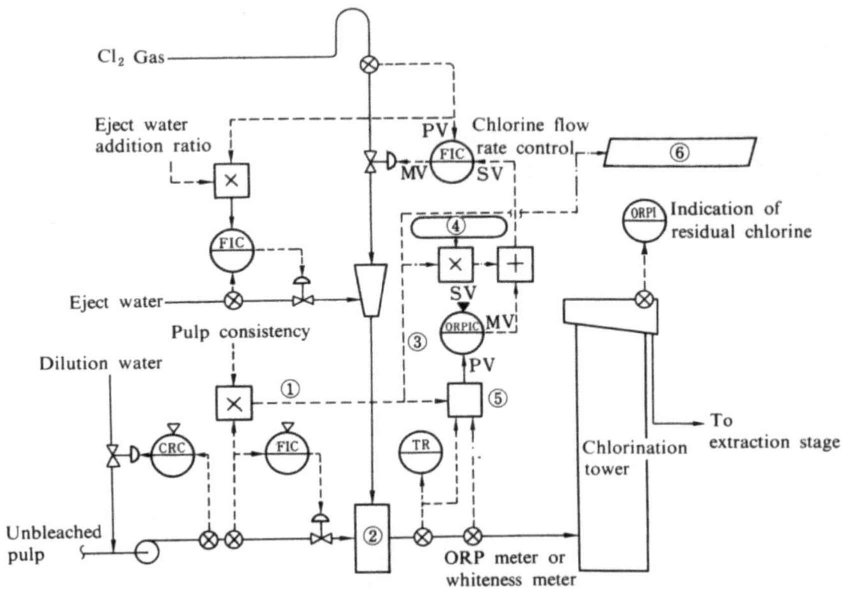

① Flow rate of bone dry pulp
② In-line mixer
③ ORP control or whiteness control
④ Chlorine addition ratio
⑤ Compensation for reaction velocity
⑥ Flow rate of bone dry pulp

Fig. 10 · 84 Control of chlorination stage

chlorine addition ratio to derive the desired value for flow control.

Once the chlorine has been added, residual chlorine is detected with a ORP meter for the purposes of feedback compensation. One problem with this method is that since the injected chlorine reacts very quickly after injection. The level of residual chlorine is indicated by the ORP meter, but flow speed and temperature fluctuations can alter the reaction speed so that errors are introduced. Thus, calculations to compensate for the time lag resulting from the detection position of the residual chlorine, as well as pulp liquor temperature, have to be factored in.

(ii) *Introduction of a whiteness gage*: The ORP meter in the chlorine feed control method indicates the inverse response of the residual chlorine to changes in the kappa number of the unbleached pulp. It is thus fairly difficult to get the right amount of chlorine to be injected. Recently, a feedback control technique has been developed to overcome this problem. It uses an optical sensor to detect pulp whiteness (whiteness gage).

(2) *Caustic soda stage* A flow diagram illustrating control of the caustic soda stage is shown in Fig. 10.85. The control objective for caustic soda feed is to maintain a consistent extraction kappa number after extraction. Because there is a high correlation between the extraction kappa number and pH value at the outlet of the caustic soda stage, pH control can be used to manipulate caustic soda feed volume.

(3) *Following the hypo stage* Control after the hypo stage is based on maintaining a constant ratio between the bone-dry weight flow of the pulp that has come through the previous two stages and the volume of

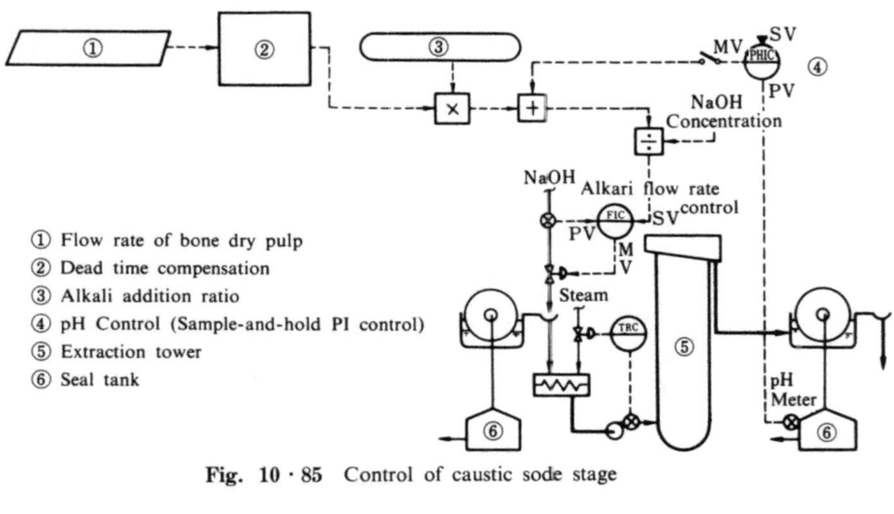

Fig. 10·85 Control of caustic sode stage

added chemicals.

As necessary, a whiteness gage is mounted at the outlet of either the hypo stage or the final stage. The detected whiteness is used to optimize the chemical flows of all the bleaching stages.

[Ⅲ] *Recovery processes instrumentation*

Chemical recovery processes include black-liquor concentration, recovery boiler, and causticizing and lime recovery processes. Here we will focus on

(1) Black liquor concentration process, and

(2) Recovery boiler process.

(a) *Black liquor concentration process*

(1) *Overview of black liquor concentration equipment* (*evaporator*) The liquid discharge from the digester to be recovered is called thin black liquor. The main constituents are cooking chemicals and lignin dissolved out of the chips. It has a soluble-solids content of 12 to 16%. Since the black liquor cannot be burned in the recovery boilers as is, it is first sent to evaporators where the liquor is concentrated under pressure to a solids content in excess of 60%.

Today, the evaporation is generally carried out in a multiple-effect evaporator with superior heat efficiency. The process flow of a multiple-effect evaporator is shown in Fig. 10.86.

(2) *Instrumentation methods*

(i) *Physical distribution control through evaporators*: As can be seen from Fig. 10.87, flow rate to the first evaporator must be controlled at a constant value coresponding to the volume of thin black liquor to be processed. Extraction of concentrated black liquor from each evaporator is

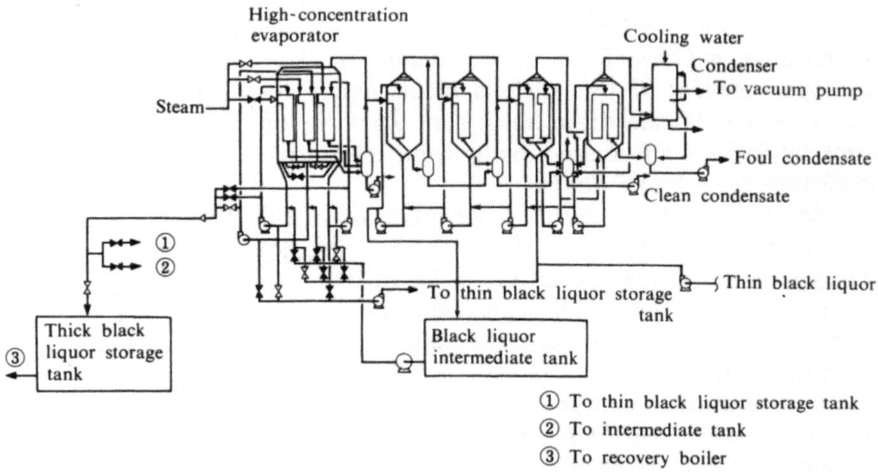

Fig. 10 · 86 Multiple-effect evaporator process

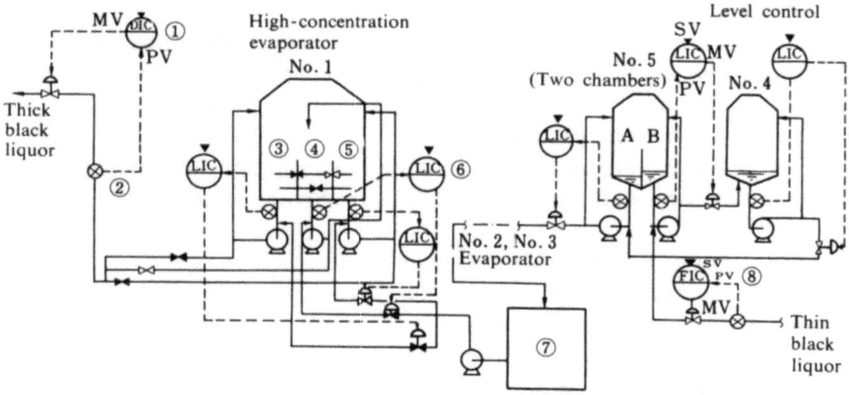

① Thick black liquor concentration control
② Black liquor concentration meter
③ A Chamber
④ B Chamber
⑤ C Chamber
⑥ Level control
⑦ Intermediate tank
⑧ Feed liquor flow control

Note : Black liquor passes through the system in the follwing
Sequence : No.5B→No.4→No.5A→No.3→No.2→No.1C→No.1B

Fig. 10・87 Physical distribution control and thick black liquor concentration control

determined by feedback control from the liquid level. An intermediate tank positioned before the high-concentration evaporator serves as a buffer; it absorbs fluctuations in the flow distribution balance. The volume drawn off from the high concentration evaporator—*i.e.*, the volume of finished thick black liquor produced—is determined by the tick black liqure concentration control (described below). The liquid level of each chamber is kept constant by feedback control of the feed volume.
(ii) *Washer-vat control*: Within the high-concentration evaporator, washing is performed in a succession of chambers, A, B, and C. Cut-over control of the various black liquor feed/extraction lines of each chamber (steam/drain lines, thin black liquor feed for washing/extraction lines) is automated using sequence control. A wash cycle takes anywhere from 3 to 5 hours.
(iii) *Thick black liquor control*: Concentration of the thick black liquor is continuously monitored with a gamma-ray density meter or optical-type consistency meter mounted at the outlet of the high-concentration evaporator. The concentration of the thick black liquor is controlled by feedback control, with the flow of drawoff extracted from the evaporator.

Immediately after cut-over between chambers, since the concentration in the wash pan falls off substantially, valve opening is controlled (as shown in Fig. 10.88) so as to minimize the duration of the transient falloff in extraction concentration. Moreover, while steady-state thick

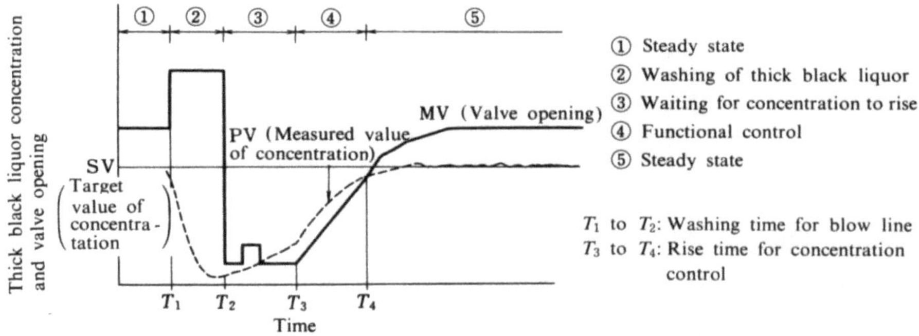

Fig. 10 · 88 Action of concentration control in transient phase

black liquor is sent to the thick black liquor tank, if the concentration falls below a standard value, the line automatically cuts over to either the intermediate tank or the thin black liquor tank as appropriate so the liquor can be recycled.

(iv) *Steam-line control*: Steam, the evaporator heat source, is supplied to the high-concentration evaporator as shown in Fig. 10.89. Steam generated from the black liquor by heat exchange is used as a heat source for the previous-stage evaporator. Then, in all succeeding downstream stages, generated steam is reused to maximize heat efficiency. The volume of steam supplied depends on the volume of black liquor fed to the evaporators. Also, where scaling is advanced (of course, every effort is taken to inhibit scaling, by such means as optimizing the wash cycle), steam pressure may increase excessively at inlets. Thus, steam pressure and steam flow override controls are sometimes implemented.

(b) *Recovery boiler process*

(1) *Process overview* While the basic structure of a recovery boiler for burning black liquor is more or less the same as a typical heavy-oil boiler, in terms of volume of steam generated (approximately 50 to 250

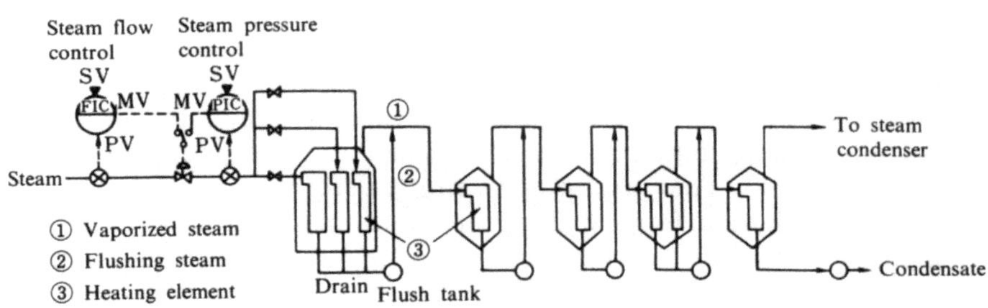

Fig. 10 · 89 Steam flow and control

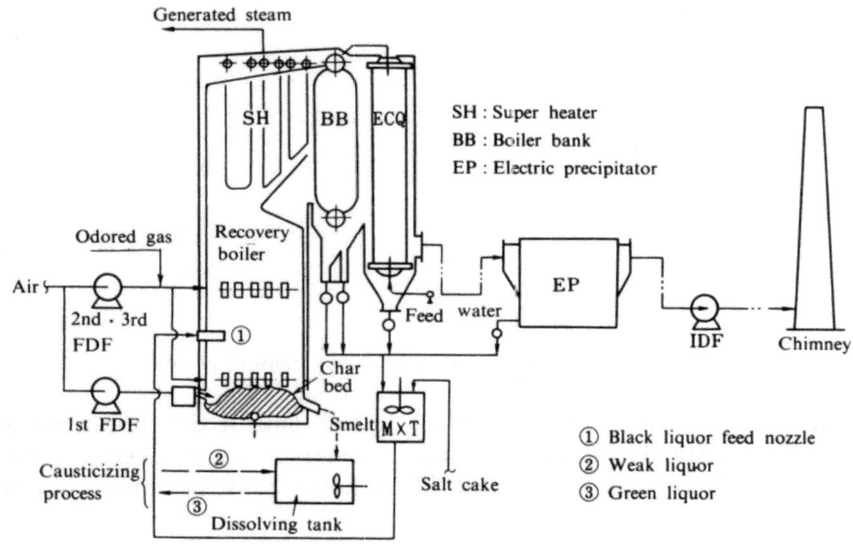

Fig. 10 · 90 Schematic of recovery boiler

t/h), it is much larger in scale. The basic configuration of a recovery boiler is shown in Fig. 10.90.

The concentrated black liquor from the evaporator is first heated and then injected into the furnace by black liquor spray-guns. As the particles of black liquor descend in the furance, water content is vaporized, they are heated, dried, thermally decomposed, and the volatile component (*i.e.*, organic) is burned. The burning operation leaves an accumulation of smelt called a char bed at the bottom of the furnace, which is continuously removed from the furnace.

The main objectives of the recovery boiler operation are to obtain the maximum energy (maximum amount of generated steam) from stable combustion, and to obtain the maximum rate of chemical recovery. The main instruments to realize these objectives are discussed below.

(2) *Instrumentation methods*

(i) *Combustion control*

i) *Air distribution ratio*: The distribution ratio of combustion air is important not only in terms of combustion efficiency, but also as an extremely important parameter influencing the reduction rate (described below). The distribution ratio is determined as a proportion of all distribution flows making up the total air flow. An effective method of finding the optimum ratio, however, is to monitor the temperature distribution inside the boiler.

ii) *Combustion control*: It is a basic premise that the role of a recovery boiler is to burn all thick black liquor produced. It is therefore not possible to adopt steady control of fuel supply as a standard for generated steam pressure (boiler master), as is usual with ordinary boilers. Moreover, because the heating value obtainable from fuels varies depending on the solid content of the black liquor, it is difficult to apply conventional air/fuel ratio control techniques. Thus, in order to manage and control the combustion, O_2 and CO concentrations in the exhaust are detected and total air flow adjusted to obtain the optimum values for those concentrations. In cases where feedback control is applied, since much dead time in the system can be attributed to boiler capacity, a control algorithm designed to cope with dead time has become necessary.

(ii) *Char bed monitoring*: The function of the char bed is recover the inorganic component in the black liquor reduced through chemical reaction at high temperature. To enhance reduction efficiency, the upper part of the char bed should be kept at the optimum temperature (about 1 050 to 1 150°C), and the optimum shape, or profile, of the char bed should be maintained.

(iii) *Soot-blower control*: Every furnace unit of a recovery boiler generates a surprising accumulation of dust, which consists mostly of salt cake. Soot blowing (sweeping with forced steam) is necessary much more frequently than in the case of ordinary boilers. Applied to a recovery boiler, soot blower are usually implemented as follows. The boiler is divided into a number of sections; boiler bank, super heater, economizer, and so on. These sections are then further subdivided into blocks, which are sequentially cleaned according to a prearranged schedule.

10.5.3 Instrumentation applied to the papermaking process

(a) Stock preparation-process instrumentation

The object of stock preparation is to improve the suitability of the pulp delivered from the previous process with chemical additives and pigments depending on the quality and end use of the paper.

Another important treatment of the stock is mechanical beating in a refiner. This improves the felting qualities of the fibers by fibrillation, causing them to bond together more readily in the papermaking process to follow.

An overview of a continuous stock preparation process is shown in Fig. 10.91.

Reflecting the objectives of preparation processing, there are three main instrumentation applications:

(1) stock mixing ratio control
(2) chemical additive/pigment mixing ratio control

(3) beating degree control (freeness control)

These controls will now be taken up for closer consideration.

(1) *Stock mixing ratio control* The purpose of this control is to deliver various stock pulps to the mixing chest in the proper predetermined proportional mix. The loop configuration for the control is shown in Fig. 10.92.

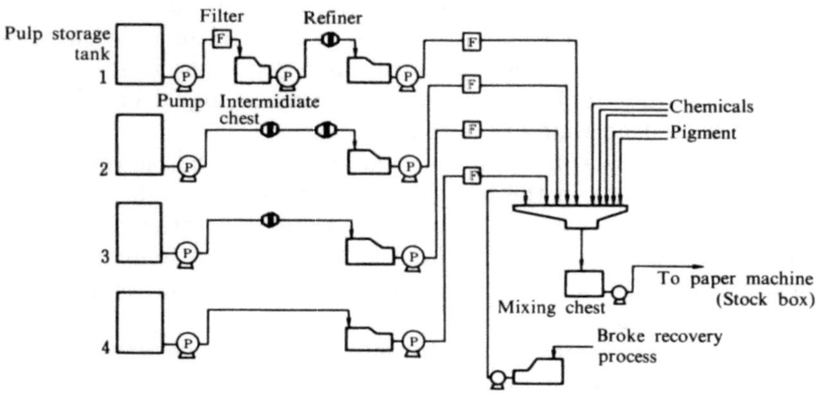

Fig. 10 · 91 Continuous stock preparation process

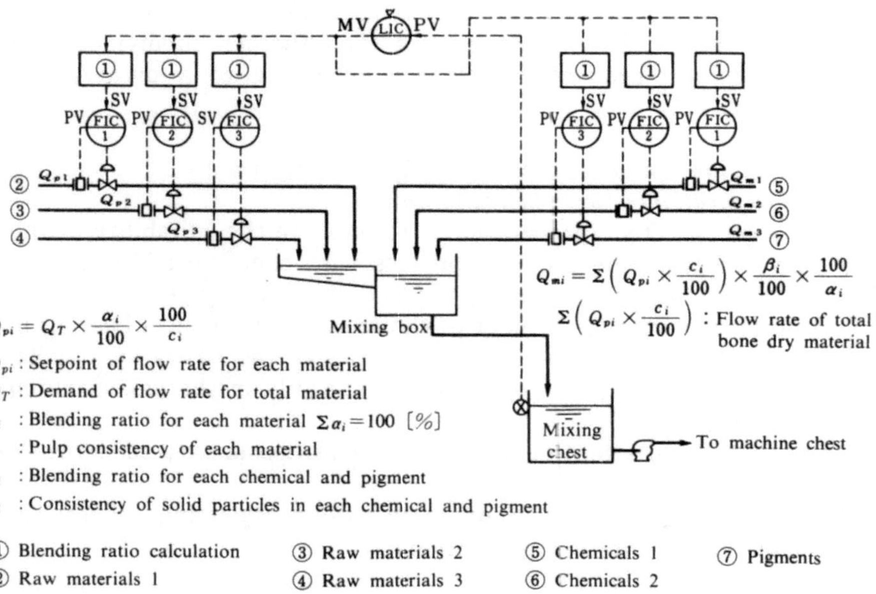

Fig. 10 · 92 Blending ratio control for materials as well as chemieals and pigments

The total stock flow into the mixing chest depends on the consumption of stock by the papermaking process. However, for the general control system, to keep the level of the mixing chest constant, the manipulated variable of the level controller is taken as the demand signal for total stock flow.

Quantities of the various pulp stocks are simply determined by multiplying the proportional factor by total required stock flow.

(2) *Chemical additive pigment mixing ratio control* The quantity of chemical additives and pigments added to the stock are figured as a proportion of total stock sent to the mixing chest (not including stock recovered from the paper-making process).

In the case of chemical additives and pigments, just as in the case of the pulp stock, consistency of solids in the stock must be taken into consideration in determining the flow setpoints.

(3) *Beating degree control* The purpose of beating degree control is to assure a uniform degree of beating (in terms of *freeness*, a measure of pulp fiber cutting and fibrillation) of the pulp in the refiner.

In a typical configuration, a refiner has two disks, of which one or both rotate and which are grooved with sharp-edged ridges. Watered pulp passes between the two disks which cuts and fibrillates the fibers. The degree of beating is varied by adjusting the clearance between disks using hydraulic pressure.

For measuring the degree of beating, the conventional method has been to use a batch-type sampling measuring device (a freeness tester, which measures the filtered water content of the watered pulp, can also be used to indirectly measure the degree of beating). This technique has drawbacks, however; the sampling time to carry out the measurement is overly long, and it is impossible to eliminate short-cycle disturbance on the basis of feedback control by means of this signal alone. A better alternative is to make use of the close correlation between motor load power and degree of beating. With this approach, disk cleanrance is adjusted by increasing the motor load power until the desired degree of beating is reached. It should be noted that when pulp is fed into the refiner, pulp flow rate and consistency changes affect the load on the motor, and hence the beating degree changes. Thus, feedforward control based on these variables is also be applied to the motor to adjust the load as appropriate. The control loop configuration is shown in Fig. 10.93.

(*b*) *Papermaking process instrumentation*

In the paper-making process, prepared stock is delivered to a paper machine where paper is formed by draining water from the stock. An overview of the process is shown in Fig. 10.94.

The stock is first diluted to around 1% in consistency, and it flows

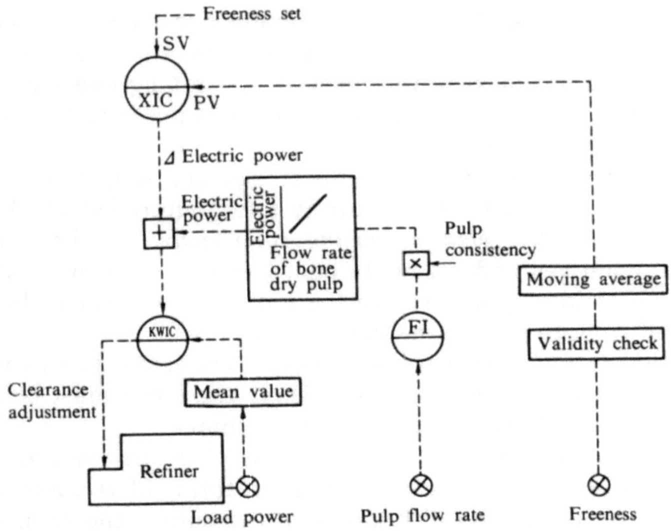

Fig. 10 · 93 Loop diagram of beating degree control

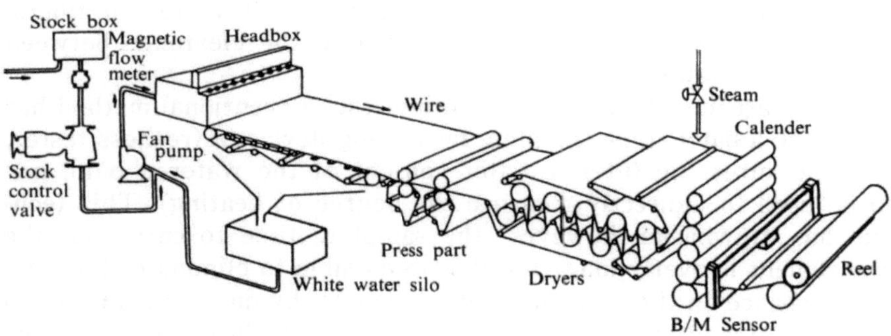

Fig. 10 · 94 Paper making process

from a headbox to the top of a moving endless belt of wire screen where it is uniformly dispersed. Drainage from the stock through the wire screen is induced both by gravity and vacuum. In the press part, more water is extracted by multiple press rolls. In the dry part, the paper is dried by steam-heated cylinders, and is finally wound into rolls with a reel device.

Here we will focus on instrumentation applied to (1) the headbox, and (2) the dry part.

(1) *Headbox instrumentation* When the headbox delivers the diluted stock on the wire, the delivery speed v_j (jet velocity) must be controlled so it is in fixed relation to the speed of the wire screen v_w in or-

der to obtain the best formation of paper (disposition of the fibers). This ratio is called the jet/wire ratio α. The three operable parameters are related as follows

$$v_j = \alpha v_w$$

The pressure (total head) yielding this jet velocity must be applied on to the headbox. This value is given by

$$\mathrm{TH} = \frac{1}{2g} v_j^2$$

where TH is total head (mAq), and g is gravitational acceleration (= 9.8×60^2 m/min^2).

Using the pressure derived by this equation to control the headbox is called total head control, and represents the single most important value for headbox instrumentation.

In total head control, pressure in the headbox is measured and controlled by the stock flow rate, or the air pressure applied. Since adjustment of the stock flow rate can be varied over a wide range and high precision is also required, a combination of rough-adjustment and fine-adjustment valves is provided. In recent years high-precision motor rpm control has become a reality. Extending this approach, and to conserve energy, the trend is toward using pump speed to control the flow rate.

(2) *Dry part instrumentation* After extraction of water in the press part, the paper web still has a water content of approximately 60%. The task of the dry part is to remove most of the remaining water down to about 6%.

In the dry part, the paper web is dried by being held in contact with cylinders that are internally heated by steam.

The steam flow configuration for a typical system is shown in Fig. 10.95.

Most paper machines are the so-called multiple-drum type, operating with anywhere from 40 to 100 cylinders. These cylinders are divided into 3 to 4 groups: a high-pressure, a middle-pressure, and a low-pressure group (facing from the reel back toward the headbox). Appropriate pressure differential is maintained between the groups (inter-group differential pressure). Steam is first supplied to the high-pressure group. After the steam condenses, the water is removed, reconverted to steam in a flashtank, and successivley supplied to the lower pressure groups.

Another common heat-recovery measure is to install a hood over the drier section. This makes it possible to effectively capture for reuse a major portion of the steam coming off the paper in the drier

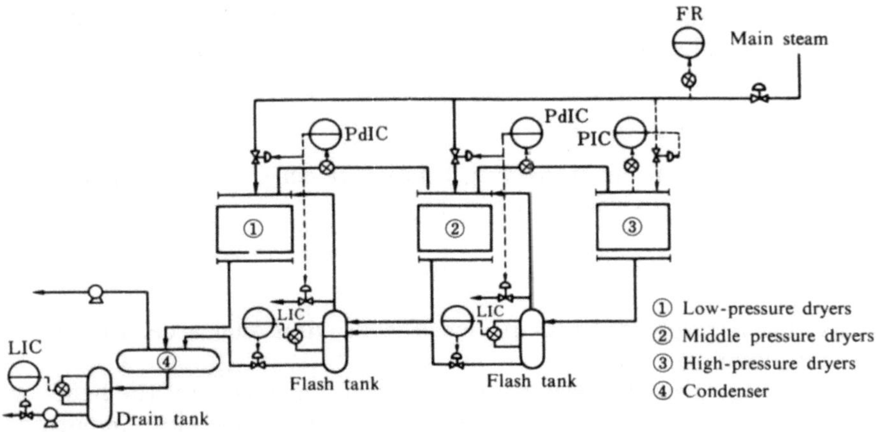

Fig. 10・95 Typical dryer steam flow system

section.

(*c*) *Web quality management and control*

There are various aspects that must be considered in web quality control:

(1) basic quality considerations such as basis weight, thickness, and dimensions related to terms of trade;

(2) Structural quality relating to such attributes of the paper fibers as dispersion, smoothness, tear strength, and tensile strength;

(3) optical quality considerations such as hue, whiteness, opacity; and

(4) absence of such problems as shrinkage, wrinkling, curling that are strongly influenced by moisture content.

Particularly important parameters in this regard are basis weight (weight per unit area 1 m^2) and moisture content. These two parameters are important not only in their own right, but also because they have such a major influence on other quality attributes. We will focus here, therefore, on the system for controlling basis weight and moisture content.

Basis weight and moisture content are measured by B/M sensors (also see Chap. 3, Sec. 3.9).

Basis weight–moisture content control is implemented in both machine and cross directions.

(1) *Machine directional basis weight control* The control object is to find the bone–dry basis weight from the basis weight and the moisture content. The final control element is a high–accuracy stock control valve (an adjustment resolution over 1/1000) which is mounted below the stock box. The control loop has the following characteristics:

(1) dead time is long (40 to 200 sec.), varying with the speed of the machine;
(2) time constant is comparatively short (10 to 50 sec.); and
(3) measured value is obtained discretely (the average measured value is used in the cross direction).

Given these characteristics, since good results cannot be obtained using a simple PID algorithm, sampled data control based on the finite time settling response method is applied, with optimization of the dead time by adjusting the speed of the machine.

Manipulated and process variable movements for the finite time settling response control method are shown in Fig. 10.96.

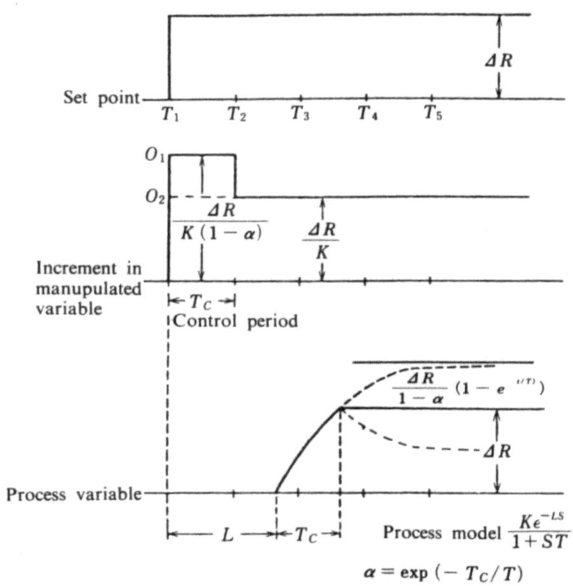

Fig. 10 · 96 Finite time settling response control

(2) *Machine directional moisture content control* This control is maintained by adjusting the pressure of the steam supplied to the driers. Characteristics of the control loop are as follows:

(1) dead time is relatively short (10 to 30 sec.);
(2) time constant is long (50 to 500 sec.);
(3) measured variables are obtained discretely; and
(4) interference is received from the basis weight (every 1% change in bone-dry basis weight produces a 0.1 to 1.0% change in moisture content).

Therefore, taking these characteristics into consideration, sampled

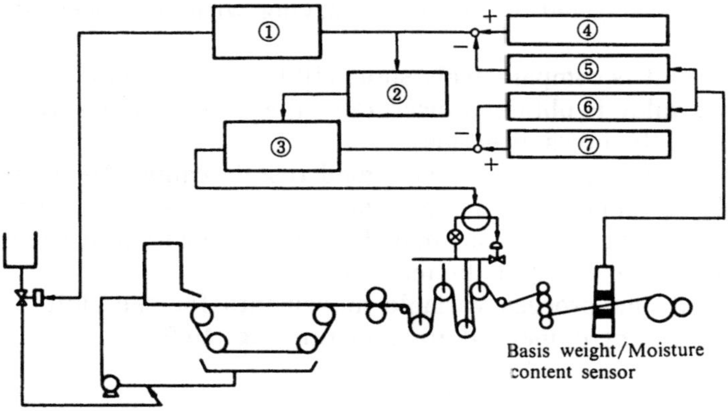

① Basis weight control algorithm
② Non-interference algorithm
③ Moisture control algorithm
④ Setpoint of bone dry basis weight
⑤ Measured value of bone dry basis weight
⑥ Measured value of moisture content
⑦ Set point of moisture content

Fig. 10・97 Basis weight/Moisture content control

data control based on bone–dry basis weight non–interference and the finite time settling response method is used. The basis weight/moisture content control is shown schematically in Fig. 10.97.

(3) *Cross directional basis weight control* The headbox delivers a jet of stock through a slice across the full width of the machine. The slice openings are spaced about 150 mm apart and are adjusted by means of slice bolts. The slice bolts are adjusted by matching transverse data detected by sensors (profile data) to a desired profile.

(4) *Cross directional moisture content control* This control is implemented with auxiliary driers arranged so the entire width of the machine is divided into zones. This control loop is simple compared to the cross directional basis weight control loop.

10.6 Waterworks Instrumentation Applications[24)–32)]

10.6.1 Overview of waterworks facilities

In broad terms, waterworks encompasses facilities for the intake and conveyance, purification, and distribution of water. Basic conditions of intake and conveyance facilities are that the planned intake volume can be regularly assured and that the quality is good. The volume of water intake is controlled by intake gates and by controlling the speed and number of pumps in operation.

Since water intake points are often located at some distance from the water purification plants, supervision and control is generally exercised from the central control room at the purification plant by means of telemeter/telecontrol (TM/TC) systems. Purification facilities include both purification processing and drainage/waste sludge processing. The former consists of arrival, settling, and filtering; while the later encompasses facilities for washing drainage, settling waste sludge, and treating sludge. The object of purification processing is to remove the various particles, microscopic organisms, and bacteria found in raw water thus making it fit for consumption as drinking water. The most common method of removing these contaminants is the chemical sedimentation-rapid filter method. Typical water treatment processing, from intake to purification to distribution, is schematically shown in Fig. 10.98.

10.6.2 Water treatment-related detectors

Detection devices commonly employed at waterworks facilities include general measuring devices for flow rate, level, and pressure; and water-quality measuring devices. This latter class of devices is used not only to supervise the quality of the water, but also to gage the effects of chemicals injected into the water. Since it would be impossible to cover the whole range of waterworks-related devices in the few pages allotted, only the most important devices and their uses will be described.

(*a*) *Water quality detectors*

The most widely used water quality devices are the turbidimeter, pH meter, residual chlorine meter, alkalinity meter, and electric conductivity meter. The points where these devices are applied and what they are designed to detect are shown in Fig. 10.99.

(*b*) *Common measurements*

Waterworks, of course, deals with water that is stored and water that is flowing. Level gages and flowmeters are thus essential to quantitatively measure water in these two basic states.

(1) *Level gages* Among the various devices available to detect water level, the most widely used are the float-type, differential-type, throw-in-type, ultrasonic-type, and electrical capacitance-type level gages. The features and uses of these various types of level gages are summarized in Table 10.7.

(2) *Flowmeters* Flowmeters are employed in processes for a variety of different reasons. The most common uses are: for the purpose of making estimates, for control (*e.g.*, to determine the quantity of chemicals to be added to the water), to create a record of operations (*i.e.*, by measuring the volume of inflow, outlfow), to keep track of transaction volume for the water-supply industry (on either the supply or delivery side). In selecting the most appropriate type of device, degree of accu-

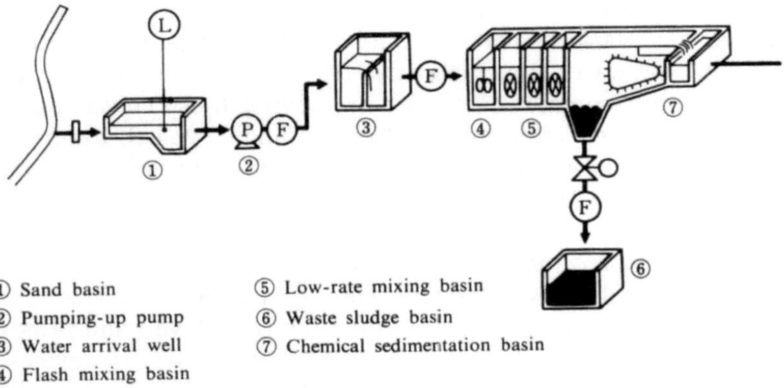

Fig. 10・98 Outline of

Table 10.7 Features and uses for various applications of miscellaneous liquid level meters

Liquid level meters	Features and uses
Float type	Simple principle and simple construction. This type is used in open tanks such as sand basins or water arrival wells. If there is any fluctuation on the liquid surface, however, a buffle tube must be installed.
Differential type	This type is used for liquid level measurement of chemicals and other storage tanks. The meter is installed at the low part of the vessel and there must be enough space to access the meter.
Throw-in type	Diaphragm is used to detect liquid pressure. Simple installation and simple maintenance.
Ultrasonic type	Certain amount of error will be introduced if there are any obstacles between the sensor and the surface of riquid or if dew and drops are produced on the sensor. This type can also be implemented to measure in a flow.
Electrical capacitance type	Offers the widest measuring range and is therefore suitable for measurement not only in tanks and basins on water purification plant, but also in deep wells. Considerable error will result if anything adheres to the electrodes.

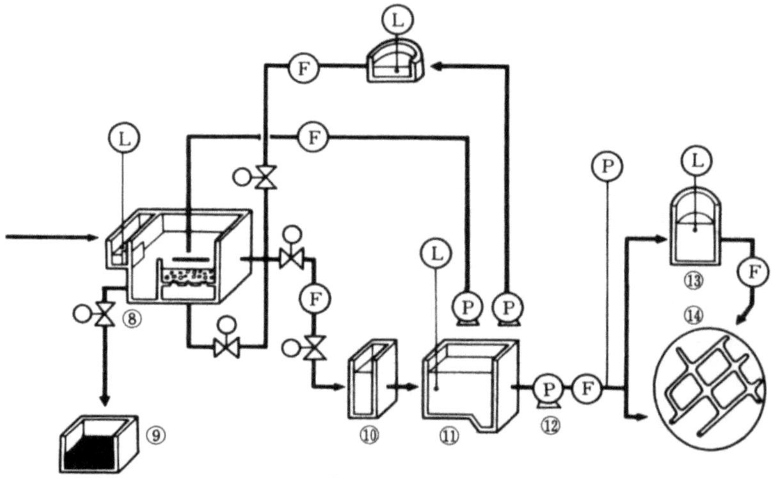

⑧ Rapid filter
⑨ Washing drainage basin
⑩ Chlorine mixing chamber
⑪ Reservoir for clean water
⑫ Water supply and distribution pump
⑬ Pump station of pressurizing for distribution reservoir
⑭ Distribution piping network

water works facilities

racy required and economy must be taken into consideration. Flow measurement methods can be broadly divided into closed–conduit and open–channel methods. Flowmeters representative of the closed–conduit method include electromagnetic, differential pressure, and ultrasonic flowmeters. For the open–channel method, there is the weir flowmeter. For waterworks applications, electromagnetic flowmeters are most widely used. Not only are electromagnetic flowmeters extremely accurate and subject to little pressure loss, they can also be applied to a wide range of pipe diameters from 2.5 to 2 600 mm. On the negative side, this type of device is generally more costly than other kinds of flowmeters; where the object is simply to obtain a rough estimate of flow, another type device would probably suffice. Typical differential pressure flowmeters include the orifice plate and venturi tube. For waterworks applications, venturi tubes are frequently used because of their very small pressure loss. For measuring filtered flow rate, ultrasonic flowmeters are sometimes used. While this type of device is probably the easiest to mount on a pipe or conduit, they cannot be applied to waste–water applications or where the flow is accompanied by foaming. Also, where current velocity fluctuations are present, say after a pump outlet has been closed or around a pipe corner, a sufficiently long section of straight pipe is necessary to prevent disturbances from being introduced.

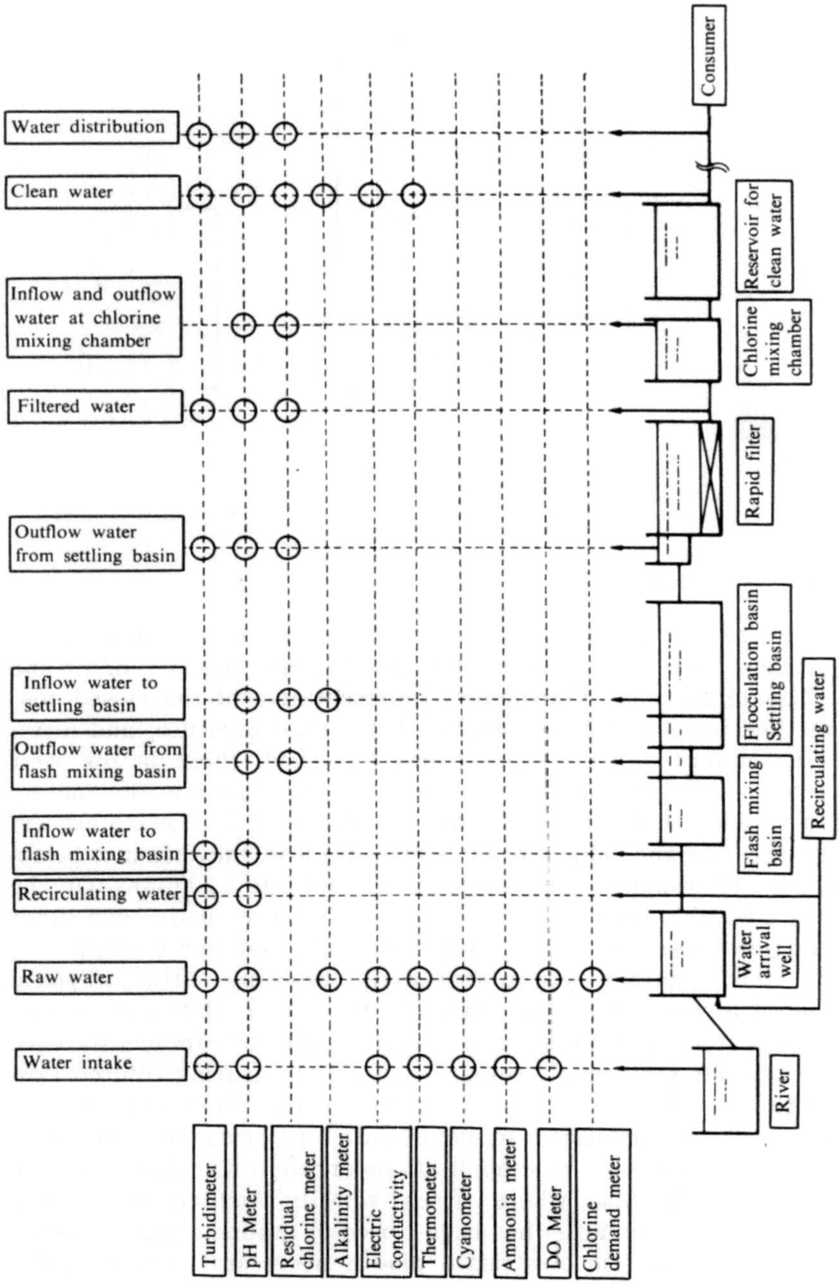

Fig. 10 · 99 Points and measurement instrumentation in a continuos water quality meter in water purification

10.6.3 Filtration equipment instrumentation

Filtration equipment is used to screen fine suspended solids that are not removed by chemical and sedimentation treatment by passing the waste stream through a bed of sand. Key instrumentation for the filter are filter rate control to maintain uniform velocity thorugh the filter, and filter washing control applied to the backwash-storage compartment that holds water used to wash the filter bed. A flow sheet of the instrumentation associated with the filter is shown in Fig. 10.100.

(*a*) *Filter rate control*

Standard filtration rates range from 120 to 150 meters/day, and must be controlled to maintain a uniform rate. The total filter rate is set by the main set station, which maintains common control over the

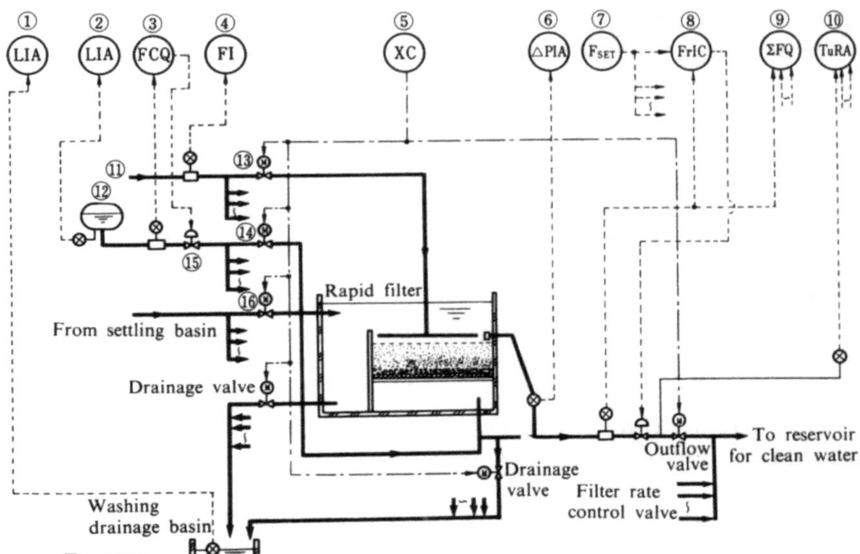

① Level indicating alarm meter on washing drainage basin
② Level indicating alarm meter on washing water tank
③ Back washing water flow integrating controller
④ Surface washing water flow meter
⑤ Filter washing control equipment (Sequence control equipment)
⑥ Filtration head loss indicating alarm meter
⑦ Main set station
⑧ Filter rate indicating controller
⑨ Total filter rate integrator
⑩ Filtered-water turbidimeter with alarm
⑪ From surface washing pump
⑫ Washing water tank
⑬ Surface washing valve
⑭ Back washing valve
⑮ Back washing water flow control valve
⑯ Inflow valve

Fig. 10 · 100 Instrumentation flow sheet of rapid filter entire group of filters in operation, and the filter rate for each controller is set automatically. The number of basins in operation and total filter rate can be set either manually according to the operator's estimate or automatically. The automatic setting method estimates water-delivery demand automatically (refer to Subsec. 10.6.8 below), then calculates balanced levels for the basins. Abruptly changing the filter rate has an adverse impact on the sand bed, which can increase the turbidity of the effluent. For this reason, rate changes are carried out after the slow start procedure.

(*b*) *Back-washing water flow control*

To remove accumulated dirt clogging the sand bed, water is injected back up through the filter bed (back washing); the resulting vigorous agitation of the sand effectively washes the bed. This backwash water is maintained at a uniform rate by flow control.

(*c*) *Filtration head-loss meter*

An important measurement item for supervising loss of filtering capacity indicates the extent that the sand bed is clogged (head loss) by filter resistance. Usually, when the value reaches 2 to 2.5 meters, filter washing is required. Filter washing is also required when the upper limit of the filter run time is reached. Two other factors have an important bearing on the wash demand: there is no basin during washing, and the drainage water in the washing drainage basin is below the prescribed level.

(*d*) *Wash sequence control*

When the above wash conditions are met, sequence control for automatic washing can be implemented. The timing chart for such an operation is shown in Fig. 10.101.

This type of filter wash sequence control and filter flow control can be automated using a microprocessor-based control device.

10.6.4 Chemical injection equipment instrumentation

(*a*) *Purposes of injections and places to be injected*

Water is treated with chemicals is to clarify the water in compliance with water quality standards prescribed in the the water works law so it is safe and fit for human consumption. Injection equipment thus has an extremely important function at waterworks facilities. The most common injection agents and the reasons they are added to the water supply are summarized in Table 10.8. The points where these agents are typically added to the purification process are shown in Fig. 10.102.

(*b*) *Injection methods*

There are three main methods of adding the chemical agents depending on the properties of the specific agent. These are: natural

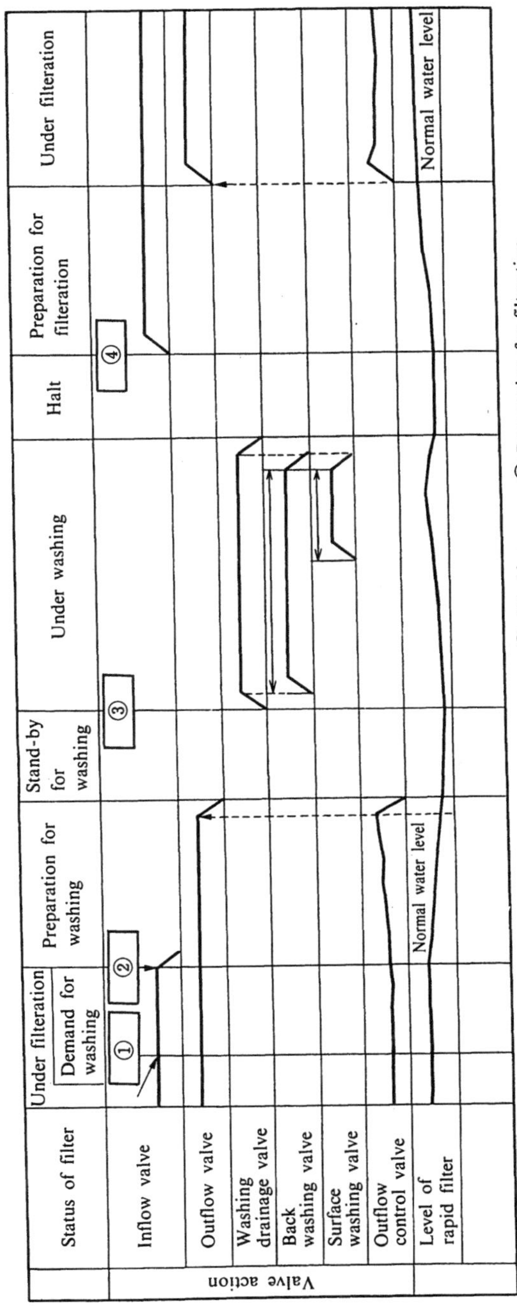

① Confirmation of alarm ② Preparation for washing ③ Washing ④ Preparation for filteration

Fig. 10 · 101 Time chart of washing sequence

Table 10.8 Injection purpose of coagulant

Injection agent	Purpose
Coagulant	Produces floc from suspended matter in raw water and promotes sedimentation as well as reducing turbidity of raw water, removal at the filter.
Coagulant aid	This is used to enhance the effect of coagulant depending on the quality and condition of raw water.
Pre-alkali	Prevents reduction of alkalinity due to injection of coagulant or pre-chlorine and raw water of low alkalinity. This aid is used to keep alkalinity or pH high to acheive good coagulation and flocculation.
Middle alkali	Maintains pH at specified value or higher to prevent corrosion of pipes.
Post-alkali	Maintains pH at specified value or higher to prevent corrosion of pipe and to maintain water quality.
Pre-chlorine Middle chlorine	Sterilizes and disinfects raw water, and removes iron, manganese, ammonia nitrogen, nitrite nitrogen and organic substance through acidification.
Post-chlorine	Chlorine injected to maintain residual chlorine (free available chlorine) at a specified value to sterilize, disinfect and prevant breeding of micro-organisms.

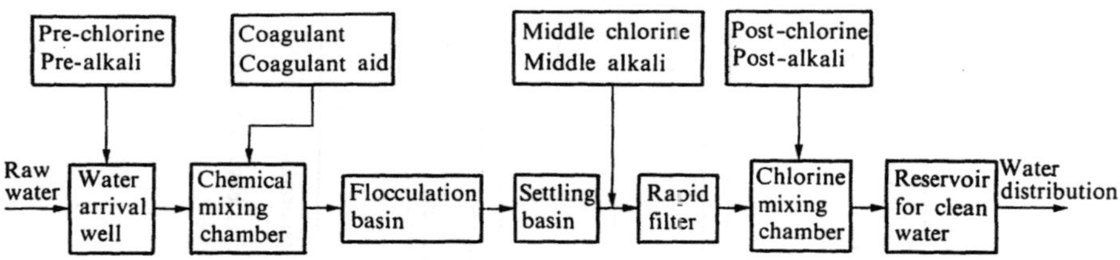

Fig. 10 · 102 Main injection agents and injection points

flow, pump force feed, and ejector.

(1) To guarantee the required pressure at the point of injection for the natural flow method, the injection tank must be placed in an elevated position. The injection dose is controlled by a flowmeter/control valve combination.

(2) With the pump force feed method, one method is to use a constant-rate pump to assure the required pressure at the injection point. Another arrangement uses a pressure pump plus a flowmeter/control valve combination.

(3) With the ejector method, chemicals are injected in powder form,

and injection pressure is provided by the ejector itself.

(*c*) *Injection dosage control*

Although the injection dose will vary to some extent depending on the quality of the water, ratio control derived by multiplying (volume of water under treatment) × (injection ratio) is the most frequent method used.

(1) *Injection control using flowmeter* (*electromagnetic*) + *control valve* The measurement range of an electro-magnetic flowmeter is 0.3 to 10 meters/s. In cases where the injection dose exceeds either the maximum or minimum values of this range, a combination of large and small systems configured with a flowmeter and a control valve is required. Modern flowmeters that incorporate a microprocessor are capable of cutting over between different flow ranges. The loop configuration for such an injection method is shown in Fig. 10.103.

(2) *Injection control with a constant-rate pump* With this method, the injection dose is controlled by varying the speed and stroke of a constant-rate pump. The injection dose is simply calculated by multiplying (speed) × (stroke). Control is effected by either increasing or decreasing the speed until the set injection dose is achieved. If the upper or lower limits of the speed are reached, the desired dosage can be

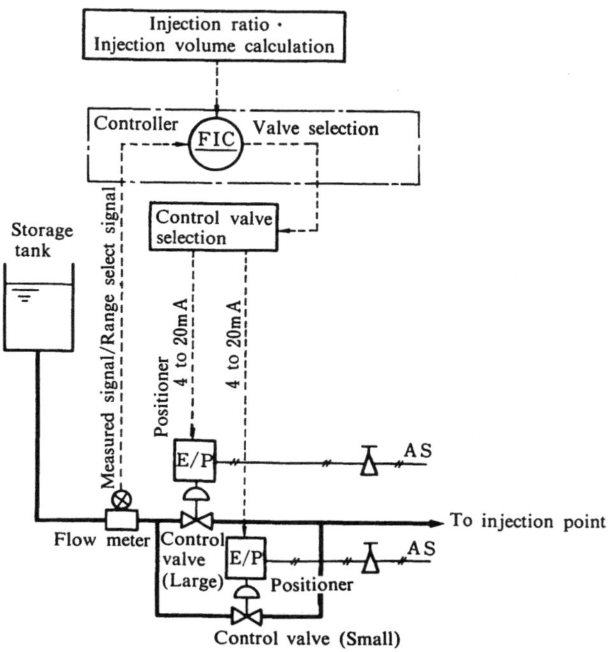

Fig. 10 · 103 Loop diagram of flow meter and control valve (Wide range control type)

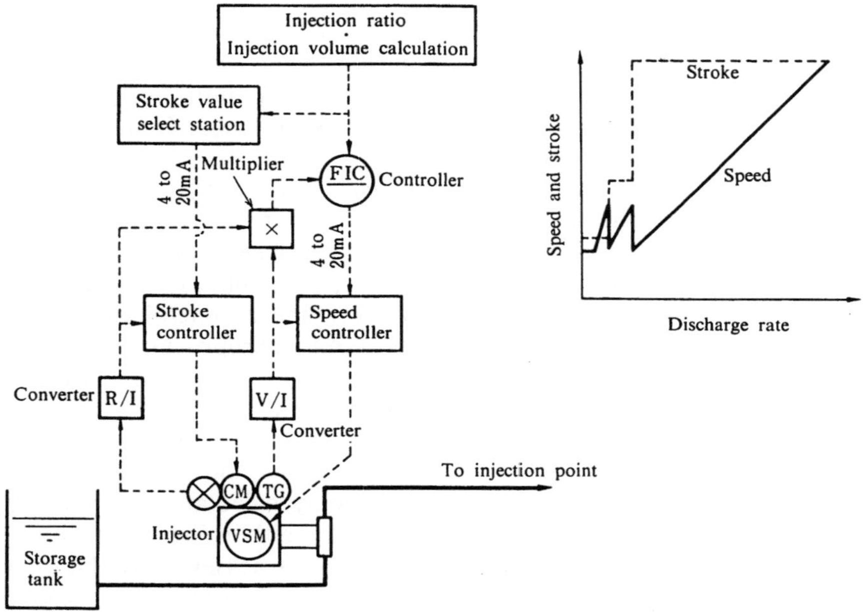

Fig. 10・104 Loop diagram of constant-rate pump (Speed and stroke control)

achieved by cutting over to the stroke and adjusting the speed. The control loop for this method is shown in Fig. 10.104.

(*d*) *Calculating the injection rate*

The injection dosage for the water under treatment is determined by the injection ratio. However, a number of methods have been developed or are being studied to enable automatic determination of the injection ratio. Here, we introduce a typical coagulant injection ratio method. The coagulant injection ratio varies depending on water quality factors and temperature of the stream being treated. To enhance the effectiveness of the coagulation and flocculation, coagulant aids and alkaline are usually added at the same time, and the pH value is kept above a prescribed level. Today, the most common method for determining the injection ratio is to conduct a jar test* (off-line). Two methods of automating this testing are as follows:

(1) Store a correlation curve of turbidity (the main turbidity elements, or parameters) and the injection ratio in the memory of an

* An apparatus for measuring the appropriate injection rate for a coagulant. The method involves gradually added a coagulant to a fixed amount of raw water while measuring the turbidity, pH, and alkalinity, and observing the effects on flocculation and suspension. Based on an overall evaluation of the results, the optimum injection ratio is determined.

automatic control device. Then, obtain the injection ratio based on this data using an on-line turbidity meter.

(2) Compile a data file of jar test injection ratios and the water quality conditions that were under test (turbidity, pH, alkalinity, electric conductivity, and water temperature). Then, by applying statistical processing to the data (*e.g.*, weighted regression analysis), a correlation is obtained between the injection ratio and the various water quality factors (elements). The correlation can be expressed as a formula or organized as a table of injection rate correlation data for use as an injection method, and stored in the memory of an automatic controller. Then, according to the reading of an on-line water quality indicator, the injection ratio is determined. One drawback of this method, however, is that the storage of the sample data needed for the statistical processing takes considerable time. This is because the quantification of floc formation, one of the measures of coagulant effectiveness, is quite difficult. This problem must be addressed before this method can be adopted on a widespread basis.

10.6.5 Instrumentation for water-supply and distribution facilities

After purification at a purification plant, water is supplied to users maintaining appropriate quality standards and water pressure. The link from the purification plant to the distribution reservoir is called the supply system, and from the reservoir to end-users is called the distribution system. A schematic of the two systems combined is shown in Fig. 10.105.

(a) Water supply flow control

Since distribution reservoirs are generally built on the highest ground available, the water supply is regulated to maintain the reservoir at a certain level by controlling the number of supply pumps that are in operation. Recently, the use of computers has had a great impact on the operation of waterworks. Computers have made it possible to analyze demand fluctuation elements in terms of time units through-

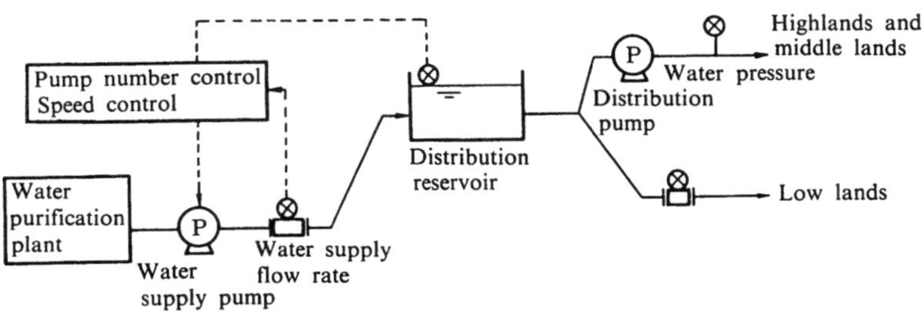

Fig. 10 · 105 Water supply and distribution system diagram out the course of a day. This data can be used to estimate water supply 24 hours ahead, which in turn, can be used to estimate the water level fluctuation in the distribution reservoir. This advanced control technique has been introduced at a number of municipal purification plants, and makes it possible to maintain the reservoir level within upper and lower permissible limits.

(*b*) *Distribution volume*

Because gravitation flow can be used to distribute water to lowland areas, no special controls are generally needed. If water pressure builds up to excess levels, presure is relieved with pressure–reducing valves. In the case of high–lying areas, water is lifted to the higher elevation by means of booster pumps. Control is effected either by adjusting pump speed or the number of pumps in operation to maintain appropriate pipe–end pressure (above 1.5 kg/cm²). This is called end pressure control. An example of end pressure control, that includes flow resistance within the pipe, is shown in Fig. 10.106.

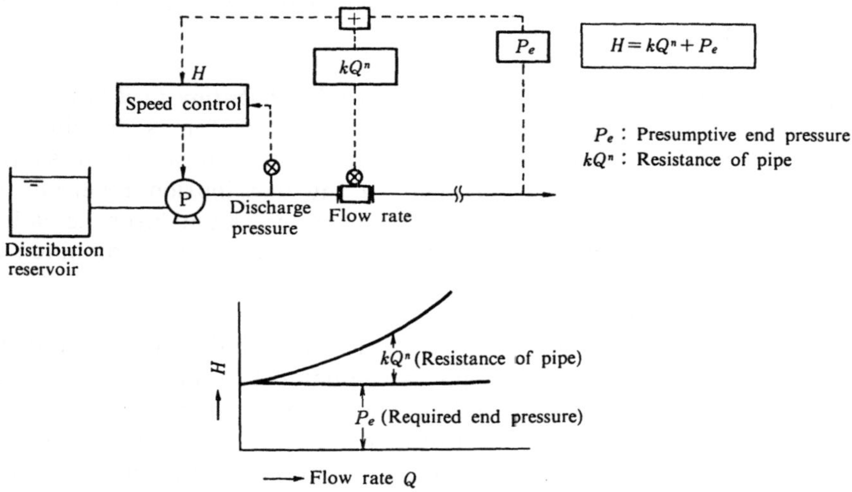

Fig. 10 · 106 An example of presumptive end pressure control in highlands

10.6.6 An integrated control system for large–scale, wide–area water works facilities

In recent years, the trend has been toward wide–area waterworks that supply water over an surprisingly dispersed geographical region. Large–scale purification plants have been constructed that sell water to multiple commuties and municipalities based on contracts individually arranged with the communities. Accordingly, more stringent management over both quantity and quality of water have come to be required. Moreover, since distances from purification plant to the local municipal reservoir on the consumer end can range from several kilometers to several tens of kilometers, unmanned facilities such as relay pumping stations and equalizing storage reservoirs must be installed at intermediate points along the way. In order to supervise and control these unmanned facilities, systems must be designed to permit centralized control from the control room at the purification plant. In short, integrated system management has become indispensable in order to maintain good quantitative balance in the supply system and to assure the maintenance of contracted supplies of water at all times. Two approaches to implementing such an integrated system are to use telemeter/telecontrol (TM/TC) over a leased circuit or to use computer control via modem. The configuration of an integrated system is shown in Fig. 10.107, and a typical control room is shown in Fig. 10.108.

10.6.7 Water distribution information management system

So far, we have focused on the segment from intake (centered on the purification plant) to the distribution reservoir—the so-called *water production process*. By contrast, the segment from distribution reservoir to each end-user over the distribution network can be likened to a *water distribution process*. The distribution takes place over a network of conduits and pipes that branches out to directly connect with end-users. Topology is an important factor that must be taken into account when laying the network. The ideal of maintaining exactly the right amount of water pressure (neither more nor less than needed) to each end-user is extremely difficult to realize. In fact, the great difficulty of network analysis is one of the main reasons computer control systems have not been widely introduced in the area of water distribution. This is in marked contrast to purification plants, where such systems have been extensively introduced for the purpose of automating processes. Here, we will focus on the network (including the distribution reservoir), and describe network analysis using a piping network calculation. Based on the network analysis, the entire water distribution process can be optimized (*i.e.*, leakage minimized by optimal adjustment of the water pressure). The analysis also makes it possible to produce a water-delivery forecast which can be used on the production-process side (*i.e.*, the purification plant) to calculate a production plan (the water supply), the most important objective for implementing the system. Briefly, the piping network calculation is carried out as follows. The actual network of pipes is replaced by a logical diagram of nodes linked by lines. The energy level (*i.e.*, water pressure) at each node, and the flow over segments between the nodes is then found.

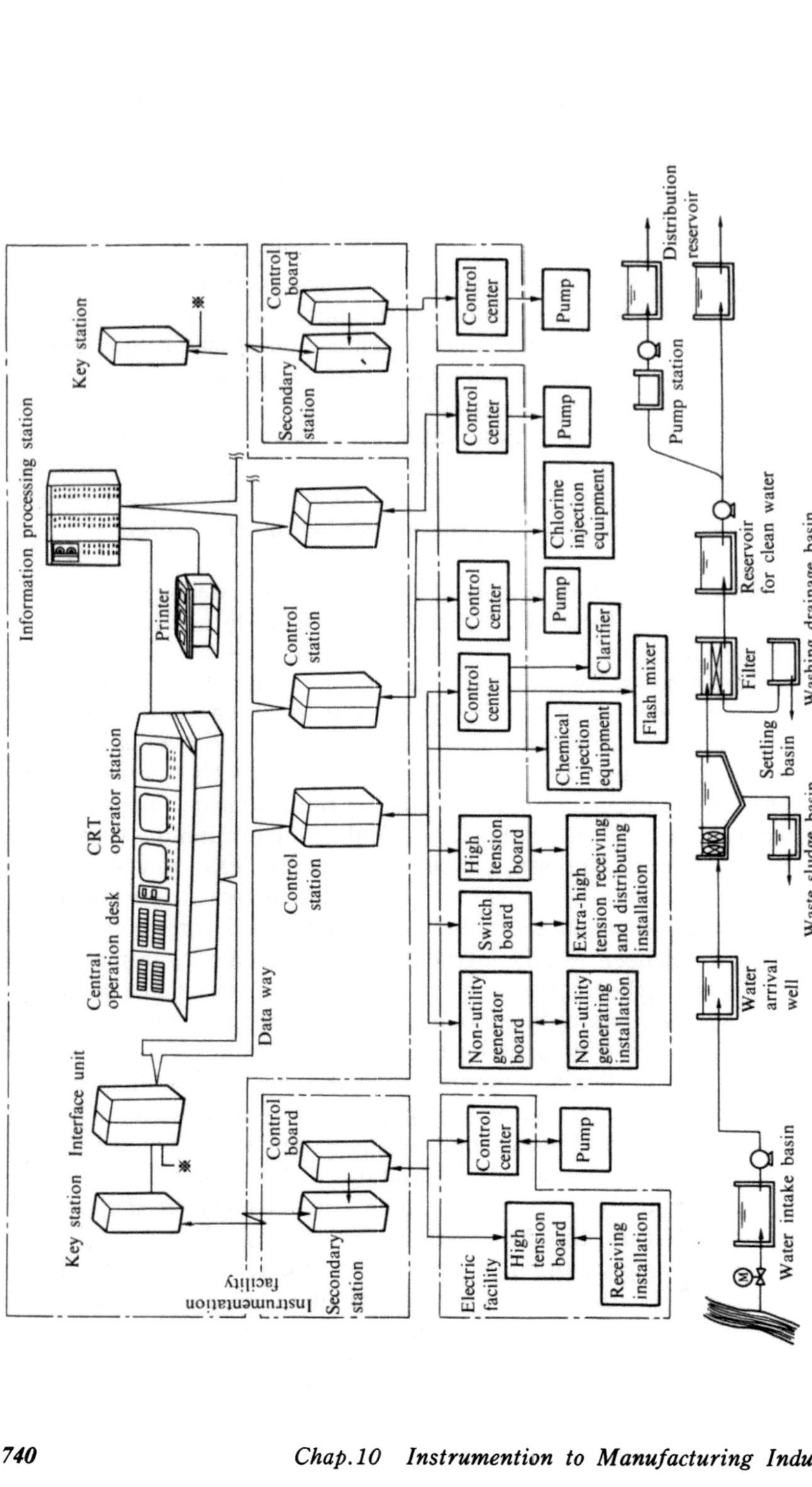

Fig. 10 · 107 Overview of wide-area water works

Fig. 10 · 108 Typical control room

Generally, the Hazen Williams formula is used, which relates a number of physical parameters of the pipe: D=pipe diameter, L=distance between nodes, C is the flow coefficient of each segment of pipe, and q is the demand volume at a node. The formula for each segment of pipe is given by

$$Q_{ij}=1\,002.7\times C_{ij}\times D_{ij}^{2.63}\times L_{ij}^{-0.54}\times(E_i-E_j)^{0.54}$$

and the formula for the continuous flow rate at a node is given by

$$\sum_i Q_{ij}+q_j=0$$

as simultaneous equations for the unknown energy levels E_i, E_j of each node. From the energy level of each node, the water pressure at those points, and the flow between nodes can be found. To apply this piping network calculation to actual networks, first the physical parameters at the pipe and node parameters are entered as a data base. Then, based on facility data (data base information) and on-line data (current water level in the reservoir, flow rate, etc.), the following can be calculated:

(1) The current state of the piping network can be estimated. That is, by knowing the water pressure of all nodes, any areas that are out of balance are apparent.

(2) Simulation can be carried out using a computer model of the piping network (*i.e.*, based on pressurizing and depressurizing elements). Simulation is useful for preevaluation of planned facilities, and is also useful for providing information on renovation and/or expansion of the piping network.

(3) Water pressure can be adjusted when there are points that need to be depressurized; the degree of opening for valves to achieve the desired water pressure can be calculated.

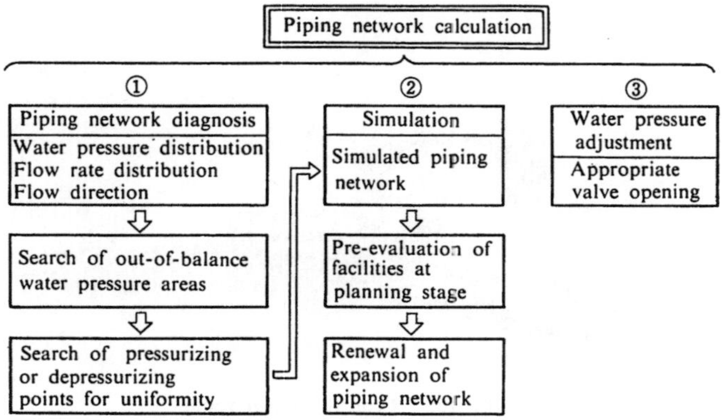

Fig. 10・109 Application of piping network calculation

Piping network applications and a typical output result of the calculation are shown as Figs. 10.109 and 10.110, respectively.

Next, we consider the method of estimating water-delivery demand. The purpose of the estimation is to forecast water-delivery in daily and time units given the great fluctuations that occur due to weather conditions and also depending on the day of the week. From the perspective of ideal plant operations, on the other hand, better efficiency can be achieved if the facility is operated under uniform load conditions. This makes it possible to operate the pumps at constant load and to stabilize the amount of chemicals injected into the water. In order to operate under constant load conditions at the plant, distribution reservoirs are constructed between the purification plant and the distribution network. The reservoir serves to absorb such fluctuations as occur, and thus performs a buffer function. To make the most effective use of this buffer function (based on upper and lower level limits of the reservoir), estimations must be made of the total volume supplied each day and the water-delivery for different times. From that data, the water-level fluctuation curve can be estimated, which makes it possible to calculate a water-delivery that will not exceed either the upper or lower level limits of the reservoir. One method of estimating the water-delivery is to estimate the load using a Kalman filter. One advantage of this method is that it is amenable to computer processing, and in fact the method has been applied at a number of municipal purification plants. An outline of plant operations based on demand forecast is shown as Fig. 10.111, and a procedure for calculating the necessary water supply is shown in Fig. 10.112. Finally, Figure 10.113 shows a method of forecasting water level, where $F_{out}(m^3/h)$ is the estimated water-delivery, $F_{in}(m^3/h)$ is the water supply, and L(m)

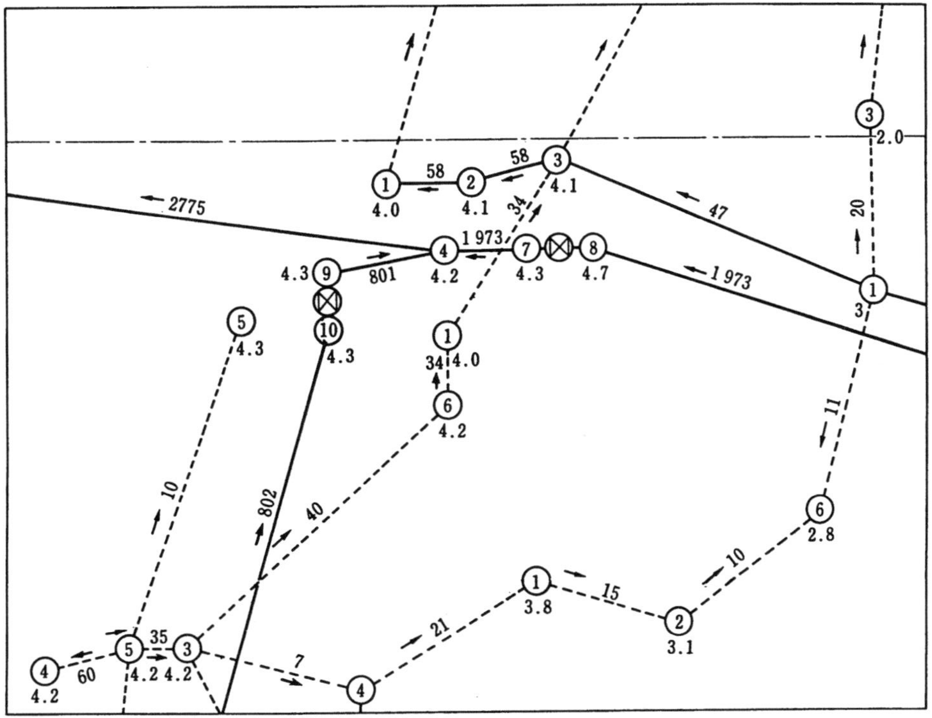

Note

1 (n) =Nodos
⊗ =Values

2 Numbers along pipe lines indicate diameter in mm (detail is omitted)

3 Numbers beside nodes = node water pressure (kgf/cm²) (The normal value is written in red)

4 Arrows indicate direction of → sign beside pipe line accompanying flow, and numbers indicate flow rate (m³/h)

Fig. 10 · 110 Example of piping network display

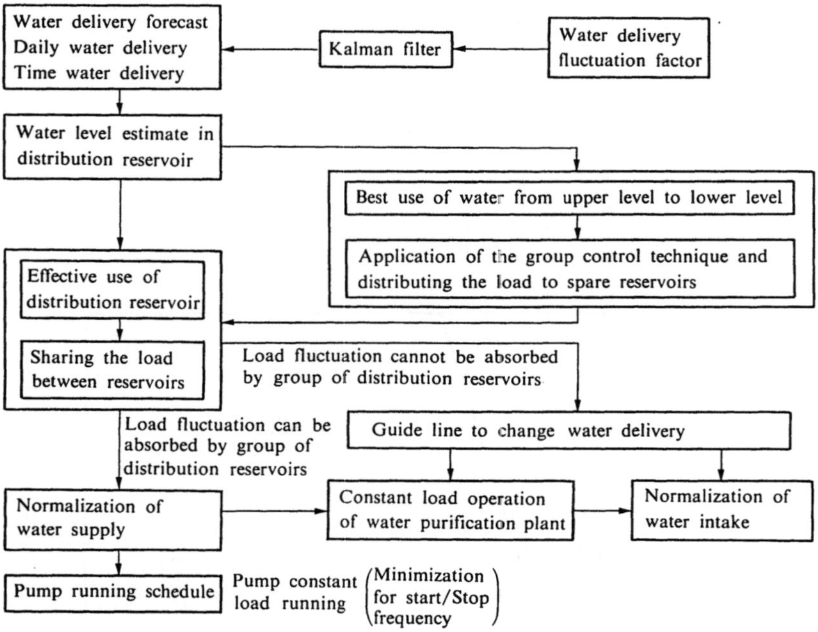

Fig. 10・111 Outline of plant operation based on demand forecast

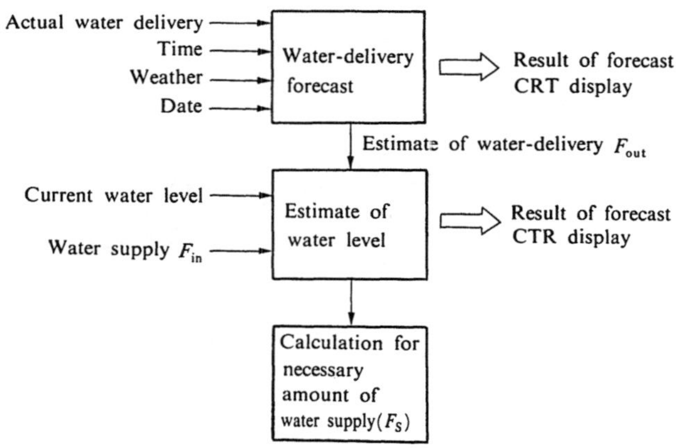

Fig. 10・112 Procedure for calculation of necessary amount of water supply

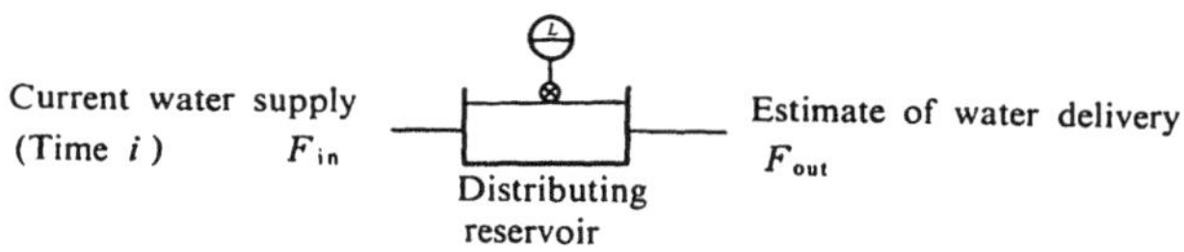

A : Equivalent area of distribution reservoir
L : Water level in distribution reservoir

General formula of water level at arbitrary time t

$$A\frac{dL}{dt} = F_{in} - F_{out}$$

$$\Downarrow$$

$$L(t) = \frac{1}{A}\left(\int_{t_0}^{t} F_{in}\,dt - \int_{t_0}^{t} F_{out}\,dt\right) + L_0$$

Where F_{in} is constant, and at time $i+n$ ($n=1\sim24$), estimated water delivery is Q_{i+n} and estimated water level is L_{i+n}

$$L_{i+1} = \frac{1}{A}(F_{in} - Q_{i-1}) + L_i$$

[L_i is the current water level at time i]

$$L_{i+2} = \frac{1}{A}(F_{in} - Q_{i-2}) + L_{i+1}$$

$$\vdots$$

$$L_{i+24} = \frac{1}{A}(F_{in} - Q_{i-24}) + L_{i-23}$$

Fig. 10 · 113 Method of estimating water level

is water level. Once the upper and lower level limits of the reservoir are calculated, then the necessary amount of water supply F_S can be determined.

10.6.8 Wastewater system overview

Wastewater systems are maintained to enhance our living environment and also to safeguard the quality of public water bodies. The wastewater system consists of a network of sewer drains, pumping stations, and a treatment plant. Pumping stations become necessary when drain pipes have to be buried beyond a certain depth.

For the past 80 or 90 years, the prevailing method of purifying large volumes of low-density waste water has been the activated-sluge method. The activated-sludge method is basically the same as the natural purification mechanism one observes taking place in rivers, only much more intensified and concentrated. Activated-sludge processing continues, even today, to be the fundamental method used in purification plants.

10.6.9 Overview of activated-sludge processes

The activated-sludge method is a biological treatment in which aeration and aerobic microbial action are used to eliminate organic pollutants. The processes involves, shown in Fig. 10.115, will be briefly

To estimate the water level at arbitrary time (t), the necessary amount of water supply is determined using upper and lower alarm forecast.

(1)

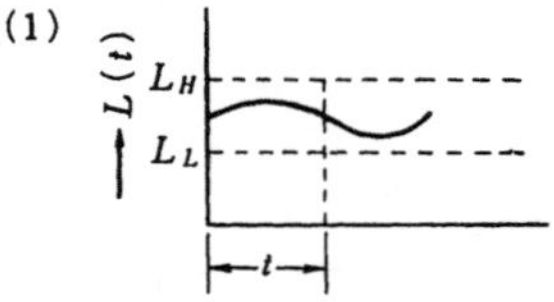

When $L_L < L(t) < L_H$ is estimated within time t,
$\therefore \ F_s = F_{in}$

(2)

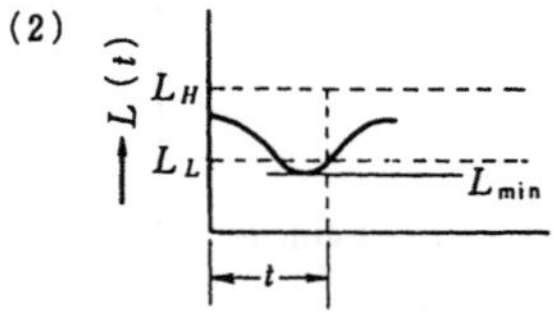

When $L_L > L(t)$ is estimated within time t,
Compensation factor $\Delta V_1 = (L_1 - L_{min}) \cdot A$ should be adopted.
$\therefore \ F_s = F_{in} + \Delta V_1 / t$

(3)

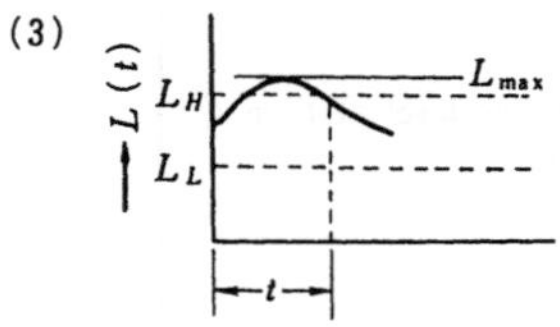

When $L(t) < L_H$ is estimated within time t,.
Compensation factor $\Delta V_2 = (L_{max} - L_H) \cdot A$ should be taken.
$\therefore \ F_s = F_{in} - \Delta V_2 / t$

(4)

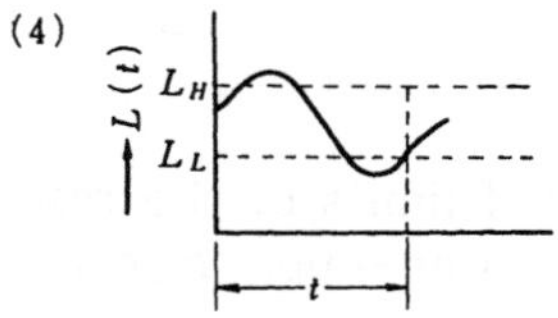

When both $L_L < L(t)$ and $L(t) > L_H$ are estimated within time t, it is considered an abnormal situation and left to operator discretion.

Fig. 10・114 Method of calculating water supply

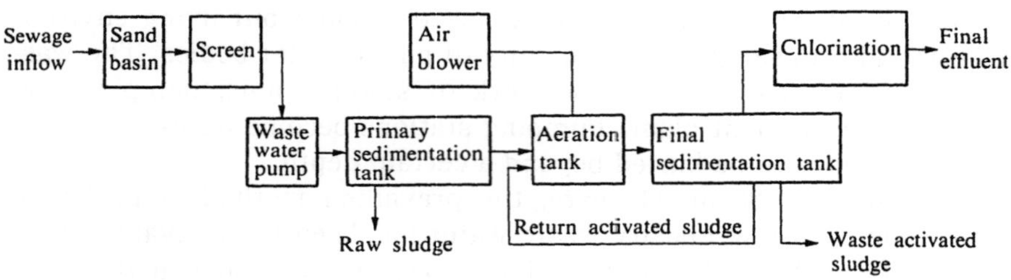

Fig. 10・115 Activated sludge process flow

described.

(*a*) *Sand basin, screen, and wastewater pump*

Gross suspended solids and trash are removed from the sewage inflow, and the stream is then pumped to the primary sedimentation tank. The sewage stream then flows under the force of gravity through the various downstream units where it is purified.

(*b*) *Primary sedimentation tank*

Suspended solid particles are removed from the liquid stream by simple gravitational settling to reduce shock loading at the following aeration tank. The sludge that accumulates in the sedimentation tank is periodically removed by pump.

(*c*) *Aeration tank, final sedimentation tank*

The sewage is churned and aerated by bubbling air through it in the aeration tank to encourage microbial action. Aerobic microorganisms in the sewage are caused to grow under favorable conditions of nutrients in the form of suspended organic matter in the waste being treated. As the organisms grow, a microbial sludge is formed having the consistency of gelatin that is known as activated sludge. Activated sludge has excellent waste-water-purifying and settling characteristics.

The clarified liquid is separated from the activated sludge in the final sedimentation tank by simple gravity settling. The sludge is then reaerated and returned to the aeration tank to contact fresh sewage. Because the activated sludge gradually accumulates as it performs its purifying role, excess sludge (waste sludge) has to be periodically removed.

(*d*) *Chlorination equipment*

The overflow from the final sedimentation tank is treated with chlorine disinfectant for safety and hygienic reasons before the effluent is discharged.

10.6.10 Wastewater treatment instrumentation

An overview of the main instruments involved in waste-water treatment is shown in Fig. 10.116. A brief description of each unit follows.

(*a*) *Sand-basin and wastewater lift-pump instrumentation*

Pump control is the main means of regulating the volume of sewage inflow. The speed of the pumps and/or the number of pumps in operation are adjusted to maintain the level of the sewage in the pump basin at a certain level. The amount of sewage inflow is measured either by an electromagnetic flowmeter or a Parshall flume mounted on the discharge side of the pump.

(*b*) *Primary sedimentation tank instrumentation*

The sludge settling on the bottom of the sedimentation tank is scrapped into sludge hoppers from which it can be removed. There are multiple hopper compartments, so the sludge removal valves is rotated from one hopper port to the next. The amount of sludge removed is generally determined using either a timer or a quantity setting device.

(*c*) *Aeration tank and blower instrumentation*

The amount of air supplied to the aeration unit is calculated to

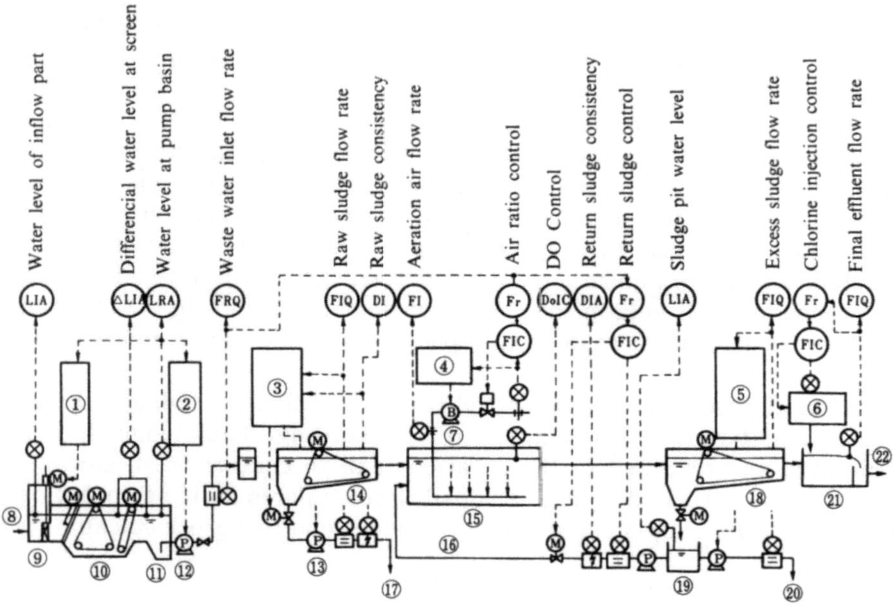

① Abnormal water level
② Pump control
③ Raw sludge pulling out sequence
④ Blower control
⑤ Excess sludge pulling out sequence
⑥ Chlorinator
⑦ Blower
⑧ Sewage inflow
⑨ Inflow gate
⑩ Sand basin
⑪ Pump basin
⑫ Waste water pump
⑬ Raw sludge pump
⑭ Primary sedimentation tank
⑮ Aeration tank
⑯ Return sludge
⑰ To sludge thickner
⑱ Final sedimentation tank
⑲ Sludge pit
⑳ To sludge thicker
㉑ Chlorine contact tank
㉒ Final effluent

Fig. 10 · 116 Instrumentation diagram of waste water treatment

maintain the dissolved oxygen concentration in the sewage at specified level. Power is also conserved by controlling this parameter. Since precision is not required in this instance, the most common approach is to supply air as a proportion of the raw sewage inflow.

Almost all of sludge that settles in the final sedimentation tank is recycled to the aeration tank. The quantity of return activated sludge is usually expressed as a percentage of the sewage waste flow, and the general practice is to use 20 to 40% of the average sewage inflow. Ratio control is used to keep this value.

(d) Final sedimentation tank instrumentaton

Settled sledge has to be scrapped and removed from the sedimentation tank on a continuous basis. Moreover, since this sludge is made up of microorganisms, it must be returned to the aeration tank as quickly as possible. Excess sludge, the incremental increase of organisms in the activated sludge, must be removed. The amount of excess to be removed is estimated by analyzing the density of the aertion tank suspended sludge. The removal of excess sludge is intermittently carried out using a quantity setting device.

(*e*) *Chlorination equipment instrumentation*

Chlorine and sodium hypochlorite are used as disinfectants. The desired dose is that maintains the coliform group bacteria count in the final effluent of less than 3 000 groups/cm^3. Ratio control is used to determine the specific chlorine dose of 2 to 4 mg/*l* of final effluent. Final effluent is generally measured with a weir-type flowmeter mounted at the outlet of the chlorination tank.

10.6.11 Sludge treatment instrumentation

Sludge produced by waste-treatment processes is first dewatered, and then hauled to a land-fill site or incinerated. Here we will briefly elaborate on the instrumentation needed to support the sludge disposal processes shown in Fig. 10.117; thickening, digestion, sludge conditioning, mechanical dewatering, and removal by truck.

(*a*) *Sludge thickener instrumentation*

Sludge produced by waste-water treatment is first thickened and then fed to the primary digestion tank. Thickener tanks are generally circular in shape, and the sludge is removed in much the same way as it is removed from the primary sedimentation tank, discussed previously.

(*b*) *Sludge digestion tank instrumentation*

In the digestion tank, the sludge gradually stabilizes and separates from the liquid while digestion gas (primarily methane) is given off in the process. This separation takes considerable time, and a number of parameters have to be supervised while it is in progress; these include the tank level, temperataure, pressure, and quantity of gas produced. The sludge and separation liquid are periodically removed from the tank. The removal, required to draw down the level of the tank when it becomes full, is under sequence control and is triggered by a timer.

(*c*) *Sludge elutriation tank instrumentation*

The sludge discharged from the digestion tank has a slimy or viscous consistency, and thus is difficult to dewater as is. Washing the sludge with water effectively removes the viscous component. Thickening is carried out in a two-stage counter-flow thickening process with the detention in each thickening unit determined by timing control. After thickening, the sludge is stored in a storage tank.

(*d*) *Coagulant mixing tank instrumentation*

Prior to dewatering, coagulants and coagulant aids are added to the sludge to facilitate the dehydration process. The chemical doses

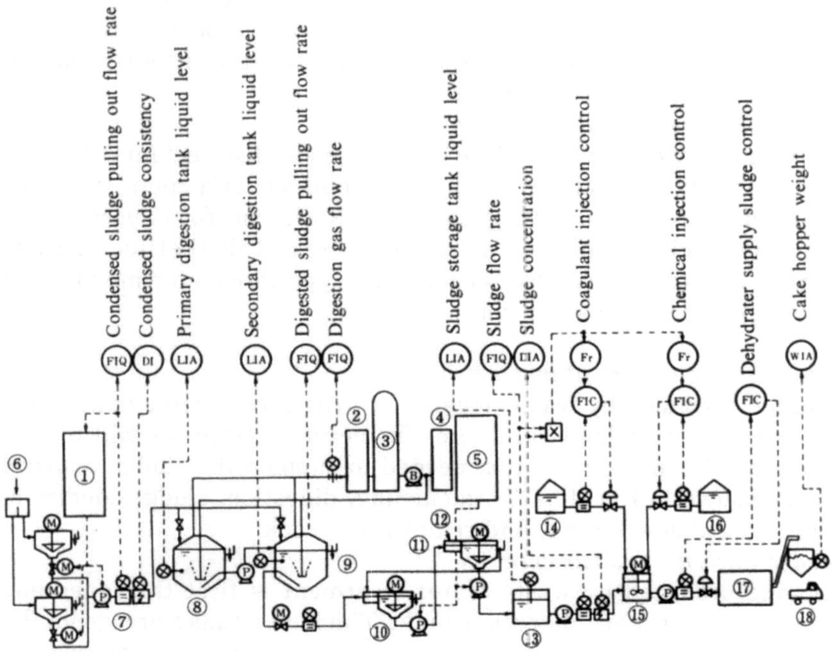

① Sludge pulling out sequence
② Desulfurizer
③ Gas holder
④ Boiler
⑤ Sludge pulling out sequence
⑥ Raw sludge and excess sludge
⑦ Sludge thickner
⑧ Primary digestion tank
⑨ Secondary digestion tank
⑩ 1st stage sludge elutriation tank
⑪ 2nd stage sludge elutriation tank
⑫ Water for washing
⑬ Sludge storage tank
⑭ Ferrous and ferric chloride
⑮ Coagulant mixing tank
⑯ Chemical
⑰ Dewatering equipment
⑱ Cake derivery

Fig. 10 · 117 Instrumentation diagram of sludge treatment

must be in proportion to the amount of solids in the sludge. By multiplying sludge flow rate times density of the sludge, the sludge solids flow rate can be calculated. Ratio control can then be applied using this value to determine the doses of the various coagulants.

Constant flow control regulates sludge feed to the dehydrator, which is set according to the capacity of the dehydrator unit. The method of flow control that is selected depends to a great extent on the type of dehydrator. After the sludge has been dewatered, it is moved by belt conveyer to the sludge cake hopper. Accumulated cake is then transported by truck to another location, say a land-fill site, or is incinerated.

10.7 *Instrumentation Applications in the Automobile Industry*

10.7.1 *Overview of automobile industry instrumentation*

The production of automobiles, a typical labor-intensive industry, is shown schematically in Fig.10.118. First, the engine and other heavy parts are cast at casting and forging factories. The pieces are machined into final shape at the machine factory, and then assembled at a engine factory. In the meantime, vehicle bodies are progressing through a succession of factories that constitutes the main line—the press factory, the body factory, the painting factory, and the assembling factory. Still other components and parts (seats, windshields, and so on) are delivered to the assembling factory by subcontractors. For the production of an automobile, all these various kinds of factories have to be systematically interrelated. Note that the production line in the automobile industry is fundamentally different from continuous processing as can be observed, for example, in the oil industry. The automobile industry is an example of non-continuous processing; each component, each body is handled as a discrete piece. Given the non-continuous nature of the automobile industry, new production technologies are constantly being introduced to realize the industry slogan "one more unit, one less yen." Here, we are referring to production technologies such robots and automatic machines which have been extensively introduced in pursuit of automation, and computers which have been introduced to aid in distributon and production management.

In this section, we will look at two instrumentation areas in the auto industry: producton management of robots and automatic machines—probably better regarded as an integrated whole—at an automobile painting factory, and instrumentation to realize storage control.

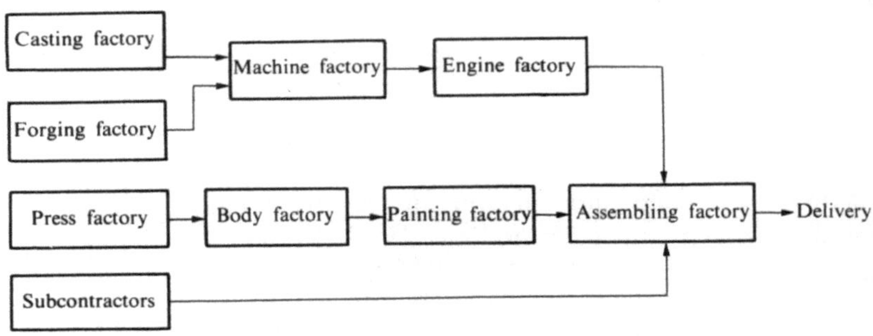

Fig. 10 · 118 Processing sequence at on automobile factory

10.7.2 Production management at a painting factory

Automobile production management seeks to manage all data pertaining to production of vehicles for the purpose of flexibly coping with modifications to the production plan. Specifically, production management monitors the following five items; (a) production instructions based on the production plan, (b) production speed (cycle time), (c) the number of car bodies in each process, (d) the location of each body (management maintained by assigning a control number to each body), (e) progress management and actual results based on tracing data (control number) updated as the body passes through processes.

(a) Painting factory process

At the painting factory, bodies received from the body factory are first put through a number of pre-treatment processes. Then, the bodies are treated for corrosion protection and multiple coats of paint are applied to provide a good visual appearance. This process flow is illustrated in Fig. 10.119.

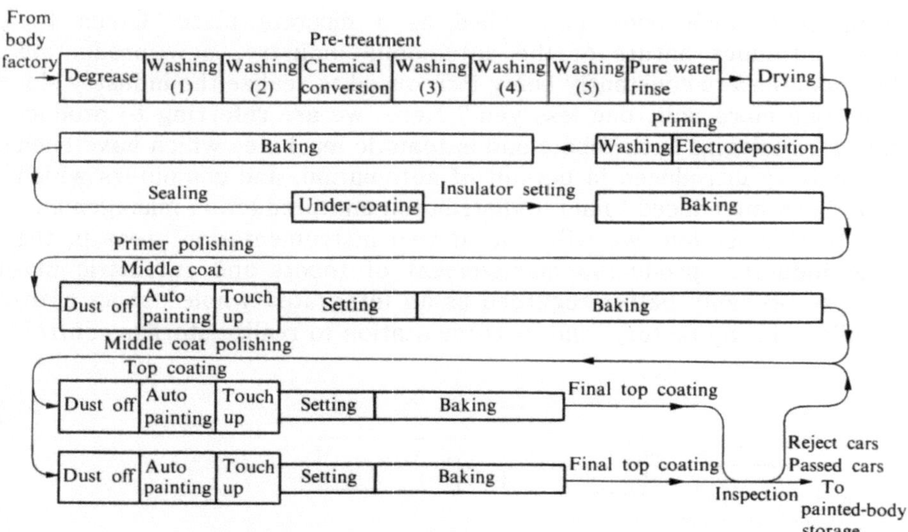

Fig. 10 · 119 Painting process and layout

After degreasing and washing in the pre-treatment process, the bodies are sent to the priming process. Next, any unevenness in the prim coat and seams in the steel sheet are smoothed out, a sealant is applied, and the bodies are sent on to the middle-coat process. After the middle coat has been applied, the bodies pass by way of the middle-coat polishing process to the final top-coat process. Here, the bodies are finished one unit at a time in accordance with purchasers' specifications to produce a durable and visually attractive external appearance.

Car bodies are moved through the processes described by chain conveyer that continues unbroken through the entire plant. There are floor-type conveyers and overhead-type conveyers in use, as shown in Fig. 10.120.

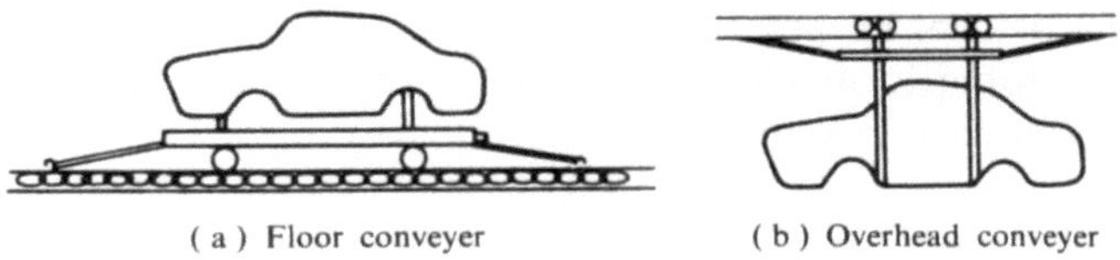

Fig. 10 · 120 Transfer by chain conveyer

(b) Production management instrumentation

We briefly touched on the contents of production management, but neglected to mention how the data is kept. A data storage system must be set up so data on the car bodies can be recorded and manipulated. In production management systems, a control number corresponding to each body is assigned. Then a file is prepared in a computer corresponding to the specifications of the specific vehicles represented by

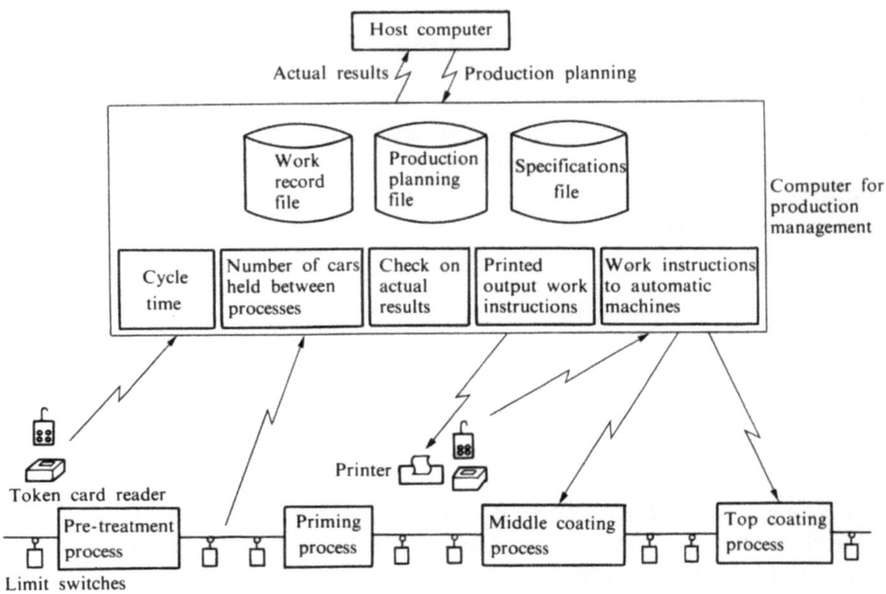

Fig. 10 · 121 Instrumentation diagram of production management

the control numbers. For the control media itself, token cards or magnetic cards are generally used.

(1) *Production instructions based on the production plan* No matter what kind of vehicle, the production sequence starts with the fabrication of the car prior to the painting processes. Once a body reaches the painting factory, the specifications file for the body is looked up using the token card control number. Then, the pertinent specifications are sent to the workers by printer, and instructions are input to the paint robots and automatic machines by contact signals. The token card is read by a token card reader.

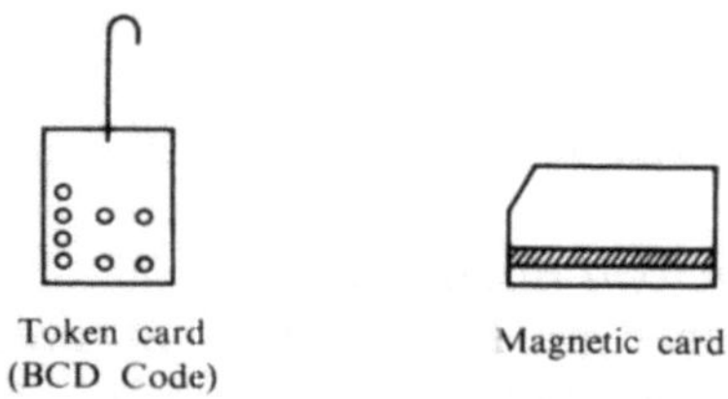

Fig. 10 · 122 Data storage media

(2) *Production speed* (*cycle time*) Speed management in auto production is exacting, and is measured in units of seconds. For example, assuming a body can be processed in 90 seconds, if the cycle time slows only slightly to 91 seconds, that means that only 791 bodies can be processes during a 20-hour operaating day versus 800 that could be processed at 90 seconds. The difference in productoin, in other words, is 9 bodies a day. Cycle time is in inverse proportion to the speed of the conveyer. Pushers (*i.e.*, hooks for latching onto the body conveyer sleds) installed at fixed intervals along the conveyer trigger limit switches. The limit-switch signals are used to measure time and monitor the cycle time.

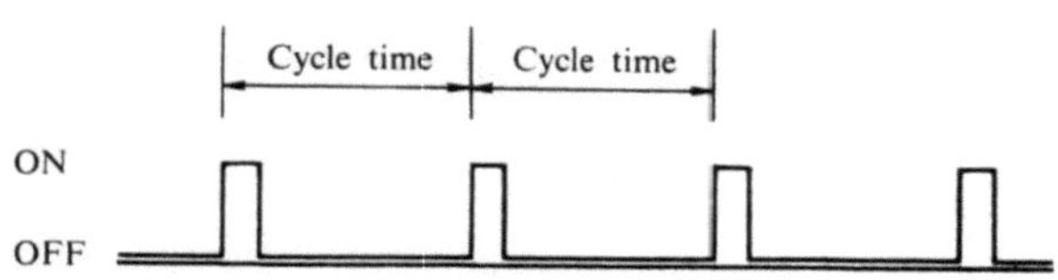

Fig. 10 · 123 Limit switch signal

(3) *Number of car bodies in each process* The car bodies trigger reset switches that are mounted at the entrances and exits of each process. By counting (incrementing or decrementing) the signals, the car bodies in each process can be accounted for.

(4) *Tracking car bodies, progress management and actual results* Buy-

ers for the cars have already been lined up even before they are produced. Thus, the production proceeds in accordance with the customers' specifications. It is necessary, therefore, to monitor the progress of the bodies so it is always known exactly which process the bodies are in. This task is managed with token card reader setpoints and limit-switch installation points, whose setpoints demarcate the individual processes. Within a given process, bodies are precisely tracked using the order in which the control numbers entered the process. This data is also used for progress management and to monitor actual results. To support such a production management system, high-performance computers are required that are capable of performing the following functions within one time cycle: (a) reading the pass signal from each setpoint limit switch, (b) outputting work instructions based on control numbers from card readers, and (c) grasping the actual results. All these requirements could not be satisfied without real-time capability and multitask processing. With the development of computers specifically designed for factory automation (FA) applications, however, it has become fairly easy to construct extremely exacting production management systems.

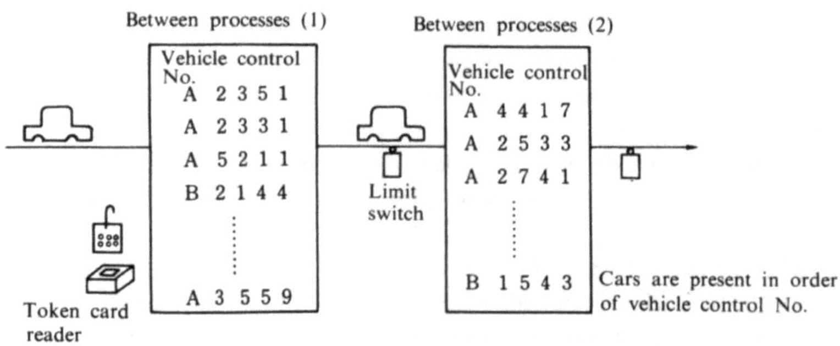

Fig. 10 · 124 Stock control No. between processes

10.7.3 Storage control

In automotive manufacturing processes, conveyer lanes are used to store car bodies between processing equipment units as well as between the painting factory and the assembling factory. This storage arrangement also functions as a buffer so that if the flow of vehicle bodies from the previous process is cut off for some reason, no adverse impact is felt at the succeeding process.

(a) Color-selection line control

The color selection line is situated between the middle-coating process and the top-coating process. The basic control function, implemented with an algorithm, is to load the vehicle into the proper col-

Fig. 10・125 Storage line

or-selection line from the middle-coating process and then unload the bodies out of the right lane for delivery to the top-coating process.

The final surface coat and color in applied to the bodies in the top-coat process. If a body out of the color selection line is inadvertently sprayed the wrong color, this creates a major inconvenience. That batch of paint has to be discarded, the wrong color paint has to be washed off the body with thinner, and the correct color then has to be loaded into the sprayer. Not only is paint wasted, but considerable time is lost. As a result, the production unit count falls, and along with it, production efficiency. To minimize the chances of this happening, the control algorithm separates vehicle bodies into separate lanes on the basis of color specifications. Then, every effort is made to send all the bodies to be painted the same color to the top-coating process in a continuous stream.

A schematic of the control functions for the color-selection line is shown in Fig. 10.126; processing flow for the system is shown in Fig. 10.127.

(1) *Production control for color-selection line* This control is activated when a vehicle body is in the stop position at the entrance of the color-selection line. The control selects which lane the body is to be sent and verifies its transfer to the selected lane.

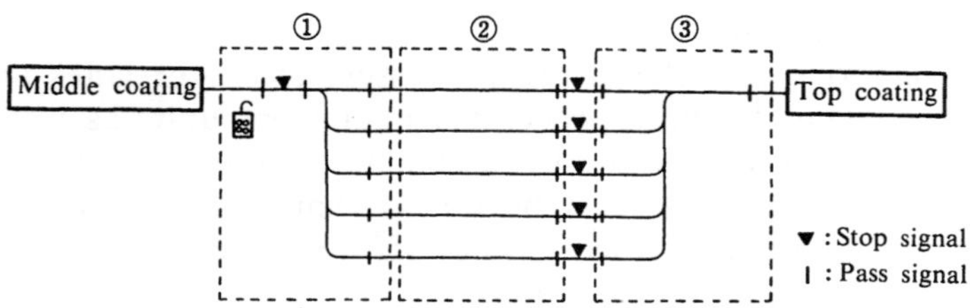

① Production control for color-selection line
② Inner control for color-selection line
③ Output production control for color-selection line

Fig. 10・126 Management functional diagram

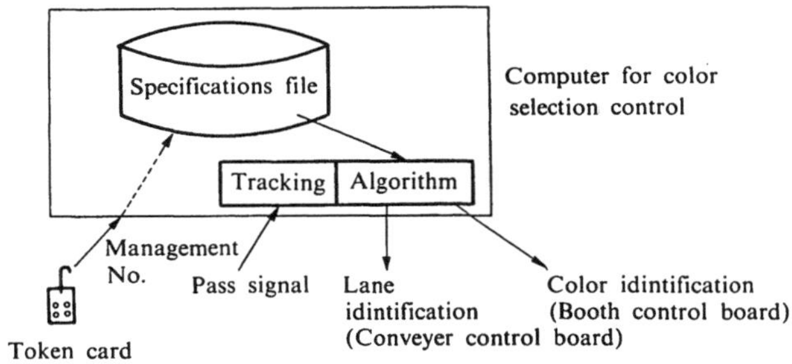

Fig. 10 · 127 Processing flow for color selection

When a vehicle body is present in the stop positon, the operator inserts the vehicle's token card in the card reader to retrieve the vehicle's control number. The control number reveals the color specifications of the body. Then, if there's a lane available where the last unit in the lane has the same color specifications as the body being processed, a pulse signal is sent identifying the lane to the conveyer control board. The stop gate then opens, and the tracking of the body to the designated lane is monitored by pass signals.

(2) *Inner control for color-selection line* This control verifies the loading and unloading of vehicle bodies to the color-selection line, and also tracks the bodies while they are in the line.

(3) *Output production control for color-selection line* This control sends bodies with the same color specifications to the top-coating process in a continuous stream. The control checks the bodies at the front of the various lanes to see if there are any to be painted the same color as the piece just sent; if there is, an unload signal is sent to the conveyer control board. When the stop barrier is opened and the workpiece transferred to the top-coat process, a pass signal detected by a sensor mounted right in front of the top-coating process sends the color identification of the body ahead to the top-coating unit.

(*b*) *PBS line control*

The Painted Body Storage (PBS) line, capable of accommodating anywhere from 100 to 200 vehicle bodies, is situated between the painting factory and the assembling factory. With the object of leveling or evening out the work on the assembly line, the PBS line control has two key functions: mixing delivery of vehicle bodies to the assembly line to achieve a good load balance on the line, and sidetracking vehicle (*i.e.*, not sending them to the assembly line) which have component(s) needed that are out of stock.

(1) *Mixed delivery to achieve load balance* Auto manufacturing involves a great deal of assembly work. In order to meet the specifications

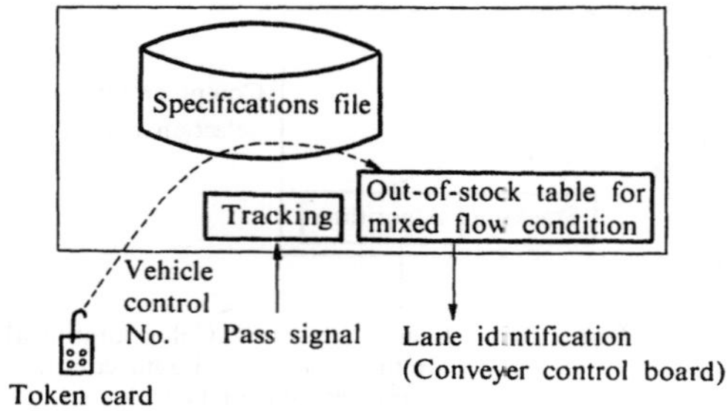

Fig. 10・128 Process flow for load balance in assembly line

for a typical vehicle, many miscellaneous parts and components are assembled in numerous processes (anywhere from 80 to 100 processes) on an assembly line. For each process, the time allowed to attach a component is strictly controlled at close to the minimum time possible for the assembly work. Under these exacting conditions, when big–load vehicle bodies come down the line, if the work isn't completed within the specified time allowed, the problem is passed on to the following downstream process. Start of assembly work on the next unit is delayed, until eventually, the line has to be shut down to catch up. To prevent this situation from developing, bodies are taken off the PBS line and delivered to the assembly line in such a way to facilite completion of the assembly work within the designated time: a heavy–load body, for example, is followed by a string of light–load bodies to even out the line load. To illustrate the mixing principle, consider five types of vehicles (A, B, C, D, and E), with the following mixing conditions:

A: can be put on the line in succession
B: 2 units in a row prohibited
C: separate by at least 2 units
D: separate by at least 3 units
E: separate by at lesst 5 units

Given these conditions, the vehicle bodies might be put on the assembly line in the following order:

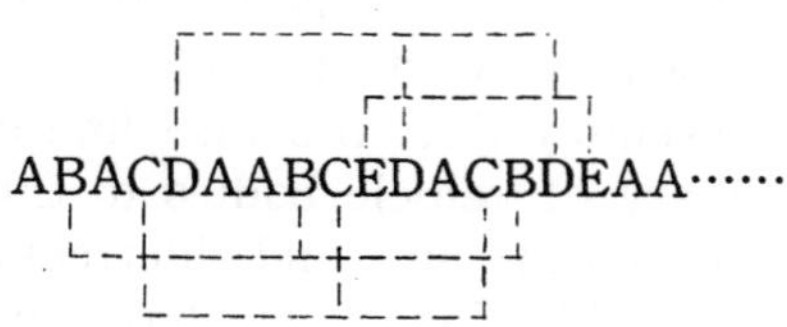

(2) *Countermeasures for missing componeuts* Components and parts manufactured by outside subcontractors (seats, tires, engines, etc.) can

normally be delivered on a just-in-time basis within two hours for a given vehicle. However, if a vehicle should come down the line and a component is not available, say because it has been held up in traffic or because the wrong part was delivered, this can create havoc on the assembly line. This situation is avoided by registering missing parts in an out-of-stock table. Then, by checking the table, bodies are held in the PBS line until the missing component is back in stock.

(3) *Processing flow* The PBS line control is implemented by means of a card reader installed at the entrance to the line; the control number read in the card reader gives access to the specifications of the vehicle. Then, transferring bodies into and out of the PBS line is done in compliance with the control conditions (mixing conditions and component availability). Bodies are tracked within the PBS line by means of pass signals.

In the auto industry, the introduction of new equipment and the revamping of manufacturing lines is practically an everyday occurrence. If lines aren't constructed around computer systems capable of flexibly responding to line conditions, then they are locked in a rigid mold, and quickly become out of date.

One final point to note is that the production management system and storage control system described both rely on the same vehicle specifications file. It would thus make sense from a systems point of view to integrate the two systems in the same computer system.

10.8 Product Control in Batch Processing[33)–37)]

The great number of new materials, bio-products, high molecular compounds that have made their appearance in recent years is testimony to the steady stream of new products that are constantly being developed based on advanced technologies. Of course, for such new products to be created, new methods of production also must be devised. In the case of large-scale chemical processes, say for the manufacture of products such as ethylene and ammonia, large-volume, continuous processing methods can be applied. The profusion of new products, however, are almost always produced in small batches, which are considerably more difficult to automate. If maximum production is the only concern, technological effort is bent toward achieving economies of scale, and continuous processing; whatever is necessary to boost volume. Once we enter the age where quality is at least as important as quantity, however, a totally different approach is called for: Inventories are reduced to the bare minimum, and only products where there is a demand are produced, and only in the quantities required. Production equipment is limited and quick modification of operating conditions is emphasized in order to expand or contract producton in respnse to changing demand. In short, this is the technology for producing small quantities of diverse products, and the methodology is flexible automation. Basic to the new production technology is the desirability of batch processing rather than continuous processing and more exacting production plans. Implementing such an approach has greatly stimulated the development of automation technology.

On the other hand, automation itself, in the form of digital control systems (DCSs), has had a profound impact. Not only can advanced control algorithms be implemented on DCSs, but even more signficant is the ease of switching back and forth between control loops and sequence operations that is possible with DCSs. Such capability has accelerated the development of DCSs much more than would have occurred otherwise. Closely adhering to the production plan, it is a simple matter with DCSs to alter control loop parameters and change the order of sequences as required for specific products. This, in essence, is the basis of flexible automation for producing diverse products in small batches.

10.8.1 Batch process recipe management

A common requirement of most small-volume batch processes is the ability to change the operating conditions for different recipes or products. However, there is great variation in detail depending on the type of batch process and other conditions. Simplified prototypical batch process configurations are shown in Fig. 10.129. Note that actual batch processes can assume many different forms depending on the product being produced; specific configurations consist of many pieces of equipment linked together in complex ways. Moreover, there are just about as many different recipe management methods as there are different processes.

Recipe management for a single equipment unit is the most basic configuration (Fig. 10.129(a)). The recipe management procedure for this case is shown in Fig. 10.130. The recipe data base is stored in an external file. When a batch is started, a command from the keyboard fetches the appropriate recipe data, and the recipe-specific control loop parameters and order of sequence operations are loaded. By specifying a recipe name, corresponding data is read from an external file and displayed on a CRT, the amounts of raw materials needed to produce the set production volume is calculated, and control loop parameters and sequence order changes are loaded.

When multiple units are configured in a parallel-line (Fig 10.129(b)), mix and charging systems such as shown in Fig.10.131 are the most common approach. Generally, the mix and charging system re-

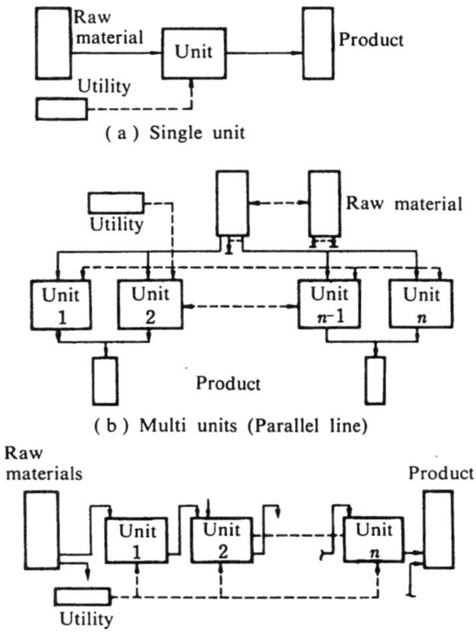

Fig. 10 · 129 Batch process configurations

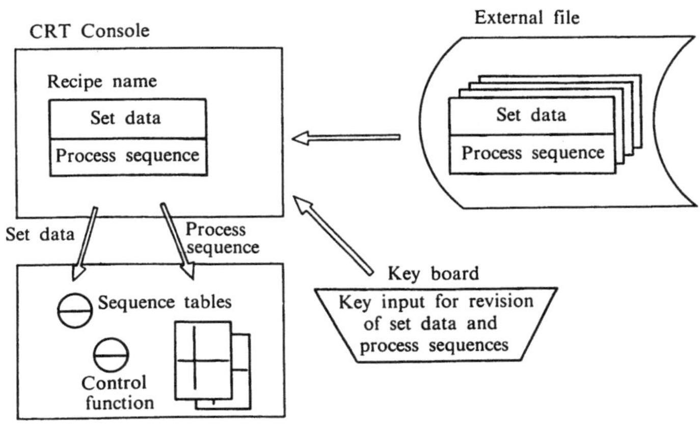

Fig. 10 · 130 Recipe management procedure

fers to the line from the raw materials intake to the mixing tank. The sequence that selects this line is often quite complex. When the recipe is selected from the mix instruction screen, the proportions of the raw materials (included in the recipe data) are brought up on the screen. If

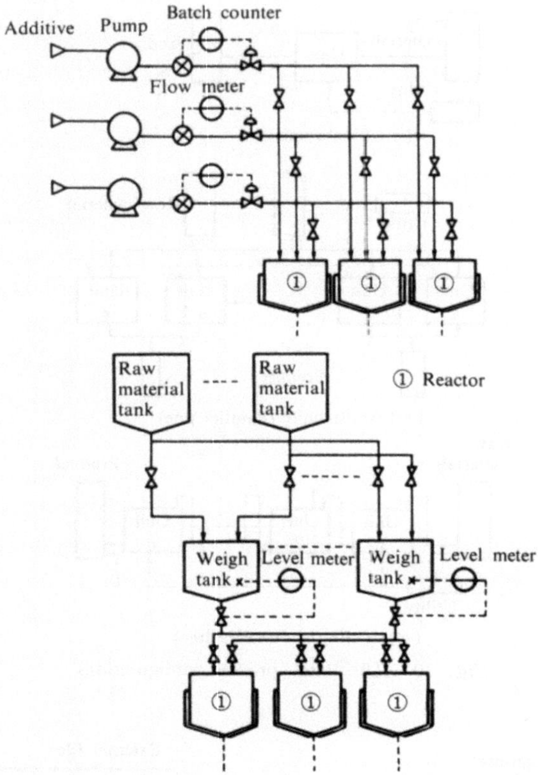

Fig. 10・131 Mix and charging system

the entire batch volume is input for the setting, proportions for all of the raw materials are calculated. Next, by entering the instruction for the mixing to start, charging from the raw materials tank begins through the line indicated by the data. In order to increase the operating efficiency of the mixing system, it is common to carry out multiple operations in parallel. When this is done, it is necessary to implement interlock between the processes.

When processing takes place over a number of stages in multiple equipment units that are configured in a series, the recipe data must encompass movement from unit to unit. Two things are required; recipe scheduling for the entire group of reactors involved, and reaction cut-over conditions for each unit. Thus, the reaction conditions for each unit must be entered into memory beforehand for each recipe. Then, when a recipe is specified from the CRT, the control loop parameters and order of sequence operations are successively changed for each unit following the path indicated by the recipe. Generally, the product flow or unit path is fixed. However, this is not necessarily the case in processes where multipurpose units are concerned; that is, where more than one reaction or operation takes place in a single unit. Combining single-function units (where only a single operation takes place) and multipurpose units, complicates the configuration. Moreover, since paths change with each recipe, sometimes the processing sequence path, or batch pass assignment from unit to unit is included in the recipe itself. Two examples illustrating this kind of arrangement are shown in Fig. 10.132.

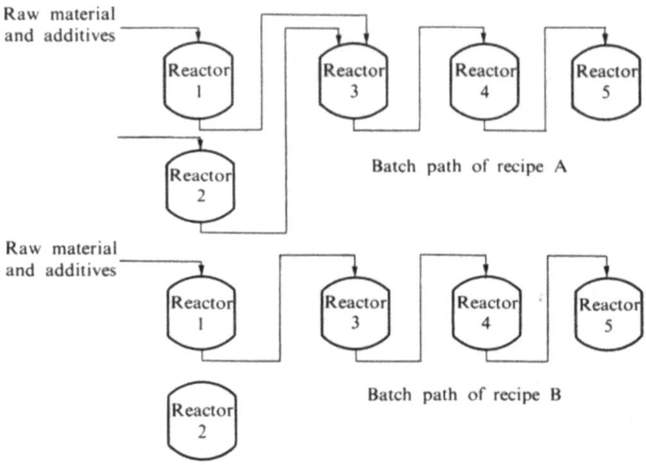

Fig. 10 · 132 Batch pass assignment of recipes

It is thus apparent that actual processes are considerably more complex than our simplified configurations in Fig. 10.129. Recipe management methods are equally complex, and show considerable variation depending on the product and the process. Nevertheless, using a basic approach to recipe management (such as illustrated in Fig. 10.130) to change control loop parameters, and the order and conditions of sequence operations, diverse products can be efficiently produced in small batches.

10.8.2 Batch process control

It would be exceedingly difficlut to extend flexible automation to diverse processes without first establishing control loop and sequence operation methods. To facilitiate changing control parameters and the order of sequence operations, a control loop and sequence operation database must be constructed for the digital control system.

First, with respect to control loops, most commonly the same approach is adopted as applied to the construction of analog modules: data is collected that is expressive of a particular modual function, and designed so it can be constructed according to some method of coupling. For example, all data pertaining to the PID module, that is, PID parameters, setpoint value, process variable, manipulated variable, high and low alarm setpoint values, loop status, etc. and input/output connection information would be lumped together in one module. In the case of cascade control, two such modules could be interconnected to form the loop. Since functions other than PID functions such as ratio, calculation, lag, and dead time are generally also implemented, all control loops can be constructed out of these modules.

Consider, for example, the reactor control shown in Fig. 10.133. For the reaction to proceed efficiently and safely, temperature control is a primary consideration. The reaction is started by applying heat, followed, at a critical moment of time just before the reaction becomes critical, by removal of the heat. As a safety precaution, the temperature is always maintained at or below some prescribed temperature. Or, there are situations where the reaction is divided into muiltiple stages, and an additive has to be introduced over several or all of the stages. For reaction temperature control in this kind of case, a control loop can be configured out of standard modules such as a program setting module and a PID module that is capable of maintaining the desired temperature pattern. A screen depicting the operation of the temperature control loop is shown in Fig. 10.134. Note that the screen also shows the program setting module and the temperature pattern in addition to the two temperature control modules. The advantage of DCSs is clearly demonstrated when its time to change the temperature pattern for a different recipe; the data in the program setting module can simply be written over to make the change.

Also, since there is some delay in the temperature response at the beginning of the reaction, a compensating element is sometimes added

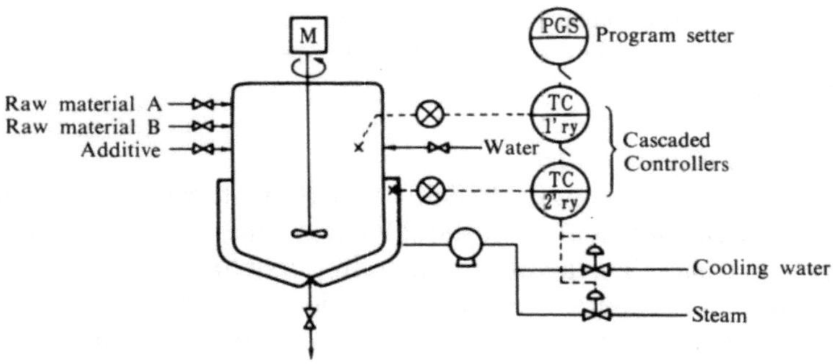

Fig. 10 · 133 Reactor control

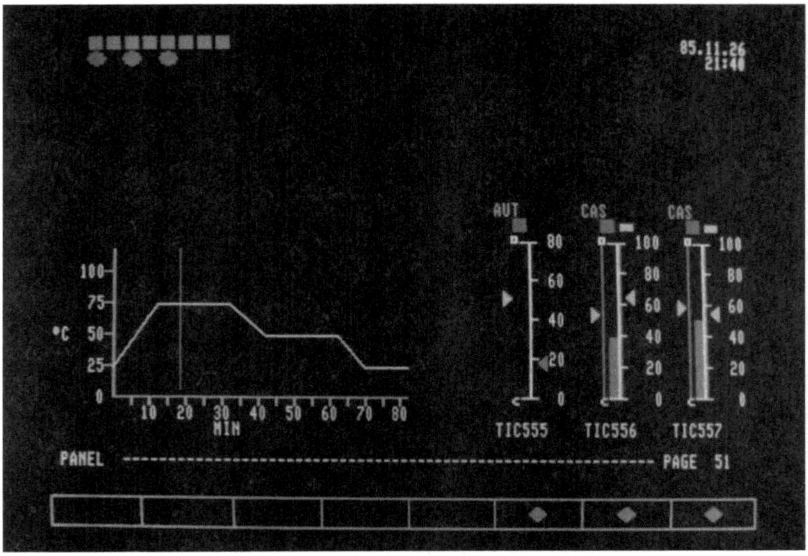

Fig. 10 · 134 Display of reactor operation

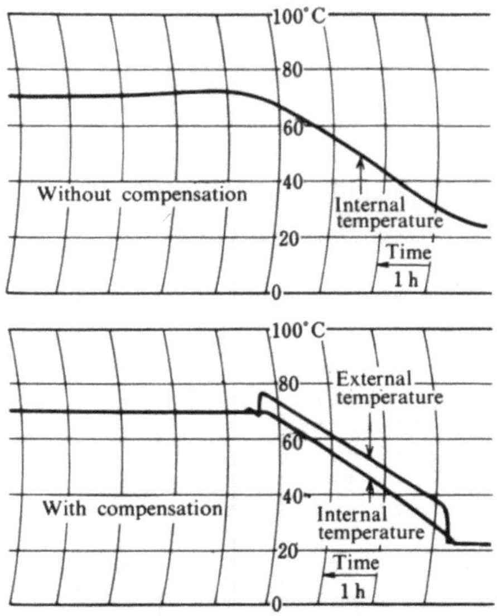

Fig. 10 · 135 Control response of reactor temperature

to assure that the raise in temperature follows the prescribed pattern. Furthermore, in the latter half of the reaction, reactant is depleted and process sensitivity declines. When this happens, it is easy for temperature deviations to occur. An appropriate means to counter this effect is conversion of temperature control sensitivity. Because the extent of compensation needed and the amount of sensitivity cut-over for temperature control are recipe dependent, these data must be prepared in advance for each recipe. Here, as we saw before, the data settings can be easily changed whenever necessary by means of a DCS. Figure 10.135 shows an example of precise temperature ramping for a particular recipe using a DCS.

Turning now to sequence operations, Figure 10.136 shows how sequence phases can be changed for different recipes. In the case of multipurpose reactors, not only operating conditions such as raw mate-

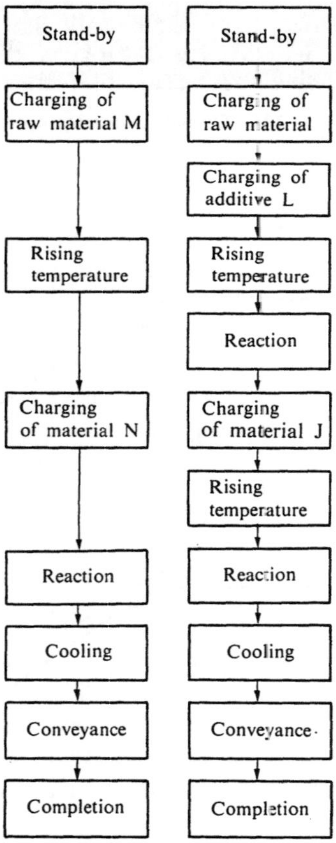

Fig. 10 · 136 Recipe-dependent sequence phases

rial and charging volumes, temperature, and time, but also sequence operation phases must be changed depending on the type of recipe. Thus, the sequence control method must be configured in such a way that the operation phases can also be rearranged in different ways according to the specific parameters involved. Moreover, each phase contains a series of detailed sequence steps: What valve is to be operated, what pump is to be activated, how control loops are to be configured, and so on. It is essential, therefore, that the detailed sequence steps within the phases are also capable of being rearranged to coordinate with the rearranged phases. In other words, the series of detailed sequence steps expressing, say, pump and valve actions, is typical of a specific phase, and the sequence of phases is stipulated in the product-specific recipe data. How finelly the detailed sequence steps are divided, or the scale of the units, will vary. Generally, however, it can be said that if the degree of freedom is large, the detailed sequence step units can be made small, and the corresponding phases will increase. Conversely, if the degree of freedom is reduced, the detailed sequence step units can be increased in size, and the corresponding number of phases can be reduced.

The relation between phases and detailed sequence steps is summarized in Fig. 10.137. So-called decision tables are shown in the same figure. When input conditions are entered in the upper tiers of the tables, then the corresponding output operations shown in the lower tiers are executed according to rules. This type of table is a useful device for describing the logical relations between input conditions and output operations. Moreover, since operatoin instructions of one decision table can be executed from another decision table, this is also an appropriate description for sequence control methods with a hierarchical structure. The step execution table is used to describe the detailed se-

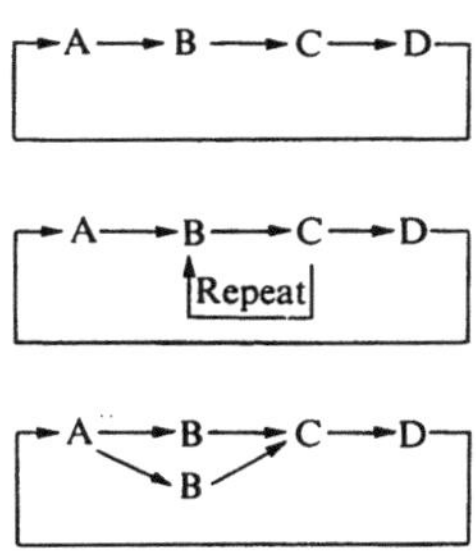

Phase control table

Step	A C D
Input	Conditions of phase transition
Output	Instruction for each phase execution

Step execution table

Step	......	......	......	......
Input	A Phase Conditions	B Phase Conditions	C Phose Conditions	D Phose Conditions
Output	A Step operations	B Step operations	C Step operations	D Step operations

Fig. 10 · 137 Sequence phase and decision table structures for batch sequence control quence steps for a single phase. These could include, for example, opening and closing valves, activating and stopping the temperature program setting module, implementing and severing control loop cascades, changing settings, changing outputs, and so on. The phase control table, on the other hand, determines what step execution table is to be executed and in what order according to the parameter conditions designated for the particular recipe.

Since indicator lamps and push buttons are generally required for batch process sequence operations, an example of what these can look like is shown in Fig. 10.138. The selector switch is set to full–auto mode when one wants the processing to proceed from phase to phase automatically. Semi–auto mode allows automatic processing of only the operations within a phase; when the phase is completed, the process stops and waits for a instruction to proceed to the next phase. Setting the selector switch to manual mode discontinues the automatic sequencing up to that point. This means that, as soon as a point is reached where it is appropriate to stop, the process does so, and subsequent operations are controlled manually. Start, stop, stepping, and reset are activated by the push buttons labeled with those functions shown in the same figure. Since this status transfer diagram can be easily modified using the decision table, the contents of the phase control table can be designed. In other words, the system has been designed so that the phases are controlled by the movement conditions of the status transfer, and the order of the phases can be rearranged by referencing the parameters.

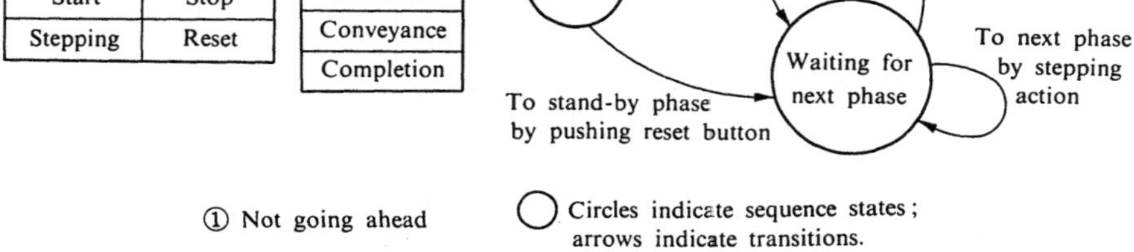

Fig. 10 · 138 Status transfer diagram of sequence and selector switches, push buttons and lamps

10.8.3 Recipe management and operation methods

In order to produce a diverse range of products by changing the batch process operating conditions, two things are required; a great volume of recipe data for each product, and a recipe data mangement function. The recipe management function is used to select the product line, display it on screen, and view the operating status to see if operating conditions need to be changed. In this section we will consider the recipe management function, and selection operation method.

It has already been mentioned that the recipe management function will differ depending on the configuration of the batch process. The first configuration we will consider is a single unit process. Through a sequence of panel operations such as shown in Fig. 10.139, the recipe to be executed is selected, data respecting the selected recipe is brought up on the screen, and instructions altering operating conditions are entered. The first panel that appears is a menu of the recipes available. A selection is made from among the available choices, and this brings the corresponding recipe data up on the screen. Since the software permits altering the data on this panel, the batch isn't started until after the recipe data has been verified. Once the batch has been started, process operating conditions proceed in accordance with the recipe data shown on the screen. Another point is that the recipe data is expressed as ratios of standard production units. In some cases the ratios are calculated when the production volume is set. In other cases, when the desired production volume cannot be ascertained from the volume of the raw materials, the ratios are obtained indirectly either by calculating the residual volume or through a volume conversion based on the concentration or purity of a weight standard chemical additive. Sometimes this recipe data is also shown on the screen.

In configurations involving multiple equipment and reactor units, a first requirement is to designate in which reactor a recipe is to be produced. This is generally handled by creating lot numbers for all of the reactors involved indicating the type of recipe and the reactor. These numbers are then displayed along with the production conditions. In some cases, the kind of recipe that can be produced in a given reactor will depend on the lines for charging raw material or additives, or on the type of reactor. In such cases, the recipe selection to be carried out in a certain reactor is made only from among those recipes that are appropriate for the reactor in question. When the product under production requires the use of a multi-unit configuration, the following information is shown: the unit in which the process starts, the sequence of succeeding units, and the recipe.

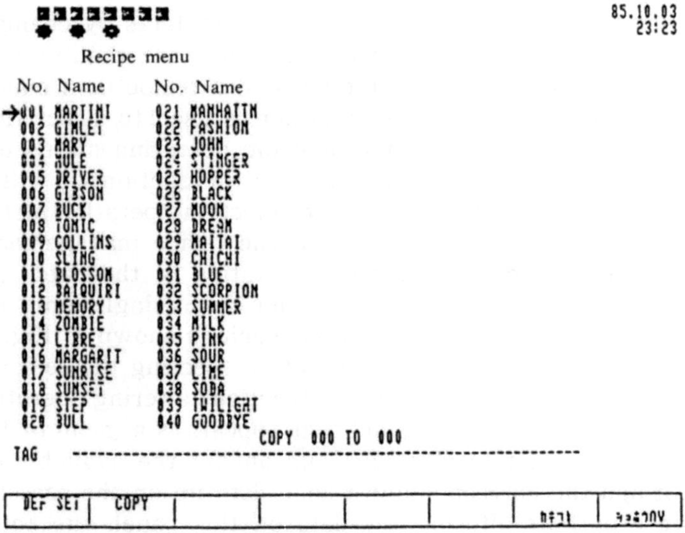

Recipe menu panel

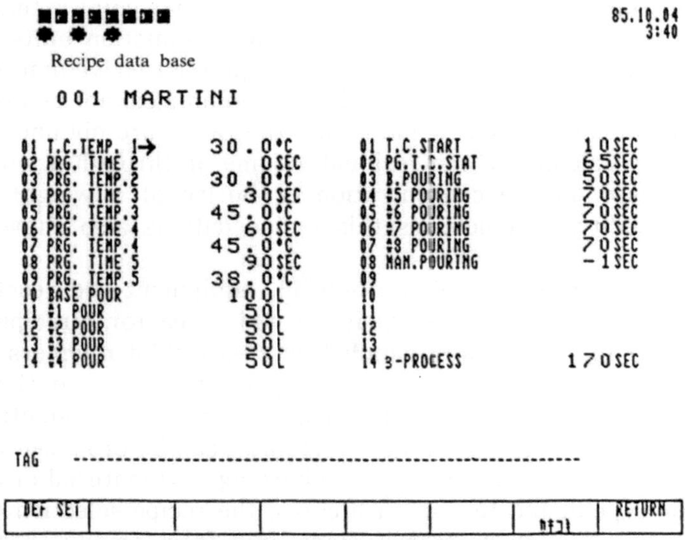

Recipe data base panel

Fig. 10 · 139 Recipe menu and data base panel

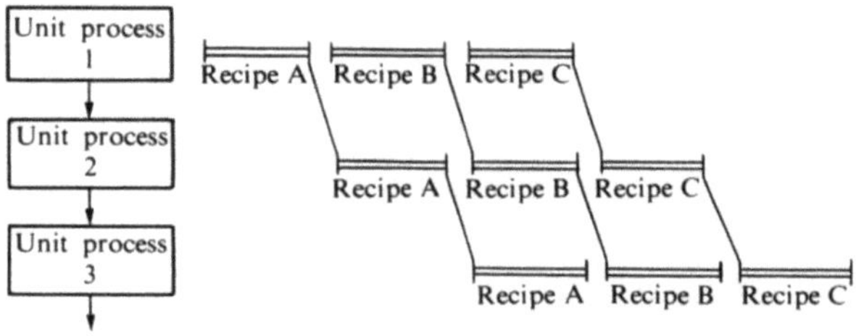

Fig. 10 · 140 Recipe management of multistage process

In cases such as illustrated in Fig. 10.140, where the equipment is arranged in a multistage configuratoin, operating conditions cannot simply cut over all at once in accordance with the recipe data. Rather, cut over must follow the sequence of the equipment units. In other words, the recipe data must be broken into segments corresponding to each unit, so that when one unit is started, only the operating conditions for that unit will be modified. Figure 10.141 illustrates how the unit sequence and recipe selection are shown on the screen. In the recipe shown in the figure, the maximum number of stages that can be designated for each sequence, eight, has been designated. When one wishes to display that recipe data, first an overview of the data for each recipe is displayed. In cases where the production involves a multistage unit configuration, the volume of recipe data generally increases. Since it is not possible to display all of the data at the same time, the data is broken up into groups. The groups, first displayed altogether on a menu screen, are segmented so as to keep all data pertaining to each stage of each unit together in one group. When a data group is indicated on the recipe overview screen, this brings up the recipe data base for a single equipment unit, such as shown for example in Fig. 10.139.

If the unit start instruction is given on the multi–unit batch selection panel, only the data group for the relevant unit is selected for modification from among the previous stage recipe data. In this case the units' path is fixed, and represents a simple, series configuration. Of course, there are possibilities: When the units are set up to serve diverse purposes, for example, the paths themselves can be changed according to the recipes. In situations such as this, the unit path is included as data in the recipe database. Then, depending on the path, the recipe name of the stage prior to that unit and the recipe data group for that unit must be searched. Figure 10.142 illustrates a recipe product cut–over method that contains a path table.

Just as process instrumentation varies widely depending on a multitude of different factors, recipe management methods also vary tre-

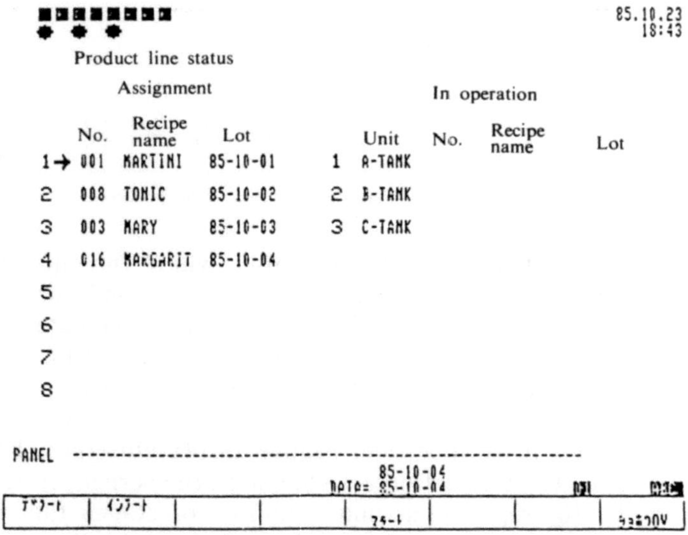

Batch assignment and current batch stage display

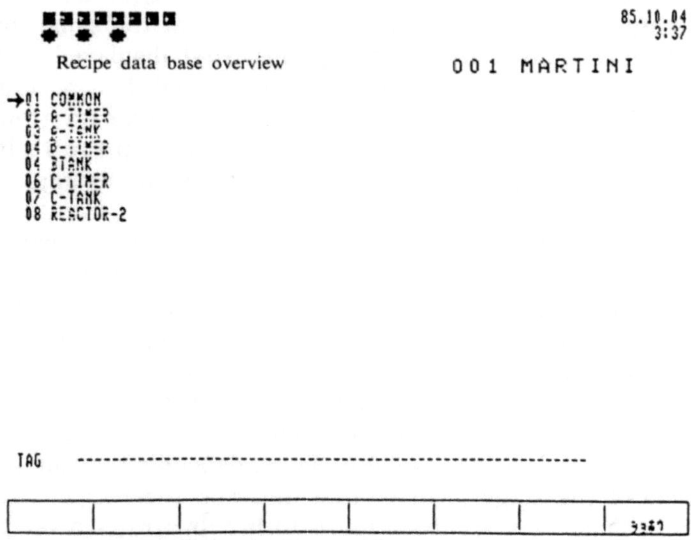

Recipe data base overview panel

Fig. 10 · 141 Batch assignment and recipe data base over view panel

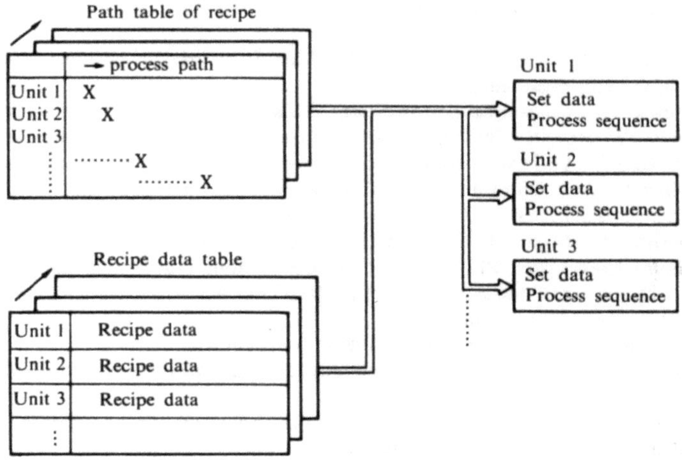

Fig. 10 · 142 Recipe management with pass assignment

mendously depending on the specific process configuration. Our approach here has been to describe prototypical schemes, or formulas, to indicate the range of different possibilities. By changing the configuration of equipment, any number of application examples could have been described. In adopting a formulaic approach, our intent was to clearly reveal the unique advantage of digital control systems—that is, flexible automation. There are other topics relating to recipe management that are also important; collection of actual results data, formation and testing of recipe data, comparison of reaction patterns, optimization of operation scheduling, and so on. With respect to digital control systems, however, it is safe to say that as these systems continue to develop and become more sophisticated, increasingly easy-to-use configurations will be realized.

REFERENCES

1) T. Ishii: Process Control Instrumentation Systems, Denki Shoin (1973) (in Japanese).
2) M. Tanaka: "Topping unit," *Petrotech*, **7**, 4 (1984) (in Japanese).
3) S. Hiramatsu, *et al.*: "Problems in instrumentation at oil industry," *Automation*, **28**,8 (1983) (in Japanese).
4) H. Hirano: MRP in Medium and Small Enterprises, Nikkan Kogyo Shinbun Sha (1982) (in Japanese).
5) T. Fujimura, *et al.*: "Instrumentation and control system for refinary off-site equipment," *Yokogawa Tech. Rep.*, **29**,3 (1985) 31–38 (in Japanese).
6) Y. Kawase, *et al.*: Oil Refinery Technology Handbook, Sangyo Tosho (1981) (in

Japanese).
7) The Iron and Steel Institute of Japan: Iron and Steel Industries in Japan (1985) (in Japanese).
8) Comprehensive Bibliography for Iron Manufacturing Machinery '80, Jukogyo Shinbunsha (in Japanese).
9) Nippon Steel Corporation: Iron Science (Making Process of Iron) (in Japanese).
10) *Seitetsu Kenkyu*, No. 308, Nippon Steel Corporation (1982) (in Japanese).
11) *Tekko Kaiho*, The Japan Iron and Steel Federation (June 1984) (in Japanese).
12) *Sumitomo Metals*, Sumitomo Metal Industries, Ltd. (in Japanese).
13) *Tetsu-to-Hagane*, **71**, 3, The Iron and Steel Institute of Japan (1985) (in Japanse).
14) *Kawasaki Steel Giho*; **14**, Kawasaki Steel Corporation (1982) (in Japanese).
15) *Seitetsu Kenkyu*; No. 313, Nippon Steel Corporation (1984) (in Japanese).
16) *The Thermal and Nuclear Power*, **29**, 6/8, Thermal and Nuclear Power Engineering Society (1978) (in Japanese).
17) K. Shirano: Zymurgy, Kodansha (1982) (in Japanese).
18) Shokuryo Kogyo: Agricultural Food-Sugar, Koseisha Kosei Kaku (1985) (in Japanese).
19) R. Machida: "Production control system on YEWCOM at sugar refinery," *Factory Automation*, **3**, 7 (1985) (in Japanese).
20) *Shigyo Times*, The Latest Pulp and Paper Technology '80 (in Japanese).
21) Y. Murakami: Japan Journal of Paper Technology (Aug. 1983) (in Japanese).
22) A. Nomoto: Japan Journal of Paper Technology (Mar. 1982) (in Japanase).
23) T. Shibata: Data Systems for Automated Production and Material Handling, Ryutsu Kenkyu Sha (Mar. 1984) (in Japanse).
24) Japan Water Works Association: Recommendation for Design of Water Works Facilities (1977) and Recommendation for Maintenance of Water Works (1982) (in Japanese).
25) S. Nagase, *et al.*: "Data base management system on city water distribution plants," *Yokogawa Tech. Rep.*, **28**,1 (1984) 8–13 (in Japanese).
26) S. Nagase: "Prediction of demand and total control systems for water supply plants," *Yokogawa Tech. Rep.*, **24**, 1, (1980) 17–22 (in Japanese).
27) H. Kamei, *et al.*: "On-line water demand predictions with Kalman filter," *Yokogawa Tech. Rep.*, **25**, 4, (1981) 36–40 (in Japanese).
28) K. Minamimura, *et al.*: "Simulation of water distribution networks," *Yokogawa Tech. Rep.* **28**,1 (1984) 14–20 (in Japanese).
29) T. Yamamoto: "Hierarchical control system for wide-area water supply network," *Yokogawa Tech. Rep.*, **28**, 1, (1984) 21–25 (in Japanese).
30) Yokogawa Electric Corporation: Application Engineering Data Chemical Injection Control and Water Quality Monitoring at Purification Plant (in Japanese).
31) Yokogawa Electric Corporation: Application Engineering Data Instrumentation on Filter Basin and Washing Control (in Japanese).
32) Japan Sewage Works Association: Recommendation and Explanation for Design of Wastwater Facilities (1984) (in Japanese).
33) K. Matsunaga: "The latest instrumentation for batch process by distributed digital control system "CENTUM"," *Yokogawa Tech. Rep.*, **23**, 3 (1979) 26–31 (in Japanese).
34) T. Hirano, *et al.*: "Distributed control system applications in batch processes and product grating," *Yokogawa Tech. Rep.*, **25**, 1, (1981) 44–52 (in Japanese).
35) K. Matsunaga: "Documenting process control sequence by decision tables," *Yokogawa Tech. Rep.*, **26**, 3, (1982) 44–50 (in Japanese).
36) T. Hirano, *et al.*: "Expediting the design of batch process control systems," *Yok-*

ogawa Tech. Rep., **29**, 3 (1985) 25–30 (in Japanese).
37) I. Miyazaki, *et al.*: "Functions and usage of CRT consol on batch process," Instrumentation & Control Eng., **29**, 6 (1986) 30–36 (in Japanese).
38) T. Kadoya: Paper Science, Chugai Sangyo Chosakai (1977) (in Japanese).

APPENDIXES

APPENDIX 1 REFERENCE THERMOELECTROMOTIVE FORCE TABLES

(Extracted from JIS C 1602-1981 which conforms to
IEC 584-1977, BS 4937-1973, DINIEC 584-1984,
ASTM E230-1983, ANSI MC96.1-1982)

Type : R   Unit : μV

Temp. (℃)	0	−10	−20	−30	−40	−50					Temp. (℃)
0	0	−51	−100	−145	−188	−226					0

Temp. (℃)	0	10	20	30	40	50	60	70	80	90	Temp. (℃)
0	0	54	111	171	232	296	363	431	501	573	0
100	647	723	800	879	959	1 041	1 124	1 208	1 294	1 380	100
200	1 468	1 557	1 647	1 738	1 830	1 923	2 017	2 111	2 207	2 303	200
300	2 400	2 498	2 596	2 695	2 795	2 896	2 997	3 099	3 201	3 304	300
400	3 407	3 511	3 616	3 721	3 826	3 933	4 039	4 146	4 254	4 362	400
500	4 471	4 580	4 689	4 799	4 910	5 021	5 132	5 244	5 356	5 469	500
600	5 582	5 696	5 810	5 925	6 040	6 155	6 272	6 388	6 505	6 623	600
700	6 741	6 860	6 979	7 098	7 218	7 339	7 460	7 582	7 703	7 826	700
800	7 949	8 072	8 196	8 320	8 445	8 570	8 696	8 822	8 949	9 076	800
900	9 203	9 331	9 460	9 589	9 718	9 848	9 978	10 109	10 240	10 371	900
1 000	10 503	10 636	10 768	10 902	11 035	11 170	11 304	11 439	11 574	11 710	1 000
1 100	11 846	11 983	12 119	12 257	12 394	12 532	12 669	12 808	12 946	13 085	1 100
1 200	13 224	13 363	13 502	13 642	13 782	13 922	14 062	14 202	14 343	14 483	1 200
1 300	14 624	14 765	14 906	15 047	15 188	15 329	15 470	15 611	15 752	15 893	1 300
1 400	16 035	16 176	16 317	16 458	16 599	16 741	16 882	17 022	17 163	17 304	1 400
1 500	17 445	17 585	17 726	17 866	18 006	18 146	18 286	18 425	18 564	18 703	1 500
1 600	18 842	18 981	19 119	19 257	19 395	19 533	19 670	19 807	19 944	20 080	1 600
1 700	20 215	20 350	20 483	20 616	20 748	20 878	21 006				1 700

Remark : Temperature of reference junction is 0℃.
When temperature of reference junction is 20℃, subtract 111 μV from the value given in the above table.

Type : S

Unit : μV

Temp. (℃)	0	−10	−20	−30	−40	−50					Temp. (℃)
0	0	−53	−103	−150	−194	−236					0

Temp. (℃)	0	10	20	30	40	50	60	70	80	90	Temp. (℃)
0	0	55	113	173	235	299	365	432	502	573	0
100	645	719	795	872	950	1 029	1 109	1 190	1 273	1 356	100
200	1 440	1 525	1 611	1 698	1 785	1 873	1 962	2 051	2 141	2 232	200
300	2 323	2 414	2 506	2 599	2 692	2 786	2 880	2 974	3 069	3 164	300
400	3 260	3 356	3 452	3 549	3 645	3 743	3 840	3 938	4 036	4 135	400
500	4 234	4 333	4 432	4 532	4 632	4 732	4 832	4 933	5 034	5 136	500
600	5 237	5 339	5 442	5 544	5 648	5 751	5 855	5 960	6 064	6 169	600
700	6 274	6 380	6 486	6 592	6 699	6 805	6 913	7 020	7 128	7 236	700
800	7 345	7 454	7 563	7 672	7 782	7 892	8 003	8 114	8 225	8 336	800
900	8 448	8 560	8 673	8 786	8 899	9 012	9 126	9 240	9 355	9 470	900
1 000	9 585	9 700	9 816	9 932	10 048	10 165	10 282	10 400	10 517	10 635	1 000
1 100	10 754	10 872	10 991	11 110	11 229	11 348	11 467	11 587	11 707	11 827	1 100
1 200	11 947	12 067	12 188	12 308	12 429	12 550	12 671	12 792	12 913	13 034	1 200
1 300	13 155	13 276	13 397	13 519	13 640	13 761	13 883	14 004	14 125	14 247	1 300
1 400	14 368	14 489	14 610	14 731	14 852	14 973	15 094	15 215	15 336	15 456	1 400
1 500	15 576	15 697	15 817	15 937	16 057	16 176	16 296	16 415	16 534	16 653	1 500
1 600	16 771	16 890	17 008	17 125	17 243	17 360	17 477	17 594	17 711	17 826	1 600
1 700	17 942	18 056	18 170	18 282	18 394	18 504	18 612				1 700

Remark : Temperature of reference junction is 0℃.
When temperature of reference junction is 20℃, subtract 113 μV from the value in the above table.

Type : K Unit : μV

Temp. (℃)	0	−10	−20	−30	−40	−50	−60	−70	−80	−90	Temp. (℃)
−200	−5 891	−6 035	−6 158	−6 262	−6 344	−6 404	−6 441	−6 458			−200
−100	−3 553	−3 852	−4 138	−4 410	−4 669	−4 912	−5 141	−5 354	−5 550	−5 730	−100
0	0	−392	−777	−1 156	−1 527	−1 889	−2 243	−2 586	−2 920	−3 242	0

Temp. (℃)	0	10	20	30	40	50	60	70	80	90	Temp. (℃)
0	0	397	798	1 203	1 611	2 022	2 436	2 850	3 266	3 681	0
100	4 095	4 508	4 919	5 327	5 733	6 137	6 539	6 939	7 338	7 737	100
200	8 137	8 537	8 938	9 341	9 745	10 151	10 560	10 969	11 381	11 793	200
300	12 207	12 623	13 039	13 456	13 874	14 292	14 712	15 132	15 552	15 974	300
400	16 395	16 818	17 241	17 664	18 088	18 513	18 938	19 363	19 788	20 214	400
500	20 640	21 066	21 493	21 919	22 346	22 772	23 198	23 624	24 050	24 476	500
600	24 902	25 327	25 751	26 176	26 599	27 022	27 445	27 867	28 288	28 709	600
700	29 128	29 547	29 965	30 383	30 799	31 214	31 629	32 042	32 455	32 866	700
800	33 277	33 686	34 095	34 502	34 909	35 314	35 718	36 121	36 524	36 925	800
900	37 325	37 724	38 122	38 519	38 915	39 310	39 703	40 096	40 488	40 879	900
1 000	41 269	41 657	42 045	42 432	42 817	43 202	43 585	43 968	44 349	44 729	1 000
1 100	45 108	45 486	45 863	46 238	46 612	46 985	47 356	47 726	48 095	48 462	1 100
1 200	48 828	49 192	49 555	49 916	50 276	50 633	50 990	51 344	51 697	52 049	1 200
1 300	52 398	52 747	53 093	53 439	53 782	54 125	54 466	54 807			1 300

Remark : Temperature of reference junction is 0℃.
When temperature of reference junction is 20℃, subtract 798 μV from the value given in the above table.

Type : E Unit : μV

Temp. (℃)	0	−10	−20	−30	−40	−50	−60	−70	−80	−90	Temp. (℃)
−200	−8 824	−9 063	−9 274	−9 455	−9 604	−9 719	−9 797	−9 835			−200
−100	−5 237	−5 680	−6 107	−6 516	−6 907	−7 279	−7 631	−7 963	−8 273	−8 561	−100
0	0	− 581	−1 151	−1 709	−2 254	−2 787	−3 306	−3 811	−4 301	−4 777	0

Temp. (℃)	0	10	20	30	40	50	60	70	80	90	Temp. (℃)
0	0	591	1 192	1 801	2 419	3 047	3 683	4 329	4 983	5 646	0
100	6 317	6 996	7 683	8 377	9 078	9 787	10 501	11 222	11 949	12 681	100
200	13 419	14 161	14 909	15 661	16 417	17 178	17 942	18 710	19 481	20 256	200
300	21 033	21 814	22 597	23 383	24 171	24 961	25 754	26 549	27 345	28 143	300
400	28 943	29 744	30 546	31 350	32 155	32 960	33 767	34 574	35 382	36 190	400
500	36 999	37 808	38 617	39 426	40 236	41 045	41 853	42 662	43 470	44 278	500
600	45 085	45 891	46 697	47 502	48 306	49 109	49 911	50 713	51 513	52 312	600
700	53 110	53 907	54 703	55 498	56 291	57 083	57 873	58 663	59 451	60 237	700
800	61 022	61 806	62 588	63 368	64 147	64 924	65 700	66 473	67 245	68 015	800
900	68 783	69 549	70 313	71 075	71 835	72 593	73 350	74 104	74 857	75 608	900
1 000	76 358										1 000

Remark : Temperature of reference junction is 0℃.
When temperature of reference junction is 20℃, subtract 1 192 μV from the value given in the above table.

Type : J　　　　Unit : μV

Temp. (℃)	0	−10	−20	−30	−40	−50	−60	−70	−80	−90	Temp. (℃)
−200	−7 890	−8 096									−200
−100	−4 632	−5 036	−5 426	−5 801	−6 159	−6 499	−6 821	−7 122	−7 402	−7 659	−100
0	0	−501	−995	−1 481	−1 960	−2 431	−2 892	−3 344	−3 785	−4 215	0
Temp. (℃)	**0**	**10**	**20**	**30**	**40**	**50**	**60**	**70**	**80**	**90**	**Temp. (℃)**
0	0	507	1 019	1 536	2 058	2 585	3 115	3 649	4 186	4 725	0
100	5 268	5 812	6 359	6 907	7 457	8 008	8 560	9 113	9 667	10 222	100
200	10 777	11 332	11 887	12 442	12 998	13 553	14 108	14 663	15 217	15 771	200
300	16 325	16 879	17 432	17 984	18 537	19 089	19 640	20 192	20 743	21 295	300
400	21 846	22 397	22 949	23 501	24 054	24 607	25 161	25 716	26 272	26 829	400
500	27 388	27 949	28 511	29 075	29 642	30 210	30 782	31 356	31 933	32 513	500
600	33 096	33 683	34 273	34 867	35 464	36 066	36 671	37 280	37 893	38 510	600
700	39 130	39 754	40 382	41 013	41 647	42 283	42 922	43 563	44 207	44 852	700
800	45 498	46 144	46 790	47 434	48 076	48 716	49 354	49 989	50 621	51 249	800
900	51 875	52 496	53 115	53 729	54 341	54 948	55 553	56 155	56 753	57 349	900
1 000	57 942	58 533	59 121	59 708	60 293	60 876	61 459	62 039	62 619	63 199	1 000
1 100	63 777	64 355	64 933	65 510	66 087	66 664	67 240	67 815	68 390	68 964	1 100
1 200	69 536										1 200

Remark : Temperature of reference junction is 0℃.
When temperature of reference junction is 20℃, subtract 1 019 μV from the value given in the above table.

Type : T　　　　Unit : μV

Temp. (℃)	0	−10	−20	−30	−40	−50	−60	−70	−80	−90	Temp. (℃)
−200	−5 603	−5 753	−5 889	−6 007	−6 105	−6 181	−6 232	−6 258			−200
−100	−3 378	−3 656	−3 923	−4 177	−4 419	−4 648	−4 865	−5 069	−5 261	−5 439	−100
0	0	−383	−757	−1 121	−1 475	−1 819	−2 152	−2 475	−2 788	−3 089	0
Temp. (℃)	**0**	**10**	**20**	**30**	**40**	**50**	**60**	**70**	**80**	**90**	**Temp. (℃)**
0	0	391	789	1 196	1 611	2 035	2 467	2 908	3 357	3 813	0
100	4 277	4 749	5 227	5 712	6 204	6 702	7 207	7 718	8 235	8 757	100
200	9 286	9 820	10 360	10 905	11 456	12 011	12 572	13 137	13 707	14 281	200
300	14 860	15 443	16 030	16 621	17 217	17 816	18 420	19 027	19 638	20 252	300
400	20 869										400

Remark : Temperature of reference junction is 0℃.
When temperature of reference junction is 20℃, subtract 789 μV from the value given in the above table.

APPENDIX 2 REFERENCE RESISTANCE VALUE OF Pt 100

(Extracted from JIS C 1604-1989 which conforms to IEC 751-1983, BS 1904-1984, DINIEC 751-1985)

Temperature ℃	0	−1	−2	−3	−4	−5	−6	−7	−8	−9	−10	Temperature ℃
−200	18.49	–	–	–	–	–	–	–	–	–	–	−200
−190	22.80	22.37	21.94	21.51	21.08	20.65	20.22	19.79	19.36	18.93	18.49	−190
−180	27.08	26.65	26.23	25.80	25.37	24.94	24.52	24.09	23.66	23.23	22.80	−180
−170	31.32	30.90	30.47	30.05	29.63	29.20	28.78	28.35	27.93	27.50	27.08	−170
−160	35.53	35.11	34.69	34.27	33.85	33.43	33.01	32.59	32.16	31.74	31.32	−160
−150	39.71	39.30	38.88	38.46	38.04	37.63	37.21	36.79	36.37	35.95	35.53	−150
−140	43.87	43.45	43.04	42.63	42.21	41.79	41.38	40.96	40.55	40.13	39.71	−140
−130	48.00	47.59	47.18	46.76	46.35	45.94	45.52	45.11	44.70	44.28	43.87	−130
−120	52.11	51.70	51.29	50.88	50.47	50.06	49.64	49.23	48.82	48.41	48.00	−120
−110	56.19	55.78	55.38	54.97	54.56	54.15	53.74	53.33	52.92	52.52	52.11	−110
−100	60.25	59.85	59.44	59.04	58.63	58.22	57.82	57.41	57.00	56.60	56.19	−100
−90	64.30	63.90	63.49	63.09	62.68	62.28	61.87	61.47	61.06	60.66	60.25	−90
−80	68.33	67.92	67.52	67.12	66.72	66.31	65.91	65.51	65.11	64.70	64.30	−80
−70	72.33	71.93	71.53	71.13	70.73	70.33	69.93	69.53	69.13	68.73	68.33	−70
−60	76.33	75.93	75.53	75.13	74.73	74.33	73.93	73.53	73.13	72.73	72.33	−60
−50	80.31	79.91	79.51	79.11	78.72	78.32	77.92	77.52	77.13	76.73	76.33	−50
−40	84.27	83.88	83.48	83.08	82.69	82.29	81.89	81.50	81.10	80.70	80.31	−40
−30	88.22	87.83	87.43	87.04	86.64	86.25	85.85	85.46	85.06	84.67	84.27	−30
−20	92.16	91.77	91.37	90.98	90.59	90.19	89.80	89.40	89.01	88.62	88.22	−20
−10	96.09	95.69	95.30	94.91	94.52	94.12	93.73	93.34	92.95	92.55	92.16	−10
0	100.00	99.61	99.22	98.83	98.44	98.04	97.65	97.26	96.87	96.48	96.09	0

Temperature ℃	0	1	2	3	4	5	6	7	8	9	10	Temperature ℃
0	100.00	100.39	100.78	101.17	101.56	101.95	102.34	102.73	103.12	103.51	103.90	0
10	103.90	104.29	104.68	105.07	105.46	105.85	106.24	106.63	107.02	107.40	107.79	10
20	107.79	108.18	108.57	108.96	109.35	109.73	110.12	110.51	110.90	111.28	111.67	20
30	111.67	112.06	112.45	112.83	113.22	113.61	113.99	114.38	114.77	115.15	115.54	30
40	115.54	115.93	116.31	116.70	117.08	117.47	117.85	118.24	118.62	119.01	119.40	40
50	119.40	119.78	120.16	120.55	120.93	121.32	121.70	122.09	122.47	122.86	123.24	50
60	123.24	123.62	124.01	124.39	124.77	125.16	125.54	125.92	126.31	126.69	127.07	60
70	127.07	127.45	127.84	128.22	128.60	128.98	129.37	129.75	130.13	130.51	130.89	70
80	130.89	131.27	131.66	132.04	132.42	132.80	133.18	133.56	133.94	134.32	134.70	80
90	134.70	135.08	135.46	135.84	136.22	136.60	136.98	137.36	137.74	138.12	138.50	90
100	138.50	138.88	139.26	139.64	140.02	140.39	140.77	141.15	141.53	141.91	142.29	100
110	142.29	142.66	143.04	143.42	143.80	144.17	144.55	144.93	145.31	145.68	146.06	110
120	146.06	146.44	146.81	147.19	147.57	147.94	148.32	148.70	149.07	149.45	149.82	120
130	149.82	150.20	150.57	150.95	151.33	151.70	152.08	152.45	152.83	153.20	153.58	130
140	153.58	153.95	154.32	154.70	155.07	155.45	155.82	156.19	156.57	156.94	157.31	140
150	157.31	157.69	158.06	158.43	158.81	159.18	159.55	159.93	160.30	160.67	161.04	150
160	161.04	161.42	161.79	162.16	162.53	162.90	163.27	163.65	164.02	164.39	164.76	160
170	164.76	165.13	165.50	165.87	166.24	166.61	166.98	167.35	167.72	168.09	168.46	170
180	168.46	168.83	169.20	169.57	169.94	170.31	170.68	171.05	171.42	171.79	172.16	180
190	172.16	172.53	172.90	173.26	173.63	174.00	174.37	174.74	175.10	175.47	175.84	190
200	175.84	176.21	176.57	176.94	177.31	177.68	178.04	178.41	178.78	179.14	179.51	200
210	179.51	179.88	180.24	180.61	180.97	181.34	181.71	182.07	182.44	182.80	183.17	210
220	183.17	183.53	183.90	184.26	184.63	184.99	185.36	185.72	186.09	186.45	186.82	220
230	186.82	187.18	187.54	187.91	188.27	188.63	189.00	189.36	189.72	190.09	190.45	230
240	190.45	190.81	191.18	191.54	191.90	192.26	192.63	192.99	193.35	193.71	194.07	240
250	194.07	194.44	194.80	195.16	195.52	195.88	196.24	196.60	196.96	197.33	197.69	250
260	197.69	198.05	198.41	198.77	199.13	199.49	199.85	200.21	200.57	200.93	201.29	260
270	201.29	201.65	202.01	202.36	202.72	203.08	203.44	203.80	204.16	204.52	204.88	270
280	204.88	205.23	205.59	205.95	206.31	206.67	207.02	207.38	207.74	208.10	208.45	280
290	208.45	208.81	209.17	209.52	209.88	210.24	210.59	210.95	211.31	211.66	212.02	290
300	212.02	212.37	212.73	213.09	213.44	213.80	214.15	214.51	214.86	215.22	215.57	300
310	215.57	215.93	216.28	216.64	216.99	217.35	217.70	218.05	218.41	218.76	219.12	310
320	219.12	219.47	219.82	220.18	220.53	220.88	221.24	221.59	221.94	222.29	222.65	320
Temperature ℃	0	1	2	3	4	5	6	7	8	9	10	Temperature ℃

Temperature ℃	0	1	2	3	4	5	6	7	8	9	10	Temperature ℃
330	222.65	223.00	223.35	223.70	224.06	224.41	224.76	225.11	225.46	225.81	226.17	330
340	226.17	226.52	226.87	227.22	227.57	227.92	228.27	228.62	228.97	229.32	229.67	340
350	229.67	230.02	230.37	230.72	231.07	231.42	231.77	232.12	232.47	232.82	233.17	350
360	233.17	233.52	233.87	234.22	234.56	234.91	235.26	235.61	235.96	236.31	236.65	360
370	236.65	237.00	237.35	237.70	238.04	238.39	238.74	239.09	239.43	239.78	240.13	370
380	240.13	240.47	240.82	241.17	241.51	241.86	242.20	242.55	242.90	243.24	243.59	380
390	243.59	243.93	244.28	244.62	244.97	245.31	245.66	246.00	246.35	246.69	247.04	390
400	247.04	247.38	247.73	248.07	248.41	248.76	249.10	249.45	249.79	250.13	250.48	400
410	250.48	250.82	251.16	251.50	251.85	252.19	252.53	252.88	253.22	253.56	253.90	410
420	253.90	254.24	254.59	254.93	255.27	255.61	255.95	256.29	256.64	256.98	257.32	420
430	257.32	257.66	258.00	258.34	258.68	259.02	259.36	259.70	260.04	260.38	260.72	430
440	260.72	261.06	261.40	261.74	262.08	262.42	262.76	263.10	263.43	263.77	264.11	440
450	264.11	264.45	264.79	265.13	265.47	265.80	266.14	266.48	266.82	267.15	267.49	450
460	267.49	267.83	268.17	268.50	268.84	269.18	269.51	269.85	270.19	270.52	270.86	460
470	270.86	271.20	271.53	271.87	272.20	272.54	272.88	273.21	273.55	273.88	274.22	470
480	274.22	274.55	274.89	275.22	275.56	275.89	276.23	276.56	276.89	277.23	277.56	480
490	277.56	277.90	278.23	278.56	278.90	279.23	279.56	279.90	280.23	280.56	280.90	490
500	280.90	281.23	281.56	281.89	282.23	282.56	282.89	283.22	283.55	283.89	284.22	500
510	284.22	284.55	284.88	285.21	285.54	285.87	286.21	286.54	286.87	287.20	287.53	510
520	287.53	287.86	288.19	288.52	288.85	289.18	289.51	289.84	290.17	290.50	290.83	520
530	290.83	291.16	291.49	291.81	292.14	292.47	292.80	293.13	293.46	293.79	294.11	530
540	294.11	294.44	294.77	295.10	295.43	295.75	296.08	296.41	296.74	297.06	297.39	540
550	297.39	297.72	298.04	298.37	298.70	299.02	299.35	299.68	300.00	300.33	300.65	550
560	300.65	300.98	301.31	301.63	301.96	302.28	302.61	302.93	303.26	303.58	303.91	560
570	303.91	304.23	304.56	304.88	305.20	305.53	305.85	306.18	306.50	306.82	307.15	570
580	307.15	307.47	307.79	308.12	308.44	308.76	309.09	309.41	309.73	310.05	310.38	580
590	310.38	310.70	311.02	311.34	311.67	311.99	312.31	312.63	312.95	313.27	313.59	590
600	313.59	313.92	314.24	314.56	314.88	315.20	315.52	315.84	316.16	316.48	316.80	600
610	316.80	317.12	317.44	317.76	318.08	318.40	318.72	319.04	319.36	319.68	319.99	610
620	319.99	320.31	320.63	320.95	321.27	321.59	321.91	322.22	322.54	322.86	323.18	620
630	323.18	323.49	323.81	324.13	324.45	324.76	325.08	325.40	325.72	326.03	326.35	630
640	326.35	326.66	326.98	327.30	327.61	327.93	328.25	328.56	328.88	329.19	329.51	640
650	329.51	329.82	330.14	330.45	330.77	331.08	331.40	331.71	332.03	332.34	332.66	650
Temperature ℃	0	1	2	3	4	5	6	7	8	9	10	Temperature ℃

The reference resistance value of Attached Table shall be calculated from the following formulas:

Range of −200℃ to 0℃: $R_t = R_o\ [1 + At + Bt^2 + C(t-100)\ t^3]$

Range of 0℃ to 650℃: $R_t = R_o\ (1 + At + Bt^2)$

where, $A = 3.908\ 02 \times 10^{-3}$ ℃$^{-1}$

$B = -5.802 \times 10^{-7}$ ℃$^{-2}$

$C = -4.273\ 5 \times 10^{-12}$ ℃$^{-4}$

Remarks: 1. R_o means a resistance value at 100 Ω and R_t at t℃

2. The relationship formulas given above are for the calculation of the reference resistance values for this Standard, and not intended for obtaining the characteristics of the individual resistance thermometer sensors.

APPENDIX 3 TABLES OF LAPLACE TRANSFORM

Table of Laplace transform pairs

No.	$f(t)$	$F(s)$	No.	$f(t)$	$F(s)$
1	$\delta(t)$	1	12	$\frac{1}{a^2}(1-\cos at)$	$\frac{1}{s(s^2+a^2)}$
2	$u(t)$	$\frac{1}{s}$	13	$\frac{1}{b-a}(e^{-at}-e^{-bt})$	$\frac{1}{(s+a)(s+b)}$
3	t	$\frac{1}{s^2}$	14	$\frac{1}{a-b}(ae^{-at}-be^{-bt})$	$\frac{s}{(s+a)(s+b)}$
4	e^{-at}	$\frac{1}{s+a}$	15	te^{-at}	$\frac{1}{(s+a)^2}$
5	$\frac{1}{a}(1-e^{-at})$	$\frac{1}{s(s+a)}$	16	$e^{-at}(1-at)$	$\frac{s}{(s+a)^2}$
6	$\frac{1}{a^2}(e^{-at}+at-1)$	$\frac{1}{s^2(s+a)}$	17	$\frac{1}{a^2}\{1-(1-at)e^{-at}\}$	$\frac{1}{s(s+a)^2}$
7	$\sin at$	$\frac{a}{s^2+a^2}$	18	$t\sin at$	$2a\frac{s}{(s^2+a^2)^2}$
8	$\cos at$	$\frac{s}{s^2+a^2}$	19	$t\cos at$	$\frac{s^2-a^2}{(s^2+a^2)^2}$
9	$\sinh t$	$\frac{a}{s^2-a^2}$	20	$e^{-at}\sin bt$	$\frac{b}{(s+a)^2+b^2}$
10	$\cosh t$	$\frac{s}{s^2-a^2}$	21	$e^{-at}\cos bt$	$\frac{s+a}{(s+a)^2+b^2}$
11	$\frac{1}{a^3}(at-\sin at)$	$\frac{1}{s^2(s^2+a^2)}$			

Laplace transform theorems

No.	$f(t)$	$F(s)$	No.	$f(t)$	$F(s)$
1	$\frac{df(t)}{dt}$	$sF(s)-f(0)$	7	$f(t-a)$	$e^{-as}F(s)$
2	$\int f(t)dt$	$\frac{1}{s}[F(s)+f^{-1}(0)]$	8	$e^{at}f(t)$	$F(s-a)$
3	$c_1f_1(t)+c_2f_2(t)$	$c_1F_1(s)+c_2F_2(s)$	9	$e^{-at}f(t)$	$F(s+a)$
4	$f(at)$	$\frac{1}{a}F\left(\frac{s}{a}\right)$	10	$\lim_{t\to 0}f(t)$	$\lim_{s\to\infty}F(s)$
5	$f(t/a)$	$aF(as)$	11	$\lim_{t\to\infty}f(t)$	$\lim_{s\to 0}F(s)$
6	$f(t+a)$	$e^{as}F(s)$	12	$\int_0^t f_1(\tau)f_2(t-\tau)d\tau$	$F_1(s)F_2(s)$

INDEX

D

E

F

G

T

MIX
Papier aus verantwortungsvollen Quellen
Paper from responsible sources
FSC® C105338

If you have any concerns about our products,
you can contact us on
ProductSafety@springernature.com

In case Publisher is established outside the EU,
the EU authorized representative is:
Springer Nature Customer Service Center GmbH
Europaplatz 3, 69115 Heidelberg, Germany

Printed by Libri Plureos GmbH
in Hamburg, Germany